Lecture Notes in Computer Science 4060

Commenced Publication in 1973
Founding and Former Series Editors:
Gerhard Goos, Juris Hartmanis, and Jan van Leeuwen

T0180721

Kokichi Futatsugi   Jean-Pierre Jouannaud
José Meseguer (Eds.)

Algebra, Meaning, and Computation

Essays Dedicated to Joseph A. Goguen
on the Occasion of His 65th Birthday

 Springer

Volume Editors

Kokichi Futatsugi
Japan Advanced Institute of Science and Technology (JAIST)
Ishikawa, Japan
E-mail: kokichi@jaist.ac.jp

Jean-Pierre Jouannaud
École Polytechnique, Laboratoire d'Informatique (LIX)
91128 Palaiseau Cedex, France
E-mail: Jean-Pierre.Jouannaud@lix.polytechnique.fr

José Meseguer
University of Illinois at Urbana-Champaign (UIUC), Computer Science Dept.
Urbana, IL 61801-2302, USA
E-mail: meseguer@cs.uiuc.edu

Library of Congress Control Number: 2006927483

CR Subject Classification (1998): F.3.1-2, F.4, F.2.2, D.1.6, D.3.1-2, I.1

LNCS Sublibrary: SL 1 – Theoretical Computer Science and General Issues

ISSN   0302-9743
ISBN-10   3-540-35462-X Springer Berlin Heidelberg New York
ISBN-13   978-3-540-35462-8 Springer Berlin Heidelberg New York

Springer is a part of Springer Science+Business Media

springer.com

© Springer-Verlag Berlin Heidelberg 2006
Printed in Germany

Typesetting: Camera-ready by author, data conversion by Boller Mediendesign
Printed on acid-free paper   SPIN: 11780274   06/3142   5 4 3 2 1 0

Joseph Goguen

Preface

Joseph Goguen is one of the most prominent computer scientists worldwide. His numerous research contributions span many topics and have changed the way we think about many concepts. Our views about data types, programming languages, software specification and verification, computational behavior, logics in computer science, semiotics, interface design, multimedia, and consciousness, to mention just some of the areas, have all been enriched in fundamental ways by his ideas.

Considering just one strand of his work, namely, the area of algebraic specifications, his ideas have been enormously influential. The concept of initiality (or co-initiality) that he introduced is now a fundamental concept in theoretical computer science applied in many subfields. The Clear formal specification language was the first language with general theory composition operations based on categorical algebra. Such generality inspired Goguen and Burstall to propose institutions as a meta-logical theory of logics, so that Clear-like languages could be defined for many logics. The OBJ language, one of the earliest and most influential executable algebraic specification languages, also incorporated the Clear ideas. Categorically based module composition operations had an enormous influence not only in formal specification, but also in software methodology: his parameterized programming methodology predates by about two decades more recent work on generic programming. These ideas, and many others that he has pioneered, reverberate through the pages of this volume, in which entire chapters are devoted to some of them. Furthermore, there are several regular scientific meetings of an international scope, including the CALCO and AMAST conferences and the WADT Workshop, dedicated to ideas either initiated or directly influenced by Joseph Goguen. There are also a number of important languages that have been influenced by his CLEAR and OBJ algebraic specification languages, including: ACT1, ML, CASL, Maude, CafeOBJ, and ELAN.

A common thread in his work is the use of abstract algebra, particularly of categorical algebra, to get at the core of each problem and formulate concepts in the most general and useful way possible. Algebraic and logical methods are then deployed to provide a rigorous account of meaning, both in computational systems and in semiotic systems. Furthermore, in areas in which social aspects are involved, a humanistic perspective is combined with mathematical and computational perspectives to do justice in a non-reductionist and critical way to a wide range of human phenomena, including phenomena arising from the use or misuse of computer systems in concrete social situations.

This Festschrift volume, published to honor Joseph Goguen on his 65th birthday, includes refereed papers by leading researchers in the different areas spanned by Joseph Goguen's work. These papers were presented at a symposium in San Diego, California, June 27-29, 2006 to honor Joseph Goguen's 65th birthday on June 28, 2006. Both the Festschrift volume and the symposium will allow the articulation of a retrospective and prospective view of a range of related research topics by key members of the research community in computer science and other fields connected with Joseph Goguen's work. We think that the papers speak for themselves and provide a wonderful overview of Joseph Goguen's enormously influential ideas in one of the best ways possible, namely, by reflecting on how they have become and are part of a vast scientific dialogue.

We feel privileged to edit this volume. For us it is a way of expressing our admiration, our gratitude, and our friendship to Joseph Goguen. The four of us worked closely together at SRI's Computer Science Laboratory designing and implementing the OBJ2 language during the 1983-4 academic year. The scientific enthusiasm, camaraderie, and friendship of that relatively short but very influential period have grown over the years and have had a great impact on our lives. We are most grateful to all the authors who responded enthusiastically to our project and have contributed an excellent collection of papers for this volume. We are also very thankful to all those, both authors and nonauthors, who have helped us in the refereeing process to achieve a well-finished scholarly volume, and to Alfred Hoffmann at Springer who has encouraged our project from its early stages and has provided valuable advice. Keith Marzullo and Briana Ronhaar at UCSD deserve very special thanks as, respectively, Local Chair of the Symposium and Main Local Coordinator. Funding from the US Office of Naval Research to partially support both this Festschrift volume and the symposium through ONR Grant N00014-06-1-0280 is also gratefully acknowledged. We are particularly grateful to Ralph Wachter at ONR, who early on encouraged our project for the Festschrift volume and the symposium. Last but not least, we warmly thank Joseph Hendrix at UIUC for his invaluable and untiring help in preparing this volume.

April 2006   Kokichi Futatsugi
                                                             Jean-Pierre Jouannaud
                                                                   José Meseguer

Table of Contents

Behavior and Formal Languages

Models, Deduction, and Computation

Bibliography of Joseph Goguen

Books Authored or Edited, Published or in Progress

1. *Theory and Practice of Software Technology*, edited with Domenico Ferrari and Mario Bolognani, North-Holland, 1983. ISBN 0–444–86647–7.
2. *Requirements Engineering: Social and Technical Issues*, edited with Marina Jirotka, Academic, 1994. ISBN 0–1238–5335–4.
3. *Algebraic Semantics of Imperative Programs*, with Grant Malcolm, MIT Press, April 1996. ISBN 0–262–07172–X.
4. *Art and the Brain*, edited book, Imprint Academic, October 1999. ISBN 0–907–84545–2.
5. *OBJ/CafeOBJ/Maude at Formal Methods '99*, edited with Kokichi Futatsugi and Josè Meseguer, Theta (Bucharest), September 1999. ISBN 973–99097–1–X.
6. *Software Engineering with OBJ: Algebraic Specification in Action*, edited with Grant Malcolm, Kluwer, April 2000. ISBN 0–7923–7757–5.
7. *Art and the Brain (Part 2)*, edited with Erik Myin, Imprint Academic, September 2000. ISBN 0–907–84512–6.
8. *Art and the Brain (Part 3)*, edited with Erik Myin, Imprint Academic, May 2004. ISBN 0–907–84598–3
9. *Theorem Proving and Algebra*, in preparation for MIT Press.
10. *Hidden Algebra*, with Grigore Roşu, in preparation.

Publications in Journals and Conference Proceedings

11. "Color Perception Using a Single Cone Type with Distributed Maximum Sensitivity," *Bulletin of Mathematical Biophysics*, Volume 26, 1964, pages 121–138.
12. "L-Fuzzy Sets," *Journal of Mathematical Analysis and Applications*, Volume 18, Number 1, 1967, pages 145–174.
13. "Categories of Fuzzy Sets," Ph.D. Dissertation, Department of Mathematics, University of California at Berkeley, 1968.
14. "The Logic of Inexact Concepts," *Synthese*, Volume 19, 1968–1969, pages 325–373.
15. "Categories of V-Sets," *Bulletin of the American Mathematical Society*, Volume 75, Number 3, 1969, pages 622–624.
16. "Mathematical Representation of Hierarchically Organized Systems," in *Global Systems Dynamics*, edited by E. Attinger, S. Karger (Basel, Switzerland) 1971, pages 112–128.
17. "Systems and Minimal Realization," *Proceedings, IEEE Conference on Decision and Control* (Miami Beach, Florida) 1972, pages 42–46.
18. "Minimal Realization of Machines in Closed Categories," *Bulletin of the American Mathematical Society*, Volume 78, Number 5, 1972, pages 777–783.
19. "Hierarchical Inexact Data Structures in Artificial Intelligence Problems," *Proceedings, Fifth Hawaii International Conference on System Sciences* (Honolulu, Hawaii) 1972, pages 345–347.
20. "The Myhill Functor, Input-Reduced Machines, and Generalized Krohn-Rhodes Theory," with Robert H. Yacobellis, *Proceedings, Fifth Princeton Conference on Information Sciences and Systems* (Princeton, New Jersey) 1972, pages 574–578.

21. "On Homomorphisms, Simulation, Correctness and Subroutines for Programs and Program Schemes," *Proceedings, 13th IEEE Symposium on Switching and Automata Theory* (College Park, Maryland) 1972, pages 52–60.

22. "Realization is Universal," *Mathematical System Theory*, Volume 6, Number 4, 1973, pages 359–374.

23. "System Theory Concepts in Computer Science," *Proceedings, Sixth Hawaii International Conference on System Sciences* (Honolulu, Hawaii) 1973, pages 77–80.

24. "The Fuzzy Tychonoff Theorem," *Journal of Mathematical Analysis and Applications*, Volume 43, 1973, pages 734–742.

25. "Categorical Foundations for General Systems Theory," in *Advances in Cybernetics and Systems Research*, edited by F. Pichler and R. Trappl, Transcripta Books (London) 1973, pages 121–130.

26. "Semantics of Computation," *Proceedings, First International Symposium on Category Theory Applied to Computation and Control* (1974 American Association for the Advancement of Science, San Francisco) University of Massachusetts at Amherst, 1974, pages 234–249. Also published in Lecture Notes in Computer Science, Volume 25, Springer, 1975, pages 151–163.

27. "Initial Algebra Semantics," with James Thatcher, *Proceedings, 15th IEEE Symposium on Switching and Automata Theory*, 1974, pages 63–77.

28. "Concept Representation in Natural and Artificial Languages: Axioms, Extensions and Applications for Fuzzy Sets," *International Journal of Man-Machine Studies*, Volume 6, 1974, pages 513–561. Reprinted in *Fuzzy Reasoning and its Applications*, edited by E. H. Mamdani and Brian Gaines, Academic, 1981, pages 67–115.

29. "On Homomorphisms, Correctness, Termination, Unfoldments and Equivalence of Flow Diagram Programs," *Journal of Computer and System Sciences*, Volume 8, Number 3, 1974, pages 333–365.

30. "Some Comments on Applying Mathematical System Theory," in *Systems Approaches and Environmental Problems*, edited by Hans W. Gottinger, Vandenhoeck and Rupert (Göttingen, Germany), 1974, pages 47–67.

31. "Factorization, Congruences, and the Decomposition of Automata and Systems," with James Thatcher, Eric Wagner, and Jesse Wright, *Mathematical Foundations of Computer Science*, Lecture Notes in Computer Science, Volume 28, Springer, 1975, pages 33–45.

32. "Objects," *International Journal of General Systems*, Volume 1, Number 4, 1975, pages 237–243.

33. "Discrete-Time Machines in Closed Monoidal Categories, I," *Journal of Computer and System Sciences*, Volume 10, Number 1, February 1975, pages 1–43.

34. "Abstract Data Types as Initial Algebras and the Correctness of Data Representations," with James Thatcher, Eric Wagner, and Jesse Wright, in *Proceedings, Conference on Computer Graphics, Pattern Recognition, and Data Structure* (Beverly Hills, California), 1975, pages 89–93.

35. "Axioms for Discrimination Information," with Lee Carlson, *IEEE Transactions on Information Theory*, September 1975, pages 572–574.

36. "On Fuzzy Robot Planning," in *Fuzzy Sets and their Applications to Cognitive and Decision Processes*, edited by Lotfi Zadeh, K. S. Fu, K. Tanaka, and M. Shimura, Academic, 1975, pages 429–448.

37. "Robust Programming Languages and the Principle of Maximum Meaningfulness," *Proceedings, Milwaukee Symposium on Automatic Computation and Control* (Milwaukee, Wisconsin), 1975, pages 87–90.

38. "Complexity of Hierarchically Organized Systems and the Structure of Musical Experiences," *International Journal of General Systems Theory*, Volume 3, Number 4, 1977, pages 237–251. Originally in *UCLA Computer Science Department Quarterly*, October 1975, pages 51–88.

39. "Some Fundamentals of Order-Algebraic Semantics," with James Thatcher, Eric Wagner, and Jesse Wright, *Proceedings, Fifth Symposium on Mathematical Foundations of Computer Science* (Gdańsk, Poland, 1976) Springer, Lecture Notes in Computer Science, Volume 45, 1976, pages 153–168.

40. "Parallel Realization of Systems, Using Factorizations and Quotients in Categories," with James Thatcher, Eric Wagner, and Jesse Wright, *Journal of the Franklin Institute*, Volume 301, Number 6, June 1976, pages 547–558.

41. "Correctness and Equivalence of Data Types," *Proceedings, Symposium on Mathematical Systems Theory* (Udine, Italy) Springer, Lecture Notes in System Theory, edited by G. Marchesini, 1976, pages 352–358.

42. "Rational Algebraic Theories and Fixed-Point Solutions," with James Thatcher, Eric Wagner, and Jesse Wright, *Proceedings, IEEE 17th Symposium on Foundations of Computer Science* (Houston, Texas), 1976, pages 147–158.

43. "Initial Algebra Semantics and Continuous Algebras," with James Thatcher, Eric Wagner, and Jesse Wright, *Journal of the Association for Computing Machinery*, Volume 24, Number 1, January 1977, pages 68–95.

44. "Abstract Errors for Abstract Data Types," in *Formal Description of Programming Concepts*, edited by Eric Neuhold, North-Holland, 1978, pages 491–522. Also in *Proceedings, IFIP Working Conference on Formal Description of Programming Concepts*, edited by Jack Dennis, MIT, 1977, pages 21.1–21.32.

45. "Putting Theories Together to Make Specifications," with Rod Burstall, *Proceedings, 5th International Joint Conference on Artificial Intelligence* (MIT, Cambridge, Massachusetts), 1977, pages 1045–1058.

46. "Correctness of Recursive Flow Diagram Programs," with José Meseguer, *Proceedings, Conference on Mathematical Foundations of Computer Science* (Tatranska Lomnica, Czechoslovakia), Springer, Lecture Notes in Computer Science, Volume 53, 1977, pages 580–595.

47. "Algebraic Specification Techniques," *UCLA Computer Science Department Quarterly*, Volume 5, Number 4, 1977, pages 53–58.

48. "The Arithmetic of Closure," with Francisco Varela, in *Journal of Cybernetics*, Volume 8, 1978. Also in *Progress in Cybernetics and Systems Research*, Volume 3, edited by R. Trappl, George Klir and L. Ricciardi, Hemisphere Publishing Co. (Washington, D.C.) 1978.

49. "A Categorical Approach to General Systems," with Susanna Ginali, in *Applied General Systems Research*, edited by George Klir, Plenum, 1978, pages 257–270.

50. "An Initial Algebra Approach to the Specification, Correctness, and Implementation of Abstract Data Types," with James Thatcher and Eric Wagner, in *Current Trends in Programming Methodology*, Volume 4, *Data Structuring*, edited by Raymond Yeh, Prentice-Hall, 1978, pages 80–149.

51. "Structure of Planning Discourse," with Charlotte Linde, *Journal of Social and Biological Structures*, Volume 1, 1978, pages 219–251.

52. "Some Design Principles and Theory for OBJ-0, a Language for Expressing and Executing Algebraic Specifications of Programs," *Proceedings, International Conference on Mathematical Studies of Information Processing* (Kyoto, Japan), Springer, Lecture Notes in Computer Science, Volume 75, 1979, pages 425–473.

53. "Fuzzy Sets at UCLA," with Efraim Shaket, *Kybernetes*, Volume 8, 1979, pages 65–66.

XVI   Bibliography of Joseph Goguen

54. "Systems and Distinctions; Duality and Complementarity," with Francisco Varela, *International Journal of General Systems*, Volume 5, 1979, pages 31–43.

55. "An Introduction to OBJ: a language for writing and testing algebraic specifications," with Joseph Tardo, *Proceedings, Conference Specification of Reliable Software* (Cambridge, Massachusetts) 1979, pages 170–189; reprinted in *Software Specification Techniques*, edited by N. Gehani and Andrew McGettrick, Addison-Wesley, 1985, pages 391–420.

56. "Algebraic Specification," in *Research Directions in Software Technology*, edited by Peter Wegner, MIT, 1979, pages 370–376.

57. "Some Ideas in Algebraic Semantics," in *Proceedings, Third IBM Symposium on Mathematical Foundations of Computer Science* (Kobe, Japan), 1979, 53 pages.

58. "Fuzzy Sets and the Social Nature of Truth," in *Advances in Fuzzy Set Theory and Applications*, edited by M. M. Gupta and Ronald Yager, North-Holland, 1979, pages 49–68.

59. "A Practical Method for Testing Algebraic Specifications," with Joseph Tardo, Norman Williamson and Maria Zamfir, *UCLA Computer Science Quarterly*, Volume 7, Number 1, 1979, pages 59–80.

60. "Thoughts on Specification, Design and Verification," *Software Engineering Notes*, Volume 5, Number 3, July 1980, pages 29–33.

61. "How to Prove Algebraic Inductive Hypotheses without Induction: With Applications to the Correctness of Data Type Implementation," *Proceedings, 5th Conference on Automated Deduction* (Les Arcs, France) edited by W. Bibel and Robert Kowalski, Springer, Lecture Notes in Computer Science, Volume 87, 1980, pages 356–373.

62. "The Semantics of Clear, a Specification Language," with Rod Burstall, in *Abstract Software Specification*, edited by Dines Bjorner (Proceedings of the 1979 Copenhagen Winter School) Lecture Notes in Computer Science, Volume 86, Springer, 1980, pages 294–332.

63. "On the Independence of Discourse Structure and Semantic Domain," with Charlotte Linde, *Proceedings, 18th Annual Meeting of the Association for Computational Linguistics, Parasession on Topics in Interactive Discourse*, (University of Pennsylvania, Philadelphia, Pennsylvania) 1980, pages 35–37.

64. "More Thoughts on Specification and Verification," *Software Engineering Notes*, Volume 6, Number 3, 1980, pages 38–41; reprinted in *Software Specification Techniques*, edited by N. Gehani and Andrew McGettrick, Addison-Wesley, 1985, pages 47–52.

65. "Algebraic Denotational Semantics using Parameterized Abstract Modules," with Kamran Parsaye-Ghomi, in *Proceedings, International Conference on Formalizing Programming Concepts* (Peniscola, Spain) edited by J. Diaz and I. Ramos, Springer, Lecture Notes in Computer Science, Volume 107, 1981, pages 292–309.

66. "An Informal Introduction to Clear, a Specification Language," with Rod Burstall, in *The Correctness Problem in Computer Science*, edited by Robert Boyer and J Moore, Academic, 1981, pages 185–213; reprinted in *Software Specification Techniques*, edited by N. Gehani and Andrew McGettrick, Addison-Wesley, 1985, pages 363–390.

67. "Completeness of Many-Sorted Equational Logic," with José Meseguer, *SIGPLAN Notices*, Volume 16, Number 7, 1981, pages 24–32. Also in *SIGPLAN Notices*, Volume 17, Number 1, 1982, pages 9–17. Extended version as Technical Report CSLI-84-15, Center for the Study of Language and Information, Stanford University, September 1984.

68. "ORDINARY Specification of KWIC Index Generation," in *Proceedings, Workshop on Program Specification*, edited by J. Staunstrup, Springer Lecture Notes in Computer Science, Volume 134, 1982, pages 114–117.

69. "ORDINARY Specification of Some Constructions in Plane Geometry," in *Proceedings, Workshop on Program Specification*, edited by J. Staunstrup, Springer, Lecture Notes in Computer Science, Volume 134, 1982, pages 31–46.

70. "Algebras, Theories and Freeness: An Introduction for Computer Scientists," with Rod Burstall, in *Theoretical Foundations of Programming Methodology* (edited by Manfred Broy and G. Schmidt) D. Reidel, 1982, pages 329–348. Also in Proceedings, 1981 Marktoberdorf Summer School, NATO Advanced Study Institute Series, Volume C91.

71. "Security Policies and Security Models," with José Meseguer, in *Proceedings, 1982 Berkeley Conference on Computer Security*, IEEE Computer Society, 1982, pages 11–20.

72. "Universal Realization, Persistent Interconnection, and Implementation of Abstract Modules," with José Meseguer, in *Proceedings, 9th International Colloquium on Automata, Languages and Programming* (Aarhus, Denmark) Springer, Lecture Notes in Computer Science, 1982, pages 265–281.

73. "Rapid Prototyping in the OBJ Executable Specification Language," in *Proceedings, Rapid Prototyping Workshop* (Columbia, Maryland) 1982. Also in *Software Engineering Notes*, ACM Special Interest Group on Software engineering, Volume 7, Number 5, 1983, pages 75–84.

74. "Programming with Parameterized Abstract Objects in OBJ," with José Meseguer and David Plaisted, in *Theory and Practice of Software Technology*, edited by Domenico Ferrari, Mario Bolognani and Joseph Goguen, North-Holland, 1983, pages 163–193.

75. "Future Directions for Software Engineering," in *Theory and Practice of Software Technology*, edited by Domenico Ferrari, Mario Bolognani and Joseph Goguen, North-Holland, 1983, pages 243–244.

76. "Correctness of Recursive Parallel Non-deterministic Flow Programs," with José Meseguer, *Journal of Computer and System Sciences*, Volume 27, Number 2, October 1983, pages 268–290.

77. "Parameterized Programming," *IEEE Transactions on Software Engineering*, Volume SE-10, Number 5, September 1984, pages 528–543; preliminary version in *Proceedings, Workshop on Reusability in Programming*, ITT, 1983, pages 138–150.

78. "Reasoning and Natural Explanation," with Charlotte Linde and James Weiner, *International Journal of Man-Machine Studies*, Volume 19, 1983, pages 521–559.

79. "Introducing Institutions," with Rod Burstall, *Logics of Programs* (Carnegie-Mellon University, June 1983), Springer, Lecture Notes in Computer Science, Volume 164, 1984, pages 221–256.

80. "Unwinding and Inference Control," with José Meseguer, *Proceedings, 1984 Symposium on Security and Privacy*, IEEE, 1984, pages 75–86.

81. "Equality, Types, Modules and Generics for Logic Programming," with José Meseguer, *Proceedings, Second International Logic Programming Conference* (Upsala, Sweden), 1984, pages 115–125; Report Number CSLI-84-5, Center for the Study of Logic and Information, Stanford University, March 1984.

82. "Some Fundamental Algebraic Tools for the Semantics of Computation, Part 1: Comma Categories, Colimits, Signatures and Theories," with Rod Burstall, *Theoretical Computer Science*, Volume 31, Number 2, 1984, pages 175–209.

83. "Some Fundamental Algebraic Tools for the Semantics of Computation, Part 2: Signed and Abstract Theories," with Rod Burstall, *Theoretical Computer Science*, Volume 31, Number 3, 1984, pages 263–295.

84. "Equality, Types, Modules and (Why Not?) Generics for Logic Programming," with José Meseguer, *Journal of Logic Programming*, Volume 1, Number 2, 1984, pages 179–210.

85. "A Full Mission Simulator Study of Aircrew Performance: The Measurement of Crew Coordination and Decisionmaking Factors and their Relationships to Flight Task Performance," with Miles Murphy, Robert J. Randle, Trieve A. Tanner, Richard M. Frankel, and Charlotte Linde, in *Proceedings, 20th Annual Conference on Manual Control*, Volume II, edited by E. James Hartzell and Sandra Hart, NASA Conference Publication 2341, 1984, pages 249–262.

86. "Crew Communication as a Factor in Aviation Accidents," with Charlotte Linde and Miles Murphy, in *Proceedings, 20th Annual Conference on Manual Control*, Volume II, edited by E. James Hartzell and Sandra Hart, NASA Conference Publication 2341, 1984, pages 217–248.

87. "Principles of OBJ2," with Kokichi Futatsugi, Jean-Pierre Jouannaud, and José Meseguer, *Proceedings, 1985 Symposium on Principles of Programming Languages*, ACM, 1985, pages 52–66.

88. "Merged Views, Closed Worlds, and Ordered Sorts: Some Novel Database Features in OBJ," *Proceedings, 1982 Workshop on Database Interfaces*, edited by Alex Borgida and Peter Buneman, University of Pennsylvania, Computer Science Department, 1985, pages 38–47.

89. "Operational Semantics for Order-Sorted Algebra," with Jean-Pierre Jouannaud and José Meseguer, in *Proceedings, 1985 ICALP*, Springer, Lecture Notes in Computer Science, 1985, pages 221–231.

90. "Completeness of Many-Sorted Equational Logic," with José Meseguer, *Houston Journal of Mathematics*, Volume 11, Number 3, 1985, pages 307–334; also appeared as Technical Report CSL-135, Computer Science Laboratory, SRI International, May 1982.

91. "Initiality, Induction and Computability," with José Meseguer, in *Algebraic Methods in Semantics*, edited by Maurice Nivat and John Reynolds, Cambridge University, Chapter 14, 1985, pages 459–541.

92. "Suggestions for Using and Organizing Libraries in Software Development," in *Proceedings, First International Conference on Supercomputing Systems*, IEEE Computer Society, 1985, pages 349–360.

93. "Reusing and Interconnecting Software Components," *IEEE Computer*, Volume 19, Number 2, special issue on "Adaptable Software and Hardware for Mission-Critical Supersystems," edited by Steven and Svetlana Kartashev, IEEE Computer Society, February 1986, pages 16–28; also in *Supercomputer Systems*, edited by Steven and Svetlana Kartashev, Elsevier, 1986; and reprinted in *Tutorial: Software Reusability*, edited by Peter Freeman, IEEE Computer Society, 1987, pages 251–263, and in *Domain Analysis and Software Systems Modelling*, Ruben Prieto-Diaz and Guillermo Arango, editors, IEEE Computer Society, 1991, pages 125–137.

94. "Eqlog: Equality, Types and Generic Modules for Logic Programming," with José Meseguer, in *Functional and Logic Programming*, edited by Doug DeGroot and Gary Lindstrom, Prentice-Hall, 1986, pages 295–363.

95. "Progress Report on the Rewrite Rule Machine," with Claude Kirchner, Sany Leinwand, José Meseguer and Timothy Winkler; *IEEE Technical Committee on Computer Architecture Newsletter*, March 1986, pages 7–21.

96. "One, None, A Hundred Thousand Specification Languages," *Information Processing 86*, Proceedings of IFIP '86, Elsevier, 1986, pages 995–1003; also, Report CSLI-87-96, Center for the Study of Language and Information, Stanford University, March 1987.

97. "Extensions and Foundations of Object-Oriented Programming," with José Meseguer, in *Proceedings, Workshop on Object-Oriented Programming* (IBM Research, Yorktown Heights, NY), *SIGPLAN Notices*, Volume 21, Number 10, October 1986, pages 153–162.

98. "A Study in the Foundations of Programming Methodology: Specifications, Institutions, Charters and Parchments," with Rod Burstall, *Proceedings, Conference on Category Theory and Computer Programming* (Guildford, Surrey, U.K.), edited by David Pitt, Samson Abramsky, Axel Poigné, and David Rydeheard, Springer, Lecture Notes in Computer Science, Volume 240, 1986, pages 313–333; also appeared as Report CSLI-86-54, Center for the Study of Language and Information, Stanford University, 1986; and as Technical Report ECS-LFCS-86-10, Laboratory for Foundations of Computer Science, Department of Computer Science, University of Edinburgh, August 1986.

99. "Models and Equality for Logical Programming," with José Meseguer, *Proceedings, TAPSOFT '87* (Pisa, Italy), edited by Hartmut Ehrig, Robert Kowalski, Giorgio Levi and Ugo Montanari, Springer, Lecture Notes in Computer Science, Volume 250, March 1987, pages 1–22; also, Report CSLI-87-91, Center for the Study of Language and Information, Stanford University, March 1987; and reprinted in *Mathematical Logic in Programming*, Mir (Moscow), 1991, pages 274–310.

100. "Parameterized Programming in OBJ2," with Kokichi Futatsugi, José Meseguer and Koji Okada, *Proceedings, 9th International Conference on Software Engineering* (Monterey, California), IEEE Computer Society, March 1987, pages 51–60.

101. "The Rewrite Rule Machine Project," Introduction to a Special Session on the Rewrite Rule Machine, in *Proceedings, Second International Supercomputing Conference, Volume I*, page 195, May 1987; also, in Technical Report SRI–CSL–87–1, Computer Science Laboratory, SRI International, May 1987.

102. "Architectural Options for the Rewrite Rule Machine," with Sany Leinwand, in *Proceedings, Second International Supercomputing Conference, Volume III*, May 1987, pages 63–70; also, in Technical Report SRI–CSL–87–1, Computer Science Laboratory, SRI International, May 1987.

103. "OBJ as a Language for Concurrent Programming," with José Meseguer, Claude Kirchner and Timothy Winkler, in *Proceedings, Second International Supercomputing Conference, Volume I*, May 1987, pages 196–198; also, in Technical Report SRI–CSL–87–1, Computer Science Laboratory, SRI International, May 1987.

104. "Simulation of Concurrent Term Rewriting," with Sany Leinwand and Timothy Winkler, in *Proceedings, Second International Supercomputing Conference, Volume I*, May 1987, pages 199–208; also, in Technical Report SRI–CSL–87–1, Computer Science Laboratory, SRI International, May 1987.

105. "Graphical Programming by Generic Example," in *Proceedings, Second International Supercomputing Conference, Volume I*, May 1987, pages 209–216; also, in Technical Report SRI–CSL–87–1, Computer Science Laboratory, SRI International, May 1987.

106. "Order-Sorted Algebra Solves the Constructor-Selector, Multiple Representation and Coercion Problems," with José Meseguer; in *Proceedings, Second Logic in Computer Science Conference*, IEEE Computer Society, June 1987, pages 18–29; also, Report CSLI-87-92, Center for the Study of Language and Information,

Stanford University, March 1987; revised version in *Information and Computation*, *103*, 1993.

107. "Final Algebras, Cosemicomputable Algebras, and Degrees of Unsolvability" with José Meseguer and Lawrence Moss, *Theoretical Computer Science, 100*, 1992, pages 267–302; preliminary version in *Proceedings*, Second Conference on Category Theory and Computer Science (University of Edinburgh, 7-9 September 1987), ed. D.H. Pitt *et al.*, Lecture Notes in Computer Science, volume 283, 1987, pages 158–181; also, Report CSLI-87-82, Center for the Study of Language and Information, Stanford University, March 1987.

108. "Unifying Functional, Object-Oriented and Relational Programming with Logical Semantics," with José Meseguer, in *Research Directions in Object-Oriented Programming*, edited by Bruce Shriver and Peter Wegner, MIT, pages 417–477, 1987; also, Report CSLI-87-93, Center for the Study of Language and Information, Stanford University, March 1987; and Technical Report SRI–CSL–87–7, Computer Science Laboratory, SRI International, July 1987, and reprinted by MIR (Moscow).

109. "Concurrent Term Rewriting as a Model of Computation," with Claude Kirchner and José Meseguer; in *Proceedings, Workshop on Graph Reduction* (Santa Fe, New Mexico), edited by Robert Keller and Joseph Fasel, Springer, Lecture Notes in Computer Science, Volume 279, October 1987, pages 53–93; also, Technical Report SRI–CSL–87–2, Computer Science Laboratory, SRI International, May 1987.

110. "Formalization in Programming Environments," with Mark Moriconi, *Computer*, Volume 20, Number 11, November 1987, pages 55–64.

111. "An Introduction to OBJ3", with Claude Kirchner, Hélène Kirchner, Aristide Mégrelis and José Meseguer, in *Proceedings*, Conference on Conditional Term Rewriting, edited by Jean-Pierre Jouannaud and Stephane Kaplan, Springer, Lecture Notes in Computer Science, Volume 308, 1988, pages 258–263.

112. "Cell and Ensemble Architecture for the Rewrite Rule Machine," with Sany Leinwand and Timothy Winkler, *Proceedings, International Conference on Fifth Generation Computer Systems* (Tokyo, Japan), ICOT, November 1988, pages 869–878.

113. "Software for the Rewrite Rule Machine," with José Meseguer, *Proceedings, International Conference on Fifth Generation Computer Systems* (Tokyo, Japan), ICOT, November 1988, pages 628–637.

114. "Modular Algebraic Specification of Some Basic Geometrical Constructions," *Artificial Intelligence*, volume 37, pages 123–153, December 1988, Special Issue on Computational Geometry, edited by Deepak Kapur and Joseph Mundy; also, Report CSLI-87-87, Center for the Study of Language and Information, Stanford University, March 1987.

115. "Rank and Status in the Cockpit: Some Linguistic Consequences of Crossed Hierarchies," with Charlotte Linde, Elisa Finnie, Susannah MacKaye and Michael Wescoat; in *Proceedings*, Conference on New Ways to Analyze Variation (NWAVE), edited by John Rickford, Stanford University, 1988, pages 300–311.

116. "OBJ as a Theorem Prover with Application to Hardware Verification," Chapter 5 in *Current Trends in Hardware Verification and Automated Theorem proving*, eds. P.S. Subramanyam and Graham Birtwhistle, Springer, 1989, pages 218–267; also Technical Report SRI–CSL–88–4R2, Computer Science Lab, SRI International, August 1988.

117. "Introducing OBJ," with José Meseguer, Timothy Winkler, Kokichi Futatsugi and Jean-Pierre Jouannaud, Technical Report SRI–CSL–88–8, Computer Science Lab, SRI International, August 1988.

118. "Principles of Parameterized Programming," in *Software Reusability, Volume I: Concepts and Models*, edited by Ted Biggerstaff and Alan Perlis, Addison-Wesley (ACM, Frontier Series), 1989, pages 159–225.

119. "What is Unification? — A Categorical View of Substitution, Equation and Solution," in *Resolution of Equations in Algebraic Structures, Volume I: Algebraic Techniques*, edited by Maurice Nivat and Hassan Aït-Kaci, Academic, 1989, pages 217–261; also Technical Report SRI–CSL–88–2R2, Computer Science Lab, SRI International, January 1988, and Center for the Study of Language and Information, Stanford University, 1988.

120. "Order-Sorted Equational Computation," with Gert Smolka, Werner Nutt and José Meseguer, in *Resolution of Equations in Algebraic Structures, Volume 2: Rewriting Techniques*, edited by Maurice Nivat and Hassan Aït-Kaci, Academic, 1989, pages 299–367; preliminary version in *Proceedings*, Colloquium on Resolving Equational Systems (MCC, Lakeway, Texas, May 1987); also appears as SEKI Report SR-87-14, Universität Kaiserslautern, December 1987.

121. "Order-Sorted Unification," with José Meseguer and Gert Smolka, *Journal of Symbolic Computation*, 8, 1990, pages 383–413; preliminary version in *Proceedings*, Colloquium on Resolving Equational Systems (MCC, Lakeway, Texas, May 1987); also, Report CSLI-87-86, Center for the Study of Language and Information, Stanford University, March 1987.

122. "Parameterized Programming and its Application to Rapid Prototyping in OBJ2," with Kokichi Futatsugi, José Meseguer and Koji Okada, in *Japanese Perspectives in Software Engineering*, edited by Yoshihiro Matsumoto and Yutaka Ohno, Addison-Wesley, 1989, pages 77–102.

123. "Hyperprogramming: A Formal Approach to Software Environments," in *Proceedings, Symposium on Formal Methods in Software Development*, Joint System Development Corporation, Tokyo, Japan, January 1990.

124. "An Algebraic Approach to Refinement," in *Proceedings, VDM'90*, edited by Dines Bjorner, C.A.R. Hoare and H. Langmaack, Springer, Lecture Notes in Computer Science, Volume 428, 1990, pages 12–28.

125. "Higher-Order Functions Considered Unnecessary for Higher-Order Programming," in *Research Topics in Functional Programming*, edited by David Turner, University of Texas at Austin Year of Programming Series, Addison-Wesley, 1990, pages 309–352; also Technical Report SRI–CSL–88–1, Computer Science Lab, SRI International, January 1988.

126. "Proving and Rewriting," in *Proceedings, Algebraic and Logic Programming*, Nancy, France, 1–3 October 1990, Springer, Lecture Notes in Computer Science Volume 463, 1990, pages 1–24.

127. "A Categorical Manifesto," in *Mathematical Structures in Computer Science*, Volume 1, Number 1, March 1991, pages 49–67; also, Programming Research Group Technical Monograph PRG–72, Oxford University Computing Lab, March 1989.

128. "A Categorial Theory of Objects as Observed Processes," with Hans-Dieter Ehrich and Amilcar Sernadas, in *Foundations of Object Oriented Languages*, Proceedings of a REX/FOOL Workshop (Noordwijkerhout, the Netherlands, May/June 1990), edited by J.W. de Bakker, W.P. de Roever and G. Rozenberg, Springer Lecture Notes in Computer Science, Volume 489, 1991, pages 203–228.

129. "OOZE: An Object Oriented Z Environment," with Antonio Alencar, in *Proceedings, 1991 European Conference on Object Oriented Programming*, edited by Pierre America, Springer Lecture Notes in Computer Science, Volume 512, 1991, pages 180–199.

130. "Types as Theories," in *Topology and Category Theory in Computer Science*, edited by George Michael Reed, Andrew William Roscoe and Ralph F. Wachter, Oxford, 1991, pages 357–390.

131. "Compiling Concurrent Rewriting onto the Rewrite Rule Machine," with Hitoshi Aida and José Meseguer, in *Conditional and Typed Rewriting Systems*, Stéphane Kaplan and Mitsuhiro Okada, editors, Lecture Notes in Computer Science, Volume 516, Springer, 1991, pages 320–332; also, Technical Report SRI–CSL–90–03, SRI International, Computer Science Laboratory, February 1990.

132. "Semantic Specifications for the Rewrite Rule Machine," in *Concurrency: Theory, Language and Architecture*, Proceedings of a U.K.-Japan Workshop (Oxford, September 1989), edited by Aki Yonezawa and Takayasu Ito, Springer, Lecture Notes in Computer Science, Volume 491, 1991, pages 216–234.

133. "FUNNEL: A CHDL with Formal Semantics," with Victoria Stavridou, Steven Eker, and Serge Aloneftis, in *Proceedings, Advanced Research Workshop on Correct Hardware Design Methodologies* (Turin, 12-14 June), IEEE Computer Society, 1991, pages 117–144.

134. "On Types and FOOPS," with David Wolfram, in *Object Oriented Databases: Analysis, Design and Construction*, Robert Meersman, William Kent and Samit Khosla (eds), (Proceedings, IFIP TC2 Conference, Windemere, UK, 2-6 July 1990) North Holland, 1991, pages 1–22.

135. "Some Fundamental Algebraic Tools for the Semantics of Computation, Part 3: Indexed Categories," with Rod Burstall and Andrzej Tarlecki, *Theoretical Computer Science 91*, 1991, pages 239–264; also, Programming Research Group Monograph PRG–77, Oxford University Computing Lab, August 1989.

136. "Institutions: Abstract Model Theory for Specification and Programming," with Rod Burstall, *Journal of the Association for Computing Machinery*, Volume 39, Number 1, January 1992, pages 95–146; also, Report ECS-LFCS-90-106, Department of Computer Science, University of Edinburgh, January 1990.

137. "Hermeneutics and Path," in *Software Development and Reality Construction*, edited by Christiane Floyd, Heinz Züllighoven, Reinhard Budde and Reinhard Keil-Slawik (papers from a conference held at Schloss Eringerfeld, Germany, October 1988), 1992, pages 39–44, Springer.

138. "The Denial of Error," in *Software Development and Reality Construction*, edited by Christiane Floyd, Heinz Züllighoven, Reinhard Budde and Reinhard Keil-Slawik (papers from a conference held at Schloss Eringerfeld, Germany, October 1988), Springer, 1992, pages 193–202.

139. "Truth and Meaning beyond Formalism," in *Software Development and Reality Construction*, edited by Christiane Floyd, Heinz Züllighoven, Reinhard Budde and Reinhard Keil-Slawik (papers from a conference held at Schloss Eringerfeld, Germany, October 1988), Springer, 1992, pages 353–362.

140. "The Dry and the Wet," in *Information Systems Concepts: Improving the Understanding*, Proceedings of IFIP Working Group 8.1 Conference (Alexandria Egypt), Elsevier North-Holland, 1992, pages 1–17; also, Technical Monograph PRG–100, March 1992, Oxford University Computing Lab.

141. "2OBJ, a Metalogical Framework based on Equational Logic," with Andrew Stevens, Keith Hobley and Hendrik Hilberdink, in *Philosophical Transactions of the Royal Society, Series A*, Volume 339, 1992, pages 69–86; also, in *Mechanized Reasoning and Hardware Design*, edited by C.A.R. Hoare and M.J.C. Gordon, Prentice-Hall, 1992, pages 69–86.

142. "Sheaf Semantics for Concurrent Interacting Objects," *Mathematical Structures in Computer Science*, Volume 2, 1992, pages 159–191; draft appears as Report CSLI-91-155, Center for the Study of Language and Information, Stanford University, June 1991.

143. "FUNNEL and 2OBJ: towards an Integrated Hardware Design Environment," with Victoria Stavridou, Andrew Stevens, Steven Eker, Serge Aloneftis and Keith Hobley, in *Theorem Provers in Circuit Design*, edited by Victoria Stavridou, Tom Melham and Raymond Boute, North-Holland, IFIP Transactions A-10, 1992, pages 197–223.

144. "A Sheaf Semantics for FOOPS Expressions," with David Wolfram, *Object-Based Concurrent Computing*, Mario Tokoro, Oscar Nierstrasz, Peter Wegner, editors, Proceedings of ECOOP'91 Workshop (Geneva, July 1991), Springer, 1992, pages 81–98.

145. "Order-Sorted Algebra I: Equational Deduction for Multiple Inheritance, Polymorphism, and Partial Operations," with José Meseguer, *Theoretical Computer Science, 105*, Number 2, November 1992, pages 217–273; also Technical Monograph PRG–80, December 1989, Oxford University Computing Lab; earlier version, Technical Report SRI–CSL–89–10, Computer Science Lab, SRI International, July 1989.

146. "OOZE," with Antonio Alencar, in *Object Orientation in Z*, edited by Susan Stepney, Rosalind Barden and David Cooper, Springer, Workshops in Computing, 1992, pages 79–94; also appeared as "Two Examples in OOZE," Technical Report PRG–TR–25–91, Programming Research Group, Oxford University Computing Lab, 1991.

147. "A Short Oxford Survey of Order Sorted Algebra," with Răzvan Diaconescu, in *Bulletin of the European Association for Computer Science*, column on Algebraic Specification, Volume 48, pages 121–133, October 1992. Also in *Current Trends in Theoretical Computer Science: Essays and Tutorials*, edited by Grigor Rozenberg and Arto Salomaa, World Scientific, 1993, pages 209–221.

148. "Simulation and Performance Estimation for the Rewrite Rule Machine," with Hitoshi Aida, Sany Leinwand, José Meseguer, Babak Taheri and Timothy Winkler, in *Proceedings of the Fourth Symposium on the Frontiers of Massively Parallel Computation* (MacLean VA, October 1992), 1992, IEEE Computer Society, pages 336–344; also, Technical Report, SRI International, Computer Science Laboratory.

149. "Techniques for Requirements Elicitation," with Charlotte Linde, in *Proceedings, Requirements Engineering '93*, edited by Stephen Fickas and Anthony Finkelstein, IEEE Computer Society, 1993, pages 152–164; reprinted in *Software Requirements Engineering, Second Edition*, ed. Richard Thayer and Merlin Dorfman, IEEE Computer Society, 1997.

150. "Social Issues in Requirements Engineering," in *Proceedings, Requirements Engineering '93*, edited by Stephen Fickas and Anthony Finkelstein, IEEE Computer Society, 1993, pages 194–195.

151. "On Notation (A Sketch of the Paper)," in *TOOLS 10: Technology of Object-Oriented Languages and Systems*, edited by Boris Magnusson, Bertrand Meyer and Jean-Francois Perrot, Prentice-Hall, 1993, pages 5–10.

152. "Order-Sorted Algebra Solves the Constructor Selector, Multiple Representation and Coercion Problems," with José Meseguer, *Information and Computation*, Volume 103, Number 1, March 1993, pages 114–158.

153. "Specification in OOZE with Examples," with Antonio Alencar, in *Object Oriented Specification Case Studies*, edited by Kevin Lano and Howard Haughton, Prentice Hall, 1993, pages 158–183; also, Technical Report PRG–TR–7–92, Programming Research Group, Oxford University Computing Lab, 1992.

154. "Logical Support for Modularisation," with Răzvan Diaconescu and Petros Stefaneas, in *Logical Environments*, edited by Gerard Huet and Gordon Plotkin (Proceedings of a Workshop held in Edinburgh, May 1991), Cambridge, 1993, pages 83–130.

155. "An Approach to Situated Adaptive Software," in *Proceedings, International Workshop on New Models of Software Architecture*, Kanazawa, Japan, November 8, 1993, pages 7–20.

156. "Introduction" to *Formal Methods in System Engineering*, edited by Peter Ryan and Chris Sennett, Springer, 1993, pages 1–5.

157. "Proof of Correctness of Object Representations," with Grant Malcolm, in *A Classical Mind: Essays in Honour of C.A.R. Hoare*, edited by A. William Roscoe, Prentice-Hall, 1994, pages 119–142.

158. "Introduction," with Marina Jirotka, to *Requirements Engineering: Social and Technical Issues*, edited with Marina Jirotka, Academic, 1994, pages 1–13.

159. "Requirements Engineering as the Reconciliation of Social and Technical Issues," in *Requirements Engineering: Social and Technical Issues*, edited with Marina Jirotka, Academic, 1994, pages 165–199.

160. "Some Suggestions for Progress in Software Analysis, Synthesis and Certification," with Luqi, *Proceedings of the Sixth International Conference on Software Engineering and Knowledge Engineering*, Jurmala, Latvia, 20–23 June, 1994, pages 501–507. An earlier version appeared as "Some Suggestions for Using Formal Methods in Software Development," in *Proceedings, Workshop on Increasing the Practical Impact of Formal Methods for Computer-Aided Software Development: Software Slicing, Merging and Integration*, Naval Postgraduate School, Monterey, California, 13–15 October 1993, pages 7–11.

161. "Towards an Algebraic Semantics for the Object Paradigm," with Răzvan Diaconescu, in *Recent Trends in Data Type Specification*, Proceedings of the 9th Workshop on Abstract Data Types (Caldes de Malevella, Spain, 28 October 1992), edited by Hartmut Ehrig and Fernando Orejas, Springer, Lecture Notes in Computer Science, Volume 785, 1994, pages 1–29.

162. "Formal Support for Software Evolution," with Luqi and Valdis Berzins, in *Proceedings, Workshop on Increasing the Practical Impact of Formal Methods for Computer-Aided Software Development: Software Evolution*, Naval Postgraduate School, Monterey, California, 7–9 September 1994, pages 10–21.

163. "An Oxford Survey of Order Sorted Algebra," with Răzvan Diaconescu, *Mathematical Structures in Computer Science*, volume 4, pages 363–392, 1994.

164. "Towards a Provably Correct Compiler for OBJ3," with Lutz Hamel, *Proceedings, Conference on Programming Language Implementation and Logic Programming* (Madrid, Spain, 14–16 September 1994), edited by Manuel Hermenegildo and Jaan Penjam, Springer, Lecture Notes in Computer Science, volume 844, pages 132–146, 1994.

165. "An Operational Semantics for FOOPS," with Paulo Borba, in *Proceedings, International Workshop on Information Systems – Correctness and Reusability, IS-CORE'94*, edited by Roel Wieringa and Remco Feenstra, Vrije Universiteit, 1994, pages 271–285; held September 1994 in Amsterdam.

166. "Module Composition and System Design for the Object Paradigm," with Adolfo Socorro, *Journal of Object Oriented Programming*, volume 7, number 9, pages 47–55, February 1995; also Technical Monograph PRG–117, January 1995, Oxford University Computing Lab.

167. "Engineering Support for Software Evolution," with Luqi, Valdis Berzins and David Dampier, in *Proceedings, Computers in Engineering Symposium*, Houston TX, 29 January–1 February 1995.

168. "Formal Methods and Social Context in Software Development," with Luqi, in *Proceedings, Sixth International Joint Conference on Theory and Practice of Software Development* (TAPSOFT 95, Aarhus, Denmark), edited by Peter Mosses, Mogens Nielsen and Michael Schwartzbach, Springer, Lecture Notes in Computer Science, Volume 915, pages 62–81, 1995.

169. "An Introduction to Category-based Equational Logic," with Răzvan Diaconescu, in *Proceeedings, AMAST 95* (Montreal), edited by V.S. Alagar and Maurice Nivat, Springer, Lecture Notes in Computer Science, volume 936, pages 91–126, 1995.

170. "Situated Adaptive Software: beyond the Object Paradigm," with Grant Malcolm, in *Proceeedings, International Symposium on New Models of Software Architecture* (Tokyo), Information-Technology Promotion Agency, pages 126–142, 1995.

171. "Parameterized Programming and Software Architecture," in *Proceedings, 1995 Monterey Workshop on Increasing the Practical Impact of Formal Methods for Computer-Aided Software Development: Specification-Based Software Architectures*, Naval Postgraduate School, Monterey, California (12–14 September 1995), pages 58–70, February 1996.

172. "An Object-Oriented Tool for Tracing Requirements," with Francisco Pinheiro, in *IEEE Software* (special issue of papers from International Conference on Requirements Engineering '96), pages 52–64, March 1996.

173. "Software Component Search," with Doan Nguyen, José Meseguer, Luqi, Du Zhang and Valdis Berzins, *Journal of Systems Integration*, Volume 6, Number 1, pages 93–134, 1996; special issue on computer-aided prototyping.

174. "Formality and Informality in Requirements Engineering," *Proceedings*, Fourth International Conference on Requirements Engineering, IEEE Computer Society, pages 102–108, April 1996. Colorado Springs CO. Keynote address.

175. "Parameterized Programming and Software Architecture," in *Proceedings*, Fourth International Conference on Software Reuse, IEEE Computer Society, pages 2–11, April 1996. Orlando FL. Keynote address.

176. "Semantics of Non-terminating Rewrite Systems using Minimal Coverings," with José Barros, in *Computer Science Logic*, Proceedings of the 9th International Workshop (Padderborn, Germany, 22–29 September 1995), edited by Hans Kleine Büning, Lecture Notes in Computer Science, volume 1092, pages 16–35, Springer, 1996; also Technical Monograph PRG–118, March 1995, Oxford University Computing Lab.

177. "An Executable Course in the Algebraic Semantics of Imperative Programs," with Grant Malcolm, in *Teaching and Learning Formal Methods*, edited by Michael Hinchey and C. Nevill Dean, Academic Press, 1996, pages 161–179.

178. "Extended Abstract of a Hidden Agenda," with Grant Malcolm, Proceedings, 1996 Conference *Intelligent Systems: A Semiotic Perspective, Volume I*, edited by John Albus, Alex Meystel and Richard Quintero, National Inst. of Standards and Technology (Gaithersberg MD, October 20–23), 1996, pages 159–167.

179. "Formal Methods: Promises and Problems," with Luqi, *IEEE Software*, January 1997, pages 73–85.

180. "Information Engineering and Industrial Relevance for International Manufacturing Enterprises," with Mohan Trivedi, *Proceedings, 1997 NSF Design and Manufacturing Grantees Conference*, January 1967, pages 601–602.

181. "Towards a Social, Ethical Theory of Information," in *Social Science, Technical Systems and Cooperative Work: Beyond the Great Divide*, edited by Geoffrey Bowker, Leigh Star, William Turner and Les Gasser, Erlbaum, 1997, pages 27–56.

182. "Extended Abstract of 'Semiotic Morphisms'," in *Intelligent Systems: A Semiotic Perspective, Volume II*, edited by John Albus, Alex Meystel and Richard Quintero, National Institute of Standards and Technology (Gaithersberg MD, 20–23 October 1996), pages 26–31.

183. "Refinement of Concurrent Object-Oriented Programs," with Paulo Borba, in *Formal Methods and Object Technology*, edited by Stephen Goldsack and Stuart Kent, Springer, 1996, chapter 11; proceedings of *BCS/FACS Workshop on Formal Aspects of Object-oriented Programming* (London, December 1993).

184. "Algebraic Semiotics, ProofWebs and Distributed Cooperative Proving," with Akira Mori and Kai Lin, in *UITP'97, User Interfaces for Theorem Provers*, ed. Yves Bertot, (Sophia Antipolis, 1–2 September 1997), INRIA (France), pages 24–34, 1997.

185. "Distributed Cooperative Formal Methods Tools," with Kai Lin, Akira Mori, Grigore Roşu and Akiyoshi Sato, *Proceedings, Automated Software Engineering* (Lake Tahoe, 3–5 November 1997), pages 55–62.

186. "Stretching First Order Equational Logic: Proofs with Partiality, Subtypes and Retracts," *Proceedings, First Order Theorem Proving '97* (Schloss Hagenberg, Austria, 27–28 October 1997), RISC-Linz Report No. 95–50, Johannes Kepler Univ. Linz, pages 78–85.

187. "Tools for Distributed Cooperative Design and Validation," with Kai Lin, Akira Mori, Grigore Roşu and Akiyoshi Sato, in *Proceedings, CafeOBJ Symposium*, Japan Institute of Science and Technology (Nomuzu, Japan, April 1998).

188. "Hidden Congruent Deduction," with Grigore Roşu, in *Proceedings, First-Order Theorem Proving — FTP'98*, edited by Ricardo Caferra and Gernot Salzer, Technische Universität Wien, pages 213–223, 1998.

189. "An Introduction to Algebraic Semiotics, with Applications to User Interface Design," edited by Chrystopher Nehaniv, *Computation for Metaphors, Analogy and Agents*, University of Aizu, Aizu–Wakamatsu, Japan, 1998, pages II: 54–79.

190. "A Hidden Herbrand Theorem," with Grant Malcolm and Tom Kemp, in *Principles of Declarative Programming*, Catuscia Palamidessi, Hugh Glaser and Karl Meinke (eds), Springer Lecture Notes in Computer Science, Volume 1490, pages 445–462, 1998. Proceedings of PLIP '98, Pisa, Italy.

191. "Tossing Algebraic Flowers down the Great Divide," in *People and Ideas of Theoretical Computer Science*, edited by Christian Calude, Springer, series in Discrete Mathematics and Theoretical Computer Science, pages 93–129, 1998.

192. "Hidden Algebra for Software Engineering," in *Combinatorics, Computation & Logic*, Proceedings of Conference on Discrete Mathematics and Theoretical Computer Science, (University of Auckland, New Zealand, 18–21 January 1999), edited by Christian Calude and Michael Dinneen, Australian Computer Science Communications, Volume 21, Number 3, Springer, pages 35–59, 1999.

193. "Signs and Representations: Semiotics for User Interface Design," with Grant Malcolm, in *Visual Representations and Interpretations*, edited by Ray Paton and Irene Nielson, Proceedings of a workshop held in Liverpool, Springer, 1999, pages 163–172.

194. "An Introduction to Algebraic Semiotics, with Application to User Interface Design," in *Computation for Metaphors, Analogy and Agents*, edited by Chrystopher Nehaniv, Springer Lecture Notes in Artificial Intelligence, Volume 1562, 1999, pages 242–291.

195. "Introduction," to special issue of *Journal of Consciousness Studies* on Art and the Brain, vol. 6, no. 6/7, June/July 1999, pages 5–14; also appears in edited book, *Art and the Brain*, Imprint Academic, October 1999.

196. "Hidden Coinduction: Behavioral Correctness Proofs for Objects," with Grant Malcolm, in *Mathematical Structures in Computer Science, 9*, number 3, pages 287–319, 1999.

197. "Hidden Algebraic Engineering," in *Semigroups and Algebraic Engineering*, edited by Chrystopher Nehaniv and Masami Ito, World Scientific, 1999, pages 17–36 (proceedings of conference "Algebraic Engineering and Semigroups," Aizu-Wakamatsu, Japan, 24–26 March 1997). Also UCSD CSE Technical Report CS97–569, December 1997.

198. "Hiding More of Hidden Algebra," with Grigore Roşu, in *FM'99 – Formal Methods '99*, Lecture Notes in Computer Sciences, Volume 1709, Proceedings of World Congress on Formal Methods (Toulouse, France), Springer, pages 1704–1719, 1999.

199. "Hiding More of Hidden Algebra," with Grigore Roşu, in *OBJ/CafeOBJ/Maude at Formal Methods '99*, edited Kokichi Futatsugi, Joseph Goguen, and Josè Meseguer, Theta (Bucharest), September 1999, pages 35–48.

200. "Social and Semiotic Analyses for Theorem Prover User Interface Design," *Formal Aspects of Computing*, volume 11, pages 272–301, 1999.

201. "A Protocol for Distributed Cooperative Work," with Grigore Roşu, in *Proceedings, Workshop on Distributed Systems* (Iasi, Romania, September 1999), in *Electronic Notes in Computer Science*, Volume 28, Elsevier, 1999, pages 1–22.

202. "Hidden Algebra and Concurrent Distributed Software," with Kai Lin and Grigore Roşu, in *Software Engineering Notes, 25*, number 1, page 51, 2000.

203. "Hidden Congruent Deduction," with Grigore Roşu, in *Automated Deduction in Classical and Non-Classical Logics*, edited by Ricardo Caferra and Gernot Salzer, Lecture Notes in Artificial Intelligence, Volume 1761, pages 252–267, 2000.

204. "On Equational Craig Interpolation," with Grigore Roşu, *Journal of Universal Computer Science*, special issue in honor of Prof. Rudeanu, edited by Christian Calude, Volume 6, No. 1, pages 194–200, February 2000. http://www.jucs.org/jucs_6_1/.

205. "An Implementation-Oriented Semantics for Module Composition," with William Tracz, in *Foundations of Component-based Systems*, edited by Gary Leavens and Murali Sitaraman, Cambridge, 2000, pages 231–263.

206. "Introduction," with Grant Malcolm, in *Software Engineering with OBJ: Algebraic Specification in Action*, book edited by Joseph Goguen and Grant Malcolm, Kluwer, 2000, pages 397–408.

207. "Introducing OBJ," with José Meseguer, Timothy Winkler, Kokichi Futatsugi and Jean-Pierre Jouannaud, in *Software Engineering with OBJ: Algebraic Specification in Action*, book edited by Joseph Goguen and Grant Malcolm, Kluwer, 2000, pages 3–167; revised and extended August 1999 version of Technical Report SRI–CSL–88–8, Computer Science Lab, SRI International, August 1988.

208. "More Higher Order Programming in OBJ3," with Grant Malcolm, in *Software Engineering with OBJ: Algebraic Specification in Action*, book edited by Joseph Goguen and Grant Malcolm, Kluwer, 2000, pages 397–408.

209. "Semantic Specification for the Rewrite Rule Machine," in *Software Engineering with OBJ: Algebraic Specification in Action*, book edited by Joseph Goguen and Grant Malcolm, Kluwer, 2000, pages 283–306.
210. "What is Art?," introduction to special issue of *Journal of Consciousness Studies*, Art and the Brain (Part 2), vol. 7, no. 8/9, September 2000, pages 7–15; also in book version, *Art and the Brain (Part 2)*, Imprint Academic, September 2000.
211. "Behavioral and Coinductive Rewriting," with Kai Lin and Grigore Roşu, in *Electronic Notes on Theoretical Computer Science*, edited by Kokichi Futatsugi, Volume 36, Elsevier, 2001. Available at http://www.elsevier.com/locate/entcs/volume36.html. Also in *Proceedings, Rewriting Logic Workshop, 2000*, edited by Kokichi Futatsugi, JAIST, Kanazawa, Japan, September 2000, pages 1–22.
212. "A Hidden Agenda," with Grant Malcolm, *Theoretical Computer Science*, volume 245, number 1, August 2000, pages 55–101, special issue on Algebraic Engineering, edited by Chrystopher Nehaniv and Masami Ito; original version, UCSD CSE Technical Report CS97-538, May 1997.
213. "An Overview of the Tatami Project," with Kai Lin, Grigore Roşu, Akira Mori and Bogdan Warinschi, in *CAFE: An Industrial-Strength Algebraic Formal Method*, edited by Kokichi Futatsugi, Ataru Nakagawa and Tetsuo Tamai, Elsevier, September 2000, pages 61–78.
214. "Circular Coinductive Rewriting," with Kai Lin and Grigore Roşu, *Proceedings, Automated Software Engineering '00*, (Grenoble), IEEE, September 2000, pages 123–131.
215. "Web-based Multimedia Support for Distributed Cooperative Software Engineering," with Kai Lin, in *Proceedings, Multimedia Software Engineering*, edited by Jeffrey Tasi (Taiwan, December 2000), IEEE, pages 25–32.
216. "Towards a Design Theory for Virtual Worlds: Algebraic semiotics and scientific visualization as a case study," in *Proceedings, Conference on Virtual Worlds and Simulation* (Phoenix AZ, 7–11 January 2001), edited by Christopher Landauer and Kirstie Bellman, Society for Modelling and Simulation, 2001, pages 298–303.
217. "Circular Coinduction," with Grigore Roşu, in *Proceedings, International Joint Conference on Automated Deduction*, Sienna, June 2001.
218. "Web-based Support for Cooperative Software Engineering," with Kai Lin, *Annals of Software Engineering, 12*, pages 167–191, 2001. Special issue on multimedia software engineering, edited by Jeffrey Tsai.
219. "A Hidden Herbrand Theorem: Combining the Object, Logic and Functional Paradigms," with Grant Malcolm and Tom Kemp, *Journal of Logic and Algebraic Programming 51*, number 1–2, 2002, pages 1–41.
220. "Institution Morphisms," with Grigore Roşu, *Formal Aspects of Computing 13*, 2002, pages 274–307; special issue in honor of Rod Burstall, edited by Don Sannella.
221. "Groundlessness, Compassion and Ethics in Management and Design," in *Proceedings, Managing as Design Workshop*, Case Western Reserve University, 2002; from a meeting hold in Cleveland, 14-15 June 2002.
222. "A Metadata Integration Assistant Generator for Heterogeneous Distributed Databases," with Young-Kwang Nam and Guilian Wang, in *Proceedings, International Conference on Ontologies, DataBases, and Applications of Semantics for Large Scale Information Systems*, Springer, Lecture Notes in Computer Science, Volume 2519, 2002, pages 1332–1344. Conference held in Irvine CA, 29–31 October 2002.
223. "Consciousness Studies," *Encyclopedia of Science and Religion, vol. 1*, J. Wentzel Vrede van Huyssteen, editor, McMillan Reference USA, 2003, pages 158–164.

224. "A Metadata Tool for Retrieval from Heterogeneous Distributed XML Documents," with Young-Kwang Nam and Guilian Wang, in *Proceedings, International Conference on Computational Science*, edited by P.M.A. Sloot and others, Springer, Lecture Notes in Computer Science, Volume 2660, 2003, pages 1020–1029. Conference held in Melbourne, Australia, 2–4 June 2003.

225. "Semiotic Morphisms, Representations, and Blending for Interface Design," in *Proceedings, AMAST Workshop on Algebraic Methods in Language Processing*, edited by Fausto Spoto, Giuseppi Scollo and Anton Nijholt, pages 1–15, 2003. Conference held in Verona, Italy, 25–27 August, 2003.

226. "Semiotics, Compassion and Value-Centered Design," in *Proceedings, Organizational Semiotics Workshop*, University of Reading, UK (11–12 July 2003).

227. "Behavioral Verification of Distributed Concurrent Systems with BOBJ," with Kai Lin. *Proceedings, Conference on Quality Software*, Dallas TX, pages 216–235, IEEE Press, November 2003.

228. "Zero, Connected, Empty: An Essay after a Cantata," with Ryoko Goguen, in *Recent Trends in Algebraic Development Techniques*, edited by Martin Wirsing, Dirk Pattinson and Rolf Hennicker, Lecture Notes in Computer Science, volume 2755, Springer, 2003, pages 127–128. Workshop held in Frauenchiemsee, Germany, 24–27 September 2002.

229. "Conditional Circular Coinductive Rewriting," with Grigore Roşu and Kai Lin, in *Recent Trends in Algebraic Development Techniques*, edited by Martin Wirsing, Dirk Pattinson and Rolf Hennicker, Lecture Notes in Computer Science, volume 2755, Springer, 2003, pages 216–232. Workshop held in Frauenchiemsee, Germany, 24–27 September 2002.

230. Review of *Intellectual Impostures*, by Alan Sokal and Jean Bricmont (second edition), in (London) *Times Higher Education Supplement, No. 1,635*, April 2004, page 26.

231. "Information Visualization and Semiotic Morphisms," with D. Fox Harrell, in *Multidisciplinary Approaches to Visual Representations and Interpretations*, ed. Grant Malcolm, Elsevier, 2004, pages 93–106.

232. "Critical Points for Interactive Schema Matching," with Guilian Wang, Young-Kwang Nam, and Kai Lin. In *Proceedings, Sixth Asia Pacific Web Conference*, edited by Jeffrey Xu yu, Xuemin Lin, Hongjun Lu and YanChun Zhang, Springer, Lecture Notes in Computer Science, volume 3007, 2004, pages 654–664. Held in Hangzhou, China, April 14–17, 2004.

233. "Sync or Swarm: Group dynamics in musical free improvisation," with David Borgo, abstract in *Proceedings, Conference on Interdisciplinary Musicology*, European Society for the Cognitive Sciences of Music, 2004, pages 52–53; extended abstract in attached CD. Held Graz, Austria, 15–18 April 2004.

234. "Editorial Introduction to Art and the Brain III," with Erik Myin, in *Journal of Consciousness Studies, 11*, no. 3/4, pages 5–8, 2004.

235. "Musical Qualia, Context, Time, and Emotion," in *Journal of Consciousness Studies, 11*, no. 3/4, pages 117–147, 2004.

236. "Composing Hidden Information Modules over Inclusive Institutions," with Grigore Roşu. In *From Object-Orientation to Formal Methods: Essays in Honor of Johan-Ole Dahl*, edited by Olaf Owe, Soren Krogdahl and Tom Lyche, Lecture Notes in Computer Science, Volume 2635, Springer, pages 96–123, 2004.

237. "Against Technological Determinism," in *Scale*, volume 1, no. 5, pages 19–22, June 2004. Available at `http://scale.ucsd.edu`.

238. "Scritical Pairs and Sunification," with Monica Marcus, in *Proceedings, Eighteenth Workshop on Unification*, edited by Michael Kohlhase, pages 110–124. 2004. Held Cork, Ireland, 4–5 July 2004, as part of Second International Joint Conference on Automated Reasoning.

239. "Data, Schema and Ontology Integration," in *Proceedings of Workshop on Combination of Logics*, edited by Walter Carnielli, Miguel Dionisio and Paulo Mateus, Center for Logic and Computation, Dept. of Mathematics, Instituto Superior Tecnico, Lisbon, Portugal, pages 21–32, July 2004.

240. "Groundlessness, Compassion, and Ethics in Management and Design," in *Managing as Design*, edited by Richard Boland and Fred Callopy, Stanford, pages 129–136, 2004.

241. "Style as a Choice of Blending Principles," with Fox Harrell, in *Style and Meaning in Language, Art, Music and Design*, edited by Shlomo Argamon, Shlomo Dubnov, and Julie Jupp, proceedings of a Symposium at the 2004 AAAI Fall Symposium Series, AAAI Press, 2004, pages 49–56 (Washington DC, October 21-24).

242. "Steps towards a Design Theory for Virtual Worlds," in *Developing Future Interactive Systems*, edited by Maria-Isabel Sánchez-Segura, Idea Publishing Group, pages 116–152, 2004.

243. "Ontology, Society, and Ontotheology," in *Formal Ontology in Information Systems*, edited by Achille Varzi and Laure Vieu, IOS Press, pages 95–103, 2004. Proceedings of FOIS'04, Torino, Italy, 4–6 November 2004.

244. "Information Visualization and Semiotic Morphisms," with Fox Harrell, in *Multidisciplinary Approaches to Visual Representations and interpretations*, edited by Grant Malcolm, Elsivier, pages 83–98, 2004.

245. "What is a Logic?," with Till Mossakowski, Razvan Diaconescu and Andrzej Tarlecki. In *Logica Universalis*, ed. Jean-Yves Beziau (Birkhauser, 2005), pages 113–133. Proceedings of First World Conference on Universal Logic, Montreaux, Switzerland.

246. "Rivers of Consciousness: The nonlinear dynamics of free jazz," with David Borgo, *Jazz Research Proceedings Yearbook*, ed. Larry Fisher, IAJE Publications, pages 46–58, 2005.

247. "Three Perspectives on Information Integration," in *Proceedings, Seminar 04391, Semantic Interoperability and Integration*, held from 20 to 24 September 2004, at Schloss Dagstuhl, Germany; available at http://drops.dagstuhl.de/opus/volltexte/2005/38.

248. "What is a Concept?", in *Proceedings, 13th International Conference on Conceptual Structures (ICCS '05)*, ed. Frithjof Dau and Marie-Laure Mugnier, Springer Lecture Notes in Artificial Intelligence, volume 3596, pages 52–77, 2005; conference held 18–22 July, 2005, Kassel, Germany.

249. "November Qualia," four short poems, in *J. Consciousness Studies 12*, No. 11, November 2005, page 73.

250. "Verifying Design with Proof Scores," with Kokichi Futatsugi and Kazuhiro Ogata, in *Proceedings, Verified Software: Theories, Tools, Experiments*, 10–13 October 2005, Zürich.

251. "Specifying, Programming and Verifying with Equational Logic," with Kai Lin, in *We Will Show Them! Essays in honour of Dov Gabbay, Vol. 2*, edited by Sergei Artemov, Howard Barringer, Artur d'Avila Garcez, Luis Lamb and John Woods, College Publications 2005, pages 1–38.

Accepted for Publication

252. "Time, Structure, and Emotion in Music," with Ryoko Goguen, to appear in *Collected University Lectures 2003–04*, Keio University, Tokyo, Japan, 2005.
253. "Data, Schema, Ontology, and Logic Integration", to appear in *Journal of the IGPL 13*, no. 6, special issue on combining logics, edited by Walter Canielli, Miguel Dionisio and Paulo Mateus, Oxford, 2006.
254. "Information Integration in Institutions," to appear in memorial volume for Jon Barwise, edited by Larry Moss, University of Indiana Press.
255. "Against Technological Determinism," to appear in *CVS: Concurrency, Versioning and Systems*, edited by Jon Phillips, 2006.
256. "Consciousness and the Decline of Cognitivism," in *Distributed Collective Practices*, UCSD, Dept. of Communication, 2002. Proceedings of a conference held 6–9 February 2002, La Jolla CA, and to appear in a journal.
257. "First Order Equational Proofs with Partiality, Subtypes and Retracts," to appear in *Journal of Symbolic Computation*.
258. "Software Architecture by Parameterized Programming," under revision for *Transactions on Software Engineering*.

Submitted for Publication

259. "Query Generation for Retrieving Data from Distributed Semistructured Documents using a Metadata Interface," with Guija Choe, Young-Kwang Nam, and Guilian Wang, submitted to *Computer Languages, Systems and Structures*, 2006.
260. "Foundations for Active Multimedia Narrative: Semiotic spaces and Structural Blending," with Fox Harrell, to appear in *Interaction Studies*.
261. "Formal verification with the OTS/CafeOBJ Method," with Kokichi D Futatsugi and Kazuhiro Ogata, in *Journal of Object Technology*, submitted; online at http://www.jot.fm/.
262. "Are Agents an Answer or a Question?"

In Preparation

263. "Style as a Choice of Blending Principles," with Fox Harrell, to appear in *Computation and Style*, edited by Shlomo Argamon, Shlomo Dubnov, and Kevin Burns, Springer, 2006.
264. "Style as a Choice of Blending Principles," with Fox Harrell, German translation of revised version to be published in special issue of *Semiotik*, devoted to papers from the 2005 annual meeting of the Deutsche Gesellshaft für Semiotik.
265. "Principles for Schema Matching Tool User Interface Design," with Guilian Wang, Vitaliy Zavesov, Kai Lin, and Young-Kwang Nam.
266. "Reality and Human Values in Mathematics."
267. "Theory and Practice of Proof Scores," with Kokichi Futatsugi and Kazuhiro Ogata.
268. "Abstract Schema Morphisms and Schema Matching Generation," with Guilian Wang, Young-Kwang Nam, and Kai Lin.
269. "A Critical Pair Theory for Rewriting Modulo Equations," with Monica Marcus.

270. "An induction Scheme in Higher Order Parameterized Programming," with Kai Lin.
271. "Morphisms and Semantics for Higher Order Parameterized Programming," with Kai Lin.
272. "Towards the Automation of Behavioral Reasoning," with Grigore Roşu.
273. "Semiotic Redesign of a Computer Language," with Kokichi Futatsugi.
274. "Models and Equality for Logical Programming," with José Meseguer, being revised for publication in *Journal of Logic Programming*; original version in *Proceedings, TAPSOFT '87* (Pisa, Italy), edited by Hartmut Ehrig, Robert Kowalski, Giorgio Levi and Ugo Montanari, Lecture Notes in Computer Science, Volume 250, Springer, March 1987, pages 1–22; also, Report CSLI-87-91, Center for the Study of Language and Information, Stanford University, March 1987, and reprinted by MIR (Moscow).
275. "Deduction with Many-Sorted Rewrite Rules," with José Meseguer; being revised for publication in *Theoretical Computer Science*; earlier version as Report CSLI-85-42, December 1985, Center for the Study of Language and Information, Stanford University.

Reports, Reviews, Abstracts, Websites, Software, Poetry, etc.

This list generally excludes items that also appear above as regular publications.

1. "Verifications with Proof Scores in CafeOBJ," with Kokichi Futatsugi and Kazuhiro Ogata, in Peter D. Mosses (Ed.) *Abstracts of Presentations of WADT 2004* (17th International Workshop on Algebraic Development Techniques Barcelona, Spain), pages 75–78, March 2004.
2. "Confessions of a Travel-worn Briefcase," poem, in *Scale 1*, No. 3, pages 12–14, April 2004. Available at scale.ucsd.edu.
3. "Critical Points for Interactive Schema Matching," with Guilian Wang, Young-Kwang Nam, and Kai Lin. Technical Report CS2004–0779, UCSD Department of Computer Science, 31 January 2004; long version of a paper of the same name to appear in *Proceedings, Sixth Asia Pacific Web Conference*, Hangzhou, China, April 14–17, 2004.
4. "Zero, Connected, Empty," words for a cantata, with music by Ryoko Amadee Goguen, in *Recent Trends in Algebraic Development Techniques*, edited by Martin Wirsing, Dirk Pattinson and Rolf Hennicker, Lecture Notes in Computer Science, volume 2755, Springer, 2003, pages 118–126. First performance, at Workshop on Algebraic Development Techniques, Frauenchiemsee, Germany, 26 September 2002. Subsequent performances in Verona, at UCSD, and in the "Powering Up / Powering Down" festival, 29 January 2004.
5. "Formal Notation for Conceptual Blending," an informal webnote, completed July 2003, available over the web at http://www.cs.ucsd.edu/users/goguen/papers/blend.html.
6. "A Scent of Skinner at Harvard," *Journal of Consciousness Studies, 10*, No. 1, pages 46–8, 2003.
7. "Perception," poem, in *Journal of Consciousness Studies 9*, No. 7, page 42, 2002.
8. The BOBJ System website, at http://www.cs.ucsd.edu/groups/tatami/bobj, 2000–2002; ftp available.

9. The Kumo System website, at `http://www.cs.ucsd.edu/groups/tatami/kumo`, 1996–2002; ftp available.
10. *Pelican, Fish, Button*, libretto for a suite from an opera in progress, with Ryoko Goguen, online at `http://www.cs.ucsd.edu/users/goguen/pelican`. Performed at UCSD Geisel Library, 10 August, 2002.
11. "The care and feeding of scenarios," paper presented at Workshop on Virtual Worlds and Simulation, San Antonio TX, 30 January 2002.
12. CSE 230, Programming Languages, website at `http://www.cs.ucsd.edu/users/goguen/courses/230`.
13. CSE 271, User Interface Design, website at `http://www.cs.ucsd.edu/users/goguen/courses/271`.
14. CSE 171, User Interface Design, website at `http://www.cs.ucsd.edu/users/goguen/courses/171`.
15. CSE 175, Social and Ethical Aspects of Technology and Science, class notes available over the web at `http://www.cs.ucsd.edu/users/goguen/courses/175`.
16. CSE 275, Social Aspects of Technology and Science, extensive class notes are available over the web at `http://www.cs.ucsd.edu/users/goguen/courses/275`.
17. "Kumo, BOBJ, and Behavioral Verification," at `http://www.cs.ucsd.edu/users/goguen/projs/tut.html`, 2002. Tutorial given 26 November 2001 at ASE'01 (Automated Software Engineering 2001), San Diego, and as lecture at Keio University, Tokyo, 18 December 2001.
18. Hidden Algebra homepage, at `/www.cs.ucsd.edu/users/goguen/projs/halg.html`, 1998–2002.
19. Algebraic Semiotics homepage, at `http://www.cs.ucsd.edu/users/goguen/projs/semio.html`, 1998–2002.
20. "What is a Proof?," on the web at `http://www.cs.ucsd.edu/users/goguen/papers/proof.html`, 2001. Online essay for user interface design courses.
21. Review of *Visual Space Perception: A Primer* by M. Hershenson, in *J. Consc. Studies 7*, no. 8/9, pages 157–60, 2000.
22. "On the Role of Algebra in Computer Science," slides for lecture at Deptartment of Mathematics, University of Lisbon, at `http://www.cs.ucsd.edu/users/goguen/pps/lisbon00.ps`, 13 July 2000.
23. "Towards the Automation of Behavioral Reasoning," slides for a lecture at a meeting of IFIP WG 1.3 (Foundations of System Specification), at Stanford University on 30 June 2000; available over the web at `http://www.cs.ucsd.edu/users/goguen/ps/has100.ps.gz`.
24. "Semiotic Morphisms," written October 1988 and revised during early 2000; available on the web at `http://www.cs.ucsd.edu/users/goguen/papers/smm.html`.
25. "Circular Coinduction," with Grigore Roşu, Dept. Computer Science and Engineering, University of California at San Diego, Technical Report CS 2000–0647, February 2000; completed October 1999.
26. "The Ethics of Databases," notes for lecture at 1999 Annual Meeting of Society for Social Studies of Science, San Diego, at `http://www.cs.ucsd.edu/users/goguen/papers/4s/4s.html`, 29 October 1999.
27. "On Notation," available at `http://www.cs.ucsd.edu/users/goguen/pps/notn.ps`, completed mid 1999.
28. The Tatami Project website, at `http://www.cs.ucsd.edu/groups/tatami`, 1996–2000.
29. "Requirements Engineering and User Interface Design," slides for lecture at Requirements Engineering Workshop, Buenes Aires, August 1999; at `http://www.cs.ucsd.edu/users/goguen/pps/wersl.ps`.

30. The UCSD Semiotic Zoo, at http://www.cs.ucsd.edu/users/goguen/zoo, 1998–2000.
31. The OBJ Family homepage, at http://www.cs.ucsd.edu/users/goguen/sys/obj.html, 1998–2000.
32. "Tatami System Motivation and Architecture," written and revised late 1998; available on the web at http://www.cs.ucsd.edu/users/goguen/ps/tata.ps.gz.
33. "Poetry, Mechanism and Consciousness," Abstract 46 in *Consciousness Research Abstracts*, Imprint Academic, 1998, page 50.
34. "Poetry, Mechanism and Consciousness," slides for lecture at the Towards a Science of Consciousness Conference, Tucson AZ, 1 May 1998, on web at http://www.cs.ucsd.edu/users/goguen/ps/t3.ps.gz.
35. "First Order Logic," a chapter from *Theorem proving and Algebra*, to be published by MIT Press; available over the web at http://www.cs.ucsd.edu/users/goguen/ps/tp/fol.ps.gz; finished September 1998.
36. Behavioral Logics Discussion Group homepage, http://www.cs.ucsd.edu/groups/tatami/behavior, begun in 2000.
37. "Semiotic Morphisms," UCSD CSE Technical Report CS97–553, August 1997.
38. "Refinement of Concurrent Object-Oriented Programs," with Paulo Borba, Programming Research Group Technical Report PRG–TR–18–95, Oxford University Computing Lab, November 1995.
39. "Algebraic Semantics of Nondeterministic Choice," with Grant Malcolm, Programming Research Group Technical Report PRG–TR–5–95, Oxford University Computing Lab, 1995.
40. "Proving Correctness of Refinement and Implementation," with Grant Malcolm, Programming Research Group Technical Monograph PRG–114, November 1994, Oxford University Computing Lab.
41. "An Operational Semantics for FOOPS," with Paulo Borba, Programming Research Group Technical Monograph PRG–115, Oxford University Computing Lab, November 1994.
42. "Research on Requirements Engineering," Centre for Requirements and Foundations, Oxford University Computing Lab, April 1993.
43. "Memories of ADJ," in *Bulletin of the European Association for Theoretical Computer Science*, Number 36, October 1989, guest column in the 'Algebraic Specification Column,' pages 96–102; also, in *Current Trends in Theoretical Computer Science: Essays and Tutorials*, edited by Grigor Rozenberg and Arto Salomaa, World Scientific, 1993, pages 76–81.
44. "Functional Object-Oriented Design," with T.H. Tse, Technical Report, Programming Research Group, Oxford University Computing Lab, 1990.
45. "Four Pieces on Error, Truth and Reality," Programming Research Group Technical Monograph PRG–89, October 1990, Oxford University Computing Lab. This includes items 127, 128 and 129 above, plus an early ancestor of item 146 and a brief preface.
46. "The Rewrite Rule Machine Project, 1988," with José Meseguer, Sany Leinwand and Timothy Winkler, Programming Research Group Technical Monograph PRG–76, August 1989, Oxford University Computing Lab.
47. "The Rewrite Rule Machine Project," with José Meseguer, Sany Leinwand, Timothy Winkler and Hitoshi Aida, Technical Report SRI–CSL–89–06, Computer Science Laboratory, SRI International, March 1989.
48. "An Abstract Machine for Fast Parallel Matching of Linear Patterns," with Ugo Montanari, Technical Report SRI–CSL–87–3, Computer Science Laboratory, SRI International, July 1987.

49. "Checklist Interruption and Resumption: A Linguistic Study," with Charlotte Linde, NASA Contractor Report 177460, NASA Contract NAS2–11052, July 1987; Structural Semantics Technical Report to NASA, Ames Research Center (Moffett Field, California).

50. "Linguistic Measures for Evaluating Flight Simulation," with Charlotte Linde, NASA Contract NAS2–11052, July 1987; Structural Semantics Technical Report to NASA, Ames Research Center (Moffett Field, California).

51. "Communication Training for Aircrews: A Review of Theoretical and Pragmatic Aspects of Training Program Design," with Charlotte Linde and Linda Devenish, NASA Contractor Report 177459, NASA Contract NAS2–12379, July 1987, Structural Semantics Final Technical Report to NASA, Ames Research Center (Moffett Field, California).

52. "Aircrew Communicative Competence: Theoretical and Pragmatic Aspects of Training Design" with Charlotte Linde and Linda Devenish, NASA Contract NAS2–11052, 1987; Structural Semantics Technical Report to NASA, Ames Research Center (Moffett Field, California).

53. Review of *Fundamentals of Algebraic Specification 1: Equations and Initial Semantics* by Hartmut Ehrig and Bernd Mahr, with José Meseguer, *SIAM Review*, Volume 28, Number 2, June 1987, pages 318–322.

54. "Some Electronic Mathematics," Abstract 87A-68-02, in *Abstracts* of papers presented to the American Mathematical Society, Volume 8, Number 7, Issue 54, page 465, American Mathematical Society, November 1987. (Abstract of lecture presented at the AMS-IMS-SIAM Joint Summer Research Conference on Category Theory and Computer Science, Boulder, Colorado, June 1987.)

55. "The Rewrite Rule Machine Project," with Claude Kirchner, Sany Leinwand, José Meseguer and Timothy Winkler, Technical Report SRI–CSL–87–1, Computer Science Laboratory, SRI International, May 1987.

56. "Remarks on Remarks on Many-Sorted Equational Logic," with José Meseguer, *Bulletin of the European Association for Theoretical Computer Science*, Volume 30, October 1986, pages 66–73; also *SIGPLAN Notices*, Volume 22, Number 4, April 1987, pages 41–48.

57. "Crew Communication as a Factor in Aviation Accidents," with Charlotte Linde and Miles Murphy, Technical Memorandum 88254, NASA Ames Research Center (Moffett Field, California), August 1986.

58. "Towards an Assessment of Japanese Fifth Generation Software Developments," with Carl Hewitt, in *Proceedings, JTECH Advanced Computer Workshop*, National Science Foundation, January 1987.

59. Abstract of "Order-Sorted Algebra: Algebraic Theory of Polymorphism," with José Meseguer, *Journal of Symbolic Logic*, Volume 51, Number 3, September 1986, pages 844–845.

60. "Models of Computation for the Rewrite Rule Machine," with Claude Kirchner and José Meseguer, SRI Technical Report, July 1986.

61. "Simulation Results for the Rewrite Rule Machine," with Sany Leinwand and Timothy Winkler, SRI Technical Report, July 1986.

62. "Architectural Options and Testbed Facilities for the Rule Machine," with Sany Leinwand and Timothy Winkler, SRI Technical Report, July 1986.

63. "Programming by Generic Example," SRI Technical Report, July 1986.

64. "Report on a Visit to Britain," *IEEE Technical Committee on Computer Architecture Newsletter*, March 1986, pages 22–25.

65. "Institutions: Abstract Model Theory for Computer Science," with Rod Burstall, Report CSLI-85-30, Center for the Study of Language and Information, Stanford University, August 1985.
66. "Optimal Structures for Multimedia Instruction," with Charlotte Linde and Tora K. Bikson, Final Report to Office of Naval Research, Psychological Sciences Division, SRI International, July 1985.
67. Review of *Fuzzy Sets and Systems: Theory and Applications* by D. DuBois and Henre Prade, *SIAM Review*, Volume 27, Number 2, June 1985, pages 270–274.
68. "Principles of OBJ2," with Kokichi Futatsugi, Jean-Pierre Jouannaud, and José Meseguer, Report CSLI-85-22, February 1985, Center for the Study of Language and Information, Stanford University; also, ICOT Technical Report TR-97, Institute for New Generation Computer Technology, Tokyo, Japan, 1984.
69. "The Structure of Aviation Procedures," with Charlotte Linde, Technical Report NAS-5, to NASA, Ames Research Center, Structural Semantics, October 1984.
70. "LIL: A Library Interconnection Language for Ada," in "Report on Ada Program Libraries Workshop," SRI International, Computer Science Lab; prepared for Ada Joint Program Office, 1984.
71. "Report on Ada Program Libraries Workshop," with Karl Levitt, SRI International, Computer Science Lab, prepared for Ada Joint Program Office; workshop held at Naval Postgraduate School (Monterey, California), November 1-3, 1984.
72. "Optimal Structures for Multimedia Instruction," with Charlotte Linde, Annual Report to Office of Naval Research, SRI International, January 1984.
73. "Transcription Conventions for Aviation Discourse," with Charlotte Linde, Technical Report to NASA Ames Research Center, Structural Semantics, July 1983.
74. "Linguistic Methodology for the Investigation of Aviation Accident Reports," with Charlotte Linde, Technical Report to NASA, Structural Semantics, January 1983; also, Contractor Report 3741, NASA, Scientific and Technical Information Branch, December 1983.
75. "Phil: A Semantic Structural Graphical Editor," with Leslie Lamport, Technical Report to Philips (Eindhoven) SRI International, September 1983.
76. "Phil: A Context-Sensitive Structured Graphical Editor," with Leslie Lamport, Technical Report to Philips (Eindhoven) SRI International, December 1981.
77. "Two ORDINARY Specifications," Technical Report CSL-128, SRI International, October 1981.
78. "Toward OBJ-1, A Study in Executable Algebraic Formal Specifications," with José Meseguer, Technical Report to Office of Naval Research, SRI International, 1981.
79. "An Ordinary Design," with Rod Burstall, draft Technical Report, SRI International, 1980.
80. "How to Prove Inductive Hypotheses without Induction," Technical Report, SRI International, 1980.
81. "CAT, a System for the Structured Elaboration of Correct Programs from Structured Specification," with Rod Burstall, SRI International Technical Report CSL-118, October 1980.
82. "Evaluating Expressions and Proving Inductive Hypothesis about Equationally Defined Data Types and Programs," Semantics and Theory of Computation Report Number 16, Computer Science Department, University of California at Los Angeles, June 1979.
83. "Order-Sorted Algebra," Semantics and Theory of Computation Report Number 14, Computer Science Department, University of California at Los Angeles, December 1978.

84. Review of *Applications of Fuzzy Sets to Systems Analysis* by Constantin Negoita and Dan Ralescu, published in *Kybernetics*, Volume 6, 1977, page 231.
85. Review of *Conversation Theory* by Gordon Pask, in *Journal of Structural Learning*.
86. Review of *Applications of Fuzzy Sets to systems Analysis* by Constantin Negoita and Dan Ralescu, in *Journal of Symbolic Logic*.
87. "OBJ–Q Preliminary User Manual," with Joseph Tardo, Semantics and Theory of Computation Report Number 10, Computer Science Department, University of California at Los Angeles, July 1977.
88. "Introducing Variables (Extended Abstract)," with Eric Wagner, Jesse Wright and James Thatcher, IBM T. J. Watson Research Center, 1977.
89. "A Junction between Computer Science and Category Theory, I: Basic Concepts and Examples (Part 2)," with James Thatcher, Eric Wagner, and Jesse Wright, report RC 5908, IBM T. J. Watson Research Center, March 1976.
90. "Cybernetics and Buddhism," unpublished manuscript, Naropa Institute (Boulder, Colorado) January 1976.
91. "Programs in Categories (Summary)," with James Thatcher, Eric Wagner, Jesse Wright, IBM T. J. Watson Research Center, August 1975.
92. "An Introduction to Categories, Algebraic Theories and Algebras," with James Thatcher, Eric Wagner and Jesse Wright, report RC 5369, IBM T. J. Watson Research Center, April 1975.
93. "Set-Theoretic Correctness Proofs," Semantics and Theory of Computation Report Number 1, Computer Science Department, University of California at Los Angeles, September 1974.
94. "Algebraic Semantics for ALGOL 68," with Daniel Berry, Modelling and Measurement Note Number 29, Computer Science Department, University of California at Los Angeles, September 1974.
95. "Factorizations, Congruences and the Decomposition of Automata and Systems," with James Thatcher, Eric Wagner, Jesse Wright, Report RC 4934, IBM T. J. Watson Research Center, May 1974.
96. "Initial Algebra Semantics (Extended Abstract)," with James Thatcher, report RC 4865, IBM T. J. Watson Research Center, May 1974.
97. "Axioms, Extensions and Applications for Fuzzy Sets: Languages and the Representation of Concepts," report RC 4547, IBM T. J. Watson Research Center, September 1973.
98. "A Junction between Computer Science and Category Theory, I: Basic Concepts and Examples (Part I)," with James Thatcher, Eric Wagner, Jesse Wright, report RC 4526, IBM T. J. Watson Research Center, September 1973.
99. "Some Comments on Data Structures," abstract of lectures given at ETH (Zürich, Switzerland) September 1973.
100. "On Homomorphisms, Correctness, Termination, Unfoldments and Equivalence of Flow Diagram Programs," Computer Science Department, University of California at Los Angeles, Report UCLA-ENG-7337, June 1973.
101. "On Mathematics in Computer Science Education," report RC 3889, IBM T. J. Watson Research Center, June 1972.
102. "Discrete-Time Machines in Closed Monoidal Categories, I," Section II B, Institute for Computer Research Quarterly Report Number 30, University of Chicago, 1970.
103. "Axioms for Discrimination Theory," with Lee Carlson, Section II A, Institute for Computer Research Quarterly Report Number 26, University of Chicago, 1970.
104. "Discrimination in Information Theory," Section III A, Institute for Computer Research Quarterly Report Number 24, University of Chicago, February 1970.

105. "Mathematical Foundations of System Theory," Section II E, Institute for Computer Research Quarterly Report Number 25, University of Chicago, May 1970.
106. "Representing Inexact Concepts," Section III A, Institute for Computer Research Quarterly Report Number 20, University of Chicago, February 1969.
107. "Categories of Fuzzy Sets: Applications of Non-Cantorian Set Theory," Technical Report to Office of Naval Research, Department of Mathematics, University of California at Berkeley, May 1968 (Ph.D. thesis).
108. "Contributions to the Mathematical Theory of Evolutionary Algorithms," with Nancy Goguen, Technical Report of Office of Naval Research, Department of Mathematics, University of California at Berkeley, September 1967.
109. "Research on a Mathematical Theory of Budgeting," with David Friedman and Kenneth Krohn, Internal Report, Krohn-Rhodes Research Institute (Berkeley, California) December 1966.
110. "Some Considerations on Evolutionary Algorithms," Technical Report to Office of Naval Research, Department of Mathematics, University of California at Berkeley, February 1966.
111. "Electrical Networks and Algebraic Topology," undergraduate thesis, Harvard University, 1963, and lecture notes, Applied Research Laboratory, Sylvania Electronic Systems (Waltham, Massachusetts) 1963.
112. "Contact Networks and Probabilistic Contact Networks," Research Report 362, Sylvania Electronic Systems (Waltham, Massachusetts) summer 1962.
113. "Reduction of Coefficient Correlation in Orthonormal Expansions by Linear Transformations," Research Note 350, Sylvania Electronic Systems (Waltham, Massachusetts) summer 1962.

Sync or Swarm:
Musical Improvisation and the Complex Dynamics
of Group Creativity

David Borgo, Ph.D.

Assistant Professor of Music
Critical Studies and Experimental Practices Program

University of California, San Diego
9500 Gilman Dr.
La Jolla, CA 92093-0326
dborgo@ucsd.edu

Abstract. This essay draws on participant observation, ethnographic interviews, phenomenological inquiry, and recent insights from the study of swarm intelligence and complex networks to illuminate the dynamics of collective musical improvisation. Throughout, it argues for a systems understanding of creativity—a view that takes seriously the notion that group creativity is not simply reducible to individual psychological processes—and it explores interconnections between the realm of musical performance, community activities, and pedagogical practices. Lastly, it offers some reflections on the ontology of art and on the role that music plays in human cognition and evolution, concluding that improvising music together allows participants and listeners to explore complex and emergent forms of social order.

1 Introduction

The nature of creativity in the arts and sciences has been of a topic of enduring human interest. But the dominant scholarly approach to the subject, until recently, has proceeded from the assumption that creativity is primarily an individual psychological process, and that the best way to investigate it is through the thoughts, emotions, and motivations of those individuals who are already thought to be gifted or innovative. In the past several decades, however, researchers have begun to focus more attention on the historical and social factors that shape and define creativity, and on its role in everyday activities and learning situations.[1] Yet despite this shift in the field towards a systems perspective, the notion that creativity operates primarily on the level of *individuals* (albeit now situated within a rich and complex environment), or that

---

[1] This shift is attributed in great part to the work of Mihaly Csikszentmihalyi [1], who has argued for a systems view of creativity. The work of sociologist Howard Becker has also been influential in this regard, as well as foundational work in sociology of knowledge (Mannheim), activity theory (Vygotsky), communities of practice (Lave and Wenger), ethnomethodology (Garfinkel), and ecological psychology (Gibson).

K. Futatsugi et al. (Eds.): Goguen Festschrift, LNCS 4060, pp. 1-24, 2006.
© Springer-Verlag Berlin Heidelberg 2006

creativity necessarily results in a creative *product,* has proved to be remarkably resilient.

The practice of improvising music together calls into question many of these assumptions.   The activity is both intrinsically collaborative and inherently ephemeral. Since roughly the middle of last century, an eclectic group of artists with diverse backgrounds in contemporary jazz and classical music–and increasingly in electronic, popular, and world music traditions as well–have pioneered an approach to improvisation that borrows freely from a panoply of musical styles and traditions and at times seems unencumbered by any overt idiomatic constraints. This musical approach, often dubbed "free improvisation," tends to devalue the two dimensions that have traditionally dominated music representation–quantized pitch and metered durations–in favor of the micro-subtleties of timbral and temporal modification and the surprising and emergent properties of collective creativity in the moment of performance.[2]

In the community of free improvisers it is not uncommon for musicians to speak of the importance of developing a "group mind" during performance. This requires, at the very least, cultivating a sense of trust or empathy among group members, and, according to some, it may also involve reaching a certain egoless state in which the actions of individuals and the group perfectly harmonize. Percussionist Adam Rudolph described his trio's approach to me this way: "We all participate in creating the musical statement of the moment. In the process of being free as a collective, you have to have selflessness to give yourself to the musical moment and not come from a place of ego."[3]

In the moment-to-moment dynamics of improvised performance it can also be difficult to separate individual contributions and intentions from those cultivated by the "group mind." Bassist Richard Davis explains: "Sometimes you might put an idea in that you think is good and nobody takes to it... And then sometimes you might put an idea in that your incentive or motivation is not to influence but it does influence."[4] Acknowledging this inherent complexity, saxophonist Evan Parker finds that:

> However much you try, in a group situation what comes out is group music and some of what comes out was not your idea, but your response to somebody else's idea... The mechanism of what is provocation and what is response–the music is based on such fast interplay, such fast reactions that it is arbitrary to say, "Did you do that because I did that? Or did I do that because you did that?" And anyway the whole thing seems to be operating at a level that involves...certainly intuition, and maybe faculties of a more paranormal nature.[5]

Research on creativity has tended to make a distinction between an ideation stage, in which the non-conscious brain produces novelty through divergent thinking, and an evaluation stage, in which the conscious mind decides which new ideas are coherent

[2] For two useful starting points on the web, covering principally the US and European scenes respectively, see www.restructures.net and www.shef.ac.uk/misc/rec/ps/efi/. See also Bailey [2].
[3] Quoted in [7], p. 80.
[4] Quoted in [6], p. 88.
[5] Quoted in [8], p. 203.

with the creative domain. From a systems perspective, however, ideation and evaluation may occur in individuals in a complex rather than a linear fashion, and during ensemble performances they may become externalized into a group process. Keith Sawyer [3], in his recent book titled *Group Creativity*, expands Mihaly Csikszentmihalyi's [4] well-known notion of "flow"–in which the skills of an individual are perfectly matched to the challenges of a task, and during which action and awareness become phenomenologically fused–to include the process of entire groups performing at their peak.[6] Group flow, according to Sawyer, can inspire individuals to play things that they would not have been able to play alone or would not have explored without the inspiration of the group. Yet as a collective and emergent property, group flow can be extremely difficult to study empirically. Sawyer describes it as an irreducible property of performing groups that cannot be reduced to psychological studies of the mental states or the subjective experiences of the individual members of the group.

Models that focus on the creativity of individuals are not wrong, but like Newtonian science, they may be inappropriate for trying to make sense of certain types of phenomena. What we need are new models operating at a different level. In the increasingly complex and interconnected world that we inhabit it is becoming apparent that structure and organization can emerge both without lead and even without seed. What happens and how it happens depends on the nature of the network.

What implications do the study of group musical performance and the study of complex network dynamics have for musical scholarship and more broadly for our understandings of human creativity? In music, networks organize not only the social world of performance (with whom you play) but also the ideascapes of creativity (by whom you are influenced and what or how you chose to create) and the dynamics of communities (how historical, cultural, and economic factors often dictate which musicians and musical ideas gain notice and prestige). Networks make communication and community possible, but they can also concentrate power and opportunities in the hands of a few. In this essay I explore the dynamics of group musical improvisation and recent insights from the study of swarm intelligence and complex networks in order to investigate some ways in which musical studies might productively grapple with the complex of factors that establish, maintain, expand, and even destroy musical communities.

2 Insect Music

"At one level, improvisation can be compared with the ultimate otherness of an ant colony or hive of bees. Perhaps it was no coincidence that in the wake of drummer John Stevens and the Spontaneous Music Ensemble, certain strands of English improvised music were known, half-disparagingly as insect music.
David Toop [9], p. 247

---

[6] Sawyer draws heavily on ethnographic work by Paul Berliner [5] and Ingrid Monson [6] for his perspective on jazz and improvisation.

Improvisation is not a revolution that pits itself against codification; it is diffuse. Like ants stripping a carcass, it works from the inside and outside of codes.
John Corbett [10], p. 237

In Euro-American art-music culture this binary [between composition and improvisation] is routinely and simplistically framed as involving the "effortless spontaneity" of improvisation, versus the careful deliberation of composition–the composer as ant, the improviser as grasshopper.
George Lewis [11], p. 38

Scientists, artists, and laypeople alike have for centuries watched in wonder as a flock of birds spontaneously takes flight and navigates in perfect harmony, or as a hive of bees throws off a collective swarm into the air. At the dawn of the twentieth century, the Belgian poet Maurice Maeterlinck wondered, "Where is 'this spirit of the hive'…where does it reside? What is it that governs here, that issues orders, foresees the future?"[7] We now know that within the swarm a half dozen or so anonymous workers scout ahead to check for possible hive locations. When they report back to the swarm, they perform an informative dance, the intensity of which corresponds to the desirability of the site they scouted. Deputy bees follow up on the more promising reports and return to either confirm or disconfirm the desirability of the new location. Although it is rare for a single bee to visit more than one potential site, through the process of compounding emphasis, the more desirable sites end up getting the most visitors. In other words, the hive chooses: the biggest crowd eventually provokes the entire swarm to dance off to its new location.

We can sense in this and other examples of complex and decentralized decision-making certain qualities that appear to inform all life. William Morton Wheeler, the founder of the field of social insects, argued as early as 1911 that an insect colony operates as a type of *superorganism*: "Like a cell or the person, it behaves as a unitary whole, maintaining its identity in space, resisting dissolution…neither a thing nor a concept, but a continual flux or process."[8] Even the sound of the swarm can fascinate human ears. For her aptly titled "Bee Project," kotoist and multimedia artist Miya Masaoka's positioned a glass-enclosed bee hive of 3,000 bees in the center of the stage and amplified, manipulated, and blended its sounds with those from a trio of improvisers, all according to the instructions in her score. Later versions of the same work have used spatialization software to twist and tilt the sound of the hive so that listeners can be sonically located within the swarm.

As the three quotes offered at the beginning of this section illustrate, there are several ways in which we might wish to locate musical connections to the swarm. Some improvised music provokes such quick reactions from players and evokes such complicated and dense soundscapes for listeners that a literal analogy to a swarm of insects may seem rather appropriate. And the ways in which individual improvisers can be heard to be "picking at" a shared body of modern techniques and sensibilities but in resolutely individualistic ways, or to be following their own creative spark while also being sensitive to and dependent on the evolving group dynamic, may

---

[7] Quoted in [12], p. 7. Maeterlink's book is available online at
http://www.eldritchpress.org/mm/b.html#toc.
[8] Quoted in [12], p. 7.

bring to mind the behavior of social insects that seem to have their own agenda while also working in ways that organize the group without supervision. Finally, the notion of "insect music" has perhaps become most associated with a type of generative compositional scheme, and often with the power of computers to create complex patterns from relatively simple materials, such that questions about the ways in which creativity may be facilitated or constrained and the ways in which cultural understandings may be reflected, reshaped, or remain concealed in this type of work become particularly important.

In addition to being an extremely skilled improviser, the English drummer John Stevens will always be remembered for his instrumental role in developing the scene at The Little Theater Club in London that nurtured many in the first generation of English free improvisers. One of his early pedagogical approaches was titled *Click Piece*, and it included little more that the instruction to play the shortest sounds on your instrument.[9] In the collective setting, however, one would gradually become aware of an emergent group sound. As David Toop [9] explains, "The piece seemed to develop with a mind of its own and almost as a by-product, the basic lessons of improvisation–how to listen and how to respond–could be learned through a careful enactment of the instructions" (pp. 242-3). Steven's *Click Piece* highlights one of the central aspects of swarm dynamics; relatively simple decentralized activities can produce dramatic, self-organizing behaviors.

In the scientific community, a growing number of researchers are exploring new ways of applying swarm intelligence (or SI) to diverse situations.[10] For instance, the foraging of ants has led to improved methods for routing telecommunications traffic in a busy network. The way in which insects cluster their dead can aid in analyzing bank data. The distributed and cooperative approach used by many social insects to transport goods and to solve navigational problems has led to new insights in the fields of robotics and artificial intelligence. And the evolving division of labor in honeybees has helped to improve the organization of factory assembly line workers and equipment. As Eric Bonabeau and Guy Théraulaz [15] see it: "The potential of swarm intelligence is enormous. It offers an alternative way of designing systems that have traditionally required centralized control and extensive preprogramming" (p.79).

Beyond these business and technological applications, however, one of the main lessons of contemplating SI is that organized behaviors can develop in decentralized ways. Can exploring and thinking about SI affect the way we make and think about music? It remains difficult for many people to envision complex systems organizing without a leader since we are often predisposed to think in terms of central control and hierarchical command. The notion that music can be organized in complex ways without a composer or conductor still leaves many scratching their heads in doubt. Scientists have also been predisposed in the past to look for chains of command, instances of clear cause and effect. But the emerging field of SI demonstrates that complex behaviors and efficient solutions can be arrived at without a leader, organized without an organizer, coordinated without a coordinator.

---

[9] Stevens titled the reverse strategy "Sustained Piece."

[10] Although this field is often presented as evolving in only the past few years, examples drawn from the world of social insects can be found in early cybernetics theory [13], pp. 156-7 and in dissipative structures as well [14], pp. 181-6.

The secret of the swarm lies in the intercommunication of its members. Through direct and indirect interactions among autonomous agents and between agents and their environment, swarm systems are able to self-organize in decentralized, robust, and flexible ways. Bonabeau, Théraulaz, and Marco Dorigo [16], a physicist, biologist, and engineer working together at the Santa Fe Institute, offer a list of four basic ingredients that through their interplay can manifest in swarm intelligence: 1) forms of positive feedback, 2) forms of negative feedback, 3) a degree of randomness or error, and finally 4) multiple interactions of multiple entities.

Positive feedback in SI can be usefully summarized as simple "rules of thumb" that promote the creation of structures: activities such as recruitment and reinforcement. Negative feedback counterbalances positive feedback and helps to stabilize the system: it may take the form of saturation, exhaustion, or competition. A certain degree of randomness or error is also crucial, since it enables the discovery of new solutions and produces fluctuations that can act as seeds from which new structures develop. Finally, SI generally requires a minimum density of mutually tolerant individuals, since individuals should be able to make use of the results of their own activities and the activities of others.

While something of a general and descriptive list, these ingredients do play important roles in collective improvisation. Through positive feedback musicians not only develop their own ideas from a kernel of inspiration, but they also work together to support the ideas of others and the evolving ensemble sound. They "recruit" others to support or sustain their own developments, or they may choose to "reinforce" the creative direction of others instead. Similar to the ways in which information about the best food source or the shortest path can be compounded among a swarm of bees or a colony of ants, positive feedback increases the ability of an improvising group to follow the more "promising" of many concurrent ideas being pursued by various members.

Negative feedback in improvisation helps to keep things interesting. By intentionally looking elsewhere for new ideas or new musical areas to explore, individuals can either signal transitions away from ensemble moments that have lingered too long or seem to be going nowhere (the feelings of saturation and exhaustion), or they can productively layer divergent sonic qualities and musical ideas together or provoke others to boost their own creativity (through a competitive element). Negative feedback helps to maintain a balance in the evolving improvisation so that one idea does not continue to amplify indefinitely (although a more static approach can produce interesting results as well).

Unexpected occurrences, in the form of randomness or error, often provide both source material and inspiration for individuals and groups to explore new sonic territory, musical techniques, and interactive strategies. Noticing and capitalizing on unexpected fluctuations as an improvisation unfolds can produce important structural cues, developments, and transitions, and it represents a particular joy of improvised music making in general. Without this third ingredient, groups of improvisers who work together over a longer period of time might become too familiar with one another's musical language and approach or might fall into regular strategies of support and counterbalance (and this of course does happen).

Finally, the notion that individuals and the group as a whole benefit from multiple interactions and perspectives is something of an axiom in ensemble forms of improvisation and in the community of improvisers. One of the particular challenges of contemporary improvisation, for both players and listeners, is to remain aware of and sensitive to the many musical gestur es and processes circulating between members of the group in the moment of performance and between members of the community as ideas circulate via recordings, impromptu meetings, and the overlapping personnel of various working groups.[11]

In much freer improvisation, the collective pattern of the group is more important than any of the individual actions heard in isolation. But this does not deny freedom to individual musicians. Saxophonist Evan Parker [17] highlights the ways in which freedom works within the collective unfolding of what might easily be termed swarm dynamics:

> The freedom is of course that since you and your response are part of the context for other people, and they have that function for you, it's very hard to unravel the knots of why anybody is doing what they do in a given context. I think it's pretty clear that you could sort of go with the flow, or you could go against the flow. And sometimes what the music really needs is for you to go with the flow, and sometimes what it really needs is for you to do something different. Or anybody, somebody, to do something different. So that's why people improvise, presumably, because they want the freedom to behave in accordance with their response to the situations. But since their response then becomes part of the new situation for the other players, it's very hard to say why a particular sequence of events unfolds in the way it does. But we get used to following the narrative of improvisational discourse...

Parker's notion that "the music" *needs* for things to happen, *needs* for musicians to do things, is a fairly common way in which improvisers speak about the process of performance. In his liner notes to the album *In Order to Survive*, bassist William Parker (no relation) expresses that, "Creative Music is any music that procreates itself as it is being played to ignite into a living entity that is bigger than the composer and player."[12]   While these comments certainly resonate with the notion of a superorganism touched on earlier, they may also highlight an additional dimension of SI research: interactions within a swarm can be both direct and indirect. The direct interactions are the obvious ones: with ants this can involve antennation or mandibular contact, food or liquid exchange, chemical contact, etc. But indirect interactions are more subtle. In SI they are referred to by the rather cumbersome term *stigmergy* (from the Greek *stigma*: sting, and *ergon*: work). Stigmergy describes the indirect interaction between individuals when one of them modifies the environment and the other responds to the new environment rather than directly to the actions of the first individual. This helps to describe the process of "incremental construction" that many social insects use to build extremely complex structures or to arrange items in ways that might at first seem arbitrary or random. And because positive feedback can produce nonlinear effects, indirect interaction can result in dramatic bifurcations when a critical point is reached: for example, some species of termites alternate

---

[11] Here we might also want to envision the creative process of each individual as a type of swarm dynamic, as the processes of ideation and evaluation can work rapidly and in complex and nonlinear ways.

[12] *Black Saint* records 12015902 (1995).

between non-coordinated and coordinated building to produce neatly arranged pillars or strips of soil pellets.

But swarm intelligence has its limits and its drawbacks. Social insects can adapt to changes in their environment, but only within a certain degree of tolerance. For instance, many social insects are able to seek out and find new food sources when an existing one is exhausted, or some species are able to reallocate labor roles if the number of required workers for a specific task dwindles, all without explicit instruction. But the "army ant syndrome" offers a compelling example of the limits to this adaptability and of swarm intelligence in general. Among army ants, when a group of foragers accidentally gets separated from the main colony, the separated workers run in a densely packed "circular mill" until they all eventually die from exhaustion. Although able to function well within the group under normal circumstances, an unpredictable perturbation of a large enough degree can destroy the colony's cohesiveness and make it impossible for the group to recover.

For a musical analogy, while sensitivity to the group is an essential component of improvised performance, to blindly base one's own playing on what others do or to simply follow the group as an overriding strategy can lead to rather inflexible and ineffective results, producing a musical "circular mill." And many improvisers, if they sense that all of the participants are following each other too carefully, will "go against the grain" or "forge out on their own" into new sonic territory; in other words, they will defy the logic of the hive mind. To return briefly to our earlier example of John Stevens's *Click Piece*, although this generative approach to collective improvisation offered an effective way to make "quite ravishing" music with a large ensemble comprising players of mixed ability and experience, to more skillful and confident musicians it quickly became an unproductive limitation. Simplifying the parameters for improvisation can be useful and even necessary for making large ensembles swarm effectively, but in the more intimate setting of a small group, arguably the preferred arrangement for the majority of free improvisation enthusiasts, a less restrictive framework is usually desired.

The cohesion of small groups can also be jeopardized by imbalances that lead to polarization. Drawing on research with decision-making among corporate boards and committees, James Surowiecki [18] identifies a few qualities that appear to factor into all intimate social settings: earlier comments are more influential; higher status people talk more and more often; and status is not always derived from knowledge/experience. Since constantly making comparisons and adjustments to others can result in an unproductive "group think," it is important for individuals to champion their own ideas in small group settings. But too much vehemence in this can lead to a completely polarized setting or to an "information cascade" when others are subsumed by a singular view or opinion. In short, deference to the ideas of others is important, but so is dissent when required.

Without a doubt there are important differences in the degrees of freedom allowed in a swarm of bees and in the sonic swarm of collective improvisation. But if interesting complexities can emerge from groupings of individuals with a limited array of communication possibilities, how much more can we expect from experienced and creative artists? J. Stephen Lansing [19], an anthropologist who also serves as external faculty at the Santa Fe Institute, wonders about complex adaptive systems in general: "What if the elements are not cells or light bulbs but agents capable of reacting with new strategies or foresight to the patterns they have helped to create?" (p. 194). Much of the current research by social scientists on complex adaptive systems is concerned with precisely this question.

The field of SI is still very much in its infancy. It is often extremely difficult for researchers to understand the inner workings of insect swarms and the variety of rules by which individuals in a swarm interact. Even in those cases when we can understand the behaviors of individuals, we may still be unable to predict or understand the dynamics of the overall system since countless other environmental factors come into play. When transposed into the realm of humans, these uncertainties only compound themselves. Discussing the business and technological applications of SI, Bonabeau and Théraulaz [15] confess that: "Although swarm-intelligence approaches have been effective at performing a number of optimization and control tasks, the systems developed have been inherently reactive and lack the necessary overview to solve problems that require in-depth reasoning techniques" (p.79). We still don't know enough about social insects, little less social humans, to be able to understand how certain group behaviors emerge and evolve.

Nevertheless, the notion that a group can have capacities and capabilities that extend beyond the scope of any of its participating members is a powerful one. In a provocative chapter titled "Hive Mind" from his book *Out of Control*, Kevin Kelly [12] points out that the hive does possess much that none of its parts possesses. Not only does swarm intelligence represent a type of distributed perception for the hive, but the hive also possesses a type of distributed memory; the average honeybee operates with a memory of six days, but the hive as a whole operates with a distributed memory of up to three months, twice as long as the lifetime of the average bee. Bonabeau et al. [16] write:

We suggest that the social insect metaphor may go beyond superficial considerations. At a time when the world is becoming so complex that no single human being can really understand it, when information (and not the lack of it) is threatening our lives, when software systems become so intractable that they can no longer be controlled, perhaps the scientific and engineering world will be more willing to consider another way of designing "intelligent" systems where autonomy, emergence and distributed functioning replace control, prepro-gramming, and centralization (p.22).

We might also hope that the music world will continue to explore ways of organizing sonic and social experiences that do not hinge on centralized notions of control. Well aware of these concerns, trombonist/composer/scholar George Lewis [20] writes in a recent essay reflecting on improvisation and the orchestra:

Orchestra performers operate as part of a network comprised not only of musicians, composers and conductors, but also administrators, foundations, critics and the media, historians, educational institutions, and much more. Each of the nodes within this network, not just those directly making music, would need to become "improvisation-aware," as part of a process of resocialization and economic restructuring that could help bring about the transformation of the orchestra that so many have envisioned.

3   A Web Without a Spider

If group improvisation may be heard in its best moments to demonstrate complex and emergent properties that are somehow greater than the sum of its parts, then investigating individuals and ensembles in isolation of the network of surrounding influences will not suffice. And as we move our gaze further into the social and historical realms, the notion that any one individual is controlling their own web of musical sounds and meanings becomes rather untenable. We need to reorient our analytical framework to take account of the dynamics that occur in ensembles as they perform together over days, weeks, months, and even years. And we need to acknowledge the ways in which influences in musical communities circulate through more than the sounds of performances and recordings; meaning is everywhere, not simply in the "sounds themselves." The networks involved include a host of social conventions and material artifacts that affect the ways in which music is made and heard: from the funding sources or media attention that a performer may receive to the casual conversations or critical reviews that a performance may provoke. While it may be fairly common to acknowledge the subtle influence that specific audiences and venues can have on performance, especially in relation to improvisation, the network of material, economic, technological, educational, and social factors at play, and the complex meanings that they generate through their interactions, are far more involved than that. In fascinating ways, this network-style organization both shapes and is shaped by the activity of all of its participants; everyone changes the state of everyone else. Although the spontaneous and surprising occurrences in improvised performance can attract our immediate attention, it is through the dynamic interplay of social, material, and sonic culture that we begin to sense the true lifeblood of the music.

Although networks have interested researchers for decades, until recently, each system tended to be treated in isolation, with little apparent reason or possible means to see if its organizational dynamics had anything in common with other networks. We are only now beginning to piece together some important qualities of, and approaches to, the study of complex dynamic networks on a broad scale. But Albert-László Barabási [21], one of the leading researchers in this still nascent field, optimistically predicts: "Network thinking is poised to invade all domains of human activity and most fields of human inquiry. It is more than another helpful perspective or tool. Networks are by their very nature the fabric of most complex systems, and nodes and links deeply infuse all strategies aimed at approaching our interlocked universe" (p. 222).

The notion of networks may bring to mind rather bare-boned models of how things are connected. To some extent this is true, since simplifying detail on one level of a network can highlight organizational similarities on another that would otherwise go unnoticed. Network models, however, are increasingly able to take account of some of the rich dynamics that occur when individual components are not only doing something–generating power, sending data, even making decisions–but also are affecting one another over time. Steven Shaviro [22] writes in his book *Connected, Or What it Means to Live in the Network Society*:

As it seems to us now, a network is a self-generating, self-organizing, self-sustaining system. It works through multiple feedback loops. These loops allow the system to monitor and modulate its own performance continually and thereby maintain a state of homeostatic equilibrium. At the same time, feedback loops induce effects of interference, amplifications, and resonance. And such effects permit the system to grow, both in size and in complexity. Beyond this, a network is always nested in a hierarchy. From the inside, it seems to be entirely self-contained, but from the outside, it turns out to be part of a still larger network (p. 10).

Music, as an inherently social practice, thrives on network organization. On perhaps the most tangible level, a musician's livelihood and creative opportunities frequently depend on the breadth and depth of one's network of social and professional contacts. But network dynamics shape the sounds, practices, and communities of music in decidedly more complex and subtle ways as well. Musicians are influenced by their years of training or apprenticeship, countless hours spent listening to music both publicly and privately, and perhaps most comprehensively (yet frequently least acknowledged) by the historical and cultural conventions of a given time and locale. The topics and techniques of music education also depend on these network-style dynamics, which inform the process of choosing canons and of exploring and imparting the intricacies of musical theory and musical aesthetics. Finally the music industry's far-reaching networks of production and distribution, and increasingly its consolidated and insular organizational practices, have the power to structure, at some degree or another, the networks of inspiration and possibility for nearly everyone who is deeply committed to music.

Yet music researchers have in the past focused the lion's share of attention on the creative work of individuals, often treating their "work" as a collection of static objects (e.g., scores or recordings) to be dissected and categorized. It is not uncommon to hear graduate students in musicology programs lamenting (or coming to terms with) the fact that they must find an increasingly obscure composer or performer on whose work to focus their "comprehensive" scholarly lens. There has, of course, been a pronounced and welcome shift in the past few decades towards a "new musicology" that takes into account the historical and cultural factors that influence not only the original production of a musical "work," but also its variable reception, taking particular notice of gender and racial constructions that may affect both of these.[13]   And there has been a marked increase in the number of scholars interested in expanding the scope of musical investigation into popular and non-Western topics as the fields of ethnomusicology and popular music studies have come into their own. But on the whole, music scholarship is only now beginning to focus attention on the organizational complexities of music rather than treat it as the provenance of a few gifted and prolific individuals.

The musical community has a vested interest in understanding network dynamics, although individuals may vary considerably in their specific expectations. Network thinking can shed light on the cultural power inequities that produce imbalances in social and economic interactions. It may also tell us much about the spread of ideas in musical communities and marketplaces under diverse historical and cultural

---

[13] For examples, see the work of Susan Mclary and Suzanne Cusick among others.

conditions. Creative musicians may hope to find in network dynamics glimpses of future directions for innovation or influence, strategies for how to avoid or disrupt network hubs and established practices in hopes of alternative community reorganization, or the means by which they might increase their own professional contacts and opportunities.

Actor-Network Theory (ANT), a sociological approach that has emerged out of science and technology studies, is geared towards embodying this very tension between the centered 'actor' on the one hand and the decentered 'network' on the other. As John Law [23], one of the field's leading researchers, remarks: "In one sense the word [actor network theory] is thus a way of performing both an elision and a difference between what Anglophones distinguish by calling 'agency' and 'structure'" (p.5).[14]   In short, ANT does not accept the notion that there is a macrosocial system on the one hand, and bits and pieces of derivative microsocial detail on the other. According to Law:

> If we do this we close off most of the interesting questions about the origins of power and organization. Instead we should start with a clean slate. For instance, we might start with interaction and assume that interaction is all that there is. Then we might ask how some kinds of interactions more or less succeed in stabilising and reproducing themselves: how it is that they overcome resistance and seem to become "macrosocial"; how it is that they seem to generate the effects such as power, fame, size, scope or organisation with which we are all familiar. This, then, is the one of the core assumptions of actor-network theory: that Napoleons are no different in kind to small-time hustlers, and IBMs to whelk-stalls. And if they are larger, then we should be studying how this comes about–how, in other words, size, power or organisation are generated.[15]

As musical traditions expand in scope and popularity, better-connected "hubs" tend to emerge. In jazz, for example, the "hubs" of Louis Armstrong, Duke Ellington, Charlie Parker, Miles Davis, and John Coltrane, among others, are impossible to ignore.   During their lifetimes these musicians were well respected and well connected (although not always early in their careers and not by everyone) and their influence has only grown since. With the spread of institutionalized jazz education and the increasing reliance of major labels on re-releasing canonical jazz recordings, the visibility and "connectedness" of these hubs may only continue to grow. For instance, in the last few years Columbia, Atlantic, and Verve have all drastically reduced their roster of living artists in favor of re-releasing older material. Even the Marsalises, perhaps the most visible jazz performers today, no longer have a major record deal. David Hajdu [24] perceptively writes in an *Atlantic Monthly* spread on Wynton: "Where the young lions saw role models and their critics saw idolatry, the record companies saw brand names–the ultimate prize of American marketing. For long established record companies with a vast archive of historic recordings, the economies were irresistible: it is far more profitable to wrap new covers around albums paid for generations ago than it is to find, record, and promote new artists" (p. 54).

---

[14] For other important work in ANT see the publications of Geoffrey C. Bowker and Susan Leigh Star.

[15] http://www.comp.lancs.ac.uk/sociology/soc054jl.html.

For an artistic tradition to remain dynamic and healthy the network dynamics that take note of history and provide hubs for a common language and style should not become too powerful. If the disparity between the hubs and the remainder becomes too great, there may be a "tipping point" beyond which communication and innovation in a tradition can suffer dramatically.[16] In the same *Atlantic Monthly* article, Jeff Levinson, the former Columbia Jazz executive, is quoted as saying: "The Frankenstein monster has turned on its creators. In paying homage to the greats, Wynton and his peers have gotten supplanted by them in the minds of the populace. They've gotten supplanted by dead people" (p. 54).[17] The disparity of attention in music seems to be regulated through the process of interaction. This can come in the direct form of collaboration between artists, but also in the indirect form of media attention, record sales, performance opportunities, and arts funding or sponsorship.

In what is perhaps its most radical move, ANT attempts to take account of the heterogeneous networks that include not only social or human dimensions, but also the material dimensions that make human and social behaviors possible. ANT explores how these heterogeneous networks come to be patterned to generate effects like organizations, inequality, and power. Joseph Goguen explains:

Actor-Network theory can be seen as a systematic way to bring out the infrastructure that is usually left out of the "heroic" accounts of scientific and technological achievements. Newton did not really act alone in creating the theory of gravitation: he needed observational data from the Astronomer Royal, John Flamsteed, he needed publication support from the Royal Society and its members (most especially Edmund Halley), he needed the geometry of Euclid, the astronomy of Kepler, the mathematics of Galileo, the rooms, lab, food, etc. at Trinity College, an assistant to work in the lab, the mystical idea of action at a distance, and more, much more.[18]

The goals of network theory are gradually shifting from describing the topology of systems to understanding the mechanisms that shape network evolution. Barabási [21] acknowledges that, "We must move beyond structure and topology and start focusing on the dynamics that take place along the links. Networks are only the skeleton of complexity, the highways for the various processes that make our world hum. To describe society we must dress the links of the social network with actual dynamical interactions between people" (p. 225).

As in a house of mirrors, the science of networks has seemingly led us to a place in which all of the details matter and, to some extent, none of them do. Since at least the work of Emile Durkheim we have known that large-scale social phenomenon–the predictable number of Parisians who commit suicide every year–can be independent of the particulars–which Parisians are actually led to kill themselves and why. And

---

[16] For a popular science treatment of the notion of a "tipping point" see Gladwell [25].

[17] For a recent example of how powerful hubs have become in jazz, the San Francisco Jazz Spring 2005 series of concerts featured no less than seven tributes to the music of John Coltrane within a month's time, including versions of his music from the albums *A Love Supreme, Ascension, Africa Brass, Crescent,* and *Interstellar Space.* There was also a concert by the Mingus Big Band and a tribute to the music of Rashaan Roland Kirk as well.

[18] http://carbon.cudenver.edu/~mryder/itc_data/ant_dff.html.

despite the enormous complexities of the Isaac Newton example described above, scientists in the modern era glean what they need to from Newton, usually without reading his original work, and they move on to more pressing concerns.

Yet the details and vagaries of a network system do seem to matter enormously. Although network theory often focuses on large-scale behaviors, these large-scale behaviors are fundamentally provoked by the ability of one individual to influence another and the notion that people can change their strategies depending on what other people are doing. Through these dynamics alone, systems can self-organize in remarkably complex ways.

In music, the practice of free improvisation is perhaps closest to this ideal of a self-organizing system. Its bottom-up style emphasizes possibilities for adaptation and emergence; it accentuates creativity-in-time and the dynamics of internal change. The structures of improvisation can also continue to be extended in boundless ways (although the system may be circumscribed, at least in part, by the abilities, materials, and experiences of those who are participating). From one perspective, improvised music is resilient to individual "mistakes" since sounds can be re-contextualized after the fact by either the original performer or others in the group. And if one musician drops out or is unable to make a performance, the system can often continue to function without major interruption, perhaps even organizing in ways that are both novel and more complex. From another perspective, however, group improvisation may be less resilient to personality conflicts or pronounced aesthetic differences between individuals. With traditional musical practices that are organized in a predominantly hierarchical manner, personality differences can often be managed in deference to the group leader, the authority of the musical score, or the professionalism of "getting the job done." Free improvisation ensembles tend to aim for a more egalitarian organization that makes them particularly susceptible to the full spectrum of both musical and so-called "extra-musical" influences.[19]

Despite its many promising qualities, improvisation is also rarely, if ever, the "optimal" means to achieve a specific musical end (although it may in fact be both a quicker and easier route to certain types of chaotic dynamics). The internal dynamics of an improvising ensemble (particularly larger groupings of musicians) can be slow to respond to change, and are, for the most part, beyond the control of any one individual. Even when things do appear to work well, it will be impossible to analyze the system's dynamics during or after the fact with absolute precision. As with other emergent forms of order, the collective dynamics of improvisation will, by definition, always transcend the full awareness of individuals. For these and other reasons, many ensembles choose to adopt certain compositional schemes or devices in order to offer some additional degrees of control over the situation. There is no guarantee, particularly in individual performances, that divergent components will find ways to self-organize effectively.[20] In general, however, the improvising music community

---

[19] For a related discussion see [26].

[20] It is interesting to note that, for a music predicated on what can be a very risky endeavor–to improvise collectively in a group setting–accounts of failure can be very difficult to locate in both the academic and trade coverage of the music. Similar to mechanical systems, we may learn as much or even more by examining occasions on which improvised performance appears to falter

demonstrates the remarkable ability to absorb the new and the diverse without disruption.

Individual ensembles will often, over time, establish their own sense of identity or coherence. The boundary that develops naturally within an ensemble is not necessarily one of personal affinity or exclusion, or one of aesthetic mandate, but rather one of trust and conviviality. Like the boundary of a storm or the membrane of a human cell, this boundary is both permeable and permanent. It defines the identity of the system but also allows for the ongoing dynamics of exchange that are necessary to maintain its existence. Of course, a certain danger may lurk for both physical and musical systems if this boundary becomes either too porous or too impermeable. If too much exchange is fostered with outside forces, the identity of a system may be put in jeopardy. Likewise, if too little exchange is allowed or encouraged, a system may decline either from reduced internal dynamics, or from its inability to continue to adapt to the changing dynamics of its environment.

Network theory tells us that very different things can be connected through surprisingly short distances. Small effects can have large causes, while at other times large disturbances may be absorbed without much notice. Although the predictive power of network theory is still an open question, it may be enough that through these perspectives and approaches we can gain a better understanding of the structure of connected systems and the way that different sorts of influences propagate through them. Duncan Watts [27], another leading voice in the field, reminds us that, "Darwin's theory of natural selection, for instance, doesn't actually predict anything. Nevertheless, it gives us enormous power to make sense of the world we observe, and therefore (if we chose) to make intelligent decisions about our place in it" (p. 302).

Although only limited work has been done on large-scale music networks to date, one study that explored the relationships between jazz musicians from 1912 to 1940 found so-called "small world" properties. By using the Red Hot Jazz Archive database on the Internet, Pablo Gleiser and Leon Danon [28] found that, on average, only 2.79 steps separated early jazz musicians from one another. Their model also captured the clustering of jazz musicians by geography, with New York and Chicago as the major hubs, and by race, due to the highly segregated nature of the music industry at the time. As in most human networks, a few individuals had very high degrees of connectivity. Guitarist Eddie Lang topped their list, with connections to 415 other musicians, while artists like Jack Teagarden, Joe Venuti, and Louis Armstrong were all in the top 10 of most connected musicians. UCSD Professor Richard Belew and I are beginning a similar project to study the network dynamics of musical communities using discographic information that will take account of more contemporary artists as well.

Through the wonders of modern network technologies we can now connect to the farthest reaches of the globe in an instant. And with more than a century of recorded music available to us, we can easily engage with sounds that are similarly removed from us, both culturally and historically. But in the age of iPods and web surfing we also experience the world in increasing isolation at the same time. Yet the resoundingly social nature of music, when viewed as performance rather than product, offers the possibility for humans to synchronize their ears, brains, and bodies in ways that may be unavailable otherwise. And improvised music's particular penchant for the emergent and unexpected may even allow us to explore and expand our own homophily parameter–the sociological tendency of like to associate with like–as familiar and less familiar sounds and people join together to find a common ground, even if only temporarily.[21]

4 Harnessing Complexity

How can these practices be nurtured, particularly within the rather serious and sedate halls of the music academy? The jazz community has traditionally stressed a type of learning that might be called in contemporary discourse embodied, situated, and distributed.[22] Not only have many performers stressed the full integration of aural, physical, and intellectual aspects of the music, but the notion that learning and development can only occur within a supportive community is seen as paramount. The Association for the Advancement of Creative Musicians (AACM) in Chicago and the Creative Music Studio (CMS) in Woodstock, NY are two of the better-known examples of this pedagogical orientation. In the standard music academy, however, the study of musical improvisation has often been shoehorned into the conventional curriculum or simply not addressed at all.

When addressed, institutionalized approaches to teaching musical improvisation have tended to stress individual facility through memorization and pre-planning, leaving little room for collective experimentation. Jonty Stockdale [29] finds that: "[I]mprovisation in jazz studies programmes is infrequently developed through a collective process, with a preference for the development of soloing facility through the absorption and imitation of pre-existing language, usage, and style. Whilst this is regarded as important for the development of a young jazz musician, matters of self-expression, individualism, and most importantly experimentation are often left to later stages, by which time exploration of free collective playing can appear unnecessary or even redundant" (p. 109).

In his account of group creativity, Keith Sawyer [3] makes an important distinction between a problem-solving and a problem-finding approach to art. Artists adopting problem-solving techniques begin with a relatively detailed plan and work to accomplish it successfully.   Those employing a problem-finding approach, by contrast, search for interesting problems as the work unfolds in an improvisatory manner. Many beginning jazz improvisers are stuck in a problem-solving mode. As pianist/composer Anthony Davis expressed to me in a recent interview: "They have been taught right and wrong–these are the notes, these are the chords, these are the arpeggios that work on a given chord. This chord happens on the 5th bar [in a blues]." But through extended listening, practicing, and playing with musicians who are more experienced, Davis finds that jazz players can move from a "dependence on articulating the form" to "using the form, realizing that [the tune structure] is the beginning of something and you have to create something else... They have to do more than just keep time, they have to articulate time... They can make melodic

---

[21] Duncan Watts's current research shows that the most searchable networks involve individuals who are neither too one-dimensional nor too scattered. As long as people have at least two dimensions along which they are able to judge their similarity to others, then small world networks are possible–people can still find short paths to remote and unfamiliar areas.

[22] For more on this topic see chapter seven in Borgo [40].

choices that are at least as strong as the melody that was there before." Even as students become more proficient, however, Davis reminds them that, "You have to get beyond your mannerisms to really come up with a musical idea as opposed to a catalog of what you do."

Problem-finding approaches are equally important when improvising in a group, since it is often impossible to determine the meaning of an action until other performers have responded to it. The particular challenge of group improvisation, then, is that each performer may have a rather different interpretation of what is going on and where the performance might be going. In other words, intersubjectivity is intrinsic to group performances. For Sawyer [3], however, "The key question about intersubjectivity in group creativity is not how performers come to share identical representations, but rather, how a coherent interaction can proceed even when they do not" (p. 9). In part, this is possible because individuals shape a performance on both denotative and metapragmatic levels; they simultaneously enact the details of a performance and negotiate their interactions together. Even if a singular meaning to performance always remains elusive, participants can shape the ways in which their various interactions unfold.

Davis stresses that it is critical that students learn the difference between listening and following: "In order to listen, you don't necessarily follow...You try to construct something that coexists or works well with something else–not necessarily this tail-wagging-the-dog thing where you just follow someone." For Davis, "Listening is knowing what someone is doing and using it in a constructive way, as opposed to mimicry, just trying to demonstrate that you are quote-unquote listening." The very notion that everything could be heard, processed, and immediately responded to during complex moments of improvised music is, by itself, far too facile. Trombonist/composer/scholar George Lewis [11] describes a type of "multi-dominance" in improvised music–an African-American aesthetic by which individuals articulate their own perspectives yet remain aware of the group dynamic, ensuring that others are able to do so as well.

Yet exactly how group flow is cultivated in improvised performances can remain rather mysterious. Describing his general approach to me, contrabassist Bertram Turetzky remarked: "One way when I play free music, I try not to think of anything. I respond or I initiate. And whatever my intuitions tell me, I go with them... Other times in free music, I play with people perhaps I don't know. And I say, well, the last one started soft and slow and got faster and then went back... So all of a sudden I start banging things and doing all kinds of stuff... For some people, I think you have to be very rational. And you perhaps have to have an idea of where you think it could go, and be the quarterback." Turetzky acknowledged that establishing a proper group rapport can be difficult "if someone has a big ego and wants to make everything compositional." When he perceives that the group flow is in jeopardy, at times he may adopt a third strategy: "If there are three of four people, maybe I'll stop a little bit and let them see what they want to do. If there is a mess, let them sort it out. Let them start something and maybe I can support them."

Certain exercises employed by improvising actors may be useful for improvising musicians. For instance, dramatist Keith Johnstone [30] believes that, "Humans are too skilled in suppressing action. All the improvisation teacher has to do is to reverse this skill and he creates very gifted improvisers. Bad improvisers block action, often with a high degree of skill. Good improvisers develop action" (p. 95).[23] Improvising actors are taught that, instead of denying or rejecting what has been previously introduced into the dramatic frame, they should accept the actions/words of others as dramatic "offers" and, in turn, add something to the dramatic frame, i.e., present a complimentary "offer," or "revoice" an existing "offer." The inherent challenge is to avoid circumscribing or over-directing the group flow. This does not, however, preclude the possibility of swiftly changing dramatic or musical directions, as the case may be, but care should be taken to do this in a way that keeps previous developments available for future moments of reference or expansion; a practice called "shelving" by improvising actors. Of course, evaluating exactly when "revoicing" or "shelving" the "offers" of others has been successful can be a tricky proposition. And the inherent complexity, polyphony, and polysemy of music can make this even more challenging. At heart, however, these exercises in improvised theater, and similar ones adopted by musicians, are designed to improve one's ability to listen and remember, so that the ongoing group development will be stimulated rather than curtailed.

Compositional schemes and strategies are often employed to help organize improvised music, either prior to, or in the moment of, performance. Deciding how or how much to organize performances, here again, becomes a tricky endeavor. John Zorn's *Cobra* may be the best-known "game piece" for improvising musicians. Making a distinction between his work and conventional notions of composition, Zorn remarked:

> In my case, when you talk about my work, my scores exist for improvisers. There are no sounds written out. It doesn't exist on a time line where you move from one point to the next. My pieces are written as a series of roles, structures, relationships among players, different roles that the players can take to get different events in the music to happen. And my concern as a composer is only dealing in the abstract with these roles like the roles of a sports game like football or basketball. You have the roles, then you pick the players to play the game and they do it. And the game is different according to who is playing, how well they are able to play...[24]

With their attentions already engaged in complex ways during performance, others worry that highly involved schemes for structuring improvisation can hinder rather than assist the natural development of the music. For instance, performer/scholar Tom Nunn [32] writes: "When improvisation plans are complicated—no matter how clear or well explained they might be—the attention of the improviser is constantly divided between the plan and the musical moment, having to remember, or look at a score, a graphic, or even a conductor. What often happens is that both the plan and the music suffer from this divided attention" (p. 162).

In a recent interview, contrabassist Mark Dresser discussed with me the challenges inherent in structuring pieces for improvisers: "Composition is often about control. You have to build [improvisation] in. I've built pieces that have been little prisons,

---

[23] For a related treatment regarding jazz improvisation, see [31].
[24] Quoted in [10], p. 233.

too. You're looking at something really specific." But he added, "It's a trip to find the balance. You try to find combinations where you have real focus and condensation, and points of real expansion. For me, it is all about being a complete musician. All of those things are interesting. At different points in the evening I try to have all of those things. Its funny, though, when you get in the composer's head it's really hard to let go of trying to control it or to create this kind of balance."

Even compositional strategies that have the sole intent of facilitating group improvisation during performance can backfire. Referring to Butch Morris's extensive system of conducted gestures designed to help organize improvised performances, Dresser commented: "I've seen the conduction thing be a disaster with people who just don't like to be controlled." Without pre-conceived strategies, however, there is an ever-present danger that improvised music will fail on its own. This danger may also increase with the size of the group. Philip Alperson [33] writes: "As the number of designing intelligences increases, the greater is the difficulty in coordinating all the parts; the twin dangers of cacophony and opacity lurk around the corner" (p. 22).

This makes those moments when group improvisation is deemed successful all the more powerful. While interviewing bassist Lisle Ellis, he confided: "A lot of improvised music I don't think is very good music. But man, when it hits, it's extraordinary! That's what I've spent my life doing–waiting for those moments when it really lines up–to find a way to have some consistency in it. Some days I think I really know how to do that and other days I think I don't have a clue." In a telling aside that highlights this balancing act of harnessing creativity, Ellis remarked, "I've got to write more stuff down. I've got to write less stuff down."

When discussing improvisation and composition, it can be particularly challenging to avoid thinking in terms of simple dichotomies while at the same time remaining leery of equally facile truisms about the music. Only with dualistic thinking, which presents two things as opposed and forces one to choose between them, are preparing for something in advance and the leap of freedom into the unforeseen viewed as antithetical or incompatible. Dresser finds that, "Within control there are lots of possibilities for freedom." And discussing his time spent as young man in classes with Muhal Richard Abrams at the AACM school, George Lewis [34] writes: "Improvisation and composition were discussed as two necessary and interacting parts of the total music-making experience, rather than essentialized as utterly different, diametrically opposed creative processes, or hierarchized with one discipline framed as being more important than the other" (p. 86). Dresser recounted a telling moment during his first tour with Anthony Braxton's quartet that resonates with this issue: "The only time that Braxton criticized the quartet, he said, 'Well, you guys are playing the music correctly, but you're just playing it correctly.' The criticism was you are being too dutiful, you're not taking a chance. That was the day that the format of the music actually changed, from being a solo-based music to an ensemble music. All of a sudden, the nature of the music became different. That moment articulated when the group came into its own."

5 Final Thoughts

Why do people tend to assume that systems are organized either by lead or by seed? In part, this is undoubtedly due to the fact that many if not most of our social institutions and artistic creations *are* organized in this way. Yet an extreme reliance on centralized organization and centralized metaphors in the past has led to a situation in which many people are unwilling or unable to imagine systems organizing in a decentralized fashion.[25] When people hear music they tend to assume a composer, a leader, or, when that music is improvised, they tend to assume that creativity emerges solely from the individual. In many cases these intuitions may be right. But one of the more encouraging aspects of much contemporary experimental music is that it is not always easy or even possible to know if a particular instance of music was or was not composed ahead of time.[26] And the generative power of computers is blurring these lines even further. Perhaps most encouraging of all, however, is the fact that creativity is increasingly being viewed as a web of network interactions operating on all scales, reflecting individual, social, cultural, and historical dimensions.

There are many compelling reasons to view artistic behavior not as some special kind of activity cut off from the rest of human behavior but rather as much an adaptation to the environment as any other human activity. Since a primary drive of human beings is to perceive the environment as comprehensible and to make successful predictions about the future, we have developed a cognitive/sensory orientation that filters out any data that is not relevant to the needs of the moment. But since such an orientation does not prepare an individual to deal with a *particular* situation but only with a *category*, or *kind*, or *class* of situations, much of the suppressed data may very well be relevant. The arts in general, and music in particular, may serve the function of breaking up entrenched orientations, weakening and frustrating our "tyrannous drive to order," so that humans are better able to deal with change, complexity, and chaos.[27]

Improvisers engage the unforeseen; they offer the experience of disorientation.[28] They look to find problems, rather than to solve them. Improvised music also reminds us that the notion of "art" is most appropriately located not in the "work" itself, but rather in the perceiver's role; a role that involves maintaining a search-behavior focused on discontinuities. Emotional affect is not intrinsic to the "work",

---

[25] Decentralization may be biological coded for ants and other social insects, but it does not seem to be as natural or automatic for humans. Or it may simply be that, because we are within the system, we remain unaware of its emergent properties, just as individual bees and ants may be unaware of their group's emergent social organization (although this hypothesis is difficult if not impossible to test). For lucid writing on this subject see [35] and [36].

[26] Although this blurring may be artistically encouraging, we still need to be aware of cultural assumptions that accompany our notions of musicking. Eddie Prévost [37] recounts an AMM performance after which a woman came up to the musicians and remarked how moved she had been by the music. Once she learned that the group had been improvising rather than playing from a memorized score, she not only doubted their artistic and intellectual integrity, but she was forced to question her own powers of discrimination. "How had it been possible for her to enjoy and admire such work when its practice had been so... primitive."

[27] For some prescient writing on this subject see [38].

[28] The Latin roots of the word improvisation are *in*-not and *provisus*-foreseen.

but rather is dependent on a successful performance of the perceiver's role; emotion is the result of a discrepancy between expectation and actuality.[29]   Perhaps most importantly, improvising music together allows participants and listeners to experience and explore complex, decentralized, interconnected, and emergent social dynamics.

Recent work in the cognitive neuroscience of music concerned with the role that music plays in human evolution and development supports this view rather well. Ian Cross [41], a leading researcher in this still nascent field, argues that music's nonefficaciousness–its general remove from immediate concerns for survival (from a strict biological perspective)–make it especially well suited to testing out aspects of social interaction, while its polysemy–its ability to producing multiple meanings– endows us with the multipurpose and adaptive cognitive capacities that make us human. In less technical language Cross writes: "[M]usic can be both a consequence free means of exploring social interaction and a 'play space' for rehearsing processes that may be necessary to achieve cognitive flexibility" (p. 51).[30]   People cooperating in a musical activity need not find the same meaning in what they do in order for the musical event to assist them in acquiring and maintaining the skill of being a member of a culture. As Cross sees it, "The singularity of the collective musical activity is not threatened by the existence of multiple simultaneous and potentially conflicting meanings" (ibid.). Through continual engagement with art–viewed as the successful performance of the perceiver's role–we may in fact be better prepared to survive and flourish in our increasingly interconnected, and therefore interdependent, world.

It is interesting to note that two of the hottest current topics for organizational design are the sciences of complexity and jazz music. Both domains emphasize adaptation, perpetual novelty, the value of variety and experimentation, and the potential of decentralized and overlapping authority in ways that are increasingly being viewed as beneficial for economic and political discourse. Robert Axelrod and Michael Cohen [43] see in the move from the industrial revolution to the information revolution a powerful shift from emphasizing discipline in organizations to emphasizing their flexible, adaptive, and dispersed nature. And Karl Weick [44], in a special issue of the journal *Organization Science* devoted to an exploration of "the jazz metaphor," finds that the music's emphasis on pitting acquired skills and pre-composed materials against unanticipated ideas or unprogrammed opportunities, options, or hazards can offset conventional organizational tendencies towards control, formalization, and routine. In a response to the heavy reliance by journal contributors on swing and bebop as the source of their jazz metaphors, Michael Zack [45] outlined ways in which free jazz might propel discourse even further into the realm of emergent, spontaneous, and mutually constructed organizational structures.

Are there lessons from improvising music that can help us to understand, or at least to cope with, the complexity of our world? Improvising music makes us aware of the power of bottom-up design, of self-organization. It operates in a network fashion,

---

[29] See Joseph Goguen's work in [39] and in the co-author chapter of [40].

[30] The notion of music as a "consequence free" activity is somewhat problematic, but it is used here in the biological sense that music, in most all cases, does not by itself do physical harm to humans. Since social interactions play an important role in our cognitive development it should also be clear that these two properties cannot be easily divorced from one another. The notion of "play" in relation to improvised music is taken up in [42].

engaging all of the participants while distributing responsibility and empowerment among them. Networks facilitate reciprocal interactions between members, fostering trust and cooperation, but they also can concentrate power in the hands of a few. Under the best of circumstances, improvising music encourages social activities that support the growth and spread of valued criteria through the network. For instance, improvisers tend to value diversity, equality, and spontaneity and often view their musical interactions as a model for appropriate social interactions. Tom Nunn [32] writes: "Free improvisers are important to the society in bringing to light some fundamental values and ideas, for example: how to get along; how to be flexible; how to be creative; how to be supportive; how to be angry; how to make do. So there is a social and political 'content' in their music that seems appropriate today, though it may not usually be overt" (p. 133).

As we continue to explore ways of improvising music, we should look for ways to assist would-be cooperators in interacting more easily and more frequently. The robustness and equity of a network system is a direct result of the range and number of interactions. We should also look to maximize participation from the fringes, rather than the core. In complex systems, a healthy fringe speeds adaptation, increases resilience, and is almost always the source of innovations. For instance, nearly every new style of American popular music has emerged from the periphery—from a localized, and often disadvantaged, community–to capture the attention of national and international audiences (at which time much of the music's original meaning may of course be sacrificed).

Fostering improvising music has the potential to overcome the inherent problems of a slow-moving traditional hierarchy, providing an effective way to handle unstructured problems, to share knowledge outside of traditional structures, and to inject local knowledge into the system. Improvising music also ensures that the cognitive models and metaphors we live by remain flexible, while it reminds us that our flexibility to learn and adapt are grounded in the bodily and the social. Without cultivating this embodied, situated, and distributed approach to music making, and without maintaining a healthy reverence for uncertainty, we can build complicated music systems, but not complex ones.

Complex systems must strike an uneasy and ever-changing balance between the exploration of new ideas or territories and the exploitation of strategies, devices, and practices that have already been integrated into the system. In other words, complex systems seek persistent disequilibrium; they avoid constancy but also restless change. Perhaps in a way similar to democracy, which along with jazz music has been a powerful symbol of liberation and resistance to oppression, improvising music teaches us to value not only cooperation, but also compromise and change. In politics, as in music, a notion of the "common good" is bound to mean different things to different individuals and groups, such that the democratic experience is one of not getting everything you want. In a similar way, the value of improvising music lies not in the outcome of a single performance, but rather it emerges over time through continued musical and social interactions. Improvising music together does not necessarily produce optimal outcomes, but the decision to improvise music together does.

References

[1]   Mihaly Csikzentmihalyi and Grant Jewell Rich. Musical improvisation: a systems approach. In R. Jeith Sawyer, editor, *Creativity in Performance*, pp. 43-66. Ablex Publishing Corporation, 1987.

[2]   Derek Bailey. *Improvisation: Its Nature and Practice in Music*, Da Capo Press, 1993.

[3]   R. Keith Sawyer. *Group Creativity: Music, Theater, Collaboration*. Lawrence Erlbaum Associates, 2003.

[4]   Mihaly Csikzentmihalyi. *Flow: The Psychology of Optimal Experience*. Dimensions, 1991.

[5]   Paul Berliner. *Thinking in Jazz: The Infinite Art of Improvisation*. University of Chicago Press, 1994.

[6]   Ingrid Monson. *Sayin' Something: Jazz Improvisation and Interaction*. University of Chicago Press, 1996.

[7]   David Borgo. The dialogics of free jazz: musical interaction in collectively improvised performance. M.A. thesis, UCLA, 1996.

[8]   John Corbett. *Extended Play: Sounding Off From John Cage to Dr. Funkenstein*. Duke University Press, 1994.

[9]   David Toop. Frame of freedom: improvisation, otherness and the limits of spontaneity. In *Undercurrents: The Hidden Wiring of Modern Music*. Continuum, 2002.

[10] John Corbett. Ephemera underscored: writing around free improvisation. In Krin Gabbard, editor, *Jazz Among the Discourses*, pp. 217-40. Duke University Press, 1995.

[11] George Lewis. Too many notes: computers, complexity and culture in voyager. *Leonardo Music Journal* 10:33-39, 2000.

[12] Kevin Kelly. *Out of Control: The New Biology of Machines, Social Systems, and the Economic World*. Addison-Wesley Publishing Company, 1994.

[13] Norbert Weiner. *Cybernetics*. MIT Press, 1961.

[14] Ilya Prigogine and Isabelle Stengers. *Order Out of Chaos*. Bantam Books, 1984.

[15] Eric Bonabeau and Guy Théraulaz. Swarm smarts. *Scientific American* (March):72-79, 2000.

[16] Eric Bonabaue, Marco Dorgio, and Guy Théraulaz. *Swarm Intelligence: From Natural to Artificial Systems*. Oxford University Press, 1999.

[17] Evan Parker. Shopping with Evan Parker. *Monastery Bulletin* (October), 2004.

[18] James Surowiecki. *The Wisdom of Crowds*. Doubleday, 2004.

[19] J. Stephen Lansing. Complex adaptive systems. *Annual Review of Anthropology* 32:183-204, 2003.

[20] George Lewis. Improvisation and the orchestra: a composer reflects. essay to accompany a performance by the American Composers Orchestra at the Improvise! Festival, April 28, 2004.

[21] Albert-László Barabási. *Linked: The New Science of Networks*. Perseus, 2002.

[22] Steven Shaviro. *Connected: Or What it Means to Live in a Network Society*. University of Minnesota Press, 2003.

[23] John Law and John Hassard. *Actor Network Theory and After*. Blackwell, 1999.

[24] David Hajdu. Wynton's blues. *The Atlantic Monthly* (March), 2003.

[25] Malcolm Gladwell. *The Tipping Point: How Little Things Can Make a Big Difference*. Back Bay Books, 2002.

[26] David Borgo. Synergy and surrealestate: the orderly-disorder of free improvisation. *Pacific Review of Ethnomusicology* 10:1-24, 2002.

[27] Duncan Watts. *Six Degrees: The Science of a Connected Age*. W.W. Norton & Co., 2003.

[28] Pablo Gleiser and Leon Danon. Community structure in jazz. *Advances in Complex Systems* 6(4):565-573, 2003.

[29] Jonty Stockdale. Reading around free improvisation. *The Source: Challenging Jazz Criticism* 1:101-114, 2004.

[30] Keith Johnstone. *Impro: Improvisation and the Theater.* Faber and Faber, 1979.

[31] Kenny Werner. *Effortless Mastery: Liberating the Master Musician Within.* Jamey Aebersold, 1996.

[32] Tom Nunn. *Wisdom of the Impulse: On the Nature of Musical Free Improvisation.* self published, (tomnunn@sirius.com), 1998.

[33] Philip Alperson. On musical improvisation. *Journal of Aesthetics and Art Criticism* 43(1):17-29, 1984.

[34] George Lewis. Teaching improvised music: an ethnographic memoir. In John Zorn, editor, *Arcana: Musicians on Music.* Granary Books, 2000.

[35] Peter Russell. *The Global Brain: Speculations on the Evolutionary Leap to Planetary Consciousness.* J.P. Tarcher, Inc., 1983.

[36] Howard Bloom. *Global Brain: The Evolution of Mass Mind from the Big Bang to the 21st Century.* John Wiley & Sons, 2000.

[37] Edwin Prévost. *Minute Particulars.* Matchless, 2004.

[38] Morse Peckham. *Man's Rage for Chaos.* Chilton Books, 1965.

[39] Joseph Goguen. Musical qualia, context, time, and emotion. *Journal of Consciousness Studies* 11(3/4):117-147, 2004.

[40] David Borgo. *Sync or Swarm: Improvising Music in a Complex Age.* Continuum, 2005.

[41] Ian Cross. Music, cognition, culture, and evolution. In Isabelle Peretz and Robert J. Zatorre, editors, *The Cognitive Neuroscience of Music,* pp. 42-56. Oxford University Press, 2003.

[42] David Borgo. The play of meaning and the meaning of play in jazz. *Journal of Consciousness Studies* 11(3/4):174-190, 2004

[43] Robert Axelrod and Michael D. Cohen. *Harnesing Complexity: Organizational Implications of a Scientific Frontier.* Basic Books, 2000.

[44] Karl Weick. Improvisation as a mindset for organizational analysis. *Organization Science* 9(5):543-555, 1998.

[45] Michael Zack. Jazz improvisation and organizing: once more from the top. *Organization Science* 11(2):227-234, 2000.

My Friend Joseph Goguen

Rod Burstall

University of Edinburgh (Retired)
4/12 Belhaven Place, Edinburgh EH10 5JN, UK

Abstract. A personal account of how Joseph Goguen and I came to
work together and of the influence that Tibetan Buddhism had on us
and on our collaboration. A brief discussion of some neurological experiments using meditators and how Goguen's work connects Buddhism,
computing, and cognition.

Mind, Heart and Meditation

Joseph Goguen has an extraordinary mind and a big heart. My friendship with
him is long and deep, and it has affected my life in two major ways since I
met him in 1974. First we worked closely together on modularity and program
specification for a dozen years or more and continued to have many conversations
about computing until my retirement in 2000. During this time I learnt a lot
from Joseph about category theory and its applications to computing. Second
he introduced me to Buddhist practice and thought under the guidance of a
Tibetan teacher, Chögyam Trungpa Rinpoche, with whom we both studied until
his death in 1987.

I will say a little about how I met him and what our motivations were, but
mainly I would like to describe a part of his life, and mine, which will be less
familiar to most readers, namely his interest in Buddhism. Later this year it
will be the thirtieth anniversary of Joseph giving me instruction in the meditation technique which he learned from Chögyam Trungpa. I am still practicing
it regularly, and now I spend some months each year in solitary or group meditation retreats. I have been thinking particularly about the connection between
Buddhist empirical knowledge and science.

Meeting and Working with Joseph

In the early seventies I was working at Edinburgh University on programming
languages and correctness proofs, and I had learned some elements of universal
algebra and category theory. When I was in the US I arranged to visit Jim
Thatcher, who had written a paper with Wright on categories and automata. It
turned out that Jim was more interested in stopping the Vietnam War than in
category theory, but both ideas were fine by me. I gave him a hundred dollars
to stop the war, and I got him a visiting fellowship to Edinburgh. One day
Jim told me that a colleague of his, a very clever mathematician named Joe
Goguen, would visit us in Edinburgh. I was quite terrified of meeting a clever

K. Futatsugi et al. (Eds.): Goguen Festschrift, LNCS 4060, pp. 25–30, 2006.

mathematician, but found to my surprise that this Joe person did not frighten me at all. So next year in 1975 when I was at a conference in Los Angeles I fixed up to stay over the weekend with him. It was a very exciting weekend, and at my invitation Joseph spent that summer in Edinburgh on a Science Research Council Visiting Fellowship. This was the start of our technical collaboration.

We were both interested in software and program correctness proofs, also in how to apply proof techniques to larger programs by imposing a modular structure on them. Joseph suggested we think about modularity of specifications, as that might be easier then thinking about programs. This led to our development of a specification language called Clear. To give it semantics we used category concepts to explain parameterised specifications and ways to combine them. It seemed that the parameter mechanisms and the ways of combining them should not depend on the particular specification language, so we came up with the categorical concept of an *institution* to abstract away from the underlying language, which might be equational logic or predicate calculus or whatever. This was done while we were visiting each other in Edinburgh and Los Angeles, two very contrasting environments. I learnt more about categories. In LA, I learned about going out for breakfast. We drew diagrams on paper napkins in the Pancake House, then in the sand on Santa Monica beach. Once when we thought we had had a good idea we danced down the street, under the suspicious eyes of a passing LAPD police car. All this was very exciting and, I believe productive. We continued to work happily together on these topics from time to time until the nineties. I also had much pleasure collaborating for some years with Joseph's talented son Healfdene, who joined my research group in Edinburgh.

Creativity and Uncertainty

Turning to the Buddhist side of Joseph's life, let me backtrack in time and explain why I had started practicing meditation before I met Joseph.

Studying physics at Cambridge University in the fifties I became fascinated by the idea of computing. After various twists and turns, teaching myself with help from friends, I wound up as a Research Fellow in Edinburgh University in the "Experimental Programming Unit". I was thinking about programming languages, artificial intelligence and lambda calculus, cheerfully staying up late at night to write and debug code — by 1966 our unit actually had a computer to ourselves, the only one in the University.

Now I found myself writing papers and taking part in workshops, part of a critical community. After a while I realized that the joy of creativity had become tinged with competition and with doubt whether people would think that I was doing well enough. So it seemed that external activity was not enough, I also had to deal with my own mind. Like many people I had long been interested in mind, indeed part of the attraction of computers was the hope that they could give us insight into the workings of our minds. So meditation was interesting for both personal and intellectual reasons. It promised an investigation of mind from the inside.

Joseph and Tibetan Buddhism

In the early seventies Joseph met and studied with an unusual man, Chögyam Trungpa Rinpoche, an accomplished and respected scholar, meditation master and teacher who fled Tibet in 1959 at the age of nineteen and wound up in the United States, via India and Britain. Trungpa was a poet, artist and practical joker who had a profound impact on many of those who met him.

What Trungpa taught was, in Western terms, somewhere between psychotherapy, philosophy, life coaching, religion and how to become a kind and open-minded person. It was the result of two and a half millennia of empirical investigation of the human mind from the inside, and it included a practical technology for training the mind. But it had been nurtured in Tibet in isolation from the rest of the world, and it was expressed in a language which was little known and hard to translate, with its own rich technical vocabulary. Trungpa opened up to Western culture, learned English and translated not just the words but also the concepts. He also developed a number of non-verbal ways of getting his message across, for example by teaching former hippies who had been attracted to his teachings in the seventies to decorously dance the Viennese waltz. He first indulged his Western students, teased them and then demanded extraordinary effort and discipline. Extraordinary effort and discipline was nothing new for Joseph: in 1975 he spent twelve weeks at a Buddhist 'Seminary' taught by Trungpa, an intensive regime of meditation and study.

It was a few weeks after this that I stayed with Joseph in Los Angeles for the first time as recounted above. He took me along to the local Buddhist centre and played me a taped talk by Trungpa. Someone asked a question about Mozart and to my surprise Trungpa seemed to admire his music — I was curious about this Tibetan guru with an appreciation of eighteenth century European music. The following year when Joseph was spending a second summer in Edinburgh I asked him to teach me the basic Buddhist meditation technique.

In 1981 Joseph attended a second twelve-week seminary, and this time my wife and I were there too (our two eldest daughters followed in later years). Over three hundred students and staff took over an off-season hotel in the Canadian Rockies. Periods of meditation, 7 a.m. to 9 p.m. with brief breaks after meals, alternated with periods of teaching, study and more meditation, running later still if Trungpa was teaching. Our family continued to practice what we had learnt, and we were very grateful to Joseph and to Trungpa Rinpoche.

We were taught practices to pacify the mind, open up our awareness and develop kindness and compassion to others. The main point was to pay moment by moment attention to what was actually happening in our minds, wandering thoughts and emotions: irritation, curiosity, regret, benevolence or whatever — touch it and let go. This was a sort of animal behaviour investigation with the animal being our own mind. Beyond this there were techniques aiming to change one's mental processes by prescribed exercises of the imagination. In particular we worked to diminish the "destructive emotions" of anger, passion/addiction, ignorance, envy and arrogance. (A selection of Trungpa's talks is available in [8].)

Since then Joseph and I have both pursued the Buddhist path under the direction Trungpa and, after his death, under his son Mipham Rinpoche. We have many times shared our ideas and experiences. So now let me sketch some ways in which this could connect with the scientific side of our lives in neuroscience, psychology of emotions or cognitive linguistics, also with Joseph's own contribution to studies of consciousness.

Connections Between Buddhist and Western Explorations of the Mind

The two and a half thousand year old culture which we call Buddhism developed psychological models for the mind and the processes of perception and action, based on internal meditative investigations and the results of many different methods of training the mind. These methods are essentially technical, complete with manuals, and not based on any kind of supernatural interventions. They are, of course, not exclusive to Buddhism, witness other Indian traditions, Sufis and Christian contemplatives.

Stephen Laberge, who has conducted experiments on lucid dreaming and compared his techniques with those of the Tibetan tradition [4], comments

> The effectiveness of a psychological technique can be tested by careful observers of the contents of consciousness without the need of technology other than a well-trained mind and a disciplined body. On contrast, testing the validity of an explanation of that technique may require the extremely sophisticated technology needed for the visualization and measurement of neural activity.

Richard Davidson at the University of Wisconsin at Madison was able to examine with an fMRI scanner and with EEG the neural activities of advanced meditators using the Tibetan methods. In a first experiment "Lama Oser" (a pseudonym), a Westerner, who has been a Tibetan monk for about thirty years, was tested using six different meditation practices, one minute each with a pause of one minute between them. Oser's brain showed clear distinctions between these different meditations and the pauses. His sharp shifts between different activities were exceptional. In the EEG tests, when meditating on compassion, his brain showed "a dramatic increase in the electrical activity known as gamma" in an area of the brain associated with "happiness, enthusiasm, high energy and alertness" [3, pp. 1–13]. In a later experiment, Davidson was able to confirm the EEG results, comparing a group of experienced meditators with a group of novice meditators [1,7].

Paul Ekman of the University of California at San Francisco, an expert on the science of emotion, tested the ability of Lama Oser and another very experienced Western meditator (each had done a total of two or three years of solitary retreats). In one test he asked them to identify "microemotions", facial emotions such as fear or contempt, which only appear for a fraction of a second and are impossible to control deliberately, showing them videotapes of flashes of one fifth or even one thirtieth of a second of the faces. They both showed ability two standard deviations above the norm, far higher than any of the five thousand other people tested, including policemen, psychiatrists and even Secret Service agents. Such a diagnostic ability for emotions would be helpful to guide students in the transformative practices of the Buddhist tradition [3, pp. 13–21, 123–131].

Turning to psychological models rather than meditation techniques, the Mahayana tradition of Buddhism emphasizes the concept of "emptiness" (Sanskrit "shunyata"). It is puzzling how this relates to Western traditions. The Madhyamaka approach tries to show the inadequacy of our conceptual system by reductio ad absurdum arguments. Some of these seem to deal with paradoxes reminiscent of Zeno's paradoxes, for example ones about movement, which have been clarified by Western work on calculus and limits. But it seems to me that these arguments should be directed, not at our mathematics or physics theories, but rather at the built-in conceptual reasoning systems which are common to all humans. These systems are of interest to cognitive science and cognitive linguistics, and Joseph has long drawn my attention to the work of George Lakoff and his associates on "metaphor". The idea here is that our conceptual models of the world start from in built sensory motor conceptual schemes, such as the idea of containment, for example "a triangle inside a square" or "the path of a movement with starting and finishing points". From these other more abstract concepts, "being in trouble" or "on the road to ruin", are derived by metaphors (mappings or morphisms). A derived concept can have its meaning determined by several such metaphorical maps. Lakoff and co-authors have applied these ideas to human understanding both in philosophy [5] and in mathematics [6]. All this is reminiscent of the early work by Joseph and myself on defining a specification language, Clear, in terms of theories and theory morphisms. In the last few years Joseph has been working on the construction of conceptual systems using "semiotic morphisms" [2] (For other references see his website http://www.cs.ucsd.edu/users/goguen/projs/semio.html).

Another connection is Joseph's work as founder and editor of the Journal of Consciousness Studies, which has fostered growing interest in this aspect of mind and published work from many disciplines by philosophers, psychologists, neuroscientists, linguists and practitioners of the ancient traditions of contemplation and meditation.

The Buddhist tradition is just one of many wisdom traditions, spiritual, psychotherapeutic and medical. We need to keep these alive. Using the analytical methods and tools of science, some elements of these traditions will be better understood, so that they can take their place as part of the global culture of accepted knowledge.

Conclusion

I hope that this personal view will have illuminated one less public side of Joseph's life journey and given some feel for how it coheres with his extensive and admirable work in computing. I count myself very fortunate to have shared some part of that journey with him.

Acknowledgements

I am deeply grateful to my teachers, notably Chögyam Trungpa Rinpoche, Sakyong Mipham Rinpoche and Ringu Tulku Rinpoche. Thanks go to José Meseguer for editorial comments and to Joe Hendrix for helping to prepare the manuscript for publication.
As for Joseph: these words do not suffice.

References

1. Davidson, R. J., Kabat-Zinn, J., Schumacher, J., Rosenkranz, M. A., Muller, D., Santorelli, S. F., Urbanowski, F., Harrington, A., Bonus, K. and Sheridan, J. F. (2003) "Alterations in brain and immune function produced by mindfulness meditation" *Psychosomatic Medicine* 65:564–570.
2. Goguen, Joseph A., Harrell, Fox (2006) "Foundations for Active Multimedia Narrative: Semiotic Spaces and structural blending" To appear in *Interaction Studies: Social Behaviour and Communication in Biological and Artificial Systems.*
3. Goleman, Daniel (2003) *Destructive Emotions: a dialogue with the Dalai Lama.* Bloomsbury UK. Bantam USA.
4. Laberge, S. (2003) "Lucid Dreaming and the Yoga of the Dream State" In *Buddhism and Science: breaking new ground.* ed. Wallace, B. Alan. Columbia University Press, New York USA and Chichester UK. 233–258.
5. Lakoff, G. and Johnson, M. (1999) *Philosophy in the Flesh: the embodied mind and its challenge to western thought.* Basic Books, New York USA.
6. Lakoff, G. and Nez, R. E. (2000) *Where Mathematics Comes From: how the embodied mind brings mathematics into being.* Basic Books, New York USA.
7. Lutz, A., Greischar, L., Rawlings, N. B., Ricard, M., Davidson, R. J. (2004). "Long-term meditators self-induce high-amplitude synchrony during mental practice". Proceedings of the National Academy of Sciences, 101: 16369–16373.
8. Trungpa, Chogyam (1999) *The Essential Chgyam Trungpa.* ed. Gimian, C., Shambhala, Boston USA.

Metalogic, Qualia, and Identity on Neptune's Great Moon: Meaning and Mathematics in the Works of Joseph A. Goguen and Samuel R. Delany

D. Fox Harrell

Department of Computer Science and Engineering
University of California, San Diego
9500 Gilman Drive, Mail Code 0404
La Jolla, CA 92093-0404
USA
fharrell@cs.ucsd.edu

Abstract. The works of Joseph A. Goguen and Samuel R. Delany address wide arrays of "big" issues in philosophy: identity and qualitative experience, semiotic representation, and the divergence between meaning in formal systems of understanding and in everyday lived experience. This essay attempts to draw out some of the parallels between the works of these two authors, in particular regarding metalogic, qualia, and identity, using illustrative examples from the works of both authors. Their works exhibit parallel dual strands: (1) a desire to rigorously and precisely map out these fundamental issues, and (2) a desire to acknowledge and embrace the ambiguities of phenomenological experience and its divergence from any formalizable theory. In the end, addressing such a wide range of issues has required both authors to develop and adopt new discourse strategies ranging from rational argumentation to mathematics, from religious and philosophical commentary to speculative (science) fiction and poetry.

1 Introduction

> *A perusal of any dozen pages from the Summa reveals Slade's formal philosophical presentation falls into three, widely differing modes. There are the closely reasoned and crystallinely lucid arguments. There are the mathematical sections in which symbols predominate over words; and what words there are, are fairly restricted to: "... therefore we can see that...," "...we can take this to stand for...," "...from following these injunctions it is evident that...," and the like. The third mode comprises those sections of richly condensed (if not impenetrable) metaphor, in language more reminiscent of the religious mystic than the philosopher of logic. For even the more informed student, it is debatable which of these last modes, mathematical or metaphorical, is the more daunting.* [8]
> – Samuel R. Delany, discussing the work of the fictitious metalogician Ashima Slade

K. Futatsugi et al. (Eds.): Goguen Festschrift, LNCS 4060, pp. 31–49, 2006.
© Springer-Verlag Berlin Heidelberg 2006

I'm afraid that the reader may have found this paper rather a long strange trip, starting from the practice of software engineering, then going to category theory, and eventually ethics, passing through topics like equational deduction, various programming and specification paradigms, semiotics, theorem proving, requirements engineering and philosophy.

From another perspective, this paper can be considered a diary from a very personal journey moving from a mathematical view of computing, through a process of questioning why it wasn't working as hoped, to a wider view that tries to integrate the technical and social dimensions of computing. This journey has required a struggle to acquire and apply a range of skills that I could never have imagined would be relevant to computer science. Always I have sought to discover things of beauty – "flowers" - and present them in a way that could benefit all beings, though of course I don't expect that very many people will share my aesthetics or my ethics. [15]
– Joseph Goguen, excerpts (slightly reordered) from an autobiographical essay tracing the trajectory of his research career

The aroma of algebraic flowers motivates this paper. Joseph Goguen has used the metaphor of flowers to describe the strivings of his own work because of the parsimonious beauty it is possible to evoke with elegant formalizations in mathematics. For him the essence of these "flowers" is rooted in compassion and a true desire to benefit humanity. Yet, Goguen's metaphor for his work is also one of loss. His autobiographical essay ".Tossing Algebraic Flowers Down the Great Divide," [14] suggests that his beautiful work is tumbling downward into a dark crevasse between technical and social scientific or humanistic disciplines, perhaps only to be discovered at an unknown time, or perhaps never.

It is not so! Goguen's algebraic flowers garland a gossamer network of bridges between diverse fields: computing, mathematics, philosophy, sociology, semiotics, narratology, and more. Though perhaps more researchers are familiar with Goguen's work on the technical side of the divide, I intend to highlight the bridge his work builds from computing and mathematics to humanistic and artistic issues. Personally, this bridge has been a profound influence on my work. My academic training is in logic, interactive media art, and computer science. In the course of these studies, I became interested in new forms of interactive narrative that take advantage of the affordances provided by computing. I came to feel that a powerful direction in interactive artwork is to allow user interaction to affect meaning with narratives, and with Professor Goguen's guidance as my advisor this intuitive direction transformed into specific goals, for example generating new metaphors or constructing narratives as users provide input. Toward this end Goguen's algebraic semiotics and his approach to user-interface design were a revelation. He is an expert mathematician dealing with semiotic issues also addressed by art theory. He is a computer scientist who espouses the importance of narrative. Underneath this all is a concern for the social, ethical applications of his work. Because he has not compromised his work toward either side of the divide, Goguen's feeling of loss regarding this work is probably due to the limited number of people on either side of the divide interested in seriously addressing the issues and methods of greatest import on the other side. I have described my own background only because I live directly in the center of the divide. For people like me, Goguen's work in these areas is of great importance both for its application and example. It can be used directly for artistic technical practices and it is an example of what is possible to achieve when combining methods from diverse fields with rigor and a careful attention to the values implicit in them. This essay is intended to convey this important aspect of Goguen's work by focusing on several particular topics in his oeuvre and contrasting them with the work of another author that has inspired me, Samuel R. Delany.

The title of this paper refers to my attempt to find sympathy in the works of these two eclectic and profound authors. The planet Neptune's largest moon is Triton, here alluding to the title of Delany's science fiction novel *Trouble on Triton: An Ambiguous Heterotopia*. The idea for the thesis of this paper was inspired by the character mentioned in the Delany quote above from that same novel. In the character Ashima Slade, using the idiosyncratic genre of "critical fiction" which allows meticulous commentary on his fictitious author, his lectures, and his theory, Delany has constructed an astounding parallel counterpart for Goguen. The parallel is astounding because of the amazing correspondence of topical concerns that exist between Delany's essay, and the content and style of his character Ashima Slade's Harbin-y Lecture *Shadows*[1] (on the topic of the "Modular Calculus," which grew out of "metalogic") [8] [10].

Goguen has never been one to shy away from "big" issues of human existence. Likewise, as a science fiction and fantasy author constructing civilizations, ancient and futuristic, in part to illuminate sociological points, Delany addresses major philosophical themes. Both are employed as university professors, Goguen in computer science and Delany in English and creative writing, yet the works of each extend well beyond their disciplinary boundaries. Indeed in the quote above Goguen expresses that his work has taken him on a journey through exotic disciplinary locales ranging from category theory to ethnomethodology, and his work also ranges to Buddhist thought and poetry and fiction writing on occasion. Similarly, Delany has commented on a wide range of concerns including semiotics, paraliterature, cultural theory, discourse analysis, gender studies, as well as producing meditations on mathematics and technology. These lists of interests of the two authors are not exhaustive, but they serve to highlight the difficulties, and pleasures for those sympathetic to deep interdisciplinary thought, in elucidating parallels in two prolific, singular authors.

There are many specific parallels in the works of Goguen and Delany. Mathematical metaphors are pervasive in Delany's oeuvre and metalogic takes a prominent role in

---

[1] Robert Elliot Fox tells us in his book *Conscientious Sorcerers* that "the title in the first lecture of the series, "Shadows," is one the Delany himself used for a speculative/critical essay. As Slade's fictitious editor tells us, Slade took the title 'from a nonfiction piece written in the twentieth century by an author of light, popular fictions." [10]

Trouble on Triton in particular. By the same token, identity and difference are major themes in Goguen's work. Often he addresses such concerns through very abstract mathematics such as the theory of institutions which allows for the comparison of logics (a type of metalogic). Though he is not as explicit about politicized social identity in the same sense as Delany, Goguen is also concerned with the relationship of these themes to everyday lived experience. This can be seen in his work on qualia. In phenomenology, philosophers use the term "qualia" to describe introspectively accessible feelings of everyday life that are irreducible to objective characteristics. [25] Goguen has carried out a set of experiments relating qualia to the issue of identity and difference. Similarly, while many artists are interested in exploring the qualitative experiences of life, Delany creates rigorous literary thought experiments that also seem to address the qualia of identity, in his case usually experiences of race, gender, sexual orientation, and similar issues of social identity. The care with which Delany constructs these detailed explorations is exemplified below in Section 2.1 as he uses the metaphor of metalogic to make very specific observations about the nature of race. Finally, both authors are brazenly concerned with mapping out meaning in all of its modularity and nuance. They are unified in this concern as they both draw upon a broad range of traditions from science, mathematics, literature, and social and cultural theories to comment upon some of the most fundamental issues we, as humans, experience in life.

The task of investigating the parallels above is quite worthwhile. It serves to highlight contributions of both Goguen and Delany that perhaps are less well-known than their main contributions to their fields, and more importantly because of the insights such an exercise provides to issues such as (1) identity and qualitative experience, (2) semiotic representation, and the (3) divergence between meaning in formal systems of understanding and everyday lived experience. These three issues are intended to focus this paper (as opposed to representing a comprehensive outline of shared concerns between the authors). This is not meant to be a complete survey of either author's work since I intend rather to highlight particularly salient parallels between them. Thus, the paper is structured as a series of two case studies followed by discussion and a conclusion.

The first case study is centered on Delany's description of "metalogic," and the "modular calculus" where appropriate, in his novel *Trouble on Triton: An Ambiguous Heterotopia*. The second case study is centered on the philosophical notion of qualia in Goguen's work in several papers [16] [19], and the theory of institutions where appropriate. [18] These case studies are unified by a concern with identity, though the starting points from which Goguen and Delany consider identity are quite different. The case studies are followed by a discussion that highlights the tension between both authors' desires to rigorously map meaning and representation (semiotic concerns), and both authors' realizations that this is a Sisyphean task when confronted with the immensity of the real world and human perception of it. The paper concludes with an account of the various discourse styles and strategies Goguen and Delany use to express their ideas – an account of the artistry of the authors. Their discourse styles can be seen as roughly fitting into the same three categories that Delany outline's for Ashima Slade's work: (1) well-reasoned rational argumentation, (2) mathematics (in Delany's case sometimes pseudomathematics used in a metaphorical way), and (3) more esoteric, artistic, or even religious/spiritual discourse.

2 Metalogic, Qualia, and Identity

2.1 Delany on Metalogic and Identity

Trouble on Triton: An Ambiguous Heterotopia is a novel that tells the story of a self-described "reasonably happy man," living in a futuristic society on Neptune's moon Triton. [3] In truth, this man, a conflicted and pompous anti-hero named Bron Helstrom, is far from satisfied. He is ill at ease with his own social identity and relationships with others. He is not a likable or sympathetic character, perhaps meant to represent the pretentiousness often brought on by experience of the privileges accompanying dominant social status. In a world where physique, gender, religion, and race are nearly instantly reconfigurable, a world at war with our own planet Earth, Bron is constantly concerned with how he presents himself externally, and with compensating for his own insecurities. Though largely a meditation on identity, the novel features a robust metaphor of mathematics to address the qualitative experience of identity and the potential for transformation of identity.

At one point early in the novel Bron Helstrom takes about seven pages, and many elaborate analogies involving colored clouds as spaces of significance, hens and a half laying eggs and a half, and the grotte between the tiles of the Taj Mahal, to provide a brief description of the field of metalogic. [6] Though in the novel's storyworld metalogic is meant to provide a rigorous theory and methodology for problem solving in the real world when rules of formal logic are inadequate, it becomes immediately clear that Delany's discussion of metalogic has the issue of identity, and especially racial identity, as a subtext.

The reader is oriented to this subtext as the character Miriamne (to whom Bron is about to pontificate on metalogic) responds to Bron's question on her preference for how she takes her coffee:

> "Black," she said from the sling chair, "as my old lady," and laughed again…
> "That's what my father always used to say." She put her hands on her knees. "My mother was from Earth – Kenya, actually; and I've been trying to live it down ever since." [5]

Bron's parents are soon to be revealed as "large, blond, diligent" and "like so many others it was embarrassing, laborers." The discussion is then, at the level of nonfictional communication between Delany and the reader [22], a commentary on the social situation of a white male, possessed of a strong sense of entitlement and oriented primarily toward class distinctions, lecturing a woman of color. This commentary plays out metaphorically and metonymically as metalogic is explained via several examples that are rich with terms that parallel racialized color such as "black," "white," "brown," "pink," "red," "tan," "colored," and "nonwhite.[2]"

Specifically, Bron begins by posing a challenge to the "beginning tenet of practically every formal logic text ever written, 'To deny P is true is to affirm P is false'." The color consciousness comes into play when Miramne responds by mentioning that she recalls "something about denying the Taj Mahal is white … is to affirm that it's not white … an idea that, just intuitively I've never felt comfortable with." Delany goes on to explicate this discomfort by having his character Bron elaborate upon metalogic, with a series of arguments using the color of the Taj Mahal as an example. This series of arguments clearly could apply as easily to a discussion of the nuances of racial identity, moving from a simplistic system of finite (binary initially: white vs. nonwhite) classification to a much more complicated system, a "parametal model of language," that stresses the metaphor to the breaking point as exemplified by the following quote:

> …he used the fanciful analogy of "meanings" like colored clouds filling up the significance space, and words as homing balloons which, when strung together in a sentence, were tugged to various specific areas in their meaning clouds by the resultant syntax vectors but, when released, would drift back more or less to where, in their cloudy ranges, they'd started out. [7]

I now present a summary of the points that Bron makes in his informal discussion of metalogic and argument against the idea that to deny P is to affirm not-P:

(1) Premise: denying the Taj Mahal is white is not to affirm that it is not white
(2) the significance of 'white' is a range of possibilities
(3) the significance of 'white' "fades imperceptibly" through grey to black and through pink to red, and even to some non-colors
(4) accepting that 'white(Taj Mahal) = F' $\square$ '$\neg$ white(Taj Mahal) = T' means placing a boundary around an area in the range of significance and to call everything in this area white and everything outside of it not-white
(5) this is already a distortion of what was already mentioned to exist, namely fading ranges of color and non-color
(6) values on the boundary line are unaccounted for
(7) objects that are piecewise white and not-white are unaccounted for, (e.g. the Taj Mahal is made of white tiles held to brown granite by tan grotte)

Notice that at this point the "Taj Mahal" in this discussion could have been substituted by "racial ambiguous individual" with no effect on Bron's argument (besides making it more socially salient or politically charged). Furthermore, we have reached a point where a solution to the problem is to describe the Taj Mahal, or racialized person, piecewise as being 'white' and also being some other discrete color signifiers. This is how archaic (really still in practice, only sometimes less overtly) systems of racial identity functioned, with any number of arbitrary discrete color

---

[2] This is strikingly reminiscent of Duke Ellington's "Black, Brown, and Beige" suite. [9]

categories often defined by quantified mixtures of identity[3]. Indeed I personally grew up well aware of the "one drop" rule that holds sway in the United States of America: any bit of "black blood" implies blackness (up to a practical limit of 1/16). It is common for individuals whose parents are identified as belonging to different racial groups to claim "biraciality," or even more finely grained subdivisions of race. DNA testing technologies [2], along with contemporary sociological theories of classification admitting the arbitrary nature of race [1], have rendered these piecewise and discrete classifications of identity obsolete. With all this in mind, I present Miriamne's response to Bron's argument so far: "Wait a second: *Part* of the Taj Mahal is white, and *part* of the Taj Mahal is brown, and *part* of the Taj Mahal is – " to which Bron responds by continuing his argument as follows:

(8) the words 'Taj Mahal' also have a range of significance

(9) the range of significance of 'Taj Mahal' is not discrete, is not unambiguous, and cannot be bounded in a simple two-dimensional model

(10) the Taj Mahal must be described in terms of continuously valued parameters, not discrete perimeters. "Language is parametal, not perimetal. Areas of significance space intermesh and fade into one another like color-clouds in a three-dimensional spectrum."

(11) thus 'logical' bounding is dangerous because it implies that boundaries can be placed around significance spaces

(12) natural language can overcome these problems and provide parametal descriptions

(13) rigorous and precise modeling of such phenomena using mathematics requires extremely advanced tools of analysis (at minimum metalogicians have simple model with seven coordinates, in practice they often use twenty-one, and even this is just an abstract model for visualization that does not fully explain the real world, i.e. "real space")

At this point, Bron's argument is not yet complete. The problem is that "significance space" has been reified. That is, it is being treated as if it exists in the real world and there is such a thing as a "real" significance space to be modeled. Delany's perspective here, as expressed through the character Bron, foreshadows recent directions in cognitive science. Bron's explanation shifts to expressing "how what-there-is manages to accomplish what-it-does," namely how the brain and sensory perception are the origins of complicated concepts such as "significance space" and other concepts in general. In short, it is almost an embodied perspective of cognition [26] (though Delany does not discuss motor operations). In this view "meaning"

---

[3] The artist Betye Saar expresses this using real historical colorized terms for black people found in popular culture and works such as those of the author Langston Hughes. Some of these are: "bright/light, cream, fair, marinee, peola, pinky/pink toes, taffy, vanilla, banana, butterscotch, café au lait, ginger, golden, honey, peaches, yella/high yella/deep yella, almond, caramel, copper, red/red bone, rusty, bark, brownie, brown sugar, cocoa brown/high brown, low brown/seal brown/tobacco brown, chocolate/chocolate drop/deep chocolate, molasses, walnut, bronze, blackie, blackbird/blackberry, black/blue black/charcoal black/coal black/dark black/deep black/lamp black/stove black, crow jane, licorice, midnight/beyond midnight, nightblack boy, tar baby."

depends upon the fact that humans exist "in a world that is inseparable from our bodies, our language, and our social history." [26]

From here Bron continues to reformulate the problem, and to describe how metalogic allows us to address it.

(14) the goals of metalogic are to delimit problems and to explore how elements in the significance space interpenetrate each other

(15) metalogical delineation of significance space means examining specific human utterances or texts (syntax vectors) to dismiss some areas from consideration

(16) the delimited area is then considered "metalogically valid"

(17) to deny "meaningfully" that the Taj Mahal is not white does not imply, but suggests, that it is some color (and not, for example, "freedom," "death," "Halley's comet," or some other thing that is not relevant)

(18) the topological representation of not-P can take any shape in the significance space, even contained within P (i.e. tangent to P at an infinite number of points – it this case it is said that it "shatters P")

(19) Summary: metalogic looks at cognitive activations triggered by linguistic parole (language as it is actually used) [24], selects a model of this in n-dimensional space, and looks at the interpenetration of truth values of relevant elements. Only in this context does it make (metalogical) sense to say that if the Taj Mahal is not white it is some other color, otherwise, the original premise is supported: denying the Taj Mahal is white is not to affirm that it is not white

The remainder of Bron's lecture merely focuses on mathematical techniques to model the significance spaces and industry protocols for doing so. So, stepping back to look at what Bron has just explained, meaning in a metalogical framework is embodied and triggered via discourse. Modeling meaning requires looking at both its cognitive basis and its relationship to language as used in practice. Mathematical modeling does not reify meaning, but it allows for precise statements to be made given an abstraction, and this abstraction may be fairly complicated with the added advantage that it can be modeled computationally in order to get closer to a precise account of the fuzzy topic of human meaning. According to Bron, regarding the issue of identity, the metalogical framework is shown to be much better than simplistic logical formalizations and their simplistic underlying assumptions.

2.2 Goguen on Identity and Qualia

Goguen is also engaged in the business of metalogic. His paper with Rod Burstall on the theory of institutions begins:

There is a population explosion among the logical systems used in Computing Science. Examples include first order logic, equational logic, Horn clause logic, higher order logic, infinitary logic, dynamic logic, intuitionistic logic, order-sorted logic, and temporal logic; moreover, there is a tendency for each theorem prover to have its own idiosyncratic logical system. We introduce the concept of

institution to formalise the informal notation of "logical system.
[18]
He notes that some "exotic" logic systems have been proposed to handle various problems ranging from program construction to natural language. The theory of institutions allows comparison between various logics, translations between results in one logic and another, and an account of the fact that "many general results used in the applications are actually *completely independent* of what underlying logic is chosen." The notion of an "institution" was introduced to "formalize the informal notion of 'logical system'," with the requirement that there be "a satisfaction relation between models and sentences which is consistent under change of notation." Thus, the use of the prefix 'meta' in the case of Goguen and Burstall is traditional in that it abstracts to a higher level of generalization than model theory, which describes only the satisfaction relationship between syntax and semantics within a logical system. The theory of institutions allows logics themselves, many different vocabularies, to be compared. It is apparent that the theory of institutions is a rigorously formulated mathematical account with practical applications and wide theoretically implications. [18]

In contrast, Delany's notion of metalogic is not 'meta' in the traditional sense, rather it is 'meta' in a socio-cultural sense. It begins by looking at formal logical reasoning and its relationship to everyday human thought and problem solving. The 'meta' level from this perspective is the issue of how "logical" reasoning and representation in cognitive, social, and cultural contexts diverges from formal logical systems. Needless to say, Delany does not present this work as rigorous mathematics (it is embedded in a science fiction novel!) and his use of the concept of a "logic" though primarily presented mathematically, is also largely meant metaphorically, without clear indication of where the boundaries between these two functions lie. This is not troublesome, however, because as seen above in Section 2.1 Delany's discussion of metalogic is multiveilant and is meant to comment upon the nuances of social identity relationships, to "ground" his novel (it is necessary in genre fiction to "mark" itself as conforming to conventions of the genre – in science fiction this is often done with detailed reference to mathematics and science) by postulating a well-thought out futuristic system of thought, and probably to explore some of his own thoughts as a philosopher and theoretician within the context of a fiction.

Goguen's work does address many overlapping issues with raised in Delany's account of metalogic, but rather than being found in Goguen's work on "metalogical" concerns (institutions), it can be found in his work on qualia and algebraic semiotics. In his paper "Time, Structure and Emotion in Music" [19], with Ryoko Goguen, it is stated that:

> In formal logic the Law of Identity is stated as "A = A" meaning that every object is equal (or identical) to itself...The Law of Identity may apply to objects of modern science or technology (e.g. numbers), but not to human experience. It appears that human senses have been optimized by evolution to find differences, in which case identity is the failure to find a significant difference.

This formulation of identity with regard to human experience also can provide commentary on sociological phenomena of identity such as prejudice, or even politically topical issues such as racial profiling and gender discrimination. It positions these practices as grounded in failures of sensory perception to account for differences (physical or cultural, nuanced or overt) between individuals that undoubtedly exist (as attested to by victims of systematic discrimination or profiling!) and implicitly states that such practices are the results of failures to respect the individuality of humans (instead relying upon inadequate and coarse systems of generalization and classification). Furthermore, Goguen emphasizes that it is not only truth values of concepts that are important, but qualitative experience in human existence. Thus, Goguen is concerned with qualia, often described informally in philosophy as "what remains when all objective features are subtracted." [19] Goguen would remark, however, that in lived experience subjective phenomena are often attributed at least as much "reality" as so-called "objective" phenomena.

Informal empirical experimentation and phenomenological analysis have moved Goguen to propose a different definition of qualia that avoids some of the vagueness of the traditional definition above. Goguen's definition is: "Qualia are the hierarchically organized constituents of conscious experience, each with a saliency and an emotional tone." To demonstrate qualia phenomena, he and Ryoko Goguen performed several musical experiments that yielded observations such as the following [19]:

(1) added notes beneath a note can change the character of a top note
(2) what comes before a note can greatly change its feeling
(3) what comes after a note can greatly change its feeling
(4) the apparent duration of a note can be changed by what comes before it
(5) repetitive phrases are expected to take a role in a larger framework, are grouped, and with extreme repetition can become seen as background noise and ignored
(6) a note can appear many times in a piece of music, but will not be interpreted merely as many instances of that note (the music is interpreted more holistically)

Clearly, though the subject matter is music, these experiments offer a strong commentary on the transitory and subjective nature of identity. It is easy to think of parallels with social identity such as: prejudices can influence dispositions from an individual toward another individual (quale 1 above), impressions of a person after meeting him or her can alter dispositions toward that person (quale 2 above), or the process of enculturation within a group can allow a shift from ignorance of social protocol to full fluency with social protocol, so that interaction becomes automatic (quale 5 above). While Goguen does not present such social experiments in his paper, probably introspection will allow the reader of this paper to agree with these phenomena. In fact, these phenomena are commonplace and not surprising at all. What is striking is that such everyday observations seem to illuminate inadequacies of common approaches to identity (prejudice and discrete classification), the limitations of "objectifying" identity, and the philosophically oft-overlooked importance of subjective experience and emotion when accounting for identity.

Since subjectivity phenomena rarely, if ever, occur in isolation, Goguen is also concerned with accounting for how qualia combine. He grounds this account in Gilles Fauconnier and Mark Turner's theory of conceptual blending from cognitive linguistics (along with Goguen's hierarchical information theory). Goguen and Goguen describe conceptual blending as the process

> ... in which relatively small, transient structures called conceptual spaces, combine or "blend" to yield a new space that may have emergent structure. Simple examples are words like "houseboat" and "roadkill," and phrases like "artificial life" and "computer virus." Blending is considered a basic human cognitive operation, invisible and effortless, but pervasive and fundamental, for example in grammar, reasoning, and combinations of text with music. [19]

Important here is the fact that conceptual blending theory has an embodied basis as discussed above in 2.1. Furthermore, Goguen has developed a theory of algebraic semiotics that uses algebraic specification from computer science to provide formal notation to describe sign systems and mappings between them that are capable of representing conceptual blends. Goguen and I have developed an algorithm that models some core aspects of conceptual blending theory. [20], [21] This means that despite the subjective nature of qualia, and the qualitative nature of identity, at least some aspects of these phenomena can be approached formally with the use of mathematics. Though Goguen is careful to claim that such work is not intended to reify the formal models (in parallel with Delany), it is clear that he seeks an account of qualia and identity that is precise and rigorous, and that corresponds with the daily realities of lived human experience.

3 Discussion

3.1 Goguen's Models and Realities

Goguen and Delany both seek rigorous accounts of social issues, and both take inspiration and ideas from logic and mathematics. Both also exhibit a tension in their work between a desire to account for social phenomena as carefully as possible, as enabled through construction of intricate models, and to acknowledge the inherent limitations of such approaches. In a very broad sense perhaps they are trying to reconcile the power of holistic accounts provided by structuralism with deeply felt postmodernist understandings of the inadequacies of such global models. The desire for rigorous modeling is exhibited as both authors offer semiotic foundations for their work.

In Goguen's algebraic semiotics the structure of complex signs, including signs in diverse media, and the blending of such structures are described using semiotic systems (also called sign systems) and semiotic morphisms. A sign system consists of [21]:

> a loose algebraic theory composed of type declarations (called sorts) and operation declarations, usually including axioms and some constants), plus a **level ordering** on sorts (having a maximum element called the **top sort**) and a **priority ordering** on the constituents at each level. Loose sorts classify the parts of signs, while data sorts classify the values of attributes of signs (e.g., color

and size). **Signs** of a certain sort are represented by terms of that sort, including but not limited to constants. Among the operations in the signature, some are **constructors**, which build new signs from given sign parts as inputs. Levels express the whole-part hierarchy of complex signs, whereas priorities express the relative importance of constructors and their arguments; social issues play an important role in determining these orderings. Conceptual spaces are the special case where there are no operations except those representing constants and relations, and there is only one sort. Many details omitted here appear in [11].

A semiotic morphism is a mapping between sign systems. One very useful type of mapping discussed above is that between information and a representation of that information. A semiotic morphism maps sorts, constructors, predicates and functions of one sign system to sorts, constructors, predicates and functions of another sign system respectively. An example of how a sign system can be represented differently via different semiotic morphisms is presented in Figure 1 [11], which depicts representations of time as reported by different types of clocks.

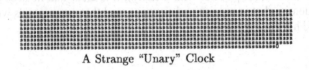

A Strange "Unary" Clock

795

A Naive Digital Clock

13 | 15

A Military Time Clock

Fig. 1. Different representations of a clock

Goguen's diagram depicts a unary clock that simply displays a character repeated a number of times equal to the number of elapsed minutes in a day, a simple digital clock that simply displays the same number of minutes in standard Arabic numerals, and a clock that displays military time. Semiotic morphisms from multiple conceptual spaces to a single conceptual space constitute a "blend."

Using a basis in conceptual blending theory and algebraic semiotics, Goguen and I have also provided an account of "style," another subjective and seemingly unformalizable topic. Still, we made modest claims that some notions of style can be captured by the principles by which concepts and signs are blended, though this is not to be seen as analogous to true, context dependent, qualitative human style. In [20], we proposed two dimensions of style (regarding computer mediated texts):

(1) Construction of formal narrative (or other) elements of media structure, at different levels of granularity. At a large grain level these elements could be

narrative clauses, or scenes of a film, at a more fine grain they could be syntactic parameters of clauses, prosody of poems, or types of shots of a film, and at the smallest grain they could include character sprites or collectible items in games, specific metaphors in poems, or icons used in a user interface.

(2) Selection of media and genres, selection of content, principles for how content elements can be combined, and controls for changing between media and genres.

Later, we even offer the following bold statement (though we mitigate both of these claims later):

Thus there are at least 12 dimensions of style in this approach, 4 at each level: choice of domain[4], content of domain, optimality principles for blending, and controls for changing domains. [20]

The point here is not the particularities of this notion of style, but rather the desire for the "cake" of a formal model of style, while being "able to eat" the facts that we do not reify this formalization and we do realize its limitations.

Indeed, in another paper we make this value very explicit [21]:

Before briefly discussing algebraic semiotics, it may be helpful to be clear about its philosophical orientation. The reason for taking special case with this is that, in Western culture, mathematical formalisms are often given a status beyond what they deserve. For example, Euclid wrote, "The laws of nature are but the mathematical thoughts of God." ... Somewhat less grandly, one might consider that conceptual spaces are somehow directly instantiated in the brain. However, the point of view of this paper is that such formalisms are constructed by researchers in the course of particular investigations, having the heuristic purpose of facilitating consideration of certain issues in that investigation.

Under this view, all theories are situated social entities, mathematical theories no less than others.

The varyingly humble and enthusiastic claims concerning the nature, and concrete applications, of algebraic semiotics illuminate what I assert is a rare attitude toward the integration of mathematics and social concern.

3.2 Delany's Models and Realities

A rare attitude, but not unique. Delany's "Informal Remarks Towards the Modular Calculus" display a similar impulse. Part one of the "remarks" consists of the body of the novel *Trouble on Triton* itself; other parts of the "remarks" are strewn throughout other novels Delany has written in a completely different genre. Thus, the literary theorist Robert Elliot Fox describes Delany's "modular calculus" as a "mapping of culture" that "embraces both science fiction and fantasy, as well as critical/confessional modes." [10]   Using the vehicle of Ashima Slade's Harbin-y

---

[4] A "domain" here refers to a collection of knowledge regarding a particular idea or theme.

Lectures, Delany provides part two of his "informal remarks toward the modular calculus [3]," discussed below.

The character Ashima Slade uses the sentence "The hammer hit a nail" to provide an example of some core concepts of the modular calculus. In summing up the modeling accomplished by that sentence Slade offers:

> We are modeling attitudes, objects, and various aspects of a relation
> between them; to do this job, we are using, among a large group of
> things and relations, various of those things and relations to stand
> for the objects, attitudes, and relations we wish to model.

Slade continues to explain that there are various ways to express the grammatical and semantic relationships evident in the sentence, and likewise there are various ways to describe the relationship between, for instance, "the three a's in the sentence." If the sentence is thought to be formed of only letters and spaces, the ways to describe the relationships that make up and describe the sentence are limited. Slade posits that if the letters in the sentence were instead made of lines in a matrix on a digital display

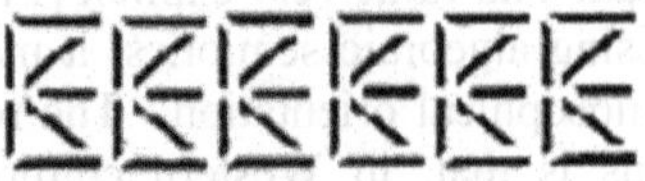

Fig. 2. Digital display flash-out from Delany's *Trouble on Triton*

(see Figure 2), the ways of describing a list of relations in the sentence would be quite different, especially considering that letters can be made in multiple forms (see Figure 3).

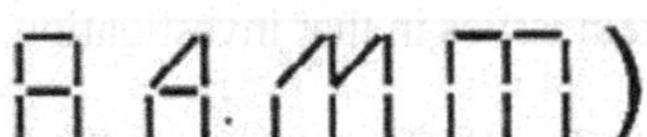

Fig. 3. Digital letter forms from Delany's *Trouble on Triton*

In explicating the modular calculus[5], Slade distinguishes between modular and non-modular descriptions. A modular description "preserves *some* of the modular properties of the sentence in a list that describes the sentence." A non-modular description "preserves *none* of the modular relations of the sentence in a list that describes the sentence." Thus, Slade asserts that the digital display is modular whereas mere letters and spaces are nonmodular. The modular calculus, then, translates between a grammar (a list of sentences about how to compose sentences – an inherently nonmodular description even if it is complete), and a modular description. Slade concludes with the following remarks about the modular calculus:

> Now the advantages of a modular description of either a modeling
> object, like a sentence, or a modeling process, like a language, are

---

[5] And distinguishing it from the "modular algebra," which sadly Delany does not have Slade explain in depth in the same essay.

obvious vis-à-vis a nonmodular description. A modular description allows us reference routes back to the elements in the situation which is being modeled. A nonmodular description is nonmodular precisely because, complete or incomplete as it may be, it destroys those reference routes: it is, in effect, a cipher.

...

The problem that still remains to the calculus, despite my work, and that will be discussed in later lectures, is the generation of formal algorithms for distinguishing incoherent modular descriptive systems from coherent modular descriptive systems. Indeed, the calculus has already given us partial descriptions of many such algorithms, as well as generating ones for determining completeness, partiality, coherence, and incoherence—processes which till now had to be considered, as in literature, matters of taste.

The parallel between the two authors' ideas described above goes far beyond the fact that both use figures depicting digital displays, Goguen and Delany share a concern for the various ways to represent a particular sign system, and the fact (following Saussure) that "signs come in systems." [11]   Both also are interested in mapping the complex ways that sign systems are composed.   But recall that Ashima Slade is naught but a character in Delany's "Informal Remarks Towards a Modular Calculus," and that the informal remarks are written in the fictional mode.   Slade's remarks and their mathematical timbre serve a metaphorical purpose (though their contents also express and reinforce that purpose) which is to express the complexities of meaning and identity formations (at the very least Delany raises many other social and philosophical issues) with fiction rather than formal modeling and the epistemological problems formalisms present.   This decision to employ a fictional mode provides an advantage outlined observation of his other character, Bron Helstrom: "Ordinary, informal, nonrigorous language overcomes all these problems, however, with a bravura, panache and elegance that leave the formal logician panting and applauding."

Like Goguen does with algebraic semiotics, Delany mitigates the modular calculus. Slade's fictitious biographer informs us that the modular calculus grew out of Slade's earlier work in metalogic.   But Bron Helstrom's lecture on metalogic was completely undermined by his unsympathetic persona.   Bron is a pompous "white" male who speaks with dominant cultural authority and in fact is filled with insecurities.   At one point he angrily berates a worker on the telephone (or some futuristic version of a telephone) whose department had mistakenly placed Miriamne, a cybralogician, in the metalogics division.   It becomes clear that Bron's performance is only displayed in the hopes of impressing Miriamne (Bron continues pretending to yell at the worker even after he is hung up on).   He exhibits an inability to relate to the woman in front of him, and is completely bewildered by his own identity, revealing the limited utility of his ability to pontificate on the subtly nuanced metalogical identity of the Taj Mahal.   And in the end, the discussion of formally modeling the color of the Taj Mahal faded out in the face of lived reality as Bron's lecture veered toward "muzzy eloquence": "...the thought struck: Somewhere in real space was the real Taj Mahal. He had never seen it: He had never been to Earth."

And the discussion of metalogic itself flashes out as Miriamne changes the subject to mention that earlier she had run into a female acquaintance that Bron was interested in. "What happened next was that his heart began to pound."

4 Conclusions

Composing this paper has been a satisfying exercise that brought into conjunction the works of two people whom I admire a great deal. This process raised important issues about topics as diverse as social identity, qualia, semiotics, and consciousness, but perhaps as importantly, an unifying aesthetic was formed. Both authors offer a type of groundless [12] work with audacity in approaching "big" issues of life. In order to locate the ambiguities and consistencies of representation and meaning, Delany and Goguen each use a diving rod that bifurcates in two seemingly opposite directions: (1) a desire to rigorously map and exploit regularities of the world(s) we inhabit, and (2) a desire to acknowledge and embrace the ambiguities of lived human experience and its divergence from any idealized theory. The feelings, sometimes tension, sometimes cool detachment, most times deep compassion, the authors evoke come in part from the subject matters of their inquiries, and in part from their methods and discourse strategies used in their explorations, meditations. I conclude with a few remarks on a final parallel between the two authors.

Delany, in a pair of quotations above, through the characters Bron Helstrom and Ashima Slade, expressed the "bravura, panache, and elegance" of informal language, and the ability of literature to formulate the modular calculus. Goguen, though cognizant of the limitations of formal methods, writes that his early formal mathematical work "may have an austere kind of beauty from its abstraction and generality," and coined the metaphor of "Tossing Algebraic Flowers Down the Great Divide" to describe his life's work in a biographical paper [14]. In the end, Goguen and Delany exhibit aesthetically motivated craftsmanship in their work. They both utilize a range of discourse styles, indeed all three that are exhibited in the fictitious work of Ashima Slade which are, once again: (1) rational argumentation, (2) logic and mathematics, and (3) more esoteric, artistic, or even religious/spiritual discourse. Samuel R. Delany's three modes can be exemplified in:

(1) the genre of critical fiction as in part two of "The Informal Remarks Towards the Modular Calculus,
(2) the exposition of metalogic,
(3) and contrasting descriptions of subcultures, both self-indulgent:

> Really, breast-bangles on a man? (even a very young man.) Just aesthetically: weren't breast bangles more or less predicated on breasts that, a) protruded and, b) bobbed...?" [4],

and acetic:

> Seven years ago, he'd actually attended a meeting of the Poor Children of the Avestal Light and Changing Secret Name; over three instruction sessions he'd learned the first of the Nintey-Seven

> Sayable   mantras/mumbles: Mimimomomizo-
> lalilamialomuelamironoriminos... [4]

along with a lyrical beauty, now sparse, now dense, in his prose style.

Joseph A. Goguen's three modes can be exemplified in:

(1) his philosophical discussion of qualia with some grounding in the work of
   Martin Heidegger and Edmund Husserl [16]

(2) a great deal of his work in mathematics, a mild example is the introduction
   of the notion of an institution:

> ...an institution consists of an abstract category
> Sign, the objects of which are signatures, a
> functor Sen: Sign $\square$ Set, and a contravariant
> functor Mod: Sign $\square$ Setop (more technically, we
> might uses classes instead of sets here).
> Satisfaction is then a parameterized relation $\models$ S
> between Mod(S) and Sen(S), such that the
> following satisfaction condition holds, for any
> signature morphism f: S $\square$ S', any S-model M, and
> any S'-sentence e:
>
> $$M \models S\ f(e)\quad iff\quad f(M) \models S'\ e$$
>
> This condition expresses the invariance of truth
> under change of notation. [18]

(3) his Buddhism based explorations of phenomenological and even
   metaphysical concerns:

> However, if Heidegger and the Buddhists are
> right, it is the possibility of non-being which
> gives beings their character of luminosity, and
> hence the nothing, i.e., shunyata, is not only prior
> to negation, but also to things.
>
> The effect of this, as Heidegger says, is to rob
> logic of its claim to supremacy, and in particular,
> to rob it of its claim to provide foundations for
> science and even for mathematics. Indeed, we
> must conclude that foundations in the sense
> sought by logicians are simply not possible. The
> judgements that we make, and in particular any
> negative judgements, are necessarily grounded in
> our being-in-the-world, and not in any pre-
> existing unshakable truths, or eternal world of
> ideal things. [17]

And finally his poetry:

> 6:41 am
>
>   Clear leaf cloud masses
>   motionlessly moving
>   past the static gray road -
>   almost too lovely to bear. [13]

Acknowledgments

Joseph Goguen made this paper possible with the gift of his work. When I was seeking a Ph.D. program, his algebraic semiotics inspired me to cross the United States of America, to move back to the city where I was raised, to work with him on combining twin streams of computation and art (especially narrative art). In his Meaning and Computation Lab, his advisorship, and his friendship, I found what I was seeking.

Samuel "Chip" Delany I have only met in passing moments, as a fan. In New York City he graciously provided me his address so that I could mail him a correspondence regarding one of his stories – my favorite short story in existence: "The Tale of Rumor and Desire" – I never could find the right words to write him. In San Diego, he offered bit of encouragement on publishing my novel. I thank Delany for forging a trail in the combination of fantasy and sociology that is my passion.

References

1. Bowker, G. C., Star, S. L.: Sorting Things Out: Classification and Its Consequences. The MIT Press, Cambridge, MA, (1999)
2. Collins, F. S.: What we do and don't know about 'race,' 'ethnicity,' genetics and health at the dawn of the genome era . In: Nature Genetics 36, S13 - S15, National Human Genome Research Institute, National Institutes of Health, Bethesda, Maryland, (2004)
3. Delany, S. R.: Trouble on Triton: An Ambiguous Heterotopia. Wesleyan University Press, Hanover, NH, (1976)
4. Ibid., 2
5. Ibid., 48
6. Ibid., 48-55
7. Ibid., 51
8. Ibid., 295-296
9. Ellington, Duke. The Duke Ellington Carnegie Hall Concerts: January 1943, Berkeley: Prestige, (1977).
10. Fox, R. E.: Conscientious Sorcerers: The Black Postmodernist Fiction of LeRoi Jones/ Amiri Baraka, Ishmael Reed, and Samuel R. Delany. Greenwood Press, New York, (1987)
11. Goguen, J.: An Introduction to Algebraic Semiotics, with Application to User Interface Design, In: Proceedings, Computation for Metaphors, Analogy and Agents, edited by Chrystopher Nehaniv. Yakamtsu, Japan, (1998)
12. Goguen, J.: Consciousness and the Decline of Cognitivism. In: Advance Papers, Second Workshop on Distributed Collective Practice. University of California, San Diego, (2002)
13. Goguen, J.: November Qualia, URL =
    <http://www.cs.ucsd.edu/users/goguen/misc/novq.html>.
14. Goguen, J.: Tossing Algebraic Flowers Down the Great Divide. In: C. S. Calude, (ed.): People and Ideas in Theoretical Computer Science, Springer, New York, (1999)
15. Ibid., 1
16. Goguen, J.A.: Musical Qualia, Context, Time and Emotion. In: J.A. Goguen and E. Myin (eds.): Journal of Consciousness Studies, Volume 11, No. 3-4, March-April, Imprint Academic, (2004)

17. Goguen, J. A.: Truth and Meaning. In: Four Pieces on Error, Truth and Reality, Technical Monograph PRG-89, Oxford University Computing Laboratory Programming Research Group, Oxford, (1990)

18. Goguen, J., Burstall, R.: Introducing Institutions. In: Logics of Programs (Carnegie-Mellon University, June 1983), Lecture Notes in Computer Science, Volume 164, Springer, (1984) 221-256

19. Goguen, J., Goguen, R.: Time, Structure and Emotion in Music, Japanese translation by Sumi Adachi to appear in book of University Lectures at Keio University, (2003-2004)

20. Goguen, J., Harrell, D. F.: Foundations for Active Multimedia Narrative: Semiotic Spaces and Structural Blending. In revision, (2006)

21. Goguen, J., Harrell, D. F.: Style as Choice of Blending Principles. In: Style and Meaning in Language, Art, Music and Design, Proceedings of a Symposium at the 2004 AAAI Fall Symposium Series, Technical Report FS-04-07, AAAI Press, Washington DC, October 21-24, (2004)

22. Jahn, M.: Narratology: A Guide to the Theory of Narrative, URL = <http://www.uni-koeln.de/~ame02/pppn.htm>, N2.3.1.

23. Saar, Betye. Colored: Consider the Rainbow, Michael Rosenfeld Gallery, New York, (2003)

24. Saussure, F.: Course in General Linguistics. Translated by Roy Harris. Duckworth, London, (1976)

25. Tye, M.: Qualia, The Stanford Encyclopedia of Philosophy (Summer 2003 Edition), Edward N. Zalta (ed.), URL = <http://plato.stanford.edu/archives/sum2003/entries/qualia/>.

26. Varela, F. J., Thompson, E., Rosch, E.: The embodied mind: Cognitive science and human experience. The MIT Press, Cambridge, MA, (1991)

Quantum Institutions

Carlos Caleiro, Paulo Mateus, Amilcar Sernadas, and Cristina Sernadas

CLC, Department of Mathematics, IST,
Av. Rovisco Pais, 1000-149 Lisbon, Portugal

Abstract. The exogenous approach to enriching any given base logic for probabilistic and quantum reasoning is brought into the realm of institutions. The theory of institutions helps in capturing the precise relationships between the logics that are obtained, and, furthermore, helps in analyzing some of the key design decisions and opens the way to make the approach more useful and, at the same time, more abstract.

1 Introduction

A new logic was proposed in [1, 2, 3] for modeling and reasoning about quantum states, embodying the relevant postulates of quantum physics (as presented, for instance, in [4]) and adopting the exogenous approach (the original models are kept). The logic was designed from the semantics upwards, starting with the key idea of adopting superpositions of classical models as the models of the quantum logic. In [5], other instances of the exogenous approach to enriching logics were presented in detail. In short, the exogenous approach is based on adopting as models of the new envisaged logic (enriched) sets of models of the given base logic without tampering with the models of the original logic. As an example assume that we want to introduce probabilities to a certain logic. Doing so, using the exogenous approach, means that we consider the possible outcomes to be the semantic structures and we assign probabilities to sets of such structures.

This novel approach to quantum logic semantics is completely different from the traditional approach [6, 7] to the problem, as initially proposed by Birkhoff and von Neumann [8], that focuses on the lattice of closed subspaces of a Hilbert space. The main drawback of Birkhoff and von Neumann's approach is that it does not yield an extension of classical logic. Our semantics has the advantage of closely guiding the design of the language around the underlying concepts of quantum physics while keeping the classical connectives and was inspired by the Kripke semantics for modal logic. The possible worlds approach was also used in [9, 10, 11, 12, 13] for probabilistic logic. Our semantics to quantum logic, although inspired by modal logic, is also completely different from the alternative Kripke semantics given to traditional quantum logics (as first proposed in [14]) still closely related to the lattice-based operations. The resulting quantum logic also incorporates probabilistic reasoning (in the style of Nilsson's calculus [9, 10]) since the postulates of quantum physics impose uncertainty on the outcome of measurements. From a quantum state (superposition of classical valuations living in a suitable Hilbert space) it is straightforward to generate a probability space

K. Futatsugi et al. (Eds.): Goguen Festschrift, LNCS 4060, pp. 50–64, 2006.

of classical valuations in order to provide the semantics for reasoning about the probabilistic measurements made on that state.

Herein, we present within the theory of institutions (a logic is identified with an institution, as originally proposed in [15, 16]), the exogenous-style construction of a quantum logic from any given base logic in order to assess how general the construction is. The construction is carried out in three main steps. Given an arbitrary institution we first build its global extension (globalization) where each model is just a set of models of the original institution. Then, we proceed with the construction of its probabilistic extension (probabilization) where each model is a probability space where the outcomes are models of the original institution. Finally, we obtain the quantum extension (quantization) of the given institution where each model is a unit vector in the Hilbert space freely generated from a set of models of the original institution. Obviously, in each step the language is enriched to take advantage and to express properties of the new models. For instance, in the globalization step, global classical connectives are added for reasoning about formulas of the original logic. The institutional perspective allows us to conclude that the first two constructions are fully general, in the sense that nothing is assumed about the given institution and also that nothing else is needed. But quantization requires some additional information (the choice of qubit formulae).

In Section 2, we briefly present the relevant notions and results of the theory of institutions. The globalization step is described in Section 3. The probabilization step is presented in Section 4. Finally, in Section 5 we carry out the quantization step of the enrichment. We conclude with an outline of further research directions.

2   Institutional Preliminaries

In this paper, as a first step towards the full understanding of the proposed approach to enriching logics, we shall adopt a variant of the original notion of institution, without morphisms between models (c.f. [17]). For simplicity we shall just call it an institution, without any further qualifiers. We denote by **Cls** the category with classes as objects and maps between classes as morphisms.

An *institution* is a tuple $I = \langle \mathbf{Sig}, \mathbf{Sen}, \mathbf{Mod}, \Vdash \rangle$ where: **Sig** is a category (of *signatures*); $\mathbf{Sen} : \mathbf{Sig} \to \mathbf{Set}$ is a (*formula*) functor; $\mathbf{Mod} : \mathbf{Sig} \to \mathbf{Cls}^{\mathrm{op}}$ is a (*model*) functor; and $\Vdash = \{\Vdash_\Sigma\}_{\Sigma \in |\mathbf{Sig}|}$ is a family of (*satisfaction*) relations $\Vdash_\Sigma \subseteq \mathbf{Mod}(\Sigma) \times \mathbf{Sen}(\Sigma)$, such that the following *satisfaction condition* holds, for every signature morphism $\sigma : \Sigma \to \Sigma'$, every formula $\varphi \in \mathbf{Sen}(\Sigma)$, and every model $m' \in \mathbf{Mod}(\Sigma')$: $\mathbf{Mod}(\sigma)(m') \Vdash_\Sigma \varphi$ iff $m' \Vdash_{\Sigma'} \mathbf{Sen}(\sigma)(\varphi)$.

As usual, given a set $\Gamma \subseteq \mathbf{Sen}(\Sigma)$ of formulas and a model $m \in \mathbf{Mod}(\Sigma)$, we will write $m \Vdash_\Sigma \Gamma$ to denote the fact that $m \Vdash_\Sigma \varphi$ for every $\varphi \in \Gamma$. Mutatis mutandis, given a set $M \subseteq \mathbf{Mod}(\Sigma)$ of models and a formula $\varphi \in \mathbf{Sen}(\Sigma)$, we will write $M \Vdash_\Sigma \varphi$ to denote the fact that $m \Vdash_\Sigma \varphi$ for every $m \in M$. Recall that I induces a family $\vDash = \{\vDash_\Sigma\}_{\Sigma \in |\mathbf{Sig}|}$ of (*entailment*) relations $\vDash_\Sigma \subseteq \mathbf{Pw}(\mathbf{Sen}(\Sigma)) \times \mathbf{Sen}(\Sigma)$ defined by $\Gamma \vDash_\Sigma \varphi$ if, for every $m \in \mathbf{Mod}(\Sigma)$, if $m \Vdash_\Sigma \Gamma$ then $m \Vdash_\Sigma \varphi$.

The notions of arrow between institutions are at least as important as the notion of institution itself. There is a rather extensive and prolific bibliography on this subject, where various meaningful notions of arrows between institutions are proposed, used, exemplified, and related with each other. A recent systematization of the field can be found in [17]. The notion of arrow that we will be using in this paper can be classified as a *comorphism* (or a *plain map* as originally named in [18], or also a *representation* as renamed in [19]). It is however a modified comorphism that maps models to sets of models, which can be explained as an instance of the general monad construction of [20]. The definition will take advantage of the usual covariant powerset endofunctor $\mathbf{Pw}$, in this case extended to classes, that is, $\mathbf{Pw} : \mathbf{Cls} \to \mathbf{Cls}$ is such that $\mathbf{Pw}(X) = 2^X$, and $\mathbf{Pw}(f : X \to X')$ maps each $Y \subseteq X$ to $f[Y] = \{f(x) : x \in Y\}$.

Definition 1. A *power-model comorphism* from institution I to institution I' is a tuple $\langle \Phi, \alpha, \beta \rangle$ where $\Phi : \mathbf{Sig} \to \mathbf{Sig}'$ is a (*signature translation*) functor; $\alpha : \mathbf{Sen} \to \mathbf{Sen}' \circ \Phi$ is a (*formula translation*) natural transformation; and $\beta : \mathbf{Mod}' \circ \Phi \to \mathbf{Pw} \circ \mathbf{Mod}$ is a (*power-model translation*) natural transformation, such that the following *coherence condition* holds, for every signature $\Sigma \in |\mathbf{Sig}|$, formula $\varphi \in \mathbf{Sen}(\Sigma)$, and model $m' \in \mathbf{Mod}'(\Phi(\Sigma))$: $\beta_\Sigma(m') \Vdash_\Sigma \varphi$ iff $m' \Vdash'_{\Phi(\Sigma)} \alpha_\Sigma(\varphi)$.

In the definition above, $\beta_\Sigma(m')$ is a set of models. Thus, the coherence condition states that $m' \Vdash'_{\Phi(\Sigma)} \alpha_\Sigma(\varphi)$ iff, for every $m \in \beta_\Sigma(m')$, $m \Vdash_\Sigma \varphi$. Clearly, the possibility that $\beta_\Sigma(m') = \emptyset$ is not excluded. In that case, m' must satisfy the translation via α of any formula whatsoever. A particularly interesting case corresponds to the situation when $\beta_\Sigma(m')$ is a singleton. If this happens for every model then we can recast the power-model natural transformation simply to $\beta : \mathbf{Mod}' \circ \Phi \to \mathbf{Mod}$, thus obtaining the usual notion of comorphism.

It is a well known fact that comorphisms preserve entailment. A further simple condition on the surjectivity of the translation of models can also guarantee the reflection of entailment. Such properties were studied in [21]. These results can easily be lifted to the level of power-model comorphisms, as stated below. (Power-model) comorphisms compose in the usual way.

Proposition 1. *Let I and I' be institutions and $\langle \Phi, \alpha, \beta \rangle : I \to I'$ a power-model comorphism. Then $\Gamma \vDash_\Sigma \varphi$ implies $\alpha_\Sigma[\Gamma] \vDash_{\Phi(\Sigma)} \alpha_\Sigma(\varphi)$. Additionally, if for each $m \in \mathbf{Mod}(\Sigma)$ there exists $m' \in \mathbf{Mod}'(\Phi(\Sigma))$ such that $\beta_\Sigma(m') = \{m\}$, then $\Gamma \vDash_\Sigma \varphi$ iff $\alpha_\Sigma[\Gamma] \vDash_{\Phi(\Sigma)} \alpha_\Sigma(\varphi)$.*

Proof. Given the power-model comorphism, assume that $\Gamma \vDash_\Sigma \varphi$. If $m' \in \mathbf{Mod}'(\Phi(\Sigma))$ is such that $m' \Vdash'_{\Phi(\Sigma)} \alpha_\Sigma[\Gamma]$ then, using the coherence condition of the power-model comorphism, we have that $\beta_\Sigma(m') \Vdash_\Sigma \Gamma$. Thus, by definition of entailment, it follows from $\Gamma \vDash_\Sigma \varphi$ that $\beta_\Sigma(m') \Vdash_\Sigma \varphi$. Using again the coherence condition, we now get $m' \Vdash'_{\Phi(\Sigma)} \alpha_\Sigma(\varphi)$. Hence, $\alpha_\Sigma[\Gamma] \vDash_{\Phi(\Sigma)} \alpha_\Sigma(\varphi)$.

Assume now that the additional surjectivity condition holds and $\alpha_\Sigma[\Gamma] \models_{\Phi(\Sigma)}$ $\alpha_\Sigma(\varphi)$. If $m \in \mathbf{Mod}(\Sigma)$ is such that $m \Vdash_\sigma \Gamma$ then $\{m\} \Vdash_\sigma \Gamma$. But we know that there exists $m' \in \mathbf{Mod}'(\Phi(\Sigma))$ such that $\beta_\Sigma(m') = \{m\}$. Thus, $\beta_\Sigma(m') \Vdash_\sigma \Gamma$ and it follows from the coherence condition of the power-model comorphism that $m' \Vdash'_{\Phi(\Sigma)} \alpha_\Sigma[\Gamma]$. Hence, by definition of entailment, it follows that $m' \Vdash'_{\Phi(\Sigma)}$ $\alpha_\Sigma(\varphi)$. Using again the coherence condition we obtain that $\beta_\Sigma(m') \Vdash_\sigma \varphi$, or equivalently, $m \Vdash_\sigma \varphi$. Therefore, $\Gamma \models_\Sigma \varphi$.   ▷

Hence, the existence of a power-model comorphism that fulfills the *surjectivity condition* stated in the second half of Proposition 1, for every signature, allows one to say that the target institution is a conservative extension of the source institution. Note that, for comorphisms, the surjectivity condition stated above simply boils down to requiring that each map $\beta_\Sigma : \mathbf{Mod}'(\Phi(\Sigma)) \to \mathbf{Mod}(\Sigma)$ is surjective. It is also a trivial task to check that the surjectivity condition is preserved by composing (power-model) comorphisms.

3   Global Institution

As a first step in our development, we aim at characterizing the exogenous enrichment of a given logic with a layer of global reasoning. For the purpose, let $I = \langle \mathbf{Sig}, \mathbf{Sen}, \mathbf{Mod}, \Vdash \rangle$ be the starting institution. We now proceed by defining the envisaged global institution I^g and then showing, by means of a power-model comorphism, that it extends I in a conservative way.

Definition 2. The *global institution* $I^g = \langle \mathbf{Sig}, \mathbf{Sen}^g, \mathbf{Mod}^g, \Vdash^g \rangle$ based on I is defined as follows:
  - $\mathbf{Sen}^g(\Sigma)$ is the least set containing $\mathbf{Sen}(\Sigma)$ such that, if $\delta, \delta_1, \delta_2 \in \mathbf{Sen}^g(\Sigma)$ then $(\boxminus \delta), (\delta_1 \sqsupset \delta_2) \in \mathbf{Sen}^g(\Sigma)$.
  - $\mathbf{Sen}^g(\sigma) = \sigma^g$ is defined inductively by: $\sigma^g(\varphi) = \mathbf{Sen}(\sigma)(\varphi)$, $\sigma^g(\boxminus \delta) = (\boxminus \sigma^g(\delta))$, and $\sigma^g(\delta_1 \sqsupset \delta_2) = (\sigma^g(\delta_1) \sqsupset \sigma^g(\delta_2))$;
  - $\mathbf{Mod}^g(\Sigma) = \{M : \emptyset \neq M \subseteq \mathbf{Mod}(\Sigma)\}$,
  - $\mathbf{Mod}^g(\sigma)(M') = \mathbf{Mod}(\sigma)[M']$;
  - $\Vdash^g_\Sigma$ is defined inductively by: $M \Vdash^g_\Sigma \varphi$ iff $M \Vdash_\Sigma \varphi$, $M \Vdash^g_\Sigma (\boxminus \delta)$ iff $M \not\Vdash^g_\Sigma \delta$, and $M \Vdash^g_\Sigma (\delta_1 \sqsupset \delta_2)$ iff $M \not\Vdash^g_\Sigma \delta_1$ or $M \Vdash^g_\Sigma \delta_2$.

Clearly, I^g is an institution. Indeed, the functoriality of $\mathbf{Sen}^g$ and $\mathbf{Mod}^g$ is straightforward. The satisfaction condition of I^g can be established by a simple induction on formulas. The only interesting case is the base case, that we analyze below, the other cases being immediate by induction hypotheses. Let $\sigma : \Sigma \to \Sigma'$ be a signature morphism, $\varphi \in \mathbf{Sen}(\Sigma)$ and $M' \in \mathbf{Mod}^g(\Sigma')$. Then, by definition of $\mathbf{Sen}^g$ and $\Vdash^g$, $M' \Vdash^g_{\Sigma'} \mathbf{Sen}^g(\sigma)(\varphi)$ iff $M' \Vdash_{\Sigma'} \mathbf{Sen}(\sigma)(\varphi)$, that is, $m' \Vdash_{\Sigma'} \mathbf{Sen}(\sigma)(\varphi)$ for every $m' \in M'$. Therefore, using the satisfaction condition of I, this is equivalent to having $\mathbf{Mod}(\sigma)(m') \Vdash_\Sigma \varphi$ for every $m' \in M'$, that is, $\mathbf{Mod}^g(\sigma)(M') \Vdash^g_\Sigma \varphi$.

In the resulting logic, the connectives $\boxminus$ and $\sqsupset$ correspond to global negation and global implication, respectively. Other connectives can be easily introduced, like global conjunction $(\delta_1 \sqcap \delta_2) \equiv \boxminus(\delta_1 \sqsupset (\boxminus \delta_2))$. If the base institution has a negation $\neg$ and an implication $\Rightarrow$, which can be understood as local, these connectives do not collapse with the global ones. For implication, for instance, we have that $\{(\varphi_1 \Rightarrow \varphi_2)\} \vDash_\Sigma^g (\varphi_1 \sqsupset \varphi_2)$, but the converse does not hold in general, given two base formulas $\varphi_1, \varphi_2 \in \mathbf{Sen}(\Sigma)$. Namely, assume that I is the institution of classical propositional logic, $\pi_1, \pi_2 \in \Sigma$ are two propositional symbols and $v_1, v_2 \in \mathbf{Mod}(\Sigma)$ are two classical valuations such that $v_1(\pi_1) = 0$, $v_1(\pi_2) = 0$, $v_2(\pi_1) = 1$ and $v_2(\pi_2) = 0$. Then, $\{v_1, v_2\} \Vdash_\Sigma^g (\pi_1 \sqsupset \pi_2)$ but $\{v_1, v_2\} \nVdash_\Sigma^g (\pi_1 \Rightarrow \pi_2)$. The logic resulting from globalizing classical propositional logic was carefully studied in [5], where a sound and complete calculus could be obtained by capitalizing on a calculus for classical logic and adding an axiomatization of the new connectives. It is an open question if the same sort of enterprise can be done in the general case. However, it seems possible to generalize the technique used there, at least if the base logic enjoys an expressibility property analogous to the disjunctive normal form of classical logic.

More interesting, at the moment, is to establish the precise relationship between the institutions I and I^g.

Proposition 2. The triple $C^g = \langle \Phi^g, \alpha^g, \beta^g \rangle$, where Φ^g is the identity functor on **Sig**; for each Σ, α_Σ^g translates $\varphi \in \mathbf{Sen}(\Sigma)$ to φ; and for each Σ, β_Σ^g translates $M \in \mathbf{Mod}^g(\Sigma)$ to M, is a power-model comorphism $C^g : I \to I^g$ and fulfills the surjectivity condition.

Proof. The naturality of α^g and β^g is straightforward. Given a signature morphism $\sigma : \Sigma \to \Sigma'$ and $\varphi \in \mathbf{Sen}(\Sigma)$, we have $\mathbf{Sen}^g(\sigma)(\alpha_\Sigma^g(\varphi)) = \mathbf{Sen}^g(\sigma)(\varphi) = \mathbf{Sen}(\sigma)(\varphi) = \alpha_{\Sigma'}^g(\mathbf{Sen}(\sigma)(\varphi))$. Similarly, given $M' \in \mathbf{Mod}^g(\Sigma')$, then we have $\mathbf{Pw}(\mathbf{Mod}(\sigma))(\beta_{\Sigma'}^g(M')) = \mathbf{Pw}(\mathbf{Mod}(\sigma))(M') = \mathbf{Mod}(\sigma)[M'] = \mathbf{Mod}^g(\sigma)(M')$ $= \beta_\Sigma^g(\mathbf{Mod}^g(\sigma)(M'))$. The coherence condition is trivial.   ▷

As a corollary, by Proposition 1, C^g shows that I^g is in fact a conservative extension of I.

4   Probability Institution

Let us now characterize the exogenous enrichment of a given logic with probabilistic reasoning. We start by introducing the essential definitions and properties of probability spaces. A *probability space* over a non-empty set Ω of *outcomes* is a pair $P = \langle \mathcal{B}, \mu \rangle$ where $\mathcal{B}$ is a Borel field over Ω, that is, $\mathcal{B} \subseteq 2^\Omega$ contains Ω and is closed for complements and countable unions; and $\mu : \mathcal{B} \to [0, 1]$ is a measure with unitary mass, that is, $\mu(M) = 1$ and $\mu(\bigcup_{i=1}^\infty B_i) = \sum_{i=1}^\infty \mu(B_i)$ if $\{B_i\}_{i=1}^\infty \subseteq \mathcal{B}$ is a family of pairwise disjoint sets.

In due course, we will need to map probability spaces along functions on their outcomes. Let $f : U \to U'$ be a function, $\Omega \subseteq U$ and $P = \langle \mathcal{B}, \mu \rangle$ a probability space over Ω. The *image of P along f* is the probability space $f(P) = \langle \mathcal{B}', \mu' \rangle$ over $\Omega' = f[\Omega]$ where $\mathcal{B}' = \{B' \subseteq \Omega' : f^{-1}(B') \cap \Omega \in \mathcal{B}\}$; and μ' is such that $\mu'(B') = \mu(f^{-1}(B') \cap \Omega)$.

Let $I = \langle \mathbf{Sig}, \mathbf{Sen}, \mathbf{Mod}, \Vdash \rangle$ be the starting institution. As before, we shall first define the envisaged probability institution I^p and then show, using power-model comorphisms, that I^p extends conservatively both I and I^g. Indeed, the whole idea is to work with sets of models of the original institution, as in the global case, but now endow them with a certain probability measure. Of course, also the linguistic resources of the logic will be augmented to allow probabilistic assertions and reasoning. For that sake, we assume fixed a set X of variables. We shall also denote by R the set of all computable real numbers (see [22]).

Definition 3. The *probability institution* $I^p = \langle \mathbf{Sig}, \mathbf{Sen}^p, \mathbf{Mod}^p, \Vdash^p \rangle$ based on I is defined as follows:

- $\mathbf{Sen}^p(\Sigma)$ is the least set containing $\mathbf{Sen}(\Sigma)$ such that: $(\boxminus \delta), (\delta_1 \sqsupset \delta_2) \in \mathbf{Sen}^p(\Sigma)$ if $\delta, \delta_1, \delta_2 \in \mathbf{Sen}^p(\Sigma)$, and $(t_1 \leq t_2) \in \mathbf{Sen}^p(\Sigma)$ if $t_1, t_2 \in T^p(\Sigma)$, where $T^p(\Sigma)$ is the least set (of *probabilistic terms*) such that: $X, R \subseteq T^p(\Sigma)$, $(\int \varphi) \in T^p(\Sigma)$ if $\varphi \in \mathbf{Sen}(\Sigma)$, and $(t_1 + t_2), (t_1 . t_2) \in T^p(\Sigma)$ if $t_1, t_2 \in T^p(\Sigma)$;
- $\mathbf{Sen}^p(\sigma) = \sigma^p$ is defined inductively by: $\sigma^p(\varphi) = \mathbf{Sen}(\sigma)(\varphi)$, $\sigma^p(\boxminus \delta) = (\boxminus \sigma^p(\delta))$, $\sigma^p(\delta_1 \sqsupset \delta_2) = (\sigma^p(\delta_1) \sqsupset \sigma^p(\delta_2))$, and $\sigma^p(t_1 \leq t_2) = (T^p(\sigma)(t_1) \leq T^p(\sigma)(t_2))$, where $T^p(\sigma)$ is inductively defined by: $T^p(\sigma)(x) = x$, $T^p(\sigma)(r) = r$, $T^p(\sigma)(\int \varphi) = (\int \mathbf{Sen}(\sigma)(\varphi))$, $T^p(\sigma)(t_1 + t_2) = (T^p(\sigma)(t_1) + T^p(\sigma)(t_2))$, and $T^p(\sigma)(t_1 . t_2) = (T^p(\sigma)(t_1) . T^p(\sigma)(t_2))$;
- $\mathbf{Mod}^p(\Sigma)$ is the class of all triples $S = \langle M, P, \rho \rangle$ where M is non-empty set subset of $\mathbf{Mod}(\Sigma)$, $P = \langle \mathcal{B}, \mu \rangle$ is a probability space over M such that $\{m \in M : m \Vdash_\Sigma \varphi\} \in \mathcal{B}$ for every $\varphi \in \mathbf{Sen}(\Sigma)$, and $\rho : X \to \mathbb{R}$ is an assignment;
- $\mathbf{Mod}^p(\sigma)(\langle M', P', \rho' \rangle) = \langle \mathbf{Mod}(\sigma)[M'], \mathbf{Mod}(\sigma)(P'), \rho' \rangle$;
- $\Vdash^p_\Sigma$ is defined inductively by: $S \Vdash^p_\Sigma \varphi$ iff $M \Vdash_\Sigma \varphi$, for $\varphi \in \mathbf{Sen}(\Sigma)$, $S \Vdash^p_\Sigma (\boxminus \delta)$ iff $S \not\Vdash^p_\Sigma \delta$, $S \Vdash^p_\Sigma (\delta_1 \sqsupset \delta_2)$ iff $S \not\Vdash^p_\Sigma \delta_1$ or $S \Vdash^p_\Sigma \delta_2$, and $S \Vdash^p_\Sigma (t_1 \leq t_2)$ iff $\llbracket t_1 \rrbracket^S \leq \llbracket t_2 \rrbracket^S$, where the denotation of probabilistic terms $\llbracket \_ \rrbracket^S : T^p(\Sigma) \to \mathbb{R}$ is defined inductively by: $\llbracket x \rrbracket^S = \rho(x)$, for $x \in X$ and $\llbracket r \rrbracket^S = r$, for $r \in R$, $\llbracket \int \varphi \rrbracket^S = \mu(\{m \in M : m \Vdash_\Sigma \varphi\})$, $\llbracket t_1 + t_2 \rrbracket^S = \llbracket t_1 \rrbracket^S + \llbracket t_2 \rrbracket^S$ and $\llbracket t_1 . t_2 \rrbracket^S = \llbracket t_1 \rrbracket^S . \llbracket t_2 \rrbracket^S$.

I^p is an institution. Indeed the functoriality of $\mathbf{Sen}^p$ is straightforward. Concerning $\mathbf{Mod}^p$, and given $\sigma : \Sigma \to \Sigma'$, just note that indeed $\langle M, P, \rho \rangle = \mathbf{Mod}^p(\sigma)(\langle M', P', \rho' \rangle) \in \mathbf{Mod}^p(\Sigma)$. Given $\varphi \in \mathbf{Sen}(\Sigma)$, $\{m \in M : m \Vdash_\Sigma \varphi\}$ is measurable, because $M = \mathbf{Mod}(\sigma)[M']$, and it is also measurable the set $\mathbf{Mod}(\sigma)^{-1}(\{m \in M : m \Vdash_\Sigma \varphi\}) \cap M' = \{m' \in M' : \mathbf{Mod}(\sigma)(m') \Vdash_\Sigma \varphi\} = \{m' \in M' : m' \Vdash_{\Sigma'} \mathbf{Sen}(\sigma)(\varphi)\}$.

The satisfaction condition of I^p can be established by a simple induction on formulas. The only interesting case is that of inequalities. Let $t \in T^p(\Sigma)$. For ease of notation let $S' = \langle M', P', \rho' \rangle$ and $P' = \langle \mathcal{B}', \mu' \rangle$, $S = \mathbf{Mod}^p(\sigma)(S') = \langle M, P, \rho \rangle$ and $P = \mathbf{Mod}(\sigma)(P') = \langle \mathcal{B}, \mu \rangle$. We need to show that $\llbracket t \rrbracket^S = \llbracket T^p(\sigma)(t) \rrbracket^{S'}$. This fact can be shown by a simple induction on terms. The interesting case concerns the terms $(\int \varphi)$. Given $\varphi \in \mathbf{Sen}(\Sigma)$, using the definitions of term denotation and of image of a probability space, the satisfaction condition of the base institution I, and the definition of term translation, along with a little set-theoretical manipulation, we have that

$$[\![\textstyle\int\varphi]\!]^S = \mu(\{m \in M : m \Vdash_\Sigma \varphi\}) =$$
$$\mu'(\mathbf{Mod}(\sigma)^{-1}(\{m \in M : m \Vdash_\Sigma \varphi\}) \cap M') =$$
$$\mu'(\{m' \in M' : \mathbf{Mod}(\sigma)(m') \Vdash_\Sigma \varphi\}) =$$
$$\mu'(\{m' \in M' : m' \Vdash_{\Sigma'} \mathbf{Sen}(\sigma)(\varphi)\}) = [\![\textstyle\int \mathbf{Sen}(\sigma)(\varphi)]\!]^{S'} = [\![T^p(\sigma)(\textstyle\int\varphi)]\!]^{S'}.$$

The term $(\int\varphi)$ denotes the probability of φ, interpreted as the probability of the models of the base institution that satisfy φ. The logic resulting from the probabilization of classical propositional logic was carefully studied in [5], where a sound and weak complete calculus could be obtained. The calculus extends the one for the globalization of classical propositional logic by exploring the interplay between the classical connectives and probability, and uses an oracle rule for reasoning with real numbers. Although the logic enjoys the deduction theorem with respect to global implication, strong completeness is out of reach simply because the logic is not compact. Take, for instance, $\Delta = \{(r \le x) : r < \frac{1}{2}\}$. Clearly, $\Delta \vDash_\Sigma^g (\frac{1}{2} \le x)$ but no finite subset of Δ does. Another interesting relevant remark is the fact that the operators $\Box$ and $\Diamond$ defined by $(\Box\varphi) \equiv (1 \le \int\varphi)$ and $(\Diamond\varphi) \equiv \boxminus(\int\varphi \le 0)$ behave as normal modalities.

In the general case depicted here, however, our aim is to establish the precise relationship between the institutions I, I^g and I^p.

Proposition 3. The triple $C^{gp} = \langle \Phi^{gp}, \alpha^{gp}, \beta^{gp} \rangle$, where Φ^{gp} the identity functor on **Sig**; for each Σ, α_Σ^{gp} translates each $\delta \in \mathbf{Sen}^g(\Sigma)$ to δ; and for each Σ, β_Σ^{gp} translates each $\langle M, P, \rho \rangle \in \mathbf{Mod}^p(\Sigma)$ to $M \in \mathbf{Mod}^g(\Sigma)$, is a comorphism and fulfills the surjectivity condition.

Proof. The naturality of α^{gp} and β^{gp} and the coherence condition are straightforward. As for surjectivity, given a non-empty $M \subseteq \mathbf{Mod}(\Sigma)$ and $m \in M$, take for instance the triple $S = \langle M, \langle \mathcal{B}, \mu \rangle, \rho \rangle$ where $\mathcal{B} = 2^M$, $\mu(B) =$ and ρ is any assignment. Then $\langle \mathcal{B}, \mu \rangle$ is a probability space over M, and $\beta_\Sigma^{gp}(S) = M$.   ▷

As a corollary, by Proposition 1 and the observations therein, C^{gp} shows that I^p is a conservative extension of I^g. By transitivity, I^p is also a conservative extension of I. Indeed, by composition, we also obtain a power-model comorphism $C^p = C^{gp} \circ C^g : I \to I^p$ that fulfills the surjectivity condition.

5   Quantum Institution

Finally, we turn our attention to the exogenous enrichment of a given logic with quantum reasoning. In order to materialize the key idea of adopting superpositions of models of the given logic as the models of the envisaged quantum logic, let us start by recalling the essential concepts of quantum systems. Let us recall the relevant postulates of quantum physics (following closely [4]) and set up some important mathematical structures.

Postulate 4 *Associated to any isolated quantum system is a Hilbert space. The state of the system is described by a unit vector $|w\rangle$ in the Hilbert space.*

For example, a quantum bit or *qubit* is associated to a Hilbert space of dimension two: a state of a qubit is a vector $\alpha_0|0\rangle + \alpha_1|1\rangle$ where $\alpha_0, \alpha_1 \in \mathbb{C}$ and $|\alpha_0|^2 + |\alpha_1|^2 = 1$. That is, the quantum state is a *superposition* of the two classical states $|0\rangle$ and $|1\rangle$ of a classical bit. Therefore, from a logical point of view, representing the qubit by a propositional constant, a *quantum valuation* is a superposition of the two classical valuations.

Postulate 5 *The Hilbert space associated to a quantum system composed of finitely many independent component systems is the tensor product of the component Hilbert spaces.*

For instance, a system composed of two independent qubits is associated to a Hilbert space of dimension four: a state of such a system is a vector $\alpha_{00}|00\rangle + \alpha_{01}|01\rangle + \alpha_{10}|10\rangle + \alpha_{11}|11\rangle$ where $\alpha_{00}, \alpha_{10}, \alpha_{01}, \alpha_{11} \in \mathbb{C}$ and $|\alpha_{00}|^2 + |\alpha_{01}|^2 + |\alpha_{10}|^2 + |\alpha_{11}|^2 = 1$. Again, representing the two qubits by two propositional constants, a *quantum valuation* is a superposition of the four classical valuations. So, the Hilbert space of the system composed of two independent qubits is indeed the tensor product of the two Hilbert spaces, each corresponding to a single qubit.

Since we want to work with an arbitrary set of qubits, we will need the following general construction. Given a nonempty set E, the *free Hilbert space* over E is $\mathcal{H}(E)$, the *inner product* space over $\mathbb{C}$ defined as follows: each element is a map $|w\rangle : E \to \mathbb{C}$ such that $\{e \in E : |w\rangle(e) \neq 0\}$ is countable, and $\sum_{e \in E} ||w\rangle(e)|^2 < \infty$; addition, scalar multiplication and inner product are defined by $|w_1\rangle + |w_2\rangle = \lambda e.\,|w_1\rangle(e) + |w_2\rangle(e)$, $\alpha|w\rangle = \lambda e.\,\alpha|w\rangle(e)$, and $\langle w_1|w_2\rangle = \sum_{e \in E} \overline{|w_1\rangle(e)}|w_2\rangle(e)$.

As usual, the inner product induces the *norm* $|||w\rangle|| = \sqrt{\langle w|w\rangle}$, which on its turn induces the *distance* $d(|w_1\rangle, |w_2\rangle) = |||w_1\rangle - |w_2\rangle||$. Since $\mathcal{H}(E)$ is complete for this distance, $\mathcal{H}(E)$ is a Hilbert space. Clearly, $\{|e\rangle : e \in E\}$ is an *orthonormal basis* of $\mathcal{H}(E)$, where $|e\rangle(e) = 1$ and $|e\rangle(e') = 0$ for every $e' \neq e$. A *unit vector* of $\mathcal{H}(E)$ is just a vector $|w\rangle \in \mathcal{H}(E)$ such that $|||w\rangle|| = 1$.

Let Q be the set of qubits in hand. If there are no dependencies between the qubits then the system is described by the Hilbert space $\mathcal{H}(2^Q)$, where 2^Q is the set of all classical valuations. However, in many cases, we will be given a finite partition $\mathcal{S} = \{Q_1, \ldots, Q_n\}$ of Q, giving rise to n independent subsystems. In the sequel, we will use $\bigcup \mathcal{S}$ to denote the set $\{\bigcup_{Q_i \in \mathcal{R}} Q_i : \mathcal{R} \subseteq \mathcal{S}\}$. Moreover, it may also be that the qubits Q_i of each isolated subsystem are also constrained and some of the classical valuations in 2^{Q_i} are impossible. Any set $V \subseteq 2^Q$ of *admissible* classical valuations induces a set of admissible classical valuations for each subsystem, that is, $V_i = \{v_i : v \in V\}$ with $v_i = v|_{Q_i}$. Analogously, we will use v_R to denote the restriction $v|_R$ of a valuation v to $R \in \bigcup \mathcal{S}$, and $V_R = \{v_R : v \in V\}$. Then, the space describing the corresponding quantum system will be the tensor product $\bigotimes_{i=1}^n \mathcal{H}(V_i)$. Still, note that although $(2^Q)_i = 2^{Q_i}$ and $2^Q = \prod_{i=1}^n 2^{Q_i}$, in general $V \subsetneq \prod_{i=1}^n V_i$. Moreover, although $\mathcal{H}(2^Q) = \bigotimes_{i=1}^{n} \mathcal{H}(2^{Q_i}) = \bigotimes_{i=1}^{n} \mathcal{H}(\prod_{i=1}^{n} 2^{Q_i})$, in general we have that $\mathcal{H}(V) \subsetneq \bigotimes_{i=1}^{n} \mathcal{H}(V_i) \subsetneq \mathcal{H}(\prod_{i=1}^{n} V_i)$.

Hence, we should only consider quantum states of $\bigotimes_{i=1}^{n} \mathcal{H}(V_i)$ that are compatible with V. Given the subspace relations stated above, we shall call a *structured quantum state* over V and $\mathcal{S}$ to a family $|w\rangle = \{|w_i\rangle\}_{i=1}^{n}$ such that each $|w_i\rangle$ is a unit vector of $\mathcal{H}(V_i)$; and $\langle v|(\bigotimes_{i=1}^{n} |w_i\rangle) = \prod_{i=1}^{n} \langle v_i|w_i\rangle = 0$ if $v \notin V$.

Note that it is easy to identify $\bigotimes_{i=1}^{n} |w_i\rangle$ with a unique unit vector in $\mathcal{H}(V)$ since all the amplitudes on valuations not in V are null. Hence, by abuse of notation, we shall also use $|w\rangle$ to denote $\bigotimes_{i=1}^{n} |w_i\rangle$.

Now, we turn our attention to the postulates concerning measurements of physical quantities.

Postulate 6 *Every measurable physical quantity of an isolated quantum system is described by an observable[1] acting on its Hilbert space.*

Postulate 7 *The possible outcomes of the measurement of a physical quantity are the eigenvalues of the corresponding observable. When the physical quantity is measured using observable A on a system in a state $|w\rangle$, the resulting outcomes are ruled by the probability space $\mathrm{Prob}_{|w\rangle}^{A} = \langle \Omega, \mathcal{B}|_{\Omega}, \mu_{|w\rangle}^{A} \rangle$ where in the case of a countable spectrum $\mu_{|w\rangle}^{A} = \lambda B. \sum_{\lambda \in \Omega} \chi_B(\lambda)|P_\lambda|w\rangle|^2$.*

For the applications we have in mind in quantum computation and information, only *logical projective measurements* are relevant. In general, the stochastic result of making a logical projective measurement of the system at a structured quantum state $|w\rangle$ determined as above is fully described by the probability space $\langle 2^V, \mu_{|w\rangle}\rangle$ over V where $\mu_{|w\rangle}(B) = \sum_{v \in B} |\langle v|w\rangle|^2$ for every $B \subseteq 2^V$.

In the sequel, we will need to be able to map quantum systems and states across qubit maps. Let $f : U \to U'$, $Q \subseteq U$ and $Q' = f[Q]$. Then, the function $f^{\bullet} : 2^{Q'} \to 2^Q$ defined by $f^{\bullet}(v')(q) = v'(f(q))$ is injective: if $f^{\bullet}(v_1') = f^{\bullet}(v_2')$ then, for each $q \in Q$, $v_1'(f(q)) = v_2'(f(q))$, which implies that $v_1' = v_2'$ since $Q' = f[Q]$. Hence, $f^{\bullet}$ establishes a bijection between any given set of classical valuations $V' \subseteq 2^{Q'}$ and $V = f^{\bullet}[V'] \subseteq 2^Q$. Therefore, $f^{\bullet}$ also establishes an isomorphism between the Hilbert spaces $\mathcal{H}(V')$ and $\mathcal{H}(V)$ obtained my mapping $|w'\rangle \in \mathcal{H}(V')$ to $|w\rangle = f^{\bullet}(|w'\rangle)$ such that $|w\rangle(f^{\bullet}(v')) = |w'\rangle(v')$. Moreover, note that every finite partition $\mathcal{S}' = \{Q_1', \dots, Q_n'\}$ induces a partition $\mathcal{S} = f^{-1}[\mathcal{S}'] = \{Q_1, \dots, Q_n\}$ of Q with each $Q_i = f^{-1}(Q_i') \cap Q$. Hence, since surjectivity guarantees that each $Q_i' = f[Q_i]$, the Hilbert space isomorphism established in the preceding paragraph by $f^{\bullet}$ also applies to the subsystems, that is, $\mathcal{H}(V_i')$ and $\mathcal{H}(V_i)$ are isomorphic.

---

[1] Recall that an observable is a Hermitian operator such that the direct sum of its eigensubspaces coincides with the underlying Hilbert space. Since the operator is Hermitian, its spectrum Ω (the set of its eigenvalues) is a subset of $\mathbb{R}$. For each $\lambda \in \Omega$, we denote the corresponding eigensubspace by E_λ and the projector onto E_λ by P_λ.

We now characterize the exogenous enrichment of a given institution I with quantum reasoning. As in the previous cases, we shall first define the envisaged quantum institution I^q and then characterize its relationship to I, as well as to the institutions previously built. To this end, qubits will be selected formulas of the original logic, that induce upon observation a probability distribution on models of the original institution. The notation I^q is a little abusive here, since the enrichment will be parameterized by a functor that chooses the qubits of interest. Hence, we consider fixed a functor $\mathbf{Qb} : \mathbf{Sig} \to \mathbf{Set}$ such that, for every signature Σ, $\mathbf{Qb}(\Sigma) \subseteq \mathbf{Sen}(\Sigma)$ and, for every signature morphism $\sigma : \Sigma \to \Sigma'$, $\mathbf{Qb}(\sigma) = \mathbf{Sen}(\sigma)|_{\mathbf{Qb}(\Sigma)}$ and $\mathbf{Sen}(\sigma)[\mathbf{Qb}(\Sigma)] = \mathbf{Qb}(\Sigma')$. Note that $\mathbf{Sen}(\sigma)$ is required to be surjective on qubits, and that this requirement is essential in the subsequent development of the I^q institution.

Clearly, models of the given institution induce classical valuations on the qubits. We denote by $V_\Sigma : \mathbf{Mod}(\Sigma) \to 2^{\mathbf{Qb}(\Sigma)}$ defined, for each qubit $\varphi \in \mathbf{Qb}(\Sigma)$, by

$$V_\Sigma(m)(\varphi) = \begin{cases} 1 \text{ if } m \Vdash_\Sigma \varphi \\ 0 \text{ otherwise} \end{cases} .$$

To fulfill the original idea of working with quantum *superpositions* of models of the original institution, we will have to restrict our attention to sets of models $M \subseteq \mathbf{Mod}(\Sigma)$ on which V_Σ is injective, that is, if $m_1, m_2 \in M$ and $m_1 \neq m_2$ then $V_\Sigma(m_1) \neq V_\Sigma(m_2)$. In this way, we have a bijection between M and $V_\Sigma[M]$.

Given $A \subseteq F \subseteq \mathbf{Qb}(\Sigma)$, we shall denote by $v_A^F \in 2^F$ the classical valuation of the qubits in F defined by $v_A^F(\varphi)$ is 1 if $\varphi \in A$ and is 0 otherwise.

The syntax of the logic will also be augmented, not only with probabilistic reasoning, but also in order to allow us to manipulate complex amplitudes and to talk about qubit independence. Hence, besides for the set X of real variables, we also assume fixed a set Z of complex variables.

Definition 8. The *quantum institution* $I^q = \langle \mathbf{Sig}, \mathbf{Sen}^q, \mathbf{Mod}^q, \Vdash^q \rangle$ based on I (and $\mathbf{Qb}$) is defined as follows:

- $\mathbf{Sen}^q(\Sigma)$ is the least set including $\mathbf{Sen}(\Sigma)$ such that: $(\boxminus \delta), (\delta_1 \sqsupseteq \delta_2) \in \mathbf{Sen}^q(\Sigma)$ if $\delta, \delta_1, \delta_2 \in \mathbf{Sen}^q(\Sigma)$; $[F] \in \mathbf{Sen}^q(\Sigma)$ if $F \subseteq \mathbf{Qb}(\Sigma)$; and $(t_1 \leq t_2) \in \mathbf{Sen}^q(\Sigma)$ if $t_1, t_2 \in T_R^q(\Sigma)$, where the sets $T_R^q(\Sigma)$ and $T_C^q(\Sigma)$ (of *real valued* and *complex valued* terms, respectively) are defined by mutual induction as follows: $X, R \subseteq T_R^q(\Sigma)$; $(\int \varphi) \in T_R^q(\Sigma)$ if $\varphi \in \mathbf{Sen}(\Sigma)$, $(t_1 + t_2), (t_1.t_2) \in T_R^q(\Sigma)$ if $t_1, t_2 \in T_R^q(\Sigma)$, and $\mathrm{Re}(u), \mathrm{Im}(u), \arg(u), |u| \in T_R^q(\Sigma)$ if $u \in T_C^q(\Sigma)$; $Z \subseteq T_C^q(\Sigma)$, $|\top)_{FA} \in T_C^q(\Sigma)$ if $A \subseteq F \subseteq \mathbf{Qb}(\Sigma)$, $(t_1 + it_2), (t_1.e^{it_2}) \in T_C^q(\Sigma)$ if $t_1, t_2 \in T_R^q(\Sigma)$, $\bar{u} \in T_C^q(\Sigma)$ if $u \in T_C^q(\Sigma)$, $(u_1 + u_2), (u_1.u_2) \in T_C^q(\Sigma)$ if $u_1, u_2 \in T_C^q(\Sigma)$, and $(\varphi \triangleright u_1; u_2) \in T_C^q(\Sigma)$ if $\varphi \in \mathbf{Sen}(\Sigma)$ and $u_1, u_2 \in T_C^q(\Sigma)$,

- $\mathbf{Sen}^q(\sigma) = \sigma^q$ is defined inductively by: $\sigma^q(\varphi) = \mathbf{Sen}(\sigma)(\varphi)$, $\sigma^q(\boxminus \delta) = (\boxminus \sigma^q(\delta))$, $\sigma^q(\delta_1 \sqsupseteq \delta_2) = (\sigma^q(\delta_1) \sqsupseteq \sigma^q(\delta_2))$, $\sigma^q([F]) = [\mathbf{Sen}(\sigma)[F]]$, and $\sigma^q(t_1 \leq t_2) = (T_R^q(\sigma)(t_1) \leq T_R^q(\sigma)(t_2))$, where $T_R^q(\sigma) = \sigma_R^q$ and $T_C^q(\sigma) = \sigma_C^q$ are defined by mutual induction: $\sigma_R^q(x) = x$, $\sigma_R^q(r) = r$, $\sigma_R^q(\int \varphi) = (\int \mathbf{Sen}(\sigma)(\varphi))$, $\sigma_R^q(t_1 + t_2) = (\sigma_R^q(t_1) + \sigma_R^q(t_2))$, $\sigma_R^q(t_1.t_2) = (\sigma_R^q(t_1).\sigma_R^q(t_2))$, $\sigma_R^q(\mathrm{Re}(u)) = \mathrm{Re}(\sigma_C^q(u))$, $\sigma_R^q(\mathrm{Im}(u)) = \mathrm{Im}(\sigma_C^q(u))$, $\sigma_R^q(\arg(u)) = \arg(\sigma_C^q(u))$, $\sigma_R^q(|u|) = $

$|\sigma_C^q(u)|$, $\sigma_C^q(z) = z$, $\sigma_C^q(|\top\rangle_{FA}) = |\top\rangle_{\mathbf{Sen}(\sigma)[F],\mathbf{Sen}(\sigma)[A]}$, $\sigma_C^q(t_1 + it_2) = (\sigma_R^q(t_1) + i\sigma_R^q(t_2))$, $\sigma_C^q(t_1.e^{it_2}) = (\sigma_R^q(t_1).e^{i\sigma_R^q(t_2)})$, $\sigma_C^q(\overline{u}) = \overline{\sigma_C^q(u)}$, $\sigma_C^q(u_1 + u_2) = (\sigma_C^q(u_1) + \sigma_C^q(u_2))$, $\sigma_C^q(u_1.u_2) = (\sigma_C^q(u_1).\sigma_C^q(u_2))$, $\sigma_C^q(\varphi \triangleright u_1; u_2) = (\mathbf{Sen}(\sigma)(\varphi) \triangleright \sigma_C^q(u_1); \sigma_C^q(u_2))$;

- $\mathbf{Mod}^q(\Sigma)$ is the class of all tuples $\langle M, \mathcal{S}, |w\rangle, \nu, \rho\rangle$ where: $\emptyset \neq M \subseteq \mathbf{Mod}(\Sigma)$ such that V_Σ is injective on M, $\mathcal{S}$ is a finite partition of $\mathbf{Qb}(\Sigma)$, $|w\rangle$ is a structured quantum state over $V_\Sigma[M]$ and $\mathcal{S}$, $\nu = \{\nu_{FA}\}_{A \subseteq F \subseteq \mathbf{Qb}(\Sigma)}$ is a family of complex numbers such that, whenever $F \in \bigcup \mathcal{S}$, $\nu_{FA} = \langle v_A^F | \bigotimes_{Q_i \subseteq F} w_i \rangle$ if $v_A^F \in V_F$, and $\nu_{FA} = 0$ if $v_A^F \notin V_F$, and ρ is an assignment such that $\rho(x) \in \mathbb{R}$ for every $x \in X$, and $\rho(z) \in \mathbb{C}$ for every $z \in Z$;

- $\mathbf{Mod}^q(\sigma)(\langle M', \mathcal{S}', |w'\rangle, \nu', \rho'\rangle) = \langle \mathbf{Mod}(\sigma)[M'], \sigma^{-1}[\mathcal{S}'], \sigma^\bullet(|w'\rangle), \nu, \rho'\rangle$ with $\sigma^\bullet = \mathbf{Sen}(\sigma)^\bullet$, $\sigma^{-1} = \mathbf{Sen}(\sigma)^{-1}$ and $\nu_{FA} = \nu'_{\mathbf{Sen}(\sigma)[F]\mathbf{Sen}(\sigma)[A]}$;

- $\Vdash_\Sigma^q$ is defined inductively by $W \Vdash_\Sigma^q \varphi$ iff $M \Vdash_\Sigma \varphi$, for $\varphi \in \mathbf{Sen}(\Sigma)$, $W \Vdash_\Sigma^q (\boxminus \delta)$ iff $W \not\Vdash_\Sigma^q \delta$, $W \Vdash_\Sigma^q (\delta_1 \sqsupset \delta_2)$ iff $W \not\Vdash_\Sigma^q \delta_1$ or $W \Vdash_\Sigma^q \delta_2$, $W \Vdash_\Sigma^q [F]$ iff $F \in \bigcup \mathcal{S}$, and $W \Vdash_\Sigma^q (t_1 \leq t_2)$ iff $[\![t_1]\!]_R^W \leq [\![t_2]\!]_R^W$, where the denotations of real terms $[\![\_]\!]_R^W : T_R^q(\Sigma) \to \mathbb{R}$ and of complex terms $[\![\_]\!]_C^W : T_C^q(\Sigma) \to \mathbb{C}$ are defined by mutual induction as follows: $[\![x]\!]_R^W = \rho(x)$, for $x \in X$, $[\![r]\!]_R^W = r$, for $r \in R$, $[\![\int \varphi]\!]_R^W = \mu_{|w\rangle}(V_\Sigma[\{m \in M : m \Vdash_\Sigma \varphi\}])$, $[\![t_1 + t_2]\!]_R^W = [\![t_1]\!]_R^W + [\![t_2]\!]_R^W$, $[\![t_1.t_2]\!]_R^W = [\![t_1]\!]_R^W.[\![t_2]\!]_R^W$, $[\![\mathrm{Re}(u)]\!]_R^W = \mathrm{Re}([\![u]\!]_C^W)$, $[\![\mathrm{Im}(u)]\!]_R^W = \mathrm{Im}([\![u]\!]_C^W)$, $[\![\arg(u)]\!]_R^W = \arg([\![u]\!]_C^W)$, and $[\![|u|]\!]_R^W = |[\![u]\!]_C^W|$, $[\![z]\!]_C^W = \rho(z)$, for $z \in Z$, $[\![|\top\rangle_{FA}]\!]_C^W = \nu_{FA}$, $[\![t_1 + it_2]\!]_C^W = [\![t_1]\!]_R^W + i[\![t_2]\!]_R^W$, $[\![t_1.e^{it_2}]\!]_C^W = [\![t_1]\!]_R^W.e^{i[\![t_2]\!]_R^W}$, $[\![\overline{u}]\!]_C^W = \overline{[\![u]\!]_C^W}$, $[\![u_1 + u_2]\!]_C^W = [\![u_1]\!]_C^W + [\![u_2]\!]_C^W$, $[\![u_1.u_2]\!]_C^W = [\![u_1]\!]_C^W.[\![u_2]\!]_C^W$, and $[\![\varphi \triangleright u_1; u_2]\!]_C^W = \begin{cases} [\![u_1]\!]_C^W & \text{if } M \Vdash_\Sigma \varphi \\ [\![u_2]\!]_C^W & \text{otherwise} \end{cases}$.

I^q is an institution. Indeed the functoriality of $\mathbf{Sen}^q$ is straightforward. Concerning $\mathbf{Mod}^q$, and given a signature morphism $\sigma : \Sigma \to \Sigma'$, note that indeed $W = \langle M, |w\rangle, \nu, \rho\rangle = \mathbf{Mod}^q(\sigma)(W') \in \mathbf{Mod}^q(\Sigma)$ if $W' = \langle M', |w'\rangle, \nu', \rho'\rangle \in \mathbf{Mod}^q(\Sigma')$. In particular, since $M = \mathbf{Mod}(\sigma)[M']$, then $\mathbf{Sen}(\sigma)^\bullet[V_{\Sigma'}[M']] = V_\Sigma[M]$ just because $V_\Sigma(\mathbf{Mod}(\sigma)(m'))(\varphi) = V_{\Sigma'}(m')(\mathbf{Sen}(\sigma)(\varphi))$ for every $m' \in \mathbf{Mod}(\Sigma')$ and $\varphi \in \mathbf{Qb}(\Sigma)$, due to the satisfaction condition of the original institution. Moreover, if $A \subseteq F \subseteq \mathbf{Qb}(\Sigma)$, and we let $F' = \mathbf{Sen}(\sigma)[F]$ and $A' = \mathbf{Sen}(\sigma)[A]$, the definition of $\nu_{FA} = \nu'_{F'A'}$ is suitable. First note that $v_{A'}^{F'} \in V_{\Sigma'}[M']$ iff $v_A^F \in V_\Sigma[M]$ just because $\mathbf{Sen}(\sigma)^\bullet(v_{A'}^{F'}) = v_A^F$. Moreover, $F \in \bigcup \mathcal{S}$ iff $F' \in \bigcup \mathcal{S}'$.

The satisfaction condition of I^q can be established by a simple induction on formulas and on terms. The only interesting cases concern independence formulas $[F]$, plus probability ($\int \varphi$) and amplitude $|\top\rangle_{FA}$ terms. In the first case, we need to show that $W \Vdash_\Sigma^q [F]$ iff $W' \Vdash_{\Sigma'}^q [F']$. The result follows immediately because $F \in \bigcup \mathcal{S}$ iff $\mathbf{Sen}(\sigma)[F] \in \bigcup \mathcal{S}'$. In the second case, we need to show that $[\![\int \varphi]\!]_R^W = [\![\int \mathbf{Sen}(\sigma)(\varphi)]\!]_R^{W'}$. Indeed, using the bijection between M and $V_\Sigma[M]$, the fact that $M = \mathbf{Mod}(\sigma)[M']$, the satisfaction condition of the institution I, and as a result the fact that $\mathbf{Sen}(\sigma)^\bullet(V_{\Sigma'}(m')) = V_\Sigma(\mathbf{Mod}(\sigma)(m'))$, we have that

$$[\![\textstyle\int\varphi]\!]^W_R = \mu_{|w\rangle}(V_\Sigma[\{m \in M : m \Vdash_\Sigma \varphi\}]) =$$
$$\textstyle\sum_{m\in M:m\Vdash_\Sigma\varphi} |\langle V_\Sigma(m)|w\rangle|^2 = \sum_{m\in M:m\Vdash_\Sigma\varphi} ||w\rangle(V_\Sigma(m))|^2 =$$
$$\textstyle\sum_{m\in M:m\Vdash_\Sigma\varphi} |\mathbf{Sen}(\sigma)^\bullet(|w'\rangle)(V_\Sigma(m))|^2 =$$
$$\textstyle\sum_{m'\in M':m'\Vdash_{\Sigma'}\mathbf{Sen}(\sigma)(\varphi)} |\mathbf{Sen}(\sigma)^\bullet(|w'\rangle)(\mathbf{Sen}(\sigma)^\bullet(V_{\Sigma'}(m')))|^2 =$$
$$\textstyle\sum_{m'\in M':m'\Vdash_{\Sigma'}\mathbf{Sen}(\sigma)(\varphi)} ||w'\rangle(V_{\Sigma'}(m'))|^2 =$$
$$\textstyle\sum_{m'\in M':m'\Vdash_{\Sigma'}\mathbf{Sen}(\sigma)(\varphi)} |\langle V_{\Sigma'}(m')|m'\rangle|^2 =$$
$$\mu_{|w'\rangle}(V_{\Sigma'}[\{m' \in M' : m' \Vdash_{\Sigma'} \mathbf{Sen}(\sigma)(\varphi)\}]) = [\![\textstyle\int\mathbf{Sen}(\sigma)(\varphi)]\!]^{W'}_R.$$

In the third case, we need to show that $[\![\top\rangle_{FA}]\!]^W_R = [\![\top\rangle_{F'A'}]\!]^W_R$. Since we already know that $F \in \bigcup \mathcal{S}$ iff $F' \in \bigcup \mathcal{S}'$, and that $v^{F'}_{A'} \in V_{\Sigma'}[M']$ iff $v^F_A \in V_\Sigma[M]$, it suffices to verify that, when it makes sense,

$$\langle v^F_A|(\bigotimes_{Q_i\subseteq F}|w_i\rangle) = \prod_{Q_i\subseteq F}\langle(v^F_A)_{Q_i}|w_i\rangle = \prod_{Q_i\subseteq F}||w_i\rangle((v^F_A)_{Q_i})|^2 =$$
$$\prod_{Q_i\subseteq F}|\mathbf{Sen}(\sigma)^\bullet(|w'_i\rangle)(\mathbf{Sen}(\sigma)^\bullet((v^{F'}_{A'})_{Q'_i}))|^2 = \prod_{Q'_i\subseteq F'}||w'_i\rangle((v^{F'}_{A'})_{Q'_i})|^2 =$$
$$\prod_{Q'_i\subseteq F'}\langle(v^{F'}_{A'})_{Q'_i}|w'_i\rangle = \langle v^{F'}_{A'}|(\bigotimes_{Q'_i\subseteq F'}|w'_i\rangle).$$

Most of the syntactic constructions introduced in I^q are self explanatory. The quantum specific constructs, besides all the operations on complex numbers, are the $[F]$ formulas and the $|\top\rangle_{FA}$ terms. Intuitively, $[F]$ holds if the qubits in F form an independent subsystem of the whole, whereas $|\top\rangle_{FA}$ evaluates, whenever it is meaningful, to the complex amplitude of the vector $|v^F_A\rangle$ in the current state of the systems. The logic resulting from the quantization of classical propositional logic was introduced and studied in [1, 2]. A sound and weak complete calculus for the logic was obtained in [3] using an iterated Henkin construction inspired by the technique in [13]. The qubits of interest in this case were the propositional symbols. Using the logic it is possible, for instance, to model and reason about quantum states corresponding to the famous case of Schrödinger's cat. The relevant attributes of the cat are **cat-in-box, cat-alive, cat-moving** being inside or outside the box, alive or dead, and moving, respectively. The following formulas constrain the state of the cat at different levels of detail:

1. [**cat-in-box, cat-alive, cat-moving**];
2. (**cat-moving** $\Rightarrow$ **cat-alive**);
3. $((\Diamond\,\textbf{cat-alive}) \sqcap (\Diamond\,(\neg\,\textbf{cat-alive})))$;
4. $(\boxminus[\textbf{cat-alive}])$;
5. $(\int\textbf{cat-alive} = \frac{1}{3})$.

Observe that the assertions are jointly consistent. They characterize the quantum states where: the qubits **cat-in-box, cat-alive, cat-moving** are not entangled with other qubits; the cat is moving only if it is alive; it is possible that the cat is alive and also that the cat is dead; the qubit **cat-alive** is entangled with the others; and the probability of observing the cat alive (after collapsing the wave function) is $\frac{1}{3}$. Our aim is now to relate I^q with I, I^g, I^p.

Proposition 4. The triple $C^{pq} = \langle \Phi^{pq}, \alpha^{pq}, \beta^{pq}\rangle$, where Φ^{pq} the identity functor on **Sig**; for each Σ, α^{pq}_Σ translates each $\delta \in \mathbf{Sen}^p(\Sigma)$ to δ; and for each Σ, β^{pq}_Σ

translates each $\langle M, \mathcal{S}, |w\rangle, \nu, \rho \rangle \in \mathbf{Mod}^q(\Sigma)$ to $\langle M, \langle 2^M, \mu \rangle, \rho|_X \rangle$ with $\mu(B) = \mu_{|w\rangle}(V_\Sigma[B])$, is a comorphism $C^{pq} : I^p \to I^q$.

Proof. The naturality of the transformation α^{pq} is straightforward. Concerning β^{pq} just note that given $W = \langle M, \mathcal{S}, |w\rangle, \nu, \rho \rangle \in \mathbf{Mod}^q(\Sigma)$, $\beta^{pq}(W)$ is well defined. The probability space $\langle 2^M, \mu \rangle$ over M is just an isomorphic copy of $\langle 2^{V_\Sigma[M]}, \mu_{|w\rangle} \rangle$ over $V_\Sigma[M]$. It is clearly a probability space, and its naturality follows easily. The coherence condition is trivial.   ▷

Note however that, in general, C^{pq} does not satisfy the surjectivity condition, and thus I^q is not a conservative extension of I^p. This happens for two essential reasons: first, the sets M of models that appear in quantum models must be in one-to-one correspondence with their induced classical valuations on the qubits; second, even for such an M, due to the independence partitions, not all probability spaces over M can be obtained from a quantum structure. Of course, by composition, we also obtain a comorphism $C^{gq} = C^{pq} \circ C^{gp} : I^g \to I^q$, and a power-model comorphism $C^q = C^{pq} \circ C^p : I \to I^q$. It is very easy to check that C^q meets the necessary surjectivity condition, and therefore I^q is still a conservative extension of I. Given Σ, and a model m of I, we just need to consider any quantum structure of the form $\langle \{m\}, \{\mathbf{Qb}(\Sigma)\}, 1|V_\sigma(m)\rangle, \nu, \rho \rangle$ with $\nu_{FA} = 1$ if $F = \mathbf{Qb}(\Sigma)$ and $v_A^F = V_\sigma(m)$, or $F = \emptyset$, and $\nu_{FA} = 0$ otherwise. On the other hand, it is easy to see that also C^{gq} will be surjective, and hence I^q a conservative extension of I^g, whenever the first of the above mentioned restrictions is trivial. That is, requiring that M is in one-to-one correspondence with its induced set of valuations should not exclude any possible set of models. For this condition to hold, it suffices to require that the qubit functor $\mathbf{Qb}$ is chosen in such a way that, for each Σ and $m_1, m_2 \in \mathbf{Mod}(\Sigma)$, if m_1 and m_2 coincide on the satisfaction of all qubits then $m_1 = m_2$. If the qubits are *representative* typically one ends up with logically equivalent models, but in many institutions it is possible to avoid having logically equivalent models. The case of classical propositional logic is paradigmatic, once we take as qubits all the propositional symbols. But similar choices are possible in many other logics. In [23] it is shown how to do this choice in any suitable finitely-valued logic. For instance, in Lukasiewicz's three-valued logic it suffices to consider as qubits all propositional symbols and negations of propositional symbols. This possibility also helps in shedding light on the usefulness of considering restricted sets of admissible valuations.

6   Conclusion

Figure 1 is the diagram of the institutions and (power-model) comorphisms we have built, where $\rightarrowtail$ is used to distinguish the arrows that guarantee a conservative extension from their source to target. Our main goal in bringing into the realm of institutions the exogenous approach to globalization, probabilization and quantization of logics was to assess how general these constructions were. The first two constructions are fully general, in the sense that nothing is assumed about the given institution and also that nothing else is needed. But quantization

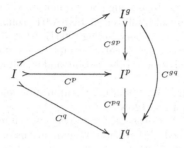

Fig. 1. Institutions and (power-model) comorphisms.

requires some additional information (the choice of qubit formulae). On the other hand, the quantum logic, as pointed out by the institutional approach, is not general enough (namely, injectivity of V_Σ on models, and surjectivity of the qubit translations). The solution seems to suggest a slight generalization of the exogenous approach towards working with multisets of models (as in Kripke structures), a promising line of further development of the approach. Furthermore, many interesting institution-theoretic questions remain open about these logics and the construction mechanisms discussed herein, like analyzing the properties of the constructions as functors on the category of institutions (or better, on some category of institutions), studying the underlying categories of models, and study their impact on the properties of the resulting categories of specifications. From a logic-theoretic point of view, the next step is to attempt at extending the completeness results in [5, 3] for a general base institution.

Acknowledgments

This work was partially supported by FCT and FEDER through POCI, namely via the QuantLog POCI/MAT/55796/2004 Project of CLC. The authors wish to express their gratitude to the regular participants in the QCI Seminar. Till Mossakowski deserves a special acknowledgment due to his prompt will to serve as an oracle on any question related to institutions and their notions of arrow. Last, but not least, we are all much indebted to Joseph Goguen for many inspirational discussions on institutions in particular, and the world at large, over the last two decades.

References

[1] Mateus, P., Sernadas, A.: Exogenous quantum logic. In Carnielli, W.A., Dionísio, F.M., Mateus, P., eds.: Proceedings of CombLog'04, Workshop on Combination of Logics: Theory and Applications, 1049-001 Lisboa, Portugal, Departamento de Matemática, Instituto Superior Técnico (2004) 141–149 Extended abstract.

[2] Mateus, P., Sernadas, A.: Reasoning about quantum systems. In Alferes, J., Leite, J., eds.: Logics in Artificial Intelligence, JELIA'04. Volume 3229 of Lecture Notes in Artificial Intelligence. Springer-Verlag (2004) 239–251

[3] Mateus, P., Sernadas, A.: Weakly complete axiomatization of exogenous quantum propositional logic. Information and Computation (in print) ArXiv math.LO/0503453.

[4] Nielsen, M.A., Chuang, I.L.: Quantum Computation and Quantum Information. Cambridge University Press (2000)

[5] Mateus, P., Sernadas, A., Sernadas, C.: Exogenous semantics approach to enriching logics. In Sica, G., ed.: Essays on the Foundations of Mathematics and Logic. Volume 1 of Advanced Studies in Mathematics and Logic. Polimetrica (2005) 165–194

[6] Foulis, D.J.: A half-century of quantum logic. What have we learned? In: Quantum Structures and the Nature of Reality. Volume 7 of Einstein Meets Magritte. Kluwer Acad. Publ. (1999) 1–36

[7] Chiara, M.L.D., Giuntini, R., Greechie, R.: Reasoning in Quantum Theory. Kluwer Academic Publishers (2004)

[8] Birkhoff, G., von Neumann, J.: The logic of quantum mechanics. Annals of Mathematics **37** (1936) 823–843

[9] Nilsson, N.J.: Probabilistic logic. Artificial Intelligence **28** (1986) 71–87

[10] Nilsson, N.J.: Probabilistic logic revisited. Artificial Intelligence **59** (1993) 39–42

[11] Bacchus, F.: Representing and Reasoning with Probabilistic Knowledge. MIT Press Series in Artificial Intelligence. MIT Press (1990)

[12] Bacchus, F.: On probability distributions over possible worlds. In: Uncertainty in Artificial Intelligence, 4. Volume 9 of Machine Intelligence and Pattern Recognition. North-Holland (1990) 217–226

[13] Fagin, R., Halpern, J.Y., Megiddo, N.: A logic for reasoning about probabilities. Information and Computation **87** (1990) 78–128

[14] Dishkant, H.: Semantics of the minimal logic of quantum mechanics. Studia Logica **30** (1972) 23–32

[15] Goguen, J., Burstall, R.: A study in the foundations of programming methodology: specifications, institutions, charters and parchments. In: Category Theory and Computer Programming. Volume 240 of LNCS. Springer (1986) 313–333

[16] Goguen, J., Burstall, R.: Institutions: abstract model theory for specification and programming. Journal of the ACM **39** (1992) 95–146

[17] Goguen, J., Roşu, G.: Institution morphisms. Formal Aspects of Computing **13** (2002) 274–307

[18] Meseguer, J.: General logics. In: Proceedings of the Logic Colloquium'87. North-Holland (1989) 275–329

[19] Tarlecki, A.: Moving between logical systems. In: Recent Trends in Data Type Specification. Volume 1130 of LNCS. Springer (1996) 478–502

[20] Mossakowski, T.: Different types of arrow between logical frameworks. In auf der Heide, F.M., Monien, B., eds.: Procs. ICALP'96. Volume 1099 of LNCS. Springer (1996) 158–169

[21] Cerioli, M., Meseguer, J.: May I borrow your logic? (Transporting logical structures along maps). Theoretical Computer Science **173** (1997) 311–347

[22] Bridges, D.S.: Computability. Volume 146 of Graduate Texts in Mathematics. Springer-Verlag (1994)

[23] Caleiro, C., Carnielli, W.A., Coniglio, M.E., Marcos, J.: Two's company: "The humbug of many logical values". In Béziau, J.Y., ed.: Logica Universalis. Birkhäuser Verlag (2005) 169–189

Jewels of Institution-Independent Model Theory

Răzvan Diaconescu

Institute of Mathematics of the Romanian Academy

Abstract. This paper is dedicated to Joseph Goguen, my beloved teacher
and friend, on the ocassion of his 65th anniversary. It is a survey of
institution-independent model theory as it stands today, the true form
of abstract model theory which is based on the concept of institution.
Institution theory was co-fathered by Joseph Goguen and Rod Burstall
in late 1970's. In the final part we discuss some philosophical roots of
institution-independent methodologies.

1 Introduction

The theory of institutions is a categorical abstract model theory which formalises
the intuitive notion of logical system, including syntax, semantics, and the satisfaction between them. Institutions constitute a model-oriented meta-theory on
logics similarly to how the theory of rings and modules constitute a meta-theory
for classical linear algebra. Another analogy can be made with universal algebra
versus groups, rings, modules, etc. By abstracting away from the realities of the
actual conventional logics, it can be noticed that institution theory comes in fact
closer to the realities of non-conventional logics.

The notion of institution arose within computing science in 1980's in response
to the population explosion of logics in use there,[1] with the ambition of doing
as much as possible at a level of abstraction independent of commitment to any
particular logic. This mathematical paradigm is called 'institution-independent'
(abbreviated *i-i*) computing science or model theory.

Since their definition by Goguen and Burstall [11,31], institutions become a
common tool in the study of algebraic specification theory and can be considered
as its most fundamental mathematical structure. It is already an algebraic specification tradition to have an institution underlying each language or system, in
which all language/system constructs and features can be rigorously explained
as mathematical entities. Most modern algebraic specification languages follow
this tradition, including CASL [2], Maude [45], or CafeOBJ [25].

---

[1] Some of them, such as first order (in many variants), second order, higher order, infinitary, Horn, equational, partial, type theoretic, intuitionistic, modal (in many variants), are well known or at least familiar to the ordinary logicians, while others such
as linear, behavioural, process, rewriting, polymorphic, coalgebraic, object-oriented,
etc. are known and used mostly in computing science.

K. Futatsugi et al. (Eds.): Goguen Festschrift, LNCS 4060, pp. 65–98, 2006.

66   Răzvan Diaconescu

An *institution* $\mathbb{I} = (\mathbb{S}ig^{\mathbb{I}}, \mathsf{Sen}^{\mathbb{I}}, \mathsf{Mod}^{\mathbb{I}}, \models^{\mathbb{I}})$ consists of

1. a category $\mathbb{S}ig^{\mathbb{I}}$, whose objects are called *signatures*,
2. a functor $\mathsf{Sen}^{\mathbb{I}} : \mathbb{S}ig^{\mathbb{I}} \to \mathbb{S}et$, giving for each signature a set whose elements are called *sentences* over that signature,
3. a functor $\mathsf{Mod}^{\mathbb{I}} : (\mathbb{S}ig^{\mathbb{I}})^{\mathrm{op}} \to \mathbb{C}at$ giving for each signature Σ a category whose objects are called Σ-*models*, and whose arrows are called Σ-(*model*) *morphisms*, and
4. a relation $\models^{\mathbb{I}}_{\Sigma} \subseteq |\mathsf{Mod}^{\mathbb{I}}(\Sigma)| \times \mathsf{Sen}^{\mathbb{I}}(\Sigma)$ for each $\Sigma \in |\mathbb{S}ig^{\mathbb{I}}|$, called Σ-*satisfaction*,

such that for each morphism $\varphi : \Sigma \to \Sigma'$ in $\mathbb{S}ig^{\mathbb{I}}$, the *satisfaction condition*

$$M' \models^{\mathbb{I}}_{\Sigma'} \mathsf{Sen}^{\mathbb{I}}(\varphi)(\rho) \quad \text{iff} \quad \mathsf{Mod}^{\mathbb{I}}(\varphi)(M') \models^{\mathbb{I}}_{\Sigma} \rho$$

holds for each $M' \in |\mathsf{Mod}^{\mathbb{I}}(\Sigma')|$ and $\rho \in \mathsf{Sen}^{\mathbb{I}}(\Sigma)$. When there is no danger of ambiguity, we may skip the superscripts from the notation of the entities of the institution, for example $\mathbb{S}ig^{\mathbb{I}}$ may be simply denoted as $\mathbb{S}ig$.

We denote the *reduct* functor $\mathsf{Mod}^{\mathbb{I}}(\varphi)$ by $\_\!\upharpoonright_\varphi$ and the sentence translation $\mathsf{Sen}^{\mathbb{I}}(\varphi)$ by $\varphi(\_)$. When $M = M'\!\upharpoonright_\varphi$ we say that M is a φ-*reduct of* M', and that M' is a φ-*expansion of* M.

I-i model theory applies to a wide variety of logics, however due to space constraints, in this paper we will discuss only examples from classical first order logic and some of its fragments and extensions.

Classical (first order) logic as institution. Let **FOL** be the institution of *many sorted first order logic with equality*. Its signatures (S, F, P) consist of a set of sort symbols S, a set F of function symbols, and a set P of relation symbols. Each function or relation symbol comes with a string of argument sorts, called *arity*, and for functions symbols, a result sort. $F_{w \to s}$ denotes the set of function symbols with arity w and sort s, and P_w the set of relation symbols with arity w.

Signature morphisms map the three components in a compatible way. Models M are first order structures interpreting each sort symbol s as a set M_s, each function symbol σ as a function M_σ from the product of the interpretations of the argument sorts to the interpretation of the result sort, and each relation symbol π as a subset M_π of the product of the interpretations of the argument sorts. Sentences are the usual first order sentences built from equational and relational atoms by iterative application of logical connectives and quantifiers. Sentence translations rename the sorts, function, and relation symbols. For each signature morphism φ, the reduct $M'\!\upharpoonright_\varphi$ of a model M' is defined by $(M'\!\upharpoonright_\varphi)_x = M'_{\varphi(x)}$ for each x sort, function, or relation symbol from the domain signature of φ. The satisfaction of sentences by models is the usual Tarskian satisfaction defined inductively on the structure of the sentences.

Without loss of generality, for the sake of simplicity of presentation, we always assume *non-empty* sorts for the models. This can be achieved in two ways. The semantic solution is to consider only models for which $M_s \neq \emptyset$ for each sort

s. The syntactical solution is to consider only signatures having at least one constant for each sort.

The institution **PL** of *propositional logic* is obtained as the sub-institution of **FOL** obtained by considering only the empty sorted signatures.

Positive first order logic, **FOL**$^+$, restricts the **FOL** sentences only to those constructed by means of $\wedge, \vee, \forall, \exists$, but not negation. Here $\vee$ and $\exists$ are no longer reducible to $\wedge$ and $\forall$ and vice versa.

An *universal Horn sentence* in **FOL** for a first order signature (S, F, P) is a sentence of the form $(\forall X)H \Rightarrow C$, where H is a finite conjunction of (relational or equational) atoms and C is a (relational or equational) atom, and $H \Rightarrow C$ is the implication of C by H. The sub-institution **HCL**, *Horn clause logic*, of **FOL** has the same signatures and models as **FOL** but only universal Horn sentences as sentences.

An *algebraic signature* (S, F) is just a **FOL** signature without relation symbols. The sub-institution of **HCL** which restricts the signatures only to the algebraic ones and the sentences to universally quantified equations is called *equational logic* and is denoted by **EQL**.

EQLN is the minimal extension of **EQL** with negation, allowing sentences obtained from atoms and negations of atoms through only one round of quantification, either universal or existential. More precisely, all sentences have the form $(QX)t_1\pi t_2$ where $Q \in \{\forall, \exists\}$ and $\pi \in \{=, \neq\}$.

Let $\forall\vee$ be the sub-institution of **FOL** determined by the *universal disjunction of atoms*.

Infinitary first order logic, **FOL**$_{\infty,\omega}$, extends **FOL** by allowing infinite conjunctions. Similarly, **HCL**$_\infty$ extends **HCL** by allowing the hypotheses H of Horn sentences $(\forall X)H \Rightarrow C$ to be infinite conjunctions of atoms. Also, $\forall\vee_\infty$ extends $\forall\vee$ by allowing infinite disjunctions of atoms.

Other examples of institutions in use in computing science include partial [10], rewriting [44], label algebra [6], higher-order [8], polymorphic [52], temporal [30], process [30], behavioural [7], coalgebraic [13], object-oriented [32] logics, and many many more...

Significance of Institution-Independent Model Theory

While the goal of i-i formal specification has been greatly accomplished in the algebraic specification literature, recently there has been significant progress towards model theory too. This responds to the feeling shared by some researchers that deep concepts and results in model theory can be reached in a significant way via institution theory. The significance of i-i model theory is manifold.

First, it fulfils the main abstract model theory ideal by providing an uniform generic approach to the model theory of various logics. This is especially relevant for areas of knowledge involving a big variety of formal logical systems, most of them unconventional. An important example comes from computing science in general, and algebraic specification in particular. Related to this, institutions also provide an ideal platform for exporting the rich and powerful body of concepts and methods developed by conventional model theory to a multitude of unconventional logics.

While conventional 'abstract' model theory of Barwise and Feferman [4,5] extends first order logic explicitly by abstracting only sentences and satisfaction and leaving signatures and models concrete and conventional, institutions axiomatise the relationship between models and sentences by leaving them abstract. Because of this lack of commitment to any particular logic, institutions can be therefore considered as *the true* form of abstract model theory, some authors even calling this 'abstract abstract model theory'...

Then, i-i model theory has a special methodological significance. The i-i top-down way of obtaining a model theoretic result, or just viewing a concept, leads to a deeper understanding which is not suffocated by the (often irrelevant) details of the actual logic and guided by structurally clean causality. A model theoretic phenomenon is thus decomposed into various layers of abstract conditions, the concepts being defined and results obtained at the *most appropriate level of abstraction*. This contrasts with the traditional bottom-up approach in which the development is done at a *given* level of abstraction. Thus concepts come naturally as presumptive features that a "logic" might exhibit or not. Hypotheses are kept as general as possible and introduced on a by-need basis. Results and proofs are modular and easy to track down, despite sometimes very deep content. Another reason for the strength of i-i methodology is that institutions provide the most complete framework for abstract model theory, emphasising the multi-signature aspect of logics by considering signature morphisms and model reducts as primary concepts.

Finally, institution theory provide an efficient framework for doing logic and model theory 'by translation or borrowing' via a general theory of mappings (homomorphisms) between institutions. For example, a certain property P which holds in an institution $\mathbb{I}'$ can be also established in another institution $\mathbb{I}$ provided that we can define a mapping $\mathbb{I} \to \mathbb{I}'$ which 'respects' P.

Apart of re-structuring known model theoretic methods, i-i model theory has already produced two classes of new concrete results. The first class is represented by model theories for a multitude of less conventional logics which did not have one properly. Out of i-i model theory, even a relatively well studied area like partial algebra gets with minimal effort (in fact almost for free!) a well developed and coherent body of advanced model theoretic concepts and results. A second class of concrete applications is constituted by new results in classical model theory obtained by institutional methods. At the moment of writing this survey, we can report interpolation and definability for numerous Birkhoff-style axiomatizable fragments of classical logic [22,49] and the elegant solution to the interpolation conjecture for many sorted logic [35]. The former results reveal a strong causality relationship between axiomatizability, on the one hand, and interpolation and definability, on the other hand. They also demount, or revise, the causal relationship between interpolation and definability. Maybe in this second class of applications we can also mention the considerably facilitated access to highly non-trivial results in classical model theory, such as Keisler-Shelah Isomorphism Theorem.

This paper is a brief journey through i-i model theory as it stands today. A full textbook on this topic is under preparation [16].

2   Basic Concepts

We assume the reader is familiar with basic notions and standard notations from category theory; e.g., see [38] for an introduction to this subject. By way of notation, $|\mathbb{C}|$ denotes the class of objects of a category $\mathbb{C}$, $\mathbb{C}(A, B)$ the set of arrows with domain A and codomain B, and composition is denoted by ";" and in diagrammatic order. The category of sets (as objects) and functions (as arrows) is denoted by $\mathbb{S}et$, and $\mathbb{C}at$ is the category of all categories.[2] The opposite of a category $\mathbb{C}$ (obtained by reversing the arrows of $\mathbb{C}$) is denoted $\mathbb{C}^{op}$.

In the following we focus on some basic institution theory concepts.

2.1   Presentations and Theories

The satisfaction relation between models and sentences determines a Galois connection between the classes of models and the sets of sentences of a signature. Let Σ be a signature in an institution $(\mathbb{S}ig, \mathsf{Sen}, \mathsf{Mod}, \models)$. Then

- for each set of Σ-sentences E, let $E^* = \{M \in |\mathsf{Mod}(\Sigma)| \mid M \models_\Sigma e$ for each $e \in E\}$, and
- for each class $\mathbb{M}$ of Σ-models, let $\mathbb{M}^* = \{e \in \mathsf{Sen}(\Sigma) \mid M \models_\Sigma e$ for each $M \in \mathbb{M}\}$.

These two functions denoted "$(\_)^*$" form what is known as a *Galois connection*. Closed classes of models $\mathbb{M} = \mathbb{M}^{**}$ are called *elementary* and closed sets of sentences $E = E^{**}$ are called *theories*.

When E and E' are sets of sentences, $E'^* \subseteq E^*$ is denoted by $E \models E'$. Two sentences e and e' of the same signature are *semantically equivalent* (denoted as $e \models\mid e'$) if they are satisfied by the same class of models, i.e., $\{e\} \models \{e'\}$ and $\{e'\} \models \{e\}$. Two models M and M' of the same signature are *elementarily equivalent* (denoted as $M \equiv M'$) if they satisfy the same set of sentences, i.e. $\{M\}^* = \{M'\}^*$. An institution is *closed under isomorphisms* when all isomorphic models are elementarily equivalent. In this paper, we will always assume that our institutions are closed under isomorphisms.

A theory E is *presented by* a set of sentences E_0 if $E_0 \subseteq E$ and $E_0 \models E$, and is *finitely presented* if there exists a finite E_0 which presents E. A *presentation morphism* $\phi \colon (\Sigma, E) \to (\Sigma', E')$ is a signature morphism such that $\phi(E) \subseteq E'^{**}$. A presentation morphism between theories is called a *theory morphism*. Notice therefore that a theory morphism $\phi \colon (\Sigma, E) \to (\Sigma', E')$ is a signature morphism

---

[2] Strictly speaking, this is only a quasi-category living in a higher set-theoretic universe.

such that $\phi(E) \subseteq E'$. It is easy to notice that under the composition of signature morphisms, presentations, respectively theory morphisms, form categories denoted $\mathbb{P}res$, respectively $\mathbb{T}h$.

Theorem 1. *[31] The forgetful functors $\mathbb{P}res \to \mathbb{S}ig$ and $\mathbb{T}h \to \mathbb{S}ig$ create limits and colimits. Consequently, in any institution, the category of its presentations/theories has whatever limits or colimits its category of signatures has.*

For example, **FOL** has all small (co)limits of signatures[3], hence it also has all (co)limits of presentations/theories.

2.2 Model Amalgamation

The model amalgamation property discussed in the following is one of the very fundamental semantical properties of logics which underlies almost all i-i model theoretic developments. It is the merit of institution theory to have discovered it.

In **FOL**, consider a model M_1 for a signature Σ_1 and a model M_2 for a signature Σ_2 such that M_1 and M_2 are 'consistent' on the intersection of the their signatures, i.e. $M_1\restriction_{\Sigma_1 \cap \Sigma_2} = M_2\restriction_{\Sigma_1 \cap \Sigma_2}$. The two models M_1 and M_2 can be 'amalgamated' to a model $M_1 \otimes M_2$ for the union of the two signatures by $(M_1 \otimes M_2)_x = (M_1)_x$ when $x \in \Sigma_1$ or $(M_1 \otimes M_2)_x = (M_2)_x$ when $x \in \Sigma_2$. Notice that this definition is correct because M_1 and M_2 are 'consistent' on $\Sigma_1 \cap \Sigma_2$, and that the amalgamation is the *unique* $(\Sigma_1 \cup \Sigma_2)$-model such that $(M_1 \otimes M_2)\restriction_{\Sigma_1} = M_1$ and $(M_1 \otimes M_2)\restriction_{\Sigma_2} = M_2$.

Such model amalgamation property can be defined in any institution by abstracting the intersection-union square of signatures to any commuting square of signatures. In any institution, a commuting square of signature morphisms

$$\begin{array}{ccc} \Sigma & \xrightarrow{\varphi_1} & \Sigma_1 \\ \varphi_2 \downarrow & & \downarrow \theta_1 \\ \Sigma_2 & \xrightarrow{\theta_2} & \Sigma' \end{array}$$

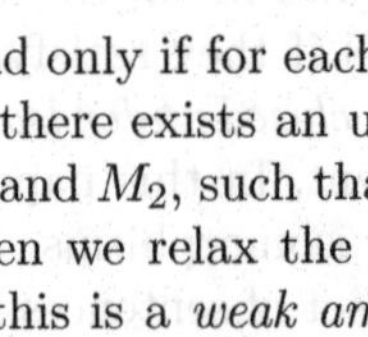

Fig. 1

is an *amalgamation square* if and only if for each Σ_1-model M_1 and a Σ_2-model M_2 such that $M_1\restriction_{\varphi_1} = M_2\restriction_{\varphi_2}$, there exists an unique Σ'-model $M_1 \otimes_{\varphi_1,\varphi_2} M_2$, called the *amalgamation* of M_1 and M_2, such that $(M_1 \otimes_{\varphi_1,\varphi_2} M_2)\restriction_{\theta_1} = M_1$ and $(M_1 \otimes_{\varphi_1,\varphi_2} M_2)\restriction_{\theta_2} = M_2$. When we relax the requirement on the uniqueness of $M_1 \otimes_{\varphi_1,\varphi_2} M_2$, we say that this is a *weak amalgamation* square. This amalgamation property is different and much more basic than some of the model amalgamation properties studied in classical model theory textbooks referring to the existence of a common elementary extension of two models of the *same signature*.

---

[3] One way to establish this is via general Grothendieck category constructions from [56].

From a categorical viewpoint, when we also involve the model homomorphisms, the model amalgamation property says that

$$\begin{array}{ccc} \mathsf{Mod}(\Sigma) & \xleftarrow{\mathsf{Mod}(\varphi_1)} & \mathsf{Mod}(\Sigma_1) \\ {\scriptstyle \mathsf{Mod}(\varphi_2)} \uparrow & & \uparrow {\scriptstyle \mathsf{Mod}(\theta_1)} \\ \mathsf{Mod}(\Sigma_2) & \xleftarrow[\mathsf{Mod}(\theta_2)]{} & \mathsf{Mod}(\Sigma') \end{array}$$

is a pullback in $\mathbb{C}at$.

At the level of arbitrary institutions model amalgamation can therefore be regarded as a limit preservation property. An institution $(\mathbb{S}ig, \mathsf{Sen}, \mathsf{Mod}, \models)$ is *semi-/directed/inductive/weakly exact* when the model functor $\mathsf{Mod} \colon \mathbb{S}ig^{\mathrm{op}} \to \mathbb{C}at$ preserves pullbacks/directed/inductive/weak[4] limits, and is simply *exact* when it preserves all small limits.

In general the many sorted institutions are exact, while the unsorted (or one-sorted) ones are only semi-exact. This is due to the fact that the initial signatures in the unsorted logics still have a sort, they are thus not initial as many sorted signatures. On the other hand the semi-exactness is not affected since pushouts of unsorted signatures are the same as pushouts of many sorted signatures.

Theorem 2. *[26] If the institution $\mathbb{I}$ is semi-exact, then the theory model functor* $\mathsf{Mod}^{\mathrm{p}} \colon (\mathbb{T}h^{\mathbb{I}})^{\mathrm{op}} \to \mathbb{C}at$ *preserves pullbacks.*[5]

This result can be of course immediately extended to other types of exactness, including full exactness.

2.3 Elementary Diagrams

The method of diagrams constitutes a traditional tool in many of the conventional first order model theory developments. Recall that the 'positive diagram' of any first order model M consists of all atoms satisfied by M in the signature extended with the elements of M. At the level of i-i model theory this is reflected as a categorical property which, in essence, formalises the idea that the class of model homomorphisms from a model M can be represented (by a natural isomorphism) as a class of models of a theory in a signature extending the original signature with syntactic entities determined by M. This can be seen as a coherence property between the semantical structure and the syntactical structure of an institution. By following the basic idea that a structure is in reality defined by its homomorphisms, the semantical structure of an actual institution is given by the model homomorphisms. On the other hand the syntactical structure of an institution is essentially determined by the atomic sentences.

---

[4] Recall [38] that a weak universal property, such as adjunction, limits, etc., is the same as the ordinary universal property except that only the existence part is required, uniqueness not being thus required.

[5] $\mathsf{Mod}^{\mathrm{p}}(\Sigma, E)$ is the full subcategory of $\mathsf{Mod}(\Sigma)$ of those models satisfying E.

An institution $(\mathbb{S}ig, \mathsf{Sen}, \mathsf{Mod}, \models)$ has *elementary diagrams* [20] if and only if for each signature Σ and each Σ-model M, there exists a signature morphism $\iota_\Sigma(M)\colon \Sigma \to \Sigma_M$, "functorial" in Σ and M, and a set E_M of Σ_M-sentences such that $\mathsf{Mod}(\Sigma_M, E_M)$ and the comma category $M/\mathsf{Mod}(\Sigma)$ are naturally isomorphic, i.e. the following diagram commutes by the isomorphism $i_{\Sigma,M}$ "natural" in Σ and M

$$
\begin{array}{ccc}
\mathsf{Mod}(\Sigma_M, E_M) & \xrightarrow{\;i_{\Sigma,M}\;} & (M/\mathsf{Mod}(\Sigma)) \\
& \searrow{\scriptstyle \mathsf{Mod}(\iota_\Sigma(M))} & \downarrow{\scriptstyle \text{forgetful}} \\
& & \mathsf{Mod}(\Sigma)
\end{array}
$$

The signature morphism $\iota_\Sigma(M)\colon \Sigma \to \Sigma_M$ is called the *elementary extension of Σ via M* and the set E_M of Σ_M-sentences is called the *elementary diagram* of the model M. For each model homomorphism $h\colon M \to N$ let N_h denote $i_{\Sigma,M}^{-1}(h)$.

The "functoriality" of ι means that for each signature morphism $\varphi\colon \Sigma \to \Sigma'$ and each Σ-model homomorphism $h\colon M \to M'\!\restriction_\varphi$, there exists a presentation morphism $\iota_\varphi(h)\colon (\Sigma_M, E_M) \to (\Sigma'_{M'}, E_{M'})$ such that

$$
\begin{array}{ccc}
\Sigma & \xrightarrow{\;\iota_\Sigma(M)\;} & \Sigma_M \\
{\scriptstyle \varphi}\downarrow & & \downarrow{\scriptstyle \iota_\varphi(h)} \\
\Sigma' & \xrightarrow[\;\iota_{\Sigma'}(M')\;]{} & \Sigma'_{M'}
\end{array}
$$

commutes.

The "naturality" of i means that for each signature morphism $\varphi\colon \Sigma \to \Sigma'$ and each Σ-model homomorphism $h\colon M \to M'\!\restriction_\varphi$ the following diagram commutes:

$$
\begin{array}{ccc}
\mathsf{Mod}(\Sigma_M, E_M) & \xrightarrow{\;i_{\Sigma,M}\;} & M/\mathsf{Mod}(\Sigma) \\
{\scriptstyle \mathsf{Mod}(\iota_\varphi(h))}\uparrow & & \uparrow{\scriptstyle h/\mathsf{Mod}(\varphi)=h;(-)\restriction_\varphi} \\
\mathsf{Mod}(\Sigma'_{M'}, E_{M'}) & \xrightarrow[\;i_{\Sigma',M'}\;]{} & M'/\mathsf{Mod}(\Sigma')
\end{array}
$$

Note that each elementary diagram (Σ_M, E_M) has an initial model $M_M = i_{\Sigma,M}^{-1}(1_M)$.

An institution with elementary diagrams ι may be denoted by $(\mathbb{S}ig, \mathsf{Sen}, \mathsf{Mod}, \models, \iota)$.

For example, classical model theory considers traditionally various kinds of model homomorphisms; each of them determine different elementary diagrams. Below we give a list of several possibilities, each of them corresponding to a specific restriction on model homomorphisms, but with the same elementary extensions. Let M be a Σ-model for a **FOL**-signature. Let $\iota_\Sigma(M)$ be the extension $\Sigma \to \Sigma_M$ adding the elements of M as new constants to Σ, and M_M be the $\iota_\Sigma(M)$-expansion of M such that $M_m = m$ for each element $m \in M$.

model homomorphisms	E_M
all	atoms in M_M^*
injective	atoms and negations of atomic equations in M_M^*
closed	atoms and negations of atomic relations in M_M^*
closed and injective	atoms and negations of atoms in M_M^*
elementary embedding	M_M^*

Recall that a **FOL**-model homomorphism $h\colon M \to N$ is *closed* when $M_\pi = h^{-1}(N_\pi)$ for each relation symbol π of the signature, and is an 'elementary embedding' when $M_M \equiv N_h$. (Notice that because $M_M \models m \neq m'$ for all $m, m' \in M$ which are different, h is also injective.)

Elementary diagrams are used in many i-i model theoretic developments. For example, in the presence of elementary diagrams, limits and colimits of models can be obtained from corresponding limits and colimits of signatures. This is an important consequence of elementary diagrams because in the actual institutions, limits, and especially colimits of models are much more difficult to establish than (co)limits of signatures.

Theorem 3. *[20] Consider and institution with elementary diagrams and initial models of presentations. Then, for each signature Σ, the category of Σ-models has $\mathbb{J}$-(co)limits whenever the category of signatures $\mathbb{S}ig$ has $\mathbb{J}$-(co)limits.*

From Theorem 3 we can immediately establish that, for any **FOL** signature its category of models has all small limits and colimits. For this, we have actually to apply Theorem 3 to the fragment of **FOL** whose sentences are just (ground) atoms. Because any Horn presentation has initial models, we can extend this argument to a stronger result: the models of any Horn theory have all small limits and colimits.

2.4   Free Models

The problem of existence of free models in institutions is often represented by the problem of existence of initial models for theories. For example, in **FOL** the largest class of theories admitting initial models is that of theories of universal Horn sentences.

At the level of an arbitrary institution $(\mathbb{S}ig, \mathsf{Sen}, \mathsf{Mod}, \models)$, a theory morphism $\varphi\colon (\Sigma, E) \to (\Sigma', E')$ is *liberal* if and only if the reduct functor $\mathsf{Mod}^{\mathrm{P}}(\varphi)\colon \mathsf{Mod}^{\mathrm{P}}(\Sigma', E') \to \mathsf{Mod}^{\mathrm{P}}(\Sigma, E)$ has a left-adjoint $(\_)^\varphi$.

Theorem 4. *[20] A semi-exact institution with elementary diagrams and pushouts of signatures is liberal when each theory has an initial model. Conversely, if the institution has initial signatures and is exact, each theory has an initial model whenever the institution is liberal.*

When we apply Theorem 4 to **FOL**, we get that **HCL** is liberal.

3   Internal Logic

Much of the i-i development of model theory relies on the possibility of defining concepts such as logical connectives, quantification, and atomic sentences internally to any institution. The main implication of this fact is that the abstract satisfaction relation between models and sentences can be decomposed at the level of arbitrary institutions into several concrete layers of satisfaction defined categorically in terms of (a simple form of) injectivity and reduction. Essentially speaking, this 'internal logic' is what gives depth to the i-i approach to model theory.

3.1   Boolean Connectives and Quantifiers

Boolean connectives. Given a signature Σ in an institution

- the Σ-sentence ρ' is a *(semantic) negation* [53] of ρ when $\rho'^* = |\mathrm{Mod}(\Sigma)| \backslash \rho^*$, and
- the Σ-sentence ρ' is the *(semantic) conjunction* [53] of the Σ-sentences ρ_1 and ρ_2 when $\rho'^* = \rho_1^* \cap \rho_2^*$.

We can easily notice that negations and conjunctions of sentences are unique modulo semantical equivalence.

An institution *has (semantic) negation* when each sentence of the institution has a negation, and *has (semantic) conjunctions* when each two sentences (of the same signature) have a conjunction. Distinguished negations are often denoted by $\neg\_$, while distinguished conjunctions by $\_ \wedge \_$.

All these can be extended in the same way to other Boolean connectives, such as disjunction ($\vee$), implication ($\Rightarrow$), equivalence ($\Leftrightarrow$), etc., and also infinitary conjunctions and disjunctions. An institution which has all semantic Boolean connectives is called a *Boolean complete institution*.

Notice that while **FOL** is Boolean complete, **EQL** and **HCL** have no semantic Boolean connectives

Quantifiers. Given a **FOL** signature morphism (S, F, P) and a set X of (new) variables (for S), any $(S, F \uplus X, P)$-sentence can be regarded as an 'open' (S, F, P)-sentence with 'unbound' variables X. When there are no unbound variables, an open sentence is just an ordinary ('closed') sentence. Recall that for any (S, F, P)-model M, $M \models (\exists X)\rho$ if and only if there exists M' an $(S, F \uplus X, P)$-expansion of M such that $M' \models \rho$.

The concept of quantification can be defined 'internally' to any institution $\mathbb{I}$ by abstracting **FOL** signature inclusions $(S, F, P) \hookrightarrow (S, F \uplus X, P)$ to any signature morphism $\chi \colon \Sigma \to \Sigma'$ in $\mathbb{I}$. Therefore at the abstract level of arbitrary institutions

- a *Σ-variable* is just a signature morphism $\chi \colon \Sigma \to \Sigma'$,
- an *(open) χ-sentence* is just a Σ'-sentence,
- a Σ-sentence ρ is a *(semantic) existential χ-quantification* [53] of a χ-sentence ρ' when $\rho^* = (\rho'^*){\upharpoonright}_\chi$; in this case we may write ρ as $(\exists\chi)\rho'$,

- a Σ-sentence ρ is a *(semantic) universal χ-quantification* [53] of a χ-sentence ρ' when $\rho \models \neg(\exists\chi)\neg\rho'$; in this case we may write ρ as $(\forall\chi)\rho'$.

For a class $\mathcal{D} \subseteq \mathbb{S}ig$ of signature morphisms, we say that the institution has universal/existential $\mathcal{D}$-*quantification* when for each $\chi\colon \Sigma \to \Sigma'$ in $\mathcal{D}$, each Σ'-sentence has a universal/existential χ-quantification.

Generally, one may consider quantification only up to what the respective concept of signature supports. For example **FOL** signatures support quantifications only up to second order, for higher order quantifications one needs to involve a different concept of signature, coding higher order types.

Based on this internal concept of variable in [21] we have introduced an internal general concept of *substitution*, which captures first-order, second-order, and higher-order substitutions in actual logics.

Finitary signature morphisms. In conventional (first order) model theory, the quantifications are finitary. At the level of abstract signature morphisms in institutions, we say that a signature morphism $\chi\colon \Sigma \to \Sigma'$ is *finitary* when for each directed diagram of Σ-models $(M_i \xrightarrow{f_{i,j}} M_j)_{(i<j)\in(I,\leq)}$ with a colimit $(M_i \xrightarrow{\mu_i} M)_{i\in I}$ and each χ-expansion M' of M there exists an index $i \in I$ and a χ-expansion μ'_i of μ_i.

3.2 Representable Signature Morphisms

Quasi-representable signature morphisms. For any **FOL** signature (S, F, P) and any set X of variables, given any (S, F, P)-model homomorphism $h\colon M \to N$, any $(S, F \uplus X, P)$-expansion M' of M determines *uniquely* a $(S, F \uplus X, P)$-expansion $h'\colon M' \to N'$ of h defined by $h' = h$ and $N'_x = h(M'_x)$ for each $x \in X$. In general, in **FOL** this property holds only for first order variables, and can be seen as an i-i generalisation of the concept of first order variable. This is important because many model theoretic results depend upon restricting quantification to first order.

In any institution, a signature morphism $\chi\colon \Sigma \to \Sigma'$ is *quasi-representable* [16] when for each Σ'-model M', the canonical functor below determined by the reduct functor $\mathsf{Mod}(\chi)$ is an isomorphism (of comma categories)

$$M'/\mathsf{Mod}(\Sigma') \cong (M'\!\restriction_\chi)/\mathsf{Mod}(\Sigma)$$

Usual 'first order' variables in actual standard institutions, but also in institutions such as $E(\textbf{FOL})$ (of **FOL** elementary embeddings) constitute examples of quasi-representable signature morphisms. However, this concept accommodates also other less conventional types of variables. For example, in the restriction of **REL** (relational logic restricting **FOL** signatures only to those without operation symbols) to strong model homomorphisms, *any* signature extension with constants and/or relation symbols is quasi-representable.

Proposition 1. *[16,19] 1. In any institution the (finitary) quasi-representable signature morphisms form a subcategory of $\mathbb{S}ig$.*

2. If the institution is semi-exact, then quasi-representable signature morphisms are stable under pushouts.

Consider a quasi-representable signature morphism $\chi\colon \Sigma \to \Sigma'$ and assume that $\mathsf{Mod}(\Sigma')$ has an initial model $0_{\Sigma'}$. We have the following canonical isomorphisms:

$$\mathsf{Mod}(\Sigma') \cong 0_{\Sigma'}/\mathsf{Mod}(\Sigma') \cong (0_{\Sigma'}{\restriction}_\chi)/\mathsf{Mod}(\Sigma)$$

This situation shows that the Σ'-models M' can be 'represented' isomorphically by Σ-model homomorphisms $M_\chi \to M'{\restriction}_\chi$, where M_χ denotes $0_{\Sigma'}{\restriction}_\chi$.

Therefore, a signature morphism $\chi\colon \Sigma \to \Sigma'$ is *representable* [19] if and only if there exists a Σ-model M_χ (called the *representation of* χ) and an isomorphism i_χ of categories such that the following diagram commutes:

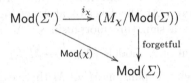

Fact 1 *A signature morphism* $\chi\colon \Sigma \to \Sigma'$ *is representable if and only if it is quasi-representable and* $\mathsf{Mod}(\Sigma')$ *has an initial model.*

Therefore, in **FOL** representable and quasi-representable signature morphisms are the same concept. For example, given a set X of variables for a **FOL** signature (S, F, P), the representation of the signature inclusion $(S, F, P) \hookrightarrow (S, F \uplus X, P)$ is given by the (free) term F-algebra $T_F(X)$. This corresponds to the fact that $(S, F \uplus X, P)$-models M are in canonical bijection with valuations of variables from X to the carrier of M, which, by the freeness of $T_F(X)$, are in canonical bijection with (S, F, P)-model homomorphisms $T_F(X) \to M$.

Proposition 2. *[14] A* **FOL***-signature morphism is representable if and only if it is bijective on sort symbols, relation symbols, and non-constant operation symbols.*

First-order substitutions can be recovered from the internal concept of substitution between representable signature morphisms; at the general level they are called *representable substitutions* [21].

3.3 Basic Sentences

Given any set of atoms (either equational or relational) E for a **FOL**-signature (S, F, P), let 0_E be the initial (S, F, P)-model satisfying E. Notice that

$$M \models E \text{ if and only if there exists a model homomorphism } 0_E \to M$$

for each (S, F, P)-model M.

Given a signature Σ in an arbitrary institution, a set E of Σ-sentences is *basic* [19] if there exists a Σ-model M_E such that for each Σ-model M, $M \models_\Sigma E$ if and only if there exists a model homomorphism $M_E \to M$.

Notice that not all sentences admitting an initial model are basic. A counterexample is given by negations of equations $t_1 \neq t_2$ in an algebraic signature (S, F).[6] On the other hand, not all basic sentences are atoms or conjunctions of atoms. For example, it can shown that **FOL** existentially quantified atoms are basic too.

When the model homomorphisms $M_E \to M$ are also unique, then we say that E is *epic basic*. We say that a sentence ρ is *(epic) basic* when $\{\rho\}$ is (epic) basic. Note that in **FOL** all atoms are epic basic.

We say that a basic set of sentences E is *finitary* if the model M_E is finitely presented in the category $\mathsf{Mod}(\Sigma)$ of Σ-models. Note that in **FOL** any finite set of atoms is finitary basic.

Proposition 3. *[16] In any institution with elementary diagrams with quasi-representable elementary extensions, the elementary diagrams are basic.*

In any institution, a *universal Horn sentence* [16] is a sentence semantically equivalent to $(\forall \chi)E \Rightarrow E'$ where $\chi \colon \Sigma \to \Sigma'$ is a quasi-representable signature morphism, E is an epic basic set of Σ'-sentences, and E' is a basic set of Σ'-sentences. A universal Horn sentence $(\forall \chi)E \Rightarrow E'$ is *finitary* when E, E' and χ are finitary. Notice that universal Horn sentences in **FOL**, as defined in the previous chapter, are the **FOL** instances of the i-i *finitary* Horn sentences.

3.4   Elementary Homomorphisms

The classical model theoretic concept of elementary embedding can be abstracted to any institution (with elementary diagrams) as follows.

First notice that in any institution with elementary diagrams, the elementary diagram of any model M has an initial model, denoted M_M. Then a model homomorphism $h \colon M \to N$ is *elementary* [34] when $N_h \models M_M^*$.

Fact 2 *For each elementary homomorphism $h \colon M \to N$, $M^* \subseteq N^*$.*

Based on the internal concept of open sentence, one may define another concept of elementary homomorphism which does not require elementary diagrams. Given a class $\mathcal{D} \subseteq \mathbb{S}ig$ of signature morphisms, a Σ-model homomorphism $h \colon A \to B$ is *$\mathcal{D}$-elementary* when $A'^* \subseteq B'^*$ for each $\mathcal{D}$-expansion $h' \colon A' \to B'$ of h.

In the actual institutions, $\mathcal{D}$ is usually taken to be the class of all signature extensions with constants. Notice that in the case of **FOL**, and in fact in all institutions with finitary sentences, elementarity with respect to signature extensions with arbitrary number of constants is equivalent to elementarity with respect to extensions adding *finite* numbers of constants. Notice that in these situations the following applies well.

---

[6] This has the term model T_F as its initial model, however it is not basic.

Proposition 4. *[34] In an weakly semi-exact institution Let $\mathcal{D}$ be a class of quasi-representable signature morphisms which is stable under pushouts. Then $\mathcal{D}$-elementary homomorphisms form a sub-institution of the original institution.*

In the case of **FOL**, when we take $\mathcal{D}$ the class of finite signature extensions with constants, this just says that **FOL** elementary embeddings form an institution.

In the presence of elementary diagrams satisfying certain 'normality' conditions (see [34] for the definition), which is very natural in actual institutions, the two notions of elementary homomorphisms coincide. This leads to another important fact: the elementary homomorphisms attached to a system of elementary diagrams bring their own system of elementary diagrams, which is in fact "more elementary" than the starting one.

Corollary 1. *[34] In any weakly semi-exact institution $\mathbb{I}$ with '$\mathcal{D}$-normal' elementary diagrams for $\mathcal{D}$ a class of quasi-representable signature morphisms which is stable under pushouts, and such that it contains all elementary extensions, the elementary homomorphisms form a (sub-)institution $E(\mathbb{I})$ which has elementary diagrams. $E(\mathbb{I})$ is called the* elementary sub-institution *of $\mathbb{I}$.*

Theorem 5 below is an i-i generalisation of famous Tarski's Elementary Chain Theorem [57] which is used for many results in classical model theory (see [12]) and shows that the closure of elementary homomorphisms under directed colimits holds when the institution either has all negations (such as **FOL**, **EQLN**), or no negation at all (such as **FOL$^+$**, **EQL**), and it may fail on intermediate cases (such as **HCL**).

Theorem 5. *[34] Assume one of the following:*
- each sentence is accessible from the basic ones by (possibly infinite) conjunctions, disjunctions, universal $\mathcal{D}$-quantifications, and finitary existential $\mathcal{D}$-quantifications, or
- the institution has negations and each sentence is accessible from the basic ones by (possibly infinite) conjunctions, negations, and finitary $\mathcal{D}$-quantifications.
Then the class of $\mathcal{D}$-elementary homomorphisms (or just elementary homomorphisms if in addition the institution has $\mathcal{D}$-normal elementary diagrams and $\mathcal{D}$-contains all elementary extensions) is closed under directed colimits.

4   Model Ultraproducts

Much of the conventional model theory can be developed through the powerful method of ultraproducts (see for example [36]). The i-i method of ultraproducts employs the following well known categorical concept of filtered products [43,1,40,41].

Let $\mathbb{C}$ be a category with small products and small directed colimits. Consider a family of objects $\{A_i\}_{i \in I}$. Each filter F over the set of indices I determines a functor $A_F \colon F \to \mathbb{C}$ such that $A_F(J \subset J') = p_{J',J} \colon \prod_{i \in J'} A_i \to \prod_{i \in J} A_i$ for each $J, J' \in F$ with $J \subset J'$, and with $p_{J',J}$ being the canonical projection.

Then the *filtered product of* $\{A_i\}_{i \in I}$ *modulo* F is the colimit $\mu \colon A_F \Rightarrow \prod_F A_i$ of the functor A_F.

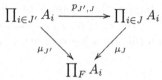

If F is an ultrafilter then the filtered product modulo F is called an *ultra-product*. When $A_i = A$ for all $i \in I$, then the filtered product is called *filtered power*. Notice that a (direct) product $\prod_{i \in I} A_i$ is the same as the filtered product $\prod_{\{\{I\}\}} A_i$.

Categorical filtered products permit the definition of filtered products of models in any institution with small products and small directed colimits for each of its categories of models. We say that an institution *has (small) products/directed colimits of models* when all its categories of models have (small) products/directed colimits.

In the case of **FOL**, model products are easy and directed colimits of models are created by the forgetful functor to (the underlying) sets because of the finiteness of the arities of the operations and relations. Alternatively, one may use the **FOL** corollary of Theorem 3. Categorical filtered products in **FOL** are the same as classical filtered products first time introduced in [39].

For a signature Σ in an institution, a Σ-sentence e is

- *preserved by $\mathcal{F}$-filtered factors* if $\prod_F A_i \models_\Sigma e$ implies $\{i \in I \mid A_i \models_\Sigma e\} \in F$, and
- *preserved by $\mathcal{F}$-filtered products* if $\{i \in I \mid A_i \models_\Sigma e\} \in F$ implies $\prod_F A_i \models_\Sigma e$,

for each filter $F \in \mathcal{F}$ over a set I and for each family $\{A_i\}_{i \in I}$ of Σ-models. A sentence is a *Łoś-sentence* [19] when is preserved by all ultrafactors and all ultraproducts.

Theorem 6. *[19] In any institution, the Łoś-sentences*
- *contain all finitary basic sentences,*
- *are closed under Boolean connectives,*
- *are closed under any finitary representable quantification, and*
- *are closed under any projectively representable quantification if the institution has epic model projections.*

An institution is a *Łoś-institution* [19] if and only if all its sentences are Łoś-sentences. For example, **FOL** is a Łoś-institution because each sentence is accessible from equations and relational atoms, which are finitary basic, by finitary representable quantifications and Boolean connectives. Instead of finitary representable quantification we may alternatively use the argument of projectively representable quantifications. This shows that the extension of **FOL** with infinitary quantifications is also a Łoś-institution. However the extension $\mathbf{FOL}_{\infty,\omega}$ of **FOL** to infinitary conjunctions is *not* a Łoś-institution.

Compactness. An institution is *m-compact* if each set of sentences is consistent when all its finite subsets have at least one model. If for each set of sentences E and each sentence e, $E \models e$ implies the existence of a finite subset $E_f \subseteq E$ such that $E_f \models e$, then we say that the institution is *compact*.

In the light of Theorem 6 the following constitutes an i-i Compactness Theorem.

Corollary 2. *Any Łoś-institution is (m-)compact.*

5   Saturated Models

Saturated models are used in many model theoretic developments (see [12]), and they can be approached naturally in an i-i framework.

Chains and $(\lambda, \mathcal{D})$-saturated models. In any category $\mathbb{C}$, for any ordinal λ, a λ-*chain* [27] is a λ-diagram $(A_i \xrightarrow{f_{i,j}} A_j)_{i<j\leq\lambda}$ such that for each limit ordinal $\zeta \leq \lambda$, $(f_{i,\zeta})_{i<\zeta}$ is the colimit of $(f_{i,j})_{i<j<\zeta}$.

For any class of arrows $\mathcal{D} \subseteq \mathbb{C}$, a $(\lambda, \mathcal{D})$-*chain* [27] is any λ-chain $(f_{i,j})_{i<j\leq\lambda}$ such that $f_{i,i+1} \in \mathcal{D}$ for each $i < \lambda$.

For each signature morphism $\chi \colon \Sigma \to \Sigma'$, a Σ-model M χ-*realizes* a set E' of Σ'-sentences, if there exists a χ-expansion M' of M which satisfies E'. It χ-*realizes* finitely E' if it realizes every finite part of E'. A Σ-model M is $(\lambda, \mathcal{D})$-*saturated* [27] for λ a cardinal and $\mathcal{D}$ a class of signature morphisms when for each ordinal $\alpha < \lambda$ and each $(\alpha, \mathcal{D})$-chain $(\Sigma_i \xrightarrow{\varphi_{i,j}} \Sigma_j)_{i<j\leq\alpha}$ with $\Sigma_0 = \Sigma$, for each $(\Sigma_\alpha \xrightarrow{\chi} \Sigma') \in \mathcal{D}$, each $\varphi_{0,\alpha}$-expansion of M χ-realizes any set of sentences if and only if it χ-realizes it finitely.

The traditional concept of λ-saturated model can be recovered from this by considering $\mathcal{D}$ to be the class of **FOL** signature extensions with a finite number of constants.

λ-small signature morphisms. A signature morphism $\Sigma \xrightarrow{\varphi} \Sigma'$ is λ-*small* [27] for a cardinal λ when for each chain $(M_i \xrightarrow{f_{i,j}} M_j)_{0\leq i<j\leq\lambda}$ of Σ-homomorphisms and each φ-expansion M' of M_λ, there exists $i < \lambda$ and $M_i' \xrightarrow{f'_{i,\lambda}} M'$ a φ-expansion of $f_{i,\lambda}$. For example, finitary signature morphisms are ω-small.

The following shows that each model can be elementarily embedded into a saturated model, thus providing an existence theorem for saturated models.

Theorem 7. *[27] 1. $M \equiv N$ if there exists a model homomorphism $M \to N$,*
   2. it has finite conjunctions and existential $\mathcal{D}$-quantifications,
   3. it has inductive colimits of signatures and is inductive-exact,
   4. for each signature Σ, the category of Σ-models has inductive colimits,

5. for each signature morphism $\Sigma \xrightarrow{\chi} \Sigma'$ and E' set of Σ'-sentences, if A realizes E' finitely then there exists a model homomorphism $A \to B$ such that B realizes E',

6. for each signature morphism $\Sigma \xrightarrow{\chi} \Sigma'$ and each Σ-model M, the class of χ-expansions of M form a set, and

7. each signature morphism from $\mathcal{D}$ is quasi-representable, the category $\mathbb{S}ig$ of signatures is $\mathcal{D}$-co-well-powered, and for each ordinal λ there exists a cardinal α such that each $(\lambda, \mathcal{D})$-chain is α-small.

Then for any cardinal λ and for each Σ-model M there exists a Σ-homomorphism $M \to N$ such that N is $(\lambda, \mathcal{D})$-saturated.

Applications of Theorem 7 considers elementary institutions. This means that in the case of **FOL**, the considered institution should be in fact the sub-institution of $E(\mathbf{FOL})$ with (arbitrarily large) signature extensions with constants as signature morphisms (in order to fulfil the inductive-exactness condition). Then it is rather easy to establish the other conditions underlying Theorem 7. The most delicate are 4., which invokes Tarski's Elementary Chain Theorem (see Theorem 5), and 5., which follows from compactness.

The uniqueness of saturated models is probably the crucial result which is used in the applications of saturated model theory. At the i-i level this requires to spell out the following rather natural property of elementary extensions.

Simple elementary diagrams. The elementary diagrams ι of an institution are *simple* [27] when for each signature Σ and all Σ-models A, B, for each $\iota_\Sigma(B)$-expansion A' of A, the following is a pushout square of signature morphisms.

$$
\begin{CD}
\Sigma @>{\iota_\Sigma(B)}>> \Sigma_B \\
@V{\iota_\Sigma(A)}VV @VV{\iota_{\Sigma_B}(A')}V \\
\Sigma_A @>>{\iota_{\iota_\Sigma(B)}(1_A)}> (\Sigma_B)_{A'}
\end{CD}
$$

It is easy to notice that in actual examples, those elementary diagrams such that their elementary extensions just add the elements of the model as new constants to its signature, like in **FOL**, are simple because the above diagram is in fact a diagram of the form

$$
\begin{CD}
\Sigma @>>> \Sigma \uplus |B| \\
@VVV @VVV \\
\Sigma \uplus |A| @>>> \Sigma \uplus |B| \uplus |A|
\end{CD}
$$

where $|A|$ and $|B|$ denote the sets of elements of (the carriers of) A and B.

Sizes of models. A $\mathcal{D}$-*size* of a model M in an institution with elementary diagrams ι is a cardinal number λ such that the elementary extension $\iota_\Sigma(M) = \varphi_{0,\lambda}$ for some $(\lambda, \mathcal{D})$-chain $(\varphi_{i,j})_{i<j\leq\lambda}$.

For example, if we take $\mathcal{D}$ to be the class of **FOL** finite extensions of signatures with constants, the $\mathcal{D}$-size of a **FOL** model M can be taken as the cardinal of its set $|M|$ of elements, i.e. $|M| = \cup_{s\in S} M_s$ where S is the set of the sorts of Σ. It can be noticed that in this case λ-saturated and $\mathcal{D}$-size λ means cardinality λ.

Theorem 8. *[27] If the institution*
  1. has pushouts and inductive colimits of signatures,
  2. is semi-exact and inductive-exact on models,
  3. has simple elementary diagrams ι,
  4. has existential $\mathcal{D}$-quantification for a (sub)category $\mathcal{D}$ of signature morphisms which is stable under pushouts,
  5. has negations and finite conjunctions, and
  6. the sentence functor preserves inductive colimits
then any two elementary equivalent $(\lambda, \mathcal{D})$-saturated Σ-models of $\mathcal{D}$-size λ are isomorphic.

In the case of **FOL**, the considered institution is just **FOL** (i.e. with the positive diagrams as (abstract) elementary diagrams). Like for Theorem 7, condition 6. holds by the finiteness of the sentences. Therefore we obtain that any two λ-saturated **FOL** models of cardinality λ are isomorphic.

The following application is an i-i generalisation of the rather famous Keisler-Shelah Theorem [12]. In the actual institutions the following conditions can be established rather easily.

Corollary 3. *[27] Consider a Łoś institution with a class $\mathcal{D}$ of signature morphisms satisfying the hypothesis of Theorem 8 and which also satisfies the following:*
  - it has finite conjunctions and existential $\mathcal{D}$-quantifications,
  - each signature morphism preserve products and directed colimits,
  - each signature morphism lifts completely ultraproducts.
Let λ be a an infinite cardinal, U a countably incomplete λ-good ultrafilter over I.
  - the cardinality of $\mathsf{Sen}(\Sigma)$ is strictly smaller than λ,
  - for each model M, if M has a $\mathcal{D}$-size λ, then each ultrapower $\prod_U M$ for an ultrafilter U over I of cardinality k, has $\mathcal{D}$-size λ^k.
  Assuming the Generalised Continuum Hypothesis, any two elementarily equivalent models have isomorphic ultrapowers (for the same ultrafilter).

Let us say that an institution has the *Keisler-Shelah property* if and only if it satisfies the conclusion of above Corollary 3.

6 Preservation and Axiomatizability

6.1 Axiomatizability by Ultraproducts

In the applications the hypotheses of the following are handled by Theorem 6.

Theorem 9. *In any institution with sentences preserved by ultraproducts that has negation and conjunction,*

- a class of models is elementary if and only if it is closed under ultraproducts and elementary equivalence,

- a class of models is finitely axiomatizable if and only if both it and its complement are elementary.

6.2   Varieties and Quasi-varieties

In classical logic it is know that in general the universal Horn sentences are essentially the most complex sentences admitting initial models in the sense that each such sentence is equivalent to a set of universal Horn sentences. It is easy to see also that Horn sentences are also preserved by (closed) sub-models are direct products. Below we show that this equivalence between the existence of initial models and the closure under direct products and submodels is independent of the actual institution.

Inclusion systems. We may use the concept of inclusion system for rephrasing the category theoretic concepts of subobjects and quotients (that are traditionally defined in terms of monics and epics).

$\langle \mathcal{I}, \mathcal{E} \rangle$ is a *inclusion system* [26][7] for a category $\mathbb{C}$ if $\mathcal{I}$ and $\mathcal{E}$ are two subcategories with $|\mathcal{I}| = |\mathcal{E}| = |\mathbb{C}|$ such that

- $\mathcal{I}$ is a partial order, and
- every arrow f in $\mathbb{C}$ can be factored uniquely as $f = e_f; i_f$ with $e_f \in \mathcal{E}$ and $i_f \in \mathcal{I}$.

The arrows of $\mathcal{I}$ are called *abstract inclusions*, and the arrows of $\mathcal{E}$ are called *abstract surjections*. The abstract surjections of some inclusion systems need not necessarily be surjective in the ordinary set-theoretic sense, take for example the inclusion system for $\mathbb{S}et$ where each function is an abstract surjection and the abstract inclusions are just the identities. An inclusion system $\langle \mathcal{I}, \mathcal{E} \rangle$ is a *epic* when all abstract surjections are epics.

In any category $\mathbb{C}$ with an inclusion system,

- A is a *subobject* of B if there exists an abstract inclusion $A \hookrightarrow B$, and
- an object B is a *quotient representation* of A if there exists an abstract surjection $A \to B$. A *quotient* of A is an isomorphism class of quotient representations.

The inclusion system is *well-powered*, respectively *co-well-powered*, if the class of subobjects, respectively quotients, of each object is a *set*.

The category of models for a **FOL**-signature (S, F, P) admits two meaningful epic inclusion systems which inherit the conventional inclusion system of the category of sets and functions. Recall that a model homomorphism $h: M \to N$ is *closed* when $M_\pi = h^{-1}(N_\pi)$ for each relation symbol $\pi \in P$, and is *strong*

---

[7] In [15] the original definition of [26] is weakened to what they called 'weak inclusion systems', which are in fact our inclusion systems.

when $h(M_\pi) = N_\pi$ for each relation symbol $\pi \in P$. Also a *submodel* M of a model N is the same with a model homomorphism $M \to N$ which is a set inclusion for each sort $s \in S$.

inclusion system	abstract inclusion	abstract surjection
closed	closed submodels	surjective homomorphisms
strong	(plain) submodels	strong surjective homomorphisms

Varieties and quasi-varieties. When $\mathbb{C}$ has small products a class of objects of $\mathbb{C}$ closed under isomorphisms
- is a *quasi-variety* if it is closed under small products and subobjects, and
- is a *variety* if it is a quasi-variety closed under quotients.

A object A of $\mathbb{C}$ is *reachable* if and only if it has no proper[8] subobjects.

The following result links the possibility of free models for theories to the quasi-variety property of the corresponding class of models. They generalise classical results from universal algebra (see [33] and [42]).

Theorem 10. *[54,20] Consider a semi-exact institution with pushouts of signatures and with elementary diagrams such that for each signature it category of models has an initial model, small products, and a co-well-powered epic inclusion system. If the class of models of each presentation is a quasi-variety, then the institution is liberal.*

The following result extends the conclusion of Theorem 10 with its opposite implication, thus obtaining an 'if and only if' characterisation of quasi-varieties.

Theorem 11. *[54,20] Consider an institution with elementary diagrams such that*
- *the category* $\mathsf{Mod}(\Sigma)$ *of* Σ-*models has an initial object* 0_Σ, *small products, and a co-well-powered epic inclusion system for each signature* Σ,
- *all model reduct functors preserve the abstract inclusions and the abstract surjections, and*
- *the model reduct functors corresponding to the elementary extensions reflect identities.*
Then each presentation has a reachable initial model if and only if the class of models of each presentation is a quasi-variety.

Under a set of appropriate conditions, the following Quasi-Variety Theorem holds in any institution.

Theorem 12. *[55,16] A class of models is a quasi-variety if and only if it is the class of models of a set of universal Horn sentences.*

The Birkhoff Variety Theorem also holds an i-i framework (under a set of appropriate conditions) when we abstract traditional 'equations' with *representable universal basic sentences* (abbreviated RUB), which are universal quantifications of basic sets of sentences by representable signature morphisms.

---

[8] Subojects which are different of A.

Theorem 13. *[16] A class of models is a variety if and only if it is the class of models of a set of RUB sentences.*

6.3   General Birkhoff Axiomatizability

In **FOL**, a finer tuned version of the Quasi-Variety Theorem 12 says that $\mathbb{M}^{**} =\overset{S_c}{\to} (P\mathbb{M})$, for each class $\mathbb{M}$ of models, where $\mathbb{M}^*$ is the set of all Horn sentences satisfied by all models of $\mathbb{M}$, $P\mathbb{M}$ is the class of all products from $\mathbb{M}$ and $\overset{S_c}{\to} \mathbb{M}$ is the class of all closed sub-models of models from $\mathbb{M}$. Similarly, if instead we consider RUB sentences, cf. Variety Theorem 13 we have that $\mathbb{M}^{**} =\overset{H_r}{\leftarrow} (\overset{S_c}{\to} (P\mathbb{M})))$, where $\overset{H_r}{\leftarrow} \mathbb{M}$ is the class of all 'quotients' of models from $\mathbb{M}$.

The i-i concept of Birkhoff-style axiomatizable closure can be captured more generally by the following concept. $(\mathbb{S}ig, \mathsf{Sen}, \mathsf{Mod}, \models, \mathcal{F}, \mathcal{B})$ is a *Birkhoff institution* [22] if and only if

- $(\mathbb{S}ig, \mathsf{Sen}, \mathsf{Mod}, \models)$ is an institution such that the category of models $\mathsf{Mod}(\Sigma)$ has small products and small directed colimits for each signature $\Sigma \in |\mathbb{S}ig|$,
- $\mathcal{F}$ is a class of filters with $\{\{*\}\} \in \mathcal{F}$, and
- $\mathcal{B}_\Sigma \subseteq |\mathsf{Mod}(\Sigma)| \times |\mathsf{Mod}(\Sigma)|$ is a reflexive binary relation for each signature $\Sigma \in |\mathbb{S}ig|$

such that

$$\mathbb{M}^{**} = \mathcal{B}_\Sigma^{-1}(\mathcal{F}\mathbb{M})$$

for each signature Σ and each class of Σ-models $\mathbb{M} \subseteq |\mathsf{Mod}(\Sigma)|$, and where $\mathcal{F}\mathbb{M}$ is the class of all $\mathcal{F}$-filtered products of models from $\mathbb{M}$.[9]

The following is a rather short list of Birkhoff institutions obtained as sub-institutions of (infinitary) **FOL** by varying the type of sentences and via various well-known axiomatizability results:

institution	$\mathcal{B}$	$\mathcal{F}$
FOL	$\equiv$	all ultrafilters
FOL	*ultraradicals*	all ultrafilters
PL	$=$	all ultrafilters
universal (quantifier-free) **FOL** sentences	$\overset{S_c}{\to}$	all ultrafilters
universal **FOL**$_{\infty,\omega}$ sentences	$\overset{S_c}{\to}$	$\{\{\{*\}\}\}$
HCL$_\infty$	$\overset{S_c}{\to}$	$\{\{I\} \mid I \text{ set}\}$
HCL	$\overset{S_c}{\to}$	all filters
universal **FOL** atoms	$\overset{H_r}{\leftarrow}; \overset{S_c}{\to}$	$\{\{I\} \mid I \text{ set}\}$
EQL	$\overset{H_r}{\leftarrow}; \overset{S_w}{\to}$	$\{\{I\} \mid I \text{ set}\}$
$\forall\vee$ (universal disjunctions of atoms)	$\overset{H_s}{\leftarrow}; \overset{S_c}{\to}$	all ultrafilters
$\forall\vee_\infty$ (univ. infinitary disj. of atoms)	$\overset{H_s}{\leftarrow}; \overset{S_c}{\to}$	$\{\{\{*\}\}\}$
$\forall\exists$ (universal-existential sentences)	*sandwiches* ([12])	all ultrafilters.

---

[9] The class of all filtered products of models modulo F for all filters $F \in \mathcal{F}$.

where H_r denote the class of surjective, H_s the class of strong surjective, H_c the class of closed surjective, S_w the class of inclusive, and S_c the class of closed inclusive model homomorphisms.

7   Interpolation

Generalised interpolation in institutions. Craig Interpolation, abbreviated CI, is classically stated as follows: if $\rho_1 \models \rho_2$ for two sentences, then there exists a sentence ρ, called the *interpolant* of ρ_1 and ρ_2, that uses logical symbols that appear both in ρ_1 and ρ_2 and such that $\rho_1 \models \rho \models \rho_2$.

An equivalent expression of the above property assumes $\rho_1 \models \rho_2$ in the *union signature* $\Sigma_1 \cup \Sigma_2$, and asks for ρ to be in the *intersection signature* $\Sigma_1 \cap \Sigma_2$, where Σ_i is the signature of ρ_i. If we naturally generalise the inclusion square

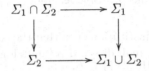

to any commuting square of signature morphisms $(\varphi_1, \varphi_2, \theta_1, \theta_2)$ like in Fig.1 and replace sentences ρ_1, ρ_2, and ρ with *sets of sentences* E_1, E_2, and E, we get the following form of CI: If $\theta_1(E_1) \models_{\Sigma'} \theta_2(E_2)$, then there exists an *interpolant* $E \subseteq \mathsf{Sen}(\Sigma)$ such that $E_1 \models_{\Sigma_1} \varphi_1(E)$ and $\varphi_2(E) \models_{\Sigma_2} E_2$. A commuting square satisfying the above property is called a *Craig Interpolation square*.

Notice that in a compact institution, if E_2 is finite, then the interpolant E can be chosen to be finite too. The immediate consequence of this fact is that in compact institutions having finite conjunctions, this CI formulation is equivalent to the more classical single sentences formulation considering single sentences rather than sets of sentences. In fact, it is the potential absence of conjunctions which motivates the generalisation from single sentences to sets of sentences.

In actual in institutions, in general, CI squares can be found among pushout squares since these constitute the accurate generalisation of intersection-union squares of signatures. While in the unsorted restriction of **FOL** *all* pushout squares have CI, this is not the case for (many sorted) **FOL**. Also, in **EQL** and **HCL**, not all pushout squares have CI. This hints that in actual institutions we should expect CI to hold not for all pushout squares, but for a restricted class of pushout squares. It is often convenient to capture such classes of CI squares by restricting independently φ_1 and φ_2 to belong to certain classes of signature morphisms. Therefore, for any classes of signature morphisms $\mathcal{L}, \mathcal{R}$, we say that the institution has the *Craig $(\mathcal{L}, \mathcal{R})$-Interpolation* [9,23] if each pushout square of signature morphism of the form is a Craig Interpolation square. The list below anticipates some of the most representatives:

institution	$\mathcal{L}$	$\mathcal{R}$	reference
unsorted **FOL**	all	all	Cor. 5 or 6
FOL	all	injective on sorts	Cor. 5 (via Thm. 15) or 6
FOL	injective on sorts	all	Cor. 5 or 6
EQL, HCL	all	injective	Cor. 4

Craig interpolation can be established in two major different ways, which have rather complementary application domains, via Birkhoff-style axiomatizability properties of institutions, or via Robinson consistency.

7.1 Interpolation Via Birkhoff-Style Axiomatizability

For a functor $C \colon I^{\mathrm{op}} \to \mathbb{C}at$, let $R = \{R_i \subseteq |C_i|^2\}_{i \in |I|}$ be a $|I|$-indexed binary relation. We say (see [22]) that an arrow $u \colon i \to i'$ in I *lifts* R if and only if for each $M' \in |C_{i'}|$ and $N \in |C_i|$, if $\langle C_u(M'), N \rangle \in R_i$ then there exists $N' \in |C_{i'}|$ such that $C_u(N') = N$ and $\langle M', N' \rangle \in R_{i'}$.

Theorem 14. *[22] In a Birkhoff institution* $(\mathbb{S}ig, \mathsf{Sen}, \mathsf{Mod}, \models, \mathcal{F}, \mathcal{B})$, *any weak amalgamation square* $(\varphi_1, \varphi_2, \theta_1, \theta_2)$ *like in Fig.1 such that*
- $\mathsf{Mod}(\varphi_1)$ *preserves products and directed colimits (of models), and*
- φ_2 *lifts* $\mathcal{B}$
is a Craig Interpolation square.

Regarding Theorem 14, CI is expected for weak amalgamation squares, which are slightly more general than pushouts squares in semi-exact institutions. The preservation of products and directed colimits by model reducts is easy in actual institutions, in fact they are created. For example, the latter holds because of the finiteness of the arities of the operation and relation symbols of the signatures. On the other hand, the lifting condition is the only interesting one which sets limits to CI in applications of Theorem 14.

Corollary 4. *[22] For universal* **FOL** *and* **FOL**$_{\infty,\omega}$ *sentences,* **HCL**, **HCL**$_\infty$, *universal* **FOL** *atoms,* **EQL**, $\forall\lor$, $\forall\lor_\infty$, *each pushout of signature morphisms* $(\varphi_1, \varphi_2, \theta_1, \theta_2)$ *like in Fig.1 with* φ_2 *injective, is a CI square.*

Interpolation via Keisler-Shelah property. In situations when the meta-Birkhoff axiomatizability is rather weak (in the sense that $\mathcal{B}$ is rather weakly defined), the lifting condition (on φ_2) can be rather hard to establish. The cost is thus shifted from the axiomatizability property to the lifting condition. A typical example is given by **FOL**, regarded as a Birkhoff institution with $\mathcal{B}$ the elementary equivalence relation $\equiv$, and $\mathcal{F}$ the class of all ultrafilters (cf. Theorem 9). However by invoking the rather powerful result that **FOL** is a Keisler-Shelah institution, [50] provides a characterisation of elementary equivalence $\equiv$ strong enough for supporting an easy applicability of Theorem 14, and which also leads to the following corollary:

Corollary 5. *In* **FOL**, *any pushout square of signature morphisms* $(\varphi_1, \varphi_2, \theta_1, \theta_2)$ *like in Fig.1 such that* φ_2 *is injective on sorts is a Craig Interpolation square.*

7.2 Interpolation Via Consistency

A set of sentences E for a signature Σ in an arbitrary institution is *consistent* if it has models, i.e. E^* is not empty. Consistency and interpolation are related by the concept of 'Robinson consistency'. A commuting square of signature morphisms $(\varphi_1, \varphi_2, \theta_1, \theta_2)$ like in Fig.1 is a *Robinson Consistency square* (abbreviated RC square) if and only if every theories $E_i \in \mathsf{Sen}(\Sigma_i)$, $i \in \{1,2\}$, with 'inter-consistent reducts', i.e. $\varphi_1^{-1}(E_1) \cup \varphi_2^{-1}(E_2)$ is consistent, have 'inter-consistent Σ'-translations', i.e. $\theta_1(E_1) \cup \theta_2(E_2)$ is consistent.

Robinson Consistency in **FOL** is classically defined only for intersection-union squares of signature morphisms, however, like for CI, this restriction is not necessary. Notice also that in some institutions, usually those supporting strong Birkhoff-style axiomatizability, such as equational logic **EQL** for example, RC is a trivial property because each set of sentences is consistent.

Theorem 15. *[51] In any institution with negation and finite conjunctions and which is compact, each commuting square of signature morphisms is a Robinson Consistency square if and only if it is a Craig Interpolation square.*

A span of signature morphisms $\Sigma_1 \xleftarrow{\varphi_1} \Sigma \xrightarrow{\varphi_2} \Sigma_2$ is said to *lift isomorphisms* [35] if for each Σ_i-models M_i, $i \in \{1,2\}$, such that $M_1\!\restriction_{\varphi_1} \cong M_2\!\restriction_{\varphi_2}$, there exists Σ_i-models N_i such that $M_i \cong N_i$ and $N_1\!\restriction_{\varphi_1} = N_2\!\restriction_{\varphi_2}$.

A commutative square of signature morphisms $(\varphi_1, \varphi_2, \theta_1, \theta_2)$ like in Fig.1 *lifts isomorphisms* if the span $\Sigma_1 \xleftarrow{\varphi_1} \Sigma \xrightarrow{\varphi_2} \Sigma_2$ lifts isomorphisms.

Theorem 16. *[35] Assume an institution such that*
1. all model homomorphisms preserve satisfaction, i.e. if $h\colon A \to B$ and then $A^* \subseteq B^*$,
2. it has pushouts of signatures and is weakly semi-exact on models,
3. it has elementary diagrams, denoted ι,
4. it has universal quantification over signature morphisms of the forms $\iota_\Sigma(h)$ and $\iota_\Sigma(M)$ for each Σ-model homomorphism $h\colon M \to N$,
5. it has ω-colimits[10] of models which are preserved by the model reduct functors.
6. it has negation and finite conjunctions, and
7. it is compact.
Then any weak amalgamation square $(\varphi_1, \varphi_2, \theta_1, \theta_2)$ like in Fig.1 (and in particular any pushout square) which lifts isomorphisms is a Robinson Consistency square (and by Theorem 15 a Craig interpolation square too).

---

[10] Here ω is the totally ordered set of the natural numbers.

In the case of classical **FOL** interpolation, the institution considered by Theorem 16 is $E(\textbf{FOL})$, the institution of the **FOL** elementary embeddings. Then, only condition 4. might need more justification, the rest being easy (for example 5. is just Tarski's Elementary Chain Theorem; see Theorem 5). Therefore, in the case of 4., if the sets of the 'empty' sorts of signatures are finite, [35] shows that quantification over $\iota_\Sigma(h)$ and $\iota_\Sigma(M)$ reduces to ordinary **FOL** quantification. Also, it is easy to see that in **FOL** a span (φ_1, φ_2) lifts isomorphisms iff either one of φ_1 or φ_2 is injective on sorts (see [35]).

Corollary 6. *The pushout of a span* $\Sigma_1 \xleftarrow{\varphi_1} \Sigma \xrightarrow{\varphi_2} \Sigma_2$ *of* **FOL** *signature morphisms such that either* φ_1 *and* φ_2 *is injective on sorts is a Craig interpolation square in* **FOL** *and* $\textbf{FOL}_{\infty,\omega}$.

8 Definability

The classical definability problem in model theory can be formulated as follows: for any **FOL**-signature (S, F, P), a new relation symbol π is 'implicitly' defined by a theory E if and only if it is 'explicitly' defined by the same theory. π is implicitly defined when the forgetful reduct $\mathsf{Mod}^{\textbf{FOL}}((S, F, P \uplus \{\pi\}), E) \to \mathsf{Mod}^{\textbf{FOL}}(S, F, P)$ is injective, which in this case can be formulated in a more syntactic but equivalent way as

$$E \cup E[\pi/\pi'] \models_{(S,F,P \uplus \{\pi,\pi'\})} (\forall X)(\pi(X) \Leftrightarrow \pi'(X))$$

for any other new relation symbol π' of the same arity and where $E[\pi/\pi']$ is the copy of E with π replaced by π', while π is explicity defined if π can be 'defined' by a $(S, F \uplus X, P)$-sentence E_π, i.e.

$$E \models_{(S,F,P \uplus \{\pi\})} (\forall X)(\pi(X) \Leftrightarrow E_\pi)$$

where X a string of variables matching the arity of π.

Generalised definability in arbitrary institutions. Definability problem can be naturally formulated at the level of abstraction of arbitrary institutions by abstracting the situation of the signature inclusion $(S, F, P) \hookrightarrow (S, F, P \uplus \{\pi\})$ to an arbitrary signature morphism.

Let $\varphi \colon \Sigma \to \Sigma'$ be a signature morphism and E' be a Σ'-theory. Then φ

- is *defined implicitly* by E' if the reduct functor $\mathsf{Mod}(\Sigma', E') \to \mathsf{Mod}(\Sigma)$ is injective, and
- is *defined (finitely) explicitly* by E' if for each signature morphism $\theta \colon \Sigma \to \Sigma_1$, and each sentence $\rho \in \mathsf{Sen}(\Sigma'_1)$, there exists a (finite) set of sentences $E_\rho \subseteq \mathsf{Sen}(\Sigma_1)$ such that

$$E' \models_{\Sigma'} (\forall \theta')(\rho \Leftrightarrow \varphi_1(E_\rho))$$

where

$$\begin{array}{ccc} \Sigma & \xrightarrow{\varphi} & \Sigma' \\ \theta \downarrow & & \downarrow \theta' \\ \Sigma_1 & \xrightarrow{\varphi_1} & \Sigma_1' \end{array}$$

is any pushout square of the span $\Sigma_1 \xleftarrow{\theta} \Sigma \xrightarrow{\varphi} \Sigma'$ of signature morphisms.

Note that E_ρ is a (finite) *set* of sentences rather than a single sentence as in the classical formulations of definability. The explicit definability says that the new part of Σ' introduced by φ can be coded only by 'symbols' of Σ. Although these formulations coincide when the institution has conjunctions, the set of sentences formulation gets the right concept of definability for institutions without conjunctions, such as **EQL**, **HCL**, etc. This situation is very similar to that of interpolation, where the concept of interpolant which is meaningful for institutions not necessarily having conjunctions is given by a set of sentences rather than by a single sentence.

One may define the concept of explicit definability such that the quantification involved is admitted by the institution by requiring θ to belong to a class $\mathcal{D}$ of signature morphisms stable under pushouts such that the institution has universal $\mathcal{D}$-quantification. Because such condition would not affect the results below, for the simplicity of presentation we prefer the unrestricted version of the explicit definability with θ *any* signature morphism.

Implicit definability contains the explicit definability. One of the most important aspects of definability theory is to establish the relationship between implicit and explicit definability. Although in classical model theory and in most of the actual institutions, explicit definability implies very easily implicit definability, the abstract model theoretic framework shows this is in fact a conditioned property holding for signature morphisms satisfying a certain condition which can be formulated by relying upon model amalgamation and elementary diagrams.

In any semi-exact institution with elementary diagrams ι, a signature morphism $\varphi \colon \Sigma \to \Sigma'$ is *tight* when for all Σ'-models M' and N' with a common φ-reduct, $M' \otimes M_M \equiv N' \otimes N_N$ implies $M' = N'$ (where $M = M'\!\upharpoonright_\varphi = N'\!\upharpoonright_\varphi = N$).

$$\begin{array}{ccc} \Sigma & \xrightarrow{\varphi} & \Sigma' \\ \iota_\Sigma(M) \downarrow & & \downarrow \theta' \\ \Sigma_M & \xrightarrow{\varphi_1} & \Sigma_1' \end{array}$$

Consider the classical situation when φ is a signature morphism in **FOL** adding one relation symbol π. Then the only possible difference between M' and N' could only be found in the difference between M'_π and N'_π. But $M'_\pi = \{X \mid M' \otimes M_M \models \pi(X)\} = \{X \mid N' \otimes N_N \models \pi(X)\} = N'_\pi$.

The situation of this example is quite symptomatic for most of the actual institutions. $M' \otimes M_M$ is just the expansion of M' interpreting the elements of M by themselves. Therefore $M' \otimes M_M \equiv N' \otimes N_N$ implies that each atom in the extended signature is satisfied either by none or by both models, which means that each symbol newly added by φ gets the same interpretation in M' and N'. This argument holds in all actual institutions in which models interpret the symbols of the signatures as sets and functions.

Corollary 7. *[49] A* **FOL** *signature morphism is tight if and only if it is surjective on sorts.*

Proposition 5. *[49] In any semi-exact institution with elementary diagrams, each tight signature morphism is defined implicitly whenever it is defined explicitly.*

For the rest of this section we focus on what is usually considered to be the 'definability problem' in model theory, i.e. the explicit contains the implicit definability. A signature morphism φ has the (finite) definability property [49] iff a theory defines φ (finitely) explicitly whenever it defines φ implicitly.

8.1 Definability Via Interpolation

Craig-Robinson interpolation. Let us strengthen the Craig interpolation property by adding to the "primary" premises E_1 a set Γ_2 (of Σ_2-sentences) as "secondary" premises. In any institution, a commuting square of signature morphisms $(\varphi_1, \varphi_2, \theta_1, \theta_2)$ like in Fig.1 is a *Craig-Robinson Interpolation square* (abbreviated *CRI* square) when for each set E_1 of Σ_1-sentences, each sets E_2 and Γ_2 of Σ_2-sentences, if $\theta_1(E_1) \cup \theta_2(\Gamma_2) \models_{\Sigma'} \theta_2(E_2)$, then there exists a set E of Σ-sentences such that $E_1 \models_{\Sigma_1} \varphi_1(E)$ and $\Gamma_2 \cup \varphi_2(E) \models_{\Sigma_2} E_2$.

We can notice easily that any CRI square is also a CI square. The following gives a sufficient condition when CI and CRI are equivalent interpolation concepts.

Proposition 6. *[29] If the institution has implications and is compact, a commuting square of signature morphisms is Craig-Robinson Interpolation square if and only if is Craig Interpolation square.*

The following can be regarded as the i-i generalisation of the Beth Definability Theorem from classical model theory.

Theorem 17. *[49] In any semi-exact (compact) institution having Craig-Robinson $(\mathcal{L}, \mathcal{R})$-interpolation for classes $\mathcal{L}$ and $\mathcal{R}$ of signature morphisms which are stable under pushouts, any signature morphism in $\mathcal{L} \cap \mathcal{R}$ has the (finite) definability property.*

By the interpolation results for **FOL** presented above (see Corollary 6) and because tight signature morphisms in **FOL** are those which are surjective on the sorts (Corollary 7), we get the following:

Corollary 8. *In* **FOL***, any signature morphism which is injective on the sorts has the finite definability property.*

Moreover, the equivalence between implicit and explicit definability holds in **FOL** *for the signature morphisms which are bijective on the sorts.*

8.2   Definability Via Axiomatizability

Definability Theorem 17 relies on Craig-Robinson interpolation, which does not hold for institutions having strong axiomatizability properties, such as **HCL** and **EQL**. In order to deal with such examples, [49] develops another definability result which relies on axiomatizability properties and which can be applied to a series of actual situations when Craig-Robinson interpolation fails.

The abstract Beth definability via axiomatizability relies on a 'lifting' condition of the signature morphism. Given a family of relations $R = \{R_\Sigma \subseteq |\mathsf{Mod}(\Sigma)| \times |\mathsf{Mod}(\Sigma)|\}_{\Sigma \in |\mathbb{S}ig|}$ indexed by the category of the signatures of an institution, a signature morphism $\varphi\colon \Sigma \to \Sigma'$ *lifts weakly* R iff for each Σ'-model M' and N', if $\langle M'{\restriction}_\varphi, N'{\restriction}_\varphi \rangle \in R_\Sigma$ then there exists P' a φ-expansion of $N'{\restriction}_\varphi$ such that $\langle M', P' \rangle \in R_{\Sigma'}$. We may recall that the first (non-weakly) lifting concept has been used by the interpolation Theorem 14. Notice that a signature morphism lifts weakly a family of relations R whenever it lifts R.

However the result below uses the lifting condition in a reverse direction than of Theorem 14.

Theorem 18. *[49] Consider a (compact) semi-exact Birkhoff institution* $(\mathbb{S}ig, \mathsf{Sen}, \mathsf{Mod}, \models, \mathcal{F}, \mathcal{B})$ *and a class* $\mathcal{S} \subseteq \mathbb{S}ig$ *of signature morphisms which is stable under pushouts and such that for each* $\varphi \in \mathcal{S}$
- φ *lifts weakly* $\mathcal{B}^{-1}$, *and*
- $\mathsf{Mod}(\varphi)$ *preserves small products and directed colimits.*
Then any signature morphism in $\mathcal{S}$ *has the (finite) definability property.*

The core technical condition which should be established in order to apply Theorem 18 is, like for Theorem 14, the lifting condition on φ. In the case of **FOL**, this leads to the following.

Corollary 9. *[49] Any* **FOL** *signature morphism which is surjective on the sort and operation symbols has the finite definability property in the institutions of the universal Horn sentences, and has the definability property in the institutions of universal sentences, of the universal infinitary sentences, and of the universal infinitary Horn sentences.*

Any **FOL** *signature morphism which is bijective on the sort and operation symbols and injective on the relation symbols has the finite definability property in the institutions of the atomic sentences and of the equations, and has the definability property in the institutions of* $\forall\lor$ *and* $\forall\lor_\infty$.

9   Other Topics

Due to space constraints, we cannot present here all important topics of today i-i model theory. Let us briefly mention here some of them which we could not develop here.

Possible worlds semantics. This development [28] refers to the treatment of modalities and their applications independently of the underlying logic. More specifically, given a base institution with model amalgamation, on the semantics side we internalise the concept of frame, and on the syntactic side we extend the existing sentences with modalities. Our concept of frame is allowed to enjoy a flexible degree of sharing which is modelled by the means of an institution morphism from the base institution to a 'domain' institution. The extension of modal sentences is based on our internal logic approach to logical connectives and quantifiers. Then on top of the satisfaction relation of the base institution, we define a modal satisfaction relation between frames and modal sentences. This generates a new 'modal' institution on top of the base institution, and due to the very mild conditions on the base institution, this 'modalisation' procedure can applied to a wide variety of actual institutions.

By employing the institution-independent method of ultraproducts [28] proves a fundamental preservation institution-independent result for modal satisfaction, that each modal sentence is preserved by ultraproducts of frames. Immediate consequences of this result includes compactness of possible worlds semantics.

Grothendieck institutions. Grothendieck institutions [18] generalise the flattening Grothendieck construction from (indexed) categories to (indexed) institutions. Regarded from a fibration theoretic angle, Grothendieck institutions are institutions for which their category of signatures is fibred. On the one hand, the actual institutions with many sorted signatures appear naturally as fibred institutions determined by the fibrations given by the functor mapping each signature to its set of sort symbols. In this sense, fibred institutions can be regarded as the reflection of the many sortedness phenomenon at the level of institution theory. On the other hand, the Grothendieck construction on institution is more adequate for modelling heterogeneous multi-logic environments. Any system of institutions which are related by institution morphisms can be flattened by the Grothendieck construction to a homogeneous institution, as has been done in the case of CafeOBJ [25] or heterogenous specification with CASL extensions [47]. In other words, this can be interpreted as putting together a system of institutions into a single institution such that their individual identities and the relationships between them are fully retained.

The Grothendieck construction on institutions can be done in two variants, using institution morphisms like in [18] or using institution comorphisms like in [46]. In the case when the institution morphisms or comorphisms correspond to adjunction situations between the categories of signatures of the institutions, the morphism-based and comorphism-based Grothendieck institutions can be shown isomorphic [47].

An important class of problems posed by the Grothendieck, or fibred, institutions is that of lifting of model-theoretic properties from the 'local' level of index institutions, or fibres, to the 'global' level of the Grothendieck, or fibred, institution. While [17] and [18] investigate the lifting of theory colimits, free models, model amalgamation, inclusion systems, [23] solves the interpolation problem for Grothendieck institutions.

Stratified institutions. They have been introduced by Marc Aiguier and Fabrice Barbier (see [3]) in order to model valuations of variables or states of models. Although it is possible to develop a great deal of model theory using this i-i technique, its biggest promise seems to be for the problem of combinations of logics, which is currently one of the most challenging problems.

Proof-theoretic aspects. Recently there has been a successful attempt to enrich institutions with proof theoretic structure [48], not by amalgamation of the often conflicting model theoretic culture of institutions and the proof theoretic culture of type theory, but by an institutional proof theory from scratch by extending categorical logic [37] to represent proof as arrows in categories of sentences.

The recent paper [24] introduces a concept of proof rules for institutions and argues that the proof systems of the actual institutions with proofs are freely generated by their presentations as systems of proof rules. It also shows that proof-theoretic quantification, an institutional refinement of the (meta-)rule of Generalisation from classical logic, can also be added freely to any proof system. By applying these universal properties, [24] is able to provide some general compactness results for proof systems and some general soundness results for institutions with proofs.

Proof systems for institutional logic emerges as a very promising new area with many interesting open questions.

10   Philosophical Roots

In this final section I would like to share with the interested readers some personal reflections about some philosophical aspects of institutions from the perspective of Tibetan Buddhism, a spiritual and philosophical tradition shared by the fathers of institution theory, Joseph Goguen and Rod Burstall, and by the author of this survey.

Institution theory is not only a mathematical theory. In fact, I think its main value resides in its unique way to approach mathematical and computing science phenomena. In my view, the institutional way can be seen as an effect of a Buddhist (trained) mind and an application of *Śunyata*, the Buddhist Mahayana perspective on reality.

The highest explanation of Śunyata has been developed by the Madhyamaka Prasangika philosophical school which had started in the great Buddhist monastic university of Nalanda about 2000 years ago. Maybe the most prominent philosophical figure of this school was Acharya Nagarjuna who wrote a series of treaties consisting mainly of very sophisticated philosophical and logical arguments supporting the doctrine of Śunyata. The Madhyamaka Prasangika philosophical viewpoint has been inherited and preserved to our days by all traditions of Tibetan Buddhism.

In brief, Śunyata means the emptiness of all phenomena, either mind or matter, of an inherent nature. All phenomena thus arise on the basis of the so-called 'co-dependent origination', which at a certain level can be thought as a very profound distributed network of interdependencies. This view avoids both extremes of eternalism (things posses an inherent nature) and of nihilism (nothing exists), hence 'Madhyamaka' translated as 'Middle Way'.

When applied to modern science, this offers a *non-essentialist* perspective. While some branches of modern science, most notably quantum physics, resonates strongly to the Madhyamaka Prasangika explanation of reality in a rather independent way (for the interested reader we recommend the recent survey [58]), i-i methodology has been directly x2influenced by this philosophical perspective.

Śunyata also means a lack of reference point, a kind of groundless. Institutions realize this in a very transparent way, because they truly transcede the idea of commitment to particular logics. Moreover, concepts such as institution (co)morphisms, which are central to institution theory, constitute efficient technical tools for understanding the immensely vast network of interdependencies between logical systems. By contrast, the original abstract model theory programme of Barwise and Feferman failed exactly because it was not based on such groundless view on logic, still having classical logic as a reference point.

The rather intensive use of category theory for institutions, at various ways and at various levels, is another illustration of the groundless aspect of institution theory. By emphasizing the relationships between objects rather than their internal structure, category theory might be the single mathematical area which realizes the principle of interdependency so close to its Buddhist meaning.

This philosophical viewpoint underlying institution theory is very intimately connected to the feeling of elegance and clarity experienced when using the i-i methodology, either in computing science or in model theory. Due to the space limitations of this paper, we leave this discussion at this point, with the promise to come back sometime with a full essay on the connections between Buddhism and i-i thinking.

Acknowledgement

To Joseph Goguen for being an ideal teacher and a close caring friend, to Joseph Goguen and Rod Burstall for inventing institutions, to the great algebraic specification community for setting high scientific and intelectual standards to our research area, to the students of Şcoala Normală Superioară Bucharest, Marius Petria, Andrei Popescu, Daniel Găină, Mihai Codescu, Traian Şerbănuţă for being bright and for their contribution first as patient students and later as researchers of i-i model theory.

References

1. Hajnal Andréka and István Németi. A general axiomatizability theorem formulated in terms of cone-injective subcategories. In B. Csakany, E. Fried, and E.T. Schmidt, editors, *Universal Algebra*, pages 13–35. North-Holland, 1981. Colloquia Mathematics Societas János Bolyai, 29.
2. Edigio Astesiano, Michel Bidoit, Hélène Kirchner, Berndt Krieg-Brückner, Peter Mosses, Don Sannella, and Andrzej Tarlecki. CASL: The common algebraic specification language. *Theoretical Computer Science*, 286(2):153–196, 2002.
3. Fabrice Barbier. *Géneralisation et préservation au travers de la combinaison des logique des résultats de théorie des modèles standards liés à la structuration des spécifications algébriques*. PhD thesis, Université Evry, 2005.
4. Jon Barwise. Axioms for abstract model theory. *Annals of Mathematical Logic*, 7:221–265, 1974.
5. Jon Barwise and Solomon Feferman. *Model-Theoretic Logics*. Springer, 1985.
6. G. Bernot, P. Le Gall, and M. Aiguier. Label algebras and exception handling. *Science of Computer and Programming*, 23:227–286, 1994.
7. Michel Bidoit and Rolf Hennicker. On the integration of the observability and reachability concepts. In *Proc. 5th Int. Conf. Foundations of Software Science and Computation Structures (FOSSACS'2002)*, volume 2303 of *Lecture Notes in Computer Science*, pages 21–36, 2002.
8. Tomasz Borzyszkowski. Higher-order logic and theorem proving for structured specifications. In Christine Choppy, Didier Bert, and Peter Mosses, editors, *Workshop on Algebraic Development Techniques 1999*, volume 1827 of *LNCS*, pages 401–418, 2000.
9. Tomasz Borzyszkowski. Logical systems for structured specifications. *Theoretical Computer Science*, 286(2):197–245, 2002.
10. Peter Burmeister. *A Model Theoretic Oriented Approach to Partial Algebras*. Akademie-Verlag, Berlin, 1986.
11. Rod Burstall and Joseph Goguen. Semantics of Clear. Unpublished notes handed out at the 1978 Symposium on Algebra and Applications, Stefan Banach Center, Warsaw, Poland, 1978.
12. C.C.Chang and H.J.Keisler. *Model Theory*. North Holland, Amsterdam, 1990.
13. Corina Cîrstea. Institutionalising many-sorted coalgebraic modal logic. In *Coalgebraic Methods in Computer Science 2002, Electronic Notes in Theoretical Computer Science*, 2002.
14. Traian Şerbănuţă. Institutional concepts in first-order logic, parameterized specification, and logic programming. Master's thesis, University of Bucharest, 2004.
15. Virgil Emil Căzănescu and Grigore Roşu. Weak inclusion systems. *Mathematical Structures in Computer Science*, 7(2):195–206, 1997.
16. Răzvan Diaconescu. *Institution-independent Model Theory*. To appear. Book draft. (Ask author for current draft at Razvan.Diaconescu@imar.ro).
17. Răzvan Diaconescu. Extra theory morphisms for institutions: logical semantics for multi-paradigm languages. *Applied Categorical Structures*, 6(4):427–453, 1998. A preliminary version appeared as JAIST Technical Report IS-RR-97-0032F in 1997.
18. Răzvan Diaconescu. Grothendieck institutions. *Applied Categorical Structures*, 10(4):383–402, 2002. Preliminary version appeared as IMAR Preprint 2-2000, ISSN 250-3638, February 2000.
19. Răzvan Diaconescu. Institution-independent ultraproducts. *Fundamenta Informaticæ*, 55(3-4):321–348, 2003.

20. Răzvan Diaconescu. Elementary diagrams in institutions. *Journal of Logic and Computation*, 14(5):651–674, 2004.
21. Răzvan Diaconescu. Herbrand theorems in arbitrary institutions. *Information Processing Letters*, 90:29–37, 2004.
22. Răzvan Diaconescu. An institution-independent proof of Craig Interpolation Theorem. *Studia Logica*, 77(1):59–79, 2004.
23. Răzvan Diaconescu. Interpolation in Grothendieck institutions. *Theoretical Computer Science*, 311:439–461, 2004.
24. Răzvan Diaconescu. Proof systems for institutional logic. *Journal of Logic and Computation*, 2006. To appear.
25. Răzvan Diaconescu and Kokichi Futatsugi. Logical foundations of CafeOBJ. *Theoretical Computer Science*, 285:289–318, 2002.
26. Răzvan Diaconescu, Joseph Goguen, and Petros Stefaneas. Logical support for modularisation. In Gerard Huet and Gordon Plotkin, editors, *Logical Environments*, pages 83–130. Cambridge, 1993. Proceedings of a Workshop held in Edinburgh, Scotland, May 1991.
27. Răzvan Diaconescu and Marius Petria. Saturated models in institutions. In preparation.
28. Răzvan Diaconescu and Petros Stefaneas. Possible worlds semantics in arbitrary institutions. Technical Report 7, Institute of Mathematics of the Romanian Academy, June 2003. ISSN 250-3638.
29. Theodosis Dimitrakos and Tom Maibaum. On a generalized modularization theorem. *Information Processing Letters*, 74:65–71, 2000.
30. J. L. Fiadeiro and J. F. Costa. Mirror, mirror in my hand: A duality between specifications and models of process behaviour. *Mathematical Structures in Computer Science*, 6(4):353–373, 1996.
31. Joseph Goguen and Rod Burstall. Institutions: Abstract model theory for specification and programming. *Journal of the Association for Computing Machinery*, 39(1):95–146, January 1992.
32. Joseph Goguen and Răzvan Diaconescu. Towards an algebraic semantics for the object paradigm. In Harmut Ehrig and Fernando Orejas, editors, *Recent Trends in Data Type Specification*, volume 785 of *Lecture Notes in Computer Science*, pages 1–34. Springer, 1994.
33. George Grätzer. *Universal Algebra*. Springer, 1979.
34. Daniel Găină and Andrei Popescu. An institution-independent generalization of Tarski's Elementary Chain Theorem. *Journal of Logic and Computation*. To appear.
35. Daniel Găină and Andrei Popescu. An institution-independent proof of Robinson consistency theorem. *Studia Logica*. To appear.
36. J.L.Bell and A.B.Slomson. *Models and Ultraproducts*. North Holland, 1969.
37. Joachim Lambek and Phil Scott. *Introduction to Higher Order Categorical Logic*. Cambridge, 1986. Cambridge Studies in Advanced Mathematics, Volume 7.
38. Saunders Mac Lane. *Categories for the Working Mathematician*. Springer, second edition, 1998.
39. J. Łoś. Quleques remarques, théorèmes et problèmes sur les classes définissables d'algèbres. In *Mathematical Interpretation of Formal Systems*, pages 98–113. North-Holland, Amsterdam, 1955.
40. Michael Makkai. Ultraproducts and categorical logic. In C.A. DiPrisco, editor, *Methods in Mathematical Logic*, volume 1130 of *Lecture Notes in Mathematics*, pages 222–309. Springer Verlag, 1985.

41. Michael Makkai. Stone duality for first order logic. *Advances in Mathematics*, 65(2):97–170, 1987.
42. A. I. Malcev. *The Metamathematics of Algebraic Systems*. North-Holland, 1971.
43. G. Matthiessen. Regular and strongly finitary structures over strongly algebroidal categories. *Canad. J. Math.*, 30:250–261, 1978.
44. José Meseguer. Conditional rewriting logic as a unified model of concurrency. *Theoretical Computer Science*, 96(1):73–155, 1992.
45. José Meseguer. A logical theory of concurrent objects and its realization in the Maude language. In Gul Agha, Peter Wegner, and Akinori Yonezawa, editors, *Research Directions in Concurrent Object-Oriented Programming*. The MIT Press, 1993.
46. Till Mossakowski. Comorphism-based Grothendieck logics. In K. Diks and W. Rytter, editors, *Mathematical foundations of computer science*, volume 2420 of *LNCS*, pages 593–604. Springer, 2002.
47. Till Mossakowski. Foundations of heterogeneous specification. In *16th Workshop on Algebraic Development Techniques 2002*, LNCS. Springer, 2003.
48. Till Mossakowski, Joseph Goguen, Răzvan Diaconescu, and Andrzej Tarlecki. What is a logic? In Jean-Yves Beziau, editor, *Logica Universalis*, pages 113–133. Birkhauser, 2005.
49. Marius Petria and Răzvan Diaconescu. Abstract Beth definability in institutions. *Journal of Symbolic Logic*, 2006/2007. To appear.
50. Andrei Popescu, Traian Şerbănuţă and Grigore Roşu. A semantic approach to interpolation. Submitted.
51. Antonio Salibra and Giuspeppe Scollo. Interpolation and compactness in categories of pre-institutions. *Mathematical Structures in Computer Science*, 6:261–286, 1996.
52. Lutz Schröder, Till Mossakowski, and Christoph Lüth. Type class polymorphism in an institutional framework. In José Fiadeiro, editor, *Recent Trends in Algebraic Development Techniques, 17th Intl. Workshop (WADT 2004)*, volume 3423 of *Lecture Notes in Computer Science*, pages 234–248. Springer, Berlin, 2004.
53. Andrzej Tarlecki. Bits and pieces of the theory of institutions. In David Pitt, Samson Abramsky, Axel Poigné, and David Rydeheard, editors, *Proceedings, Summer Workshop on Category Theory and Computer Programming*, pages 334–360. Springer, 1986. Lecture Notes in Computer Science, Volume 240.
54. Andrzej Tarlecki. On the existence of free models in abstract algebraic institutions. *Theoretical Computer Science*, 37:269–304, 1986. Preliminary version, University of Edinburgh, Computer Science Department, Report CSR-165-84, 1984.
55. Andrzej Tarlecki. Quasi-varieties in abstract algebraic institutions. *Journal of Computer and System Sciences*, 33(3):333–360, 1986. Original version, University of Edinburgh, Report CSR-173-84.
56. Andrzej Tarlecki, Rod Burstall, and Joseph Goguen. Some fundamental algebraic tools for the semantics of computation, part 3: Indexed categories. *Theoretical Computer Science*, 91:239–264, 1991. Also, Monograph PRG–77, August 1989, Programming Research Group, Oxford University.
57. Alfred Tarski and R.Vaught. Arithmetical extensions of relational systems. *Compositio Mathematicæ*, 13:81–102, 1957.
58. His Holiness the XIVth Dalai Lama. *The Universe in a Single Atom*. Wisdom Publications, 2005.

Semantic Web Languages – Towards an Institutional Perspective*

Dorel Lucanu[1], Yuan Fang Li[2], and Jin Song Dong[2]

[1] Faculty of Computer Science
"A.I.Cuza" University
Iaşi, Romania
dlucanu@info.uaic.ro
[2] School of Computing
National University of Singapore, Singapore
{liyf,dongjs}@comp.nus.edu.sg

Abstract. The Semantic Web (SW) is viewed as the next generation of the Web that enables intelligent software agents to process and aggregate data autonomously. Ontology languages provide basic vocabularies to semantically markup data on the SW. We have witnessed an increase of numbers of SW languages in the last years. These languages, such as RDF, RDF Schema (RDFS), the OWL suite of languages, the OWL⁻ suite, SWRL, are based on different semantics, such as the RDFS-based, description logic-based, Datalog-based semantics. The relationship among the various semantics poses a challenge for the SW community for making the languages interoperable. Institutions provide a means of reasoning about software specifications regardless of the logical system. This makes it an ideal candidate to represent and reason about the various languages in the Semantic Web. In this paper, we construct institutions for the SW languages and use institution morphisms to relate them. We show that RDF framework together with the RDF serializations of SW languages form an indexed institution. This allows the use of Grothendieck institutions to combine Web ontologies described in various languages.

1 Introduction

The family of Semantic Web (SW) languages increased very much in the last years and we guess it will continue to increase in the future. This is somehow surprising for SW community and it contradicts the initial intentions of the SW creators. But it is a reality and we have to live with it. This increase refers specially to the languages describing Web ontologies. Here are several examples: OWL with its three dialects (Lite, DL, and Full) [20], SWRL [15], SWRL FOL [21], DLP [12], OWL⁻ with its three dialects (Lite⁻, DL⁻, Full⁻) [5], WRL [1], and the list does not finish here. For these languages, different definitions for their semantics were proposed in the literature:

---

* This work is partially supported by Singapore MOE project Rigorous Design Methods and Tools for Intelligent Autonomous (R-252-000-201-112) and NUS EERSS Program. The second author would like to thank Singapore Millennium Foundation (SMF) for the financial support.

K. Futatsugi et al. (Eds.): Goguen Festschrift, LNCS 4060, pp. 99–123, 2006.

model-theoretic semantics [20,13], RDF based semantics [1,20], first-order logic based semantics [15,21], frame logic semantics [5], Datalog semantics [5], Z semantics [8,18], and so on. This gives rise to some confusions and debates about the meaning of the hierarchy of SW languages as it has been illustrated in the well-known Tim Berners-Lee's "Semantic Web Stack" diagram (Fig. 1).

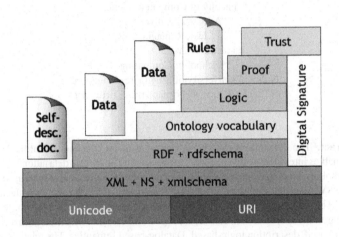

Fig. 1. The Semantic Web stack of languages

In this paper we use the institution theory in order to investigate the exact relationships among these languages.

The notion of institutions [10] was introduced to formalize the concept of "logical systems". Institutions provide a means of reasoning about software specifications regardless of the logical system. Hence, it serves as a natural candidate to study the relationship among the various SW languages, as they are based on different logical systems (semantics).

In this paper, we investigate the relationship among languages RDF [17], RDF Schema [4], OWL suite [20], and OWL$^-$ suite [5] by defining their respective institutions and relating these institutions using morphisms or comorphisms. A main advantage is a better understanding of the semantical relationship among the various SW languages. Here we focus only on the RDF triple-based semantics. We show that RDF framework (RDF and RDF Schema) together with RDF serializations of SW languages form an indexed institution, and hence the whole framework can be organized as a Grothendieck institution [7]. An interesting fact is that the construction of the indexed institution is based on a diagram of RDF theories. We define a method of constructing institutions starting from theories and then we extend it to diagrams of categories and indexed institutions. We believe that we answer in this way the question regarding the layering of SW languages [22]. Semantically, the "stack" of SW languages depicted by Berners-Lee is an indexed institution. This indexed institution produces a Grothendieck institution which offers a formal framework for combining ontologies written in various languages. In this way, all SW languages can "live" together.

The rest of the paper is organized as follows. In Section 2, we briefly present the background information on SW languages and institutions. In Section 3, we define the institutions of (bare) RDF and RDF Schema languages. These institutions are used as the basis on which one of the semantics for SW languages is constructed using a method presented in Section 2. Section 4 is devoted to the construction of the institutions defining SW languages. In Section 5, we construct an indexed institution based on a diagram of RDF theories and we show that the RDF-based semantics of SW languages can be defined as institution comorphisms from these languages to the indexed institution. Section 6 concludes the paper and discusses future work directions.

Acknowledgment

This paper is dedicated, warmly and respectfully, to Professor Joseph Goguen on the occasion of his 65th birthday. Joseph has determinative contributions in promoting Algebra from the status of an abstract notation to that of a practical specification language, widely used today in software engineering, and promoting logical systems as the "institution of the specification languages". The research and teaching activity of the first author is definitely guided by Joseph's papers on these two issues. We wish him success and happiness in his future.

2 Preliminaries

2.1 Semantic Web Languages

The Semantic Web is a vision as the new generation of the current World Wide Web in which information is semantically marked-up so that intelligent software agents can autonomously understand, process and aggregate data. This ability is realized through the development of a "stack" of languages, as depicted by Berners-Lee in Fig. 1.

Based on mature technologies such as XML, Unicode and URI (Uniform Resource Identifier), The Resource Description Framework (RDF) [17] is the foundation of later languages in the SW. RDF is a model of metadata defining a mechanism for describing resources that makes no assumptions about a particular application domain. It provides a simple way to make *statements* about Web resources. An RDF document is a collection of *triples*: statements of the form ⟨*subject*, *predicate*, *object*⟩, where *subject* is the resource we are interested in, *predicate* specifies the property or characteristic of the subject and *object* states the value of the property. This is the basic structure of the subsequent ontology languages. RDF also defines vocabularies for constructing containers such bags, sequences and lists.

RDF Schema [4] provides additional vocabularies for describing RDF documents. It defines semantical entities such as *Resource*, *Class*, *Property*, *Literal* and various properties about these entities, such as *subClassOf*, *domain*, *range*, etc. In RDF Schema, *Resource* is the universe of description. It can be further categorized as classes, properties, datatypes or literals. With these semantical constructs, RDF Schema can be regarded as the basic ontology language.

The Web resources are represented by full URIs, consisting of a prefix, representing a namespace, and a name representing the actual resource that is being described. In its full form, the prefix and the resource name are separated by a #. In shorthand form, the prefix can be represented by a shorter name and it is separated from the actual name by a colon (:), as the following example shows. After a resource has been introduced by an rdf:ID construct (in shorthand form of the URI), it can be subsequently accessed and augmented by the rdf:about constructs. When there is no possibility of confusion, the prefix can be omitted (but not the separator #).

Example 1. The following RDF fragment defines an RDFS class *Carnivore*, which is a sub class of *Animal*.

```
<rdfs:Class rdf:ID="Animal"/>
<rdfs:Class rdf:ID="Carnivore">
   <rdfs:subClassOf rdf:resource=#Animal"/>
</rdfs:Class>
```

In this example, the namespace is the URI http://ex.com/animals. The full URI for the class Animal is http://ex.com/animals#Animal.

The ability to organize and categorize domain knowledge is a necessity for software agents to process and aggregate Web resources. Domain knowledge is usually organized as inter-related conceptual entities in a hierarchy. The RDF language is not expressive enough to tackle such complexity.

In 2003, W3C published a new ontology language, the Web Ontology Language (OWL) [20]. Based on description logics and RDF Schema, the OWL suite consists of three sublanguages: Lite, DL and Full, with increasing expressiveness. The three sublanguages are meant for user groups with different requirements of expressiveness and decidability. OWL Lite and DL are decidable whereas OWL Full is generally not.

By saying that an ontology language is decidable, it actually means that the core reasoning problems, namely, concept subsumption, concept/ontology satisfiability and instantiation checking, are decidable [16].

One of the major extensions of OWL over RDF Schema is the ability to define restrictions using existing classes and properties. By using restrictions, new classes can be built incrementally. In OWL, conceptual entities are organized as classes in hierarchies. Individuals are grouped under classes and are called instances of the classes. Classes, properties and individuals can be related by properties.

Example 2. The following OWL fragment shows the definition of an object property *eats* and a class *carnivore*, which is further defined as an animal that only eats animals. This is achieved through the use of an allValuesFrom restriction in OWL.

```
<owl:ObjectProperty rdf:ID="eats"/>
<owl:Class rdf:about="#Carnivore">
   <rdfs:subClassOf>
      <owl:Restriction><owl:allValuesFrom>
         <owl:Class rdf:resource="#Animal"/>
      </owl:allValuesFrom>
```

```
    <owl:onProperty>
       <owl:ObjectProperty rdf:resource=
            "http://ex.com/animals#eats"/>
       </owl:onProperty>
    </owl:Restriction></rdfs:subClassOf>
</owl:Class>
```

The class *Carnivore* is defined to be a sub class of an OWL restriction that defines an anonymous class which only *eats Animals*.

The OWL⁻ [5] suite of languages, namely Lite⁻, DL⁻ and Full⁻, is a restricted variant of OWL languages. OWL Lite⁻ and DL⁻ are strict subsets of OWL Lite and DL respectively and they can be directly translated into Datalog. According to [5], the main advantages of the OWL⁻ include the following. Firstly, by translating OWL⁻ to Datalog, highly efficient deductive database querying capabilities can be used; Secondly, rules extension and query languages can be easily implemented on top of OWL⁻.

In order to expand the expressiveness of SW languages, several rules extensions have been proposed. SWRL [15] is a direct *extension* of OWL DL that incorporates Horn-style rules. Among other things, it supports (universally quantified) variables and built-in predicates/ functions for various data types.

On the contrary, the Web Rules Language (WRL) [1] is a rule-based ontology language. Based on deductive databases and logic programming, it is designed to be complementary to OWL, which is strong at checking subsumption relationships among concepts. WRL focuses on checking instance data and the specification of and reasoning about arbitrary rules. Moreover, WRL assumes a "Closed World Assumption", whereas OWL and SWRL assume an "Open World Assumption".

2.2 Institutions

Institutions supply a uniform way for structuring the theories in various logical systems. Many logical systems have been proved to be institutions. Recent research showed that institutions are useful in designing tools supporting verification over multiple logics. The basic reference for institutions is [10]. A comprehensive overview on institutions and their applications can be found in [6]. A well structured approach of the various institution morphisms and many other recent constructions can be found in [11]. A recent application of institutions in formalizing the information integration is given in [9]. The Grothendieck institution construction we use in this paper follows the line from [7]. The institutions use intensively category theory; we recommend [2] for a detailed presentation of categories and their applications in computer science.

In this section we recall the main definitions for institutions and we introduce two new constructions. The first is simple and it generalizes the notion of theoroidal comorphism by allowing to encode sentences from the source institution by conjunctions of sentences in the target institution. The second is more complex and is used to construct indexed institutions starting from diagrams of semantically constrained theories from a basic institution. We use this construction to define the indexed institutions based on RDF triples corresponding to SW languages.

An *institution* is a quadruple $\Im = (\mathsf{Sign}, \mathsf{Mod}, \mathsf{sen}, \models)$, where Sign is a category whose objects are called *signatures*, $\mathsf{Mod} : \mathsf{Sign}^{op} \to \mathsf{Cat}$ is a functor which associates with each signature Σ a category whose objects are called Σ-*models*, sen is a functor $\mathsf{sen} : \mathsf{Sign} \to \mathsf{Set}$ which associates with each signature Σ a set whose elements are called Σ-*sentences*, and $\models$ is a function which associates with each signature Σ a binary relation $\models_\Sigma \subseteq |\mathsf{Mod}(\Sigma)| \times \mathsf{sen}(\Sigma)$, called *satisfaction relation*, such that for each signature morphism $\phi : \Sigma \to \Sigma'$ the *satisfaction condition*

$$\mathsf{Mod}(\phi^{op})(M') \models_\Sigma \varphi \Leftrightarrow M' \models_{\Sigma'} \phi(\varphi)$$

holds for each model $M' \in \mathsf{Mod}(\Sigma')$ and each sentence $\varphi \in \mathsf{sen}(\Sigma)$. The functor sen abstracts the way the sentences are constructed from signatures (vocabularies). The functor Mod is defined over the opposite category Sign^{op} because a "translation between vocabularies" $\phi : \Sigma \to \Sigma'$ defines a forgetful functor $\mathsf{Mod}(\phi^{op}) : \mathsf{Mod}(\Sigma') \to \mathsf{Mod}(\Sigma)$ such that for each Σ'-model M', $\mathsf{Mod}(\phi^{op})(M')$ is M' viewed as a Σ-model. The satisfaction condition may be read as "M' satisfies the ϕ-translation of φ iff M', viewed as a Σ-model, satisfies φ", i.e., the meaning of φ is not changed by the translation ϕ.

We often use $\mathsf{Sign}(\Im)$, $\mathsf{Mod}(\Im)$, $\mathsf{sen}(\Im)$, $\models_\Im$ to denote the components of the institution $\Im$. If $\phi : \Sigma \to \Sigma'$ is a signature morphism, then the Σ-model $\mathsf{Mod}(\phi^{op})(M')$ is also denoted by $M'\!\restriction_\phi$ and we call it *the ϕ-reduct of M'*. We also often write $\phi(\varphi)$ for $\mathsf{Mod}(\phi)(\varphi)$.

If F is a set of Σ-sentences and M a Σ-model, then $M \models \mathsf{F}$ denotes the fact that M satisfies all the sentences in F. Let $\mathsf{F}^\bullet$ denote the set $\mathsf{F}^\bullet = \{\varphi \mid (\forall M \text{ a } \Sigma \text{ model})M \models_\Sigma \mathsf{F} \Rightarrow M \models_\Sigma \varphi\}$. A sentence φ is *semantical consequence* of F, we write $\mathsf{F} \models_\Sigma \varphi$, iff $\varphi \in \mathsf{F}^\bullet$.

A specification (presentation) is a way to represent the properties of a system independent of model (= implementation). Formally, a *specification* is a pair (Σ, F), where Σ is a signature and F is a set of Σ-sentences. A (Σ, F)-*model* is a Σ-model M such that $M \models_\Sigma \mathsf{F}$. We sometimes write $(\Sigma, \mathsf{F}) \models \varphi$ for $\mathsf{F} \models_\Sigma \varphi$. A *specification morphism* from (Σ, F) to (Σ', F') is a signature morphism $\phi : \Sigma \to \Sigma'$ such that $\phi(\mathsf{F}) \subseteq \mathsf{F}'^\bullet$. We denote by Spec the category of the specifications. A *theory* is a specification (Σ, F) with $\mathsf{F} = \mathsf{F}^\bullet$; the full subcategory of theories in Spec is denoted by Th. The inclusion functor $U : \mathsf{Th} \to \mathsf{Spec}$ is an equivalence of categories, having a left-adjoint-left-inverse $F : \mathsf{Spec} \to \mathsf{Th}$ given by $F(\Sigma, \mathsf{F}) = (\Sigma, \mathsf{F}^\bullet)$ on objects and identity on morphisms.

Given an institution $\Im = (\mathsf{Sign}, \mathsf{Mod}, \mathsf{sen}, \models)$, the *theoroidal institution* $\Im^{th}$ of $\Im$ is the institution $\Im^{th} = (\mathsf{Th}, \mathsf{Mod}^{th}, \mathsf{sen}^{th}, \models^{th})$, where Mod^{th} is the extension of Mod to theories, sen^{th} is $\mathsf{sign};\mathsf{sen}$ with $\mathsf{sign} : \mathsf{Th} \to \mathsf{Sign}$ the functor which forgets the sentences of a theory, and $\models^{th} = |\mathsf{sign}|;\models$.

Let $\Im = (\mathsf{Sign}, \mathsf{Mod}, \mathsf{sen}, \models)$ and $\Im' = (\mathsf{Sign}', \mathsf{Mod}', \mathsf{sen}', \models')$ be two institutions. An *institution morphism* $(\Phi, \beta, \alpha) : \Im \to \Im'$ consists of:

1. a functor $\Phi : \mathsf{Sign} \to \mathsf{Sign}'$,
2. a natural transformation $\beta : \mathsf{Mod} \Rightarrow \Phi^{op};\mathsf{Mod}'$, i.e., a natural family of functors $\beta_\Sigma : \mathsf{Mod}(\Sigma) \to \mathsf{Mod}'(\Phi(\Sigma))$, and
3. a natural transformation $\alpha : \Phi;\mathsf{sen}' \Rightarrow \mathsf{sen}$, i.e., a natural family of functions $\alpha_\Sigma : \mathsf{sen}'(\Phi(\Sigma)) \to \mathsf{sen}(\Sigma)$,

such that the following satisfaction condition holds:

$$M \models_\Sigma \alpha_\Sigma(\varphi') \text{ iff } \beta_\Sigma(M) \models'_{\Phi(\Sigma)} \varphi'$$

for any Σ-model M in $\mathfrak{I}$ and $\Phi(\Sigma)$-sentence φ' in $\mathfrak{I}'$. Usually, the institution morphisms are used to express the embedding relationship. An example of institution morphism is $(\Phi, \beta, \alpha) : \mathfrak{I}^{th} \to \mathfrak{I}$ which express the embedding of $\mathfrak{I}$ in $\mathfrak{I}^{th}$. $\Phi : \text{Th} \to \text{Sign}$ is given by $\Phi(\Sigma, F) = \Sigma$, $\beta : \text{Mod}^{th} \Rightarrow \Phi^{op}; \text{Mod}$ is defined such that $\beta_{(\Sigma, F)}$ is the identity, and $\alpha : \Phi; \text{sen} \Rightarrow \text{sen}^{th}$ is defined such that $\alpha_{(\Sigma, F)}$ is the identity.

An *institution comorphism* $(\Phi, \beta, \alpha) : \mathfrak{I} \to \mathfrak{I}'$ consists of:

1. a functor $\Phi : \text{Sign} \to \text{Sign}'$,
2. a natural transformation $\beta : \Phi^{op}; \text{Mod}' \Rightarrow \text{Mod}$, i.e., a natural family of functors $\beta_\Sigma : \text{Mod}'(\Phi(\Sigma)) \to \text{Mod}(\Sigma)$, and
3. a natural transformation $\alpha : \text{sen} \Rightarrow \Phi; \text{sen}'$, i.e., a natural family of functions $\alpha_\Sigma : \text{sen}(\Sigma)) \to \text{sen}'(\Phi(\Sigma))$,

such that the following satisfaction condition holds:

$$\beta_\Sigma(M') \models_\Sigma \varphi \text{ iff } M' \models_{\Phi(\Sigma)} \alpha_{\Sigma'}(\varphi)$$

for any $\Phi(\Sigma)$-model M' in $\mathfrak{I}$ and Σ-sentence φ in $\mathfrak{I}$. If β_Σ is surjective for each signature Σ, then we say that (Φ, β, α) is *conservative*. Usually, the institution comorphisms are used to express the representation (encoding) relationship. A simple example of comorphism is $(\Phi, \beta, \alpha) : \mathfrak{I} \to \mathfrak{I}^{th}$, where $\Phi : \text{Sign} \to \text{Th}$ is given by $\Phi(\Sigma) = (\Sigma, \emptyset)$, $\beta : \Phi; \text{Mod}^{th} \Rightarrow \text{Mod}$ is defined such that β_Σ is the identity, and $\alpha : \text{sen} \Rightarrow \Phi; \text{sen}^{th}$ is defined such that α_Σ is the identity.

In many practical examples, we have to represent (encode) a sentence from the source institution with a conjunction of sentences from the target institution. A simple example is the representation of the equivalence $\varphi \leftrightarrow \varphi'$ by the conjunction of two Horn rules: $\varphi \to \varphi' \wedge \varphi' \to \varphi$. Hence the following construction. The *conjunction extension* of $\mathfrak{I}$ is the institution $\mathfrak{I}^\wedge = (\text{Sign}, \text{Mod}, \text{sen}^\wedge, \models^\wedge)$, where $\text{sen}^\wedge(\Sigma) = \text{sen}(\Sigma) \cup \{\varphi_1 \wedge \cdots \wedge \varphi_k \mid \varphi_1, \ldots, \varphi_k \in \text{sen}(\Sigma)\}$, $M \models_\Sigma^\wedge \varphi$ iff $M \models_\Sigma \varphi$ for all $\varphi \in \text{sen}(\Sigma)$, and $M \models_\Sigma^\wedge \varphi_1 \wedge \cdots \wedge \varphi_k$ iff $M \models_\Sigma \varphi_i$ for $i = 1, \ldots, k$. There is an institution morphism $(\Phi, \beta, \alpha) : \mathfrak{I}^\wedge \to \mathfrak{I}$ expressing the embedding of $\mathfrak{I}$ in $\mathfrak{I}^\wedge$. This embedding can also be represented by a comorphism from $\mathfrak{I}$ to $\mathfrak{I}^\wedge$.

An *indexed category* is a functor $G : I^{op} \to \text{Cat}$, where I is a category of indices. The *Grothendieck category* $G^\#$ of an indexed category $G : I^{op} \to \text{Cat}$ has pairs $\langle i, \Sigma \rangle$, with i an object in I and Σ an object in $G(i)$, as objects, and $\langle u, \varphi \rangle : \langle i, \Sigma \rangle \to \langle i', \Sigma' \rangle$, with $u : i \to i'$ an arrow in I and $\varphi : \Sigma \to G(u)(\Sigma')$ an arrow in $G(i)$, as arrows.

The *Grothendieck institution* $\mathfrak{I}^\#$ of an indexed institution $\mathfrak{I} : I^{op} \to \text{Ins}$ has

1. the Grothendieck category $\text{Sign}^\#$ as its category of signatures, where $\text{Sign} : I^{op} \to \text{Cat}$ is the indexed category of signatures of $\mathfrak{I}$;
2. $\text{Mod}^\# : (\text{Sign}^\#)^{op} \to \text{Cat}$ as its model functor, where $\text{Mod}^\# \langle i, \Sigma \rangle = \text{Mod}^i(\Sigma)$ and $\text{Mod}^\# \langle u, \varphi \rangle = \beta_{\Sigma'}^u; \text{Mod}^i(\varphi)$;
3. $\text{sen}^\# : \text{Sign}^\# \to \text{Set}$ as its sentence functor, where $\text{sen}^\# \langle i, \Sigma \rangle = \text{sen}^i(\Sigma)$; and
4. $M \models_{\langle i, \Sigma \rangle}^\# \varphi$ iff $M \models_\Sigma^i \varphi$ for all $i \in |I|$, $\Sigma \in |\text{Sign}^i|$, $M \in |\text{Mod}^\#(i, \Sigma)|$, and $\varphi \in \text{sen}^\#(i, \Sigma)$;

where $\Im(i) = (\text{Sign}^i, \text{Mod}^i, \text{sen}^i, \models^i)$ for each index $i \in |I|$ and $\Im(u) = (\phi^u, \beta^u, \alpha^u)$ for each index morphism $u \in I$.

We show how a theory (Σ_0, F_0) and a model constraint can define an institution $\widehat{(\Sigma_0, F_0)}$. A *model constraint* is a map $[\![\_]\!]_c$ which associates a subcategory $[\![\Sigma, F]\!]_c \subseteq \text{Mod}^{th}(\Sigma, F)$ with each theory (Σ, F), such that $M'\restriction_\phi \in [\![\Sigma, F]\!]_c$ for all $\phi : (\Sigma, F) \to (\Sigma', F')$ and $M' \in [\![\Sigma', F']\!]_c$. Moreover, a model constraint implies in fact a *semantical extension* in the following sense. $[\![\Sigma, F]\!]_c \subseteq \text{Mod}^{th}(\Sigma, F)$ implies $[\![\Sigma, F]\!]_c^\bullet \supseteq \text{Mod}^{th}(\Sigma, F)^\bullet$, where $\mathcal{M}^\bullet$ denotes the set of sentences satisfied by all models in $\mathcal{M}$. In other words, in the presence of model constraints we can prove more properties. The constraints defined in [10] are a particular case of model constraints when the subcategory can be syntactically represented. The institution $\widehat{(\Sigma_0, F_0)}$ is defined as follows:

1. the category of signatures is the comma category $(\Sigma_0, F_0) \downarrow \text{Th}$, where the objects are theory morphisms $f : (\Sigma_0, F_0) \to (\Sigma, F)$, and the arrows $\phi : f \to f'$ are consisting of theory morphisms $\phi : (\Sigma, F) \to (\Sigma', F')$ such that $f; \phi = f'$,
2. the model functor $\text{Mod}\widehat{(\Sigma_0, F_0)}$ maps each signature $f : (\Sigma_0, F_0) \to (\Sigma, F)$ into the subcategory $[\![\Sigma, F]\!]_c$,
3. the sentence functor $\text{sen}\widehat{(\Sigma_0, F_0)}$ maps a signature $f : (\Sigma_0, F_0) \to (\Sigma, F)$ into the set of Σ-sentences,
4. the satisfaction relation is defined by $M \models_f \varphi$ iff $M \models_\Sigma \varphi$.

Note that the model constraint is required only for theories (Σ, F) for that there exists a theory morphism $f : (\Sigma_0, F_0) \to (\Sigma, F)$. We extend the above construction to diagrams of theories and indexed institutions. Let $D : I \to \text{Th}$ be a diagram of theories and $([\![\_]\!]_i \mid i \in |I|)$ a model constraint such that if $u : i \to j$ is an arrow in I, then $M'\restriction_u \in [\![D(i)]\!]_i$ for each $M' \in [\![D(j)]\!]_j$. We denote $D(i)$ by (Σ_i, F_i), $i \in |I|$. If $u : j \to i$ is an arrow in I^{op}, then there is an institution morphism $(\Phi, \beta, \alpha) : \widehat{(\Sigma_j, F_j)} \to \widehat{(\Sigma_i, F_i)}$, where

1. Φ maps a signature $f : (\Sigma_j, F_j) \to (\Sigma, F)$ into the signature $D(u); f : (\Sigma_i, F_i) \to (\Sigma, F)$;
2. $\beta : \text{Mod}\widehat{(\Sigma_j, F_j)} \to \Phi; \widehat{(\Sigma_i, F_i)}$ is as follows: if $f : (\Sigma_j, F_j) \to (\Sigma, E)$ is a signature in $\widehat{(\Sigma_j, F_j)}$, then β_f is the identity because $M\restriction_f \restriction_{D(u)}$ is a Σ_i-model by functoriality of $\text{Mod}(\Im)$ and by the fact that $D(u)$ and f are theory morphisms;
3. $\alpha : \Phi; \text{sen}\widehat{(\Sigma_j, F_j)} \to \widehat{(\Sigma_i, F_i)}$ is as follows: if $f : (\Sigma_j, F_j) \to (\Sigma, E)$ is a signature in $\widehat{(\Sigma_j, F_j)}$, then $\alpha_{\Phi(f)}$ is identity.

The diagram $D : I \to Th$ together with the model constraint $([\![\_]\!]_i \mid i \in |I|)$ produces an indexed institution $\mathcal{D} : I^{op} \to \text{Ins}$, where Ins is the category of institutions and the arrows are institution morphisms. We can define now the Grothendieck institution $\mathcal{D}^\#$, where

1. the category of signatures $\text{Sign}(\mathcal{D}^\#)$ is the Grothendieck construction $\text{Sign}(\mathcal{D})^\#$ corresponding to $\text{Sign}(\mathcal{D}) : I^{op} \to \text{Cat}$, which maps each index i into $\text{Sign}\widehat{(\Sigma_i, F_i)}$;

2. the model functor $\mathsf{Mod}(\mathcal{D}^\#) : \mathsf{Sign}(\mathcal{D}^\#) \to \mathsf{Cat}$ is given by:

   $\mathsf{Mod}(\mathcal{D}^\#)(\langle i, f : (\Sigma_i, \mathsf{F}_i) \to (\Sigma, \mathsf{F}) \rangle)$ is $\mathsf{Mod}(\widehat{\Sigma_i, \mathsf{F}_i})(f)$ (that is equal to $[\![\Sigma, \mathsf{F}]\!]_i$),
   and if $\langle u, \phi \rangle : \langle i, f{:}(\Sigma_i, \mathsf{F}_i) \to (\Sigma, \mathsf{F}) \rangle \to \langle j, f'{:}(\Sigma_j, \mathsf{F}_j) \to (\Sigma', \mathsf{F}') \rangle$, then
   $\mathsf{Mod}(\mathcal{D}^\#)(\langle u, \phi \rangle) = \beta_{f'}(u); \mathsf{Mod}(\widehat{\Sigma_i, \mathsf{F}_i})(\phi)$, $(\Phi, \beta, \alpha) : \widehat{(\Sigma_j, \mathsf{F}_j)} \to \widehat{(\Sigma_i, \mathsf{F}_i)}$;

3. the sentence functor $\mathsf{sen}(\mathcal{D}^\#) : \mathsf{Sign}(\mathcal{D}^\#) \to \mathsf{Set}$ is given by:

   $\mathsf{sen}(\mathcal{D}^\#)(\langle i, f : (\Sigma_i, \mathsf{F}_i) \to (\Sigma, \mathsf{F}) \rangle)$ is $\mathsf{sen}(\widehat{\Sigma_i, \mathsf{F}_i})(f)$ (that is equal to $\mathsf{sen}(\Im)(\Sigma)$),
   and if $\langle u, \phi \rangle : \langle i, f : (\Sigma_i, \mathsf{F}_i) \to \Sigma, \mathsf{F}) \rangle \to \langle j, f' : (\Sigma_j, \mathsf{F}_j) \to \Sigma', \mathsf{F}') \rangle$, then
   $\mathsf{sen}(\mathcal{D}^\#)(\langle u, \phi \rangle) = \mathsf{sen}(\widehat{\Sigma_i, \mathsf{F}_i})(\phi); \alpha_f(u)$, where $(\Phi, \beta, \alpha) : \widehat{(\Sigma_j, \mathsf{F}_j)} \to \widehat{(\Sigma_i, \mathsf{F}_i)}$;

4. if $f : (\Sigma_i, \mathsf{F}_i) \to (\Sigma, \mathsf{F})$, $M \in \mathsf{Mod}(\mathcal{D}^\#)(\langle i, f \rangle)$ and $\varphi \in \mathsf{sen}(\mathcal{D}^\#)(\langle i, f \rangle)$, then
   $M \models_{\langle i, f \rangle} \varphi$ iff $M \models_f \varphi$.

This construction will be used to formalize the RDF triple-based logics underlying SW languages. For instance, it is useful to combine ontologies described in various SW languages.

3 RDF and RDF Schema Logics

In this section, we define the institutions for the languages RDF and RDF Schema. The construction of these institutions is divided into three steps. Firstly, we construct a bare-bone institution for RDF logic, capturing only the very essential concepts in RDF, namely the resource references and the triples format. This logic then serves as the basis on which the institutions of the actual RDF and RDF Schema are constructed. In turn, the institutions defined in this section serve to define the RDF serialization of ontology languages defined in Section 4.

3.1 Bare RDF Logic $\widehat{\mathrm{BRDF}}$

As introduced in Section 2.1, the Resource Description Framework (RDF) is the foundation language of the Semantic Web and all upper layer languages are based on it. Hence, they are all based on the syntax defined in RDF, which is, the *triples* format. Together with the use of URI for resource referencing, these two features of the RDF language are common to all other languages. Hence, we extract them from RDF language and define an institution, the bare RDF logic $\widehat{\mathrm{BRDF}}$.

Example 3. Since resource references are the only signatures in $\widehat{\mathrm{BRDF}}$, any triple will be part of the sentences. As $\widehat{\mathrm{BRDF}}$ is a bare-bone RDF institution, it does not define the XML serialization presented in the previous two examples. Therefore, we will use the informal syntax in this example. Note that the separator # is replaced by a : in this notation. The following triple is a legal sentence in $\widehat{\mathrm{BRDF}}$, stating that carnivores eat animals.

    (animals:Carnivore, animals:eats, animals:Animal)

The Bare RDF logic $\widehat{\text{BRDF}}$ is a bare-bone institution with resource references as the only signatures. The sentences are triples. $\widehat{\text{BRDF}}$ is not expressive at all. We use it as a basis upon which we develop the Grothendieck institution of the triples-based logics underlying SW languages.

A signature RR in $\widehat{\text{BRDF}}$ is a set of resource references. A signature morphism $\phi :$ $RR \rightarrow RR'$ is an arrow in Set. The RR-sentences are triples of the form (sn, pn, on), where $sn, pn, on \in RR$. Usually, sn is for subject name, pn is for property (predicate) name, and on is for object name. RR-models $\mathbb{I}$ are tuples $\mathbb{I} = (R_{\mathbb{I}}, P_{\mathbb{I}}, S_{\mathbb{I}}, \text{ext}_{\mathbb{I}})$, where $R_{\mathbb{I}}$ is a set of resources, $P_{\mathbb{I}}$ is a subset of $R_{\mathbb{I}}$ ($P_{\mathbb{I}} \subseteq R_{\mathbb{I}}$) - the set of properties, $S_{\mathbb{I}} :$ $RR \rightarrow R_{\mathbb{I}}$ is a mapping function that maps each resource reference to some resource, and $\text{ext}_{\mathbb{I}} : P_{\mathbb{I}} \rightarrow \mathcal{P}(R_{\mathbb{I}} \times R_{\mathbb{I}})$ is an extension function mapping each property to a set of pairs of resources that it relates. An RR-homomorphism $h : \mathbb{I} \rightarrow \mathbb{I}'$ between two RR-models is a function $h : R_{\mathbb{I}} \rightarrow R_{\mathbb{I}'}$ such that $h(P_{\mathbb{I}}) \subseteq P_{\mathbb{I}'}$, $S_{\mathbb{I}}; h = S_{\mathbb{I}'}$, and $\text{ext}_{\mathbb{I}}; \mathcal{P}(h \times h) = h|_{P_{\mathbb{I}}}; \text{ext}_{\mathbb{I}'}$. The satisfaction is defined as follows:

$$\mathbb{I} \models_{RR} (sn, pn, on) \text{ iff } (S_{\mathbb{I}}(sn), S_{\mathbb{I}}(on)) \in \text{ext}_{\mathbb{I}}(S_{\mathbb{I}}(pn)),$$

that (sn, pn, on) is satisfied if and only if the pair consisting of the resources associated with the subject name sn and the object name on is in the extension of pn.

In order to simplify the notation, we often write $\text{ext}_{\mathbb{I}}(pn)$ instead of $\text{ext}_{\mathbb{I}}(S_{\mathbb{I}}(pn))$.

3.2   RDF Logic $\widehat{\text{RDF}}$

The RDF logic $\widehat{\text{RDF}}$ is constructed with $\widehat{\text{BRDF}}$ as the basis. The addition in $\widehat{\text{RDF}}$ is the built-in vocabularies of the RDF language and the semantics of these language constructs. Hence, as shown below, we denote the signatures of $\widehat{\text{RDF}}$ using theories, which consist of these built-in vocabularies and sentences giving them semantics. We also add some weak model constraints. More precisely, $\widehat{\text{RDF}}$ is defined using the construction we defined in Section 2.2 starting from a theory RDF and a model constraint $[\![-]\!]_{RDF}$.

The *RDF theory* is RDF $= (\text{RDFVoc}, T_{\text{RDF}})$, where the RDF vocabulary RDFVoc includes the following items:

```
rdf:type, rdf:Property, rdf:value,
rdf:Statement, rdf:subject, rdf:predicate, rdf:object,
rdf:List, rdf:first, rdf:rest, rdf:nil,
rdf:Seq, rdf:Bag, rdf:Alt, rdf:_1 rdf:_2 ...
```

and T_{RDF} consists of triples expressing properties of the vocabulary symbols:

```
(rdf:type, rdf:type, rdf:Property),
(rdf:subject, rdf:type, rdf:Property),
(rdf:predicate, rdf:type, rdf:Property),
(rdf:object, rdf:type, rdf:Property),
(rdf:value, rdf:type, rdf:Property),
(rdf:first, rdf:type, rdf:Property),
(rdf:rest, rdf:type, rdf:Property),
(rdf:nil, rdf:type, rdf:List),
(rdf:_1, rdf:type, rdf:Property),
(rdf:_2, rdf:type, rdf:Property),
...
```

Note that the above vocabularies such as rdf:type are all shorthands of legal URIs, as described in Section 2. All the above triples are self explanatory. For instance, the triple (rdf:value, rdf:type, rdf:Property) states that rdf:value is a property.

We suppose that there is a given set R_{RDF} of RDF resources and a function S_{RDF} : RDFVoc $\rightarrow R_{RDF}$ which associates a resource with each RDF symbol. It is easy to see that R_{RDF} and S_{RDF} can be extended to an RDF-model $\mathbb{RDF}$.

For each theory such that there is a theory morphism f : RDF $\rightarrow$ (RR, T), we consider the model constraint $[\![RR, T]\!]_{RDF}$ as consisting of those (RR, T)-models $\mathbb{I}$ such that

- $R_{\mathbb{I}}$ includes R_{RDF} and the restriction of $S_{\mathbb{I}}$ to RDFVoc coincides with S_{RDF},
- if $p \in P_{\mathbb{I}}$, then $(p, S_{\mathbb{I}}(\text{rdf:Property})) \in \text{ext}_{\mathbb{I}}(\text{rdf:type})$.

Since f is a theory morphism, the restriction of $\mathbb{I}$ to RDFVoc is an RDF-model. We denote by $\widehat{RDF}$ the institution defined by the theory RDF together with the model constraint $[\![-]\!]_{RDF}$ using the method presented in Section 2.2.

If we denote by $\widehat{(\emptyset, \emptyset)}$ the institution defined by the theory $(\emptyset, \emptyset)$ and the model constraint $[\![RR, T]\!]_{\emptyset} = \text{Mod}(\widehat{BRDF})^{\text{th}}(RR, T)$, then $\widehat{BRDF}^{\text{th}}$ is isomorphic to $\widehat{(\emptyset, \emptyset)}$ and we have the institution morphisms $\widehat{RDF} \rightarrow \widehat{BRDF}^{\text{th}} \rightarrow \widehat{BRDF}$.

3.3   The RDF Schema Logic $\widehat{RDFS}$

RDF Schema defines additional language constructs for the RDF language. It expands the expressiveness of RDF by introducing the concept of universe of resources (rdfs:-Resource), the classification mechanism (rdfs:Class) and a set of properties that relate them (rdfs:subClassOf, rdfs:domain, rdfs:range). Hence, it is natural for the RDFS institution $\widehat{RDFS}$ to be developed on top of $\widehat{RDF}$, with some more model constraints added to capture the semantics of RDFS language constructs.

Example 4. Example 1 defines the sub class relationship between two RDF Schema classes. In the shorthand, it can be represented in the informal syntax as follows.

(animals:Carnivore, rdfs:subClassOf, animals:Animal)

The *RDF Schema theory* RDFS = (RDFSVoc, $T_{RDFSVoc}$) is composed of the RDF Schema vocabulary RDFSVoc including RDFVoc together with

```
rdfs:domain, rdfs:range, rdfs:Resource,
rdfs:Literal, rdfs:Datatype, rdfs:Class,
rdfs:subClassOf, rdfs:subPropertyOf, rdfs:member,
rdfs:Container, rdfs:ContainerMembershipProperty
```

and the sentences T_{RDFS} including T_{RDFS} together the triples setting the properties of the new symbols (for the whole list of triples see [13]):

```
(rdf:type, rdfs:domain, rdfs:Resource),
(rdfs:domain, rdfs:domain, rdf:Property),
(rdfs:range, rdfs:domain, rdf:Property),
```

```
(rdfs:subPropertyOf, rdfs:domain, rdf:Property),
(rdfs:subClassOf, rdfs:domain, rdfs:Class),
 ...
(rdf:type, rdfs:range, rdfs:Class),
(rdfs:domain, rdfs:range, rdfs:Class),
(rdfs:range, rdfs:range, rdfs:Class),
(rdfs:subPropertyOf, rdfs:range, rdf:Property),
(rdfs:subClassOf, rdfs:range, rdfs:Class),
 ...
```

We suppose that there is a given set R_{RDFS} of RDF Schema resources and a function S_{RDFS} : RDFSVoc $\rightarrow R_{\text{RDFS}}$ which associates a resource with each RDF Schema symbol and that satisfies $S_{\text{RDFS}}|_{\text{RDFVoc}} = S_{\text{RDF}}$.

For each theory such that there is a theory morphisms f : RDFS $\rightarrow (RR, T)$, we define the model constraint $[\![RR, T]\!]_{\text{RDFS}}$ obtained by strengthening $[\![RR, T]\!]_{\text{RDF}}$ with the following conditions. If $\mathbb{I} \in [\![RR, T]\!]_{\text{RDFS}}$, then:

- $R_{\mathbb{I}}$ includes R_{RDFS} and the restriction of $S_{\mathbb{I}}$ to RDFSVoc coincides with S_{RDFS}
- $\text{ext}_{\mathbb{I}}(\text{rdfs:Resource}) = R_{\mathbb{I}}$
- $(\forall\, x, y, u, v : R_{\mathbb{I}})(x, y) \in \text{ext}_{\mathbb{I}}(\text{rdfs:domain}) \wedge (u, v) \in \text{ext}_{\mathbb{I}}(x) \Rightarrow$
  $u \in \text{ext}_{\mathbb{I}}(y)$
- $(\forall\, x, y, u, v : R_{\mathbb{I}})(x, y) \in \text{ext}_{\mathbb{I}}(\text{rdfs:range}) \wedge (u, v) \in \text{ext}_{\mathbb{I}}(x) \Rightarrow$
  $v \in \text{ext}_{\mathbb{I}}(y)$
- $(\forall\, x, y : R_{\mathbb{I}})(x, y) \in \text{ext}_{\mathbb{I}}(\text{rdfs:subClassOf}) \Rightarrow \text{ext}_{\mathbb{I}}(x) \subseteq \text{ext}_{\mathbb{I}}(y)$
- $(\forall\, x : \text{ext}_{\mathbb{I}}(\text{rdf:Class}))(x, S_{\mathbb{I}}(\text{rdfs:Resource})) \in \text{ext}_{\mathbb{I}}(\text{rdfs:subClassOf}))$
- $(\forall\, x, y : R_{\mathbb{I}})(x, y) \in \text{ext}_{\mathbb{I}}(\text{rdfs:subPropertyOf}) \Rightarrow \text{ext}_{\mathbb{I}}(x) \subseteq \text{ext}_{\mathbb{I}}(y)$
- $(\forall\, x : \text{ext}_{\mathbb{I}}(\text{rdfs:ContainerMembershipProperty}))$
  $(x, S_{\mathbb{I}}(\text{rdfs:member})) \in \text{ext}_{\mathbb{I}}(\text{rdfs:subPropertyOf})$

In other words, $[\![\_]\!]_{\text{RDFS}}$ gives the intended semantics to syntactic constructions such as domain, range, subClassOf, subPropertyOf, etc.

We denote by $\widehat{\text{RDFS}}$ the institution such that $\widehat{\text{RDFS}} \rightarrow \widehat{\text{RDF}}$ is the indexed institution produced by the diagram RDF $\rightarrow$ RDFS together with the model constraint $[\![\_]\!]_{\text{RDFS}}$. We have the theory morphisms (inclusions) $(\emptyset, \emptyset) \rightarrow$ RDF $\rightarrow$ RDFS and $[\![\_]\!]_{\emptyset} \subseteq [\![\_]\!]_{RDF} \subseteq [\![\_]\!]_{RDFS}$. We can formalize now the logics underlying RDF framework as the Grothendieck institution defined by the indexed institution:

$$\widehat{\text{RDFS}} \longrightarrow \widehat{\text{RDF}} \longrightarrow \widehat{\text{BRDF}} \overset{th}{\underset{co}{\rightleftarrows}} \widehat{\text{BRDF}}$$

4   Semantic Web Logics

A number of ontology languages have been proposed in the past years. These include the OWL suite of languages, the OWL$^-$ suite of languages, OWL Flight, etc. They are all based on RDF and RDFS but imposes different restrictions on the usage of RDF(S) language constructs. Hence, their expressiveness is different. In this section, we construct institutions in an RDF(S)-independent way for some of these languages and inter-relate them using institution morphisms. Then we relate them to RDF(S) by exhibiting the comorphisms defining the RDF serializations. These institutions are incrementally constructed using the same pattern. Therefore we present more details only for the first (smallest) one.

4.1 OWL Lite$^-$ Logic $\widehat{\mathrm{OWLLite}}^-$

OWL Lite$^-$ [5] is a proper subset of OWL Lite (see the next subsection) that can be translated in Datalog. It is obtained from OWL Lite by removing those features considered hard to reason about. OWL Lite$^-$ is the lightest dialect of SW languages and therefore we start with it. We denote by $\widehat{\mathrm{OWLLite}}^-$ the institution of the ontology language OWL Lite$^-$.

Example 5. ·The class subsumption relationship is allowed in OWL Lite$^-$ as long as neither of the classes is either top ($\top$, the super class of all classes) or bottom ($\bot$, the sub class of all classes, i.e., the empty class). Moreover, allValuesFrom restrictions like that mentioned in Example 2 is also allowed in OWL Lite$^-$. Hence, Example 2 is also an OWL Lite$^-$ fragment.

The signatures of $\widehat{\mathrm{OWLLite}}^-$ are triples $\Sigma = (CN, PN, IN)$, where CN is a set of class names, PN is a set of individual property names, and IN is a set of individual names. We assume that CN, PN, and IN are pairwise disjoint. A signature morphism $\phi : \Sigma \rightarrow \Sigma'$ is a function $\phi : CN \cup PN \cup IN \rightarrow CN' \cup PN' \cup IN'$ such that $\phi(CN) \subseteq CN', \phi(PN) \subseteq PN', \phi(IN) \subseteq IN'$, where $\Sigma' = (CN', PN', IN')$.

A Σ-model $\mathbb{I}$ consists of $(R_{\mathbb{I}}, S_{\mathbb{I}}, \mathrm{ext}_{\mathbb{I}})$, where $R_{\mathbb{I}}$ is a set of resources, $S_{\mathbb{I}} : CN \cup PN \cup IN \rightarrow R_{\mathbb{I}}$ is a map such that $S_{\mathbb{I}}(CN), S_{\mathbb{I}}(PN)$ and $S_{\mathbb{I}}(IN)$ are pairwise disjoint, and $\mathrm{ext}_{\mathbb{I}}$ is a map associating a subset of $R_{\mathbb{I}}$ with each class name $cn \in CN$, and a subset of $R_{\mathbb{I}} \times R_{\mathbb{I}}$ with each property name pn. A Σ-homomorphism $h : \mathbb{I} \rightarrow \mathbb{I}'$ between two Σ-models is a function $h : R_{\mathbb{I}} \rightarrow R_{\mathbb{I}'}$ such that $S_{\mathbb{I}}; h = S_{\mathbb{I}'}, \mathrm{ext}_{\mathbb{I}}|_{CN}; \mathcal{P}(h) = \mathrm{ext}_{\mathbb{I}'}|_{CN}$, and $\mathrm{ext}_{\mathbb{I}}|_{PN}; \mathcal{P}(h \times h) = \mathrm{ext}_{\mathbb{I}'}|_{PN}$.

For class expressions and Σ-sentences we use a more compact notation:

$$\mathcal{R}es ::= \mathtt{restriction}(pn\ \mathtt{allValuesFrom}\ cn)\ |$$
$$\mathtt{restriction}(pn\ \mathtt{minCardinality}(0))$$
$$\mathcal{C} ::= cn\ |\ \mathcal{R}es$$
$$\mathcal{S} ::= \mathtt{Class}(cn\ \mathtt{partial}\ \mathcal{C}_1 \ldots \mathcal{C}_k)\ |\ \mathtt{Class}(cn\ \mathtt{complete}\ cn_1 \ldots cn_k)\ |$$
$$\mathtt{EquivalentClasses}(cn_1 \ldots cn_k)\ |$$
$$\mathtt{ObjectProperty}(pn\ \mathtt{super}(pn_1) \ldots \mathtt{super}(pn_k))\ |$$
$$pn.\mathtt{domain}(cn_1 \ldots cn_k)\ |\ pn.\mathtt{range}(cn_1 \ldots cn_k)\ |\ pn.\mathtt{inverseOf}(pn_1)\ |$$
$$pn.\mathtt{Symmetric}\ |\ pn.\mathtt{Transitive}\ |$$
$$\mathtt{SubProperty}(pn_1\ pn_2)\ |\ \mathtt{EquivalentProperties}(pn_1 \ldots pn_k)\ |$$
$$\mathtt{Individual}(in\ \mathtt{type}(cn_1) \ldots \mathtt{type}(cn_k))\ |\ in.\mathtt{value}(pn\ in_1)$$

The semantics of expressions is given by:

$$\text{ext}_{\mathbb{I}}(\texttt{restriction}(pn \texttt{ allValuesFrom } cn)) =$$
$$\{x \mid (\forall y)(x, y) \in \text{ext}_{\mathbb{I}}(pn) \Rightarrow y \in \text{ext}_{\mathbb{I}}(cn)$$
$$\text{ext}_{\mathbb{I}}(\texttt{restriction}(pn \texttt{ minCardinality}(0))) =$$
$$\{x \mid \#(\{y \mid (x, y) \in \text{ext}_{\mathbb{I}}(pn)\}) \geq 0\}$$

The satisfaction relation between OWL Lite$^-$ Σ-models $\mathbb{I}$ and OWL Lite$^-$ Σ-sentences is defined as it is intuitively suggested by syntax. For instance, we have:

$\mathbb{I} \models_{\Sigma} \texttt{Class}(cn \texttt{ partial } \mathcal{C}_1 \ldots \mathcal{C}_k)$ iff $\text{ext}_{\mathbb{I}}(cn) \subseteq \text{ext}_{\mathbb{I}}(cn_1) \cap \cdots \cap \text{ext}_{\mathbb{I}}(cn_k)$

$\mathbb{I} \models_{\Sigma} \texttt{ObjectProperty}(pn \texttt{ super}(pn_1) \ldots \texttt{super}(pn_k))$ iff
    $\text{ext}_{\mathbb{I}}(pn) \subseteq \text{ext}_{\mathbb{I}}(pn_1) \cap \cdots \cap \text{ext}_{\mathbb{I}}(pn_k)$

$\mathbb{I} \models_{\Sigma} pn.\texttt{domain}(cn_1 \ldots cn_k)$ iff $\text{dom ext}_{\mathbb{I}}(pn) \subseteq \text{ext}_{\mathbb{I}}(cn_1) \cap \cdots \cap \text{ext}_{\mathbb{I}}(cn_k)$

$\mathbb{I} \models_{\Sigma} \texttt{SubProperty}(pn_1 \ pn_2)$ iff $\text{ext}_{\mathbb{I}}(pn_1) \subseteq \text{ext}_{\mathbb{I}}(pn_2)$

$\mathbb{I} \models_{\Sigma} \texttt{Individual}(in \texttt{ type}(cn_1) \ldots \texttt{type}(cn_k))$ iff $S_{\mathbb{I}}(in) \in \text{ext}_{\mathbb{I}}(cn_1) \cap \cdots \cap \text{ext}_{\mathbb{I}}(cn_k)$

$\mathbb{I} \models_{\Sigma} in.\texttt{value}(pn \ in_1)$ iff $(S_{\mathbb{I}}(in), S_{\mathbb{I}}(in_1)) \in \text{ext}_{\mathbb{I}}(pn)$

$\ldots$

4.2 OWL Lite Logic $\widetilde{\textsf{OWLLite}}$

OWL Lite is the least expressive species of the OWL suite. It is obtained by imposing some constraints on OWL Full. These constraints include, for example, that the sets of classes, properties and individuals are mutually disjoint; that min and max cardinality restrictions can only be applied on numbers 0 and 1; that value restrictions such as `allValuesFrom` and `someValuesFrom` can only be applied to named classes. Compared with OWL Lite$^-$, OWL Lite is more expressive since it removes some constraints that are imposed on the latter. The details are discussed in the following. We denote by $\widetilde{\textsf{OWLLite}}$ the institution of OWL Lite.

Example 6. For example, OWL Lite$^-$ does not support the relationship between OWL individuals, namely `sameAs` and `differentFrom`, whereas these features are present in OWL Lite. Suppose that we have two URI references for carnivores `car1` and `car2`, which are actually referring to the same animal. We use the following code fragment to represent this piece of knowledge:

```
<animals:Carnivore rdf:ID="car1"/>
<animals:Carnivore rdf:ID="car2"/>
<animals rdf:about="#car1>
   <owl:sameAs rdf:resource="http://ex.com/animals#car2/>
</animals>
```

$\widetilde{\textsf{OWLLite}}$ is obtained from $\widetilde{\textsf{OWLLite}}^-$ by replacing the definition of expressions with

$$\mathcal{R}es ::= \texttt{restriction}(pn \texttt{ allValuesFrom } cn) \mid$$
$$\texttt{restriction}(pn \texttt{ someValuesFrom } cn) \mid$$
$$\texttt{restriction}(pn \texttt{ minCardinality}(n)) \mid$$
$$\texttt{restriction}(pn \texttt{ maxCardinality}(n))$$
$$\mathcal{C} ::= cn \mid \texttt{owl:Thing} \mid \texttt{owl:Nothing} \mid \mathcal{R}es$$

where $n \in \{0, 1\}$, and adding the following sentences:

pn.Functional | pn.InverseFunctional |
SameIndividual($in_1, \ldots, in_k$) | DifferentIndividuals($in_1, \ldots, in_k$)

The semantics of the new expressions is as follows:

$\text{ext}_{\mathbb{I}}(\text{owl:Thing})$ is a subset of $R_{\mathbb{I}}$ s.t. $(\forall \, cn \in CN)\text{ext}_{\mathbb{I}}(cn) \subseteq \text{ext}_{\mathbb{I}}(\text{owl:Thing})$,
$\text{ext}_{\mathbb{I}}(\text{owl:Nothing}) = \emptyset$,
$\text{ext}_{\mathbb{I}}(\text{restriction}(pn \, \text{someValuesFrom} \, cn)) =$
$\quad \{x \mid (\exists \, y)(x, y) \in \text{ext}_{\mathbb{I}}(pn) \land y \in \text{ext}_{\mathbb{I}}(cn)\}$,
$\text{ext}_{\mathbb{I}}(\text{restriction}(pn \, \text{minCardinality}(1) \, cn)) =$
$\quad \{x \mid \#\{y \mid (x, y) \in \text{ext}_{\mathbb{I}}(pn)\} \geqslant 1\}$,
$\text{ext}_{\mathbb{I}}(\text{restriction}(pn \, \text{maxCardinality}(1) \, cn)) =$
$\quad \{x \mid \#\{y \mid (x, y) \in \text{ext}_{\mathbb{I}}(pn)\} \leqslant 1\}$.

The satisfaction of the new sentences is intuitive and straightforward and we omit its formal definition.

Proposition 1. *There is a conservative comorphism from* $\widehat{\text{OWLLite}}^{-}$ *to* $\widehat{\text{OWLLite}}$.

4.3   OWL DL⁻ Logic $\widehat{\text{OWLDesLog}}^{-}$

OWL DL⁻ [5] is an extension of OWL Lite⁻ and a subset of OWL DL (see the next subsection) which can also be translated in Datalog. We denote by $\widehat{\text{OWLDesLog}}$ the institution of OWL DL⁻. Compared to OWL Lite⁻, OWL DL⁻ allows additional language constructs value, someValuesFrom and oneOf, albeit that the latter two are only allowed as the first argument of subClassOf (left hand side).

Example 7. In OWL DL⁻, the value restriction not present in OWL Lite⁻ is allowed in OWL DL⁻. This restriction constructs a class that for a given property, each of whose instances must have (among others) a particular individual as the value mapped by this property. Suppose that we want to model the fact that the ancestor of all humans is Adam (among all his/her other ancestors), assuming that we have defined an individual Adam and a property hasAncestor. Here is the definition in OWL DL⁻.

```
<owl:Class rdf:ID="Human">
   <rdfs:subClassOf><owl:Restriction>
      <owl:onProperty rdf:resource="#hasAncestor"/>
      <owl:hasValue rdf:resource="#Adam"/>
   </owl:Restriction></rdfs:subClassOf>
</owl:Class>
```

$\widetilde{\text{OWLDesLog}}^-$ is obtained from $\widetilde{\text{OWLLite}}^-$ by replacing the definition of expressions with

$$\mathcal{C} ::= cn \mid \mathcal{R}es \mid \texttt{intersectionOf}(\mathcal{C}_1, \dots, \mathcal{C}_n)$$
$$\mathcal{L}hs\_D ::= \mathcal{C} \mid \mathcal{L}hs\_\mathcal{R}es \mid \texttt{unionOf}(\mathcal{L}hs\_D, \dots, \mathcal{L}hs\_D) \mid \texttt{oneOf}(in_1, \dots, in_k)$$
$$\mathcal{R}hs\_D ::= \mathcal{C} \mid \mathcal{R}hs\_\mathcal{R}es$$
$$\mathcal{R}es ::= \texttt{restriction}(pn \, \texttt{value}(in))$$
$$\mathcal{L}hs\_\mathcal{R}es ::= \mathcal{R}es \mid \texttt{restriction}(pn \, \texttt{someValuesFrom}(\mathcal{L}hs\_D))$$
$$\mid \texttt{restriction}(pn \, \texttt{minCardinality}(1))$$
$$\mathcal{R}hs\_\mathcal{R}es ::= \mathcal{R}es \mid \texttt{restriction}(pn \, \texttt{allValuesFrom}(\mathcal{R}hs\_D))$$

and replacing the class-related sentences with the following:

$$\texttt{Class}\,(cn \, \texttt{partial} \, \mathcal{R}hs\_D) \mid \texttt{Class}\,(cn \, \texttt{complete} \, \mathcal{C}) \mid$$
$$\texttt{EquivalentClass}(\mathcal{C}_1, \dots, \mathcal{C}_n) \mid \texttt{subClassOf}(\mathcal{L}hs\_D, \mathcal{R}hs\_D)$$

Note the use of class expressions instead of named classes. $\mathcal{L}hs\_D$ and $\mathcal{R}hs\_D$ represent class descriptions that can only appear in the left hand side and right hand side of the $\texttt{subClassOf}$ sentence, respectively.

The semantics of the value restriction is $\text{ext}_\mathbb{I}(\texttt{restriction}(pn \, \texttt{value} \, in)) = \{x \mid (x, S_\mathbb{I}(in)) \in \text{ext}_\mathbb{I}(pn)\}$. The semantics of the other expressions and the satisfaction relation for the new sentences are defined as expected.

Proposition 2. *There is a conservative comorphism from* $\widetilde{\text{OWLLite}}^-$ *to* $\widetilde{\text{OWLDesLog}}^-$.

4.4   OWL DL Logic $\widetilde{\text{OWLDesLog}}$

OWL DL is the main ontology language of the OWL suite. It is more expressive than OWL Lite yet still decidable. It relaxes some constraints imposed on OWL Lite and allows more language constructs to describe classes and properties. Still, classes, properties and individuals are mutually disjoint in OWL DL. We denote by $\widetilde{\text{OWLDesLog}}$ the institution of OWL DL. Compared to OWL DL$^-$, OWL DL adds a number of language features, such as enumerated class, disjointness classes, functional property, etc.

Example 8. The class $\texttt{Continents}$ defines the continents of the Earth, namely Africa, Antarctica, Asia, Australia, Europe, North America and South America. As this class only contains these 7 instances, it is natural to use an enumeration to define it.

```
<owl:Class rdf:ID="Continents">
   <owl:oneOf rdf:parseType="Collection">
      <owl:Thing rdf:about="#Africa"/>
      <owl:Thing rdf:about="#Antarctica"/>
      <owl:Thing rdf:about="#Asia"/>
      <owl:Thing rdf:about="#Australia"/>
      <owl:Thing rdf:about="#Europe"/>
```

```
      <owl:Thing rdf:about="#North America"/>
      <owl:Thing rdf:about="#South America"/>
   </owl:oneOf>
</owl:Class>
```

OWLDesLog is obtained from OWLLite replacing the definition of expressions with

$$\mathcal{R}es ::= \texttt{restriction}(pn\ \texttt{value}\ in)\ |$$
$$\texttt{restriction}(pn\ \texttt{allValuesFrom}\ \mathcal{C})\ |$$
$$\texttt{restriction}(pn\ \texttt{someValuesFrom}\ \mathcal{C})\ |$$
$$\texttt{restriction}(pn\ \texttt{minCardinality(n)})\ |$$
$$\texttt{restriction}(pn\ \texttt{maxCardinality(n)})\ |$$
$$\mathcal{C} ::= cn\ |\ \texttt{owl:Thing}\ |\ \texttt{owl:Nothing}\ |\ \mathcal{R}es$$
$$\texttt{intersectionOf}(\mathcal{C}_1,\ldots,\mathcal{C}_k)\ |\ \texttt{unionOf}(\mathcal{C}_1,\ldots,\mathcal{C}_k)\ |$$
$$\texttt{complementOf}(\mathcal{C})\ |\ \texttt{oneOf}(in_1,\ldots,in_k)$$

and adding the following sentences:

$$\texttt{EnumeratedClass}(cn\ in_1,\ldots,in_k)\ |\ \texttt{SubClassOf}(\mathcal{C}_1,\mathcal{C}_2)\ |$$
$$\texttt{DisjointClasses}(\mathcal{C}_1,\ldots,\mathcal{C}_k)\ |\ \texttt{EquivalentClasses}(\mathcal{C}_1,\ldots,\mathcal{C}_k)\ |$$
$$pn.\texttt{domain}(\mathcal{C}_1\ldots\mathcal{C}_k)\ |\ pn.\texttt{range}(\mathcal{C}_1\ldots\mathcal{C}_k)\ |$$
$$\texttt{Individual}(in\ \texttt{type}(\mathcal{C}_1),\ldots,\texttt{type}(\mathcal{C}_k))\ |\ in.\texttt{value}(pn_1\ in_1)$$

The semantics of the new expressions and the satisfaction relation for the new sentences are defined as expected.

Proposition 3. *a) There is a conservative comorphism from OWLLite to OWLDesLog.*

b) There is a conservative comorphism from OWLDesLog⁻ to OWLDesLog.

4.5   OWL Full Logic OWLFull

OWL Full adds a number of features on top of OWL DL and also removes some restrictions. The vocabulary no longer needs to be separated. This means an identifier can denote a class, an individual and/or a property at the same time.

Let X be a name denoting a class and a property in the same ontology $\Sigma = ((CN, PN, IN), \mathsf{F})$, i.e., $X \in CN \cap PN$, and let $\mathbb{I}$ be a (Σ, F)-model. For the moment, we denote with $X{:}CN$ the occurrence of X as a class and with $X{:}PN$ the occurrence of X as a property. We have $S_{\mathbb{I}}(X{:}CN) = S_{\mathbb{I}}(X{:}PN) = S_{\mathbb{I}}(X)$, $\text{ext}_{\mathbb{I}}(X{:}CN) \subseteq R_{\mathbb{I}}$, and $\text{ext}_{\mathbb{I}}(X{:}PN) \subseteq R_{\mathbb{I}} \times R_{\mathbb{I}}$. Since X denotes just one entity, we relate the two sets by means of a bijection $rdef_{\mathbb{I}}(X) : \text{ext}_{\mathbb{I}}(X{:}PN) \rightarrow \text{ext}_{\mathbb{I}}(X{:}CN)$. We may think that $rdef_{\mathbb{I}}(X)(r_1, r_2)$ is the URL address where the pair (r_1, r_2) is defined as an instance of $\text{ext}_{\mathbb{I}}(X{:}PN)$. If X denotes a class (property) and an individual, then its meaning as an individual is given by $S_{\mathbb{I}}(X)$ and its meaning as class (property) is $\text{ext}_{\mathbb{I}}(X)$.

Also, keywords of the language can be used in place of classes, properties and individuals, and restrictions. For instance, we may assume that subClassOf and `subPropertyOf` are in PN. Then for X, Y denoting both classes and properties, we have `subClassOf`(X, Y) iff `SubPropertyOf`(X, Y); this is semantically expressed by $(S_{\mathbb{I}}(X{:}CN), S_{\mathbb{I}}(Y{:}CN)) \in \text{ext}_{\mathbb{I}}(\texttt{SubClassOf})$ iff $(S_{\mathbb{I}}(X{:}PN), S_{\mathbb{I}}(Y{:}PN)) \in \text{ext}_{\mathbb{I}}(\texttt{SubPropertyOf})$.

We skip the formal definition of OWL Full here. The new features are added in a similar way to other languages. In the definition of the signatures we remove the restriction as the sets CN, PN, IN to be pairwise disjoint. The corresponding restriction from the definition of models is also removed.

The original definition of OWL Full [20] is given directly over RDF Schema. Here we refer an RDF independent definition for OWL Full. The fact that OWL Full is built over RDF Schema is given by the following result.

Proposition 4. *There is a conservative comorphism from* $\widehat{\text{RDFS}}$ *to* $\widehat{\text{OWLFull}}^{\text{th}}$.

It is easy to see that we cannot embed $\widehat{\text{RDFS}}$ in $\widehat{\text{OWLDesLog}}$. For instance, triples like $(\texttt{rdf:type}, \texttt{rdf:type}, \texttt{rdf:Property})$ cannot be encoded in $\widehat{\text{OWLDesLog}}$ but can be expressed as a sentence in $\widehat{\text{OWLFull}}$.

If (Σ, F) is an ontology in OWL DL, then, syntactically, it is also an ontology in OWL Full. However, the class of (Σ, F)-models in $\widehat{\text{OWLFull}}$ is richer than that of (Σ, F)-models in $\widehat{\text{OWLDesLog}}$. The reason is that in OWL Full we removed some model constraints which are present in OWL DL. Hence we have the following result:

Proposition 5. *There is not an embedding comorphism from* $\widehat{\text{OWLDesLog}}$ *to* $\widehat{\text{OWLFull}}$.

The theorem above has a drastic consequence: it could be unsound to use OWL Full reasoners for OWL DL ontologies. This is refereed in literature as inappropriate layering [5].

The relationships between SW logics are expressed by the following diagram:

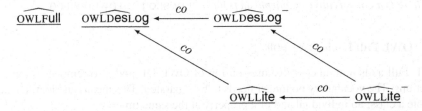

Between $\widehat{\text{OWLFull}}$ and $\widehat{\text{OWLDesLog}}$ we can define only a "syntactical comorphism" consisting of an inclusion functor $\text{Sign}(\widehat{\text{OWLDesLog}}) \rightarrow \Phi : \text{Sign}(\widehat{\text{OWLFull}})$ and a natural transformation $\alpha : \Phi; \text{sen}(\widehat{\text{OWLDesLog}}) \rightarrow \text{sen}(\widehat{\text{OWLFull}})$.

5   RDF Serialization of Semantic Web Logics

In this section, we define the RDF serialization of the SW languages discussed in the previous section. In terms of institution theory, an RDF serialization is a comorphism (encoding) from the source language to an institution built over an RDF theory as in

Section 2.2. Since the corresponding theories are related by morphisms, we get an indexed institution. This approach results in a much clearer understanding of the relationship among the various languages, as is shown at the end of the section.

5.1 RDF Serialization of OWL Lite$^-$

We define the theory OWLLM = (OWLLMVoc, T_{OWLLM}), where OWLLMVoc is RDFSVoc together with an enumerable set of anonymous names and the symbols

```
owl:allValuesFrom, owl:Class, owl:equivalentClass,
owl:equivalentProperty, owl:hasValue, owl:inverseOf,
owl:minCardinality, owl:ObjectProperty,
owl:SymmetricProperty, owl:TransitiveProperty,
owl:Restriction, owl:onProperty,
owl:allValuesFrom, owl:hasValue, owl:minCardinality
```

and T_{OWLLM} is defined as $T_{\text{OWLLM}} = T_{\text{RDFS}}\quad\cup$

```
{
  (owl:Class, rdfs:subClassOf, rdf:Class),
  (owl:allValuesFrom, rdf:type, rdf:Property),
  (owl:allValuesFrom, rdfs:domain, rdf:Property),
  (owl:equivalentProperties, rdf:type, rdf:Property),
  (owl:equivalentProperties, rdfs:domain, rdf:Property),
  (owl:equivalentProperties, rdfs:subPropertyOf,
                                    rdfs:subPropertyOf),
  (owl:ObjectProperty, rdf:type, rdfs:Class),
  (owl:inverseOf, rdf:type, rdf:Property),
  ...
}
```

The anonymous names are used in translating OWL sentences into conjunctions of triples [20]. As for RDF and RDF Schema, we suppose that there is a given set R_{OWLLM} of OWL Lite$^-$ resources and a function $S_{\text{OWLLM}} : \text{OWLLMVoc} \to R_{\text{OWLLM}}$ which associates a resource to each OWL Lite$^-$ symbol, and satisfies $S_{\text{OWLLM}}|_{\text{RDFSVoc}} = S_{\text{RDFS}}$.

For each theory such that there is a theory morphisms $f : \text{OWLLM} \to (RR, T)$, we define the model constraint $[\![RR, T]\!]_{\text{OWLLM}}$ obtained by strengthening $[\![RR, T]\!]_{\text{RDFS}}$ with the following conditions. If $\mathbb{I} \in [\![RR, T]\!]_{\text{OWLLM}}$, then:

- $R_{\mathbb{I}}$ includes R_{OWLLM} and the restriction of $S_{\mathbb{I}}$ to OWLLMVoc coincides with S_{OWLLM}.
- $\text{ext}_{\mathbb{I}}(\text{owl:Class}) \cap \text{ext}_{\mathbb{I}}(\text{owl:ObjectProperty}) = \emptyset$
- $(\forall x, y)(x, \text{owl:Class}) \in \text{ext}_{\mathbb{I}}(\text{rdf:type}) \wedge (y, x) \in \text{ext}_{\mathbb{I}}(\text{rdf:type}) \Rightarrow$
  $((y, \text{owl:Class}) \notin \text{ext}_{\mathbb{I}}(\text{rdf:type})) \wedge$
  $((y, \text{rdf:Property}) \notin \text{ext}_{\mathbb{I}}(\text{rdf:type}))$
- $(\forall x, y)(x, y) \in \text{ext}_{\mathbb{I}}(\text{owl:equivalentClass}) \Rightarrow$
  $(x, S_{\mathbb{I}}(\text{owl:Class})) \in \text{ext}_{\mathbb{I}}(\text{rdf:type}) \wedge$
  $(y, S_{\mathbb{I}}(\text{owl:Class})) \in \text{ext}_{\mathbb{I}}(\text{rdf:type}) \wedge \qquad \text{ext}_{\mathbb{I}}(x) = \text{ext}_{\mathbb{I}}(y)$
- $(\forall u, w, v)(w, S_{\mathbb{I}}(\text{owl:Restriction})) \in \text{ext}_{\mathbb{I}}(\text{rdf:type}) \wedge$
  $(w, u) \in \text{ext}_{\mathbb{I}}(\text{owl:onProperty}) \wedge (w, v) \in \text{ext}_{\mathbb{I}}(\text{xowl:allValuesFrom}) \Rightarrow$
  $\text{ext}_{\mathbb{I}}(w) = \{x \mid (x, y) \in \text{ext}_{\mathbb{I}}(u) \Rightarrow y \in \text{ext}_{\mathbb{I}}(v)\}$

- $(\forall\, u, v, w)(w, S_{\mathbb{I}}(\texttt{owl:Restriction})) \in \text{ext}_{\mathbb{I}}(\texttt{rdf:type}) \wedge$
  $(w, u) \in \text{ext}_{\mathbb{I}}(\texttt{owl:onProperty}) \wedge (w, v) \in \text{ext}_{\mathbb{I}}(\texttt{owl:hasValue}) \Rightarrow$
  $\text{ext}_{\mathbb{I}}(w) = \{x \mid (x, v) \in \text{ext}_{\mathbb{I}}(u)\}$
- $(\forall\, u, w, y)(w, S_{\mathbb{I}}(\texttt{owl:Restriction})) \in \text{ext}_{\mathbb{I}}(\texttt{rdf:type}) \wedge$
  $(w, u) \in \text{ext}_{\mathbb{I}}(\texttt{owl:onProperty}) \wedge (w, 0) \in \text{ext}_{\mathbb{I}}(\texttt{owl:minCardinality}) \Rightarrow$
  $\text{ext}_{\mathbb{I}}(w) = \{x \mid \#(\{(x, y) \in \text{ext}_{\mathbb{I}}(u)\}) \geq 0$
- $(\forall\, v, w, x, y)(x, y) \in \text{ext}_{\mathbb{I}}(\texttt{owl:inverseOf}) \Rightarrow$
  $(x, S_{\mathbb{I}}(\texttt{owl:ObjectProperty})) \in \text{ext}_{\mathbb{I}}(\texttt{rdf:type}) \wedge$
  $(y, S_{\mathbb{I}}(\texttt{owl:ObjectProperty})) \in \text{ext}_{\mathbb{I}}(\texttt{rdf:type}) \wedge$
  $(w, v) \in \text{ext}_{\mathbb{I}}(x) \Leftrightarrow (v, w) \in \text{ext}_{\mathbb{I}}(y)$
- $(\forall\, u, x, y)(u, S_{\mathbb{I}}(\texttt{owl:SymmetricProperty})) \in \text{ext}_{\mathbb{I}}(\texttt{rdf:type}) \wedge$
  $(x, y) \in \text{ext}_{\mathbb{I}}(u) \Rightarrow (y, x) \in \text{ext}_{\mathbb{I}}(u)$
- $(\forall\, u, x, y, z)(u, S_{\mathbb{I}}(\texttt{owl:TransitiveProperty})) \in \text{ext}_{\mathbb{I}}(\texttt{rdf:type}) \wedge$
  $(x, y) \in \text{ext}_{\mathbb{I}}(x) \wedge (y, z) \in \text{ext}_{\mathbb{I}}(x) \Rightarrow (x, z) \in \text{ext}_{\mathbb{I}}(x)$
- $(\forall\, x, y)(x, y) \in \text{ext}_{\mathbb{I}}(\texttt{owl:equivalentProperty}) \Rightarrow$
  $(x, S_{\mathbb{I}}(\texttt{rdf:Property})) \in \text{ext}_{\mathbb{I}}(\texttt{rdf:type}) \wedge$
  $(y, S_{\mathbb{I}}(\texttt{rdf:Property})) \in \text{ext}_{\mathbb{I}}(\texttt{rdf:type}) \wedge \text{ext}_{\mathbb{I}}(x) = \text{ext}_{\mathbb{I}}(y)$

The second and the third conditions say that the vocabulary is separated: a resource cannot be a class, an individual and/or a property at the same time. The last three conditions give the intended meaning of the symmetric property, transitive property, and equivalent property, respectively. The other conditions give semantics to the new syntactical constructions. We denote by $\widehat{\texttt{OWLLM}}$ the institution generated by the theory OWLLM and the model constraint $[\![\_]\!]_{\text{OWLLM}}$ using the method presented in Section 2.2.

Proposition 6. *There is a conservative comorphism* (Φ, β, α) *from* $\widehat{\texttt{OWLLite}}^{-}$ *to* $(\widehat{\texttt{OWLLM}}^{\text{th}})^{\wedge}$.

Here is a brief description of the comorphism given by Proposition 6.

$\Phi : \text{Sign}(\widehat{\texttt{OWLLite}}^{-}) \to \text{Sign}((\widehat{\texttt{OWLLM}}^{\text{th}})^{\wedge})$ is defined as follows. If $\Sigma = (CN, PN, IN)$ is an OWL Lite^{-} signature, then $\Phi(\Sigma)$ is $\Sigma : \texttt{OWLLM} \to (RR, T)$, where $RR = \texttt{OWLLMVoc} \cup CN \cup PN \cup IN$, and T includes $T_{\texttt{OWLLM}}$ together with:
a triple $(cn, \texttt{rdf:type}, \texttt{owl:Class})$ for each $cn \in CN$,
a triple $(pn, \texttt{rdf:type}, \texttt{owl:ObjectProperty})$ for each $pn \in PN$, and
a triple $(in, \texttt{rdf:type}, \texttt{rdf:Resource})$ for each $in \in IN$.
$\beta : \Phi; \text{Mod}((\widehat{\texttt{OWLLM}}^{\text{th}})^{\wedge}) \Rightarrow \text{Mod}(\widehat{\texttt{OWLLite}}^{-})$ is defined as follows. If $\mathbb{I}'$ is a $\Phi(\Sigma)$-model, then $\beta_{\Sigma}(\mathbb{I}') = \mathbb{I}$, where $R_{\mathbb{I}} = R_{\mathbb{I}'}$, $S_{\mathbb{I}}(name) = S_{\mathbb{I}'}(name)$ for each $name \in CN \cup PN \cup IN$, $\text{ext}_{\mathbb{I}}(pn) = \text{ext}_{\mathbb{I}'}(pn)$, and
$\text{ext}_{\mathbb{I}}(cn) = \{x \mid (x, S_{\mathbb{I}'}(cn)) \in \text{ext}_{\mathbb{I}'}(\texttt{rdf:type})\}$.
$\alpha : \text{sen}(\widehat{\texttt{OWLLite}}^{-}) \Rightarrow \Phi; \text{sen}((\widehat{\texttt{OWLLM}}^{\text{th}})^{\wedge})$ is given such that α_{Σ} associates with each OWL Lite^{-} syntactical construction a set (conjunction) of triples similar to that defined in [20]. If Σ is an OWL Lite^{-}-signature and M a Σ-model, then M can be extended to $\Phi(\Sigma)$-model by giving semantics to symbols in OWLLMVoc according to triples in $T_{\texttt{OWLLM}}$.

5.2 RDF Serialization of OWL Lite

We define the theory OWLL = (OWLLVoc, T_{OWLL}), where OWLLVoc is OWLLMVoc together with

```
owl:Thing, owl:Nothing, owl:FunctionalProperty,
owl:SameIndividual, DifferentIndividuals, owl:someValues,
owl:maxCardinality
```

and T_{OWLL} is defined as $T_{\text{OWLL}} = T_{\text{OWLLM}}$ $\cup$

```
{
    (owl:Thing, rdf:type, rdfs:Class),
    (owl:Nothing, rdf:type, rdfs:Class),
    (owl:FunctionalProperty, rdf:type, rdfs:Class),
    (owl:InverseFunctionalProperty, rdf:type, rdfs:Class),
    (owl:InverseFunctionalProperty, rdfs:subClassOf,
                                        owl:ObjectProperty),
    (owl:sameAs, rdf:type, rdf:Property),
    . . .
}
```

As for OWL Lite⁻, we suppose that there is given a set R_{OWLL} of RDF Schema resources and a function $S_{\text{OWLL}} : \text{OWLL} \to R_{\text{OWLL}}$ which associates a resource to each OWL Lite symbol, and satisfies $S_{\text{OWLL}}|_{\text{OWLLMVoc}} = S_{\text{OWLLM}}$.

For each theory such that there is a theory morphisms $f : \text{OWLL} \to (RR, T)$ we define the model constraint $[\![RR, T]\!]_{\text{OWLL}}$ obtained by strengthening $[\![RR, T]\!]_{\text{OWLLM}}$ with the following conditions. If $\mathbb{I} \in [\![RR, T]\!]_{\text{OWLL}}$, then:

– $R_{\mathbb{I}}$ includes R_{OWLL} and the restriction of $S_{\mathbb{I}}$ to OWLLVoc coincides with S_{OWLL}.
– $\mathbb{I}$ satisfies the restrictions expressing the intended meaning of the new features.

We denote by $\widehat{\text{OWLL}}$ the institution generated by the theory OWLL and the model constraint $[\![\_]\!]_{\text{OWLL}}$ using the method presented in Section 2.2.

Theorem 1. *There is a conservative comorphism from $\widehat{\text{OWLLite}}$ to $(\widehat{\text{OWLL}}^{\text{th}})^{\wedge}$.*

5.3 RDF Serialization of OWL DL⁻

We define the theory OWLDLM = (OWLDLMVoc, T_{OWLDLM}), where the vocabulary OWLDLMVoc is OWLLMVoc together with

```
owl:SubClassOf, owl:intersectionOf, owl:unionOf, owl:oneOf,
owl:someValues, owl:hasValue
```

and T_{OWLDLM} is defined as $T_{\text{OWLDLM}} = T_{\text{OWLLM}}$ $\cup$

```
{
    (owl:intersectionOf, rdf:type, rdf:Property),
    (owl:unionOf, rdf:type, rdf:Property),
    (owl:oneOf, rdf:type, rdf:Property),
    (owl:oneOf, rdfs:range, rdf:List),
    . . .
}
```

As usual, we suppose that there is a given set R_{OWLDLM} of RDF Schema resources and a function $S_{\text{OWLDLM}} : \text{OWLDLM} \to R_{\text{OWLDLM}}$ which associates a resource to each OWL DL$^-$ symbol, and satisfies $S_{\text{OWLDLM}}|_{\text{OWLLMVoc}} = S_{\text{OWLLM}}$.

For each theory such that there is a theory morphism $f : \text{OWLDLM} \to (RR, T)$ we define the model constraint $[\![RR, T]\!]_{\text{OWLDLM}}$ obtained by strengthening $[\![RR, T]\!]_{\text{OWLLM}}$ with the following conditions. If $\mathbb{I} \in [\![RR, T]\!]_{\text{OWLDLM}}$, then:

- $R_{\mathbb{I}}$ includes R_{OWLDLM} and the restriction of $S_{\mathbb{I}}$ to OWLDLMVoc is S_{OWLDLM}.
- $\mathbb{I}$ satisfies the restrictions expressing the intended meaning of the new features.

We denote by $\widehat{\text{OWLDLM}}$ the institution generated by the theory OWLDLM and the model constraint $[\![\_]\!]_{\text{OWLDLM}}$ using the method presented in Section 2.2.

Theorem 2. *There is a conservative comorphism from* $\widehat{\text{OWLDesLog}}^-$ *to* $(\widehat{\text{OWLDLM}}^{\text{th}})^{\wedge}$.

5.4   RDF Serialization of OWL DL

We define the theory $\text{OWLDL} = (\text{OWLDLVoc}, T_{\text{OWLDL}})$, where OWLDLVoc is OWLLVoc together with

```
owl:DepricatedClass, owl:DisjointClasses, owl:SubClassOf,
owl:Functional,  owl:InverseFunctional, owl:Transitive,
owl:SameIndividual, DifferentIndividuals, owl:someValues,
owl:Thing, owl:Nothing,
owl:intersectionOf, owl:unionOf, owl:complementOf,
owl:oneOf, owl:someValues, owl:hasValue, owl:maxCardinality
```

and T_{OWLDL} is defined as $T_{\text{OWLDL}} = T_{\text{OWLL}} \quad \cup$

```
{
  (owl:intersectionOf, rdf:type, rdf:Property),
  (owl:intersectionOf, rdfs:domain, owl:Class),
  (owl:equivalentClass, rdf:type, rdf:Property),
  (owl:disjointWith, rdf:type, rdf:Property),
  . . .
}
```

As usual, we suppose that there is given a set R_{OWLDL} of RDF Schema resources and a function $S_{\text{OWLDL}} : \text{OWLDL} \to R_{\text{OWLDL}}$ which associates a resource to each OWL DL symbol, and satisfies $S_{\text{OWLDLM}}|_{\text{OWLLVoc}} = S_{\text{OWLL}}$.

For each theory such that there is a morphisms $f : \text{OWLDL} \to (RR, T)$ we define the model constraint $[\![RR, T]\!]_{\text{OWLDL}}$ obtained by strengthening $[\![RR, T]\!]_{\text{OWLL}}$ with the following conditions. If $\mathbb{I} \in [\![RR, T]\!]_{\text{OWLDL}}$, then:

- $R_{\mathbb{I}}$ includes R_{OWLDL} and the restriction of $S_{\mathbb{I}}$ to OWLDLVoc coincides with S_{OWLDL}.
- $\mathbb{I}$ satisfies the restrictions expressing the intended meaning of the new features.

We denote by $\widehat{\text{OWLDL}}$ the institution generated by the theory OWLDL and the model constraint $[\![\_]\!]_{\text{OWLDL}}$ using the method presented in Section 2.2.

Theorem 3. *There is a conservative comorphism from* $\widehat{\text{OWLDesLog}}$ *to* $(\widehat{\text{OWLDL}}^{\text{th}})^{\wedge}$.

5.5   RDF Serialization of OWL Full

The theory $\mathtt{OWLF} = (\mathtt{OWLFVoc}, T_{\mathtt{OWLF}})$ is defined as follows. The vocabulary $\mathtt{OWLFVoc}$ includes $\mathtt{OWLDLVoc}$ and the symbols corresponding to the new features. Similarly, $T_{\mathtt{OWLF}}$ includes $T_{\mathtt{OWLDL}}$ together with triples restricting the use of the new symbols as intended and triples expressing the equality of the parts of the OWL universe with their analogues in RDF Schema.

R_{OWLF} and S_{OWLF} are defined as usual. The model constraint $[\![\_]\!]_{\mathrm{OWLF}}$ includes:

- The restriction corresponding to R_{OWLF} and S_{OWLF}.
- Restrictions expressing the intended meaning of all the features.
- Restrictions that force the parts of the OWL universe to be the same as their analogues in RDF.

The vocabulary separation restriction is not included in $[\![\_]\!]_{\mathrm{OWLF}}$.

Theorem 4. *There is a conservative comorphism from* $\widetilde{\mathtt{OWLFull}}$ *to* $(\widetilde{\mathtt{OWLF}}^{\mathrm{th}})^{\wedge}$.

5.6   Summing Up

All institutions we defined in this paper and their relationships are represented in Figure 2. The lower side includes the RDF indexed institution and it gives the semantics for RDF layer in the Berners-Lee's stack. It is worth to note that all arrows are institution morphisms; hence we may define the semantics of the layer as being the Grothendieck institution defined by this indexed institution. The upper side includes the institutions corresponding to the SW languages and their relationships expressed as comorphisms. The Grothendieck institution defined by this indexed institution gives the semantics for ontology layer. The semantics of the layering of web ontology languages on the RDF framework is given by a comorphism of Grothendieck institutions. Note that the embedding of RDF Schema in OWL Full is not a component of this comorphism.

6   Conclusion

The multitude of languages causes certain confusion in the Semantic Web community as they are based on different formalisms (description logics, Datalog, RDF Schema, etc.). A careful and thorough investigation of the relationship among the various languages will certainly reveal subtle differences among them.

Institutions and institution morphisms were developed to capture the notion of "logical systems" and relate software systems regardless of the underlying logical system. Hence, it is natural to use institutions to represent the various Semantic Web languages (including RDF and RDF Schema) and study their relationship using institution morphism.

In this paper, based on RDF(S), we define indexed institutions for RDF framework layer and ontology layer. An overall relationship among all these languages can be seen in Fig. 2. The figure shows that the institution approach can precisely capture the

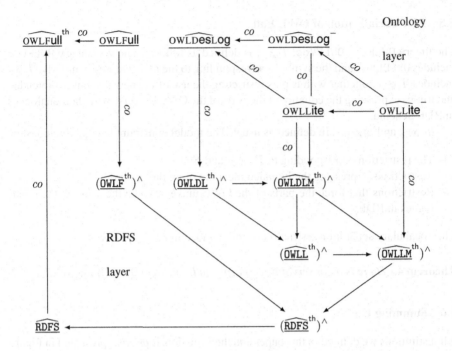

Fig. 2. RDF serialization

relationship among the various languages. The work presented in this paper opens up a new practical application domain for the institutions theory.

One future work direction is to further investigate the relationship of various ontology languages with regard to their respective underlying logical systems. Languages such as the OWL suite (OWL Lite, DL and Full) are based on description logics and they assume an "Open World Assumption". On the other hand, languages such as OWL Flight and WRL are based on logic programming and they assume a "Closed World Assumption". The interoperability of these kinds of languages has been intensively discussed but is still an open question. We believe institution theory can help to clarify this issue by establishing links at the logical level. It is also of interest to investigate the properties of the indexed institution like theory colimits, liberality, exactness, inclusions, and how the design of tools for SW can benefit from these properties.

References

1. J. Angele, H. Boley, J. de Brujin, D. Fensel, P. Hitzler, M. Kifer, R. Krummenacher, H. Lausen, A. Polleres, R.Studer. Web Rule Language (WRL). Version 1.0, 2005. http://www.wsmo.org/wsml/wrl/wrl.html.
2. M. Barr and Ch. Wells. *Category Theory for Computing Science*. Les Publications CRM, Montreal, third edition, 1999
3. T. Berners-Lee, J. Hendler, and O. Lassila. The Semantic Web. *Scientific American*, May 2001.

4. D. Brickley and R.V. Guha (editors). Resource Description Framework (RDF) Schema Specification 1.0. http://www.w3.org/TR/rdf-schema/, February 2004.

5. J. de Brujin, H. Lausen, and D. Fensel. OWL⁻. Deliverable D20.1v0.2, WSML, 2004. http://www.wsmo.org/TR/d20/d20.1/v0.2/.

6. R. Diaconescu. *Institution-independent Model Theory*. To appear. http://www.imar.ro/~diacon/.

7. R. Diaconescu. Grothendieck institutions. *Applied Categorical Structures*, 10:383–402, 2002.

8. J. S. Dong, C. H. Lee, Y. F. Li, and H. Wang. Verifying DAML+OIL and beyond in Z/EVES. In *Proceedings of 26th International Conference on Software Engineering (ICSE'04)*, pages 201–210, Edinburgh, Scotland, May 2004.

9. J. Goguen. Information Integration in Institutions, 2004. To appear in avolume dedicated to Jon Barwise, edited by Larry Moss.

10. J. Goguen and R. Burstall. Institutions: Abstract Model Theory for Specification and Programming. *Journal of the Association for Computing Machinery*, 39(1):95–146, 1992.

11. J. Goguen and G. Roşu. Institution Morphisms. *Formal Aspects of Computing*, 13, pages 274-307, 2002.

12. B.N. Grosof, I. Horrocks, R. Volz, St. Decker. Description Logic Programs: Combining Logic Programs with Description Logic. In *Proc. of the Twelfth International World Wide Web Conference*, pages 48-57, ACM, 2003.

13. P. Hayes. RDF Semantics. http://www.w3.org/TR/rdf-mt/, February 2004

14. I. Horrocks and P. Patel-Schneider. A proposal for an OWL rules language. In Proc. of the Thirteenth International World Wide Web Conference (WWW 2004), ACM, 2004.

15. I. Horrocks, P. Patel-Schneider, H. Boley, S. Tabet, B. Grosof, and M. Dean. SWRL: A semantic web rule language combining OWL and RuleML. http://www.daml.org/2004/11/fol/rules-all.html, November 2004.

16. I. Horrocks, P. Patel-Schneider, and F. van Harmelen. From $\mathcal{SHIQ}$ and RDF to OWL: The making of a web ontology language. *J. of Web Semantics*, 1(1):7–26, 2003.

17. G. Klyne, J. Carroll (editors). Resource Description Framework (RDF): Concepts and Abstract Syntax. W3C Recommendation, 2004. http://www.w3.org/TR/rdf-concepts/.

18. D. Lucanu, Y. F. Li, and J. S. Dong. Web Ontology Verification and Analysis in the Z Framework. Technical Report TR 05-01, University "Alexandru Ioan Cuza" of Iaşi, Romania, January 2005. http://thor.info.uaic.ro/~tr/tr05-01.ps.

19. B. McBride. Jena: Implementing the RDF Model and Syntax Specification. In 2nd Int'l Semantic Web Workshop, 2001.
http://www.hpl.hp.com/personal/bwm/papers/20001221-paper/.

20. P. Patel-Schneider, P. Hayes, and I. Horrocks (editors). OWL Web Ontology Semantics and Abstract Syntax. http://www.w3.org/TR/2004/REC-owl-semantics-20040210/, 2004.

21. P. Patel-Schneider. A proposal for a SWRL extension to forst-order logic. http://www.daml.org/2004/11/fol/proposal, November 2004.

22. P. Patel-Schneider, and D. Fensel, Layering the Semantic Web: Problems and Directions. In *First International Semantic Web Conference (ISWC2002)*, Sardinia, Italy". citeseer.ist.psu.edu/article/patel-schneider02layering.html, 2002.

23. B.C. Pierce. *Basic Category Theory for Computer Science*. MIT, 1991.

Institutional 2-cells
and Grothendieck Institutions

Till Mossakowski

DFKI Lab Bremen and Dept. of Computer Science, University of Bremen, Germany

Abstract. We propose to use Grothendieck institutions based on 2-categorical diagrams as a basis for heterogeneous specification. We prove a number of results about colimits and (some weak variants of) exactness. This framework can also be used for obtaining proof systems for heterogeneous theories involving institution semi-morphisms.

1 Introduction

"There is a population explosion among the logical systems used in computer science. Examples include first order logic, equational logic, Horn clause logic, higher order logic, infinitary logic, dynamic logic, intuitionistic logic, order-sorted logic, and temporal logic; moreover, there is a tendency for each theorem prover to have its own idiosyncratic logical system. We introduce the concept of *institution* to formalize the informal notion of 'logical system'." [10]

This famous quote from Joseph Goguen's and Rod Burstall's seminal paper introducing institutions lead, in its consequences, also to the introduction of Grothendieck institutions by Răzvan Diaconescu [5], which provide the semantic basis for heterogeneous specifications, i.e. the involvement of a multitude of logical systems within a single specification.

While the properties of Grothendieck institutions and their interaction with colimits, exactness, liberality, Craig interpolation etc. is well-studied now (cf. the forthcoming book [4]), the present theory of Grothendieck institutions still does not answer certain practical problems. During the development of the heterogeneous tool set (HETS) [15,17], a parsing, static analysis and proof management tool for heterogeneous specifications, we have encountered the following problems:

- often there is a plethora of possible translations between two given institutions, making choice difficult for the user;
- often premises for theorems about Grothendieck institutions do not hold for some of the institution involved — however, failure of a premise just for one institution usually destroys applicability of a theorem;
- also, the premises needed for institution (co)morphisms do not hold in all cases;

K. Futatsugi et al. (Eds.): Goguen Festschrift, LNCS 4060, pp. 124–149, 2006.
© Springer-Verlag Berlin Heidelberg 2006

– finally, this means that the applicability of theorem proving for structured specifications [2] is limited for Grothendieck institutions, and hence for heterogeneous specifications.

We introduce two ideas that may help solving these problems: the use of institutional 2-cells, and the weakening of exactness properties to quasi-exactness. We prove a number of properties of these and discuss examples. Proofs can be found in the appendix.

2 Institutions

Let $\mathcal{CAT}$ be the category of categories and functors.[1]

Definition 1. *An* institution $I = (\mathbf{Sign}, \mathbf{Sen}, \mathbf{Mod}, \models)$ *consists of*

– *a category* $\mathbf{Sign}$ *of signatures,*
– *a functor* $\mathbf{Sen}\colon \mathbf{Sign} \longrightarrow \mathbf{Set}$ *giving, for each signature* Σ, *the set of sentences* $\mathbf{Sen}(\Sigma)$, *and for each signature morphism* $\sigma\colon \Sigma \longrightarrow \Sigma'$, *the sentence translation map* $\mathbf{Sen}(\sigma)\colon \mathbf{Sen}(\Sigma) \longrightarrow \mathbf{Sen}(\Sigma')$, *where often* $\mathbf{Sen}(\sigma)(\varphi)$ *is written as* $\sigma(\varphi)$,
– *a functor* $\mathbf{Mod}\colon (\mathbf{Sign})^{op} \longrightarrow \mathcal{CAT}$ *giving, for each signature* Σ, *the category of* models $\mathbf{Mod}(\Sigma)$, *and for each signature morphism* $\sigma\colon \Sigma \longrightarrow \Sigma'$, *the reduct functor* $\mathbf{Mod}(\sigma)\colon \mathbf{Mod}(\Sigma') \longrightarrow \mathbf{Mod}(\Sigma)$, *where often* $\mathbf{Mod}(\sigma)(M')$ *is written as* $M'|_{\sigma}$,
– *a satisfaction relation* $\models_{\Sigma} \subseteq |\mathbf{Mod}(\Sigma)| \times \mathbf{Sen}(\Sigma)$ *for each* $\Sigma \in |\mathbf{Sign}|$,

such that for each $\sigma\colon \Sigma \longrightarrow \Sigma'$ *in* $\mathbf{Sign}$ *the following* satisfaction condition *holds:*

$$M' \models_{\Sigma'} \sigma(\varphi) \Leftrightarrow M'|_{\sigma} \models_{\Sigma} \varphi$$

for each $M' \in |\mathbf{Mod}(\Sigma')|$ *and* $\varphi \in \mathbf{Sen}(\Sigma)$. □

Institutions can alternatively, and more succinctly, be characterized as functors into a certain category of "twisted relations" [10], called "rooms" in [9]:
   An *institution room* $(S, \mathcal{M}, \models)$ consists of

– a set of S of *sentences,*
– a category $\mathcal{M}$ of *models,* and
– a satisfaction relation $\models \subseteq |\mathcal{M}| \times S$.

Rooms are connected via corridors (which model change of notation within one logic, as well as translations between logics).
   An *institution corridor* $(\alpha, \beta)\colon (S_1, \mathcal{M}_1, \models_1) \longrightarrow (S_2, \mathcal{M}_2, \models_2)$ consists of

– a sentence translation function $\alpha\colon S_1 \longrightarrow S_2$, and
– a model reduction functor $\beta\colon \mathcal{M}_2 \longrightarrow \mathcal{M}_1$, such that

---

[1] Strictly speaking, $\mathcal{CAT}$ is not a category but only a so-called quasicategory, which is a category that lives in a higher set-theoretic universe [11].

$$M_2 \models_2 \alpha(\varphi_1) \Leftrightarrow \beta(M_2) \models_1 \varphi_1$$

holds for each $M_2 \in |\mathcal{M}_2|$ and each $\varphi_1 \in S_1$ (*satisfaction condition*).

Now, an institution can equivalently be defined to be just a functor $I\colon \mathbf{Sign} \longrightarrow \mathbf{InsRoom}$ (where $\mathbf{Sign}$ is the category of signatures).

Example 2. The institution $FOL^=$ of many-sorted first-order logic with equality. Signatures are many-sorted first-order signatures, i.e. many-sorted algebraic signatures enriched with predicate symbols with arities. Signature morphisms map signature symbols in a coherent way. Models are many-sorted first-order structures, and model morphisms are standard algebra homomorphisms that preserve the holding of predicates. Model (morphism) reduction is done by renaming model (morphism) components. Sentences are first-order formulas, and sentence translation means replacement of the translated symbols. Satisfaction is the usual satisfaction of a first-order sentence in a first-order structure. □

Example 3. The institution $Eq^=$ of equational logic is the restriction of $FOL^=$ to signatures without predicates, and (universally quantified) equations as the only sentences. □

Example 4. The institution $PFOL^=$ of partial first-order logic with equality. Signatures are many-sorted first-order signatures enriched by partial function symbols. Models are many-sorted partial first-order structures. Sentences are first-order formulas containing existential equations, strong equations, definedness statements and predicate applications as atomic formulas. Satisfaction is defined using total valuations of variables, while valuation of terms is partial due to the existence of partial functions. An existential equation holds if both sides are defined and equal, whereas a strong equation also holds if both sides are undefined. A definedness statement holds if the term is defined. A predicate application holds if the terms contained in it are defined, and the corresponding tuple of values is in the interpretation of the predicate. This is extended to first-order formulas as usual. Moreover, signature morphisms, model reductions and sentence translations are defined like in $FOL^=$. □

Example 5. The CASL institution extends $PFOL^=$ with subsorting and induction (for datatypes), see [14,3] for details. CASL has, among others, a modal logic extension MODALCASL [15] and a coalgebraic extension CoCASL [18]. □

Example 6. There is an institution $PLNG$ of a programming language [21]. It is built over an algebra of built-in data types and operations of a programming language. Signatures are given as function (functional procedure) headings; sentences are function bodies; and models are maps that for each function symbol, assign a computation (either diverging, or yielding a result) to any sequence of actual parameters. A model satisfies a sentence iff it assigns to each sequence of parameters the computation of the function body as given by the sentence.

Hence, sentences determine particular functions in the model uniquely. Finally, signature morphisms, model reductions and sentence translations are defined similarly to those in $FOL^=$. □

Institution morphisms [10,7] relate two given institutions. A typical situation is that an institution morphism expresses the fact that a "larger" institution *is built upon* a "smaller" institution by *projecting* the "larger" institution onto the "smaller" one. Dually, institution comorphisms [7] typically express that an institution is included, or *encoded* into another one.

Using the notation of institutions as functors, given institutions $I_1: \mathbf{Sign}_1 \longrightarrow \mathbf{InsRoom}$ and $I_2: \mathbf{Sign}_2 \longrightarrow \mathbf{InsRoom}$, an *institution morphism* $(\Psi, \mu): I_1 \longrightarrow I_2$ consists of a functor $\Psi: \mathbf{Sign}_1 \longrightarrow \mathbf{Sign}_2$ and a natural transformation $\mu: I_2 \circ \Psi \longrightarrow I_1$. (Alternatively, we split μ into two natural transformations, denoted by α and β). By contrast, an *institution comorphism* $(\Phi, \rho): I_1 \longrightarrow I_2$ consists of a functor $\Phi: \mathbf{Sign}_1 \longrightarrow \mathbf{Sign}_2$ and a natural transformation $\rho: I_1 \longrightarrow I_2 \circ \Psi$.

Together with obvious identities and composition, this gives us the category **Ins** (**CoIns**) of institutions and institution (co)morphisms. An institution *semi-(co)morphism* is like an institution (co)morphism, but without the sentence translation component (and hence also without the satisfaction condition).

Example 7. There is an institution morphism going from first-order logic with equality to equational logic. A first-order signature is translated to an algebraic signature by just forgetting the set of predicate symbols; similarly, a first-order model is turned into an algebra by forgetting the predicates. Sentence translation is just inclusion of equations into first-order sentences. □

Example 8. There is an institution semi-morphism *toCASL* from PLNG to CASL [21]. It extracts an algebraic signature with partial operations out of a PLNG-signature by adding the signature of built-in data types and operations of the programming language. For any function declared, any PLNG-model determines its computations on given arguments, from which we can extract a partial function that maps any sequence of arguments to the result of the computation (if any). These are used to expand the built-in algebra of data types and operations of the programming language with an interpretation for the extra function names in the signature obtained. □

Example 9. There is an institution comorphism going from equational logic to first-order logic with equality. An algebraic signature is translated to a first-order signature by just taking the set of predicate symbols to be empty. Sentence translation is just inclusion of equations into first-order sentences. A first-order model with empty set of predicates is translated by just considering it as an algebra. □

Example 10. Similarly, there are obvious inclusion comorphisms from CASL to MODALCASL and COCASL, see [15]. □

Example 11. Define an institution comorphism going from partial first-order logic with equality to first-order logic with equality as follows: A partial first-order signature is translated to a total one by encoding each partial function symbol as a total one, plus a (new) unary predicate D ("definedness") and a (new) function symbol $\perp$ ("undefined") for each sort (this means that $\perp$ and D are heavily overloaded). Furthermore, we add axioms[2] stating that D does not hold on $\perp$, and that (encoded) total functions preserve ("totality") and reflect ("strictness") D, while partial functions only reflect D (and the holding of predicates implies D to hold on the arguments). Sentence translation is done by replacing all partial function symbols by the total functions symbols encoding them, replacing strong equations $t = u$ by $(D(t) \vee D(u)) \Rightarrow t = u$, existence equations by conjunctions of the equation and the definedness (using D) of one of the sides of the equation, replacing definedness with D, and leaving predicate symbols as they are. For a given total model of the translated signature, we just take as carriers of the partial model the interpretations of the definedness predicates in the total model, while the total functions are restricted to these new carriers, yielding partial functions.   □

3   Institution (Co)Morphism Modifications

A typical experience with using the heterogeneous tool set [15,17] is the following: for some specification, you want to prove a theorem, and hence want to see a list of its possible translations (along (co)comorphisms) into tool-supported institutions. Now even with a small diagram of institutions, the list can become quite large, because also composites should be shown (see Fig. 1 for a menu of such translations). Now such lists generally bear a lot of redundancy, since two different translation paths, though differing as (co)morphisms, lead to essentially same results, as the following example shows:

Example 12. There are two ways to go from equational logic to first-order logic: one is the obvious subinstitution comorphism ρ_1 from Example 9, the other one is the composition ρ_2 of the obvious subinstitution comorphism from equational logic to partial first-order logic composed with the encoding of partial first-order logic into first-order logic from Example 11.[3] These comorphisms are different: ρ_2 adds some (superfluous) coding of partiality. Yet, for e.g. the purpose of re-using proof tools, ρ_1 and ρ_2 are essentially the same.

In this context, the notion of *modification* helps.

In order to ensure that the difference between two translations really is inessential, a crucial property of modifications is that they do not lead to identifications of different sentence or model translation maps. Hence, we strengthen the original notion from [5] to *discrete* modifications:

---

[2] Hence, strictly speaking, this comorphism is a so-called simple theoroidal one, see [19] for details.

[3] Actually, since the latter is a simple theoroidal comorphism, we should take both to end in FOL^{th}, the institution of FOL-theories.

Fig. 1. Dozens of translation possibilities for a CASL theory in HETS (from a logic graph without comorphism modifications; using modifications, the number of possible translations can be greatly reduced).

Definition 13. *Given institution morphisms* $(\Psi, \mu)\colon I_1 \longrightarrow I_2$ *and* $(\Psi', \mu')\colon I_1 \longrightarrow I_2$, *a* discrete institution morphism modification $\theta\colon (\Psi, \mu) \longrightarrow (\Psi', \mu')$ *is just a natural transformation* $\theta\colon \Psi \longrightarrow \Psi'$ *such that* $\mu = \mu' \circ (I_2 \cdot \theta)$. *Similarly, given institution comorphisms* $(\Phi, \rho)\colon I_1 \longrightarrow I_2$ *and* $(\Phi', \rho')\colon I_1 \longrightarrow I_2$, *a* discrete institution comorphism modification $\theta\colon (\Phi, \rho) \longrightarrow (\Phi', \rho')$ *is a natural transformation* $\theta\colon \Phi \longrightarrow \Phi'$ *such that* $(I_2 \cdot \theta) \circ \rho = \rho'$.

Together with obvious identities and compositions, modifications can serve as 2-cells, leading to 2-categories **Ins** *and* **CoIns**. □

In [5,4], a weaker notion of institution morphism modification has been introduced, involving an additional natural transformation on the side of the models. We have not found this extra generality of practical use and hence work with the above stronger notion of discrete modification. However, since we will not use any non-discrete modification, we will omit the qualification of being discrete henceforth.

Example 14. Consider the comorphisms from Example 12.

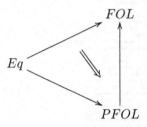

The comorphism modification $\theta\colon \rho_1 \longrightarrow \rho_2$ is just the pointwise inclusion of an algebraic signature viewed as first-order signature into the theory coding a partial variant of that signature. □

Modifications also interplay with amalgamation:

Definition 15. *Let* $\rho = (\Phi, \alpha, \beta)\colon I_1 \longrightarrow I_2$, $\rho_1 = (\Phi_1, \alpha_1, \beta_1)\colon I_1 \longrightarrow J$ *and* $\rho_2 = (\Phi_2, \alpha_2, \beta_2)\colon I_2 \longrightarrow J$ *be three comorphisms. A lax triangle*

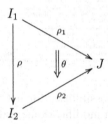

of institution comorphisms and modifications is called (weakly) *amalgamable, if*

$$
\begin{array}{ccc}
\mathbf{Mod}^{I_1}(\Sigma) & \xleftarrow{\ (\beta_1)_\Sigma\ } & \mathbf{Mod}^{J}(\Phi_1(\Sigma)) \\
{\scriptstyle \beta_\Sigma}\big\uparrow & & \big\uparrow {\scriptstyle \mathbf{Mod}^{J}(\theta_\Sigma)} \\
\mathbf{Mod}^{I_2}(\Sigma) & \xleftarrow{\ (\beta_2)_\Sigma\ } & \mathbf{Mod}^{J}(\Phi_2(\Sigma))
\end{array}
$$

is a (weak) *pullback for each signature* $\Sigma \in |\mathbf{Sign}^{I}|$. □

4   Colimits in Hom-Categories

As a first result about the 2-categorical structure of **CoIns**, we examine colimits in the Hom-categories, which play a rôle for some results about the Grothendieck construction (see Prop. 22 below):

Proposition 16. *Given two institutions* I *and* J, *if* J *has pushouts of signatures, then the Hom-category* **CoIns**(I, J) *has pushouts as well. This generalizes to arbitrary non-empty colimits of connected diagrams.* □

Note that initial objects in Hom-categories $\mathbf{CoIns}(I, J)$ generally do not exist: an initial comorphism from I to J would have to translate I-sentences to J-sentences over the initial signature, thereby losing any specific reference to the signature, which generally destroys the satisfaction condition.

The dual situation is better for initial objects:

Proposition 17. *Given two institutions I and J, if J has an initial signature with empty set of sentences and terminal model category, then the Hom-category* $\mathbf{Ins}(I, J)$ *has an initial object.*   □

However, pushouts in $\mathbf{Ins}(I, J)$ seem to exist only under rather strong additional assumptions.

We hence prefer to work with comorphisms in the sequel.

5   Comorphism-Based Grothendieck Institutions

Grothendieck institutions have been introduced by Diaconescu [5] as a foundation for heterogeneous specification. The basic data for comorphism-based heterogeneous specification is a graph of institutions, comorphisms and modifications. Remember from Sect. 1 that the modifications are needed because we want to express that certain compositions of comorphisms are the same. This means that we need to specify both compositions and modifications. We hence arrive at the following:

Definition 18. *Given an index 2-category Ind, a 2-indexed coinstitution is a 2-functor $\mathcal{I}\colon Ind^* \longrightarrow \mathbf{CoIns}$[4] into the 2-category of institutions, institution comorphisms and institution comorphism modifications.*   □

A 2-indexed coinstitution can be flattened, using the so-called *Grothendieck construction*. The basic idea here is that all signatures of all institutions are put side by side, and a signature morphism in this large realm of signatures consists of an intra-logic signature morphism plus an inter-logic translation (along some logic comorphism). The other components (sentences, models, satisfaction) are then defined in a straightforward way.

The Grothendieck construction for indexed institutions has been described in [5]; we develop its dual here [13]. In an indexed coinstitution $\mathcal{I}$, we use the notation $\mathcal{I}^i = (\mathbf{Sign}^i, \mathbf{Sen}^i, \mathbf{Mod}^i, \models^i)$ for $\mathcal{I}(i)$, (Φ^d, ρ^d) for the comorphism $\mathcal{I}(d)$, and $\mathcal{I}^u$ for the modification $\mathcal{I}(u)$.

Definition 19. *Given a 2-indexed coinstitution $\mathcal{I}\colon Ind^* \longrightarrow \mathbf{CoIns}$, define the Grothendieck institution $\mathcal{I}^\#$ as follows:*

- *signatures in $\mathcal{I}^\#$ are pairs (i, Σ), where $i \in |Ind|$ and Σ a signature in $\mathcal{I}^i$,*
- *signature morphisms $(d, \sigma)\colon (i, \Sigma_1) \longrightarrow (j, \Sigma_2)$ consist of a morphism $d\colon j \longrightarrow i \in Ind$ and a signature morphism $\sigma\colon \Phi^d(\Sigma_1) \longrightarrow \Sigma_2$ in $\mathcal{I}^j$,*
- *composition is given by $(d_2, \sigma_2) \circ (d_1, \sigma_1) = (d_1 \circ d_2, \sigma_2 \circ \Phi^{d_2}(\sigma_1))$,*
- *$\mathcal{I}^\#(i, \Sigma) = \mathcal{I}^i(\Sigma)$, and $\mathcal{I}^\#(d, \sigma) = \mathcal{I}^i(\Sigma_1) \xrightarrow{\rho^d} \mathcal{I}^j(\Phi^d(\Sigma_1)) \xrightarrow{\mathcal{I}^j(\sigma)} \mathcal{I}^j(\Sigma_2)$.*

□

---

[4] Ind^* is the 2-categorical dual of Ind, where both 1-cells and 2-cells are reversed.

That is, the room $\mathcal{I}^{\#}(i, \Sigma)$ (consisting of sentences, models and satisfaction) for a Grothendieck signature (i, Σ) is defined component-wise, while the corridor for a Grothendieck signature morphism is obtained by composing the corridor given by the inter-institution comorphism with that given by the intra-institution signature morphism. We also denote the Grothendieck institution by $(\mathbf{Sign}^{\#}, \mathbf{Sen}^{\#}, \mathbf{Mod}^{\#}, \models^{\#})$.

While the comorphism based Grothendieck construction nearly satisfies all of our needs, one problem remains. Sometimes, the Grothendieck construction makes too many distinctions between signature morphisms (cf. Fig. 1). Therefore, we use the institution comorphism modifications to obtain a congruence on Grothendieck signature morphisms: the congruence is generated by

$$(d', \mathcal{I}_{\Sigma}^{u} \colon \Phi^{d'}(\Sigma) \longrightarrow \Phi^{d}(\Sigma)) \equiv (d, id \colon \Phi^{d}(\Sigma) \longrightarrow \Phi^{d}(\Sigma)) \tag{1}$$

relating morphisms from (i, Σ) to $(j, \Phi^{d}(\Sigma))$, for $\Sigma \in |\mathbf{Sign}^{i}|$, $d, d' \colon j \longrightarrow i \in Ind$, and $u \colon d \Rightarrow d' \in Ind$. We will later examine what is really added by the congruence closure. But first, let us state the following crucial property:

Proposition 20. $\equiv$ *is contained in the kernel of* $\mathcal{I}^{\#}$ *(considered as a functor).*
□

Let $q^{\mathcal{I}} \colon \mathbf{Sign}^{\#} \longrightarrow \mathbf{Sign}^{\#}/\equiv$ be the quotient functor induced by $\equiv$ (see [12] for the definition of quotient category). Note that it is the identity on objects. We easily obtain that the functor $\mathcal{I}^{\#}$ factors through the quotient category $\mathbf{Sign}^{\#}/\equiv$:

Corollary 21. $\mathcal{I}^{\#} \colon \mathbf{Sign}^{\#} \longrightarrow \mathbf{InsRoom}$ *leads to a quotient Grothendieck institution* $\mathcal{I}^{\#}/\equiv \colon \mathbf{Sign}^{\#}/\equiv \longrightarrow \mathbf{InsRoom}$. □

By abuse of notation, we denote $\mathcal{I}^{\#}/\equiv$ by $(\mathbf{Sign}^{\#}/\equiv, \mathbf{Sen}^{\#}, \mathbf{Mod}^{\#}, \models^{\#})$.

When considering e.g. the comorphism going from partial first-order logic $PFOL^{=}$ to first-order logic $FOL^{=}$, and the composite comorphism going from $PFOL^{=}$ to CASL and then to $FOL^{=}$, we end up in different comorphisms, which are however related by a comorphism modification. The above identification process in the Grothendieck institution now tells us that it does not matter which way we choose.

In some cases, the congruence $\equiv$ can be described succinctly:

Proposition 22. *Assume that* Ind^{*} *has cocones for diagrams of 2-cells of shape* $\bullet \Longrightarrow \bullet \Longleftarrow \bullet$ *that are mapped to pushouts of 2-cells in* $\mathbf{CoIns}$. *Then the congruence* $\equiv$ *defined above is explicitly given by*

$$(d_1, \sigma \circ \mathcal{I}_{\Sigma}^{u_1}) \equiv (d_2, \sigma \circ \mathcal{I}_{\Sigma}^{u_2})$$

for $\Sigma \in |\mathbf{Sign}^{i}|$, $d, d_1, d_2 \colon j \longrightarrow i \in Ind$, $\sigma \colon \Phi^{d}(\Sigma) \longrightarrow \Sigma' \in \mathbf{Sign}^{j}$ *and* $u_1 \colon d \Rightarrow d_1, u_2 \colon d \Rightarrow d_2 \in Ind$. □

Note that according to Prop. 16, under relatively mild assumptions, pushouts of 2-cells in **CoIns** exist. Hence, the assumption of Prop. 22 that Ind^* has cocones for diagrams of 2-cells of shape $\bullet \Longrightarrow \bullet \Longleftarrow \bullet$ that are mapped to pushouts of 2-cells in **CoIns** is quite realistic. In particular, it is possible to add suitable cocones to Hom-categories in Ind^* and interpret these as pushouts in **CoIns**.

6 Amalgamation and Exactness

The amalgamation property (called 'exactness' in [6]) is a major technical assumption in the study of specification semantics [20] and is important in many respects. It allows the computation of normal forms for specifications [1,2], and it is a prerequisite for good behaviour w.r.t. parameterization, conservative extensions [6] and proof systems [16].

Definition 23. *A cocone for a diagram in* **Sign** *is called* (weakly) amalgamable *if it is mapped to a (weak) limit under* **Mod**. *I (or* **Mod***) admits* (finite) (weak) amalgamation *if (finite) colimit cocones are (weakly) amalgamable, i.e. if* **Mod** *maps (finite) colimits to (weak) limits. This property is also called (weak) exactness, while (weak) semi-exactness is its restriction to pushout diagrams.* □

More generally, given a diagram $D: J \longrightarrow \mathbf{Sign}^I$, a family of models $(M_j)_{j \in |J|}$ is called *D-consistent* if $M_k|_{D(\delta)} = M_j$ for each $\delta: j \longrightarrow k \in J$. A cocone $(\Sigma, (\mu_j)_{j \in |J|})$ over the diagram in $D: J \longrightarrow \mathbf{Sign}^I$ is called *weakly amalgamable* if for each D-consistent family of models $(M_j)_{j \in |J|}$, there is a Σ-model M with $M|_{\mu_j} = M_j$ $(j \in |J|)$. If this model is unique, the cocone is called *amalgamable*.

Proposition 24. *An institution admits (weak) amalgamation iff each colimiting cocone in the category of signatures is (weakly) amalgamable.* □

A further weakening just requires the existence of weakly amalgamable cocones:

Definition 25. *Call an institution I* quasi-exact *if for each diagram* $D: J \longrightarrow \mathbf{Sign}^I$, *there is some weakly amalgamable cocone over D.* Quasi-semi-exactness *is the restriction of this notion to diagrams of shape* $\bullet \longleftarrow \bullet \longrightarrow \bullet$.

The importance of this definition lies in the fact that it

1. interacts quite nicely with heterogeneous specification (the property holds for Grothendieck institutions under very mild and practically feasible assumptions), and it
2. is a prerequisite for the (soundness and completeness of the) proof calculus of development graphs [15,16].

The theory of amalgamation and exactness in Grothendieck institutions for indexed institutions has been developed by Diaconescu [5]. Actually, the corresponding theory for indexed coinstitutions turns out to be much simpler [13].

Theorem 26. *Let* $\mathcal{I}: Ind^{op} \longrightarrow \mathbf{CoIns}$ *be an indexed coinstitution and* K *be some small category such that*

1. *Ind is K-complete,*
2. Φ^d *is K-cocontinuous for each* $d: i \longrightarrow j \in Ind$, *and*
3. *the indexed category of signatures of* $\mathcal{I}$ *is locally K-cocomplete (the latter meaning that* $\mathbf{Sign}^i$ *is K-cocomplete for each* $i \in |Ind|$).

Then the signature category $\mathbf{Sign}^{\#}$ *of the Grothendieck institution has K-colimits.*

□

We cannot expect that this result directly carriers over to the quotient Grothendieck institution, since quotients of categories generally do not interact well with colimits. However, we can say something provided that we work with a quotient of the index category Ind:

Proposition and Definition 27 Given a 2-category Ind, the relation of being in the same connected component of a Hom-category defines a congruence $\equiv$ on the objects of the Hom-categories, i.e. the morphisms of Ind. $Ind/\equiv$ is the corresponding quotient 1-category.

□

Lemma 28. *Given a 2-indexed coinstitution* $\mathcal{I}: Ind^* \longrightarrow \mathbf{CoIns}$, *if* $(d_2, \sigma_1) \equiv (d_1, \sigma_2)$ *in* $\mathbf{Sign}^{\#}$, *then* $d_1 \equiv d_2$.

□

Proposition 29. *Assume that* Ind^* *has cocones for diagrams of 2-cells of shape* $\bullet \Longrightarrow \bullet \Longleftarrow \bullet$ *that are mapped to pushouts of 2-cells in* $\mathbf{CoIns}$. *Then the congruence* $\equiv$ *in* Ind *defined above is explicitly given by* $d_1: i \longrightarrow j \equiv d_2: i \longrightarrow j$ *iff there exist* $d: i \longrightarrow j \in Ind$ *and* $u_1: d \longrightarrow d_1, u_2: d \longrightarrow d_2 \in Ind$.

□

Theorem 30. *Let* $\mathcal{I}: Ind^* \longrightarrow \mathbf{CoIns}$ *be a 2-indexed coinstitution such that*

1. $Ind/\equiv$ *is K-complete for some small category* K,
2. *each connected component (considered as a subcategory) of a Hom-category* $Ind(i, j)$ *has a distinguished canonical weakly terminal object, such that these canonical objects are stable under composition,*
3. $(d, \sigma_1) \equiv (d, \sigma_2)$ *in* $\mathbf{Sign}^{\#}$ *implies* $\sigma_1 = \sigma_2$,
4. Φ^d *is K-cocontinuous for each* $d: i \longrightarrow j \in Ind$, *and*
5. *the indexed category of signatures of* $\mathcal{I}$ *is locally K-cocomplete.*

Then the signature category $\mathbf{Sign}^{\#}/\equiv$ *of the quotient Grothendieck institution has K-colimits. (Note that assumptions 2 and 3 are vacuous in case of discrete Hom-categories; we then get Theorem 26 as a special case.)*

□

By contravariance of $\mathcal{I}$, assumption 2 of the above proposition means that if institution comorphisms are linked by modifications, there is always a "smallest" comorphism that can be embedded into the other ones. This is quite realistic in practice. However, it is not so realistic to assume that these smallest co-morphisms are stable under composition. For example, the composition of the smallest embedding of $FOL^=$ into CASL with the smallest embedding of CASL into second-order logic will give not given the smallest embedding of $FOL^=$ into second-order logic, but rather a more complex one.

Assumption 3 basically means that the congruence does not identify signature morphisms within one institution, i.e. that each signature category $\mathbf{Sign}^i$ is faithfully embedded into $\mathbf{Sign}^\#/\equiv$. This assumption is a reasonable and desirable property in practice. We record this explicity:

Proposition 31. $emb^i \colon \mathbf{Sign}^i \longrightarrow \mathbf{Sign}^\#/\equiv$ *is an embedding preserving colimits under the assumptions of Theorem 30.*   □

Let us now come to exactness. We extend the notion of semi-exactness to comorphisms and to the indexed case. An institution comorphism (Φ, α, β) is called (weakly) exact, if the naturality squares for β are (weak) pullbacks. An 2-indexed coinstitution $\mathcal{I}\colon Ind^* \longrightarrow \mathbf{CoIns}$ is called *(weakly) locally semi-exact*, if each institution I^i is (weakly) semi-exact ($i \in |Ind|$). Assuming that equivalence classes of 2-cells have canonical weakly terminal objects, $\mathcal{I}$ is called *(weakly) semi-exact* if for each pullback in $Ind/\equiv$

$$
\begin{array}{ccc}
i & \xleftarrow{\;[d_1]\;} & j1 \\
{\scriptstyle[d_2]}\big\uparrow & & \big\uparrow{\scriptstyle[e_1]} \\
j2 & \xleftarrow{\;[e_2]\;} & k
\end{array}
$$

the square

$$
\begin{array}{ccc}
\mathbf{Mod}^i(\Sigma) & \xleftarrow{\qquad \beta_\Sigma^{d_1} \qquad} & \mathbf{Mod}^{j1}(\Phi^{d_1}(\Sigma)) \\
{\scriptstyle\beta_\Sigma^{d_2}}\big\uparrow & & \big\uparrow{\scriptstyle\beta_\Sigma^{e_1}} \\
\mathbf{Mod}^{j2}(\Phi^{d_2}(\Sigma)) & \xleftarrow{\;\beta_\Sigma^{e_2}\;} & \mathbf{Mod}^k(\Phi^{e_1}(\Phi^{d_1}(\Sigma))) = \mathbf{Mod}^k(\Phi^{e_2}(\Phi^{d_2}(\Sigma)))
\end{array}
$$

is a (weak) pullback for each signature Σ in $\mathbf{Sign}^i$, where canonical weakly terminal representatives are used.[5]

Theorem 32. *Assume that the 2-indexed coinstitution* $\mathcal{I}\colon Ind^* \longrightarrow \mathbf{CoIns}$ *fulfills the assumptions of Theorem 30. Then the quotient Grothendieck institution* $\mathcal{I}^\#/\equiv$ *is (weakly) semi-exact if and only if*

1. $\mathcal{I}$ *is (weakly) locally semi-exact,*
2. $\mathcal{I}$ *is (weakly) semi-exact, and*
3. *for all canonical weakly terminal* $d\colon i \longrightarrow j \in Ind$, *in* $\mathcal{I}^d$ *is (weakly) exact.*

□

---

[5] It might be useful to weaken these notions in the way such that model morphisms are ignored.

Theorems 26, 30 and 32 already provide a good theoretical basis for hetero-geneous specification. However, in some cases, these theorems are not general enough: Given a diagram $J \to Ind$, its limit must be the index of some insti-tution that can serve to encode (via comorphisms) all the institutions indexed by the diagram. While the existence of such an institution may not be a prob-lem (e.g. higher-order logic often serves as such a "universal" logic for coding other logics), the uniqueness condition imposed by the limit property is more problematic. This means that any two such "universal" institutions must have isomorphic indices and hence be isomorphic themselves. This might work well is some circumstances, but may not desirable in others: after all, a number of non-isomorphic logics, such as classical higher-order logic, the calculus of con-structions and rewriting logic have been proposed as such a "universal" logic.[6]

A related problem[7] is that the assumptions of Theorem 32 are too strong to be met for all practical examples. E.g. the CASL institution is not weakly semi-exact, and its encoding into $HOL^=$ [14] is neither exact, nor does it have a cocontinuous signature translation.

We hence now generalize the previous results by replacing weak exactness with quasi-exactness, i.e. amalgamable colimits with weakly amalgamable co-cones, and thereby dropping the uniqueness requirement. Hence, several non-isomorphic "universal" institutions may coexist peacefully with our approach, and also non-exact institutions and comorphisms may be included in the indexed coinstitution serving as basis for heterogeneous specification.

We first extend Def. 25 to indexed coinstitutions:

Definition 33. *An indexed coinstitution* $\mathcal{I}\colon Ind^{op} \longrightarrow \mathbf{CoIns}$ *is called* locally quasi-exact, *if each institution* $\mathcal{I}^i$ *is quasi-exact* ($i \in |Ind|$). *It is called* quasi-exact, *if for each diagram* $D\colon J \longrightarrow Ind$, *there is some cone* $(l, (d_j)_{j\in|J|})$ *over* D *whose image under* $\mathcal{I}$ *is weakly amalgamable.* Quasi-semi-exactness *is the restriction of these notions to diagrams of shape* $\bullet \longleftarrow \bullet \longrightarrow \bullet$. $\square$

However, for the index level, even quasi-exactness may be too strong. Con-sider the diagram

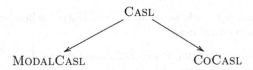

How do we obtain a weakly amalgamable cocone? A simple way is to use the embedding of MODALCASL into CASL and compose it with the inclusion of CASL into COCASL:

---

[6] This problem can possibly be circumvented by formally adjoining limits to the index category, which are then interpreted using Grothendieck institutions over subdia-grams. However, this would add considerable complexity to the construction.

[7] This problem already has been noted by Diaconescu [5] for his more special version of Theorem 32; see [13] why we consider it to be more special.

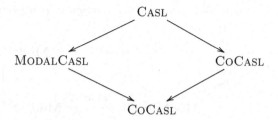

but the resulting square does not even commute.[8] The reason is that on the way from CASL to COCASL via MODALCASL, MODALCASL adds an implicit set of worlds, which is made explicit by the embedding of MODALCASL into CASL.[9] To obtain a commuting square, we would need to have a comorphism from COCASL to itself which adds an explicit set of worlds. However, this solution is rather inelegant, since it means that any (present of future) extension of CASL without possible world semantics (e.g. for HASCASL), we need a similar comorphism.

We hence prefer to split the square into two lax triangles:

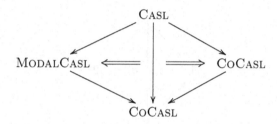

and indeed, the square is weakly amalgamable in the following sense:

Definition 34. *Given a 2-indexed coinstitution* $\mathcal{I}: Ind^* \longrightarrow \mathbf{CoIns}$, *a square consisting of two lax triangles of index morphisms*

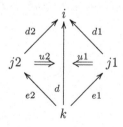

---

[8] Of course, we could also embed everything into HOL, which would not cause any relevant change to the subsequent discussion.

[9] See [15] for the reason why the set of worlds cannot be omitted even for models of signatures without modalities.

is called (weakly) amalgamable, if the following diagram is a (weak) pullback

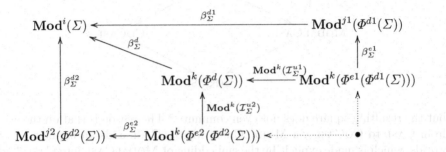

where the lower right square is a pullback. That is, each pair consisting of a $\Phi^{d2}(\Sigma)$- and a $\Phi^{d1}(\Sigma)$-model with the same Σ-reduct is (weakly) amalgamable to a pair consisting of a $\Phi^{e2}(\Phi^{d2}(\Sigma))$- and a $\Phi^{e1}(\Phi^{d1}(\Sigma))$-model having the same $\Phi^{d}(\Sigma)$-reduct.

$\mathcal{I}$ *is called* lax-quasi-exact, *if each for pair of arrows* $j1 \xrightarrow{d_1} i \xleftarrow{d_2} j2$ *in Ind, there is some square*

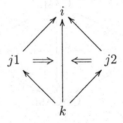

consisting of a weakly amalgamable square of lax triangles, such that additionally $\mathcal{I}^k$ is quasi-semi-exact.   □

Note that this property is different from (and indeed, incomparable to) amalgamability of the individual lax triangles:

Definition 35. *Given a 2-indexed coinstitution $\mathcal{I}: Ind^* \longrightarrow \mathbf{CoIns}$, a lax triangle of index morphisms*

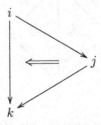

is called (weakly) amalgamable, if $\mathcal{I}$ maps it to a (weakly) amalgamable lax triangle in the sense of Definition 15.   □

Theorem 36. *For a 2-indexed coinstitution* $\mathcal{I}\colon Ind^* \longrightarrow \mathbf{CoIns}$, *assume that*

- $\mathcal{I}$ *is lax-quasi-exact, and*
- *all institution comorphisms in* $\mathcal{I}$ *are weakly exact.*

Then $\mathcal{I}^{\#}/\equiv$ *is quasi-semi-exact.*   □

Call a diagram *acyclic (connected)* if the graph underlying its index category is acyclic (connected) when the identity arrows are deleted.

Corollary 37. *Let* $\mathcal{I}$ *satisfy the assumptions of Theorem 36. Then* $\mathcal{I}^{\#}/\equiv$ *admits weak amalgamation of finite acyclic connected diagrams.*   □

As stated above, the importance of these results lies in the fact that quasi-(semi-)exactness is a prerequisite for the (soundness and completeness of the) proof calculus of development graphs [15,16]. Due to lack of space, we cannot go into the details here. Instead, we provide a simple application of a typical situation of a view (or a refinement) involving hiding, illustrating a simple application of the rule *Theorem-Hide-Shift* from the calculus of [15,16].

Proposition 38. *In an institution, let a span of theories*

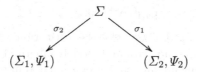

be given. Then the refinement statement

$$\mathbf{Mod}(\sigma_1)^{-1}(\mathbf{Mod}(\sigma_2)(|\mathbf{Mod}(\Sigma_2, \Psi_2)|)) \subseteq |\mathbf{Mod}(\Sigma_1, \Psi_1)|$$

follows from (and, hence can be reduced to) the statement

$$\mathbf{Mod}(\Sigma_3, \theta_2(\Psi_2)) \subseteq \mathbf{Mod}(\Sigma_3, \theta_1(\Psi_1))$$

provided that

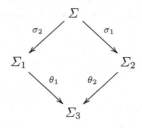

is a weakly amalgamable square.   □

7   From Specifications to Programs

Consider a specification *SortSpec* of sorting written in CASL (let it have signature Σ_S), and a sorting program *SortProg* written in PLNG (let it have signature Σ_P). We can use the institution semi-morphism $toCASL\colon PLNG \longrightarrow$ CASL from example 8 to express that *SortProg* is an implementation of *SortSpec*. Let (Φ, β) be *toCASL* decomposed in its signature and model translation component. Then the property that we need to express is

$$\beta_{\Sigma_P}(\mathbf{Mod}^{PLNG}(SortProg)) \subseteq \mathbf{Mod}^{\mathrm{CASL}}(SortSpec)$$

assuming that $\Phi(\Sigma_P) = \Sigma_S$ (if needed, we can ensure this property by massaging the CASL specification appropriately).

Now the question arises how to prove this property. It would be easy if *toCASL* could be extended to an institution morphism; however, there is no hope to translate CASL formulas into programs. However, we can split the semi-morphism $toCASL = (\Phi, \beta)$ into a span of comorphisms

$$PLNG \xleftarrow{\;\;toCASL^-\;\;} \mathrm{CASL} \circ \Phi \xrightarrow{\;\;toCASL^+\;\;} \mathrm{CASL}$$

as follows:

$$\begin{array}{ccccc}
\mathbf{Sign}^{PLNG} & \xleftarrow{\;\;id\;\;} & \mathbf{Sign}^{PLNG} & \xrightarrow{\;\;\Phi\;\;} & \mathbf{Sign}^{\mathrm{CASL}} \\[4pt]
\mathbf{Sen}^{PLNG} & \xleftarrow{\;\;incl\;\;} & \emptyset & \xrightarrow{\;\;incl\;\;} & \mathbf{Sen}^{\mathrm{CASL}} \circ \Phi \\[4pt]
\mathbf{Mod}^{PLNG} & \xrightarrow{\;\;\beta\;\;} & \mathbf{Mod}^{\mathrm{CASL}} \circ \Phi^{op} & \xleftarrow{\;\;id\;\;} & \mathbf{Mod}^{\mathrm{CASL}} \circ \Phi^{op}
\end{array}$$

Here, the "middle" institution CASL$\circ\Phi$ is the institution with signature category inherited from PLNG, no sentences, and models inherited from CASL via Φ.

Our refinement statement can now be reformulated in terms of comorphisms:

$$(\beta_{\Sigma_P}^{toCASL^+})^{-1}(\beta_{\Sigma_P}^{toCASL^-}(\mathbf{Mod}^{PLNG}(SortProg))) \subseteq \mathbf{Mod}^{\mathrm{CASL}}(SortSpec)$$

We can regard this in a suitable Grothendieck institution; then it has exactly the form of the statement in Prop. 38. We hence can reformulate the statement, provided that we have quasi-semi-exactness. By Theorem 36, we need lax-quasi-exactness of the indexed coinstitution. The essential ingredient to find a square of two weakly amalgamable lax triangles for the span $PLNG \xleftarrow{\;\;toCASL^-\;\;} \mathrm{CASL} \circ \Phi \xrightarrow{\;\;toCASL^+\;\;} \mathrm{CASL}$. But this can e.g. be given by coding of both CASL and PLNG into a common logic such as higher order logic (indexing institutions and comorphisms by themselves):

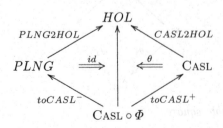

By Theorem 36, this lead to a weakly amalgamable square in the Grothendieck institution:

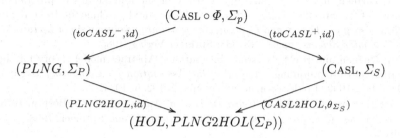

By Prop. 38, our refinement statement can now be reformulated as follows:

$$\mathbf{Mod}^{HOL}(PLNG2HOL(SortProg)) \subseteq \mathbf{Mod}^{HOL}(\theta(CASL2HOL(SortSpec)))$$

which is amount to proving, in HOL,

$$PLNG2HOL(SortProg) \vdash \theta(CASL2HOL(SortSpec)).$$

An implementation of this machinery for the case PLNG=Haskell is under way, to become part of the Heterogeneous Tool Set HETS [15,17].

Acknowledgments There are a number of colleagues who have introduced me into the field and who always are open for interesting discussions and collaborations; here I shall name only Joseph Goguen, Răzvan Diaconescu and Andrzej Tarlecki. Andrzej Tarlecki made very valuable comments on a draft.

This work has been supported by the *Deutsche Forschungsgemeinschaft* under Grants KR 1191/5-1 and KR 1191/5-2.

References

1. J. Bergstra, J. Heering, and P. Klint. Module Algebra. *J. ACM*, 37(2):335–372, 1990.
2. T. Borzyszkowski. Generalized interpolation in CASL. *Information Processing Letters*, 76/1-2:19–24, 2000.
3. CoFI (The Common Framework Initiative). CASL *Reference Manual*. LNCS Vol. 2960 (IFIP Series). Springer, 2004.
4. R. Diaconescu. *Institution-independent Model Theory*. To appear. Book draft. (Ask author for a current draft.).
5. R. Diaconescu. Grothendieck institutions. *Applied categorical structures*, 10:383–402, 2002.
6. R. Diaconescu, J. Goguen, and P. Stefaneas. Logical support for modularisation. In G. Huet and G. Plotkin, editors, *Proceedings of a Workshop on Logical Frameworks*, 1991.
7. J. Goguen and G. Roşu. Institution morphisms. *Formal aspects of computing*, 13:274–307, 2002.

8. J. A. Goguen and R. M. Burstall. Introducing institutions. volume 164 of *Lecture Notes in Computer Science*, pages 221–256. Springer Verlag, 1984.
9. J. A. Goguen and R. M. Burstall. A study in the foundations of programming methodology: Specifications, institutions, charters and parchments. In D. P. et al., editor, *Category Theory and Computer Programming*, volume 240 of *Lecture Notes in Computer Science*, pages 313–333. Springer Verlag, 1985.
10. J. A. Goguen and R. M. Burstall. Institutions: Abstract model theory for specification and programming. *Journal of the Association for Computing Machinery*, 39:95–146, 1992. Predecessor in: LNCS 164, 221–256, 1984.
11. H. Herrlich and G. Strecker. *Category Theory*. Allyn and Bacon, Boston, 1973.
12. S. Mac Lane. *Categories for the Working Mathematician*. Springer, 1998. Second edition.
13. T. Mossakowski. Comorphism-based Grothendieck logics. In K. Diks and W. Rytter, editors, *Mathematical foundations of computer science*, volume 2420 of *LNCS*, pages 593–604. Springer, 2002.
14. T. Mossakowski. Relating CASL with other specification languages: the institution level. *Theoretical Computer Science*, 286:367–475, 2002.
15. T. Mossakowski. Heterogeneous specification and the heterogeneous tool set. Habilitation thesis, University of Bremen, 2005.
16. T. Mossakowski, S. Autexier, and D. Hutter. Development graphs – proof management for structured specifications. *Journal of Logic and Algebraic Programming*, 67(1-2):114–145, 2006.
17. T. Mossakowski, C. Maeder, K. Lüttich, and S. Wölfl. The heterogeneous tool set. Submitted for publication.
18. T. Mossakowski, L. Schröder, M. Roggenbach, and H. Reichel. Algebraic-co-algebraic specification in CoCASL. *Journal of Logic and Algebraic Programming*, 67(1-2):146–197, 2006.
19. G. Roşu and J. Goguen. Composing hidden information modules over inclusive institutions, 2004.
20. D. Sannella and A. Tarlecki. Specifications in an arbitrary institution. *Information and Computation*, 76:165–210, 1988.
21. A. Tarlecki. Moving between logical systems. In M. Haveraaen, O. Owe, and O.-J. Dahl, editors, *Recent Trends in Data Type Specifications. 11th Workshop on Specification of Abstract Data Types*, volume 1130 of *Lecture Notes in Computer Science*, pages 478–502. Springer Verlag, 1996.
22. A. Tarlecki, R. M. Burstall, and J. A. Goguen. Some fundamentals algebraic tools for the semantics of computation. Part 3: Indexed categories. *Theoretical Computer Science*, 91:239–264, 1991.

A   Proofs of the Theorems

Proof of Prop. 16. Given comorphisms $(\Phi_i, \rho_i): I \longrightarrow J$ $(i = 1, 2, 3)$ and a span of modifications

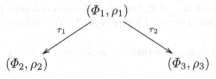

construct the signature component $\Phi(\Sigma)$ of the resulting comorphism as the pushout

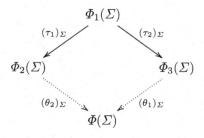

By the universal property of the pushout, this extends to a functor $\Phi: \mathbf{Sign}^I \longrightarrow \mathbf{Sign}^J$ such that $\theta_1: \Phi_3 \longrightarrow \Phi$ and $\theta_2: \Phi_2 \longrightarrow \Phi$ become natural transformations.

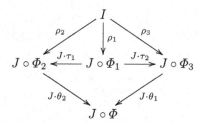

We can then define room component of the pushout comorphism $\rho: I \longrightarrow J \circ \Phi$ to be $J \cdot \theta_2 \circ \rho_2 = J \cdot \theta_1 \circ \rho_3$, and the cocone consisting of $\theta_1: (\Phi_3, \rho_3) \Longrightarrow (\Phi, \rho)$ and $\theta_2: (\Phi_2, \rho_2) \Longrightarrow (\Phi, \rho)$ is easily seen to satisfy the universal property of a pushout.

The proof for coproducts, coequalizers or arbitrary non-empty colimits of connected diagrams is very similar.   □

Proof of Prop. 17: The initial institution morphism $(\Phi, \mu): I \longrightarrow J$ is defined by letting $\Phi(\Sigma)$ be the initial signature, and μ_Σ consist of the empty map of sentences and the unique functor into the terminal model category.   □

Proof of Prop. 20: By the definition of comorphism modification, $(\mathcal{I}^j \cdot \mathcal{I}^u) \circ \rho^{d'} = \rho^d$. But this just means that equivalent signature morphisms induce the same corridors.   □

Proof of Prop. 22: It is easy to see that the above relation is contain in the relation generated by (1): just apply (1) twice. It remains to show that the above relation is a congruence. Reflexivity and symmetry are clear. Concerning transitivity, assume that

$$(d_1, \sigma_1 \circ \mathcal{I}_\Sigma^{u_1}) \equiv (d_3, \sigma_1 \circ \mathcal{I}_\Sigma^{u_2}) = (d_3, \sigma_2 \circ \mathcal{I}_\Sigma^{u_3}) \equiv (d_5, \sigma_2 \circ \mathcal{I}_\Sigma^{u_4}),$$

the first relation being witnessed by $u_1 : d_2 \Rightarrow d_1, u_2 : d_2 \Rightarrow d_3$, and the second by by $u_3 : d_4 \Rightarrow d_3, u_4 : d_4 \Rightarrow d_5$. Take the pullback in $Ind(j,i)$ of the two spans

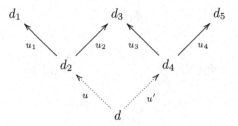

By the construction of pushouts of 2-cells in **CoIns** (see Prop.16), the middle square in

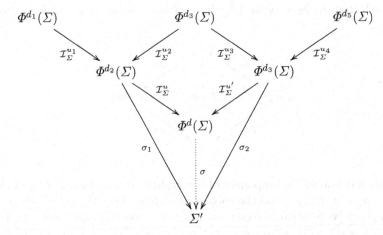

is a pushout, and the mediating morphism σ leads to the desired form

$$(d_1, \sigma_1 \circ \mathcal{I}_\Sigma^{u_1}) = (d_1, \sigma \circ \mathcal{I}_\Sigma^{u_1 \circ u}) \equiv (d_5, \sigma \circ \mathcal{I}_\Sigma^{u_4 \circ u'}) = (d_5, \sigma_2 \circ \mathcal{I}_\Sigma^{u_4}).$$

Concerning composition, assume that

$$(d_1, \sigma \circ \mathcal{I}_\Sigma^{u_1}) \equiv (d_2, \sigma \circ \mathcal{I}_\Sigma^{u_2})$$

via $u_1 : d \Rightarrow d_1, u_2 : d \Rightarrow d_2$, and

$$(e_1, \tau \circ \mathcal{I}_{\Sigma'}^{v_1}) \equiv (e_2, \tau \circ \mathcal{I}_{\Sigma'}^{v_2})$$

via $v_1 : e \Rightarrow e_1, v_2 : e \Rightarrow e_2$. Then for $k = 1, 2$,

$$
\begin{aligned}
&(e_k, \sigma \circ \mathcal{I}_\Sigma^{u_k}) \circ (d_k, \tau \circ \mathcal{I}_{\Sigma'}^{v_k}) \\
&= (d_k \circ e_k, \sigma \circ \mathcal{I}_\Sigma^{u_k} \circ \Phi^{e_k}(\tau) \circ \Phi^{e_k}(\mathcal{I}_{\Sigma'}^{v_k})) &&\text{(def. Grothendieck composition)} \\
&= (d_k \circ e_k, \sigma \circ \Phi^{e_k}(\tau) \circ \Phi^{e_k}(\mathcal{I}_{\Sigma'}^{v_k}) \circ \mathcal{I}_{\Phi^{e_k}(\Sigma')}^{u_k}) &&\text{(naturality of } \mathcal{I}^{u_k}) \\
&= (d_k \circ e_k, \sigma \circ \Phi^{e_k}(\tau) \circ \mathcal{I}_{\Sigma'}^{v_k \cdot u_k}) &&\text{(functoriality of } \mathcal{I})
\end{aligned}
$$

which shows that we arrive at the desired form.   □

Proof of Thm. 26: Apply Theorem 1 of [22] with $C_i = \mathbf{Sign}^i$ and $C_m = \Phi^m$. Note that $\mathbf{Sign}^\#$ is then $Flat(C^{op})^{op}$.   □

Proof of Lemma 28: Easy induction over the definition of $(d_1, \sigma_1) \equiv (d_2, \sigma_2)$.   □

Proof of Prop. 29: Analogous to the proof of Prop. 22.   □

Proof of Thm. 30: The proof idea follows that of Theorem 1 in [22], the necessary modifications being caused by the congruences. By assumption 2, we can always choose representatives $d \in Ind$ of congruences classes $[d] \in Ind/\equiv$ in such a way that d is a canonical weakly terminal object. Similarly, we can always choose representatives (d, σ) of congruence classes $[(d, \sigma)]$ in $\mathbf{Sign}^\#/\equiv$ in such a way that d is the canonical weakly terminal object in its connected component: given an arbitrary $(d, \sigma: \Phi^d(\Sigma) \longrightarrow \Sigma')$ in $\mathbf{Sign}^\#$, let $u: d \Longrightarrow t$ be a 2-cell into the canonical weakly terminal object. Then $(t, \sigma \circ \mathcal{I}^u_\Sigma)$ is equivalent to (d, σ).

Given a diagram $D: K \longrightarrow \mathbf{Sign}^\#/\equiv$, we introduce the notation (i_k, Σ_k) for $D(k)$ $(k \in |K|)$ and $[(d_m, \sigma_m)]: (i_k, \Sigma_k) \longrightarrow (i_{k'}, \Sigma_{k'})$ for $D(m)$ $(m: k \longrightarrow k' \in K)$. Let $\bar{D}: K \longrightarrow Ind/\equiv$ be the projection of D to the first component; by Lemma 28 this is a well-defined diagram in $Ind/\equiv$. By assumption 1, $\bar{D}$ has a limit $([m_k]: i \longrightarrow i_k)_{k \in |K|}$.

Let the diagram $G: K \longrightarrow \mathbf{Sign}^i$ be defined by

$$G(k) = \Phi^{m_k}(\Sigma_k) \ (k \in |K|)$$
$$G(m) = \Phi^{m_k}(\sigma_m) \ (m: k' \longrightarrow k \in K)$$

Note that m_k is chosen to be canonical weakly terminal in $[m_k]$. By assumption 5, G has a colimit $(\sigma_k: G(k) \longrightarrow \Sigma)_{k \in |K|}$. We show that $([(m_k, \sigma_k)]: (i_k, \Sigma_k) \longrightarrow (i, \Sigma))_{k \in |K|}$ is a colimit of D.

Since equality implies congruence, $([(m_k, \sigma_k)])_{k \in |K|}$ is a cocone of D. Let $([(n_k, \theta_k)]: (i_k, \Sigma_k) \longrightarrow (i', \Sigma'))_{k \in |K|}$ be another cocone. By Lemma 28, $([n_k]: i' \longrightarrow i_k)_{k \in |K|}$ is a cocone for $\bar{D}$. Hence there is a unique $[d]: i' \longrightarrow i$ with $[m_k] \circ [d] = [n_k]$. Since we choose representatives canonically in a way closed under composition, $m_k \circ d = n_k$.

By assumption 4, $(\Phi^d(\sigma_k))_{k \in |K|}$ is a colimit of $\Phi^d \circ G$. Note that the source of $\Phi^d(\sigma_k)$ is $\Phi^d(G(k)) = \Phi^d(\Phi^{m_k}(\Sigma_k)) = \Phi^{n_k}(\Sigma_k)$. By the cocone property of $([(n_k, \theta_k)])_{k \in |K|}$, $(n_k, \theta_k) \equiv (d_m \circ n_{k'}, \theta_{k'} \circ \Phi^{n_{k'}}(\sigma_m))$ for $m: k \longrightarrow k' \in K$. By the assumption of weakly terminal canonical representatives, $n_k = d_m \circ n_{k'}$. By assumption 3, $\theta_k = \theta_{k'} \circ \Phi^{n_{k'}}(\sigma_m)$. This shows that $(\theta_k: \Phi^{n_k}(\Sigma_k) \longrightarrow \Sigma')_{k \in |K|}$ is a cocone for $\Phi^d \circ G$. Hence, there is a unique $\tau: \Phi^d(\Sigma) \longrightarrow \Sigma'$ with $\tau \circ \Phi^d(\sigma_k) = \theta_k$. Then $[(d, \tau)]: (i, \Sigma) \longrightarrow (i', \Sigma')$ is a unique morphism in $\mathbf{Sign}^\#/\equiv$ such that $[(d, \tau)] \circ [(m_k, \sigma_k)] = [(n_k, \theta_k)]$.   □

Proof of Prop. 31: Clearly, emb^i is injective on objects. Faithfulness follows from assumption 3. Preservation of colimits can be seen by inspecting the construction of the proof of Theorem 30: if the indices are all i, then the colimit is just that in $\mathbf{Sign}^i$.   □

Proof of Thm. 32: "Only if", 1: Following Prop. 2 in [5], it is easy to see that for each $i \in |Ind|$, the model functor $\mathbf{Mod}^i$ is the restriction $\mathbf{Mod}^{\#}(i, \_)$ of the model functor of the Grothendieck institution to the subcategory $\mathbf{Sign}^i$ of the Grothendieck signature category $\mathbf{Sign}^{\#}/\equiv$.

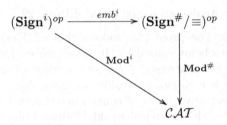

By Prop. 31, the canonical injection $emb^i \colon \mathbf{Sign}^i \longrightarrow \mathbf{Sign}^{\#}$ preserves colimits, hence $\mathbf{Mod}^i$ takes pushouts to (weak) pullbacks because $\mathbf{Mod}^{\#}$ does so.

"Only if", 2: Given a pullback in $Ind/\equiv$

$$
\begin{array}{ccc}
i & \xleftarrow{\;[d_1]\;} & j1 \\
{\scriptstyle[d_2]}\big\uparrow & & \big\uparrow{\scriptstyle[e_1]} \\
j2 & \xleftarrow{\;[e_2]\;} & k
\end{array}
$$

choose d_1, d_2, e_1, e_2 canonically. By the construction of colimits in Theorem 30, for any signature Σ in $\mathbf{Sign}^i$,

$$
\begin{array}{ccc}
(i, \Sigma) & \xrightarrow{\;[(d_1, id)]\;} & (j1, \Phi^{d_1}(\Sigma)) \\
{\scriptstyle[(d_2, id)]}\big\downarrow & & \big\downarrow{\scriptstyle[(e_1, id)]} \\
(j2, \Phi^{d_2}(\Sigma)) & \xrightarrow{\;[(e_2, id)]\;} & (k, \Phi^{e_1}(\Phi^{d_1}(\Sigma))) = (k, \Phi^{e_2}(\Phi^{d_2}(\Sigma)))
\end{array}
$$

is a pushout in $\mathbf{Sign}^{\#}/\equiv$ and is therefore mapped to a (weak) pullback by the model functor. This gives exactly the desired property.

"Only if", 3: Let $d \colon j \longrightarrow i$ by canonical and $\sigma \colon \Sigma_1 \longrightarrow \Sigma_2$ a signature morphism in $\mathbf{Sign}^i$. By the construction of colimits in Theorem 30,

$$
\begin{array}{ccc}
(i, \Sigma_1) & \xrightarrow{\;[(id, \sigma)]\;} & (i, \Sigma_2) \\
{\scriptstyle[(d, id)]}\big\downarrow & & \big\downarrow{\scriptstyle[(d, id)]} \\
(j, \Phi^d(\Sigma_1)) & \xrightarrow{\;[(id, \Phi^d(\sigma))]\;} & (j, \Phi^d(\Sigma_2))
\end{array}
$$

is a pushout in $\mathbf{Sign}^{\#}/\equiv$ and is therefore mapped to a (weak) pullback by the model functor. Again, this gives exactly the desired property.

"If": Consider an arbitrary pushout in $\mathbf{Sign}^{\#}/\equiv$

$$
\begin{array}{ccc}
(i, \Sigma_0) & \xrightarrow{[(d_1,\sigma_1)]} & (j_1, \Sigma_1) \\
\downarrow{\scriptstyle [(d_2,\sigma_2)]} & & \downarrow{\scriptstyle [(e_1,\theta_1)]} \\
(j_2, \Sigma_2) & \xrightarrow{[(e_2,\theta_2)]} & (k, \Sigma')
\end{array}
$$

and assume that representatives are chosen canonically. By the construction of colimits in Theorem 30, the above pushout can be expressed as the following composition of four pushout squares:

$$
\begin{array}{ccccc}
(i, \Sigma_0) & \xrightarrow{[(d_1,id)]} & (j_1, \Phi^{d_1}(\Sigma_0)) & \xrightarrow{[(id,\sigma_1)]} & (j_1, \Sigma_1) \\
\downarrow{\scriptstyle [(d_2,id)]} & & \downarrow{\scriptstyle [(e_1,id)]} & & \downarrow{\scriptstyle [(e_1,id)]} \\
(j_2, \Phi^{d_2}(\Sigma_0)) \xrightarrow{[(e_2,id)]} (k, \Phi^{e_1}(\Phi^{d_1}(\Sigma_0))) & = & (k, \Phi^{e_2}(\Phi^{d_2}(\Sigma_0))) \xrightarrow{[(id,\Phi^{e_1}\sigma_1)]} (k, \Phi^{e_1}(\Sigma_1)) \\
\downarrow{\scriptstyle [(id,\sigma_2)]} & & \downarrow{\scriptstyle [(id,\Phi^{e_2}\sigma_2)]} & & \downarrow{\scriptstyle [(id,\theta_1)]} \\
(j_2, \Sigma_2) & \xrightarrow{[(e_2,id)]} & (k, \Phi^{e_2}(\Sigma_2)) & \xrightarrow{[(id,\theta_2)]} & (k, \Sigma')
\end{array}
$$

Now the model functor of the quotient Grothendieck institution maps the upper left pushout to a (weak) pullback because the 2-indexed coinstitution is (weakly) semi-exact, maps the lower right pushout to a (weak) pullback because the 2-indexed coinstitution is (weakly) locally semi-exact, and maps the remaining two squares to (weak) pullbacks because the comorphisms for canonical index morphisms are (weakly) exact. Since (weak) pullback squares compose, the result follows.   □

Proof of Thm. 36:

Let a diagram $(j_1, \Sigma_1) \xleftarrow{(d_1,\sigma_1)} (i, \Sigma) \xrightarrow{(d_2,\sigma_2)} (j_2, \Sigma_2)$ in $\mathbf{Sign}^{\#}$ be given. Let

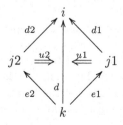

be a weakly amalgamable square of two lax triangles with $\mathcal{I}^k$ quasi-semi-exact. By the latter property, there are θ_1, θ_2 such that

$$
\begin{array}{ccc}
\Phi^d(\Sigma) \xrightarrow{\;\mathcal{I}_\Sigma^{u1}\;} \Phi^{e1}(\Phi^{d1}(\Sigma)) \xrightarrow{\;\Phi^{e1}\sigma_1\;} \Phi^{e1}(\Sigma_1) \\
\downarrow{\scriptstyle \mathcal{I}_\Sigma^{u2}} \\
\Phi^{e2}(\Phi^{d2}(\Sigma)) \qquad\qquad\qquad\qquad\quad \theta_1 \\
\downarrow{\scriptstyle \Phi^{e2}\sigma_2} \\
\Phi^{e2}(\Sigma_2) \xdashrightarrow{\qquad\quad\theta_2\qquad\quad} \Sigma'
\end{array}
$$

is a weakly amalgamable square, which leads to weak amalgamability of the lower right square in

$$
\begin{array}{ccccc}
(i,\Sigma) & \xrightarrow{(d1,id)} & (j1,\Phi^{d1}(\Sigma)) & \xrightarrow{(id,\sigma_1)} & (j1,\Sigma_1) \\
\big\downarrow{\scriptstyle (d2,id)} \searrow{\scriptstyle (d,id)} & & \big\downarrow{\scriptstyle (e1,id)} & & \big\downarrow{\scriptstyle (e1,id)} \\
 & (k,\Phi^d(\Sigma)) \xrightarrow{(id,\mathcal{I}_\Sigma^{u1})} (k,\Phi^{e1}(\Phi^{d1}(\Sigma))) \xrightarrow{(id,\Phi^{e1}(\sigma_1))} (k,\Phi^{e1}(\Sigma_1)) \\
 & \big\downarrow{\scriptstyle (id,\mathcal{I}_\Sigma^{u2})} & & & \\
(j2,\Phi^{d2}(\Sigma)) \xrightarrow{(e2,id)} (k,\Phi^{e2}(\Phi^{d2}(\Sigma))) & & & (id,\theta_1) \\
\big\downarrow{\scriptstyle (id,\sigma_2)} & \big\downarrow{\scriptstyle (id,\Phi^{e2}(\sigma_2))} & & \\
(j2,\Sigma_2) \xrightarrow{(e2,id)} (k,\Phi^{e2}(\Sigma_2)) & \xdashrightarrow{\quad(id,\theta_2)\quad} & (k,\Sigma')
\end{array}
$$

The upper right and lower left squares are weakly amalgamable by weak exactness of $\mathcal{I}^{e1}$ and $\mathcal{I}^{e2}$. The pair of the remaining two squares is jointly weakly amalgamable since it is induced by a weakly amalgamable square of two lax triangles (and note that squares in $\mathbf{Sign}^{\#}/\equiv$ induced by lax triangles in Ind commute by definition of $\equiv$). Since weakly amalgamable squares can be pasted together, we get a weakly amalgamable cocone for the original diagram. $\square$

Proof of Corollary 37: In the sequel, we will use terms like "connected", "maximal", "lower bound" for small categories, when we really mean the pre-order obtained from the category by collapsing the hom-sets into singletons. A maximal element in a pre-order is an element which is equivalent to any element above it.

Let $D\colon J \longrightarrow \mathbf{Sign}^{\#}$ be a connected diagram and let Max be the set of maximal nodes in J. We successively construct new diagrams out of J. Take two nodes in Max that have a common lower bound (if two such nodes do not exist, the diagram is not connected). By Theorem 36, there is a weak amalgamating cocone for the sub-diagram consisting of the two maximal nodes and the lower bound (together with the arrows from the lower bound into the maximal nodes).

Extend the diagram with the cocone. The diagram thus obtained now has a set of maximal nodes whose size is decreased by one. By iterating this construction, we get a diagram with one maximal node. The maximal node then is just the tip of a weakly amalgamating cocone for the original diagram.   □

Proof of Prop. 38:

A model $M_1 \in |\mathbf{Mod}(\sigma_1)^{-1}(\mathbf{Mod}(\sigma_2(\mathbf{Mod}(\Sigma_2, \Psi_2))))|$ is nothing but a pair (M_1, M_2) of models $M_1 \in |\mathbf{Mod}(\Sigma_1)|$, $M_2 \in |\mathbf{Mod}(\Sigma_2, \Psi_2)|$ with common reduct to Σ. This pair can be amalgamated to a model $M_3 \in |\mathbf{Mod}(\Sigma_3)|$. Since $M_3|_{\theta_2} = M_2$, by the satisfaction condition, $M_3 \models_{\Sigma_3} \theta_2(\Psi_2)$. By the assumption, also $M_3 \models_{\Sigma_3} \theta_1(\Psi_1)$. But this means $M_1 = M_3|_{\theta_1} \models_{\Sigma_1} \Psi_1$.   □

Some Varieties of Equational Logic
(Extended Abstract)*

Gordon Plotkin

LFCS, School of Informatics, University of Edinburgh, UK.

The application of ideas from universal algebra to computer science has long been a major theme of Joseph Goguen's research, perhaps even *the* major theme. One strand of this work concerns algebraic datatypes. Recently there has been some interest in what one may call algebraic *computation* types. As we will show, these are also given by equational theories, if one only understands the notion of equational logic in somewhat broader senses than usual.

One moral of our work is that, suitably considered, equational logic is not tied to the usual first-order syntax of terms and equations. Standard equational logic has proved a useful tool in several branches of computer science, see, for example, the RTA conference series [9] and textbooks, such as [1]. Perhaps the possibilities for richer varieties of equational logic discussed here will lead to further applications.

We begin with an explanation of computation types. Starting around 1989, Eugenio Moggi introduced the idea of monadic notions of computation [11,12] with the idea that, for appropriately chosen monads T on, e.g., **Set**, the category of sets, one thinks of $T(X)$ as the type of computations of an element of X. For example, for side-effects one takes the monad $T_S(X) =_{\mathrm{def}} (S \times X)^S$ where S is the set of states. Below, we take $S =_{\mathrm{def}} V^{\mathrm{Loc}}$ where V is a countably infinite set of values such as the natural numbers, and Loc is a finite set of locations. See [2] for a recent exposition of Moggi's ideas, particularly emphasising the connections with functional programming, where the monadic approach has been very influential.

As is well known, equational theories give rise to free algebra monads. For example the free semilattice monad arises from the theory of a binary operation $\cup$ subject to the axioms of associativity, commutativity and idempotence, where the last is the equation $x \cup x = x$. The induced monad $T_N(X)$ is the collection of all non-empty finite subsets of X. In general, the equational theories with operations of finite arity induce exactly those monads which have finite rank, see, e.g., [19].

In denotational semantics one typically employs a category of ordered structures, such as ω-**Cpo**, the category of ω-cpos, which are partial orders with lubs of increasing ω-chains, and with morphisms those monotonic functions preserving the ω-lubs. An ω-**Cpo**-semilattice is a semilattice in ω-**Cpo**, that is an ω-cpo together with a continuous binary function satisfying the semilattice axioms;

---

* This work has been done with the support of EPSRC grant GR/S86372/01 and a Royal Society-Wolfson research merit award.

K. Futatsugi et al. (Eds.): Goguen Festschrift, LNCS 4060, pp. 150–156, 2006.
© Springer-Verlag Berlin Heidelberg 2006

the free ω-**Cpo**-semilattice monad is (a generalisation of) the convex powerdomain monad, originally defined only on a subcategory [5]. There are also lower, or Hoare, and upper, or Smyth, powerdomain monads; these are obtained by adding an additional axiom, viz:

$$x \leq x \cup y$$

for the lower powerdomain, and:

$$x \geq x \cup y$$

for the upper one. Note that these are inequations rather than equations.

This idea was carried further in [15] where similar characterisations were noted for other important monads arising in Moggi's approach, such as those for exceptions, state, input/output, probabilistic nondeterminism and nontermination. One of the main contributions there was an axiomatisation of the state monad employing families of operations of finite or countably infinite arity, as follows. For each location l one assumes given an operation symbol:

$$\text{lookup}_l$$

of arity the countably infinite set V (it is convenient to allow any set to be an arity, not just a cardinal) and for each each location l and value v one assumes given a unary operation symbol:

$$\text{update}_{l,v}$$

The idea is that a term of the form $\text{lookup}_l(\ldots t_v \ldots)$ denotes the computation which looks up the contents of l in the current state and, if this is v, then proceeds according to the computation denoted by the v-th argument, t_v. Similarly a term of the form $\text{update}_{l,v}(t)$ denotes the computation which first updates the contents of the location l to v and then proceeds according to the computation denoted by t.

These ideas have been elaborated into what may be termed the algebraic theory of notions of computation, where the operations and equations are primary and determine the monads. The computational importance of the operations is that it is they that give rise to the effects at hand [16]. Applications include the operational semantics of effects [14], their modular combination [7] and, prospectively, a general logic of effects [17]; see [18] for a survey.

The examples demonstrate that the algebraic theory of computation would benefit from a wider means of expression than is provided by standard equational theories: one also needs to consider parameterization, operations of countable, i.e., denumerable, arity and inequations. As we will see, a unifying rôle is played by Lawvere theories: each such kind of 'equational' theory corresponds to a kind of Lawvere theory, possibly enriched or countable rather than finitary, as standard.

Parameterization This occurs naturally in mathematics, for example in the notion of a vector space over a given field $\mathbb{F}$. There one has the axiom:

$$\lambda(x + y) = \lambda x + \lambda y$$

which involves both field elements and vectors. To treat the notion as an equational theory in the standard sense, one would introduce a unary operation of 'multiplication by λ' for each field element λ and the axiom would be rendered as a family of equations, with one for each field element. We will instead treat the axiom as a single *parametric* equation, with λ a variable ranging over the field and with multiplication by a field element treated as a parametric unary operation on the vector space.

One can go further and allow 'side-conditions,' involving only the parameter variables. For example, in the case of state, treating update as a unary operation parametric over locations and values, one has the following parametric equation:

$$\text{update}_{l,v}(\text{update}_{l',v'}(x)) = \text{update}_{l',v'}(\text{update}_{l,v}(x)) \ (\text{if } l \neq l')$$

which has the side condition that $l \neq l'$; the equation states that the order in which one updates distinct locations does not matter.

Such parametric equational theories abbreviate ordinary equational theories, but, by allowing a schema to be replaced by a parametric equation with side conditions, may enable finitary axiomatisation and consequent direct computer implementation. Formally one assumes given an interpretation $\mathfrak{A}$ of a many-sorted first-order signature, the *parameter* signature; for the equational part one further assumes given a *parametric* signature where the operation symbols are assigned a given list of sorts from the parameter signature as well as the usual natural number. There is then a natural notion of parametric term where the parameters are given by standard first-order terms over the parameter signature and so of parametric equation:

$$t = u \ (\varphi)$$

with side condition φ written in first-order logic with equality over the parameter signature. A collection of such equations abbreviate, as indicated above, a standard equational theory over a derived signature.

There is a natural system for deriving these parametric equations from a given collection Th of first-order formulas with equality over the parameter signature, together with another given collection Eqn of parametric equations; the system includes first-order logic with equality for the parameter spaces and equational logic for the parametric equations. One can define whether a parametric equation is a semantic consequence of Th and Eqn relative to the fixed interpretation $\mathfrak{A}$, but, unfortunately, taking Th to be the theory of $\mathfrak{A}$, completeness need not hold. It may, however, hold in particular cases: one such is that of vector spaces mentioned above taking the standard 'ring signature' for the many-sorted first-order signature. On the other hand, fixing Th and Eqn, one can show completeness, if by validity one means with respect to *all* models of Th.

Infinitary operations One can treat operations of countable arity using the evident natural notions of countable equational theory and countable Lawvere theory; the induced monads are those of countable rank. Here is an example of a schema of infinitary equations involving the operation of looking up the contents of a location:

$$\text{lookup}_l(\ldots \text{update}_{l,v}(x)\ldots) = x$$

The equation states that if a location is looked up and then updated with the value found, then that is equivalent to doing nothing.

However it would again be preferable to have a finitary syntax, now for operations of countably infinite arity. To that end, we employ binding on variables of the arity sort, here val (standing for V); the term-forming construction for lookup is then:

$$\text{lookup}_a(v:\text{val}.t)$$

where a is a parameter term of sort loc (standing for Loc) and t is a parametric term given the environment $v:\text{val}$. With this, the above infinitary schema can be written as the following finitary 'equation':

$$\text{lookup}_l(v:\text{val}.\text{update}_{l,v}(x)) = x$$

We consider next the following infinitary equation scheme:

$$\text{lookup}_l(\ldots \text{update}_{l',v'}(x_v)\ldots) = \text{update}_{l',v'}(\text{lookup}_l(\ldots x_v \ldots)) \ (\text{if } l \neq l')$$

which states that the operations of looking up one location and updating another commute. Notice that it employs a family x_v of variables. If we introduce the notion of a parametric variable (ranging over a suitable collection of functions) this infinitary equation scheme can also be rendered in a finitary fashion:

$$\text{lookup}_l(v:\text{val}.\text{update}_{l',v'}(x_v)) = \text{update}_{l',v'}(\text{lookup}_l(v:\text{val}.x_v)) \ (\text{if } l \neq l')$$

These two ideas of binding and parametric variables suffice to write down all the parameterized, possibly infinitary, equation schemes for global state given in [15] finitarily.

In the general formalism, we again begin with an interpretation $\mathfrak{A}$ of a parameter signature, as above, except that we assume also given a subcollection of the sorts, called the *arity* sorts. In the parametric signature an operation symbol has m parameter arguments of given parameter sorts, and n argument positions, with the ith being abstracted on k_i arity sorts. A collection of parametric equations abbreviates a countable equational theory, provided that the arity sorts are interpreted by countable sets.

One can then give a logic following the previous lines. An immediate question is whether the logic is complete for global state, where for the many-sorted first-order signature one would take the two sorts, loc and val, and constants for all the elements of Loc, with the evident interpretation using V and Loc. We would also like to know whether we have completeness relative to all interpretations of a given theory, as we do in the simpler case, considered above, of finitary operations. Positive answers to such questions would demonstrate that valid uniform infinitary equations have uniform proofs.

Inequations These are a natural generalisation of equations and there is an evident notion of inequational, or ordered, equational logic over operations of finite arity, which has a straightforward completeness theorem using posets rather than sets [3]. The resulting ordered equational theories correspond to ordered Lawvere theories, in the sense of [23,3]. These are not the same as the **Pos**-enriched Lawvere theories of [19], as the latter allow all finite posets as arities of operations, not just the discrete ones. However they are the same as the **Pos**-enriched Lawvere theories of [10], equivalently the **Pos**-enriched discrete Lawvere theories of [20]. There is a natural generalisation to countable inequational logic, and the inequational theories of this logic correspond to the discrete countable **Pos**-theories (the countable case is the main one considered in [20]). In general discrete **V**-theories of a given rank freely induce **V**-theories of that rank, in the sense of [19], and the latter induce the **V**-monads of the same rank; not all such monads arise from discrete theories.

Parameterization, now over given posets, is again an expressive convenience, and there are inequational versions of the two equational deductive systems considered above: one for parametric inequations and the other with finitary syntax for infinitary operations. For the parameter interpretation $\mathfrak{A}$ it is natural to work with enriched first-order structures, which we take to mean here that sorts are interpreted by posets, operations by monotonic functions and relations by subsets; one then naturally works with first-order logic with inequations $a \leq b$, rather than equations, to express parameter conditions. One evidently requires arity sorts to be interpreted by countable discrete partial orders to obtain discrete countable **Pos**-theories from a collection of parametric inequations.

Turning to ω-**Cpo**-enrichment, one can consider discrete finitary or countable ω-**Cpo**-theories. Here parameterization is more than an expressive convenience: it enables one to implicitly write down equations involving sups of increasing chains. One can still work with simple inequations, but rather than finitary or countably infinitary operation symbols, one takes families of such, parameterized over a collection of parameter ω-cpos. They are to be interpreted by functions which are continuous over the parameter ω-cpos as well as the algebra ω-cpo. A natural example is provided by d-cones, which arise when considering powerdomains for mixed ordinary and probabilistic nondeterminism [22]. These are the ω-**Cpo**-semimodules over the semiring $\overline{\mathbb{R}}_+$, which latter is the ω-cpo of the nonnegative reals extended with a point at infinity, and endowed with the natural semiring structure [13].

Collections of such inequations induce the discrete finitary or countable ω-**Cpo**-theories, according to the arities of the operation symbols allowed. However there is a question as to what is the appropriate inequational logic. It may be best to introduce an explicit infinitary syntax for sups of increasing ω-sequences, but then sup-terms would only be well-formed if one could prove the sequence was increasing, and that would mean a mutual recursion between the definitions of proofs and well-formed terms. It remains to investigate such a system.

The next question is to what extent one can achieve a useful finitary system. One can clearly investigate analogues of the methods used above to handle parameterization and operations of countably infinite arity. But it is far from clear what to do about the sup-terms. Perhaps one can restrict to considering only least fixed-points and work with a combination of the above ideas and the μ-calculus, for which, and associated logical and categorical results, see [4,8,21].

Whatever the difficulties are with finding the right logic, it is at least the case that the combination of parameterization, binding constructions and inequations, interpreted over ω-**Cpo**, is enough to express all the theories of computation types so far considered over that category. We should admit, however, that this is not quite enough to account for all the computation types so far considered. One difficult case is that of the continuations monad. However one can argue that there the types should not be treated as algebraic since the natural operations are not even of the right type to be algebraic operations, and, further, the monad does not have a rank [6].

A more interesting case is that of *local* state, as opposed to the above *global* state, where one can declare new locations. This was treated using a monad over a presheaf category in [15]. The monad was specified by equations, but they involved a mixture of linear and ordinary operations, with the linear structure coming from the Day tensor on the presheaf category. This example feels as if it should be treatable within an algebraic framework, but we do not see the proper notions of Lawvere theory or equational theory. Finally there is also the possibility of employing other semantic categories in place of ω-**Cpo** for the algebraic computational types; we content ourselves here with the remark that for reasonable such categories, one would expect the relevant free algebras still to exist.

References

1. F. Baader & T. Nipkow, *Term Rewriting and All That*, Cambridge University Press, 1998.
2. N. Benton, J. Hughes & E. Moggi, Monads and Effects, *Proc. APPSEM 2000*, LNCS, **2395**, 42–122, Springer-Verlag, 2002.
3. S. L. Bloom, Varieties of Ordered Algebras, *J. Comput. Syst. Sci.*, **13**(2), 200–212, 1976.
4. S. L. Bloom & Z. Ésik, *Iteration Theories: The Equational Logic of Iterative Processes*, EATCS Monographs on Theoretical Computer Science, Springer-Verlag, 1993.
5. M. Hennessy & G. D. Plotkin, Full Abstraction for a Simple Parallel Programming Language, *Proc. 8th. MFCS* (ed. J. Becvár), LNCS, **74**, 108–120, Springer-Verlag, 1979.
6. J. M. E. Hyland, P. B. Levy, G. D. Plotkin & A. J. Power, Combining Continuations with Other Effects, *Proc. 4th. ACM SIGPLAN Continuations Workshop*, 45–47, University of Birmingham Report CSR-04-1, 2004.
7. J. M. E. Hyland, G. D. Plotkin & A. J. Power, Combining Effects: Sum and Tensor, *Theoretical Computer Science*, to appear, 2006.

8. A. J. C. Hurkens, M. McArthur, Y. N. Moschovakis, L. S. Moss & G. T. Whitney, The Logic Of Recursive Equations, *JSL*, **63**(2), 451–478, 1998.

9. J-P. Jouannaud, Twenty Years Later, *Proc. 16th. RTA* (ed. J. Giesl), LNCS, **3467**, 368–375, Springer-Verlag, 2005.

10. J. Meseguer, Varieties of Chain-Complete Algebras, *JPAA*, **19**, 347–383, 1980.

11. E. Moggi, Computational Lambda-Calculus and Monads, *Proc. 4th. LICS*, 14–23, IEEE Press, 1989.

12. E. Moggi, Notions of Computation and Monads, *Inf. & Comp.*, **93**(1), 55–92, 1991.

13. G. D. Plotkin, A Domain-Theoretic Banach-Alaoglu Theorem, Festschrift for Klaus Keimel, *MSCS*, **16**, 1–13, 2006.

14. G. D. Plotkin & A. J. Power, Adequacy for Algebraic Effects, *Proc. 4th. FOSSACS* (eds. F. Honsell & M. Miculan), LNCS, **2030**, 1–24, Springer-Verlag, 2001.

15. G. D. Plotkin & A. J. Power, Notions of Computation Determine Monads, *Proc. 5th. FOSSACS*, LNCS, **2303**, 342–356, Springer-Verlag, 2002.

16. G. D. Plotkin & A. J. Power, Algebraic Operations and Generic Effects, *Applied Categorical Structures*, **11**(1), 69–94, 2003.

17. G. D. Plotkin & A. J. Power, Logic for Computational Effects: Work in Progress, *Proc. 6th. IWFM* (eds. J. M. Morris, B. Aziz & F. Oehl), Electronic workshops in computing, BCS, 2003.

18. G. D. Plotkin & A. J. Power, Computational Effects and Operations: an Overview, *Proc. Domains VI*, ENTCS, **73**, 149–163, Elsevier, 2004.

19. A. J. Power, Enriched Lawvere Theories, *Theory and Applications of Categories*, **6**, 83–93, 2000.

20. A. J. Power, Discrete Lawvere Theories, *Proc. 1st. CALCO* (eds. J. L. Fiadeiro, N. Harman, M. Roggenbach & J. J. M. M. Rutten), LNCS, **3629**, 348–363, Springer-Verlag, 2005.

21. A. Simpson & G. Plotkin, Complete Axioms for Categorical Fixed-point Operators, *Proc. 15th. LICS*, 30–41, IEEE Press, 2000.

22. R. Tix, K. Keimel and G. Plotkin, *Semantic Domains for Combining Probability and Non-Determinism*, ENTCS, Vol. 129, pp. 1–104, Amsterdam: Elsevier, 2005.

23. E. G. Wagner, J. B. Wright, J. A. Goguen & J. W. Thatcher, Some Fundamentals of Order-Algebraic Semantics, *Proc. 5th. MFCS* (ed. A. W. Mazurkiewicz), LNCS, **45**, 153–168, Springer-Verlag, 1976.

Complete Categorical Deduction
for Satisfaction as Injectivity

Grigore Roşu

Department of Computer Science
University of Illinois at Urbana-Champaign, USA

Abstract. Birkhoff (quasi-)variety categorical axiomatizability results
have fascinated many scientists by their elegance, simplicity and generality. The key factor leading to their generality is that equations, conditional or not, can be regarded as special morphisms or arrows in a
special category, where their satisfaction becomes injectivity, a simple
and abstract categorical concept. A natural and challenging next step is
to investigate complete deduction within the same general and elegant
framework. We present a categorical deduction system for equations as
arrows and show that, under appropriate finiteness requirements, it is
complete for satisfaction as injectivity. A straightforward instantiation
of our results yields complete deduction for several equational logics, in
which conditional equations can be derived as well at no additional cost,
as opposed to the typical method using the theorems of constants and
of deduction. At our knowledge, this is a new result in equational logics.

1   Introduction

Equational logic is an important paradigm in computer science. It admits complete deduction and is efficiently mechanizable by rewriting: CafeOBJ [15],
Maude [12] and Elan [9] are equational specification and verification systems
in the OBJ [21] family that can perform millions and tens of millions of rewrites
per second on standard PC platforms. It is expressive: Bergstra and Tucker [5,6]
showed that any computable data type can be characterized by means of a finite
equational specification, and Goguen and Malcolm [17], Wand [41], Broy, Wirsing and Pepper [11], and many others showed that equational logic is essentially
strong enough to easily describe virtually all traditional programming language
features. It has simple semantic models: its models are algebras, straightforward
and intuitive structures. We suggest Goguen and Malcolm [19] and Padawitz
and Wirsing [31] as good references for many-sorted equational logic, its completeness, as well as applications to computer science.

There are many variants and generalizations of equational logics, ranging
from unsorted [7] to many-sorted [19,31], to partial [32], to order-sorted [20,40],
to membership [27,10], to local [13], to hidden [18,34] equational logics, and
so on. A major challenge is to develop a uniform common framework for all
these variants, that allows one to formulate and prove at least some of their
important properties, such as Birkhoff axiomatizability, complete deduction and

K. Futatsugi et al. (Eds.): Goguen Festschrift, LNCS 4060, pp. 157–172, 2006.
© Springer-Verlag Berlin Heidelberg 2006

Craig interpolation. Whether this is possible or not is open, but what is certain is the existence of elegant categorical equational variants by Banaschewski and Herrlich [4], Andréka, Németi and Sain [2,3,30], Adámek and Rosický [1] and many others, in which equations are viewed as epimorphisms and their satisfaction as injectivity, and that these allow very general treatments of variety and quasi-variety results. We also adopt this categorical view in the present paper.

To emphasize the simplicity and generality of this approach, we mention that everything happens within only one category, denoted by $\mathcal{C}$ in this paper, which has a factorization system $\langle \mathcal{E}, \mathcal{M} \rangle$. The objects of $\mathcal{C}$ are viewed as models and the morphisms in $\mathcal{E}$, which for simplicity will be called *equations*, are viewed as sentences[1]. In order to define our sound (w.r.t. injectivity) four rule inference system for arrows in $\mathcal{E}$, $\mathcal{C}$ is required to additionally have pushouts and enough $\mathcal{E}$-projectives. To show it complete, $\mathcal{C}$ also needs to have directed colimits and to be $\mathcal{E}$-co-well-powered, and some appropriate notions of finiteness for arrows in $\mathcal{E}$ need to be introduced. A related variant by Diaconescu [14], called category-based equational logic, considers equations as pairs of arrows, one for each term, and then gives a set of deduction rules that resembles that of equational logics.

The present paper is part of our efforts to develop a unifying, categorical framework for axiomatizability, deduction and interpolation for equational and coequational logics. In [37] it is shown that the difference w.r.t. injectivity between epimorphisms of free/projective sources and epimorphisms of any sources is exactly as the difference w.r.t. usual satisfaction between unconditional and conditional equations, that is, the first define varieties while the second define quasi-varieties. In [33,36], equational axiomatizability for hidden equational logic and coalgebra is investigated, and in [38] a categorical generalization of equational interpolation is given. The closest to the present paper is [35], where we also present a complete four rule inference system for equations as epics, but limited to unconditional axioms. In the present paper, due to crucial developments of finiteness concepts and results, especially Proposition 3, we non-trivially extend the results in [35] by eliminating the admittedly frustrating limitation to unconditional axioms, putting thus an end to our quest for complete deduction when satisfaction is injectivity. We show that a four rule inference system for epics is complete provided that all the axioms have finite conditions and the equation to be derived is finite. An interesting characteristic of our deduction system is that it is also complete for conditional equations, and that those can be derived the same way as the unconditional ones. We are not aware of any similar result for any equational paradigm in the literature until [35], where a version of it, restricted to unconditional axioms, was presented.

Section 2 recalls some categorical concepts and introduces our notational conventions. Section 3 revises factorization systems. Section 4 shows how equations, both unconditional and conditional, are equivalent to surjective morphisms and their satisfaction to injectivity; clarifying examples are presented. Section

---

[1] If one thinks that equations should be *regular epimorphisms* then one can read so instead of "epimorphism." Our results hold for any epimorphisms, so a restriction to regular epimorphisms would be technically artificial and less general.

5 introduces our four rule inference system for arrows and shows how it works on various examples. Finiteness concepts and results are explored in Section 6, which are necessary in Section 7 to show the completeness result. The last section concludes the paper and presents challenges for further research.

2 Preliminaries

The reader is assumed familiar with basic concepts of category theory [26,23] and equational logics [7,8,31,19]. In this section we introduce our notations and conventions, and recall some less frequent notions. Given a category $\mathcal{C}$, let $|\mathcal{C}|$ denote its class of objects; we use diagrammatic order for composition of morphisms, i.e., if $f\colon A \to B$ and $g\colon B \to C$ then $f;g\colon A \to C$. If the source or the target of a morphism is not important in a certain context, then we replace it by a bullet to avoid inventing new letters; for example, $f\colon A \to \bullet$. In situations where there are more bullet objects, they may be different. If $f\colon A \to B$ and $g\colon A \to C$ have a pushout then we let $f^g\colon C \to \bullet$ and $g^f\colon B \to \bullet$ denote the opposite arrows, up to isomorphism, of f and g in that pushout.

Given a class of morphisms $\mathcal{E}$ in a category $\mathcal{C}$, $P \in |\mathcal{C}|$ is called $\mathcal{E}$-**projective** iff for any $e\colon \bullet \to X$ in $\mathcal{E}$ and any $h\colon P \to X$, there is a g s.t. $g; e = h$. $\mathcal{C}$ **has enough $\mathcal{E}$-projectives** iff for each object $X \in |\mathcal{C}|$ there is some $\mathcal{E}$-projective object P_X and a morphism $e_X\colon P_X \to X$ in $\mathcal{E}$. It is known that any set is $\mathcal{E}$-projective where $\mathcal{E}$ consists of all the surjective functions, that free algebras are $\mathcal{E}$-projective where $\mathcal{E}$ is the class of surjective morphisms, and that the category of algebras has enough $\mathcal{E}$-projectives (for an algebra X, one can take P_X to be the free algebra over the elements in X seen as variables). Dually, I is $\mathcal{E}$-**injective** iff for any $e\colon X \to \bullet$ and any $h\colon X \to I$, there is a g s.t. $e; g = h$. $\mathcal{C}$ is called $\mathcal{E}$-**co-well-powered** iff for any $X \in |\mathcal{C}|$ and any *class* $\mathcal{D}$ of morphisms in $\mathcal{E}$ of source X, there is a *set* $\mathcal{D}' \subseteq \mathcal{D}$ such that each morphism in $\mathcal{D}$ is isomorphic to some morphism in $\mathcal{D}'$; we often call $\mathcal{D}'$ a **representative set** of $\mathcal{D}$.

If X is an object in a category $\mathcal{E}$, then $X \downarrow \mathcal{E}$ is the comma category containing morphisms $e, e', ...\colon X \to \bullet$ in $\mathcal{E}$ as objects and morphisms $h \in \mathcal{E}$ such that $e; h = e'$ as morphisms. Notice that if $\mathcal{E}$ contains only epimorphisms then there is at most one morphism between any two objects in $X \downarrow \mathcal{E}$. The intuition in our framework for the the objects $e, e', ...\colon X \to \bullet$ in the comma category $X \downarrow \mathcal{E}$ will be that of equations over the same source (variables, condition).

3 Factorization Systems

The idea to form subobjects by factoring each morphism f as $e; m$, where e is an epic and m is a mono, seems to go back to Grothendieck [22] in 1957, and was intensively used by Isbell [24], Lambek [25], Mitchell [28], and many others. Lambek was probably the first to explicitly state a diagonal-fill-in property in 1966 [25], called also "orthogonality" by Freyd and Kelly in [16]. One of the first formal definition of a factorization system that we are aware of was given by Herrlich and Strecker [23] in 1973, under the name *factorizable category*, and a comprehensive study of factorization systems, containing different equivalent definitions, was done by Németi [29] in 1982.

Definition 1. *A* **factorization system** *of a category* C *is a pair* $\langle \mathcal{E}, \mathcal{M} \rangle$, *s.t.:*

- $\mathcal{E}$ *and* $\mathcal{M}$ *are subcategories of epics and monics, respectively, in* C,
- *all isomorphisms in* C *are both in* $\mathcal{E}$ *and* $\mathcal{M}$, *and*
- *each morphism* f *in* C *can be factored as* $e; m$ *with* $e \in \mathcal{E}$ *and* $m \in \mathcal{M}$ *"uniquely up to isomorphism", that is, if* $f = e'; m'$ *is another factorization of* f *then there is a unique isomorphism* α *such that* $e; \alpha = e'$ *and* $\alpha; m' = m$.

The following are important properties of factorization systems:

Proposition 1. *Let* $\langle \mathcal{E}, \mathcal{M} \rangle$ *be a factorization system for* C, *and let* $e \in \mathcal{E}$ *and* $f \in C$ *be morphisms having the same source. Then*

1. **Diagonal-fill-in.** *If* $f; m = e; g$ *then there is a "unique up to isomorphism"* $h \in C$ *such that* $e; h = f$ *and* $h; m = g$, *and*
2. **Pushout.** *If the pushout of* e *and* f *exists then* $e^f \in \mathcal{E}$.

For the rest of the paper, suppose that $\langle \mathcal{E}, \mathcal{M} \rangle$ is a factorization system for a category C. The proof of the following proposition, which intuitively shows conditions under which "equations can be put together," can be found in [35]:

Proposition 2. *If* $X \in |C|$ *and* C *has colimits then* $X \downarrow \mathcal{E}$ *has colimits.*

When C is $\mathcal{E}$-co-well-powered, colimits in $X \downarrow \mathcal{E}$ also exist for large diagrams $\mathcal{D}$ (whose nodes form a class): one takes the colimit of a representative set of $\mathcal{D}$.

Definition 2. *We let* $(\{\gamma_i\}_{i \in I}, e_{\mathcal{D}} \colon X \to X_{\mathcal{D}})$ *denote the colimit of* $\mathcal{D} \subseteq X \downarrow \mathcal{E}$, *and use* $e_1 \cup e_2$ *instead of* $e_{\mathcal{D}}$ *if* $\mathcal{D}$ *consists of only* $e_1 \colon X \to \bullet$ *and* $e_2 \colon X \to \bullet$.

4   Equations as Epimorphisms

As advocated by Banaschewski and Herrlich [4], by Andréka, Németi and Sain [2,30], and by many others including the author [37,35], equations can be regarded as epimorphisms and their satisfaction as injectivity. Readers with different background bases can find/have different explanations or intuitions for these relationships. We next informally give our version which seems closest in spirit to the subsequent results, together with some examples inspired from group theory.

An unconditional equation e over variables $x, y, \ldots$ is nothing but a binary relation R_e (containing only one pair) on the term algebra $T(x, y, \ldots)$. This relation generates a congruence C_e, which further generates a surjective morphism of free source $s_e \colon T(x, y, z, \ldots) \to T(x, y, \ldots)/C_e$. An algebra satisfies e iff it is $\{s_e\}$-injective. Conversely, the kernel K_s of a surjective morphism of free source $s \colon T(x, y, \ldots) \to \bullet$ is nothing but a set of equations quantified by $x, y, \ldots$, and an algebra is $\{s\}$-injective iff satisfies K_s. It is often more convenient to work with sets of equations rather than with individual equations, as perhaps best illustrated by Craig interpolation results that do not hold for individual equations but do hold for sets of equations [39,38]. In this paper, by equation we also mean a set of individual equations over the same variables, so there is a one-to-one correspondence between equations and epimorphisms of free sources.

Example 1. Let Σ be the unsorted signature consisting of a constant 1, a unary operation $\overline{(\_)}$ and a binary operation $\_\_$, and let us consider the equations $(\forall x)\ x1 = x$, $(\forall x)\ x\overline{x} = 1$, and $(\forall x, y, z)\ x(yz) = (xy)z$. In our notation, these equations correspond to the following three epimorphisms:

$\quad$ axiom$_1$: $T_\Sigma(x) \to \bullet \qquad$ generated by $(x1, x)$,
$\quad$ axiom$_2$: $T_\Sigma(x) \to \bullet \qquad$ generated by $(x\overline{x}, 1)$,
$\quad$ axiom$_3$: $T_\Sigma(x, y, z) \to \bullet$ generated by $(x(yz), (xy)z)$,

where $T_\Sigma(x)$ and $T_\Sigma(x, y, z)$ are the Σ-term algebras over the variable x and over the variables x, y, z, respectively, and an epimorphism $e \colon T_\Sigma(x, y, ...) \to \bullet$ is *generated* by a binary relation R of terms iff e is the natural surjection $T_\Sigma(x, y, ...) \to T_\Sigma(x, y, ...)/_R$ that maps each term to its congruence class. Notice that we could have also merged the first two epics into the epic axiom$_1$ $\cup$ axiom$_2$: $T_\Sigma(x) \to T_\Sigma(x)/_{\{(x1,x),(x\overline{x},1)\}}$. It is known that the algebras satisfying the three equations above are exactly the groups, i.e., the left unit and left inverse equations can be proved from the above. We will focus on these proofs in the next section.

What is less known is that conditional equations can also be viewed as epics and their satisfaction as injectivity. This is explained in detail in [35]. Intuitively, one first factors the term algebra by the condition and then takes the epic generated by the equivalence classes of the conclusion.

Example 2. The conditional equation $(\forall x)\ x = 1$ **if** $xx = 1$ on groups (see Example 1), is in our notation equivalent to the epic

$\quad$ axiom$_4$: $T_\Sigma(x)/_{(xx,1)} \to \bullet$ generated by $(x, 1)$,

where, for simplicity, we have identified equivalence classes with some representatives: $(x, 1)$ should normally be $(\hat{x}, \hat{1})$. A group satisfies this new axiom iff it has no proper square roots of unity iff it is $\{\text{axiom}_4\}$-injective.

In theoretical efforts, it is often technically more easily to abstract freeness by projectivity. We have shown in [37] that there is essentially no difference between projective and free sources of epimorphisms with respect to axiomatizability, and that free objects are usually projective in almost any category. The results in this paper also hold for both situations, but we only discuss projective sources.

For the rest of the paper we assume that $\mathcal{C}$, besides its factorization system $\langle \mathcal{E}, \mathcal{M} \rangle$, also has enough $\mathcal{E}$-projectives. Moreover, for each object $X \in |\mathcal{C}|$ we fix an arbitrary $\mathcal{E}$-projective object P_X and an arbitrary morphism $e_X \colon P_X \to X$ in $\mathcal{E}$. If $\mathcal{C}$ is the category of algebras over some signature and X is the quotient of a free algebra by some congruence, then P_X is usually taken to be the free algebra and e_X to map each term to its congruence class.

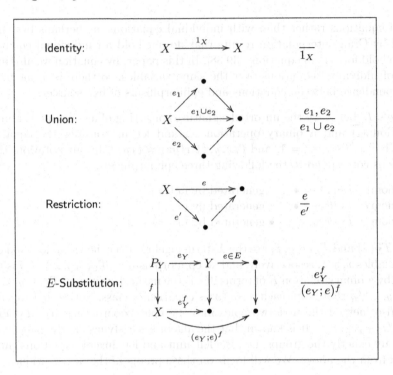

Fig. 1. Categorical inference rules.

Definition 3. *We call the morphisms in $\mathcal{E}$* **equations**. *If $e\colon X \to \bullet$ is an equation then $e_X\colon P_X \to X$ is called its* **condition**. *If $X = P_X$ then e is called* **unconditional**. *An object A in $\mathcal{C}$* **satisfies** *the equation $e\colon X \to \bullet$, written $A \models e$, if and only if A is $\{e\}$-injective. $\models$ trivially extends to sets of equations.*

5 Sound Deduction

In this section we give four inference rules for equations as arrows as defined in the previous section, show that they are sound and give some examples. The first three rules also appeared in [35]. The fourth rule appeared in an over-simplified form in [35] because conditional axioms were not allowed there.

In this section we assume that $\mathcal{C}$, besides a factorization system $\langle \mathcal{E}, \mathcal{M} \rangle$ and enough $\mathcal{E}$-projectives, also has pushouts.

Definition 4. *Given a set of equations E, let $\vdash$ denote the* **derivation relation** *generated by the rules in Fig. 1, where E-Substitution is a class of rules, one for each $f\colon P_Y \to X$. If the source of e is X and $E \vdash e$ then e is called an* **X-derivation** *of E. Let $\mathcal{D}_X(E)$ denote the full subcategory of $X \downarrow \mathcal{E}$ of X-derivations of E.*

Note that $E \vdash e$ for each $e \in E$ since one can take $f = e_Y$ in E-Substitution, and also that $\mathcal{D}_X(E)$ can be a class in general because E can be a class. Since equations in E were allowed to have only $\mathcal{E}$-projective sources in [35], E-Substitution was a simple pushout there, for which reason it was called E-Pushout.

Theorem 1. <u>Soundness.</u> $E \vdash e$ *implies* $E \models e$.

Proof. The soundness of the first three rules is easy; we only show the soundness of E-Substitution. Let us assume that $E \models e_Y^f$, let A be any object such that $A \models E$, and let $g \colon X \to A$ be any morphism.

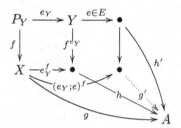

Since $A \models e_Y^f$, there is a morphism h like in the diagram above, such that $e_Y^f ; h = g$. Further, since $A \models e$ there is a morphism h' such that $e; h' = f^{e_Y}; h$. Hence $f; g = (e_Y; e); h'$, so by the pushout property there is a morphism g' such that $(e_Y; e)^f; g' = g$. Therefore, $A \models (e_Y; e)^f$, i.e., $E \models (e_Y; e)^f$.

Example 3. We show that the three arrows $E = \{\text{axiom}_1, \text{axiom}_2, \text{axiom}_3\}$ defined in Example 1 define indeed the groups, that is, that the remaining arrows $g_1 \colon T_\Sigma(x) \to \bullet$ generated by $(1x, x)$, and $g_2 \colon T_\Sigma(x) \to \bullet$ generated by $(\overline{x}x, 1)$, stating the left-unit and left-inverse axioms of groups, can be derived from E. The table in Figure 2 shows a possible proof, where the first column shows or gives names to newly inferred arrows, the second shows a set of generators of the kernel of the new arrow (a dash "-" means that the set of generators is obvious, so we do not write it to save space), and the third column shows the inference rule used to derive the new arrow (identity is omitted).

To derive e_1, for example, one applies the substitution rule for $e = \text{axiom}_1$ where $f \colon T_\Sigma(x) \to T_\Sigma(x)$ takes x to $\overline{x}x$, using tacitly the identity rule on $1_{T_\Sigma(x)}$:

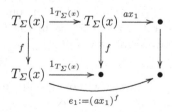

We showed in 29 inference steps that the three axioms define groups. The careful reader may have noticed that we have used unnecessarily many Restriction steps. Indeed, if one does all the substitutions first, followed by all the unions, and then by reductions, then one can prove the above in only 19 steps.

	Generated by $(\_,\_)$	Inference rule
e_1	$(\overline{x}x)1,\ \overline{x}x$	Substitution : axiom$_1$
e_2	$(\overline{x}x)\overline{\overline{x}x},\ 1$	Substitution : axiom$_2$
e_3	$(\overline{x}x)((\overline{x}x)\overline{\overline{x}x}),\ ((\overline{x}x)(\overline{x}x))\overline{\overline{x}x}$	Substitution : axiom$_3$
e_4	–	Union : $e_1 \cup e_2$
e_5	–	Union : $e_3 \cup e_4$
e_6	$((\overline{x}x)(\overline{x}x))\overline{\overline{x}x},\ \overline{x}x$	Restriction : e_5
e_7	$\overline{x}1,\ \overline{x}$	Substitution : axiom$_1$
e_8	$\overline{x}(1x),\ (\overline{x}1)x$	Substitution : axiom$_3$
e_9	–	Union : $e_7 \cup e_8$
e_{10}	$\overline{x}(1x),\ \overline{x}x$	Restriction : e_9
e_{11}	$x\overline{x},\ 1$	Substitution : axiom$_1$
e_{12}	–	Union : $e_{10} \cup e_{11}$
e_{13}	$\overline{x}((x\overline{x})x),\ \overline{x}x$	Restriction : e_{12}
e_{14}	$x(\overline{x}x),\ (x\overline{x})x$	Substitution : axiom$_3$
e_{15}	–	Union : $e_{13} \cup e_{14}$
e_{16}	$\overline{x}(x(\overline{x}x)),\ \overline{x}x$	Restriction : e_{15}
e_{17}	$\overline{x}(x(\overline{x}x)),\ (\overline{x}x)(\overline{x}x)$	Substitution : axiom$_3$
e_{18}	–	Union : $e_{16} \cup e_{17}$
e_{19}	$(\overline{x}x)(\overline{x}x),\ \overline{x}x$	Restriction : e_{18}
e_{20}	–	Union : $e_2 \cup e_{19}$
e_{21}	$((\overline{x}x)(\overline{x}x)\overline{\overline{x}x},\ 1$	Restriction : e_{20}
e_{22}	–	Union : $e_6 \cup e_{21}$
g_2	$\overline{x}x,\ 1$	Restriction : e_{22}
e_{23}	–	Union : $e_{14} \cup g_2$
e_{24}	$x1,\ (x\overline{x})x$	Restriction : e_{23}
e_{25}	$x1,\ x$	Substitution : axiom$_1$
e_{26}	–	Union : $e_{24} \cup e_{25}$
e_{27}	–	Union : $e_{11} \cup e_{26}$
g_1	$1x,\ x$	Restriction : e_{27}

Fig. 2. Deriving the remaining group properties.

As mentioned before, a benefit of our deduction system is that one can also directly infer conditional equations.

Example 4. In the same context as in Example 3, one can infer the conditional equation $(\forall x)\ x = \overline{x}$ if $xx = 1$, which in our notation is the arrow $g_3: T_\Sigma(x)/_{(xx,1)} \to \bullet$ generated by $(x,\overline{x})$ as in the table in Figure 3. Note that e_{30} was possible since its source was $T_\Sigma(x)/_{(xx,1)}$.

Example 5. In the context of groups without square roots of unity in Example 2, where $E = \{\text{axiom}_1, \text{axiom}_2, \text{axiom}_3, \text{axiom}_4\}$, we can derive the conditional equation $(\forall x,y)\ x = y$ if $x\overline{y} = y\overline{x}$, which in our notation is the morphism $g_4: T_\Sigma(x,y)/_{(x\overline{y},y\overline{x})} \to \bullet$ generated by (x,y). To apply substitution on axiom$_4$, with $f: T_\Sigma(x) \to T_\Sigma(x,y)/_{(x\overline{y},y\overline{x})}$ taking x to $x\overline{y}$, where $Y = T_\Sigma(x)/_{(xx,1)}$ and $X = T_\Sigma(x,y)/_{(x\overline{y},y\overline{x})}$ and where e_Y^f is generated by $((x\overline{y})(x\overline{y}),1)$ and

$T_\Sigma(x)/_{(xx,1)} \to \bullet$	Generated by $(\_,\_)$	Inference rule
$e_1,..,e_{22},g_2,..,g_1$	same as before	same as before
e_{28}	$\overline{x}(xx),(\overline{x}x)x$	Substitution : axiom$_3$
e_{29}	$-$	Union : $g_2 \cup e_{28}$
e_{30}	$1x,\overline{x}1$	Restriction : e_{29}
e_{31}	$-$	Union : $e_7 \cup e_{30}$
e_{32}	$-$	Union : $g_1 \cup e_{31}$
g_3	$x,\overline{x}$	Restriction : e_{32}

Fig. 3. Inferring a conditional property.

$T_\Sigma(x)/_{(x\overline{y},y\overline{x})} \to \bullet$	Generated by $(\_,\_)$	Inference rule
$e_1,...,e_{22},g_2,...,g_1$	same as before	same as before
e_{28}	$(x\overline{y})(x\overline{y}),((x\overline{y})x)\overline{y}$	Substitution : axiom$_3$
e_{29}	$(x\overline{y})(x\overline{y}),((y\overline{x})x)\overline{y}$	Restriction : e_{28}
e_{30}	$y(\overline{x}x),(y\overline{x})x$	Substitution : axiom$_3$
e_{31}	$-$	Union : $e_{29} \cup e_{30}$
e_{32}	$-$	Union : $g_2 \cup e_{31}$
e_{33}	$(x\overline{y})(x\overline{y}),(y1)\overline{y}$	Restriction : e_{32}
e_{34}	$y1,y$	Substitution : axiom$_1$
e_{35}	$y\overline{y},1$	Substitution : axiom$_2$
e_Y^f	$(x\overline{y})(x\overline{y}),1$	Restriction : e_{35}
$(e_Y;\text{axiom}_4)^f$	$x\overline{y},1$	Substitution : axiom$_4$
e_{36}	$x(\overline{y}y),(x\overline{y})y$	Substitution : axiom$_3$
e_{37}	$\overline{y}y,1$	similar to g_2
e_{38}	$1y,y$	similar to g_1
e_{39}	$-$	Union : $(e_Y;\text{axiom}_4)^f \cup e_{36}$
e_{40}	$-$	Union : $e_{37} \cup e_{39}$
e_{41}	$-$	Union : $e_{38} \cup e_{40}$
g_4	x,y	Restriction : e_{41}

Fig. 4. xyz

$(e_Y;\text{axiom}_4)^f$ by $(x\overline{y},1)$, we first derive e_Y^f like in Figure 4. The diagram below shows the relevant morphisms involved in this proof:

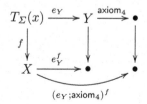

6 Finiteness

Since derivation of arrows involves a finite number of steps, one cannot expect any deduction system to be complete without some form of finiteness requirements. In this section we first recall the usual categorical concept dedicated to finiteness, then instantiate it to our framework, and then add one more requirement to factorization systems that makes them deal with finiteness smoothly.

A nonempty partially ordered set $(\mathcal{I}, \leq)$ is *directed* provided that each pair of elements has an upper bound. A *directed colimit* in a category $\mathcal{K}$ is a colimit of a diagram $D\colon (\mathcal{I}, \leq) \to \mathcal{K}$, where $(\mathcal{I}, \leq)$ is a directed poset (regarded as a category). An object K of a category $\mathcal{K}$ is *finitely presentable* provided that its hom-functor $Hom(K, \_)\colon \mathcal{K} \to \mathbf{Set}$ preserves directed colimits. It is easy to see that K is finitely presentable iff for each directed colimit $(\{\gamma_i\colon D(i) \to C\}_{i\in|\mathcal{I}|}, C)$ and each morphism $f\colon K \to C$, there is an $i \in |\mathcal{I}|$ and a unique morphism $f_i\colon K \to D(i)$ such that $f_i; \gamma_i = f$.

There are many examples of finitely presentable objects, such as finite sets and posets, finite graphs and automata, finite and discrete topological spaces, algebras presented by finitely many generators and finitely many equations, etc. We refer the interested reader to [1] for many more examples, as well as interesting properties of finitely presentable objects. What is relevant to our paper is that a surjective morphism $e\colon X \to \bullet$ of algebras is finitely presentable in the comma category of surjective morphisms of source X iff its kernel, regarded as a subalgebra of $X \times X$, is finitely generated; in our setting, where equations are surjective morphisms, that means that e stands for a finite set of equations.

Definition 5. *Equation* $e\colon X \to \bullet$ *is* **finite** *iff it is finitely presentable in* $X \downarrow \mathcal{E}$.

If $\mathcal{D} \subseteq X \downarrow \mathcal{E}$ is a finite diagram of finite equations, then with the notation in Definition 2, by Proposition 1.3 in [1] it follows that $e_\mathcal{D}$ is also finite. In particular, $e_1 \cup e_2$ is finite whenever e_1 and e_2 are finite, so finiteness is preserved by union. We next give conditions under which finiteness is also preserved by pushout. Given a morphism $m\colon X' \to X$ in $\mathcal{M}$, one can build "up to an isomorphism" a functor $\mathcal{F}_m\colon X \downarrow \mathcal{E} \to X' \downarrow \mathcal{E}$ as follows: for each $e\colon X \to \bullet$, let $\mathcal{F}_m(e)\colon X' \to \bullet$ be the epic by which $m; e$ factorizes, and for each $e_1\colon X \to X_1$, $e_2\colon X \to X_2$ and $\gamma\colon X_1 \to X_2$ with $e_1; \gamma = e_2$, let $\mathcal{F}_m(\gamma)$ be the unique "up to isomorphism" morphism given by the diagonal-fill-in property applied to the diagram $\mathcal{F}_m(e_2); m_2 = \mathcal{F}_m(e_1); (m_1; \gamma)$, where $m; e_1$ factors through $\mathcal{F}_m(e_1); m_1$ and $m; e_2$ factors through $\mathcal{F}_m(e_2); m_2$, like in the diagram below:

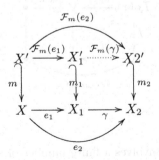

$\mathcal{F}_m(e)$ should be thought of as the restriction of e to X'. Interestingly, $\mathcal{F}_m$ does not preserve colimits in general. For example, if $\mathcal{C}$ is the category of sets then one can take $X = \{a_1, a_2, a_3\}$, $X' = \{a_1, a_3\}$, and $e_1 : X \to \bullet$ and $e_2 : X \to \bullet$ such that $e_1(a_1) = e_1(a_2)$ and $e_2(a_2) = e_2(a_3)$, respectively, and note that $(e_1 \cup e_2)(a_1) = (e_1 \cup e_2)(a_3)$, while $(\mathcal{F}_m(e_1) \cup \mathcal{F}_m(e_2))(a_1) \neq (\mathcal{F}_m(e_1) \cup \mathcal{F}_m(e_2))(a_3)$, where m is the inclusion $X' \subset X$. However, $\mathcal{F}_m$ does preserve directed colimits both for sets and algebras. The proof is relatively easy but takes much space, so we let it as an exercise to the interested reader (Hint: work with kernels instead of epics). With the notation above,

Definition 6. *The factorization system $\langle \mathcal{E}, \mathcal{M} \rangle$ is* **reasonable** *provided that $\mathcal{F}_m$ preserves directed colimits for each $m \in \mathcal{M}$.*

The following important property can be shown:

Proposition 3. *In the context of Proposition 1, if $\langle \mathcal{E}, \mathcal{M} \rangle$ is reasonable and e is finite, then e^f is finite.*

Proof. Due to factorization, it suffices to show the result separately for $f \in \mathcal{E}$ and for $f \in \mathcal{M}$. Let $\mathcal{D} \subseteq X \downarrow \mathcal{E}$ be a directed diagram and let h be a morphism such that $e^f ; h = e_{\mathcal{D}}$ (see Definition 2).

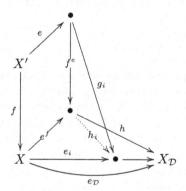

If $f \in \mathcal{E}$ then note that $f; \mathcal{D}$ is also a directed diagram and that $e_{f;\mathcal{D}} = f; e_{\mathcal{D}}$. Since e is finite and since $e; (f^e; h) = e_{f;\mathcal{D}}$, there is some $e_i : X \to \bullet$ in $\mathcal{D}$ such that $f; e_i$ factors through e, i.e., there is some morphism g_i such that $f; e_i = e; g_i$. By the pushout property of e and f, it follows that there is some morphism h_i such that $e^f ; h_i = e_i$ and $f^e; h_i = g_i$. Hence, e_i factors through e^f, so e^f is finite. If $f \in \mathcal{M}$ then, since $\langle \mathcal{E}, \mathcal{M} \rangle$ is reasonable, there is some morphism $m \in \mathcal{M}$ with $f; e_{\mathcal{D}} = e_{\mathcal{F}_f(\mathcal{D})}; m$.

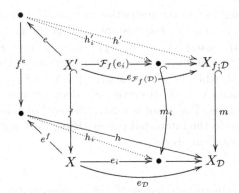

Then $e; (f^e; h) = e_{\mathcal{F}_f(\mathcal{D})}; m$, so by the diagonal-fill-in property there is a morphism h' such that $e; h' = e_{\mathcal{F}_f(\mathcal{D})}$ and $h'; m = f^e; h$. Since e is finite and $\mathcal{F}_f(\mathcal{D})$ is directed, there is some $e_i \in \mathcal{D}$ such that $\mathcal{F}_f(e_i)$ factors through e, so there is an h'_i with $\mathcal{F}_f(e_i) = e; h'_i$. Therefore, $e; (h'_i; m_i) = f; e_i$, where $m_i \in \mathcal{M}$ is such that $f; e_i$ factors as $\mathcal{F}_f(e_i); m_i$, so by the pushout property there is some morphism h_i such that $e^f; h_i = e_i$ and $f^e; h_i = h'_i; m_i$. Hence e_i factors through e^f, so e^f is also finite.

7   Completeness

In this section we fix the following

> Framework: A category $\mathcal{C}$ that
> - admits a reasonable factorization system $(\mathcal{E}, \mathcal{M})$,
> - has enough $\mathcal{E}$-projectives,
> - is $\mathcal{E}$-co-well-powered,
> - has colimits[2],

and show that, under appropriate finiteness conditions, the four rules presented in the previous section are complete wrt satisfaction as injectivity.

The usual notion of closure under inference rules is extended to classes of epics; in particular, $\mathcal{D} \subseteq X \downarrow \mathcal{E}$ is closed under E-substitution iff for any $e: Y \to \bullet$ in E and any $f: P_Y \to X$, if e_Y^f is in $\mathcal{D}$ then so is $(e_Y; e)^f$. Notice that $\mathcal{D}_X(E)$ is closed under all the four inference rules, so it is non-empty (because of closure under Identity) and directed (due to closure under Union). If $\mathcal{D}_X(E)$ is not a set then, since $\mathcal{C}$ is $\mathcal{E}$-co-well-powered, it can be replaced by some representative set that it includes, so we can let $e_{\mathcal{D}_X(E)}: X \to X_{\mathcal{D}_X(E)}$ denote its colimit object, as usual (see Definition 2). Then, with the notation in Definition 2,

Theorem 2. *If E contains only equations of finite conditions, then*

1. $X_{\mathcal{D}} \models E$ for any non-empty directed diagram $\mathcal{D} \subseteq X \downarrow \mathcal{E}$ closed under Restriction and E-Substitution;

---

[2] Actually only directed colimits and certain pushouts are needed.

2. *For any equation e of source X, $E \models e$ iff $X_{\mathcal{D}_X(E)} \models e$;*

3. **Completeness.** *$E \models e$ implies $E \vdash e$ whenever e is finite.*

Proof. 1. Let $e\colon Y \to \bullet$ be any equation in E and let $g\colon Y \to X_{\mathcal{D}}$ be a morphism. Since P_Y is $\mathcal{E}$-projective and since $e_{\mathcal{D}} \in \mathcal{E}$, there is a morphism $f\colon P_Y \to X$ such that $e_Y; g = f; e_{\mathcal{D}}$.

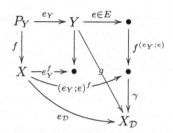

Since e_Y^f is an arrow in the pushout of f and e_Y, $e_{\mathcal{D}}$ factors through e_Y^f, and since e_Y^f is finite (Proposition 3) and $\mathcal{D}$ is directed and non-empty, there is an e' in $\mathcal{D}$ which factors through e_Y^f. It follows then that $e_Y^f \in \mathcal{D}$ because $\mathcal{D}$ is closed under Restriction, and further that $(e_Y; e)^f \in \mathcal{D}$ because $\mathcal{D}$ is closed under E-Substitution. Thus there is a morphism γ such that $(e_Y; e)^f; \gamma = e_{\mathcal{D}}$. Notice that $e_Y; (e; f^{(e_Y;e)}; \gamma) = f; (e_Y; e)^f; \gamma = f; e_{\mathcal{D}} = e_Y; g$, so $e; (f^{(e_Y;e)}; \gamma) = g$ because e_Y is an epimorphism. Therefore, $X_{\mathcal{D}} \models e$.

2. If $E \models e$ then by *1.*, noticing that $\mathcal{D}_X(E)$ is closed under Restriction and E-Substitution and is directed (because it is closed under Union) and non-empty (because it is closed under Identity), it follows that $X_{\mathcal{D}_X(E)} \models e$. Conversely, if $X_{\mathcal{D}(E)} \models e$ then there is an e' such that $e; e' = e_{\mathcal{D}_X(E)}$. Let $A \models E$ and let $h\colon X \to A$. Since $A \models \mathcal{D}_X(E)$, for each $e_j \in \mathcal{D}_X(E)$ there is a β_j such that $e_j; \beta_j = h$. Then A together with the morphisms β form a cocone in $\mathcal{C}$ for $\mathcal{D}_X(E)$, so there is a unique $g\colon X_{\mathcal{D}(E)} \to A$ such that $\gamma_j; g = \beta_j$ for all $e_j \in \mathcal{D}_X(E)$. It follows then that $e; (e'; g) = e_{\mathcal{D}_X(E)}; g = e_j; \gamma_j; g = e_j; \beta_j = h$, that is, $A \models e$.

3. $X_{\mathcal{D}_X(E)} \models e$ by *2.*, so there is an e' such that $e; e' = e_{\mathcal{D}_X(E)}$. Since e is finite and since $\mathcal{D}_X(E)$ is non-empty, there is an e_j in $\mathcal{D}_X(E)$ which factors through e. Since $E \vdash e_j$, by Restriction it follows that $E \vdash e$.

Therefore, under reasonable and necessary finiteness conditions, the four rule inference system can be used to derive any arrow e which is injectively satisfied by all objects satisfying E. On the one hand, this can be regarded as a purely categorical characterizing result, independently from logics. On the other hand, instantiated to equational logics it gives an inference system which can derive any conditional equational semantical consequence *directly*. For example, the Identity rules corresponds to reflexivity; the Union rule corresponds to closures under transitivity and congruence closures over conclusions of conditional equations, assuming that they have (provably) the same hypotheses (note that closure under symmetry is implicit, because kernels or morphisms are symmetric binary relations); the Restriction allows one to retain only a part of the conclusion of a conditional equation, in case one proved more than needed; finally, the E-Substitution rule corresponds as expected to substitution, but note that it can also derive conditional equations (when X is not projective).

8   Conclusion and Future Work

We presented a four rule categorical deduction system for a categorical abstraction of equational logics, in which equations are regarded as epimorphisms and their satisfaction as injectivity. We showed that under reasonable finiteness conditions, the four rule deduction system is complete. The research presented in this paper is part of a project aiming at developing a categorical framework in which axiomatizability, complete deduction and interpolation can be treated uniformly. Birkhoff variety and quasi-variety results for equations regarded as epics and for satisfaction regarded as injectivity are known and considered folklore among category theorists. The results in this paper show that there is also complete deduction within this framework. We are not aware of other similar categorical completeness results in the literature, except previous work by the author [35] where only unconditional axioms where supported and some interesting results by Diaconescu [14] within his category-based equational logic, where equations were regarded as parallel pairs of arrows and his five inference rules were the typical ones for equational deduction.

There is much challenging research to be done. Can the Craig-like interpolation results in [38] be instantiated to the categorical equational logic framework presented in this paper? Can the present results be dualized, hereby obtaining complete deduction for some variant of modal or coalgebraic logics? Would it be possible to implement the four rules and thus develop an arrow-based, perhaps graphical, equational reasoning engine?

Dedication. The author dedicates this paper to his former PhD adviser, Joseph Goguen, to whom he warmly thanks for all his teachings and unforgettable time spent at the University of California at San Diego. The author is also grateful to Joseph Goguen for his enthusiasm in categorical approaches to equational logics, and in particular for his encouragements in writing this material up.

References

1. J. Adámek and J. Rosický. *Locally Presentable and Accessible Categories.* Cambridge University Press, 1994. London Math. Society. Lecture Note Series 189.
2. H. Andréka and I. Németi. A general axiomatizability theorem formulated in terms of cone-injective subcategories. In B. Csakany, E. Fried, and E. Schmidt, editors, *Universal Algebra*, pages 13–35. North-Holland, 1981. Colloquia Mathematics Societas János Bolyai, 29.
3. H. Andréka and I. Németi. Generalization of the concept of variety and quasi-variety to partial algebras through category theory. In *Dissertationes Mathematicae*, volume 204. Polish Scientific Publishers, 1983.

4. B. Banaschewski and H. Herrlich. Subcategories defined by implications. *Houston Journal Mathematics*, 2:149–171, 1976.
5. J. Bergstra and J. Tucker. Characterization of computable data types by means of a finite equational specification method. In J. de Bakker and J. van Leeuwen, editors, *Automata, Languages and Programming, Seventh Colloquium*, pages 76–90. Springer, 1980. Lecture Notes in Computer Science, Volume 81.
6. J. Bergstra and J. V. Tucker. Equational specifications, complete term rewriting systems, and computable and semicomputable algebras. *Journal of the Association for Computing Machinery*, 42(6):1194–1230, 1995.
7. G. Birkhoff. On the structure of abstract algebras. *Proceedings of the Cambridge Philosophical Society*, 31:433–454, 1935.
8. G. Birkhoff and J. Lipson. Heterogenous algebras. *Journal of Combinatorial Theory*, 8:115–133, 1970.
9. P. Borovanský, H. Cîrstea, H. Dubois, C. Kirchner, H. Kirchner, P.-E. Moreau, C. Ringeissen, and M. Vittek. Elan. User manual; http://www.loria.fr/equipes/protheo/SOFTWARES/ELAN.
10. A. Bouhoula, J.-P. Jouannaud, and J. Meseguer. Specification and proof in membership equational logic. *Theoretical Computer Science*, 236:35–132, 2000.
11. M. Broy, M. Wirsing, and P. Pepper. On the algebraic definition of programming languages. *ACM Trans. on Prog. Lang. and Systems*, 9(1):54–99, Jan. 1987.
12. M. Clavel, S. Eker, P. Lincoln, and J. Meseguer. Principles of Maude. In J. Meseguer, editor, *Proceedings, First International Workshop on Rewriting Logic and its Applications*. Elsevier Science, 1996. Volume 4, *Electronic Notes in Theoretical Computer Science*.
13. V. Căzănescu. Local equational logic. In Z. Esik, editor, *Proceedings, 9th International Conference on Fundamentals of Computation Theory FCT'93*, pages 162–170. Springer-Verlag, 1993. Lecture Notes in Computer Science, Volume 710.
14. R. Diaconescu. *Category-based Semantics for Equational and Constraint Logic Programming*. PhD thesis, University of Oxford, 1994.
15. R. Diaconescu and K. Futatsugi. *CafeOBJ Report: The Language, Proof Techniques, and Methodologies for Object-Oriented Algebraic Specification*. World Scientific, 1998. AMAST Series in Computing, volume 6.
16. P. Freyd and G. Kelly. Categories of continuous functors. *Journal of Pure and Applied Algebra*, 2:169–191, 1972.
17. J. Goguen and G. Malcolm. *Alg. Semantics of Imperative Programs*. MIT, 1996.
18. J. Goguen and G. Malcolm. A hidden agenda. *Theoretical Computer Science*, 245(1):55–101, August 2000.
19. J. Goguen and J. Meseguer. Completeness of many-sorted equational logic. *Houston Journal of Mathematics*, 11(3):307–334, 1985. Preliminary versions have appeared in: *SIGPLAN Notices*, July 1981, Volume 16, Number 7, pages 24–37; SRI Report CSL-135, May 1982; and Report CSLI-84-15, Stanford, Sep. 1984.
20. J. Goguen and J. Meseguer. Order-sorted algebra I: Equational deduction for multiple inheritance, overloading, exceptions and partial operations. *Theoretical Computer Science*, 105(2):217–273, 1992.
21. J. Goguen, T. Winkler, J. Meseguer, K. Futatsugi, and J.-P. Jouannaud. Introducing OBJ. In J. Goguen and G. Malcolm, editors, *Software Engineering with OBJ: algebraic specification in action*, pages 3–167. Kluwer, 2000.
22. A. Grothendieck. Sur quelques points d'algèbre homologique. *Tôhoku Mathematical Journal*, 2:119–221, 1957.
23. H. Herrlich and G. Strecker. *Category Theory*. Allyn and Bacon, 1973.

24. J. R. Isbell. Subobjects, adequacy, completeness and categories of algebras. *Rozprawy Matematyczne*, 36:1–33, 1964.
25. J. Lambek. *Completions of Categories*. Springer-Verlag, 1966. Lecture Notes in Mathematics, Volume 24.
26. S. M. Lane. *Categories for the Working Mathematician*. Springer, 1971.
27. J. Meseguer. Membership algebra as a logical framework for equational specification. In *Proceedings, WADT'97*, volume 1376 of *LNCS*, pages 18–61. Springer, 1998.
28. B. Mitchell. *Theory of categories*. Academic Press, New York, 1965.
29. I. Németi. On notions of factorization systems and their applications to cone-injective subcategories. *Periodica Mathematica Hungarica*, 13(3):229–335, 1982.
30. I. Németi and I. Sain. Cone-implicational subcategories and some Birkhoff-type theorems. In B. Csakany, E. Fried, and E. Schmidt, editors, *Universal Algebra*, pages 535–578. North-Holland, 1981. Colloquia Mathematics Societas János Bolyai, 29.
31. P. Padawitz and M. Wirsing. Completeness of many-sorted equational logic revisited. *Bulletin of the European Association for Theoretical Computer Science*, 24:88–94, Oct. 1984.
32. H. Reichel. *Initial Computability, Algebraic Specifications, and Partial Algebras*. Oxford University Press, 1987.
33. G. Roşu. A Birkhoff-like axiomatizability result for hidden algebra and coalgebra. In B. Jacobs, L. Moss, H. Reichel, and J. Rutten, editors, *Proceedings of the First Workshop on Coalgebraic Methods in Computer Science (CMCS'98), Lisbon, Portugal, March 1998*, volume 11 of *Electronic Notes in Theoretical Computer Science*, pages 179–196. Elsevier Science, 1998.
34. G. Roşu. *Hidden Logic*. PhD thesis, University of California at San Diego, 2000. http://ase.arc.nasa.gov/grosu/phd-thesis.ps.
35. G. Roşu. Complete categorical equational deduction. In L. Fribourg, editor, *Proceedings of Computer Science Logic (CSL'01)*, volume 2142 of *Lecture Notes in Computer Science*, pages 528–538. Springer, 2001.
36. G. Roşu. Equational axiomatizability for coalgebra. *Theoretical Computer Science*, 260(1-2):229–247, 2001.
37. G. Roşu. Axiomatizability in inclusive equational logics. *Mathematical Structures in Computer Science*, to appear. http://ase.arc.nasa.gov/grosu/iel.ps.
38. G. Roşu and J. Goguen. On equational Craig interpolation. *Journal of Universal Computer Science*, 6(1):194–200, 2000.
39. P. H. Rodenburg. A simple algebraic proof of the equational interpolation theorem. *Algebra Universalis*, 28:48–51, 1991.
40. G. Smolka, W. Nutt, J. Goguen, and J. Meseguer. Order-sorted equational computation. In M. Nivat and H. Aït-Kaci, editors, *Resolution of Equations in Algebraic Structures, Volume 2: Rewriting Techniques*, pages 299–367. Academic, 1989.
41. M. Wand. First-order identities as a defining language. *Acta Informatica*, 14:337–357, 1980.

Extension Morphisms for CommUnity

Nazareno Aguirre[1], Tom Maibaum[2], and Paulo Alencar[3]

[1] Departamento de Computación, FCEFQyN, Universidad Nacional de Río Cuarto,
Ruta 36 Km. 601, Río Cuarto (5800), Córdoba, Argentina,
naguirre@dc.exa.unrc.edu.ar
[2] Department of Computing & Software, McMaster University,
1280 Main St. West, Hamilton, Ontario, Canada L8S 4K1,
tom@maibaum.org
[3] School of Computer Science, University of Waterloo,
200 University Avenue West, Waterloo, Ontario, Canada N2L 3G1,
palencar@csg.uwaterloo.ca

Abstract. Superpositions are useful relationships between programs or components in component based approaches to software development. We study the application of invasive superposition morphisms between components in the architecture design language CommUnity. This kind of morphism allows us to characterise component extension relationships, and in particular, serves an important purpose for enhancing components to implement certain *aspects*, in the sense of aspect oriented software development. We show how this kind of morphism combines with regulative superposition and refinement morphisms, on which CommUnity relies, and illustrate the need and usefulness of extension morphisms for the implementation of aspects, in particular, certain fault tolerance related aspects, by means of a case study.

1 Introduction

The demand for adequate methodologies for modularising designs and development is increasing rapidly, due to the inherent complexities of modern software systems. Of course, these modularisation methodologies do not affect only the final implementation stages, but they also have an impact on earlier stages of software development processes. Thus, it is generally accepted that, for the modularisation to be effective (and persistent, and resistant to evolution), it needs to be applied from the start, at the level of specification or modelling of systems. Modularising, or structuring, specifications has important benefits. It allows one to divide the specifications into manageable parts, and to evaluate the consequences of our architectural design decisions prior to the implementation of the system. Moreover, it also favours the reuse of parts of the resulting implementations, and their adaptations and extensions for new application domains.

In the area of critical systems, specification languages are required to have a precise meaning (since formal semantics is crucial for eliminating ambiguities in specifications, and for developing tools for verification), and therefore specifications tend to be much longer than those of informal frameworks. Thus,

K. Futatsugi et al. (Eds.): Goguen Festschrift, LNCS 4060, pp. 173–193, 2006.

mechanisms for structuring or modularising specifications and designs are especially important for formal specification languages, as they help to make the specification and verification activities scalable. There exist many formal specification languages which put an emphasis on the way systems are built out of components (e.g., those reported in [19,9,20,18]). CommUnity is one of these languages; it is a formal program design language which puts special emphasis on ways of composing abstract designs of components to form designs of systems [6,5]. CommUnity is based on Unity [3] and IP [8], and its foundations lie in the categorical approach to systems design [10]. Its mechanisms for composing specifications have a formal interpretation in terms of category theoretic constructions [5,6]. Moreover, CommUnity's composition mechanisms combine nicely with a sophisticated notion of refinement, which involves separate concepts of action blocking and action progress.

We are particularly interested in CommUnity because, in our view, its design composition mechanisms make it suitable for the specification and combination (or "weaving") of *aspects*, in the aspect oriented software development sense [7]. Moreover, its rather abstract designs for components allow us to deal with aspects at a design level, in contrast to most of the work on aspects, which concerns implementation related stages (e.g., [14]). Some evidence of the adequacy of CommUnity as a design language for aspects relies on the possibility of defining *higher-order connectors* [16]. As shown in [16], a wide variety of aspects (e.g., fault tolerance, security, monitoring, compression, etc) can be superimposed on existing CommUnity architectures, by building "stacks" of more and more complex connectors between components.

Higher-order connectors provide a very convenient way of enhancing the behaviour of an architecture of component designs, by the superimposition of aspects. A crucial characteristic of CommUnity, which makes this possible, is the *complete externalisation of the definition of interaction* between components (a feature also exhibited by various other architecture description languages). The component coordination mechanism of CommUnity reduces the coupling between components to a minimum, and makes it feasible to superimpose behaviour (related to aspects) on existing systems via superposition and refinement of components. However, higher-order connectors are not powerful enough for defining various kinds of aspects, since some of these, as we will show, require extensions of the components as well as in the connectors. Thus, we are forced to consider another kind of superposition, known as *invasive superposition* [12], which allows us to define *extensions* of components. By combining extension with regulative superposition and refinement, we believe that we obtain a powerful framework in which we can define architectures, and enhance their behaviours by superimposing behaviour through aspects defined in terms of component extension and higher-order connectors. Having the possibility of extending components also provides us with a way of *balancing* the distribution of extended behaviour among connectors and components, which would otherwise be put exclusively on the connector side. This problem has also arisen in the context of object oriented design and programming, attempting to define various forms of inheritance, resulting in the proposals attempting to characterise the concept of substitutability [15,21]. We believe that this proposal provides a more solid foundation for substitutivity, one that is better structured and more amenable to analysis. We propose a definition of extension in CommUnity, partly motivated by the definitions and proof obligations used to define the structuring mechanisms in B [1,4], that justifies the notion of substitutivity and provides a structuring principle for augmenting components by breaking encapsulation of the component. (Perhaps this should be considered a contradiction in terms!)

We show how extension morphisms combine with the superposition and refinement morphisms already present in CommUnity. We will also illustrate the need and usefulness of extension morphisms for the implementation of aspects, by means of a case study, based on a simple sender/receiver architecture communicating via an unreliable channel, which is then enhanced with some typical aspects, imposing a standard fault tolerance mechanism.

2   The Architecture Design Language CommUnity

In this section, we introduce the reader to the CommUnity design language and its main features, by means of an example. The computational units of a system are specified in CommUnity through *designs*. Designs are abstract programs, in the sense that they describe a *class* of programs (more precisely, the class of all the programs one might obtain from the design by refinement), rather than a single program [23,5].

Before describing in some detail the refinement and composition mechanisms of CommUnity, let us describe the main constituents of a CommUnity design. Let us first assume that we have a fixed set $\mathcal{ADT} = \langle \Sigma_{\mathcal{ADT}}, \Phi_{\mathcal{ADT}} \rangle$ of datatypes, specified as usual via a first-order specification. A CommUnity design is composed of:

- A set V of *channels*, typed with sorts in $\mathcal{ADT}$. V is partitioned into three subsets V_{in}, V_{prv} and V_{out}, corresponding to input, private and output channels, respectively. Input channels are the ones controlled, from the point of view of the component, by the environment. Private and output channels are the local channels of the component. The difference between these is that output channels can be read by the environment, whereas private channels cannot.
- A first-order sentence $Init(V)$, describing the initial states of the design[4].
- A set Γ of actions, partitioned into private actions Γ_{prv} and public actions Γ_{pub}. Each action $g \in \Gamma$ is of the form:

$$g[D(g)] : L(g), U(g) \to R(g)$$

where $D(g) \subseteq V_{prv} \cup V_{out}$ is the (write) *frame* of g (the local channels that g modifies), $L(g)$ and $U(g)$ are two first-order sentences such that $\Phi_{\mathcal{ADT}} \vdash$

---

[4] Some versions of CommUnity, such as the one presented in [17], do not include an initialisation constraint.

$U(g) \Rightarrow L(g)$, called the lower and upper bound guards, respectively, and $R(g)$ is a first-order sentence $\alpha(V \cup D(g)')$, indicating how the action g modifies the values of the variables in its frame. ($D(g)$ is a set of channels and $D(g)'$ is the corresponding set of "primed" versions of the channels in $D(g)$, representing the new values of the channels after the execution of the action g.)

The two guards $L(g)$ and $U(g)$ associated with an action g are related to refinement, in the sense that the actual guard of an action g_r implementing the abstract action g, must lie between $L(g)$ and $U(g)$. As explained in [17], the negation of $L(g)$ establishes a blocking condition ($L(g)$ can be seen as a lower bound on the actual guard of an action implementing g), whereas $U(g)$ establishes a progress condition (i.e., an upper bound on the actual guard of an action implementing g, in the sense that it implies the enabling condition of an action implementing g).

Of course, $R(g)$ might not uniquely determine values for the variables $D(g)'$. As explained in [17], $R(g)$ is typically composed of a conjunction of implications *pre* $\Rightarrow$ *post*, where *pre* is a precondition and *post* defines a multiple assignment.

To clarify the definition of CommUnity designs, let us suppose that we would like to model the unreliable communication between a sender and a receiver. We will abstract away from the actual contents of messages between these components, and represent them simply by an integer, identifying particular messages. Then, a sender is a simple CommUnity design composed of:

- An output channel `msg: int`, representing the current message of the sender.
- A private channel `rts: bool` ("ready to send"), indicating whether the sender is ready to send the current message or not (messages need to be produced before sending them).
- An action `send`, which, if the sender is ready to send (indicated by the boolean variable above), then goes back to a "ready to produce" state (characterised by the `rts` variable being false).
- An action `prod`, that, if the sender is in a "ready to produce" state, increments by one the `msg` variable (i.e., generates a new message to be sent) and moves to a "ready to send" state.

The CommUnity design corresponding to this component is shown in Figure 1.

In Fig. 1, the actions of the design have a single guard, meaning that their lower and upper bound guards coincide. We will illustrate refinement through more abstract designs below. An important point to notice in the sender design is the way it communicates with the environment through the `send` action. This action does not make a call to an external action, as one might expect; it will be the responsibility of other components to "extract" the value of the output variable `msg`, by synchronising other actions with the `send` action of the sender. This will become clearer later on, when we build architectures and describe in more detail the model of interaction between components in CommUnity.

To complete the picture, let us introduce some further designs. One is a simple component with a single integer typed output variable, used for communication

```
Design Sender
out
    msg: int
prv
    rts: bool
init
    msg=0 ∧ rts=false
do
    prod[msg,rts]: ¬rts → rts'=true ∧ msg'=msg+1
[]  send[rts]: rts → rts'=false
```

Fig. 1. A CommUnity design for a simple sender component.

and for modelling the loss of messages (Figure 2). The other one is a receiver component, somewhat similar in structure to the sender, but with an input variable instead of an output one, and a boolean channel rtr (ready to receive) instead of rts (Figure 3). To complete the picture, let us introduce some further designs. One is a simple component with a single integer typed output variable, used for communication and for modelling the loss of messages (Figure 2). The other one is a receiver component, somewhat similar in structure to the sender, but with an input variable instead of an output one, and a boolean channel rtr (ready to receive) instead of rts. To complete the picture, let us introduce some further designs. One is a simple component with a single integer typed output variable, used for communication and for modelling the loss of messages (Figure 2). The other one is a receiver component, somewhat similar in structure to the sender, but with an input variable instead of an output one, and a boolean channel rtr (ready to receive) instead of rts.

```
Design Communication_Medium
in
    in_msg: int
out
    out_msg: int
prv
    rts: bool
init
    out_msg=0 ∧ rts=false
do
    transmit[out_msg,rts]: ¬rts → out_msg'=in_msg ∧ rts'=true
[]  lose[]: ¬rts → true
[]  send[rts]: rts → rts'=false
```

Fig. 2. A CommUnity design for an unreliable communication medium.

```
Design Receiver
in
    msg: int
out
    curr_msg: int
local
    rtr: bool
init
    curr_msg=0 ∧ rtr=true
do
        rec[rtr,curr_msg]: rtr → rtr'=false ∧ curr_msg'=msg
[] prv cons[rtr]: ¬rtr → rtr'=true
```

Fig. 3. A CommUnity design for a receiver component.

2.1   Refinement Morphisms

Refinement morphisms constitute an important relationship between CommU-
nity designs. Not only do these morphisms allow one to establish "realisation"
relationships, indicating that a component is a more refined or concrete version
of another one, but they also serve an important purpose for parameter instantiation. In particular, refinement morphisms are essential for the implementation
of *higher-order connectors* [16].

We will not give a fully detailed description of refinement morphisms here. We
refer the interested reader to [6,23,16,17,5] for a detailed account of refinement
in CommUnity.

A *refinement morphism* $\sigma : P_1 \to P_2$ between designs $P_1 = (V_1, \Gamma_1)$ and
$P_2 = (V_2, \Gamma_2)$ consists of a total function $\sigma_{ch} : V_1 \to V_2$ and a partial function
$\sigma_{ac} : \Gamma_2 \to \Gamma_1$ such that:

- σ_{ch} preserves the sorts and kinds (output, input or private) of channels;
  moreover, σ_{ch} is *injective* on input and output channels,
- σ_{ac} maps shared actions to shared actions and private actions to private
  actions; moreover, every shared action in Γ_1 has at least one corresponding
  action in Γ_2 (via σ_{ac}^{-1}),
- the initialisation condition is strengthened through the refinement, i.e.,
  $\Phi_{ADT} \vdash Init_{P_2} \Rightarrow \underline{\sigma}(Init_{P_1})$,
- every action $g \in \Gamma_2$ whose frame $D_2(g)$ includes a channel $\sigma_{ch}(v)$, with
  $v \in V_1$, is mapped to an action $\sigma_{ac}(g)$ whose frame $D_1(\sigma_{ac}(g))$ includes v,
- if an action $g \in \Gamma_2$ is mapped to an action $\sigma_{ac}(g)$, then $\Phi_{ADT} \vdash L_2(g) \Rightarrow$
  $\underline{\sigma}(L_1(\sigma_{ac}(g)))$ and $\Phi_{ADT} \vdash R_2(g) \Rightarrow \underline{\sigma}(R_1(\sigma_{ac}(g)))$,
- for every action $g \in \Gamma_1$, $\Phi_{ADT} \vdash \underline{\sigma}(U_1(g)) \Rightarrow \bigvee_{h \in \sigma^{-1}(g)} U_2(h)$.

As specified by these conditions, the interval determined by the lower and
upper bound guards can be reduced through refinement, and the assignments can
be strengthened. The interface of a component design, determined by the output
and input channels and shared actions, is *preserved* along refinement morphisms,

and the new actions that can be defined in a refinement are not allowed to modify the channels originating in the abstract component. Essentially, one can refine a component by making its actions more detailed and less underspecified (cleverly characterised by the reduction of the guard interval and the strenghthening of the assignments), and possibly adding more detail to the component, in the form of further channels or actions [17].

As an example of refinement, consider the more abstract version of the sender design shown in Figure 4. Notice that the assignment associated with prod is more abstract or liberal than the assignment of the same action in the Sender design. Also, the lower bound guards of both actions are equivalent to those of the corresponding actions in Sender, but the upper bound guards are *strengthened* to false. Clearly, Abstract_Sender is a more abstract version of the Sender (or, equivalently, Sender is a refinement of Abstract_Sender), and it is not difficult to prove that there exists a refinement morphism between these designs. In fact, Abstract_Sender is also a refinement of the Communication_Medium component (where the abstract prod operation corresponds to the operations lose and transmit), although it is less evident than in the first case.

```
Design Abstract_Sender
out
    msg: int
prv
    rts: bool
init
    msg=0 ∧ rts=false
do
    prod[msg,rts]: ¬rts, false → rts'=true ∧ msg'∈int
[]  send[rts]: rts, false → rts'=false
```

Fig. 4. A more abstract CommUnity design for a sender component.

2.2   Component Composition

In order to build a system out of the above components, we need a mechanism for composition. The mechanism for composing designs in Community is based on action synchronisation and the "connection" of output channels to input channels (shared memory). Basically, we need to connect the sender and receiver through the unreliable medium. This can be achieved by:

- identifying the output channel msg of the sender with the input channel in_msg of the medium,
- identifying the input channel msg of the receiver with the output channel out_msg of the medium,

– synchronising the action `send` of the sender with actions `transmit` and `lose` of the medium,
– synchronising the action `send` of the medium with action `rec` of the receiver.

The resulting architecture can be graphically depicted as shown in Figure 5. In this diagram, the architecture is shown using the CommUnity Workbench [24] graphical notation, where boxes represent designs, with its channels and actions[5], and lines represent the interactions ("cables" in the sense of [17]), indicating how input channels are connected to output channels, and which actions are synchronised. Notice that, in particular, action `send` of the sender is connected to two different actions of the medium; this requires that, in the resulting system, there will be two different actions corresponding to (or "invoking") the `send` action in the sender, one that is synchronised with `transmit` and another one that is synchronised with `lose`. This allows us to model very easily the fact that, sometimes, the sent message is lost (when the action `send-lose` is executed), without using further channels in the communication medium.

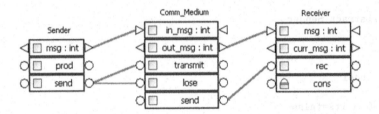

Fig. 5. A graphical view of the architecture of the system.

Semantics of Architectures. CommUnity designs have an operational semantics based on (labelled) transition systems. Architectural configurations, of the kind shown in Fig. 5, also have a precise semantics; they are interpreted as categorical diagrams, representing the architecture [17]. The category has designs as objects and the morphisms are *superposition relationships*. A superposition morphism between two designs A and B captures, in a formal way, the fact that B *contains* A, and uses it while respecting the encapsulation of A (regulative superposition). The interesting fact is that the joint behaviour of the system can be obtained by taking the *colimit* of the categorical diagram corresponding to the architecture [5,6]. Therefore, one can obtain a single design (the colimit object), capturing the behaviour of the whole system.

---

[5] Private actions are not displayed by the CommUnity Workbench, although we decided to show these actions, conveniently annotated, in the diagrams.

More formally, a *superposition morphism* $\sigma : P_1 \to P_2$ between designs $P_1 = (V_1, \Gamma_1)$ and $P_2 = (V_2, \Gamma_2)$ consists of a total function $\sigma_{ch} : V_1 \to V_2$ and a partial function $\sigma_{ac} : \Gamma_2 \to \Gamma_1$ such that:

- σ_{ch} preserves the sorts of channels; private and output channels must be mapped to channels of the same kind, but input channels can be mapped to output channels,
- σ_{ac} maps shared actions to shared actions and private actions to private actions,
- the initialisation condition is strengthened through the superposition, i.e., $\Phi_{\mathcal{ADT}} \vdash Init_{P_2} \Rightarrow \underline{\sigma}(Init_{P_1})$,
- every action $g \in \Gamma_2$ whose frame $D_2(g)$ includes a channel $\sigma_{ch}(v)$, with $v \in V_1$, is mapped to an action $\sigma_{ac}(g)$ whose frame $D_1(\sigma_{ac}(g))$ includes v,
- if an action $g \in \Gamma_2$ is mapped to an action $\sigma_{ac}(g)$, then $\Phi_{\mathcal{ADT}} \vdash L_2(g) \Rightarrow \underline{\sigma}(L_1(\sigma_{ac}(g)))$, $\Phi_{\mathcal{ADT}} \vdash R_2(g) \Rightarrow \underline{\sigma}(R_1(\sigma_{ac}(g)))$, and $\Phi_{\mathcal{ADT}} \vdash U_2(g) \Rightarrow \underline{\sigma}(U_1(\sigma_{ac}(g)))$.

As for refinement morphisms, superposition morphisms allow assignments to be strengthened, but not weakened. Intuitively, P_2 enhances the behaviour of P_1 via the superposition of additional behaviour, described in other components (and synchronised with P_1). So, the actions of the augmented component P_2 "using" corresponding actions in P_1 do at least what the actions of P_1 originally did. Since actions in P_2 should use the corresponding actions in P_1 within enabledness bounds, the lower bound guards of actions in P_1 must be strengthened when superposed in actions of P_2. Notice however that, as opposed to the case of refinement morphisms, upper bound guards can be strengthened, but not weakened; as explained in [17], this is a key difference between refinement and superposition, and reflects the fact that "all the components that participate in the execution of a joint action have to give their permission for the action to occur." (cf. p. 9 of [17]).

3   Component Extension in CommUnity

In this section we describe the main contribution of this paper, namely, a new kind of morphism between components for CommUnity. This kind of morphism, that we call *extension morphism*, enables us to establish extension relationships between components (of the kind defined by inheritance in object orientation), and is of a different nature, compared to the already existing refinement and superposition morphisms of CommUnity.

In order to illustrate the need for extension morphisms, let us consider the following case. Suppose that, for the existing system of communicating sender and receiver, we would like to superimpose behaviour related to the *monitoring* of the received messages. As explained in [16], this is possible to achieve, in an elegant and structured way, by using higher-order connectors. Essentially, an abstract monitoring structure is defined; this structure is composed of various abstract designs, used for characterising roles of the architecture, like sender, receiver and monitor, and others necessary for the implementation of the "observed connector". These abstract designs are interconnected as shown in Figure 6. We will not describe these designs in detail, and refer the reader to [16], where a detailed description of this higher-order connector is given. However, it is important to mention that Abstract_Sender (which is given in Fig. 4) and Abstract_Receiver can be refined by, essentially, any pair of components providing the basic functionality for sending and receiving messages. Then, this higher-order connector is plugged into the existing architecture, through refinement, to obtain the resulting architecture of Figure 7. It is important to notice the difference between Figures 6 and 7; Fig. 6 describes the (abstract, non instantiated) higher-order connector for monitoring, whereas Fig. 7 described the *instantiation* of this higher-order connector (see how Abstract_Sender, Abstract_Receiver and Abstract_Monitor have been instantiated by Communication_Medium, Receiver and Simple_Monitor, respectively). The reader might observe that the actual monitor that we are using, described in Figure 8, simply counts the number of messages received by the receiver component. Notice that the guard of the monitor must be as weak as possible (i.e., true), to avoid interfering with the behaviour of the monitored operations.

In [16], several aspects are characterised and superimposed by using this same technique.

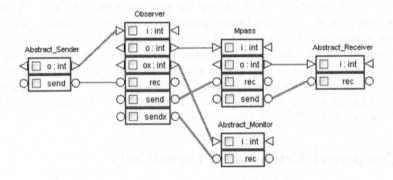

Fig. 6. A higher-order connector for monitoring.

Now suppose that we would like to superimpose a "resend message" mechanism on the architecture, in order to make the communication reliable. We can capture the loss of a "packet" through a monitor, instead of using it simply for counting the messages, as we did before. However, for the sender to reset and start sending the message again, we need to replace it with a slightly more sophisticated sender component, namely one with a reset operation, such as the one shown in Figure 9. Notice that RES_Sender cannot be obtained from Sender by superposition, since it is clear that the new reset operation modifies a channel originating in Sender. RES_Sender cannot be obtained through the refinement of

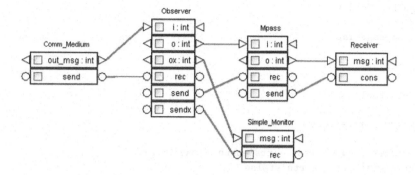

Fig. 7. Communication enhanced with a monitoring system.

```
Design Simple_Monitor
in
     msg: int
prv
     counter: int
init
     counter=0
do
     rec[counter]: true  →  counter'=counter+1
```

Fig. 8. A simple monitor to count received messages.

Sender either, since clearly its **reset** action, which modifies channels originating in **Sender**, should be mapped to a corresponding action in this design, but it does not respect any of the original assignments of actions **prod** and **send**, so it cannot be mapped to any of these.

However, there exists a clear relationship between the original **Sender** component and the new **RES_Sender**: the state of the original is extended, and more operations are provided (which might modify the channels of the original component), but the effect of the original actions is maintained. This relationship is a special case of what is called *invasive superposition* [12].

Invasive superposition has already been recognised as a possible relationship between CommUnity designs in [6]; moreover, therein it has been shown that CommUnity designs and invasive superpositions constitute a category. However, not much attention has been paid to invasive superposition for the architectural modelling of systems in CommUnity so far. Although not in the context of CommUnity, some researchers have employed various kinds of superpositions for defining architectures of components and augmenting their behaviours, particularly the work in [11]. Here, we propose the use of invasive superposition for characterising component extension in CommUnity.

184   Nazareno Aguirre, Tom Maibaum, and Paulo Alencar

```
Design RES_Sender
in
    lost-msg: int
out
    msg: int
prv
    rts: bool
init
    msg=0 ∧ rts=false
do
    prod[msg,rts]: ¬rts → rts'=true ∧ msg'=msg+1
[]  send[rts]: rts → rts'=false
[]  reset[msg,rts]: true → rts'=true ∧ msg'=lost-msg
```

Fig. 9. A CommUnity design for a sender component with a *reset* capability.

A distinguishing property typically associated with sound component extension is what is normally known as the *substitutability principle* [15]. This principle requires, in concordance with the now highly regarded "design by contract" approach [21], that if a component P_2 extends another component P_1, then one must be able to replace P_1 by P_2, and the "clients" of the original component must not perceive the difference. In other words, component P_2 should behave exactly as P_1, when put in a context where P_1 was expected. It is our aim to characterise such extensions through the definition of extension morphisms below.

Definition 1. *An extension morphism* $\sigma : P_1 \rightarrow P_2$ *between designs* $P_1 = (V_1, \Gamma_1)$ *and* $P_2 = (V_2, \Gamma_2)$ *consists of a total function* $\sigma_{ch} : V_1 \rightarrow V_2$ *and a partial mapping* $\sigma_{ac} : \Gamma_2 \rightarrow \Gamma_1$ *such that:*

- σ_{ch} *is injective and* σ_{ac} *is surjective,*
- σ_{ch} *preserves the sorts and kinds of channels,*
- σ_{ac} *maps shared actions to shared actions and private actions to private actions,*
- *there exists a formula* α, *using only variables that are contained in* $(V_2 - \sigma_{ch}(V_1))$, *and such that* $\Phi_{ADT} \vdash \exists \overline{v} : \alpha(\overline{v})$ *and* $\Phi_{ADT} \vdash Init_{P_2} \Leftrightarrow \underline{\sigma}(Init_{P_1}) \wedge \alpha$,
- *for every* $g \in \Gamma_2$ *such that* $\sigma_{ac}(g)$ *is defined, and for every* $v \in V_1$, *if* $\sigma_{ch}(v) \in D_2(g)$, *then* $v \in D_1(\sigma_{ac}(g))$,
- *if an action* $g \in \Gamma_2$ *is mapped to an action* $\sigma_{ac}(g)$, *then* $\Phi_{ADT} \vdash \underline{\sigma}(L_1(\sigma_{ac}(g))) \Rightarrow L_2(g)$ *and* $\Phi_{ADT} \vdash \underline{\sigma}(U_1(\sigma_{ac}(g))) \Rightarrow U_2(g)$,
- *for every* $g \in \Gamma_2$ *such that* $\sigma_{ac}(g)$ *is defined, there exists a formula* α, *using only primed variables that are contained in* $(V_2 - \sigma_{ch}(V_1))'$, *such that* $\Phi_{ADT} \vdash \underline{\sigma}(L_1(\sigma_{ac}(g))) \Rightarrow (R_2(g) \Leftrightarrow \underline{\sigma}(R_1(\sigma_{ac}(g))) \wedge \alpha)$ *and* $\Phi_{ADT} \vdash \exists \overline{v} : \alpha(\overline{v})$, *where* $\overline{v}$ *represents the primed variables of* α.

The first condition for extension morphisms requires all actions of the original component to be mapped to actions in the extended one, and the preservation of all the channels of the original component. In particular, it is not allowed for several channels to be mapped to a single channel in the extended component. (Notice that if this was allowed, then the extended component might not be "plugged" into architectures where the original component could be "plugged", due to insufficient "ports" in the interface.) The second and third conditions above require the types and kinds of channels and actions to be preserved. The fourth condition allows the initialisation to be strengthened when a component is extended, but respecting the initialisation of the channels of the original component, and via realisable assignments for the new variables. The fifth condition indicates that "old actions" of the extended component can modify new variables, but the only old variables these can modify are the ones they already modified in the original component (in other words, frames can be expanded only with new channels). The sixth condition establishes that both the lower and upper bound guards can be weakened, but not strengthened. Finally, the last condition establishes that the actions corresponding to those of the original component must preserve the assignments to old variables, if the lower bound guard of the original component is satisfied; this provides the extension with some freedom, to decide how the action might modify old and new variables when executed under circumstances where the original action could not be executed. Again, it is required for the assignments for new variables to be "realisable".

Going back to our example, notice that RES_Sender is indeed an extension of Sender, where the associated extension morphism $\sigma = \langle \sigma_{ch}, \sigma_{ac} \rangle$ is composed of the identity mappings σ_{ch} and σ_{ac} on channels and actions, respectively. It is clear that these mappings are injective and surjective, respectively, and that sorts and kinds of channels are preserved by σ_{ch}, and the visibility constraints on actions are preserved by σ_{ac}. Moreover, since the initialisation and the write frames, guards and assignments of actions send and prod are not modified in the extension, the last four conditions in the definition of extension morphisms are trivially met.

Notice that extension morphisms are *invasive*, in the sense that new actions in the extended component are allowed to modify variables of the original component. However, extension morphisms differ from invasive superposition morphisms, as formalised in [6] in various ways. In particular, guards are weakened in extension morphisms, whereas these are strengthened in invasive superposition morphisms. Moreover, our allowed forms of assignment and initialisation strengthening are more restricted than those of invasive superposition morphisms.

It is not difficult to prove the following theorem, showing that, as for other morphisms in CommUnity, designs and extension morphisms constitute a category.

Theorem 1. *The structure composed of CommUnity designs and extension morphisms constitutes a category, where the composition of two morphisms σ_1 and σ_2 is defined in terms of the composition of the corresponding channel and action mappings of σ_1 and σ_2.*

Proof. The proof can be straightforwardly reduced to proving that the composition of extension morphisms is an extension morphism (the remaining points to prove that the proposed structure is a category are straightforward). So, let $\sigma_1 : P_1 \to P_2$ and $\sigma_2 : P_2 \to P_3$ be extension morphisms. The composition $\sigma_1; \sigma_2$ is defined by the composition of the corresponding mappings of these morphisms.

Let us prove each of the restrictions concerning the definition of extension morphism.

- First, $\sigma_{1_{ch}}; \sigma_{2_{ch}}$ must be injective, and $\sigma_{2_{ac}}; \sigma_{1_{ac}}$ must be surjective; this is easy to show, since the composition of injective mappings is injective, and the composition of surjective mappings is surjective.
- It is clear that since both $\sigma_{1_{ch}}$ and $\sigma_{2_{ch}}$ preserve the sorts and kinds of channels, so does the composition $\sigma_{1_{ch}}; \sigma_{2_{ch}}$.
- We have as hypotheses that there exist two formulas α_1 and α_2, referring to variables in $(V_2 - \sigma_{1_{ch}}(V_1))$ and $(V_3 - \sigma_{2_{ch}}(V_2))$ respectively, and such that $\Phi_{ADT} \vdash Init_{P_2} \Leftrightarrow \underline{\sigma_1}(Init_{P_1}) \wedge \alpha_1$ and $\Phi_{ADT} \vdash Init_{P_3} \Leftrightarrow \underline{\sigma_2}(Init_{P_2}) \wedge \alpha_2$; moreover, both these formulas are "satisfiable", in the sense that $\Phi_{ADT} \vdash \exists \overline{v_1} : \alpha_1(\overline{v_1})$ and $\Phi_{ADT} \vdash \exists \overline{v_2} : \alpha_2(\overline{v_2})$. We must show that there exists a formula α_3, using only variables that are contained in $(V_3 - \sigma_{1_{ch}}; \sigma_{2_{ch}}(V_1))$, such that $\Phi_{ADT} \vdash \exists \overline{v_3} : \alpha_3(\overline{v_3})$ and $\Phi_{ADT} \vdash Init_{P_3} \Leftrightarrow \underline{\sigma_1; \sigma_2}(Init_{P_1}) \wedge \alpha_3$. We propose $\alpha_3 \hat{=} \sigma_{2_{ch}}(\alpha_1) \wedge \alpha_2$.
  - The fact that α_3 refers only to variables in $V_3 - \sigma_{1_{ch}}; \sigma_{2_{ch}}(V_1))$ is obvious.
  - Let us prove that α_3 is satisfiable. First, since α_1 is satisfiable, so is $\sigma_{2_{ch}}(\alpha_1)$ (satisfiability is preserved under injective language translation). Second, it is easy to see that $\sigma_{2_{ch}}(\alpha_1)$ and α_2 refer to disjoint sets of variables; therefore (and since the only free variables allowed in initialisation conditions are the ones corresponding to channels), the satisfiability of the conjunction $\sigma_{2_{ch}}(\alpha_1) \wedge \alpha_2$ is guaranteed.
  - Let us now prove that $\Phi_{ADT} \vdash Init_{P_3} \Leftrightarrow \underline{\sigma_1; \sigma_2}(Init_{P_1}) \wedge \alpha_3$. We know that $\Phi_{ADT} \vdash Init_{P_3} \Leftrightarrow \underline{\sigma_2}(Init_{P_2}) \wedge \alpha_2$, and that $\Phi_{ADT} \vdash Init_{P_2} \Leftrightarrow \underline{\sigma_1}(Init_{P_1}) \wedge \alpha_1$. Combining these two hypotheses, we straightforwardly get that $\Phi_{ADT} \vdash Init_{P_3} \Leftrightarrow \underline{\sigma_2}(\underline{\sigma_1}(Init_{P_1}) \wedge \alpha_1) \wedge \alpha_2$, which leads us to $\Phi_{ADT} \vdash Init_{P_3} \Leftrightarrow (\underline{\sigma_2}(\underline{\sigma_1}(Init_{P_1})) \wedge \underline{\sigma_2}(\alpha_1)) \wedge \alpha_2$, as we wanted.
- We have to prove that for every $g \in \Gamma_3$ such that $\sigma_{2_{ac}}; \sigma_{1_{ac}}(g)$ is defined, and for every $v \in V_1$, if $\sigma_{1_{ch}}; \sigma_{2_{ch}}(v) \in D_3(g)$, then $v \in D_1(\sigma_{1_{ac}}; \sigma_{2_{ac}}(g))$. This is straightforward, thanks to our hypotheses regarding frame preservation of morphisms σ_1 and σ_2.
- To prove that the composition of the morphisms σ_1 and σ_2 weakens both the lower and the upper bound guards is also straightforward.
- We have as hypotheses that:
  - for every $g \in \Gamma_2$ such that $\sigma_{1_{ac}}(g)$ is defined, there exists a formula α_1 whose referring primed variables are contained in $(V_2' - \sigma_{1_{ch}}(V_1)')$ such that: $\Phi_{ADT} \vdash \exists \overline{v_1} : \alpha_1(\overline{v_1})$ and $\Phi_{ADT} \vdash \underline{\sigma}(L_1(\sigma_{1_{ac}}(g))) \Rightarrow (R_2(g) \Leftrightarrow \underline{\sigma_1}(R_1(\sigma_{1_{ac}}(g))) \wedge \alpha_1)$,
  - for every $g \in \Gamma_3$ such that $\sigma_{2_{ac}}(g)$ is defined, there exists a formula α_2 whose referring primed variables are contained in $(V_3' - \sigma_{2_{ch}}(V_2)')$ such

that: $\Phi_{\mathcal{ADT}} \vdash \exists \overline{v_2} : \alpha_2(\overline{v_2})$ and $\Phi_{\mathcal{ADT}} \vdash \underline{\sigma}(L_2(\sigma_{2_{ac}}(g))) \Rightarrow (R_3(g) \Leftrightarrow \sigma_2(R_2(\sigma_{2_{ac}}(g))) \wedge \alpha_2)$.

Let $g \in \Gamma_3$ such that $\sigma_{2_{ac}}; \sigma_{1_{ac}}(g)$ is defined. We have to find a formula α_3 whose referring primed variables are contained in $(V_3' - \sigma_{1_{ch}}; \sigma_{2_{ch}}(V_1)')$ such that: $\Phi_{\mathcal{ADT}} \vdash \exists \overline{v_3} : \alpha_3(\overline{v_3})$ and $\Phi_{\mathcal{ADT}} \vdash \underline{\sigma}(L_1(\sigma_{2_{ac}}; \sigma_{1_{ac}}(g))) \Rightarrow (R_3(g) \Leftrightarrow \underline{\sigma_1}; \sigma_2(R_1(\sigma_{2_{ac}}; \sigma_{1_{ac}}(g))) \wedge \alpha_3)$. We propose $\alpha_3 \hat{=} \sigma_{2_{ch}}(\alpha_1) \wedge \alpha_2$. The justification of the "satisfiability" of α_3 is justified, as for the case of the initialisation, by the fact that both $\sigma_2(\alpha_1)$ and α_2 are "satisfiable", and they refer to disjoint sets of variables. Proving that $\Phi_{\mathcal{ADT}} \vdash \underline{\sigma}(L_1(\sigma_{2_{ac}}; \sigma_{1_{ac}}(g))) \Rightarrow (R_3(g) \Leftrightarrow \underline{\sigma_1}; \sigma_2(R_1(\sigma_{2_{ac}}; \sigma_{1_{ac}}(g))) \wedge \alpha_3)$ is also simple; having in mind that $\underline{\sigma}(L_1(\overline{\sigma_{2_{ac}}; \sigma_{1_{ac}}}(g)))$ is stronger than $\underline{\sigma}(L_2(\sigma_{2_{ac}}(g)))$ and $L_3(g)$, we can "expand" $R_3(g)$ into $\underline{\sigma_2}(R_2(\sigma_{2_{ac}}(g))) \wedge \alpha_2)$, and this into $(\underline{\sigma_1}; \sigma_2(R_1(\sigma_{2_{ac}}; \sigma_{1_{ac}}(g)) \wedge \alpha_1) \wedge \alpha_2)$, obtaining what we wanted.

The rationale behind the definition of extension morphisms is the characterisation of the substitutability principle (a property that can be shown to fail for invasive superposition as defined in [6]). The following result shows that, if there exists an extension morphism σ between two designs P_1 and P_2 (and this extension is realisable), then all behaviours exhibited by P_1 are also exhibited by P_2. Since superposition morphisms, used as a representation of "clientship" (strictly, the existence of a superposition morphism between two designs indicates that the first is part of the second, as a component is part of a system when the first is used by the system), restrict the behaviours of superposed components, it is guaranteed that all behaviours exhibited by a component when this becomes part of a system will also be exhibited by an extension of this component, if replaced by the first one in the system. Of course, one can also obtain *more behaviours*, resulting from the explicit use of new actions of the component. But if none of the new actions are used, then the extended component behaves exactly as the original one.

Theorem 2. *Let P_1 and P_2 be two CommUnity designs, and $\sigma : P_1 \to P_2$ an extension morphism between these designs. Then, every run of P_1 can be embedded in a corresponding run of P_2.*

Proof. For this theorem, we consider a semantics based on runs, i.e., infinite sequences of interpretations such that they all coincide on the interpretation for $\mathcal{ADT}$, the first interpretation in the sequence satisfies the initial condition and any pair of consecutive interpretations in the sequence either only differ in the interpretation of input variables (stuttering), or they are in the "'consequence" relation for one of the actions of the component.

Let $P1$ and P_2 be two CommUnity designs, and $\sigma : P_1 \to P_2$ an extension morphism between these designs. Let $s = s_0, s_1, s_2, \ldots$ be a run for P_1. We will inductively construct a sequence $s' = s_0', s_1', s_2', \ldots$ which is a run for P_2, and such that, for all i, $(s_i')|_{\sigma_{ch}(V_1)} \equiv s_i$, i.e., the reduct of each s_i' to the symbols originating in P_1 coincides with the interpretation s_i.

- Base case. The initialisation of P_2 is of the form $\underline{\sigma}(Init_{P_1}) \wedge \alpha$, with α a formula satisfying $\Phi_{\mathcal{ADT}} \vdash \exists \overline{v} : \alpha(\overline{v})$, and whose variables are "new variables", in the sense that they differ from those appearing in the initialisation of P_1. Then, there exists an interpretation I_α of the variables in α that makes it true. We define s_0' as the extension of the interpretation s_0, appropriately translated via σ, with the interpretation I_α for the remaining variables. Clearly, this interpretation satisfies the initial condition of P_2, and its reduct to the language of P_1 coincides with s_0.

- Inductive step. Assuming that we have already constructed a prefix $s' = s_0', s_1', s_2', \ldots, s_i'$ of the run s', we build the interpretation s_{i+1}' in the following way. We know that s_{i+1} is in one of the following two cases:
  - s_{i+1} is reached from s_i via stuttering. In such a case, we define $s_{i+1}' \hat{=} s_i'$, and clearly, by inductive hypothesis, we have that the reduct of s_{i+1}' to the variables of P_1 coincides with s_{i+1}, and (s_i', s_{i+1}') are in the "stuttering relationship".
  - there exists some action $g \in \Gamma_1$ such that (s_i, s_{i+1}) are in the consequence relationship corresponding to g. If this is the case, notice that, under the "stronger guard" $L_1(g)$, the assignment of an action g_2 in $\sigma_{ac}^{-1}(g)$ (which is nonempty, since σ_{ac} is surjective) is of the form $\underline{\sigma}(R_1(\sigma_{ac}(g))) \wedge \alpha)$, for a formula α referring only to the primed versions of new variables. Since we know that $\Phi_{\mathcal{ADT}} \vdash \exists \overline{v} : \alpha(\overline{v})$, there exists an interpretation I_α of the variables in α that makes is true. We define s_{i+1}' as the extension of the interpretation s_{i+1}, with symbols appropriately translated via σ, with I_α for the interpretation of the remaining variables. It is straightforward to see that (s_i', s_{i+1}') are in the consequence relation of g_2, and that the reduct of s_{i+1}' to the variables originating in P_1 coincides with s_{i+1}.

3.1 Replacing Components by Extensions in Configurations

The intention of extension morphisms is to characterise component extension, respecting the substitutability principle. One might then expect that, if a component C can be "plugged" into an architecture of components, then we should be able to plug an extension C' in the architecture, instead of C. Due to the restrictions for valid extension, it can be guaranteed that a design in a well formed diagram can be replaced with an extension of it, preserving the wellformedness of the diagram (although it is necessary to consider an "open system" semantics, since extensions might introduce new input variables, which would be "disconnected" after the replacement). Moreover, we can also prove that the colimit of the new diagram (where a component was replaced by an extension of it) is actually an extension of the colimit of the original diagram. This basically means that the joint behaviour of the original system is *augmented* by the extension of a component, but never restricted (i.e., the resulting system exhibits all the behaviours of the original one, and normally also more behaviours).

We are not in a similar situation when combining extension and refinement. As we mentioned, refinement plays an important role in the implementation of higher-order connectors, since it allows us to "instantiate" roles with actual components. Roles, as the Abstract_Sender example, specify the minimum requirements that have to be satisfied in order to be able to plug components using a higher-order connector. Notice that, in particular, the interval determined by the guards of actions of the role has to be preserved or reduced by the actual parameter, i.e., the component with which the role is instantiated. Consider, for instance, the case of the `Abstract_Sender`. As we mentioned, the design `Sender` refines this more abstract `Abstract_Sender`, and therefore we can instantiate the abstract sender with the concrete one. Moreover, `RES_Sender`, an extension of `Sender`, also refines `Abstract_Sender`, so it also can instantiate this role. However, since extensions weaken both guards, it is not difficult to realise that, if a component B refines a component A, and B' is an extension of B, then it is not necessarily the case that B' also refines A. With respect to configurations of systems, this means that, when replacing components by corresponding extensions, one might lose the possibility of applying or using some higher-order connectors.

Although this might seem an unwanted restriction, it is actually rather natural. The conditions imposed by roles of a higher-order connector are a kind of "rely-guarantee" assumptions. When extending a component we might lose some properties the role requires for the component.

4   An Example Using Extension

Let us go back to our example of communicating components via an unreliable channel. As we explained in previous sections, we would like to superpose behaviour on the existing architecture, to make the communication reliable by implementing a reset in the communication when packets are lost. The mechanism we used was very simple, and required a "reset" operation on the sender, which, as we discussed, can be achieved by component extension. In order to complete the enhanced architecture to implement the reset acknowledgement mechanism, we need a monitor that, if it detects a missing packet, issues a call for reset. The idea is that, if a message is not what the monitor expected (characterised by the `msg-exp`), then it will go to a "reset" cycle, and wait to see if the expected packet arrives. If the expected packet arrives, then the component will start waiting for the next packet. Notice that, for the sake of simplicity, we assume that the communication between the monitor and the extended sender is reliable. The monitor used for this task is shown in Figure 10. The final architecture for the system is shown in Figure 11.

Notice that, since the superposed monitor is *spectative*, we can guarantee that, if the augmented system works without the need for reset in the communication, i.e., no messages are lost, then its behaviour is exactly the same as the one of the original architecture with unreliable communication.

5   Related Work

The original work on CommUnity took its inspiration from languages like Unity [3] and IP [8] and on related software engineering research [12] using superimposition/superposition as structuring principles. Recently, research by Katz and

```
Design RES_Monitor
in
     msg: int
out
     msg-rst: int
prv
     msg-exp: int
     w: bool
init
     msg-rst=0 ∧ msg-exp=0 ∧ w=true
do
  rec1[msg-exp]: w ∧ msg-exp=msg  → msg-exp'=msg-exp+1
  rec2[msg-exp,msg-rst,w]: w ∧ msg-exp≠msg →
                          msg-exp'=msg-exp ∧ msg-rst'=msg-exp ∧ w'=false
  rec3[msg-exp]: ¬w → msg-exp'=msg-exp
  res[w]: ¬w → w'=true
```

Fig. 10. A monitor for detecting lost packets.

his collaborators has recognised the usefulness of superimposition as a way of characterising aspects [13,11,22]. Especially in [11], there is a recognition of the same principles we espouse in this paper, namely that aspects should be characterised and applied at the architectural level of software development. Aspects are seen as patterns to be applied to underlying architectures (which may already have been modified by the application of previous concerns), based on specifications of the aspects. These specifications include descriptions of components and connectors used to define the aspect, as well as "dummy" components defining required services in order to be able to apply the aspect. The relationships and structuring mechanisms and the instantiation of the "dummy" components are explained in terms of superimpositions.

The motivation for our research is very similar, we want to lift the treatment of aspects to the architectural level and view the application of aspects to the design of some underlying system as the application of a transformation defined by the aspect design to the underlying architecture, resulting in an augmented architecture. The application of various aspects can be seen as the application of a sequence of transformations to the underlying architecture (see [2]). This raises concerns analogous to those discussed in [11]. In order to develop this framework, we found it necessary to come to a better understanding of invasive superpositions in the context of CommUnity. In particular, we needed to characterise a structured use of invasive superpositions, which allows arbitrary changes breaking encapsulation of the component being superimposed. As noted earlier, this problem has also arisen in the context of object oriented design and programming, resulting in the various proposals attempting to characterise the concept of substitutivity ([15]). We believe that this proposal provides a more solid foun-

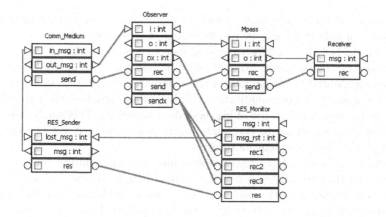

Fig. 11. The architecture of the system, with the reset mechanism.

dation for substitutivity, one that is better structured and more amenable to analysis.

Of course, the work reported in [16,17] is related to our work, both because it is based on CommUnity and because it recognises that the concept of higher order connector (a kind of parameterised connector that can be applied to other connectors to obtain more sophisticated connectors) can be used to characterise certain aspects. Again the emphasis is on using the specification of the aspect, as a higher order connector, to transform an existing architectural pattern in order to apply the aspect. As we demonstrate in this paper, some interesting aspects cannot be characterised in terms of this mechanism alone and it is necessary to consider transformations that apply uniformly to connectors and to the components they connect. Furthermore, some of the transformations require the use of invasive superpositions, as in the main example used in this paper. This is a subject that has received very little scrutiny in the CommUnity literature.

6   Conclusion

We have studied a special kind of invasive superposition for the characterisation of extensions between designs in the CommUnity architecture design language. This kind of morphism, that we have defined with special concern regarding the substitutability principle [15] (an essential property associated with sound component extension), allows us to complement the refinement and (regulative) superposition morphisms of CommUnity, and obtain a suitable formal framework to characterise certain *aspects*, in the sense of aspect oriented software development. We have argued that some useful aspects require extensions on the components, as well as in the connectors, and therefore the introduced extension morphisms are necessary. Also, having the possibility of extending components provides us a way of balancing the distribution of augmented behaviour in the connectors and the components, which would otherwise be put exclusively on the connector side (typically by means of higher-order connectors).

We illustrated the need for extension morphisms by means of a simple case study based on the communication of two components via an unreliable channel. We then augmented the behaviour of this original system with a fault tolerance aspect for making the communication reliable, which required the extension of components, as well as the use of higher-order connectors. This small case study also allowed us to illustrate the relationships and combined use of extension, superposition and refinement morphisms.

As we mentioned before, This problem has also arisen in the context of object oriented design and programming, attempting to define various forms of inheritance, resulting in the proposals attempting to characterise the concept of substitutability [15,21]. We believe that this proposal provides a more solid foundation for substitutivity, one that is better structured and more amenable to analysis. The definition of extension in CommUnity that we introduced has been partly motivated by the definitions and proof obligations used to define the structuring mechanisms in B [1,4], that justifies the notion of substitutivity and provides a structuring principle for augmenting components by breaking the encapsulation of the component.

References

1. J.-R. Abrial, *The B-Book, Assigning Programs to Meanings*, Cambridge University Press, 1996.
2. N. Aguirre, T. Maibaum and P. Alencar, *Abstract Design with Aspects*, submitted, 2005.
3. K. Chandy, J. Misra, *Parallel Program Design - A Foundation*, Addison-Wesley, 1988.
4. T. Dimitrakos, J. Bicarregui, B. Matthews and T. Maibaum, *Compositional Structuring in the B-Method: A Logical Viewpoint of the Static Context*, in Proceedings of the International Conference of B and Z Users ZB2000, York, United Kingdom, LNCS, Springer-Verlag, 2000.
5. J. Fiadeiro, *Categories for Software Engineering*, Springer, 2004.
6. J. Fiadeiro and T. Maibaum, *Categorical Semantics of Parallel Program Design*, in Science of Computer Programming 28(2-3), Elsevier, 1997.
7. R. Filman, T. Elrad, S. Clarke and M. Aksit, *Aspect-Oriented Software Development*, Addison-Wesley, 2004.
8. N. Francez and I. Forman, *Interacting Processes*, Addison-Wesley, 1996.
9. D. Garlan and R. Monroe and D. Wile, *ACME: An Architecture Description Interchange Language*, in Proceedings of CASCON'97, Toronto, Ontario, 1997.
10. J. Goguen, *Categorical Foundations for General System Theory*, in F.Pichler and R.Trappl (eds), Adavaces in Cybernetics anda Systems Research, Transcripta Books, pages 121-130, 1973.
11. M. Katara and S. Katz, *Architectural Views of Aspects*, in Proceedings of International Conference on Aspect-Oriented Software Design AOSD 2003, 2003.
12. S. Katz, *A Superimposition Control Construct for Distributed Systems*, ACM Transactions on Programming Languages and Systems, 15:337-356, 1993.

13. S. Katz and J. Gil, *Aspects and Superimpositions*, ECOOP Workshop on Aspect Oriented Programming, 1999.
14. G. Kiczales, *An overview of AspectJ*, in Proceedings of the European Conference on Object-Oriented Programming ECOOP 2001, Lecture Notes in Computer Science, Budapest, Hungary, Springer-Verlag, 2001.
15. B. Liskov and J. Wing, *A Behavioral Notion of Subtyping*, ACM Transactions on Programming Languages and Systems, Vol 16, No 6, ACM Press, November 1994.
16. A. Lopes, M. Wermelinger and J. Fiadeiro, *Higher-Order Architectural Connectors*, ACM Transactions on Software Engineering and Methodology, vol. 12 n. 1, 2003.
17. A. Lopes and J. Fiadeiro, *Superposition: Composition vs. Refinement of Non-Deterministic, Action-Based Systems*, in Formal Aspects of Computing, Vol. 16, N. 1, Springer-Verlag, 2004.
18. D. Luckham, J. Kenney, L. Augustin, J. Vera, D. Bryan and W. Mann, *Specification and Analysis of System Architecture Using Rapide*, IEEE Transactions on Software Engineering, Special Issue on Software Architecture, 21(4), 1995.
19. J. Magee, N. Dulay, S. Eisenbach and J. Kramer, *Specifying Distributed Software Architectures*, in Proceedings of the 5th European Software Engineering Conference (ESEC '95), Sitges, Spain, Lecture Notes in Computer Science, Springer-Verlag, 1995.
20. N. Medvidovic, P. Oreizy, J. Robbins and R. Taylor, *Using Object-Oriented Typing to Support Architectural Design in the C2 Style*, in Proceedings of the ACM SIG-SOFT '96 Fourth Symposium on the Foundations of Software Engineering, ACM SIGSOFT, San Francisco, CA, 1996.
21. B. Meyer, *Object-Oriented Software Construction*, Second Edition, Prentice Hall, 2000.
22. M. Sihman and S. Katz, *Superimpositions and Aspect-Oriented Programming*, BCS Computer Journal, vol. 46, 2003.
23. M. Wermelinger, A. Lopes and J. Fiadeiro, *A Graph Based Architectural (Re)configuration Language*, in ESEC/FSE'01, V.Gruhn (ed), ACM Press, 2001.
24. M. Wermelinger, C. Oliveira, *The CommUnity Workbench*, In Proc. of the 24th Intl. Conf. on Software Engineering, page 713. ACM Press, 2002.

Non-intrusive Formal Methods and Strategic Rewriting for a Chemical Application

Oana Andrei[1], Liliana Ibanescu[2], and Hélène Kirchner[3]

[1] INRIA - LORIA
[2] ULP Strasbourg
[3] CNRS - LORIA
Campus scientifique BP 239
F-54506 Vandoeuvre-lès-Nancy Cedex, France
First.Last@loria.fr

Abstract. The concept of formal islands allows adding to existing programming languages, formal features that can be compiled later on into the host language itself, therefore inducing no dependency on the formal language. We illustrate this approach with the TOM system that provides matching, normalization and strategic rewriting, and we give a formal island implementation for the simulation of a chemical reactor.

1 Introduction

Concerned by the crucial need for improvement of existing software in their logic, algorithmic, security and maintenance qualities, formal methods are more and more used in the software design process. Usually they come into play both at the design and verification levels either for formal specification or high-level programming. But this approach does not take into account existing software, while billions of code lines are executed every day. This might be one of the reasons why formal methods did not yet fully succeed at the industrial level.

Among many formal method approaches, algebraic techniques providing a clear semantics for signatures and rewrite rules are used in high-level languages and environments like ASF+SDF [25], Maude [9], CafeOBJ [11], or ELAN [6,18] which have been designed on these concepts. These rule-based systems have gained considerable interest with the development of efficient compilers. However, when programs are developed in these languages, they can hardly interact with programs written in another language like C or Java.

The work presented here proposes an alternative reconciling the use of algebraic formal features with the widely used object-oriented language Java. This is possible through the *Formal Islands* approach developed in the Protheo project team since a few years [21]. A formal island is a piece of code introducing formal features. These new features are anchored in terms of the available functionalities of the host language. Once compiled, these features are translated into pure host language constructs preserving the behavior of the program. The formal island concept is implemented through the software system TOM [3] which

K. Futatsugi et al. (Eds.): Goguen Festschrift, LNCS 4060, pp. 194–215, 2006.

is built upon the concepts of rules and strategic rewriting. TOM is a good language for programming by pattern-matching, and it is particularly well-suited for programming various transformations on trees/terms or XML data-structures. Moreover, its compiler has been designed with the TOM language.

The approach and the use of TOM are illustrated in this paper with a specific example: we apply strategic rewriting to model a chemical reactor by means of a formal island implementation. The considered problem is the automated generation of reaction mechanisms: a set of molecules and a list of generic elementary reactions (reaction patterns) are given as input to a generator that produces the list of all possible elementary reactions according to a specific reactor dynamics. The solution of this problem consists of generating all possible reactions and collecting all products starting from a small set of reactants. We are therefore interested only in the qualitative aspects of this problem.

A number of software systems [5,14,22,30] have been developed for the automated generation of reaction mechanisms [10,24]. As far as literature says, these systems are implemented using traditional programming languages, employing rather ad-hoc data structures and procedures for the representation and transformations of molecules (e.g. Boolean adjacency matrices and matrices transformations). Furthermore, existing systems are limited, sometimes by their implementation technology, to acyclic species, or mono-cyclic species, whereas combustion mechanisms often involve aromatic species, which are polycyclic.

In the GasEl project [7,8,16] we already have explored the use of rule-based systems and strategies for the problem of automated generation of kinetics mechanisms [24,10] in the whole context of its use by chemists and industrial partners. In GasEl the representation of chemical species uses the notion of molecular graphs, encoded by a term structure called GasEl term [7] which is inspired by the linear notation SMILES [31]. The graph isomorphism test is based on the Unique SMILES algorithm [31] which provides a unique notation for each chemical structure regardless of the many possible equivalent description of the structure that might be input; the order of this algorithm is $N^2 log_2 N$ where N is the number of atoms in the structure. Reactions patterns are encoded by a set of conditional rewriting rules on GasEl terms. The molecular graph rewriting relation is simulated by a rewriting relation on equivalence classes of terms [8]. The control of the chemical reactions chaining (i.e. the reactor dynamics) is described using a strategy language [7]. GasEl prototype is implemented in ELAN [6,18], encoding a set of nine reaction patterns. Qualitative validations have been performed with chemists [16].

The formal background of strategic rewriting is quite relevant for the considered problem: (i) chemical reactions are naturally expressed by chemists themselves using conditional rules; (ii) matching power associated with rewriting allows retrieving patterns in chemical species; (iii) defining the control on rules is essential for designing automated mechanisms generators in a flexible way and for controlling combinatorial explosion. This gives the possibility to the chemist of activating and deactivating reactions patterns, and of tuning their application during each stage. The main technical difficulty with ELAN implementation consisted in the encoding of reaction patterns on GasEl terms that correctly simulates the corresponding transformation on molecular graphs. The TOM implementation provides another approach to this problem, while keeping the same molecular graph rewriting relation, and preserving the same chemical principles and hypotheses as in GasEl.

The paper is structured as follows. Section 2 presents the formal island concept that will be further illustrated in the sequel. The TOM system is briefly described in Section 3 and the main language constructions needed to understand the considered application are introduced. Section 4 is devoted to the chemical example and explains what kind of reactor is modelled. Section 5 addresses the formal island implementation of the chemical reactor and details the different steps performed to achieve the Java implementation. Finally Section 6 draws some conclusions and perspectives for future work.

2   Formal Islands

Since several years, we have been strongly concerned with the feasibility of strategic rewriting as a practical programming paradigm [1,18]. The development of efficient compilation concepts and techniques took an important place in the language support design. The results presented in [20] led to a quite efficient implementation and thus demonstrated the practicality of the paradigm.

Making strategic rewriting easily available in many programming languages was the main concern that led to the emergence of *formal island*. This concept provides a general way to make formal methods, and in particular matching and rewriting, widely available.

We use the notions of *formal island* and *anchoring* to extend an existing language with formal capabilities. A formal island is a piece of code introducing formal features, while anchoring means to describe these new features in terms of the available functionalities of the host language. Once compiled, these features are translated into pure host language constructs, allowing us to say that the formal islands are *not intrusive* with respect to the behavior of the application.

In the following we review the definitions of representation functions and formal anchor for the unsorted case.

In order to precisely define these notions, we recall a few concepts of first order term algebra needed here [17]. A *signature* $\mathcal{F}$ is a set of function symbols, each one associated to a natural number by the arity function, $ar : \mathcal{F} \rightarrow \mathbb{N}$. $\mathcal{F}_n$ is the set of function symbols of arity n, $\mathcal{F}_n = \{f \in \mathcal{F} \mid ar(f) = n\}$. $\mathcal{T}(\mathcal{F}, \mathcal{X})$ is the set of *terms* from a given finite set $\mathcal{F}$ of function symbols and a denumerable set $\mathcal{X}$ of variable symbols. A *position* within a term is represented as a sequence ω of positive integers describing the path from the root of the term to the root of the subterm at that position, denoted by $t_{|\omega}$. $Symb(t)$ is a partial function from $\mathcal{T}(\mathcal{F}, \mathcal{X})$ to $\mathcal{F}$ which associates to each term t its root symbol $f \in \mathcal{F}$. The set of variables occurring in a term t is denoted by $Var(t)$. If $Var(t)$ is empty, t is called a *ground term* and $\mathcal{T}(\mathcal{F})$ is the set of ground terms. We write $t_1 = t_2$ when t_1 and t_2 are syntactically equal.

Definition 1. ([19]) *Given a tuple composed of a signature $\mathcal{F}$, a set of variables $\mathcal{X}$, booleans $\mathbb{B}$ and integers $\mathbb{N}$, given sets of host language constructs $\Omega_{\mathcal{F}}$, $\Omega_{\mathcal{X}}$, $\Omega_{\mathcal{T}}$, $\Omega_{\mathbb{B}}$, and $\Omega_{\mathbb{N}}$, we consider a family of representation functions $\ulcorner\,\urcorner$ that map:*

- *function symbols $f \in \mathcal{F}$ to elements of $\Omega_{\mathcal{F}}$, denoted $\ulcorner f \urcorner$,*
- *variables $v \in \mathcal{X}$ to elements of $\Omega_{\mathcal{X}}$, denoted $\ulcorner v \urcorner$,*
- *ground terms $t \in \mathcal{T}(\mathcal{F})$ to elements of $\Omega_{\mathcal{T}}$, denoted $\ulcorner t \urcorner$,*
- *booleans $b \in \mathbb{B} = \{\top, \bot\}$ to elements of $\Omega_{\mathbb{B}}$, denoted $\ulcorner b \urcorner$,*
- *natural numbers $n \in \mathbb{N}$ to elements of $\Omega_{\mathbb{N}}$, denoted $\ulcorner n \urcorner$.*

Definition 2. ([19]) *Given a tuple $\langle \mathcal{F}, \mathcal{X}, \mathcal{T}(\mathcal{F}), \mathbb{B}, \mathbb{N} \rangle$ and the operations* eq: $\Omega_{\mathcal{T}} \times \Omega_{\mathcal{T}} \to \Omega_{\mathbb{B}}$, is_fsym: $\Omega_{\mathcal{T}} \times \Omega_{\mathcal{F}} \to \Omega_{\mathbb{B}}$, *and* subterm$_f$: $\Omega_{\mathcal{T}} \times \Omega_{\mathbb{N}} \to \Omega_{\mathcal{T}}$ *($f \in \mathcal{F}$), a representation function $\ulcorner\,\urcorner$ is a formal anchor if it preserves the structural properties of $\mathcal{T}(\mathcal{F})$ in $\ulcorner \mathcal{T}(\mathcal{F}) \urcorner$ by the semantics of* eq, is_fsym, *and* subterm$_f$:

$$\forall\ t, t_1, t_2 \in \mathcal{T}(\mathcal{F}), \forall f \in \mathcal{F}, \forall i \in [1..ar(f)]\ :$$

$$\mathsf{eq}(\ulcorner t_1 \urcorner, \ulcorner t_2 \urcorner) \equiv \ulcorner t_1 = t_2 \urcorner$$
$$\mathsf{is\_fsym}(\ulcorner t \urcorner, \ulcorner f \urcorner) \equiv \ulcorner Symb(t) = f \urcorner$$
$$\mathsf{subterm}_f(\ulcorner t \urcorner, \ulcorner i \urcorner) \equiv \ulcorner t_{|i} \urcorner\ if\ Symb(t) = f$$

We illustrate the concept of formal anchor with a small example from [19]:

Example 1. In C or Java like languages, the notation of *term* can be implemented by a record *(sym:integer, sub:array of term)*, where the first slot *(sym)* denotes the top symbol, and the second slot *(sub)* corresponds to the subterms. It is easy to check that the following definitions of eq, is_fsym, and subterm$_f$ (where = denotes an atomic equality) provide a formal anchor for $\mathcal{T}(\mathcal{F})$:

$$\mathsf{eq}(t_1, t_2) \triangleq t_1.sym = t_2.sym\ \wedge\ \forall i \in [1..ar(t_1.sym)],$$
$$\mathsf{eq}(t_1.sub[i], t_2.sub[i])$$
$$\mathsf{is\_fsym}(t, f) \triangleq t.sym = f$$
$$\mathsf{subterm}_f(t, i) \triangleq t.sub[i]\ if\ t.sym = f \wedge i \in [1..ar(f)]$$

3   TOM

TOM is an implementation of the idea of formal island [21]. TOM [3] provides matching, normalization, and strategic rewriting in Java, C, and Caml [21,15]. In particular, we have used Java for developing the chemical application described in this paper. In each of the three instances, matching and rewriting primitives can be combined with constructs of the programming language, then compiled to the host language, using similar techniques as for compiling ELAN. The normal forms provided by rewriting are available to get conciseness and expressiveness in programs written in the host language. Moreover one can prove that these sets of rewrite rules have useful properties like termination or confluence. Once the programmer has used rewriting to specify functionalities and to prove properties, the compilation dissolves this formal island in the existing code. The TOM constructs are non-intrusive because their use induces no dependence: once compiled, a TOM program contains no more trace of the rewriting and matching statements that were used to build it.

Basically, a TOM program is a list of blocks, where each block is either a TOM construct, or a sequence of characters. The idea is that after transformation, the sequence of characters merged with the compiled TOM constructs becomes a valid host language program having the same behavior as the initial TOM program.

The main construct, %match, is similar to the match primitive found in functional languages: given an object (called *subject*) and a list of *patterns-actions*, the match primitive selects the first pattern that matches the subject and performs the associated action. The subject against which we match can be any object, but in practice, this object is usually a tree-based data-structure, also called a term in the algebraic programming community. The match construct may be seen as an extension of the classical switch/case construct. The main difference is that the discrimination occurs on a term and not on atomic values like characters or integers: the patterns are used to discriminate and retrieve information from an algebraic data structure.

In addition to %match TOM provides the %rule construct which allows describing rewrite rule systems. This construct supports conditional rewrite rules as well as rules with matching conditions (as in ELAN or ASF+SDF). By default, TOM rules provide a leftmost innermost normalization strategy which computes normal forms in an efficient way. It is of course possible to combine these features with more complex strategies, like generic traversal strategies, to describe more complex or generic transformations. When understanding all the possibilities offered by TOM, this general purpose system becomes as powerful and expressive as many specific rewrite rule based programming languages.

Another construct of TOM is the backquote ('). This construct is used for building an algebraic term or to retrieve the value of a TOM variable (a variable instantiated by pattern-matching).

The %vas construct allows the user to define a many-sorted signature. This construct is replaced at compile time by the content of the generated formal anchor.

Other available constructs like %typeterm, %typelist, and %op which define the formal anchor between signature formalism and concrete implementations (Java classes) allow performing pattern matching against any data structure.

In order to make easier the use of TOM, two tools were developed: ApiGen and Vas [3]. ApiGen is a system which takes a many-sorted signature as input, and generates both a concrete implementation for the abstract data-type (for example Java classes), and a formal anchor for TOM. Vas is a preprocessor for ApiGen which provides a human-readable syntax definition formalism inspired from SDF. These two systems are useful for manipulating Abstract Syntax Trees since they offer an efficient implementation based on ATerms [26] which supports maximal memory sharing, strong static typing, as well as parsers and pretty-printers.

TOM provides a library inspired by ELAN, Stratego [27], and JJTraveler [29], which allows us to easily define various kinds of traversal strategies. Figure 1 provides an algebraic view of elementary strategy constructors, and defines their evaluation using the application operator @. We note that, according to the definition, if c is a constant operator then $All(s)@(c)$ returns c, while $One(s)@(c)$ returns $failure$. In this context, the application of a strategy to a term can fail. In Java, the failure is implemented by an exception (`VisitFailure`).

$$
\begin{aligned}
&Identity@(t) && => t \\
&Fail@(t) && => failure \\
&Sequence(s_1, s_2)@(t) && => failure \text{ if } s_1@(t) \text{ fails} \\
& && \quad s_2@(t') \text{ if } s_1@(t) => t' \\
&Choice(s_1, s_2)@(t) && => t' \text{ if } s_1@(t) => t' \\
& && \quad s_2@(t) \text{ if } s_1@(t) \text{ fails} \\
&All(s)@(f(t_1, ..., t_n)) && => f(t'_1, ..., t'_n) \text{ if } s@(t_1) => t'_1, ..., s@(t_n) => t'_n \\
& && \quad failure \text{ if there exists } i \text{ such that } s_i@(t_i) \text{ fails} \\
&One(s)@(f(t_1, ..., t_n)) && => f(t_1, ..., t'_i, ..., t_n) \text{ if } s@(t_i) => t'_i \\
& && \quad failure \text{ if } s@(t_1) \text{ fails}, ..., s@(t_n) \text{ fails} \\
&Omega(i, s)@(f(t_1, ..., t_n)) && => f(t_1, ..., t'_i, ..., t_n) \text{ if } s@(t_i) => t'_i \\
& && \quad failure \text{ if } s@(t_i) \text{ fails}
\end{aligned}
$$

Fig. 1. Strategy constructors

These strategy constructors are the key-component that can be used to define more complex strategies. In order to define recursive strategies, the μ abstractor was introduced. This allows giving a name to the current strategy, which can be referenced later. Using strategy operators and the μ abstractor, new strategies can be defined [28] as illustrated by Figure 2.

$$
\begin{aligned}
&Try(s) && = Choice(s, Identity) \\
&Repeat(s) && = \mu x.Choice(Sequence(s, x), Identity()) \\
&BottomUp(s) && = \mu x.Sequence(All(x), s)) \\
&TopDown(s) && = \mu x.Sequence(s, All(x))) \\
&Innermost(s) && = \mu x.Sequence(All(x), Try(Sequence(s, x)))
\end{aligned}
$$

Fig. 2. Examples of strategies

The Try strategy never fails: it tries to apply the strategy s; if it succeeds, the result is returned, otherwise, the $Identity$ strategy is applied, and the subject is not modified.

The $Repeat$ strategy applies the strategy s as many times as possible, until a failure occurs. The last unfailing result is returned.

The strategy $BottomUp$ tries to apply the strategy s to all nodes, starting from the leaves. Note that the application of s should not fail, otherwise the whole strategy also fails.

The *TopDown* strategy tries to apply the strategy s to all nodes, starting from the root. It fails if the application of s fails at least once.

The strategy *Innermost* tries to apply s as many times as possible, starting from the leaves. This construct is useful to compute normal forms.

4   Strategic Rewriting for a Chemical Reactor

The purpose of an automated generator of detailed kinetic mechanisms is to take as input one or more hydrocarbon molecules and the reaction conditions, and to give as output a reaction model, i.e. the list of applied reactions. We are interested only in exhaustive generation of chemical reactions, therefore we consider only the qualitative aspects of the model; the quantitative or probabilistic features are treated separately by the chemists. For this kind of modeling the two dimensional model of molecules is sufficient.

In this section we present the model used for the representation of chemical species, the reaction pattern we considered, and the reactor dynamics.

4.1   Molecular Graphs

We now describe formally the chemical model we want to implement.

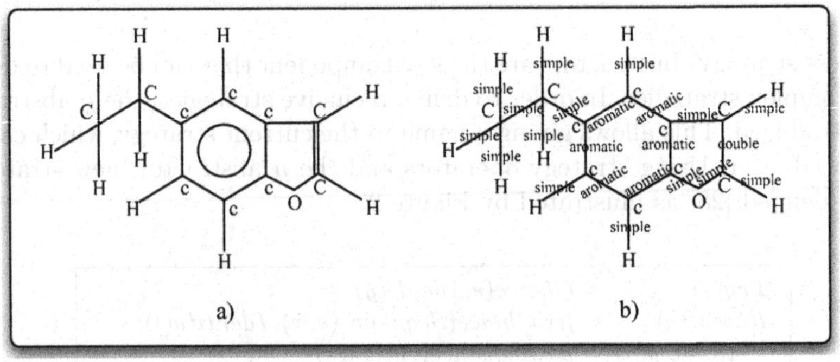

Fig. 3. Molecular graphs

A molecular graph [12] is a vertex-labelled and edge-labelled graph, where each vertex is labelled with an atom and each edge graphically suggests the bond type or is explicitly labelled with the bond type, as illustrated in Figure 3. A chemical reaction is expressed as a rewriting rule for molecular graphs. Figure 4 gives an example of a chemical reaction.

4.2   Rules for Decorated Labelled Graphs

In the so-called primary mechanism, a set of nine reaction patterns is applied to an initial mixture of molecules. A complete description of the involved reaction

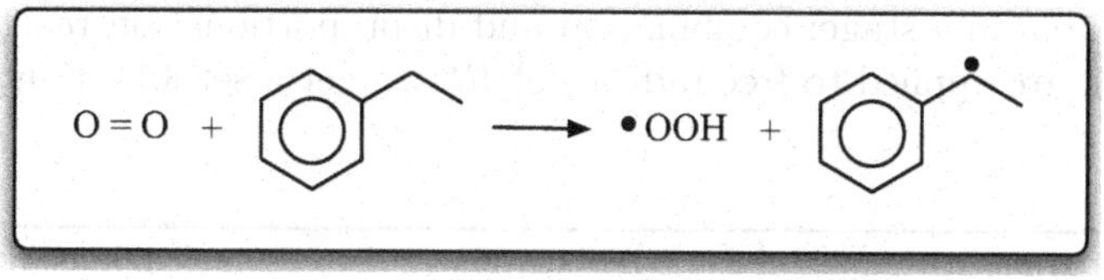

Fig. 4. Bimolecular initiation for ethylbenzene

patterns is out of the scope of this paper, but the chemistry-like presentation from Figure 5 gives the flavor of the transformations needed to be encoded.

Name	Description					
ui	$x - y$	$\longrightarrow$	$x\bullet$	$+$	$\bullet y$	
bi	$O = O$	$+$	$H - x$	$\longrightarrow$	$\bullet OOH$	$+$	$\bullet x$
ipso	$\bullet H$	$+$	$Ar - x$	$\longrightarrow$	$H - Ar$	$+$	$\bullet x$
me	$\bullet\beta$	$+$	$H - x$	$\longrightarrow$	$\beta - H$	$+$	$\bullet x$
bs	$\bullet x - y - z$	$\longrightarrow$	$x = y$	$+$	$\bullet z$	
ox	$O = O$	$+$	$H - x - y\bullet$	$\longrightarrow$	$\bullet OOH$	$+$	$x = y$
co.O.	$\bullet O \bullet$	$+$	$\bullet x$	$\longrightarrow$	$\bullet O - x$	
co	$\bullet x$	$+$	$\bullet y$	$\longrightarrow$	$x - y$	
di	$\bullet x$	$+$	$H - y - z\bullet$	$\longrightarrow$	$x - H$	$+$	$y = z$

Fig. 5. Reaction patterns of primary mechanism: patterns involve simple $(-)$ or double $(=)$ bonds, free radicals $(\bullet x)$, specific atoms $(\mathbf{O}, \mathbf{H})$; variables x, y, z can be instantiated by any reactants

Every reaction pattern is also guarded by "chemical filters", i.e. chemical conditions of applications, not mentioned here, even if several of them are currently implemented: they include considerations on the number of atoms in involved molecules or free radicals, the type of radicals or the type of bonds, etc. Some of them are discussed in [10].

4.3   Primary Mechanism

The primary mechanism can be described as the result of three stages (see Figure 6):

1. The *initiation* stage: unimolecular and bimolecular initiation reactions, (ui) and (bi), are applied to initial reactants, i.e. to the initial mixture of molecules. Let $RS_1 = RS$ be the set of all reactants that can be obtained.
2. The *propagation* stage: the set of reactions, (ipso), (me), (bs), (ox), and (co.O.), are applied to all reactants in RS_i to obtain a new set RS_{i+1} of reactants. The reactants from RS_i are then added to RS_{i+1}. This step is iterated until no new reactant is generated.

202   Oana Andrei, Liliana Ibanescu, and Hélène Kirchner

3. The *termination* stage: combination and disproportionation reactions, (co) and (di), are applied to free radicals of RS_i to get a set RS' of molecules.

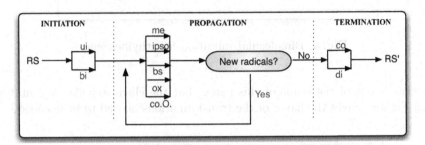

Fig. 6. Primary mechanism

The initial mixture of molecules, RS, is a finite set of reactants. Working only on the qualitative aspects of this chemical problem, we are not interested in the quantity or concentration of each reactant; hence, for each element in the current set of reactants we consider to have an infinite supply.

The set of reaction rules R is partitioned into three sets R_i, R_p, and R_t where $R_i = \{(\text{ui}), (\text{bi})\}$, $R_p = \{(\text{me}), (\text{ipso}), (\text{bs}), (\text{ox}), (\text{co.O.})\}$, and $R_t = \{(\text{co}), (\text{di})\}$.

For expository reasons we consider here that all reactions have the generic form $m_1 + m_2 \to m_1' + m_2'$, where at most one reactant in each side of the rule can be a "dummy" reactant which is always present in the set of reactants.

```
𝒫(S) UNIT (R : 𝒫(R), P₁ : 𝒫(S) [, P₂ : 𝒫(S)])
  begin
    P' := ∅;
    while(¬terminate()) do
    (m₁, m₂) := select(P₁ [, P₂]);
      for all (m₁ + m₂ → m₁' + m₂') ∈ R
          P' := insert(P', m₁', m₂');
      fi
    od
    return P'
  end
```

Fig. 7. The UNIT algorithm

The algorithms for the reactor dynamics for each stage have a common part, which we call UNIT (Figure 7), parametrized by a set of reaction rules and one or two input sets of reactants. $select(P_1)$ returns randomly each time a new pair of reactants from P_1 without removing them from P_1. As expected, $select(P_1, P_2)$ returns randomly a new pair of reactants, first from P_1 and the second from P_2, without removing the reactants from the two sets. $insert(P, m_1', m_2')$ adds the two reaction products m_1' and m_2' to P if they are not already in P. The function $terminate()$ returns $false$ as long as there are reactants that can interact by means of rules from R.

$\mathcal{P}(S)$ AlgInit $(P_0 : \mathcal{P}(S))$
  begin
    return $P_0 \cup \text{UNIT}(R_i, P_0)$
  end

$\mathcal{P}(S)$ AlgTermin $(P_0 : \mathcal{P}(S))$
  begin
    return $P_0 \cup \text{UNIT}(R_t, P_0)$
  end

$\mathcal{P}(S)$ AlgPropag $(P_0 : \mathcal{P}(S))$
  begin
    $i := 0$;
    $P'' := \emptyset$;
    repeat
      $P'' := P'' \cup P_i$;
      $P_{i+1} := \text{UNIT}(R_p, P'', P_i) \setminus P''$;
      $i := i + 1$;
    until $P_i = \emptyset$;
    return P'';
  end

Fig. 8. The stage algorithms

Now the algorithms for the three stages can be written in a rather uniform way as given in Figure 8.

We consider that the three stages of the reactor are executed sequentially due to chemical hypothesis. Therefore the reactor dynamics is described in Figure 9.

$$AlgTermin(AlgPropag(AlgInit(P_0)))$$

Fig. 9. The reactor dynamics

5 A Formal Island Implementation of the Primary Mechanism

We present in this section how the primary mechanism is implemented in TOM using the formal island principle.

The TOM implementation involves four steps, in order to design:

1. An algebraic view of molecular graphs, as a set of terms on a convenient signature.
2. A representation mapping that establishes a correspondence between algebraic terms and Java objects. (This is the formal anchor.)
3. Reaction rules implemented with match constructs: the left-hand side consists of a TOM term, while the right-hand side is a mixture of Java code and TOM constructs.

4. Strategies for applying the reaction rules within each stage, and the chaining of stages.

In the following subsections, we develop each of these steps.

5.1   Molecular Graphs Viewed as Algebraic Terms

A molecular graph (see Figure 3) is encoded by a term, as proposed in the linear notation SMILES presented in [31]. Representing graphs as terms is a choice design: terms provide an intermediate structure between graphs and their representation by adjacency lists which appears to be well suited to the patterns specific to our application.

We briefly recall the principles of this representation. Molecules are represented as hydrogen-suppressed molecular graphs (hydrogen atoms are not represented) with atom-labelled vertices and bond-labelled edges. If the hydrogen-suppressed molecular graph has cycles, it can be transformed into a tree by applying the following rule to every cycle: arbitrarily choose one simple- or aromatic-labelled edge of the cycle, delete the edge, and add a fresh digit and the label of the edge to the labels of the formerly adjacent vertices. This corresponds to a spanning tree of the molecular graph. A vertex is chosen as root, and the tree is represented in a (semi)parenthesized preorder traversal (the parentheses are omitted for the right-most child of each vertex). Moreover, an aromatic cycle is represented by lower case letters, and the aromatic and simple bonds are not represented.

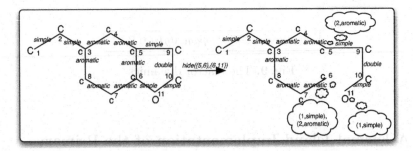

Fig. 10. From a cyclic molecular graph to an acyclic decorated molecular graph

In the first molecular graph from Figure 10 two edges are transformed into *implicit* edges: (i) edge {6,11} labelled with *simple* is hidden and encoded by labels (1, *simple*) on vertices 6 and 11; (ii) edge {5,6} labelled with *aromatic* is hidden and encoded by labels (2, *aromatic*) on vertices 5 and 6. The aromaticity of a bond is propagated to its end vertices which are labelled by lower case letters in the SMILES notation, and by upper case letters prefixed by **ar** in the signature. For example, if the vertex number 1 is chosen as root, a linear notation is CCc(ccc12)cc2C=CO1; if the root is the vertex number 3 another notation is C(CC)(ccc12)cc2C=CO1.

The user must provide as input for this prototype a list of molecules in the SMILES notation. The associated TOM terms are quite heavy to handle, hence the user does not need to deal with them. Moreover, the user can use a Java based software provided by Chemaxon, called Marvin [2], which allows editing and visualizing molecules on a web page: by simply drawing a molecule, one gets its SMILES notation.

The syntax for the TOM terms encoding decorated molecular graphs is given by Figure 11. The operator lab constructs a label composed of an integer and a bond type encoding an implicit edge, while the operator symb constructs a label for vertices composed of an atom name and a list of labels.

```
sorts Atom Bond Label LabelList Symbol Reactant ReactantList
abstract syntax
  C   -> Atom
  arC -> Atom
  O   -> Atom
  H   -> Atom
  e   -> Atom
  none    -> Bond
  simple -> Bond
  double -> Bond
  triple -> Bond
  arom    -> Bond
  lab(no:int, bond:Bond) -> Label
  concLab( Label* ) -> LabelList
  symb(atom:Atom, labels:LabelList) -> Symbol
  rct(bond:Bond,symbol:Symbol,rctList:ReactantList) -> Reactant
  conc( Reactant* ) -> ReactantList
```

Fig. 11. The signature for TOM terms

We represent a decorated molecular tree as a term of sort Reactant as follows:

− a leaf v is a term of sort Reactant,

```
rct(b, symb(a, concLab(labs*)), conc())
```

where a encodes the label of the leaf (an atom symbol), b encodes the label of the edge connecting v with his father, and labs* is a possibly empty list of pairs of integers and bond types representing the associated set of broken cycle labels;
− an internal vertex is a term of sort Reactant,

```
rct(b, symb(a, concLab(labs*)), conc(rcts*))
```

where rcts* encodes the list of its term-like represented children;
− the root has a dummy bond label, none, for uniformity reasons.

Operation symbols like `conc` above represent variadic associative operators that construct a list from its arguments (that can be empty).

We consider that a radical point is an atom of valence 1 labelled by e (for electron). For efficiency reasons, we consider all free radicals (such as $\bullet x$ in Figure 5) to have tree representations where the electron is the root.

The signatures for GasEl terms and TOM terms are slightly different, but the principles for building the terms are the same. The differences rise from restricting TOM signatures to many-sorted ones, while in ELAN one can use order-sorted signatures. The operation symbols in TOM are given in prefix notation and are always explicit.

5.2   Mapping Construction

In order to define necessary abstract data-types, we use the signature definition mechanism (`%typeterm`, `%typelist`, `%op`, etc.) provided by TOM.

For example, given a Java class `Reactant`, we can define the following algebraic mapping for it:

```
%typeterm Reactant {
    implement { Reactant }
    equals(t1, t2) { t1.equals(t2) }
}
```

where the class `Reactant` has the following structure:

```
class Reactant {
    private Bond bond;
    private Symbol symbol;
    private ArrayList rctlist;
    public Reactant(Bond bond, Symbol symbol, ArrayList rctlist) {...}
    ....
}
```

We can define the following constructor for the `Reactant` type:

```
%op Reactant rct(bond:Bond, symbol:Symbol, radlist:ReactantList) {
    is_fsym(t)                  { t instanceof Reactant }
    get_slot(bond,t)            { t.getBond() }
    get_slot(symbol,t)          { t.getSymbol() }
    get_slot(rctlist,t)         { t.getRctlist() }
    make(bond,symbol,radlist) { new Reactant(bond, symbol, radlist) }
}
```

In fact, this algebraic operation is a mapping from algebraic terms to Java objects that preserves the structural properties of `Reactant` sorted terms for `Reactant` Java instances, i.e. is a *formal anchor*. Let us remind that the formal anchor is determined by the semantics of three mappings: eq, is_fsym, subterm. The construct `%typeterm` contains the definition of *eq* which is `equals`. The other two mapping definitions are given by means of the `%op` construct for the operation symbol `rct`: the mapping is_fsym(t, rct) is implemented by the construct `is_fsym(t)`, while the mapping subterm(t, i) is implemented by three constructs `get_slot` for retrieving each of the three arguments of `rct`.

Instead of explicitly building this mapping, we can use the two external tools developed together with TOM, Vas and ApiGen, to generate Java files implementing the signature. In this way, we take advantage of the ATerm library and the VisitableVisitor design pattern which are automatically implemented by the generated classes. The memory sharing is very important for the implementation of reactants because the terms encoding them have in general many common subterms, while the Visitor pattern is necessary for doing term traversals.

The construct `%vas` allows defining a Vas grammar in a .tom file:

```
%vas {
  module data
   imports ...
   public
   sorts Atom Bond CLabel CLabelList Symbol Reactant ReactantList ...
   abstract syntax
    ...
}
```

Considering the signature described by Figure 11, after running Vas , some standard directories are generated containing all classes that make up the API for the signature. At the root level, the directory contains several standard classes and the mapping for TOM (`data.tom`). The subdirectory `types` contains abstract base classes for each sort defined in the signature, and one subdirectory per sort that contains concrete classes for each operator of this co-arity.

The TOM implementation uses a specialized version of the Visitor design pattern, the VisitableVisitor pattern, based on the *visitor combinators* concept introduced in [29] which allows composition and full tree traversal control. The basic visitor combinators are inspired by the strategy primitives of Stratego which are presented in Figure 1 (except the *Omega* strategy). The Java classes generated for the algebraic operations defined within a Vas construct implement the `Visitable` interface. On one side, the built-in or user defined traversal strategies are visitable as algebraic terms; on the other side they define `visit_Sort` and `visit_ValueSort_OperationSymbol` methods necessary for visiting algebraic operations.

5.3   Reaction Rules

The reaction rules have the form:

$$r : t_1[+ t_2] \rightarrow t'_1 + t'_2 \text{ if } cond$$

(where the elements between square brackets are optional), and we implement them using a match construct according to the following schema:

```
%match(Reactant subject1 [, Reactant subject2]) {
    t₁ [, t₂] → {
                if(cond) return pair(σ(t'₁), σ(t'₂));
    }
}
```

where the argument of match is the term we want to rewrite (the reactant), and σ is the substitution resulting from the matching process. Let us notice that only the implementations of termination rules have two reactants in their left-hand sides.

For all types of reaction rules, we define a base class **ChemicalRule** which encloses the common features of all reaction rules. For each reaction application, we determine the reaction products and its *degeneration* (how many times the reaction can be applied in different parts of reactants with equal results).

In GasEl one of the implementation difficulties was to have exhaustive application of a reaction rule on one or two reactants. Since the reaction rules are encoded in ELAN as named strategies which can be applied only at the top of a term, exhaustive application in GasEl is achieved by generating all tree-like visions of an acyclic decorated molecular graph (a vision is obtained by choosing a root on a spanning tree).

In TOM this problem is handled in an elegant way by using the strategy *Omega* (Figure 1). Given a term t and a rewrite rule $r : t_1 \rightarrow t_2$, the *Omega* strategy provides the following features:

- we can apply a topdown (or other traversal) strategy for solving the matching problem $t_1 \ll t$; successful matches give rise to a family of substitutions $\{\sigma_i\}_{i \in \mathbb{N}}$;
- for each match solution i, the position ω_i in t where the pattern matched can be retrieved as a Java object by means of the static method **getPosition()** of the class **MuTraveler**;
- for a position ω_i, the subterm $t_{|\omega_i}$ is returned by the method **getSubterm()**;
- for a position ω_i, the term resulting from t after applying the rewriting rule r, i.e. $\sigma_i(t)$ is computed using the method **getReplace($\sigma_i(t_2)$)**.

This is, up to our knowledge, an original feature that provides full control for applying a rewriting rule and allows a wide range of applications. In particular this is quite convenient for applying a reaction rule.

From the implementation point of view, there are two classes of reaction rules: the first class consists of the reactions (ui), (bi), (me), and (ipso) corresponding to an implementation by topdown traversal of a term in search for a reaction pattern, while the second class consists of the rest of the reactions for which the pattern (with the radical point) is always searched at the root. We illustrate these two types of implementation with the following two examples.

Example 2. [Bimolecular initiation reaction] The generic reaction is:

$$\mathbf{O} = \mathbf{O} \quad + \quad \mathbf{H} - x \quad \longrightarrow \quad \bullet\mathbf{OOH} \quad + \quad \bullet x$$

and an application is illustrated in Figure 4. The result of applying the (bi) reaction rule on a term `subject` is implemented by means of the following code:

```
if( !containsElectron(subject) && (nC(subject) > 1)) {
    VisitableVisitor birule = new BiRule();
    'TopDown(Try((birule))).visit(subject);
    this.setResultList(birule.getResultList());
}
```

First we test if the reactant does not contain a radical point (encoded as an electron), and if it contains at least two carbon atoms. If the test is successful, then we apply in a topdown manner a rule, instance of the class `BiRule`.

For every subterm of sort `Reactant`, during the top-down traversal of the subject of the reaction, the following method of the object `birule` is applied:

```
public Reactant visit_Reactant(Reactant arg) throws VisitFailure {
    Reactant r1, r2;
    int n;
    %match(Reactant arg) {
        rct(b, symb(C(), concLab(labs*)), conc(rcts*)) -> {
            n = nH(arg);
            if( n >= 1) {
                Position pos = MuTraveler.getPosition(this);
                r1 = insertElectron(pos.getSubterm().visit(globalSubject));
                r2 = hangE(pos.getReplace(r1).visit(globalSubject));
                addMPack('mpack(n, pack(ctRcts.eoo, ctRcts.seoo),
                                  pack(r2, usmiles(r2)))));
            }
        }
    }
    return 'Fail().visit(arg);
}
```

The variable `globalSubject` is set to the value of the term participating to the reaction. We search within the term for a non-aromatic carbon atom which has at least one hydrogen bound by examining all subterms of sort `Reactant`. nH computes the number of hydrogen atoms connected to the **C** atom.

We attach an electron to the found carbon atom, we insert the new term in the context, and then we twist the term by means of **hangE** such that the node labelled by e becomes the root in the corresponding molecular tree in order to preserve the chosen representation of free radicals.

A term of sort **Pack** represents a pair composed of a **Reactant** term and its SMILES form computed with the algorithm presented in [31]; eoo is a constant term corresponding to •OO, while seoo is the canonical form of eoo; n is the degeneration of the reaction. The method **addMPack** adds an element consisting of a pair of **Pack**-sorted terms with the multiplicity n to a private list of this type; this list represents the result of the exhaustive application of a particular reaction rule.

Example 3. [Beta-scission reaction with no cycle breaking] The generic reaction is:

$$\bullet x - y - z \quad \longrightarrow \quad x = y \quad + \quad \bullet z$$

This reaction rule described by subgraphs is easily translated in a rule over trees (as we can see schematically in Figure 12) which is matched at the top of a term (because the electron is always placed in the root).

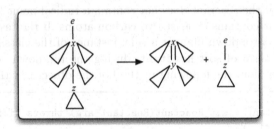

Fig. 12. Beta-scission on terms

5.4 Reactor Strategy

We present in this section the implementation of the reactor dynamics formally described by the algorithms in Figures 8 and 9. We implement the function UNIT given in Figure 7 by means of the visitor class **UnitRule** with a private member consisting of an array of chemical rules:

```
class UnitRule extends data.dataVisitableFwd {
  private Object rules[];
  public UnitRule(Object rules[]) {
    super('Fail());
    this.rules = rules;
  }
  public PairPackList visit_PairPackList(PairPackList arg) { ... }
  public PackList visit_PackList(PackList arg)  { ... }
}
```

UnitRule can be used as a rule with a particular behavior on terms of sorts
PairPackList and PackList. Each of the visit_Sort methods contains applications of the rules passed as arguments on lists of reactants.

The initiation stage described by $UNIT(R_i)$ given by Figure 8 is implemented as follows:

```
ChemicalRule initRules[] =
            {new UICCRule(), new UICHRule(), new BiRule()};
VisitableVisitor init_unit  = new UnitRule(initRules);
plist = 'Try(init_unit).visit(plist);
```

where plist from the right-hand side is the input list of chemical reactants
(the initial set of reactants), while plist from the left-hand side contains the
products obtained from the initiation stage together with the input reactants.

For the propagation stage, chemical hypotheses impose to apply the reactions (me) and (ipso) only on the products resulted from the initiation stage.
Therefore we describe the propagation stage by means of the strategy
$UNIT(R_p); repeat(UNIT(R_p - \{(me), (ipso)\}))$, and we implement it as follows:

```
ChemicalRule propagRules1[] = {new MeRule(), new IpsoRule(),
   new BSCCRule(), new BSCHRule(), new OxRule(), new CombeOeRule()};
VisitableVisitor propag_unit1 = new UnitRule(propagRules1);
tmplist = 'Try(propag_unit1).visit(plist);
tmplist = diff(tmplist, plist);
plist = appendLists(plist, tmplist);
pairlist = 'pair(plist, tmplist);
ChemicalRule propagRules2[] = {new BSCCRule(), new BSCHRule(),
                        new OxRule(), new CombeOeRule() };
VisitableVisitor propag_unit2 = new UnitRule(propagRules2);
pairlist = 'RepeatId(Try(propag_unit2)).visit(pairlist);
plist = getFirstList(pairlist);
```

First we put the reaction products from all propagation rules in tmplist,
then we select only the free radicals not already in the input list, and put them
together with the initial reactants. We make a pair of lists with the first element
consisting of all reactants, and the second element consisting of the list of new
free radicals, and we provide it as input for the strategy that applies the chemical
rules from the array propagRules2. The application of this strategy ends when
the list of new free radicals is empty. The result of the propagation stage consists
of the list of all products concatenated with the list of input reactants.

The termination stage described by $UNIT(R_t)$ is implemented in TOM as
follows:

```
ChemicalRule terminRules[] = {new CoRule(), new DiRule()};
VisitableVisitor termin_unit  = new UnitRule(terminRules);
plist = 'Try(termin_unit).visit(plist);
```

For a given list of input molecules, this prototype writes in a file the chemical products for each stage as well as the elementary reactions that took place during the entire mechanism.

6   Conclusion

The first output of this work is a new prototype of a chemical reactor. First results revealed good properties with respect to chemical validations of the model. In all but one cases, this prototype is faster (less than 13 seconds) than GasEl. Moreover, for non cyclic molecules with 16 carbon atoms and a big number of simple bonds, this implementation in TOM is up to 9 times faster than GasEl. The execution times for the two prototypes have been compared on all examples validated with chemists and presented in [16]. For the most complex molecule tested (JP10) not completely handled in [16], the prototype in TOM was able to terminate with 1165 generated reactions in 139 minutes. A complete comparison between the GasEl prototype and the current implementation in TOM is not trivial due to notation and implementation differences, and out of the scope of this paper.

It may be worth noticing that the rule-based approach on graph structures has also been studied in the modelling of signal transduction networks [13] and metabolic pathways [23] in the domains of biological systems and protein interactions. Our model of chemical reactor seems to be easily adaptable to these domains.

Our second concern was to explore the formal island concept and methodology on a significant example. The objective of the formal island approach to extend the expressivity of the host language with higher-level constructs at design time is well-illustrated in this example. From this point of view, the TOM implementation appeared to be quite convenient to implement chemical rules with conditions and actions expressed in the Java host language. On the other hand, control was expressed with a high-level language of strategies which makes now possible to reason about formal properties, especially the termination property of each phase [4]. This illustrates the idea to perform formal proof on the formal island constructions.

A further idea would be to implement a new version of the TOM compiler able to perform graph rewriting. Representing cyclic structures in TOM is not too difficult but matching and rewriting have to be adapted to this context. Indeed this capability would open new application areas.

A long-term objective of the formal island approach is to certify the implementation of the formal island compilation into the host language. A first step in this direction has been presented in [19] to generate proof obligations for the compilation of matching. A similar concern is underway for rewriting and strategies.

Further improvements of the formal island approach is to anchor other language extensions, especially modules and parameters, while improving the capacity of the compiler to generate verification requirements related to properties to be checked.

Acknowledgements: We sincerely thank Olivier Bournez, Pierre-Etienne Moreau and Antoine Reilles for helpful remarks and interactions on this work, and the Protheo group for scientific and financial support. This work has been partially funded by an INRIA international internship program.

References

1. Elan web site. http://elan.loria.fr.
2. Marvin: A tool for Molecule Drawing and Visualization.
   http://www.chemaxon.com/marvin/.
3. Tom web site. http://tom.loria.fr.
4. O. Andrei. Term graph and chemical rewriting. Internship report, LORIA, Protheo Team, Nancy, France, September 2005.
5. E. S. Blurock. Reaction: System for Modeling Chemical Reactions. *Journal of Chemical Information and Computer Science*, 35:607–616, 1995.
6. P. Borovanský, C. Kirchner, H. Kirchner, P.-E. Moreau, and C. Ringeissen. An Overview of ELAN. In C. Kirchner and H. Kirchner, editors, *Proceedings of the Second International Workshop on Rewriting Logic and Applications*, volume 15, http://www.elsevier.nl/locate/entcs/volume15.html, Pont-à-Mousson (France), September 1998. Electronic Notes in Theoretical Computer Science. Rapport LORIA 98-R-316.
7. O. Bournez, G.-M. Côme, V. Conraud, H. Kirchner, and L. Ibǎnescu. A Rule-Based Approach for Automated Generation of Kinetic Chemical Mechanisms. In R. Nieuwenhuis, editor, *Proceedings of the 14th International Conference on Rewriting Techniques and Applications, RTA 2003, Valencia, Spain, June 9-11, 2003*, volume 2706 of *Lecture Notes in Computer Science*, pages 30–45. Springer, 2003.
8. O. Bournez, L. Ibanescu, and H. Kirchner. From Chemical Rules to Term Rewriting. In *6th International Workshop on Rule-Based Programming*, To appear in ENTCS series, Nara, Japan, April 2005.
9. M. Clavel, F. Durán, S. Eker, P. Lincoln, N. Martí-Oliet, J. Meseguer, and J. F. Quesada. Maude: Specification and Programming in Rewriting Logic. *Theoretical Computer Science*, (285):187–243, August 2002.
10. G.-M. Côme. *Gas-Phase Thermal Reactions. Chemical Engineering Kinetics.* Kluwer Academic Publishers, 2001.
11. R. Diaconescu and K. Futatsugi. An overview of CafeOBJ. In C. Kirchner and H. Kirchner, editors, *Electronic Notes in Theoretical Computer Science*, volume 15. Elsevier, 2000.
12. J. Dugundji and I. Ugi. An Algebraic Model of Constitutional Chemistry as a Basis for Chemical Computer Programs. *Topics in Current Chemistry*, 39:19–64, 1973.

13. J. R. Faeder, M. L. Blinov, and W. S. Hlavacek. Graphical rule-based representation of signal-transduction networks. In H. Haddad, L. M. Liebrock, A. Omicini, and R. L. Wainwright, editors, *Proceedings of the 2005 ACM Symposium on Applied Computing (SAC), Santa Fe, New Mexico, USA, March 13-17, 2005*, pages 133–140. ACM, 2005.
14. J. M. Grenda, I. Androulakis, A. M. Dean, and W. H. Green. Application of Computational Kinetic Mechanism Generation to Model the Autocatalytic Pyrolysis of Methane. *Industrial & Engineering Chemistry Research*, 42:1000–1010, 2003.
15. J. Guyon, P.-E. Moreau, and A. Reilles. An integrated development environment for pattern matching programming. In B. Barry and O. de Moor, editors, *Proceedings of the 2nd eclipse Technology eXchange workshop, eTX'2004 (Barcelona, Spain)*, Barcelona (Spain), 2004. Electronic Notes in Theoretical Computer Science.
16. L. Ibanescu. *Programmation par règles et stratégies pour la génération automatique de mécanismes de combustion d'hydrocarbures polycycliques*. Thèse de Doctorat d'Université, Institut National Polytechnique de Lorraine, Nancy, France, June 2004.
17. C. Kirchner and H. Kirchner. Rewriting, solving, proving. A preliminary version of a book available at `www.loria.fr/~ckirchne/rsp.ps.gz`, 1999.
18. C. Kirchner and H. Kirchner. Rule-based programming and proving: the ELAN experience outcomes. In *Proceedings of the Ninth Asian Computing Science Conference ASIAN'04*, volume 3371, pages 363–379, Chiang Mai, Thailand, December 2004. Lecture Notes in Computer Science.
19. C. Kirchner, P.-E. Moreau, and A. Reilles. Formal validation of pattern matching code. In *PPDP '05: Proceedings of the 7th ACM SIGPLAN international conference on Principles and practice of declarative programming*, pages 187–197, New York, NY, USA, 2005. ACM Press.
20. H. Kirchner and P.-E. Moreau. Promoting rewriting to a programming language: A compiler for non-deterministic rewrite programs in associative-commutative theories. *Journal of Functional Programming*, 11(2):207–251, 2001.
21. P.-E. Moreau, C. Ringeissen, and M. Vittek. A Pattern Matching Compiler for Multiple Target Languages. In G. Hedin, editor, *12th Conference on Compiler Construction, Warsaw (Poland)*, volume 2622 of *LNCS*, pages 61–76. Springer-Verlag, May 2003.
22. E. Ranzi, T. Faravelli, P. Gaffuri, and A. Sogaro. Low-Temperature Combustion: Automatic Generation of Primary Oxidation Reactions and Lumping Procedures. *Combustion and Flame*, 102:179–192, 1995.
23. F. Roselló and G. Valiente. Analysis of metabolic pathways by graph transformation. In H. E. et al., editor, *2nd International Conference on Graph Transformation - ICGT'04, Roma, Italy*, volume 3256 of *Lecture Notes in Computer Science*, pages 70 – 82. Springer, 2004.
24. A. S. Tomlin, T. Turányi, and M. J. Pilling. *Mathematical Tools for the Construction, Investigation and Reduction of Combustion Mechanisms*, volume 35 of *Comprehensive Chemical Kinetics*, chapter 4, pages 293–437. Elsevier, Amsterdam, 1997.
25. M. van den Brand, A. van Deursen, J. Heering, H. A. de Jong, M. de Jonge, T. Kuipers, P. Klint, L. Moonen, P. A. Olivier, J. Scheerder, J. J. Vinju, E. Visser, and J. Visser. The ASF+SDF Meta-environment: A Component-Based Language Development Environment. In *Computational Complexity*, pages 365–370, 2001.
26. M. G. J. van den Brand, H. A. de Jong, and P. Olivier. Efficient annotated terms. Technical report, University of Amsterdam, 2000. SEN-R0003, ISSN 1386-369X.

27. E. Visser. Stratego: A Language for Program Transformation based on Rewriting Strategies. System Description of Stratego 0.5. In A. Middeldorp, editor, *Rewriting Techniques and Applications (RTA'01)*, volume 2051 of *Lecture Notes in Computer Science*, pages 357–361. Springer-Verlag, May 2001.

28. E. Visser, Z.-e.-A. Benaissa, and A. Tolmach. Building program optimizers with rewriting strategies. *ACM SIGPLAN Notices*, 34(1):13–26, January 1999. Proceedings of the International Conference on Functional Programming (ICFP'98).

29. J. Visser. Visitor combination and traversal control. In *ACM Conference on Object-Oriented Programming, Systems, Languages, and Applications - OOPSLA'01, Tampa Bay, Florida, USA*, volume 36(11) of *ACM SIGPLAN Notices*, pages 270–282, 2001.

30. V. Warth, F. Battin-Leclerc, R. Fournet, P.-A. Glaude, G.-M. Côme, and G. Scacchi. Computer Based Generation of Reaction Mechanisms for Gas-Phase Oxidation. *Computers and Chemistry*, 24:541–560, 2000.

31. D. Weininger, A. Weininger, and J. L. Weininger. SMILES. 2. Algorithm for Generation of Unique SMILES Notation. *Journal of Chemical Information and Computer Science*, 29:97–101, 1989.

From OBJ to ML to Coq

Jacek Chrząszcz[1] * and Jean-Pierre Jouannaud[2] **

[1] Institute of Informatics, Warsaw University, ul. Banacha 2, Warsaw
http://www.mimuw.edu.pl/~chrzaszc
[2] École Polytechnique, LIX, CNRS UMR 7161, F-91400 Palaiseau
http://www.lix.polytechnique.fr/Labo/Jean-Pierre.Jouannaud/

This work is dedicated to our colleague Joseph Goguen, who was extremely influential in the design of modern programming languages.

1 Introduction

Rigorous program development is notoriously difficult because it involves many aspects, among which specification, programming, verification, code reuse, maintenance, and version management. Besides, these various tasks are interdependent, requiring going back and forth between them. In this paper, we are interested in certain language features and in languages which help make the user's life easier for developing programs satisfying their specifications.

Our interest focuses on three implemented specification/programming languages, OBJ [14,18], ML[27] and Coq [10], which have played an important historical role in the process of coming up with better languages. And indeed, both OBJ and ML had many successors or dialects, among which OBJ3 [20], Cafe-OBJ [28], Maude [9] and ELAN [2] for OBJ, and SML [23], CAML [30] and OCaml [29] among others for ML. Coq has evolved with many different versions keeping the same name, following the evolution of type theory from the calculus of constructions [11] to the extended calculus of constructions [22] and the development of the theory of inductive types from Martin-Löf's type theory [25,26] to the calculus of inductive constructions [12,31]. Other proof assistants based on a similar historical development include Lego [21], Alf [24] and Agda/Alfa [1]. Coq remains the most mature and widely used of them all.

We explain briefly in the introduction what important properties are shared by these three languages, and how OBJ has been influential in such a way that many important characteristics of ML and Coq were already present in OBJ, sometimes in disguise. In what sense can these three languages be considered as specification languages, or programming languages, or proof development systems is another important aspect we are interested in.

The user does not like doing things twice. Writing a specification in one language before coding it in another language is more than a challenge: it is helpless. The coding part must be automated as is the case in all three languages we are interested in. This automation obeys the same principle: forgetting the non-executable subpart of the specification or of its proof.

* Partly supported by Polish KBN Grant 3 T11C 002 27
** Project LogiCal, Pôle Commun de Recherche en Informatique du Plateau de Saclay, CNRS, École Polytechnique, INRIA, Université Paris-Sud.

K. Futatsugi et al. (Eds.): Goguen Festschrift, LNCS 4060, pp. 216–234, 2006.
© Springer-Verlag Berlin Heidelberg 2006

A specification is nothing but a logical property of the form $\forall \overline{x}.P(\overline{x}) \rightarrow Q(\overline{x})$, where $\overline{x}$ is the vector of data, $P(\overline{x})$ is the assumption, and $Q(\overline{x})$ is the conclusion. Therefore, the specification/programming language must contain (possibly via an encoding) a mechanism for expressing properties, as well as one for expressing computations and, possibly, a last one for expressing proofs. In ML, the specification part is simple enough to be inferred automatically by the system from the user's functional program: this is called type inference. The program then satisfies this (extremely poor) specification without requiring any further proof. In OBJ, specifications are algebraic, that is, conditional equations giving meaning to the various functions and predicates introduced by the user, and are executable via rewriting. Showing that the rewrite program implements the specification requires several checks (confluence and termination) left to the user. Proving properties of an OBJ specification can be done in the language itself by using reflection, this has been done in Maude and Elan, as well as in CafeOBJ — to a limited extent. Coq uses higher-order intuitionistic logic as a specification language, and includes the possibility to carry out the development of a (constructive) proof of the specification by using a tactic language which generates a Coq term representing the proof. A functional program meeting the specification can then be extracted automatically from that proof by erasing all subepressions without computational content which differ from the others by their type.

The old paradigm that the same program piece can be used several times in a bigger program with different data has led to a first notion of abstraction, giving rise to the notion of function, or subprogram. The idea that a program operating upon certain data should not depend upon the way they are actually represented has led to the notion of abstract data type. The same paradigm applied to groups of functions or subprograms achieving some well-defined task, processing some well-defined data, has led to the notion of module. Abstracting over modules themselves has lead to the notion of functor. All three languages have pioneered the design of modules and functors in their respective areas, not to speak about abstract types, and OBJ has been very influential in this matter.

Object-orientation is a different, important abstraction mechanism that is not part of our three languages, and indeed, the first two have been extended so as to include object-oriented features. We will not say more about object orientation, although OCaml, a dialect of ML, played an important role in popularizing object orientation among the community of functional programmers.

Among the programming tasks that should be eased by a good language choice, only the last one, version management, is not taken care of at all by our three languages. Some of the others tasks are better taken care of by OBJ or by ML or by Coq. In particular, the verification principles behind these languages differ in the expressivity of their underlying specification language. In OBJ, typing looks very elementary, since OBJ static types are checked in linear time by a bottom-up tree automaton. But OBJ types are not all static, requiring some runtime type-checking as well. In ML, static typing is more advanced, with a polymorphic type discipline for which types can be inferred by an exponential (but practically linear) algorithm. In Coq, types are arbitrary formulas of higher-order (intuitionistic) logic which can be checked in finite (but indefinite) time, and cannot be inferred in general. This typing system generalizes both OBJ's and ML's typing as we will see. Verification can also be achieved by model checking or testing. Both are lacking in OBJ, ML and Coq, but can of course be made available as tactics in OBJ's successors and Coq.

The quest for the ideal programming language will continue until a satisfactory language is designed that internalizes features still taken care of by the user or by the programming environment.

2   The Three Languages

2.1   OBJ

In their first landmark paper on CLEAR, Rod Burstall and Joseph Goguen introduced the brand new bright idea that specifying a program required a specific language able to reflect the structure of the problem itself [6]. Following the ADJ group [16,15], they advocated for an algebraic specification language based on equationnal logic, together with a module system in which logical theories could be specified. This was the birth of CLEAR, later developped more formally in [7]. To our knowledge, CLEAR was the very first specification language. CLEAR was algebraic, using many-sorted algebras with error-sorts, an approach later revised to yield OBJ's order-sorted algebras. CLEAR had parameterized modules and theories, but no functors and was not implemented, although one can consider that the first implementation of OBJ by Joseph Tardo [17], a student of Joseph Goguen at UCLA, was indeed an implementation of CLEAR. A second more advanced implementation was then written by David Plaisted when visiting Joseph Goguen at SRI in 1982, which included associative-commutative rewriting.

OBJ2 was the third implementation of OBJ. It was developped in 1984, when Kokichi Futatsugi and the second author visited Joseph Goguen and and José Meseguer at SRI for one year. OBJ2 was the first algebraic specification language based on a fragment of a Horn logic built on the equality predicate and finitely many membership predicates called subsorts [14]. The many novel features of OBJ2 included a flexible user-defined syntax, defining subsorts by Horn sentences, rapid prototyping via rewriting modulo associativity, commutativity, identity and their combinations, parameterized modules and functors. OBJ2 was followed by OBJ3 [20], an improved implementation developped by Claude and Hélène Kirchner whose postdoctoral visit closely followed their advisor's. Full Horn logic is available in the Maude language [9,3], one of OBJ's successors developped by José Meseguer and his collaborators.

An OBJ program is a collection of modules followed by queries. A module is either an *object* or a *theory*. A module has a name, which we always write with capital letters. Objects are made of two parts: a *signature* made of basic types called sorts, and of constructors and (defined) operators for these sorts; the meaning of the operators and of the subsorts is given by (executable) Horn clauses (called *equalities* or *sort constraints* depending on the predicate heading the positive atom). We will also use the name of *membership* for sort constraint, as in Maude. In general, the *principal sort* of a module bears the same name as the module itself, but the first letter only is capitalized. Semantically, objects are initial algebras, implemented via the computation of normal forms: the meaning of the defined operators must be given by a convergent set of conditional rewrite rules (possibly modulo associativity, commutativity and the like). A theory is much like an object except that it is not executable: its (*loose*) semantics is given by the class of all algebras that satisfy the arbitrary first-order logical sentences specifying its properties. The definition of an object or theory can use other objects or theories. The keywords: *using* allows to import a module without ensuring any property of the imported module which must therefore be copied; *protecting* ensures that the imported module is not modified, making copying unnecessary; *extending* stands in-between, since new values can be added in sorts, but old values cannot be made equal unless they were equal beforehand. *Parameterization* is one more way for importing a module. If T is a theory, parameterizing a module M by an abstract module X satisfying T will allow using the symbols defined in T in order to build M, possibly by using qualification as a disambiguation mechanism. The parameterized module M can later be instantiated by an actual A provided A satisfies the axioms of T. Asserting a module property is done by a *view*, which is the third kind of entity in OBJ. The construction of the instantiated module may also involve some copying.

OBJ has a much more powerful mechanism for defining types than it appears. Besides its basic types called *sorts*, like $\mathbb{N}$ and $List$, it also has type constructors: if the module $LIST$ is parameterized by an abstract module X assumed to satisfy the theory T, then any type $List(Elt)$ exists potentially, provided Elt is the sort of a module satisfying T. This allows to build the types $List(\mathbb{N})$ as well as $List(List(\mathbb{N}))$, therefore providing with some form of polymorphism. However, these types can only be used if the corresponding module instances $LIST[NAT]$ and $LIST[LIST[NAT]]$ are explicitly constructed. The same mechanism provides with dependent types like bounded lists of length n, where n can be a parameter of sort $\mathbb{N}$ defined via a theory. It also has arbitrary first-order Horn sentences as types, written $t : s'$ if A, where A is an arbitrary conjunction of equations and memberships built from the variables in t. OBJ's subsort declaration is a static restriction of this mechanism. So, OBJ's type system was quite strong at the time OBJ was implemented, and has even some Curry-Howard flavour. In retrospect, theories themselves can be seen as types for modules, and a view becomes then an assertion that a module has some theory as type.

OBJ's types, however, only serve specification purposes. Unlike modern functional programming languages like ML, typing is not really internalized in OBJ: property checking is left to the user's responsibility. Still, a limited amount of type-checking is done. For example, the left-hand and right-hand side of an equality must have the same sort. And the expression occurring in the head of a membership must have a sort whose asserted sort must be a subsort.

OBJ specifications are assumed to satisfy a few other properties, all left to the user. For example, the set of rules in a module is supposed terminating and confluent, and the operators should be completely defined. Maude provides support for checking these properties.

2.2   ML

ML was the first functional programming language in which specifications were given (actually, inferred) as types, another novel bright idea from the late seventies due to

Robin Milner [27]. ML has a powerful higher-order module system, an efficient execution model via separate compilation, and a primitive verification mechanism via type inference.

An ML program is a collection of *modules*. A module is either a *structure*, which corresponds to an OBJ non-parametric object, or a *functor* which corresponds to a parametric object. Contrary to the latter, ML functors can be higher order, i.e. they can be parametrized by a module which itself is parametrized. Specification of a functor parameter is given by a *module type*. This can either be a *signature*, corresponding to an OBJ theory, or a functor type. Contrary to OBJ theories, values cannot be specified by equations, but types can.

Another difference is the lack of views in the ML module system. Since subtyping is implicit, a functor *F*, expecting an argument of type *SIG*, can be applied to all modules *M*, whose principal module type *MSIG* is a subtype of *SIG*. Using type inference, the principal module type can be computed efficiently and since subtyping is an extension of inclusion, views are not necessary. On the other hand, the OBJ views can also be used to rename components of an object, which in ML can only be done via a functor.

The important feature of OBJ that is missing in ML is theory extension via keywords *extending* and *using*. Because equational specification of values is lacking in ML, signature inclusion, present in most ML implementations, is much weaker than its OBJ counterpart, hence cannot be seen as a substitute. Indeed, theory extension can be used as another means of parametrisation: assume one declares a function *f* of some type in a theory *A* and one then uses it in a subsequent equational specification of some function *g*; in a theory *B* extending *A*, one can then provide equations defining *f*, therefore completing the specifications of *g* at the same time. In fact, the specification of *g* is parametrized by *f*. Similar ideas are currently being investigate by the ML community with the so called mixins [4,19].

2.3 Coq

In the mid-eighties, following the path initiated by Curry, Howard, Girard and De Bruijn, Thierry Coquand and Gérard Huet made another important step with the beautiful Calculus of Constructions [11], in which types are arbitrary sentences of higher-order intuitionistic logic. This calculus was the start of the language Coq, a proof assistant including a full functional programming language as an executable subset. Coq has a powerful higher-order module system with cut elimination semantics studied and implemented by the first author [8], at that time a phd-student of the second author, a primitive execution model via rewriting and an efficient execution model via compilation. It also includes a sophisticated proof search engine via tactics (and a tactic language), a secure proof checker based on type checking, and an extraction mechanism towards modular ML code. Here, it must be stressed that the module system is used to structure first specifications, then proofs, and finally the programs extracted from proofs. The latter is of course facilitated by the fact that the module systems of Coq and ML are essentially the same.

The logical formalism implemented in Coq is based on the calculus of inductive constructions [12,31]. The terms in Coq are of two sorts: calculable Set and logical

Prop[3]. Values are typed by types, which are typed by the sort Set (for example 0 : nat and nat : Set). The second sort, Prop, is a type of logical formulas, which in turn are types of their proofs (formula, whose proof is e.g. fun x ⇒ x). In type theory with dependent types these two worlds interleave, but it is nevertheless possible to use this dichotomy in order to extract the computable content of a proof, by deleting all its (logical) subterms of sort Prop.

The general structure of a Coq development is the same as that of an ML program. The main difference lies in logical parts: axioms in specifications and theorems in implementations. While in ML code precise specifications are usually written informally as comments and correctness is based on trusting the programmer, in Coq one can write specifications as logical formulas, and then carry out the proof that the specification is satisfied.

3 Example

To compare the modular features of the three languages, we shall study a simple sorting algorithm using an abstract priority queue. We also provide a naive implementation of the priority queue and show how the abstract algorithm can be composed with the given implementation. The obtained algorithm and data structure remain parameterized with respect to the element ordering, which can itself be instantiated later on.

Priority queues are data structures implementing the following functionalities: creation of an empty queue, insertion of an element into the queue and extraction of the minimal element from the queue. They can be realized very efficiently imperatively (Fibonacci heaps, binomial heaps, etc) but efficient functional implementations also exist (see e.g. [5]).

Using a priority queue, one can implement the following sorting algorithm: insert all element into the queue and then extract them one by one. Several apparently different sorting algorithms can be seen as instances of this abstract schema using a particular implementation of a priority queue: selection sort uses unsorted lists, insertion sort uses sorted lists and heapsort uses heaps.

This example, despite being so small and simple, illustrates quite well the modular features of our three languages and how they evolved from OBJ to ML and Coq. We show how a specification and an implementation of a data structure look like, how an implementation of the data structure can be composed with an abstract algorithm, and how the resulting concrete but parametric algorithm can be instantiated and used in a program.

Our example shows the advantages of each approach: in OBJ one can write very concise equational specifications, in ML specifications are very brief (and imprecise) but implementations are very efficient, and Coq allows one to formally specify and prove correctness of a data structure or algorithm. The comparison between ML and Coq further shows how much work is needed to formally specify and verify a piece of code.

We will give the actual code of the example in the presentation.

---

[3] There are other sorts in Coq, namely the predicative hierarchy of $Type_i$, $i \in \mathbb{N}$, called universes [22], but we do not use them in this paper.

4 Priority Queues in OBJ

We will take the liberty to exploit the full power of Maude and use its syntax when appropriate, to ease the understanding. Using OBJ instead would sometimes require some irrelevant detour.

Specification of an ordered type, pairs, queues and priority queues.

We define successively trivial theories with a distinguished sort, pairs, totally ordered sets, queues and priority queues. Being part of any OBJ specification, the predefined module BOOL has one sort, Bool, two (truth) values, true and false, and the usual Boolean connectives as operations. In all examples, italics are used to identify OBJ keywords. All sentences are terminated by a dot for parsing purposes. Underscores are used to indicate arguments of operators which use a mixfix syntax.

```
th   TRIV is
sort Elt .
endt
```

The theory TRIV requires the existence of (at least) one sort, named Elt.

```
obj  PAIR[X :: TRIV, Y :: TRIV] is
sort Pair .
op   pair : Elt.X Elt.Y -> Pair .
op   1st : Pair -> Elt.X .
op   2nd : Pair -> Elt.Y .
var  E : Elt.X .
var  E' : Elt.Y .
eq   1st(pair(E, E')) == E .
eq   2nd(pair(E, E')) == E' .
endo
```

The parameterized object PAIR builds upon two formal objects X and Y satisfying TRIV, which acts as a binder for the sort names Elt.X and Elt.Y, therefore providing for the polymorphic sort constructor pair. Note the use of qualification for disambiguating between the two instances of TRIV. The symbol == is used for equations in theories and for rules in objects. It is also used for the built-in equality available at all sorts. Similarly, : s is the built-in membership predicate available at sort s. In the equations, the variables E, E' and E'' are universally quantified by the binding declaration var.

```
th   TOSET[X :: TRIV] is protecting BOOL .
op   _ ≤ _: Elt Elt -> Bool .
var  E E' E'' : Elt .
E    E ≤ E == true .
eq   E == E' if E ≤ E' and E' ≤ E .
eq   E ≤ E'' == true if E ≤ E' and E' ≤ E'' .
eq   E ≤ E' or E' ≤ E == true .
endt
```

The theory TOSET uses the module BOOL with the keyword *protecting* implying two important properties: no new element of sort Bool can exist in the semantics (for any two elements e, e' of sort X, e≤e' must be equal to either true or false), and no two elements of sort Bool that were semantically different in BOOL can be equated in TOSET.

th   QUEUE[X :: TRIV] *is protecting* BOOL .
sorts   NeQueue Queue .
subsorts Elt < NeQueue < Queue .
op   empty : Queue .
op   get : NeQueue -> Elt .
op   rest : NeQueue -> Queue .
op   insert : Elt Queue -> NeQueue .
op   eq : Queue Queue -> Bool .
var   Q : NeQueue .
eq   eq(empty, empty) == true .
eq   eq(insert(E, Q), empty) == false .
eq   eq(insert(E, Q), insert(E', Q') ==
            (E == E') and eq(Q, Q') .
eq   eq(insert(get(Q), rest(Q)), Q) == true .
endt

In the theory of queues, the declaration NeQueue < Queue implies that get and rest are total on their domain. An alternative is

var Q : NeQueue .
mb Q : Queue .

th   PRIOQUE[X :: TRIV, Y :: POSET[X]] *is extending*
       PAIR[X, QUEUE[X]] .
op   extract : NeQueue -> Pair .
op   ≤ : Elt Queue -> Bool .
var Q : NeQueue .
var E, E' : Elt .
eq   E ≤ nil == true .
eq   E ≤ insert(E', Q) == E ≤.Y E' and E ≤ Q .
eq   extract(insert(E, Q)) == pair(E, Q) *if* E ≤ Q .
eq   extract(insert(E, Q)) == pair(1st(extract(Q)),
       insert(E, 2nd(extract(Q)))) *if* E ≤ Q == false .
endt

Note how models of PRIOQUE alternate loose interpretations (of TRIV, QUEUE and PRIOQUE) with initial interpretations (of PAIR and BOOL). The role of the PAIR is to provide a polymorphic pairing construct.

Specification of an abstract sorting algorithm based on priority queues.

```
th        LIST[X :: TRIV] is protecting BOOL .
sorts     NeList List .
subsorts  Elt < NeList < List .
op        nil : List .
op        _ _ : List List -> List [assoc id : nil] .
op        head : NeList -> Elt .
op        tail : NeList -> List .
var       E E' : Elt .
var       L L' : List .
eq        head(E L) == E .
eq        tail(E L) == L .
mb        L L' : NeList if L : NeList or L' : NeList .
endt
```

```
th        ORDLIST[X :: TRIV, Y :: POSET[X],
          Z :: LIST[X]] is
sorts     NeOList OList .
subsorts  NeOList < OList < List .
subsorts  NeOList < NeList .
op        sorted : List -> Bool .
op        sort : List -> OList .
var       L L' L'' : List .
var       E E' : Elt .
eq        sorted(nil) == true .
eq        sorted(E) == true .
eq        sorted(E E' L) == E ≤ E' and sorted(E' L) .
mb        nil : OList .
mb        L : NeOList if sorted(L) and L : NeList .
eq        sort(L E L' E' L'') == sort(L E' L' E L'') .
eq        sort(L) == L if sorted(L) .
endt
```

Note the subtle use of associativity and identity of concatenation in specifying `sort` and `sorted`.

```
obj       SORT[X :: TRIV, Y :: POSET[X], Z :: PRIOQUE[X, Y]] is
op        sort : Queue -> OList .
var       Q : NeQueue .
eq        sort(empty) == nil .
eq        sort(Q) == 1st(extract(Q)) sort(2nd(extract(Q))) .
endo
```

Concrete algorithms for sorting elements of an ordered set.

```
view QLIST[X :: TRIV] of LIST[X] as QUEUE[X] .
sort  Queue to List .
sort  NeQueue to NeList .
op    empty to nil
op    get to head .
op    rest to tail .
op    insert to _ _ .
endv
```

This kind of typing assertion implies proof obligations to be checked by the user. Here, the equation given for `insert`, `get` and `rest` must be verified for their interpretation in `LIST`. We now construct specific priority queues as views to instantiate the abstract sorting algorithm.

```
view PRIOQUE1[X :: TRIV, Y :: POSET[X]] of
     PAIR[X, QLIST[X]] as PRIOQUE[X, Y] .
var  L L' : Queue .
var  E : Elt .
op   extract(L E L') to pair(E, L L')
     if  E ≤ L and E ≤ L' .
op   insert(E, L) to E L .
endv

view PRIOQUE2[X :: TRIV, Y :: POSET[X]] of
     PAIR[X, ORDLIST[X, QLIST[X]]] as PRIOQUE[X, Y] .
var  L : NeOList .
var  L' : OList .
var  E E' E'' : Elt .
op   extract(L) to pair(head(L), tail(L)) .
op   insert : Elt List -> NeList .
eq   insert(E, nil) == E .
eq   insert(E, E') == E E' if E ≤ E' .
eq   insert(E, L E' E'' L') == L E' E E'' L'
     if E' ≤ E and E ≤ E'' .
endv
```

The module `SORT[X, Y, PRIOQUE1[X, Y]]` and the module `SORT[X, Y, PRIOQUE2[X, Y]]` both inherit a sorting algorithm still parameterized by `X`, a set, and `Y`, an order on that set. Applying further to, for example, the built-in module `NAT` of natural numbers having the usual ordering on natural numbers, will generate objects in which we can run the obtained sorting algorithms.

5   Priority Queues in ML

The ML version of our example is given in the Caml [29] dialect. It is divided into four parts: the definition of all needed signatures, a simple implementation of priority queues as unsorted lists `ListPQ`, an implementation of sorting by an abstract priority queue `PQSort` and composition of both implementations into a sorting module `Sort`.

The first file contains the signatures of an ordered type (consisting of a type and an ordering function), a priority queue and a sorting algorithm. The latter two declare a submodule E defining the ordering.

```
module type OrderedType =
  sig
    type t
      (* The type of elements *)
    val compare : t → t → int
      (* compare a b is smaller than 0 if a is smaller than b, 0 if a=b, and is
      larger than 0 if a is larger than b *)
  end
module type PrioQueSig =
  sig
    module E : OrderedType
      (* The type and ordering of the elements of the queue *)
    type t
      (* The type of priority queues *)
    (* Operations: *)
    val create : t
    val insert : E.t → t → t
    val extract : t → t * E.t
      (* raises Not_found if the queue is empty *)
  end
module type SortSig =
  sig
    module E : OrderedType
      (* The type and ordering of the elements to sort *)
    val sort : E.t list → E.t list
      (* The sorting function *)
  end
```

The second file contains the definition of a priority queue based on unordered lists. We skip the (straightforward) implementation here, the only interesting thing is the functor's header:

```
module ListPQ (O: OrderedType)
  : PrioQueSig with module E=O
```

which says that the module `ListPQ` is a functor, taking an order O as parameter and returning a priority queue where the ordering is the same as in O. Note that since the output signature of this functor is given, its users will only have access to types and functions specified in this signature. Other types and functions are treated as local and implementation specific and therefore they will be inaccessible.

The third element is the abstract algorithm, whose implementation is also trivial. Again the interesting part is the functor's header, which can have two possible forms. The first one is the following:

```
module PQSort1 (O: OrderedType)
                (PQ: PrioQueSig with module E=O)
 : SortSig with module E=O
```

Now, in order to obtain the final sorting algorithm one can do it in OCaml in the following way:

```
module Sort1 (O: OrderedType)
 : SortSig with module E=O
 = PQSort1(O)(ListPQ(O))
```

The module's output signature is the signature of sorting with respect to the argument ordering. Its implementation is simply the composition of existing algorithms, all this under the abstraction with respect to the argument ordering.

There is also a second way of writing the header of the abstract priority queue sorting algorithm:

```
module type PQFunctSig
 = functor (O': OrderedType)
            → PrioQueSig with module E=O'

module PQSort2 (O: OrderedType) (PQF: PQFunctSig)
 : SortSig with module E=O
```

The above code fragment consists of two parts: first the functor type is defined, which corresponds exactly to the specification of ListPQ. Then the sorting algorithm is presented as a higher-order functor, i.e. a functor which itself takes a functor as a parameter. Of course, the first line of PQSort2 is the application of PQF to O in order to get the priority queue PQ, and from this point on the code of both functors is identical.

Higher-order functors are not available in OBJ.

In order to obtain the final sorting algorithm, one applies PQSort2 to ListPQ:

```
module Sort2 (O: OrderedType)
 : SortSig with module E=O
 = PQSort2(O)(ListPQ)
```

The first approach to composing modules is more general than the second, because one does not necessarily have to use a generic priority queue functor. Consequently the use of data structures specialized to a given type is possible (e.g. if a set of values is finite a priority queue can be based on counting elements).

On the other hand, the higher-order functor may correspond better to the intended way the programmer wishes to use a given part of code in the whole program. This is exactly our case, since we want to compose PQSort with the generic ListPQ *functor*.

Of course it is possible to get the advantages of both approaches: write the most general specification, as in PQSort1, and then wrap it in a higher-order functor, presenting the intentions of the programmer:

```
module PQSort2' (O: OrderedType) (PQF: PQFunctSig)
  : SortSig with module E=O
  = PQSort1(O)(PQF(O)).
```

6 Priority Queues in Coq

The structure of the Coq development is the same as in ML, but the signatures now contain formal specifications, and structures contain proofs of desired properties.

The first file, as in ML, contains the definition of all needed signatures. The signatures are preceded by the definition of the type of a three-value proof-carrying comparison: the type comparison t < = a b is for example inhabited by Lt p, where p is a proof of the property a < b.

Inductive *comparison* $(X : Set)$ $(lt\ eq : X \to X \to Prop)$ $(x\ y : X) : Set :=$
   | $Lt : lt\ x\ y \to comparison\ X\ lt\ eq\ x\ y$
   | $Eq : eq\ x\ y \to comparison\ X\ lt\ eq\ x\ y$
   | $Gt : lt\ y\ x \to comparison\ X\ lt\ eq\ x\ y.$

Module Type *OrderedType.*

   Parameter $t : Set.$

   Parameter $eq : t \to t \to Prop.$
   Parameter $lt : t \to t \to Prop.$

   Parameter $compare : \forall\ x\ y : t, comparison\ t\ lt\ eq\ x\ y.$

   Axiom $eq\_refl : \forall\ x : t, eq\ x\ x.$
   Axiom $eq\_sym : \forall\ x\ y : t, eq\ x\ y \to eq\ y\ x.$
   Axiom $eq\_trans : \forall\ x\ y\ z : t, eq\ x\ y \to eq\ y\ z \to eq\ x\ z.$

   Axiom $lt\_trans : \forall\ x\ y\ z : t, lt\ x\ y \to lt\ y\ z \to lt\ x\ z.$
   Axiom $lt\_not\_eq : \forall\ x\ y : t, lt\ x\ y \to \neg\ eq\ x\ y.$

   Hint Immediate $eq\_sym.$
   Hint Resolve $eq\_refl\ eq\_trans\ lt\_not\_eq\ lt\_trans.$
End *OrderedType.*

Module Type *PrioQueSig.*

   (* Declarations *)
   Declare Module $E : OrderedType.$

   Parameter $t : Set.$

   Parameter $create : t.$
   Parameter $insert : t \to E.t \to t.$
   Parameter $extract : t \to option\ (t \times E.t).$

   (* Specification - auxiliary functions and predicates *)

Parameter *number* : *t* → *E.t* → *nat* .

Definition *empty q* : *Prop* := ∀ *x, number q x* = 0.

(* Queues are similar iff *q1* = *q2* + {*x*} *)
Definition *similar* (*q1 q2* : *t*) (*x* : *E.t*) : *Prop* :=
   (∀ *y* : *E.t*, ¬ *E.eq x y* → *number q1 y* = *number q2 y*)
   ∧ (∀ *y* : *E.t*, *E.eq x y* → *number q1 y* = *S* (*number q2 y*)).

(* Specification of operations *)

Axiom *create_empty* : *empty create*.

Axiom *insert_similar* :
   ∀ (*q* : *t*) (*x* : *E.t*), *similar* (*insert q x*) *q x*.

Axiom *extract_similar* :
   ∀ (*q q2* : *t*) (*x* : *E.t*),
      extract q = *Some* (*q2, x*) → *similar q q2 x*.

Axiom *extract_minimal* :
   ∀ (*q q2* : *t*) (*x y* : *E.t*),
      extract q = *Some* (*q2, x*) → *E.lt y x* → *number q y* = 0.

Axiom *extract_empty_none* :
   ∀ *q* : *t, extract q* = *None* → *empty q*.

End *PrioQueSig*.

Module Type *SortSig*.

   Declare Module *E* : *OrderedType*.

   Parameter *sort* : *list E.t* → *list E.t*.

   Definition *le e1 e2* := *E.lt e1 e2* ∨ *E.eq e1 e2*.
   Axiom *sort_sorted* : ∀ *l* : *list E.t, Sorting.sort le* (*sort l*).

   Axiom *eq_dec* : ∀ *e1 e2* : *E.t*, {*E.eq e1 e2*} + { ¬ *E.eq e1 e2*}.
   Axiom *sort_permut* :
      ∀ *l* : *list E.t, Permutation.permutation E.eq eq_dec l* (*sort l*).

End *SortSig*.

The signature OrderedType, taken from [13], contains the same calculable elements as its ML counterpart, but is constructed differently. Its main elements are the type and the equality and ordering predicates (i.e. logical elements). The function compare is only an addition to the predicates. Instead of an int, the compare function returns an element of the comparison type defined earlier, i.e. the ordering decision together with the proof that the decision is right.

   Apart from this, the OrderedType signature contains axioms specifying the properties of ordering and equality and hints to instrument automatic tactics, trying to prove properties concerned with the order. The latter element is of course not part of the type theory.

   The priority queue signature is also divided into two parts: declarations and specifications. The declarations contain the same elements as in ML with the only exception of the extract function, which returns an option type, i.e. Some value if the queue is not empty and None otherwise (instead of raising an exception). Note, however, that in order to specify the queue operations one must declare additional functions, counting the number of occurrences of a given element in the queue. Based on this function, two predicates empty and similar can easily be defined in order to write the purely logical axioms specifying how create, insert and extract work.

The signature of a sorting algorithm is simply an extension of its ML counterpart by the logical axioms, saying that the list resulting from sorting is sorted and is a permutation of the input list. The Sorting.sort and Permutation.permutation predicates from the Coq standard library need additional elements such as less than or equal predicate le or equality decidability property eq_dec.

In the second file, the header of ListPQ is the following:

Module *ListPQ* (*O*: *OrderedType*) <: (*PrioQueSig* with Module *E*:=*O*).

The difference between the ML and Coq versions of this functor is the way the resulting module type is declared. The Coq syntax Module *M* <: *SIG* means that the type checker should check that the principal signature of *M* is included in *SIG* and the users of *M* are allowed to use all the information inferred in its principal signature. We say that this module type annotation is transparent as opposed to the opaque one that was used in the ML version. The fact the transparent annotation is used is only important for evaluation of programs inside Coq, such as Eval compute in *(sort l)*, see below. Thanks to transparency the reduction mechanism can *see* the definitions of all functions and evaluate them. For typechecking reasons the opaque module type annotations would be equally good.

In Coq, we also have two possibilities of writing the *PQSort* functor. The header of the first order one is as follows:

Module *PQSort1* (*O*: *OrderedType*)
                 (*PQ*: *PrioQueSig* with Module *E* := *O*)
  <: *SortSig* with Module *E* := *O*.

Unfortunately, due to the requirement that functors are applied only to *names* of modules, and the lack of local module bindings, the composition of *PQSort1* and *ListPQ* is somewhat lengthy:

Module *Sort1* (*O*: *OrderedType*) <: (*SortSig* with Module *E*:=*O*).
    Module *ListPQ_O* := *ListPQ O*.
    Module *PQSort_O* := *PQSort1 O ListPQ_O*.
    (* Include PQSort_O. *)
    Module *E* := *PQSort_O.E*.
    Definition *sort* := *PQSort_O.sort*.
    Definition *le* := *PQSort_O.le*.
    Definition *sort_sorted* := *PQSort_O.sort_sorted*.
    Definition *eq_dec* := *PQSort_O.eq_dec*.
    Definition *sort_permut* := *PQSort_O.sort_permut*.
End *Sort1*.

Now we can apply the functor to an example module NatOrder and test the sorting!

Module *NatSort1* <: (*SortSig* with Module *E:=NatOrder*)
:= *Sort1 NatOrder*.
Eval compute *in* (*NatSort1.sort* (4::5::1::2::nil)).

The higher-order way of writing *PQSort*

Module Type *PQFunctSig* (*O' : OrderedType*)
:= *PrioQueSig* with Module *E := O'*.

Module *PQSort2* (*O: OrderedType*) (*PQF: PQFunctSig*)
<: *SortSig* with Module *E := O*.

starting with the creation of the priority queue for *O*:

   Module *PQ := PQF O*.
leads to a much simpler composition code:

Module *Sort2* (*O: OrderedType*)
<: *SortSig* with Module *E:=O*
:= *PQSort2 O ListPQ*.

Unfortunately, due to a certain weakness of the Coq module system with respect to transparency of higher-order functors, the instances of the *PQSort2* functor cannot be evaluated inside Coq. However, the ML code extracted from both functors can of course be evaluated without any problems.

    To summarize, it is interesting to compare the size of ML and Coq code. It follows that Coq signatures with specifications by logical formulas are about 2-3 times longer than their commented ML counterparts. Unfortunately, the implementations, which in Coq contain proofs of required properties, are about 10-20 times longer than the corresponding ML code.

7   Conclusion

We have presented three languages which integrate specification and implementation. With the simple example of an abstract sorting algorithm based on a priority queue, we demonstrate how each of the three languages can be used for programming in the large by writing specifications, implementations and by composing abstract components. In particular, we want to stress that parameterization should be available for all kinds of modules.

    We have seen that the most important concepts of the OBJ modules are still present in more recent systems such as ML and Coq. Indeed, OBJ objects correspond to structures, parametric objects to functors and OBJ theories to signatures. Only the parametric OBJ theories do not have direct representatives in the ML and Coq module systems, but abstract signatures can easily be refined to concrete ones using the "with" notation. On the other hand, higher-order modules are lacking in OBJ. Although they are not much used in practice, our example shows their adequacy to describing dependencies on other parametric components.

Concerning the ability of these languages to specify and implement software components, OBJ lies somewhere between ML and Coq. In ML, specifications are simply given as types for functions, and execution is based on an efficient call-by-value evaluation strategy. In OBJ, one can write first-order equational and membership specifications that are executable via an efficient built-in associative-commutative rewriting mechanism guided by user-defined strategies. In Coq, the specification language is higher-order predicate logic, which is by far the most expressive of the three. This makes it possible to write a specification, implement it, prove that the implementation is correct, run the implementation inside Coq and even extract the program into an executable ML code. Some of these steps may of course involve complex, lengthy machine computations.

The question arises of which language is best suited for fast prototyping. If no verification is needed, the answer would probably be ML. Separating signatures from their actual implementation is just very neat, and allows a two steps development methodology which does not require much interaction between these two phases unless there are major design errors. Because OBJ modules provide at the same time with an interface and logical requirements for the interface, specification and coding are no more clearly separated. The development process becomes more complicated, going back and forth between different pieces of the code. A comparison with Coq is more difficult, since Coq gives you a lot more: while it is possible in OBJ to forget about the proof obligations generated when typing modules, this is not the case with Coq. A consequence is that every change requires tedious adjustments of the proofs.

Acknowledgments: We thank Andrzej Gąsienica-Samek and Tomasz Stachowicz for their help with the Coq development, Pierre-Yves Strub for checking preliminary versions of the OBJ development in Maude, and the referee for many valuable comments.

References

1. The Agda proof assistant. http://www.cs.chalmers.se/~catarina/agda/.
2. Peter Borovanský, Claude Kirchner, Hélène Kirchner, Pierre-Etienne Moreau, and Marian Vittek. ELAN: A logical framework based on computational systems. In J. Meseguer, editor, *1st International Workshop on Rewriting Logic and its Applications, Electronic Notes in Theoretical Computer Science 4*, 1996.
3. Adel Bouhoula, Jean-Pierre Jouannaud, and José Meseguer. Specification and proof in membership equational logic. *Theoretical Computer Science*, 236:35–132, 1999.
4. Gilad Bracha. *The Programming Language Jigsaw: Mixins, Modularity and Multiple Inheritance.* PhD thesis, Dept. of Computer Science, University of Utah, 1992.
5. Gerth Stolting Brodal and Chris Okasaki. Optimal purely functional priority queues. *Journal of Functional Programming*, 6(6):839–857, 1996.
6. Rod M. Burstall and Joseph A. Goguen. Putting theories together to make specifications. In *Proc. 5th International Joint Conference of Artificial Intelligence, Cambridge Massachusetts*, pages 1045–1058, Edinburgh University, 1977.
7. Rod M. Burstall and Joseph A. Goguen. The semantics of CLEAR, a specification language. In *1979 Copenhagen Winter School on Abstract Software Specification*, volume 86 of *LNCS*. Springer-Verlag, 1980.

8. Jacek Chrząszcz. Modules in Coq are and will be correct. In Stefano Berardi, Mario Coppo, and Ferruccio Damiani, editors, *Types for Proofs and Programs, International Workshop, TYPES 2003, Torino, Italy, April 30 - May 4, 2003, Revised Selected Papers*, volume 3085 of *LNCS*, pages 130–146. Springer, 2004.

9. Manuel Clavel, Steven Eker, Patrick Lincoln, and José Meseguer. Principles of Maude. In J. Meseguer, editor, *1st International Workshop on Rewriting Logic and its Applications, Electronic Notes in Theoretical Computer Science 4*, 1996.

10. The Coq proof assistant. http://coq.inria.fr/.

11. Thierry Coquand and Gérard Huet. The calculus of constructions. *Information and Computation*, 76:95–120, February 1988.

12. Thierry Coquand and Christine Paulin-Mohring. Inductively defined types. In P. Martin-Löf and G. Mints, editors, *COLOG-88: International conference on computer logic*, volume 417 of *LNCS*. Springer-Verlag, 1990.

13. Jean-Christophe Filliâtre and Pierre Letouzey. Functors for Proofs and Programs. In *European Symposium on Programming*, volume 2986 of *LNCS*, pages 370–384, Barcelona, Spain, April 2004. Springer-Verlag.

14. Kokichi Futatsugi, Joseph A. Goguen, Jean-Pierre Jouannaud, and José Meseguer. Principles of OBJ2. In *Proc. 12th ACM Symp. on Principles of Programming Languages, New Orleans*, 1985.

15. J. A. Goguen, J. W. Thatcher, and E. G. Wagner. An initial algebra approach to the specification, correctness and implementation of abstract data types. In *Current Trends in Programming Methodology, vol. 4*, pages 80–149. Prentice Hall, 1978.

16. J. A. Goguen, J. W. Thatcher, E. W. Wagner, and J. B. Wright. Initial algebra semantics and continuous algebra. *Journal of the ACM*, 24(1):68–95, January 1977.

17. Joseph A. Goguen and Joseph J. Tardo. An introduction to obj, a language for writing and testing formal algebraic specifications. In *Specification of Reliable Software Conference*, pages 170–189, April 1979.

18. Joseph A. Goguen, Timothy Winkler, José Meseguer, Kokichi Futatsugi, and Jean-Pierre Jouannaud. *Applications of Algebraic Specifications Using OBJ*, chapter Introducing OBJ*. Cambridge University Press, 1993. D. Coleman, R. Gallimore and J. A. Goguen, eds.

19. Tom Hirschowitz and Xavier Leroy. Mixin modules in a call-by-value setting. In D. Le Métayer, editor, *Programming Languages and Systems, ESOP'2002*, volume 2305 of *LNCS*, pages 6–20. Springer-Verlag, 2002.

20. Claude Kirchner, Hélène Kirchner, and José Meseguer. Operational semantics of OBJ3. In *15th International Conference on Automata, Languages and Programming*, volume 317 of *LNCS*, pages 287–301. Springer-Verlag, 1988.

21. The LEGO proof assistant. http://www.dcs.ed.ac.uk/home/lego/.

22. Zhaohui Luo. ECC an Extended Calculus of Constructions. In *4th Symposium on Logic in Computer Science*, Pacific Grove, California, 1989.

23. David MacQueen. Theory and practice of higher-order type systems or the Standard ML type system. Copy of Transparencies.

24. Lena Magnusson and Bengt Nordström. The alf proof editor and its proof engine. In H. Barendregt and T. Nipkow, editors, *Types for Proofs and Programs*, volume 806 of *LNCS*, pages 213–237. Springer-Verlag, 1993.

25. Per Martin-Löf. An intuitionistic theory of types: Predicative part. In H. E. Rose and J. C. Sheperdson, editors, *Logic Colloquium '73*, volume 80 of *Studies in Logic*, pages 73–118. North-Holland, 1975.

26. Per Martin-Löf. *Intuitionistic Type Theory*. Biblioplois, Napoli, 1984. Notes of Giovanni Sambin on a series of lectues given in Padova.

27. Robert Milner. A theory of type polymorphism programming. *Journal of Computer and System Sciences*, 17, 1978.

28. Shin Nakajima and Kokichi Futatsugi. An object-oriented modeling method for algebraic specifications in Cafe OBJ. In *19th International Conference on Software Engineering*, pages 34–44. ACM Press, 1997.
29. The Objective Caml language. http://caml.inria.fr/.
30. Pierre Weis et al. The CAML reference manual. Rapport de Recherche 121, INRIA, 1990.
31. Benjamin Werner. *Méta-théorie du Calcul des Constructions Inductives*. PhD thesis, Univ. Paris VII, 1994.

Weak Adhesive High-Level Replacement Categories and Systems: A Unifying Framework for Graph and Petri Net Transformations

Hartmut Ehrig and Ulrike Prange

Technical University of Berlin, Germany
ehrig|uprange@cs.tu-berlin.de

Abstract. Adhesive high-level replacement (HLR) systems have been recently introduced as a new categorical framework for graph tranformation in the double pushout (DPO) approach. They combine the well-known concept of HLR systems with the concept of adhesive categories introduced by Lack and Sobociński.

While graphs, typed graphs, attributed graphs and several other variants of graphs together with corresponding morphisms are adhesive HLR categories, such that the categorical framework of adhesive HLR systems can be applied, this has been claimed also for Petri nets. In this paper we show that this claim is wrong for place/transition nets and algebraic high-level nets, although several results of the theory for adhesive HLR systems are known to be true for the corresponding Petri net transformation systems.

In fact, we are able to define a weaker version of adhesive HLR categories, called weak adhesive HLR categories, which is still sufficient to show all the results known for adhesive HLR systems. This concept includes not only all kinds of graphs mentioned above, but also place/transition nets, algebraic high-level nets and several other kinds of Petri nets. For this reason weak adhesive HLR systems can be seen as a unifying framework for graph and Petri net transformations.

1 Introduction

The use of categorical techniques for unifying frameworks in Computer Science has a long tradition. In the early 1970ies the concept of closed monoidal categories was proposed by Goguen in [1] as a unifying framework for different kinds of deterministic automata. An extension of this framework to nondeterministic and stochastic automata using pseudo-closed categories was presented in [2]. Other important examples are the unifying frameworks of institutions and specification frames respectively. This first framework is based on a categorical treatment of signatures, models and sentences introduced by Goguen and Burstall [3], and the second one in [4] combines directly signatures and sentences to specifications. In both cases we obtain a unifying framework for all kinds of algebraic and logical specification techniques.

K. Futatsugi et al. (Eds.): Goguen Festschrift, LNCS 4060, pp. 235–251, 2006.

Most recently the unifying framework of adhesive high-level replacement (HLR) systems for different kinds of graph transformation systems has been introduced in [5, 6]. The corresponding concept of adhesive HLR categories integrates those of HLR categories in [7] and adhesive categories by Lack and Sobociński [8], which was later extended to quasi-adhesive categories [9]. The concept of adhesive categories requires the existence of pushouts along monomorphisms and pullbacks, and the property that pushouts along monomorphisms are van Kampen (VK) squares. Roughly spoken the last property means that pushouts are stable under pullbacks and vice versa pullbacks are stable under combined pushouts and pullbacks. In the case of adhesive HLR categories the class of all monomorphisms is replaced by a subclass $\mathcal{M}$ of monomorphisms closed under composition and decomposition and the existence of all pullbacks by pullbacks along $\mathcal{M}$-morphisms. In [5, 6] it is shown that there is a unifying framework of adhesive HLR systems for graph transformation systems based on the double pushout (DPO) approach [10] concerning a large variety of different graph concepts, like labeled graphs, typed graphs, attributed graphs, typed attributed graphs and hypergraphs. The key idea is to show that adhesive HLR categories satisfy a number of different properties, called HLR properties, which are used in [7] to prove important results like the Local Church-Rosser Theorem, the Parallelism Theorem and the Concurrency Theorem. This was first shown for adhesive categories in [8] for the class $\mathcal{M}$ of all monomorphisms and later extended to adhesive HLR categories in [5, 6] and to quasiadhesive categories in [9], where $\mathcal{M}$ is the class of all regular monomorphisms.

The idea to apply the DPO approach to Petri nets was first considered for place/transition nets in [7] and for algebraic high-level nets in [11]. In [5] we have claimed that the category (**PTNets**, $\mathcal{M}$) of place/transition nets with the class $\mathcal{M}$ of all injective morphisms is an adhesive HLR category in order to apply the general theory of adhesive HLR systems also to place/transition nets. Unfortunately this claim is wrong as we show in this paper. The reason is that **PTNets** has general pullbacks, but pullbacks in general cannot be constructed componentwise in **Sets**. However, pullbacks along monomorphisms in **PTNets** can be constructed componentwise in **Sets**. This is the key idea to weaken the concept of adhesive HLR categories using weak VK squares, such that (**PTNets**, $\mathcal{M}$) is a weak adhesive HLR category, and nevertheless this weaker concept still allows to verify the HLR properties used in [7, 5, 6] to prove under some additional assumptions the following main results:
1. Local Church-Rosser Theorem,
2. Parallelism Theorem,
3. Concurrency Theorem,
4. Embedding and Extension Theorem,
5. Local Confluence Theorem - Critical Pair Lemma.

In this paper we show for elementary nets, place/transition nets and algebraic high-level nets that they are weak adhesive HLR categories for a suitable class of morphisms. In [5, 6] we have shown already that adhesive HLR categories satisfy the HLR properties to prove the main results stated above. In this paper we show that this is already true for weak adhesive HLR categories. This implies that the main results are also true for different kinds of Petri net transformation systems including elementary, place/transition and algebraic high-level nets. Note, that in contrast to the "classical" theory of Petri nets and systems based on the token game, where the structure of the nets remains unchanged, the theory of Petri net transformations allows not only the token game, but also to change the structure of the nets. In this sense weak adhesive HLR categories can be seen as a unifying framework not only for graph but also for Petri net transformations.

This paper is organized as follows:
In Section 2 we review adhesive and adhesive HLR categories as introduced in [8] and [5]. In Section 3 we extend these concepts to weak adhesive HLR categories and systems. This is the basis to define Petri net transformation systems as an instance of weak adhesive HLR systems in Section 4.

Acknowledgement

This paper is a contribution to a special issue of Springer LNCS as a Festschrift on the occasion of the 65th birthday of J.A. Goguen. We are glad to announce that the important contributions of J.A. Goguen concerning categorical unifying frameworks for different areas in Computer Science have mainly influenced the school in Berlin leading to similar ones and also to the new unifying framework for graph transformation published recently and for Petri net transformation presented in this paper.

2   Review of Adhesive and Adhesive HLR Categories

The intuitive idea of adhesive categories are categories with suitable pushouts and pullbacks which are compatible with each other. More precisely the definition is based on so-called van Kampen squares.

The idea of a van Kampen (VK) square is that of a pushout which is stable under pullbacks, and vice versa that pullbacks are stable under combined pushouts and pullbacks. The name van Kampen derives from the relationship between these squares and the Van Kampen Theorem in topology (see [12]).

Definition 1 (van Kampen square). *A pushout (1) is a van Kampen square, if for any commutative cube (2) with (1) in the bottom and the back faces being pullbacks holds: the top face is a pushout iff the front faces are pullbacks.*

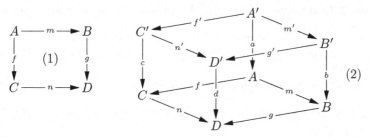

It might be expected that at least in the category **Sets** of sets and functions each pushout is a van Kampen square. Unfortunately this is not true (see Ex. 1). But at least pushouts along monomorphisms (injective functions) are VK squares (see [8, 9]).

Fact 1 (VK squares in Sets). *In* **Sets**, *each pushout along a monomorphism is a VK square. Pushout (1) is called a pushout along a monomorphism, if m (or symmetrically f) is a monomorphism.*

Example 1 (VK squares in **Sets***).* In the following diagram on the left hand side a VK square along an injective function in **Sets** is shown. All morphisms are inclusions, or 0 and 1 are mapped to $*$ and 3 to 2.

Arbitrary pushouts are stable under pullbacks in **Sets**. That means, one direction of the VK square property is also valid for arbitrary morphisms. But the other direction is not necessarily fulfilled. The cube on the right hand side is such a counterexample for arbitrary functions: all faces commute, the bottom and the top are pushouts and the back faces are pullbacks. But obviously the front faces are no pullbacks, therefore the pushout in the bottom fails to be a VK square.

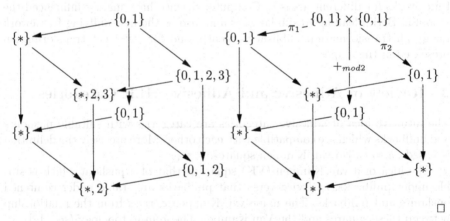

□

In the following definition of adhesive categories only those VK squares of Def. 1 are considered where m is a monomorphism. According to Lack and Sobociński [8] we define

Definition 2 (adhesive category). *A category* **C** *is an* adhesive category, *if*

1. **C** *has pushouts along monomorphisms (i.e. pushouts, where at least one of the given morphisms is a monomorphism),*
2. **C** *has pullbacks,*
3. *pushouts along monomorphisms are VK squares.*

Let us first consider some basic examples and counterexamples for adhesive categories (see [8]).

Fact 2 (Sets, Graphs, Graphs$_{TG}$ as adhesive categories). *The categories* **Sets** *of sets and functions,* **Graphs** *of graphs and graph morphisms and* **Graphs$_{TG}$** *of typed graphs and typed graph morphisms are adhesive categories.*

Counterexample 2 (non-adhesive categories). For example, the category **Posets** of partially ordered sets and the category **Top** of topological spaces and continuous functions are not adhesive categories. In the following diagram a cube in **Posets** is shown that fails to be a van Kampen square. The bottom is a pushout with injective functions (monomorphisms) and all lateral faces are pullbacks, but the top square is no pushout in **Posets**. The proper pushout over the corresponding morphisms is the square (1).

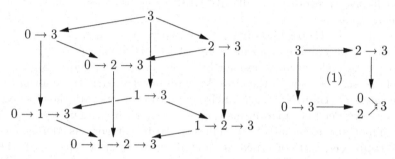

□

Remark 1. In [9] Lack and Sobociński have also introduced a variant of adhesive categories, called quasiadhesive categories, where the class of monomorphisms in Def. 2 is replaced by regular monomorphisms. A monomorphism is called regular, if it is the equalizer of two morphisms. For adhesive and also for quasiadhesive categories Lack and Sobociński have shown, that all the HLR properties, shown for adhesive HLR categories in Thm. 2 below, are valid. This allows to prove several important results of graph transformation systems in the framework of adhesive and also of quasiadhesive categories. On the other hand adhesive and also quasiadhesive categories are special cases of adhesive HLR categories $(\mathbf{C}, \mathcal{M})$ (see Def. 3 below), where the class $\mathcal{M}$ is specialized to the class of all monos and of all regular monos respectively.

The main difference between adhesive HLR categories and adhesive categories is that a distinguished class $\mathcal{M}$ of monomorphisms is considered instead of all monomorphisms, so that only pushouts along $\mathcal{M}$-morphisms have to be VK squares. Moreover, only pullbacks along $\mathcal{M}$-morphisms and not over arbitrary morphisms are required (see [5, 6]).

Definition 3 (adhesive HLR category). *A category* **C** *with a morphism class* $\mathcal{M}$ *is called an* adhesive HLR category, *if*

1. $\mathcal{M}$ *is a class of monomorphisms closed under isomorphisms, composition* $(f : A \to B \in \mathcal{M}, g : B \to C \in \mathcal{M} \Rightarrow g \circ f \in \mathcal{M})$ *and decomposition* $(g \circ f \in \mathcal{M}, g \in \mathcal{M} \Rightarrow f \in \mathcal{M})$,

2. **C** *has pushouts and pullbacks along* $\mathcal{M}$*-morphisms and* $\mathcal{M}$*-morphisms are closed under pushouts and pullbacks,*
3. *pushouts in* **C** *along* $\mathcal{M}$*-morphisms are VK squares.*

Remark 2. $\mathcal{M}$-morphisms are closed under pushouts if, for a pushout (1) in Def. 1, $m \in \mathcal{M}$ implies that $n \in \mathcal{M}$. Analogously, $\mathcal{M}$-morphisms are closed under pullbacks if, for a pullback (1), $n \in \mathcal{M}$ implies that $m \in \mathcal{M}$.

Example 3 (adhesive HLR categories).

- All adhesive categories are adhesive HLR categories for the class $\mathcal{M}$ of all monomorphisms.
- The category (**HyperGraphs**, $\mathcal{M}$) of hypergraphs with the class $\mathcal{M}$ of injective hypergraph morphisms is an adhesive HLR category.
- Another example for an adhesive HLR category is the category (**Sig**, $\mathcal{M}$) of algebraic signatures with the class $\mathcal{M}$ of all injective signature morphisms.
- The category (**ElemNets**, $\mathcal{M}$) of elementary Petri nets with the class $\mathcal{M}$ of all injective Petri net morphisms is an adhesive HLR category (see Fact 3).
- An important example of an adhesive HLR category is the category (**AGraphs**$_{\mathbf{ATG}}$, $\mathcal{M}$) of typed attributed graphs with a type graph ATG and the class $\mathcal{M}$ of all injective morphisms with isomorphisms on the data part.   □

Counterexample 4 (non-adhesive HLR categories). The categories (**PTNets**, $\mathcal{M}$) of place/transition nets and (**Spec**, $\mathcal{M}$) of algebraic specifications, where $\mathcal{M}$ is the class of all the corresponding monomorphisms, fail to be adhesive HLR categories (see Ex. 6).   □

3   Weak Adhesive HLR Categories and Systems

As pointed out in Counterex. 4 the category (**PTNets**, $\mathcal{M}$) of place/transition nets with the class $\mathcal{M}$ of all monomorphisms fails to be an adhesive HLR category. For this reason we introduce now a slightly weaker version, called weak adhesive HLR category.

For a weak adhesive HLR category we only soften item 3 in Def. 3, so that only special cubes are considered for the VK square property.

Definition 4 (weak adhesive HLR category). *A category* **C** *with a morphism class* $\mathcal{M}$ *is called a* weak adhesive HLR category, *if*

1. $\mathcal{M}$ *is a class of monomorphisms closed under isomorphisms, composition and decomposition,*
2. **C** *has pushouts and pullbacks along* $\mathcal{M}$*-morphisms and* $\mathcal{M}$*-morphisms are closed under pushouts and pullbacks,*

3. *pushouts in* $\mathbf{C}$ *along* $\mathcal{M}$*-morphisms are weak VK squares, i.e. the VK square property holds for all commutative cubes with* $m \in \mathcal{M}$ *and (* $f \in \mathcal{M}$ *or* $b, c, d \in \mathcal{M}$*) (see Def. 1).*

Remark 3. For the weak version of the VK square property it is sufficient to require $f \in \mathcal{M}$ or $b, c, d \in \mathcal{M}$. In both cases this makes sure that the pullback squares in the cube are pullbacks along $\mathcal{M}$-morphisms.

Example 5 (weak adhesive HLR categories).

- All adhesive HLR categories are weak adhesive HLR categories.
- The category (**PTNets**, $\mathcal{M}$) of place/transition nets with the class $\mathcal{M}$ of all monomorphisms is a weak adhesive HLR category (see Fact 4).
- Similarly the category **AHLNets(SP, A)** of algebraic high-level nets with fixed specification SP and algebra A considered with the class $\mathcal{M}$ of injective morphisms is a weak adhesive HLR category (see Fact 5).
- An interesting example of high-level structures, which are not graph-like, are algebraic specifications (see [13]). The category (**Spec**, $\mathcal{M}_{strict}$) of algebraic specifications with the class $\mathcal{M}_{strict}$ of all strict injective specification morphisms is a weak adhesive HLR category.   □

Similar to adhesive HLR categories also weak adhesive HLR categories are closed under product, slice, coslice, functor and comma category constructions. That means we can construct new weak adhesive HLR categories from given ones.

Theorem 1 (construction of weak adhesive HLR categories). *Weak adhesive HLR categories can be constructed as follows:*

1. *If* $(\mathbf{C}, \mathcal{M}_1)$ *and* $(\mathbf{D}, \mathcal{M}_2)$ *are weak adhesive HLR categories, then the product category* $(\mathbf{C} \times \mathbf{D}, \mathcal{M}_1 \times \mathcal{M}_2)$ *is a weak adhesive HLR category.*
2. *If* $(\mathbf{C}, \mathcal{M})$ *is a weak adhesive HLR category, so are the slice category* $(\mathbf{C} \backslash X, \mathcal{M} \cap \mathbf{C} \backslash X)$ *and the coslice category* $(X \backslash \mathbf{C}, \mathcal{M} \cap X \backslash \mathbf{C})$ *for any object* X *in* $\mathbf{C}$.
3. *If* $(\mathbf{C}, \mathcal{M})$ *is a weak adhesive HLR category, then for every category* $\mathbf{X}$ *the functor category* $([\mathbf{X}, \mathbf{C}], \mathcal{M}$*-functor transformations) is a weak adhesive HLR category. An* $\mathcal{M}$*-functor transformation is a natural transformation* $t : F \rightarrow G$ *where all morphisms* $t_X : F(X) \rightarrow G(X)$ *are in* $\mathcal{M}$.
4. *If* $(\mathbf{A}, \mathcal{M}_1)$ *and* $(\mathbf{B}, \mathcal{M}_2)$ *are weak adhesive HLR categories and* $F : \mathbf{A} \rightarrow \mathbf{C}$, $G : \mathbf{B} \rightarrow \mathbf{C}$ *are functors, where* F *preserves pushouts along* $\mathcal{M}_1$*-morphisms and* G *preserves pullbacks (along* $\mathcal{M}_2$*-morphisms), then the comma category* $(ComCat(F, G; \mathcal{I}), \mathcal{M})$ *with* $\mathcal{M} = (\mathcal{M}_1 \times \mathcal{M}_2) \cap Mor_{ComCat(F,G;\mathcal{I})}$ *is a weak adhesive HLR category* .

In the following theorem we show several important properties for weak adhesive HLR categories, which are essential to prove the main results in Cor. 1. These properties have been required as HLR properties in [7] to show some of the main results for HLR systems. In [8], it was shown already that these HLR properties are valid for adhesive categories. They were extended to adhesive HLR categories in [5], and now also for weak adhesive HLR categories using almost the same proofs.

Theorem 2 (properties of weak adhesive HLR categories). *Given a weak adhesive HLR category* $(\mathbf{C}, \mathcal{M})$, *then the following properties hold:*

1. Pushouts along $\mathcal{M}$-morphisms are pullbacks: *Given the following pushout (1) with* $k \in \mathcal{M}$, *then (1) is also a pullback.*
2. $\mathcal{M}$ pushout-pullback decomposition lemma: *Given the following commutative diagram with (1)+(2) being a pushout, (2) a pullback,* $w \in \mathcal{M}$ *and* $(l \in \mathcal{M}$ *or* $u \in \mathcal{M})$. *Then (1) and (2) are pushouts and also pullbacks.*
3. Cube pushout-pullback lemma: *Given the following commutative cube (3), where all morphisms in the top and in the bottom are in* $\mathcal{M}$, *the top is a pullback and the front faces are pushouts. Then we have: the bottom is a pullback iff the back faces of the cube are pushouts.*

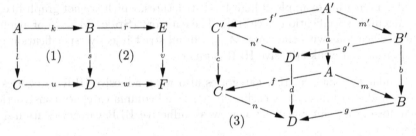

4. Uniqueness of pushout complements: *Given* $k : A \to B \in \mathcal{M}$ *and* $s : B \to D$ *then there is up to isomorphism at most one* C *with* $l : A \to C$ *and* $u : C \to D$ *such that (1) is a pushout.*

Now we are able to generalize graph transformation systems, grammars and languages in the sense of [10] based on the category **Graphs** to weak adhesive HLR categories, which was already done for HLR, adhesive and adhesive HLR categories in [7], [8] and [5] respectively.

In general, a weak adhesive HLR system is based on productions, also called rules, that describe in an abstract way how objects in this system can be transformed. An application of a production is called direct transformation and describes how an object is actually changed by the production. A sequence of these applications yields a transformation.

Definition 5 (production and transformation). *Given a weak adhesive HLR category* $(\mathbf{C}, \mathcal{M})$, *a production* $p = (L \xleftarrow{l} K \xrightarrow{r} R)$ *(also called rule) consists of three objects* L, K *and* R *called left hand side, gluing object and right hand side respectively, and morphisms* $l : K \to L$, $r : K \to R$ *with* $l, r \in \mathcal{M}$.

Given a production $p = (L \xleftarrow{l} K \xrightarrow{r} R)$ *and an object* G *with a morphism* $m : L \to G$, *called match. A direct transformation* $G \xRightarrow{p,m} H$ *from* G *to an object* H *is given by the following diagram, where (1) and (2) are pushouts.*

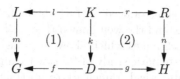

A sequence $G_0 \Rightarrow G_1 \Rightarrow ... \Rightarrow G_n$ of direct transformations is called a trans-formation and is denoted as $G_0 \overset{*}{\Rightarrow} G_n$. For $n = 0$, we have the identical trans-formation $G_0 \overset{id}{\Rightarrow} G_0$, i.e. $f = g = id_{G_0}$. Moreover, we allow for $n = 0$ also isomorphisms $G_0 \cong G_0'$, because pushouts and hence also direct transformations are only unique up to isomorphism.

Definition 6 (weak adhesive HLR system, grammar and language). *A weak adhesive HLR system $AHS = (\mathbf{C}, \mathcal{M}, P)$ consists of a weak adhesive HLR category $(\mathbf{C}, \mathcal{M})$ and a set of productions P.*

A weak adhesive HLR grammar $AHG = (AHS, S)$ is a weak adhesive HLR system together with a distinguished start object S.

The language L of a weak adhesive HLR grammar is defined by $L = \{G \mid \exists$ transformation $S \overset{*}{\Rightarrow} G\}$.

In [5, 6] it is shown that the HLR properties stated in Thm. 2 together with binary coproducts compatible with $\mathcal{M}$ are sufficient to prove the following main results for adhesive HLR systems. Hence we also have the following main results for weak adhesive HLR systems which are stated explicitly in [7] for HLR systems and in [5, 6] for adhesive HLR systems.

Corollary 1 (main results for weak adhesive HLR systems). *Given a weak adhesive HLR system with binary coproducts compatible with $\mathcal{M}$ (i.e. $f, g \in \mathcal{M} \Rightarrow f + g \in \mathcal{M}$), then we have the following results:*

1. *Local Church-Rosser Theorem,*
2. *Parallelism Theorem,*
3. *Concurrency Theorem.*

The Local Church-Rosser Theorem allows one to apply two graph transfor-mations $G \Rightarrow H_1$ via p_1 and $G \Rightarrow H_2$ via p_2 in an arbitrary order leading to the same result H, provided that they are parallel independent. In this case they can also be applied in parallel, leading to a parallel graph transformation $G \Rightarrow H$ via the parallel production $p_1 + p_2$. This second main result is called the Paral-lelism Theorem. The Concurrency Theorem is concerned with the simultanous execution of causally dependent transformations.

4   Petri Net Transformation Systems

Petri net transformation systems have been first introduced in [7] for the case of low-level nets and in [11] for high-level nets using the algebraic presentation of Petri nets as monoids as introduced in [14]. The main idea of Petri net transformation systems is to extend the well-known theory of Petri nets based on the token game by general techniques which allow to change also the net structure of Petri nets. In [15], a systematic study of Petri net transformation systems has been presented in the categorical framework of abstract Petri nets, which can be instantiated to different kinds of low-level and high-level Petri nets. In this chapter we show that the category (**ElemNets**, $\mathcal{M}$) of elementary Petri nets is an adhesive HLR category (see Fact 3) and that the categories (**PTNets**, $\mathcal{M}$) of place/transition nets and (**AHLNets(SP, A)**, $\mathcal{M}$) of algebraic high-level nets over (SP, A) are weak adhesive HLR categories (see Fact 4 and 5). The corresponding instantiations of weak adhesive HLR systems lead to different kinds of Petri net transformation systems.

In the following we present a simple grammar $ENGG$ (elementary net graph grammar) for elementary Petri nets, which allows to generate all elementary nets. The start net S of $ENGG$ is empty. We have a production $addPlace$ to create a new place p and productions $addTrans(n, m)$ for $n, m \in \mathbb{N}$ to create a transition with n input and m output places.

$addPlace$:

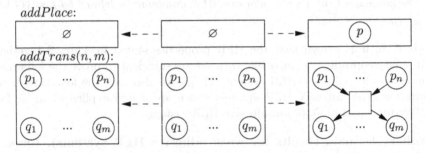

$addTrans(n, m)$:

The grammar $ENGG$ can be modified to a grammar $PTGG$ (place/transition net graph grammar) for place/transition nets if we replace the productions $addTrans(n, m)$ by productions $addTrans(n, m)(i_1, ..., i_n, o_1, ..., o_m)$, where $i_1, ..., i_n$ resp. $o_1, ..., o_m$ correspond to the arc weights of the input places $p_1, ..., p_n$ resp. the output places $q_1, ..., q_m$.

Definition 7 (elementary Petri net). *An elementary Petri net is given by* $N = (P, T, pre, post : T \to \mathcal{P}(P))$ *with a set P of places, T of transitions and pre- and post-domain functions $pre, post : T \to \mathcal{P}(P)$, where $\mathcal{P}(P)$ is the power set of P. A morphism $f : N \to N'$ in **ElemNets** is given by $f = (f_P : P \to P', f_T : T \to T')$ compatible with the pre- and post-domain function, i.e. $pre' \circ f_T = \mathcal{P}(f_P) \circ pre$ and $post' \circ f_T = \mathcal{P}(f_P) \circ post$.*

Fact 3 (elementary Petri nets as adhesive HLR category). *The category (**ElemNets**, $\mathcal{M}$) of elementary Petri nets is an adhesive HLR category, where $\mathcal{M}$ is the class of all injective morphisms.*

Proof idea. The category **ElemNets** is isomorphic to the comma category $ComCat(ID_{\mathbf{Sets}}, \mathcal{P}; \mathcal{I})$, where $\mathcal{P} : \mathbf{Sets} \to \mathbf{Sets}$ is the power set functor and

$\mathcal{I} = \{1, 2\}$. According to Thm. 1.4 it suffices to note that $\mathcal{P} : \mathbf{Sets} \to \mathbf{Sets}$ preserves pullbacks using the fact that $(\mathbf{Sets}, \mathcal{M})$ is an adhesive HLR category. $\Box$

Definition 8 (place/transition net). *According to [14] a place/transition net $N = (P, T, pre, post : T \to P^{\oplus})$ is given by a set P of places, a set T of transitions, as well as pre- and post-domain functions $pre, post : T \to P^{\oplus}$, where $P^{\oplus}$ is the free commutative monoid over P. A morphism $f : N \to N'$ in* **PTNets** *is given by $f = (f_P : P \to P', f_T : T \to T')$ compatible with the pre- and post-domain functions, i.e. $pre' \circ f_T = f_P^{\oplus} \circ pre$ and $post' \circ f_T = f_P^{\oplus} \circ post$.*

Fact 4 (place/transition nets as weak adhesive HLR category). *The category* **(PTNets**, $\mathcal{M})$ *of place/transition nets is a weak adhesive HLR category, but not an adhesive HLR category, if $\mathcal{M}$ is the class of all injective morphisms.*

Proof idea. The category **PTNets** is isomorphic to the comma category $ComCat(ID_{\mathbf{Sets}}, \square^{\oplus}; \mathcal{I})$ with $\mathcal{I} = \{1, 2\}$, where $\square^{\oplus} : \mathbf{Sets} \to \mathbf{Sets}$ is the free commutative monoid functor. According to Thm. 1.4 it suffices to note $\square^{\oplus} : \mathbf{Sets} \to \mathbf{Sets}$ preserves pullbacks along injective morphisms using the fact that $(\mathbf{Sets}, \mathcal{M})$ is a weak adhesive HLR category. This implies that **(PTNets**, $\mathcal{M})$ is a weak adhesive HLR category.

It remains to show that **(PTNets**, $\mathcal{M})$ is not an adhesive HLR category. This is due to the fact, that $\square^{\oplus} : \mathbf{Sets} \to \mathbf{Sets}$ does not preserve general pullbacks. This would imply that pullbacks in **PTNets** are constructed componentwise for places and transitions. In fact, in Ex. 6 we present a non-injective pullback in **PTNets**, where the transition component is not a pullback in **Sets**, and a cube which violates the VK properties of adhesive HLR categories. $\Box$

Example 6 (non-VK square in **PTNets***).* The square (1) in Fig. 1 with non-injective morphisms g_1, g_2, p_1, p_2 is a pullback in the category **PTNets**, where the transition component is not a pullback in **Sets**. In the cube in Fig. 1 the bottom square is a pushout in **PTNets** along an injective morphism $m \in \mathcal{M}$, all side squares are pullbacks, but the top square is no pushout in **PTNets**. Hence we have a counterexample for the VK property. $\Box$

In the following we combine algebraic specifications with Petri nets leading to algebraic high-level (AHL) nets (see [11]). For simplicity we fix the corresponding algebraic specification SP and the SP-algebra A. For the more general case, where also morphisms between different specifications and algebras are allowed, we refer to [11]. Under suitable restrictions for the morphisms we also obtain a weak adhesive HLR category in the more general case (see [15] for HLR properties of high-level abstract Petri nets).

Intuitively, an AHL net is a Petri net, where ordinary, uniform tokens are replaced by data elements from the given algebra. Firing a transition t means to remove some data elements from the input places and add some data elements, computed by term evaluation, to the output places of t. There could be also some

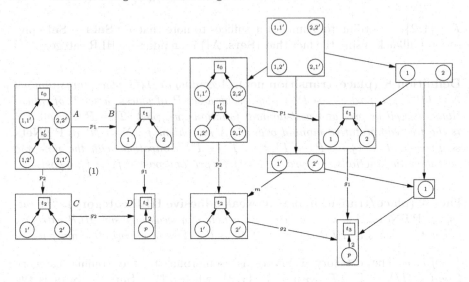

Fig. 1. A pullback and a non-VK square in **PTNets**

firing conditions to restrict the firing behaviour of a transition. In addition, a typing of the places restricts the data elements which could be put on each place to that of a certain type.

Definition 9 (AHL net). *An AHL net over* (SP, A), *where* $SP = (SIG, E, X)$ *has additional variables* X *and* $SIG = (S, OP)$, *is given by* $N = (SP, P, T,$ $pre, post, cond, type, A)$ *with sets* P *and* T *of places and transitions,* $pre, post : T \rightarrow (T_{SIG}(X) \otimes P)^{\oplus}$ *as pre- and post-domain functions,* $cond : T \rightarrow \mathcal{P}_{fin}(Eqns(SIG, X))$ *assigning to each* $t \in T$ *a finite set* $cond(t)$ *of equations over* SIG *and* X, *type* : $P \rightarrow S$ *a type function and* A *an* SP-*algebra.* *Note that* $T_{SIG}(X)$ *is the* SIG-*term algebra with variables* X, $(T_{SIG}(X) \otimes P) =$ $\{(term, p) \mid term \in T_{SIG}(X)_{type(p)}, p \in P\}$ *and* $\square^{\oplus}$ *is the free commutative monoid functor. A morphism* $f : N \rightarrow N'$ *in* **AHLNets(SP, A)** *is given by a pair of functions* $f = (f_P : P \rightarrow P', f_T : T \rightarrow T')$ *which are compatible with the pre, post, cond and type functions as shown below.*

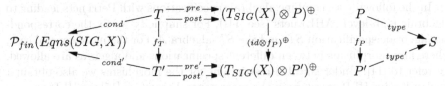

Fact 5 (AHL nets as weak adhesive HLR category). *Given an algebraic specification* SP *and an* SP-*algebra* A, *the category* $(\textbf{AHLNets}(SP, A), \mathcal{M})$ *of algebraic high-level nets over* (SP, A) *is a weak adhesive HLR category.* $\mathcal{M}$ *is the class of all injective morphisms* f, *i.e.* f_P *and* f_T *are injective.*

Proof idea. According to the fact that (SP, A) is fixed the construction of push-outs and pullbacks in **AHLNets(SP, A)** is essentially the same as in **PTNets**, which is already a weak adhesive HLR category. We can apply the idea of comma categories $ComCat(F, G; \mathcal{I})$, where in our case the source functor of the operations $pre, post, cond, type$ is always the identity $ID_{\mathbf{Sets}}$, and the target functors are $(T_{SIG}(X) \otimes \_)^\oplus$: **Sets** → **Sets** and two constant functors. In fact $(T_{SIG}(X) \otimes \_)$: **Sets** → **Sets**, the constant functors and $\Box^\oplus$: **Sets** → **Sets** preserve pullbacks along injective functions. This implies that also $(T_{SIG}(X) \otimes \_)^\oplus$: **Sets** → **Sets** preserves pullbacks along injective functions, which is sufficient to verify the properties of a weak adhesive HLR category.   □

Corollary 2 (main results for Petri net transformation systems). *The results stated in Cor. 1 are valid for Petri net transformation systems based on the following categories:*

1. *(**PTNets**, $\mathcal{M}$) (see Fact 4),*
2. *(**ElemNets**, $\mathcal{M}$) (see Fact 3),*
3. *(**AHLNets**, $\mathcal{M}$) (see Fact 5).*

Example 7 (place/transition net transformation). We present an example of a place/transition net transformation system from [16], where a communication network is created and analyzed w.r.t. lifeness and safety properties. Here we only consider the construction using Petri net transformations. The system is composed of 3 components: a buffer, a printer and a communication unit depicted in Fig. 2. The behaviour of the buffer and the printer are obvious from the

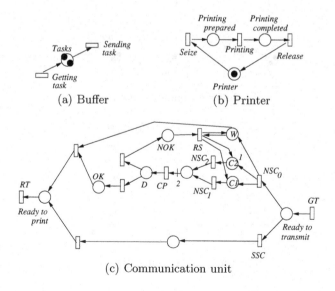

(a) Buffer   (b) Printer

(c) Communication unit

Fig. 2. Components of the system

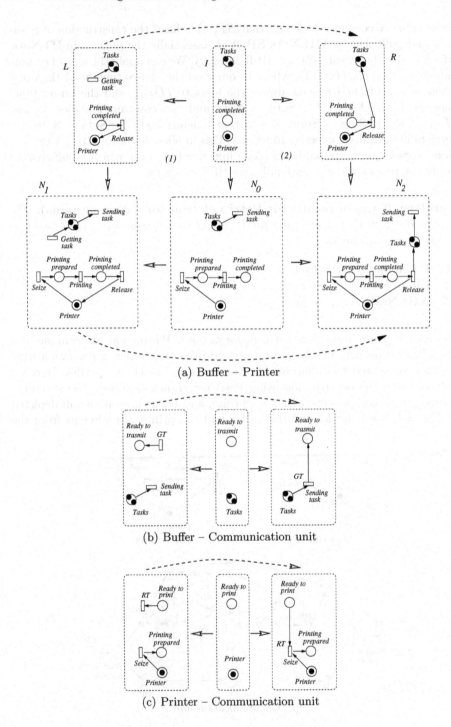

(a) Buffer – Printer

(b) Buffer – Communication unit

(c) Printer – Communication unit

Fig. 3. Interconnection of components

figure. The communication unit can send a message through a secure (SSC) or non-secure (NSC) channel. Using the NSC channel a message may become corrupted, therefore two copies of the message are sent, which are compared by the receiving subunit D. If both copies differ (NOK), then the transmission has to be repeated, otherwise (OK) it ends.

Petri net transformations are used to connect these three components. In the top row of Fig. 3(a) the production to connect buffer and printer is depicted. Fig. 3(a) shows the whole Petri net transformation as the application of this production to the components buffer and printer. In Fig. 3(b) and Fig. 3(c) the corresponding productions for connecting the communication unit with buffer and printer are shown respectively. Applying all three productions leads to the communication network depicted in Fig. 4.

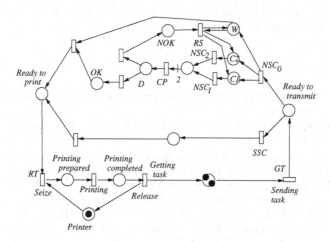

Fig. 4. Resulting communication network

5   Conclusion

In this paper we have shown how to extend adhesive HLR categories and systems - recently introduced as a new categorical framework for graph transformation in [5, 6] - to weak adhesive HLR categories and systems in order to be suitable also as a unifying framework for Petri net transformations. It is interesting to note that all the results for HLR systems based on adhesive HLR categories are still valid under the weaker assumptions of weak adhesive HLR categories. But we might need the stronger assumptions for results based on general pullback constructions as considered in [8, 9]

Especially we have shown in this paper that the category (**PTNets**, $\mathcal{M}$) of place/transition nets with the class $\mathcal{M}$ of all monomorphisms is not an adhesive HLR category, but a weak adhesive HLR category. This is sufficient to show

that the following main results of graph transformation systems are also valid
for Petri net transformation systems:
1. Local Church-Rosser Theorem
2. Parallelism Theorem
3. Concurrency Theorem
We conjecture that also the following results
4. Embedding and Extension Theorem
5. Local Confluence Theorem
stated explicitly in [5, 6] for adhesive HLR systems are valid for our Petri
net transformation systems considered above. The Embedding and Extension
Theorem allows us to embed transformations into larger contexts, and with the
Local Confluence Theorem we are able to show local confluence of transformation
systems on the basis of the confluence of critical pairs. As additional properties
we need a suitable $\mathcal{E}'$-$\mathcal{M}'$ pair factorization and initial pushouts for Petri nets
which have been shown for graphs already in [5, 6].

References

[1] Goguen, J.: Discrete-Time Machines in Closed Monoidal Categories. Bull. AMS **78** (1972) 777–783

[2] Ehrig, H., Kiermeier, K.D., Kreowski, H., Kühnel, W.: Universal Theory of Automata. Teubner (1974)

[3] Goguen, J., Burstall, R.: Introducing Institutions. In Clarke, E., Kozen, D., eds.: Logic of Programs. Volume 164 of LNCS., Springer (1984) 221–256

[4] Ehrig, H., Mahr, B.: Fundamentals of Algebraic Specification 2: Module Specifications and Constraints. Volume 21 of EATCS. Springer (1990)

[5] Ehrig, H., Habel, A., Padberg, J., Prange, U.: Adhesive High-Level Replacement Categories and Systems. In Ehrig, H., Engels, G., Parisi-Presicce, F., Rozenberg, G., eds.: Proceedings of ICGT 2004. Volume 3256 of LNCS., Springer (2004) 144–160

[6] Ehrig, H., Habel, A., Padberg, J., Prange, U.: Adhesive High-Level Replacement Systems: A New Categorical Framework for Graph Transformation. Fundamenta Informaticae (2005)

[7] Ehrig, H., Habel, A., Kreowski, H.J., Parisi-Presicce, F.: Parallelism and Concurrency in High-Level Replacement Systems. Math. Struct. in Comp. Science **1** (1991) 361–404

[8] Lack, S., Sobociński, P.: Adhesive Categories. In: Proc. of FOSSACS '04. Volume 2987 of LNCS. Springer (2004) 273–288

[9] Lack, S., Sobociński, P.: Adhesive and Quasiadhesive Categories. Theoretical Informatics and Applications **39**(3) (2005) 511–546

[10] Ehrig, H.: Introduction to the Algebraic Theory of Graph Grammars (A Survey). In: Graph Grammars and their Application to Computer Science and Biology. Volume 73 of LNCS. Springer (1979) 1–69

[11] Padberg, J., Ehrig, H., Ribeiro, L.: Algebraic High-Level Net Transformation Systems. MSCS **2** (1995) 217–256

[12] Brown, R., Janelidze, G.: Van Kampen Theorems for Categories of Covering Morphisms in Lextensive Categories. Journal of Pure and Applied Algebra **119** (1997) 255–263

[13] Ehrig, H., Mahr, B.: Fundamentals of Algebraic Specification 1: Equations and Initial Semantics. Volume 6 of EATCS. Springer (1985)

[14] Meseguer, J., Montanari, U.: Petri Nets are Monoids. Information and Computation **88**(2) (1990) 105–155

[15] Padberg, J.: Abstract Petri Nets: A Uniform Approach and Rule-Based Refinement. PhD thesis, TU Berlin (1996)

[16] Braatz, B., Ehrig, H., Urbášek, M.: Petri Net Transformations in the Petri Net Baukasten. In Ehrig, H., Reisig, W., Rozenberg, G., Weber, H., eds.: Petri Net Technology for Communication-Based Systems. Volume 2472 of LNCS., Springer (2003) 37–65

From OBJ to Maude and Beyond

José Meseguer

University of Illinois at Urbana-Champaign, USA

Dedicated to Joseph Goguen on his 65th Birthday

Abstract. The OBJ algebraic specification language and its Eqlog and
FOOPS multiparadigm extensions are revisited from the perspective of
the Maude language design. A common thread is the quest for ever more
expressive computational logics, on which executable formal specifications of increasingly broader classes of systems can be based. Several
recent extensions, beyond Maude itself, are also discussed.

1 Introduction

Joseph and I met for the first time in San Francisco on February 25, 1977 at
the First (and last!) International Symposium on *Category Theory Applied to
Computation and Control* [80]. We wrote our first paper together in 1977 [57].
We worked very closely together at SRI from 1980 to 1988, when the bulk of
our joint published work appeared, and, after his departure to Oxford and his
subsequent return to San Diego, we have continued collaborating in various ways.
In honoring him as a friend, colleague, and mentor of those early years, I want
to reflect on some great things we did together at SRI from the perspective of
how they have influenced the work that other colleagues and I have done on
Maude in the 1990s and in the present decade. Since Maude itself is evolving
and expanding in different directions, my reflections, will not only look at the
past, but will also try to sketch what those directions, leading beyond Maude
itself, look like. My views are necessarily subjective and partial, and my memory
too; but that does not prevent me from trying to recollect things as best as I
can, and from taking full responsibility for my own words and actions.

One common thread of our joint work at SRI was the OBJ language. Joseph
and I worked on OBJ1 with David Plaisted [63], and then, in the *annus mirabilis*
1983–84, with Kokichi Futatsugi and Jean-Pierre Jouannaud on OBJ2 [47]. Then
came OBJ3 [53], the most ambitious and far-reaching language design and implementation on which we worked with Claude and Hélène Kirchner, Patrick
Lincoln, Aristide Mégrelis, and Timothy Winkler. A long paper combining in
some way the OBJ2 and OBJ3 ideas appeared later [65], within an entire book
dedicated to the OBJ experience [66]. I try to explain in this paper how not only
OBJ, but also the Eqlog [59] and FOOPS [61] multiparadigm extensions of OBJ,
on which Joseph and I also worked together at SRI, have influenced Maude. But
to make better sense of all this, I think that it may be worthwhile to first present
my own perspective on the specification language design *challenges* that we have

K. Futatsugi et al. (Eds.): Goguen Festschrift, LNCS 4060, pp. 252–280, 2006.

been trying to meet all along, and which have motivated the design of each of these languages.

1.1   System Specification Vs. Property Specification

In discussing different uses of logic in computer science, considerable confusion can arise from lack of relevant distinctions. One that I have repeatedly found useful to clarify some key issues is the distinction between *system specification* and *property specification*. In a system specification we are after an unambiguous specification of a given system and how it actually *works*. In its most useful form, a system specification is *executable* and therefore provides an *executable model* of the system. Such specifications are enormously useful, since a system design can then be *tested* and *analyzed* in various ways, and it is possible to *refine*, sometimes even automatically, such an executable model into an actual system implementation.

By contrast, when specifying *properties* of a system we are not necessarily after an executable model of our system. Instead, we *assume it*, as either already given or to be developed later, and specify such properties in a typically nonexecutable manner: for example in first-order logic, higher-order logic, or some temporal logic. That is, the properties we specify have an *intended model*, namely the system design captured by a system specification, and we are interested in *verifying* by different methods that the intended model *satisfies* the properties stated in our property specification.

1.2   System Specification in Computational Logics

The above distinction brings us to the heart of a real problem: how can we *formally*, that is using logical and mathematical methods, verify a property if the system specification we have is *informal*, that is, if it does not precisely define a *mathematical model* of our system? This is indeed a genuine problem. Having a formal grammar is a necessary but insufficient condition: we also need a formal *semantics*. This is where the rub comes with system specifications based on conventional programming languages. For some such languages nobody has managed so far to give a complete formal semantics and therefore the only unambiguous "specifications" of some languages are their different compilers, which may exhibit different behaviors. Here is where computational logics can render an invaluable service. A computational logic can either:

1. be used as a *declarative programming language* with a *precise mathematical semantics* to *directly* express system specifications; or
2. be used to *give a precise mathematical semantics* to a conventional programming language, so that a system specified by a program in such a language will *indirectly* acquire a precise mathematical meaning in the computational logic.

I have not yet defined what I mean by a *computational logic*. The simplest practical answer is: a logic that you can implement as a programming language. That is, you can define and implement a programming language whose programs are exactly *theories* in the given logic and whose program execution is *logical deduction*. You then call such a language a *declarative* programming language. The point is that from the earliest times of computability theory, logical formalisms and mathematical definitions of computability have gone hand in hand. For example, Herbrand-Gödel computable functions are defined by *equational* theories; and Church computability is defined in terms of the lambda calculus. Over time, this has given rise to various declarative programming languages. For example, pure Prolog is a declarative programming language associated to Horn logic; pure ML and Haskell are declarative programming languages associated to the typed lambda calculus; OBJ is a declarative programming language based on order-sorted equational logic; and Maude is a declarative programming language based on rewriting logic.

One can always blur the above distinctions, but this is not very helpful. For example, there is always the Quixotic and amusing possibility of declaring that *everything is a logic!*, including, say, C++, thus arriving at a toothless notion of "logic". The opportunities for confusion and obscurantism are indeed endless; but such verbal games are for the most part a waste of time. Furthermore, it is possible to give *meta-logical* requirements for declarative programming languages that cut through silly verbal games of this kind: Joseph and I gave such requirements in terms of institutions in [60]; and I gave more detailed requirements in terms of general logics in [85].

1.3   The Quest for More Expressive Computational Logics

A lot of water has gone under the bridges since the 1930s. Founding computation on a theory of recursive functions was a great achievement at its time and is still very useful today; but it is clearly a limited theory, and we know it. There is, for example, no meaningful way of thinking of internet computations as definable by recursive functions. Massive changes in the nature of computing and emergence of entirely new applications do not make older computational logics and declarative languages incorrect or useless; but they can make them limited, relegated to specific *niches*. If a wider, more general applicability beyond such niches is desired, computational logics are typically in need of either generalization or replacement. One good example is functional programming, which is of course a very elegant and powerful way of programming *functional* applications. It is certainly possible to add bells and whistles to a functional language, for example by grafting a process calculus on top of it, so as to make it suitable for nonfunctional applications such as distributed computing. But what is the *logic* of such a centaur? The fact that it can be given a semantics, just as Java can, proves nothing, since the real issue is whether the resulting language remains declarative in the precise sense of programs being theories in a logic, for a decent meta-theoretic notion of logic, and computation being deduction in such a logic.

Therefore, to preserve the declarative nature of a language, when extending it to cover new application areas, one should think primarily of how its underlying logic can be extended, and only secondarily about the extended syntax: declarative language design is primarily a task of *logic design*. The design space is therefore the space, in fact the *category*, of logics. But there are tight design constraints and tradeoffs that require good judgment. Not all logics are computational; and having a recursively enumerable set of deducible formulas is *not* a sufficient condition: first-order logic has that, but it is hopeless as a programming language. The logic has to remain *lean and mean* in order to allow efficient implementations as a programming language, and not just as a theorem prover. Yet, the whole point of an extension is to make the logic more expressive. How to achieve both goals in an optimal way is the challenge.

OBJ and its extensions are a good case in point. As algebraic specification/equational programming languages, OBJ2 [47] and OBJ3 [53,65] were arguably the most expressive such languages in the 1980s. But they were, by the very nature of their underlying order-sorted equational logic [62] and their associated operational semantics [52,70], *functional languages*. Extending OBJ in a *multiparadigm* way was a task that Joseph and I undertook in the mid 1980s, resulting in two new language designs: Eqlog [59], and FOOPS [61]. Eqlog unified functional/equational programming and Horn-logic programming; its logic design task was to embed order-sorted equational logic and Horn logic without equality into a suitable Horn logic with equality [60]. FOOPS unified equational/functional programming, Horn-logic programming, and object-oriented programming. Although an underlying model-theoretic semantics was given, based on algebraic data types with hidden sorts and behavioral equivalence between them in the sense of [58,94], FOOPS fell short of having an underlying logic with modules as theories and computation as deduction. This was remedied later, by theoretical developments presenting various proposals for a hidden or "observational" equational logic [50,51,56,55,122,64,68,11,10,9,115,116,117,30,120]. In hindsight, one can view CafeOBJ [46], BOBJ [54] and BMaude [96] as full-blooded declarative languages that achieve in a more satisfactory way many of the FOOPS goals.

1.4   Rewriting Logic and Maude

With rewriting logic [87,88,13] and Maude [86,90,18,19], several of us undertook the task of unifying within a single declarative language: (i) equational/functional programming; (ii) object-oriented programming; and (iii) concurrent/distributed programming. That (iv) Horn-logic programming was also naturally embeddable in this framework was clear from the early stages of this project [89,90], but at the operational semantics level this required a generalization of narrowing that was achieved later [132,133]. Three more insights emerged over time as part of different research collaborations: (v) that real-time and hybrid systems could be naturally specified in rewriting logic [108]; (vi) that higher-order type theory was naturally embeddable in rewriting logic [130]; and (vii) that probabilistic systems were likewise expressible in a natural probabilistic extension of rewriting logic and could be simulated within rewriting logic itself [73,4]. In spite of being multiparadigm in all the above (i)–(vii) ways, rewriting logic remains remarkably lean and mean: it is a very simple formalism and, thanks to Steven Eker, has a very high-performance Maude implementation. Modules are indeed theories in the logic, and nothing more. Computation *is* deduction, and theories have initial models [88,13], which give semantics to modules and support inductive reasoning. Furthermore, operational properties such as termination can be usefully formulated and verified by adopting this logical/deductive viewpoint [36,79]

2   From Order-Sorted to Membership Equational Logic

Rewriting logic contains membership equational logic [92] as a sublogic. In Maude's language design this is reflected in its sublanguage of *functional modules*, for equational theories with initial semantics, and of *functional theories* for equational theories with "loose" semantics. Therefore, in relating OBJ and Maude the first task at hand is relating their corresponding equational logics.

One key reason why OBJ2 and OBJ3 were so expressive was their order-sorted type theory. That one should use types to make any reasonable sense of algebraic specifications goes without saying. But the problem with many-sorted equational logic is that it does not deal well with partiality. Many simple operations, such as selectors in data structures or just simple arithmetic operations, are *partial*. To the embarrassment of many-sorted specifications, simple trade examples, such as the perennial stacks or the rational numbers, cannot be given elegant many-sorted specifications: the top of the empty stack or division by zero raise their ugly heads and require ugly ad-hoc solutions.

The appeal of order-sorted equational logic [62] is that, by allowing the expressive power of subtypes (subsorts), many partial functions become total on appropriate subsorts. Furthermore, function symbols can be subsort overloaded, which is very convenient in practice. But there are limits to the kind of partiality expressible by typing means alone, which are those available in order-sorted algebra. When the definedness of a function depends on *semantic conditions* such as, for example, the fact that for the concatenation of two paths in a graph to be defined the target node of the first must coincide with the source node of the second, order-sorted equational logic is not enough. This was understood early on, and led to formulating notions of unconditional [49] or conditional [52] *sort constraints*; but how to extend order-sorted equational logic so as to fully account for conditional sort constraints remained an open question.

The appeal of membership equational logic (MEL) [92] is that it gives a full account of partiality, and even a systematic, functorial way of relating partial and total specifications [92,95]. Furthermore, as shown in [92], it embeds in a conservative way the "right" version of order-sorted equational logic, one that solves several anomalies, including the lack of pushouts of theory morphisms, in the version given in [62]. But does membership equational logic remain lean and mean? The relevant facts are that it: (i) has a well-developed operational semantics by rewriting (see the systematic study [12], which also deals with many other automated deduction techniques); (ii) enjoys a high-performance Maude implementation; (iii) is a quite simple logic; and (iv) has initial and free models [92], on which inductive proof methods and inductive theorem proving tools can be based [12,20]. From these facts it seems fair to conclude that the answer is definitely yes.

In summary, therefore, we can view OBJ3 as a *sublanguage* of Maude's functional sublanguage. The generalization from OBJ3 to Maude is further stressed by the fact that Maude supports order-sorted notation as convenient syntactic sugar for membership equational logic axioms. In membership equational logic atomic propositions are either equations $t = t'$, or memberships $t : s$, stating that term t has sort s. A subsort declaration $s < s'$ is then just syntactic sugar for a conditional axiom $x : s \Rightarrow x : s'$. Similarly, an order-sorted operator declaration $f : s_1 \ldots s_n \longrightarrow s$ is syntactic sugar for the conditional axiom $x_1 : s_1 \wedge \ldots \wedge x_n : s_n \Rightarrow f(x_1, \ldots, x_n) : s$.

A membership equational theory is a pair (Σ, H) with Σ a signature specifying the kinds, sorts, and function symbols, and with H a set of Horn clauses involving both equations and memberships. Kinds classify potentially meaningful expressions, and sorts within a kind classify actually defined expressions. Terms having a kind but not a sort correspond to undefined or error expressions. For example, $2/0$ is in the *Number* kind but has no sort. For execution purposes we typically impose some requirements on such a theory. First of all, its Horn clauses H may be decomposed as a union $E \cup A$, with A a set of equations that we will reason *modulo* (for example, A may include associativity, commutativity and/or identity axioms for some of the operators in Σ). Second, the remaining Horn clauses E are typically required to be Church-Rosser[1] modulo A, so that we can use the conditional equations in E as equational rewrite rules modulo A. Third, for some applications it is useful to make the equational rewriting relation[2] *context-sensitive* [76,77]. This can be accomplished by specifying a function $\mu : \Sigma \longrightarrow \mathbb{N}^*$ assigning to each function symbol $f \in \Sigma$ (with, say, n arguments) a list $\mu(f) = i_1 \ldots i_k$ of *argument positions*, with $1 \leq i_j \leq n$, which must be fully evaluated (up to the context-sensitive equational reduction strategy specified by μ) in the order specified by the list $i_1 \ldots i_k$ before applying any equations whose lefthand sides have f as their top symbol. For example, for $f = \mathit{if\_then\_else\_fi}$ we may give $\mu(f) = \{1\}$, meaning that the first argument must be fully evaluated before the equations for $\mathit{if\_then\_else\_fi}$ are

---

[1] See [12] for a detailed study of equational rewriting concepts and proof techniques for MEL theories.

[2] As we shall see, in a rewrite theory $\mathcal{R}$ rewriting can happen at two levels: (1) *equational rewriting* with (possibly conditional) equations E; and (2) *non-equational rewriting* with (possibly conditional) rewrite rules R. These two kinds of rewriting *are different*. Therefore, to avoid confusion I will qualify rewriting with equations as *equational rewriting*.

applied[3]. Therefore, for execution purposes we can specify a membership equational theory as a triple $(\Sigma, E \cup A, \mu)$, with A the axioms we rewrite modulo, and with μ the map specifying the context-sensitive equational reduction strategy. A Maude functional module is then, essentially, a specification of the form fmod $(\Sigma, E \cup A, \mu)$ endfm.

3 Rewriting Logic: From OBJ to Maude

As already mentioned, the whole point of rewriting logic [87,88,13] and Maude [86,90,18,19] was to unify within a single logic and associated declarative language: (i) equational/functional programming; (ii) object-oriented programming; and (iii) concurrent/distributed programming. For this unification, a purely equational/functional framework would be clearly unsuitable[4] The challenge therefore was to find a lean and mean superlogic of equational logic in which this unification could take place.

A related challenge was to make some sense of the quite diverse menagerie of concurrency models that were around, often competing with each other as the "right" model of concurrency. A key strategy in this competition game was to produce, sometimes quite complicated, translations from other models, adduced as proof of universality of the proposed model. Implicit in this strategy was the belief that, given enough time, the right model, capable of expressing all the relevant concurrency concepts would emerge. This search for the Holy Grail of concurrency is certainly a chivalrous one; but I find serious grounds for being skeptical about its success. The main difficulty is that concurrency encompasses a very wide range of phenomena: there are concurrent functional programs, concurrent grammars, dataflow networks, actors, Petri nets of various ilks and colors, various synchronous and asynchronous process calculi, neural networks, and so on. Although translations between some of these models are possible, the fact that in this way some concurrency features can simulate others, perhaps in a complex way, is not particularly helpful.

In my view, what was missing was a *computational logic for concurrency* that could serve as a *semantic framework* in which different concurrency models could

---

[3] As in OBJ2–3, in Maude maps μ specifying context-sensitive equational reduction strategies are called *evaluation strategies* [47,40,18], and $\mu(f) = i_1 \ldots i_k$ is specified with the strat keyword followed by the string $(i_1 \ldots i_k \ 0)$, with 0 indicating evaluation at the top of the function symbol f. For an in-depth study of the relationship between OBJ/Maude evaluation strategies and context-sensitive rewriting see [76,77].

[4] The key point is that concurrency and nondeterminism cannot be *directly modeled* in an equational/functional framework, which typically assumes determinism in the form of a Church-Rosser property. Therefore, one needs special devices to model some concurrency aspects *indirectly*. Two good examples of indirectly modeling concurrency within a purely functional framework are the ACL2 semantics of the JVM using a scheduler [101], and the use of lazy data structures in Haskell to analyze cryptographic protocols [7].

be naturally unified without requiring any translations. That is, in a logic one can define quite different *theories* which have associated *models*. The logic then allows one to understand in a unified way all such models as models in the same logic; but there is plenty of room for diversity between them. Furthermore, once we understand that a logical framework of this kind can give us an enormous range of possibilities for naturally expressing different concurrency phenomena, we realize that we can have a general *framework* without in any way needing a general *model*, whatever that means.

Is rewriting logic a suitable general framework in exactly this sense? The answer is necessarily an empirical one, and can never be claimed to be definitive. But the amount of positive evidence gathered up to now, thanks to the research of different people and covering indeed a very wide range of concurrency models, is in my view very strong. The key point is the naturalness and directness with which different concurrency models can be expressed as rewrite theories. It is not a matter of complicated *encodings*: typically the original representations of a model and those of its associated rewrite theory are isomorphic. Since all this is a matter carefully documented in many papers and in several rewriting logic surveys, I will not go over the, indeed quite large, body of work backing the view that rewriting logic is a very expressive general framework for concurrency. I refer the reader to the survey paper [82]; and for an explanation of how rewriting logic unifies and improves upon other semantic frameworks such as algebraic semantics and structural operational semantics (SOS) to the more recent papers [97,98].

3.1 Rewrite Theories: Their Execution and Formal Analysis

A *rewrite theory* is a tuple $\mathcal{R} = (\Sigma, E \cup A, \mu, R, \phi)$, with: (1) $(\Sigma, E \cup A, \mu)$ a membership equational theory with "modulo" axioms A and context-sensitive equational reduction strategy μ; (2) R a set of *labeled conditional rewrite rules* of the general form

$$r : (\forall X)\, t \longrightarrow t' \ \text{ if } \ (\bigwedge_i u_i = u_i') \wedge (\bigwedge_j v_j : s_j) \wedge (\bigwedge_l w_l \longrightarrow w_l') \qquad (1)$$

where the variables appearing in all terms are among those in X, terms in each rewrite or equation have the same kind, and in each membership $v_j : s_j$ the term v_j has kind $[s_j]$; and (3) $\phi : \Sigma \longrightarrow \mathcal{P}(\mathbb{N})$ a mapping assigning to each function symbol $f \in \Sigma$ (with, say, n arguments) a set $\phi(f) = \{i_1, \ldots, i_k\}$, $1 \leq i_1 < \ldots < i_k \leq n$ of *frozen argument positions*[5] under which it is forbidden to perform any rewrites.

Intuitively, $\mathcal{R}$ specifies a *concurrent system*, whose states are elements of the initial algebra $T_{\Sigma/E \cup A}$ specified by $(\Sigma, E \cup A)$, and whose *concurrent transitions* are specified by the rules R, subject to the frozenness requirements imposed by ϕ.

---

[5] In Maude, $\phi(f) = \{i_1, \ldots, i_k\}$ is specified by declaring f with the `frozen` attribute, followed by the string $(i_1 \ \ldots \ i_k)$. Although originated by a quite different motivation, frozen operators have some similarities with notions such as "non-coherent operators" in CafeOBJ [46], and "non-congruent" operators in BOBJ [54].

The frozenness information is important in practice to forbid certain rewritings. For example, when defining the rewriting semantics of a process calculus, one may wish to require that in prefix expressions $\alpha.P$ the operator $\_.\_$ is *frozen in the second argument*, that is, $\phi(\_.\_) = \{2\}$, so that P cannot be rewritten under a prefix. Note that a rewrite theory $\mathcal{R} = (\Sigma, E \cup A, \mu, \phi, R)$ specifies two kinds of *context-sensitive* rewriting requirements: (1) equational rewriting with E modulo A is made context-sensitive by μ; and (2) non-equational rewriting with R is made context-sensitive by ϕ. But the maps μ and ϕ impose *different types* of context-sensitive requirements: (1) $\mu(f)$ specifies a list of arguments where we *are allowed* to rewrite with equations in E; and (2) $\phi(f)$ specifies arguments where we *are forbidden* to rewrite with the rules R. The maps μ and ϕ substantially increase the expressive power of rewriting logic, because various order-of-evaluation and context-sensitive requirements, which would have to be specified by explicit rules in a formalism like *SOS*, become implicit and are encapsulated in μ and ϕ.

For execution purposes a rewrite theory $\mathcal{R} = (\Sigma, E \cup A, \mu, R, \phi)$ should satisfy some basic requirements that are assumed to hold for Maude *system modules*. Such modules are specifications of the form mod $(\Sigma, E \cup A, \mu, R, \phi)$ endm. First, in the membership equational theory $(\Sigma, E \cup A, \mu)$, E should be ground Church-Rosser modulo A – for A a set of equational axioms for which matching modulo A is decidable – and ground terminating modulo A, up to the context-sensitive strategy μ^6. Second, the rules R should be *coherent* with E modulo A [136]; intuitively, this means that, to get the effect of rewriting in equivalence classes modulo $E \cup A$, we can always first simplify a term with the equations in E to its canonical form modulo A, and then rewrite with a rule in R. Finally, the rules in R should be *admissible* [18], meaning that in a conditional rewrite rule of the form (1), besides the variables appearing in t there can be extra variables in t', provided that they also appear in the condition and that they can all be *incrementally instantiated* by either matching a pattern in a "matching equation" or performing breadth first search in a rewrite condition (see [18] for a detailed description of admissible equations and rules).

Computation in Maude is then deduction with the inference rules of rewriting logic (see [13]) that are efficiently implemented by the Maude engine under the above executability assumptions. Specifically, equivalence classes modulo $E \cup A$ are represented by their unique canonical forms modulo A. That is, Maude performs equational rewriting to reach a canonical form with the equations in E modulo A by means of the reduce command. This is entirely analogous to OBJ's reduce command for equational specifications, but applies now to more general theories. It also supports two variants of fair rewriting with the rules R modulo A which, in combination with equational rewriting and under the coherence assumption, achieves the effect of rewriting with R in $(E \cup A)$-equivalence classes. These two commands are the rule-fair rewrite command; and the rule and po-

---

[6] μ-termination is a weaker requirement than termination [77]; the interactions between context-sensitive rewriting and the Church-Rosser property are somewhat subtle [75,78].

sition fair `frewrite` command which, for object-based systems (see Section 3.3) is also object and message fair. Furthermore, the context-sensitive requirements provided by μ and ϕ are always respected. Since the rules R need not be confluent and may be highly nondeterministic, the `rewrite` and `frewrite` commands give just *one* execution path among many others. This is still very useful for execution and simulation purposes, but for analysis purposes Maude's `search` command supports a systematic breadth-first exploration of all rewrite paths until states matching a specified pattern and satisfying specified semantic conditions are reached. For example, we may want to know whether the concurrent system specified by our rewrite theory satisfies a given invariant (say, is deadlock-free). We can then search for a reachable state satisfying the *negation* of the given invariant. Within the practical limitations of time and memory, the *search* command then gives us a semi-decision procedure for the failure of such invariants, regardless of the in general infinite number of reachable states of our systems. Furthermore, for systems whose sets of reachable states are *finite*, Maude also provides a *decision procedure* for the satisfaction of linear-time temporal logic (LTL) properties. This is achieved through its built-in MODEL-CHECKER module which, in the experiments that we have evaluated [41,42], performs explicit-state on-the-fly model checking of LTL formulas with efficiency comparable to that of the SPIN model checker [69].

3.2   Module Algebra: The Power of Reflective Thinking

One of the most powerful features of OBJ2 and OBJ3 was the possibility of defining *parameterized modules* having semantic requirements for their instantiation specified in the form of *parameter theories*. Such modules could then be instantiated by means of *views* (theory interpretations) in the typical pushout construction way of Clear [14]. They could also be *renamed*, and instantiations and renamings could be composed in very expressive *module expressions* (see [47,65]). This supported a very powerful discipline of *parameterized programming* that inspired similar mechanisms in ML and in module interconnection languages such as LILEANNA [135]. In hindsight, however, there were two limitations. The first was that it took in practice a long time (several years of hard work) to properly implement this part of the language. Indeed, it proved to be the most complex and sophisticated component of OBJ3's LISP-based implementation. The second limitation, much less apparent to us at the time, was that OBJ's module algebra, while very powerful, was a *closed* algebra, in the sense of offering a fixed repertoire of theory operations. Of course, one could have imagined other operations, but this would have required both a new metatheory and big implementation efforts.

An important breakthrough at the theoretical level was the formulation of a general axiomatic notion of reflective logic by Manuel Clavel and myself in [23], followed by a series of papers, a Ph.D. thesis, and a book, showing that several conditional and unconditional versions of rewriting logic, as well as membership equational logic and many-sorted Horn logic with equality, are indeed reflective [24,15,16,25,26]. Intuitively, a logic is reflective if it can represent its metalevel at the object level in a sound and coherent way. Specifically, rewriting logic can represent its own theories and their deductions by having a finitely presented rewrite theory $\mathcal{U}$ that is *universal*, in the sense that for any finitely presented rewrite theory $\mathcal{R}$ (including $\mathcal{U}$ itself) we have the following equivalence

$$\mathcal{R} \vdash t \to t' \;\; \Leftrightarrow \;\; \mathcal{U} \vdash \langle \overline{\mathcal{R}}, \overline{t} \rangle \to \langle \overline{\mathcal{R}}, \overline{t'} \rangle,$$

where $\overline{\mathcal{R}}$ and $\overline{t}$ are terms representing $\mathcal{R}$ and t as data elements of $\mathcal{U}$. Since $\mathcal{U}$ is representable in itself, we can achieve a "reflective tower" with an arbitrary number of levels of reflection.

Reflection is a very powerful property: it allows defining rewriting strategies by means of metalevel theories that extend $\mathcal{U}$ and guide the application of the rules in a given object-level theory $\mathcal{R}$ [24,16,83]; it is efficiently supported in the Maude implementation by means of *descent functions* [17], implemented in the built-in META-LEVEL module; it can be used to build a variety of theorem proving and theory transformation tools [15,20,21,27]; and it can also be used to prove metalogical properties about families of theories in rewriting logic, and about other logics represented in the rewriting logic (meta-)logical framework [5,22,6].

From the module algebra point of view, the key advantage is that the universal theory $\mathcal{U}$, and the META-LEVEL module that implements key descent functions for it, have a sort Module whose terms represent finitary rewrite theories. This means that theories become *data* that can be manipulated *within* the logic in a declarative way. Similar sorts, defining data types for parameterized modules and for views, can likewise be easily defined in extensions of the META-LEVEL module. In this way, Francisco Durán and I showed that many powerful theory composition operations endowing Maude with a module algebra can be defined within the logic [37,32,39]. Furthermore, the module algebra so defined now becomes easily *extensible*. For example, the notion of parameterized module, and the way in which module instantiation can be defined does not necessarily have to follow a pushout-like pattern. Different forms of parameterization, understood as new metalevel functions, can be easily defined. For instance, it is very easy to define in the Full Maude extension of Maude a TUPLE(n) module that for each nonzero natural number n provides a parameterized module of n-tuples [32]. Indeed, reflection has allowed considerable flexibility in easily defining and experimenting with different module composition operations before implementing some of them in the underlying Core Maude system, as has been recently done in Maude 2.2. Furthermore, Full Maude itself has been an excellent basis for building other Maude extensions such as Real-Time Maude (see Section 4.1), a strategy language for Maude [83], and the Maude termination tool (MTT) [36].

More generally, reflection has made it quite easy to build an environment of formal analysis tools for Maude. Such tools, by their very nature, manipulate and analyze rewrite theories. By reflection, a rewrite theory $\mathcal{R}$ becomes a *term* $\overline{\mathcal{R}}$ in the universal theory, which can be efficiently manipulated by the descent functions in the META-LEVEL module. As a consequence, Maude formal tools have a reflective design and are built in Maude as suitable extensions of the META-LEVEL module. They include the following:

- the Maude Church-Rosser Checker, and Knuth-Bendix and Coherence Completion tools [20,38,34,33]
- the Full Maude module composition tool [32,39]
- the Maude Predicate Abstraction tool [118]
- the Maude Inductive Theorem Prover (ITP) [16,20,27]
- the Real-Time Maude tool [109] (discussed in Section 4.1)
- the Maude Sufficient Completeness Checker (SCC) [67]
- the Maude Termination Tool (MTT) [36].

3.3   Object-Oriented Modules

A declarative treatment of the object paradigm was also a key goal from the very beginning of rewriting logic [86], and was more fully realized as part of Maude's language design in [90]. Of course, since concurrent programming was also a key goal, the point was to have a declarative way to specify and program *concurrent object systems*. This declarative approach, by using subsort overloading and proposing a key distinction between *class inheritance* and *module inheritance* solved also an old chestnut in concurrent object-oriented programming, namely the so-called *inheritance anomaly* [91].

The essential idea is extremely simple. We view the state of a concurrent object system as a "soup" of objects and messages. Mathematically, such a soup is modeled as a multiset, built up from the objects and the messages by means of a multiset union operator that is associative and commutative and has the empty multiset as its identity element. Concurrent *interactions* between objects, and between objects and messages, are then described by means of rewrite rules that transform a fragment of such a soup into a new fragment. By rewriting logic's congruence rule [88], many such rewrites can of course take place concurrently within the soup. Rules whose lefthand sides involve a single object and at most one message are called *asynchronous* and essentially correspond to the Actor model of computation [3,1]. Rules whose lefthand sides involve more than one object are called *synchronous*, because such objects have to come together synchronously in order for the interaction to take place.

More generally, the soup describing the distributed state of an object system needs not be "flat" but may instead be a "soup of soups" with arbitrary nesting depth. For example, the Internet is a network of networks and a soup of soups in exactly this sense. This structuring is very useful, for example for security and management/monitoring purposes. Carolyn Talcott and I modeled this in rewriting logic by means of our "Russian dolls" model of concurrent object reflection [100]. The "dolls" in question are meta-objects, which may contain in their belly a whole soup of other (meta-) objects, and so on "all the way down." In this way, all kinds of mechanisms for concurrent meta-object reflection can be naturally axiomatized, programmed, and reasoned about [100]. The Russian dolls model is also useful in clarifying the relationship between *object-oriented* reflection and *logical* reflection in the sense of Section 3.2. Some object-oriented reflection mechanisms do not need logical reflection: the hierarchical nesting of dolls (meta-object nesting) is enough to express them. But more powerful concurrent object reflection mechanisms may use both the nesting of dolls and logical reflection. For example, the mobility features of Mobile Maude [35] use both meta-object reflection and logical reflection.

In Maude, concurrent object systems are specified in *object-oriented modules* [90,37,32,18]. Such modules provide syntactic sugar supporting all the usual object-oriented concepts: objects, object attributes, messages, object classes, and multiple class inheritance. Furthermore, they can be parameterized with parameter theories just like any other Maude module. Semantically, all this useful syntactic sugar can be stripped away, so that a Maude object-oriented module is semantically equivalent to an ordinary rewrite theory, that is, to a corresponding Maude *system module* into which it can be desugared. Operationally, however, knowledge of the existence of objects and messages within a multiset representing a distributed object state is used by Maude's `frewrite` command to support a rule, position, and object and message *fair* rewriting strategy. In conjunction with Maude 2.2's built-in internet sockets feature [19], this provides a very simple and elegant way of doing declarative internet programming in Maude, because there is no need whatsoever for writing any complicated thread scheduling code, which is typically needed when a conventional language is used.

4 Beyond Maude

How general and expressive is rewriting logic? The best way to find out is by pushing its limits. What follows is a progress report on how, through several research collaborations, some of us have been extending rewriting logic and its range of applications beyond those of Maude itself so as to encompass: (i) real-time and hybrid systems; (ii) probabilistic systems; (iii) deduction with logical variables; (iv) higher-order specifications; and (v) behavioral specifications.

4.1 Real-Time Maude

In many reactive and distributed systems, real-time properties are essential to their design and correctness. Therefore, the question of how systems with real-time features can be best specified, analyzed, and proved correct in the semantic framework of rewriting logic is an important one. This question has been investigated by several authors from two perspectives. On the one hand, an extension of rewriting logic called *timed rewriting logic* has been investigated, and has been applied to some examples and specification languages [71,105,125]. On the other hand, Peter Ölveczky and I have found a simple way to express real-time and hybrid system specifications *directly* in rewriting logic [106,108]. Such specifications are called *real-time rewrite theories* and have rules of the form

$$r : \{t\} \overset{\delta}{\longrightarrow} \{t'\} \ \ if \ \ C$$

with δ a term denoting the *duration* of the transition (where the time domain can be chosen to be either discrete or dense), $\{t\}$ representing the *whole* state of a system, encapsulated with {_}, and C an equational condition. Peter Ölveczky and I have shown that, by making the clock an explicit part of the state, these theories can be *desugared* into semantically equivalent ordinary rewrite theories [106,108,109]. That is, in the desugared version we can model the state of a real-time or hybrid system as a pair $(\{t\}, \tau)$, with $\{t\}$ the current state, and with τ the current global clock time. Then the above rule becomes desugared as

$$r : (\{t\}, \tau) \longrightarrow (\{t'\}, \tau + \delta) \quad \textit{if} \quad C$$

Rewrite rules can then be either *instantaneous rules*, that take no time and only change some part of the state t, or *tick rules*, that advance the global time of the system according to some time expression δ and may also change the state t. When time is dense, tick rules may be *nondeterministic*, in the sense that the time δ advanced by the rule is not uniquely determined, but is instead a parametric expression (however, this time parameter is typically subjected to some equational condition C). In such cases, tick rules need a *time sampling strategy* to choose suitable values for time advance. Besides being able to show that a wide range of known real-time models (including, for example, timed automata, hybrid automata, timed Petri nets, and timed object-oriented systems) can be naturally expressed in a direct way in rewriting logic (see [108]), an important advantage of our approach is that one can use an existing implementation of rewriting logic to execute and analyze real-time specifications. Because of some technical subtleties, this seems difficult for the alternative of timed rewriting logic, although a mapping into our framework does exist [108].

Real-Time Maude [102,107,109,110] is a specification language and a formal tool built in Maude by reflection. It provides special syntax to specify real-time systems, and offers a range of formal analysis capabilities. The Real-Time Maude 2.1 tool [109,112] systematically exploits the underlying Maude efficient rewriting, search, and LTL model checking capabilities to both execute and formally analyze real-time specifications. Reflection is crucially exploited in the Real-Time Maude 2.1 implementation. On the one hand, Real-Time Maude specifications are internally desugared into ordinary Maude specifications by transforming their meta-representations. On the other, reflection is also used for execution and analysis purposes. The point is that the desired modes of execution and the formal properties to be analyzed have real-time aspects with no clear counterpart at the Maude level. To faithfully support these real-time aspects a *reflective transformational approach* is adopted: the original real-time theory and query (for either execution or analysis) are *simultaneously* transformed into a semantically equivalent pair of a Maude rewrite theory and a Maude query [109,112]. One important concern about the search and model checking analyses thus performed by Real-Time Maude is their *completeness*. Note that not all state-time pairs are visited, but only those allowed by the given time sampling strategy. For dense time it is even impossible to visit *all* times. Fortunately, under simple conditions on the specification, that are indeed satisfied by almost all examples that have been analyzed in Real-Time Maude, the analyses are indeed complete: if the tool finds no counterexamples, the given property holds [111].

In practice, Real-Time Maude executions and analyses are quite efficient. They allow scaling up to highly nontrivial specifications and case studies. In fact, both the naturalness of Real-Time Maude to specify large nontrivial real-time applications (particularly for distributed object-oriented real-time systems) and its effectiveness in simulating and analyzing the formal properties of such systems have been demonstrated in a number of substantial case studies, including: (1) the AER/NCA suite of active network protocols [102,104,113]; (2) the NORM multicast protocol [74]; (3) the OGDC wireless sensor network algorithm [134,114]; and (4) the CASH adaptive scheduling algorithm [103]. Real-Time Maude is freely available from http://www.ifi.uio.no/RealTimeMaude. It is a mature and quite efficient tool, and its source code, a tool manual, examples, case studies, and papers are all available in its web page.

4.2   PMaude and SHYMaude

Many systems are probabilistic in nature. This can be due either to the uncertainty of the environment in which they must operate, such as message losses and other failures in an unreliable environment, or to the probabilistic nature of some of their algorithms, or to both. In general, particularly for distributed systems, both probabilistic and nondeterministic aspects may coexist, in the sense that different transitions may take place nondeterministically, but the outcomes of some of those transitions may be probabilistic in nature. To specify systems of this kind, rewrite theories have been generalized to *probabilistic rewrite theories* in [72,73,4]. Rules in such theories are *probabilistic rewrite rules* of the form

$$r : t(\boldsymbol{x}) \rightarrow t'(\boldsymbol{x}, \boldsymbol{y}) \ \ if \ \ C(\boldsymbol{x}) \ \ with \ \ probability \ \ \boldsymbol{y} := \pi_r(\boldsymbol{x})$$

where the first thing to observe is that the term t' has new variables $\boldsymbol{y}$ disjoint from the variables $\boldsymbol{x}$ appearing in t. Therefore, such a rule is *nondeterministic*; that is, the fact that we have a matching substitution θ for the variables $\boldsymbol{x}$ such that $\theta(C)$ holds does not uniquely determine the next state fragment: there can be many different choices for the next state, depending on how we instantiate the extra variables $\boldsymbol{y}$ in t'. In fact, we can denote the different such next states by expressions of the form $t'(\theta(\boldsymbol{x}), \rho(\boldsymbol{y}))$, where θ is fixed as the given matching substitution, but ρ ranges along all the possible substitutions for the new variables $\boldsymbol{y}$. The probabilistic nature of the rule is expressed by the notation: *with probability* $\boldsymbol{y} := \pi_r(\boldsymbol{x})$, where $\pi_r(\boldsymbol{x})$ is a probability distribution *which may depend on the matching substitution* θ. We then choose the values for $\boldsymbol{y}$, that is, the substitution ρ, probabilistically according to the distribution $\pi_r(\theta(\boldsymbol{x}))$.

The fact that the probability distribution may depend on the substitution θ can be illustrated by means of a simple example. Consider a battery-operated clock. We may represent the state of the clock as a term clock(T,C), with T a natural number denoting the time, and C a positive real denoting the amount of battery charge. Each time the clock ticks, the time is increased by one unit, and the battery charge slightly decreases; however, the lower the battery charge, the greater the chance that the clock will stop, going into a state of the form `broken(T,C')`. We can model this system by means of the probabilistic rewrite rule

```
rl [tick]: clock(T,C) => if B then clock(s(T),C - (C / 1000))
                         else broken(T,C - (C / 1000))
                         fi
           with probability B := BERNOULLI(C / 1000) .
```

that is, the probability of the clock breaking down instead of ticking normally *depends on the battery charge*, which is here represented by the battery-dependent bias of the coin in a Bernoulli trial. Note that here the new variable on the rule's righthand side is the Boolean variable B, corresponding to the result of tossing the biased coin. As shown in [72], probabilistic rewrite theories can express a wide range of models of probabilistic systems, including continuous-time Markov chains [131], probabilistic non-deterministic systems [119,123], and generalized semi-Markov processes [48]; they can also naturally express probabilistic object-based distributed systems [73,4], including real-time ones. Yet another class of probabilistic models that can be simulated by probabilistic rewrite theories is the class of object-based stochastic hybrid systems discussed in [99].

The PMaude language [73,4] is an experimental specification language whose modules are probabilistic rewrite theories. Note that, due to their nondeterminism, probabilistic rewrite rules *are not directly executable*. However, probabilistic systems specified in PMaude can be *simulated* in Maude. As explained in [4,93], this is accomplished by transforming a PMaude specification into a corresponding Maude specification in which actual values for the new variables appearing in the righthand side of a probabilistic rewrite rule are obtained by *sampling* the corresponding probability distribution functions using standard techniques based on random number generation and Maude's built-in `COUNTER` and `RANDOM` modules.

In general, provided that sampling for the probability distributions used in a PMaude module is supported in the underlying infrastructure, we can associate to it a corresponding Maude module. We can then use this associated Maude module to perform Monte Carlo simulations of the probabilistic systems thus specified. As explained in [4], provided all nondeterminism has been eliminated from the original PMaude module[7], we can then use the results of such Monte Carlo simulations to perform a *statistical model checking analysis* of the given system to verify certain properties. For example, for a PMaude specification of a

---

[7] The point is that, as explained above, in general, given a probabilistic rewrite theory and a term t describing a given state, there can be several different rewrites, perhaps with different rules, at different positions, and with different matching substitutions, that can be applied to t. Therefore, the choice of rule, position, and substitution is *nondeterministic*. To eliminate all nondeterminism, at most one rule at exactly one position and with a unique substitution should be applicable to any term t. As explained in [4], for many systems, including probabilistic real-time object-oriented systems, this can be naturally achieved, essentially by scheduling events at real-valued times that are all different, because we sample a continuous probability distribution on the real numbers.

TCP/IP protocol variant that is resistant to Denial of Service (DoS) attacks, we may wish to establish that, even if an attacker controls 90% of the network bandwidth, it is still possible for the protocol to establish a connection in less than 30 seconds with 99% probability. Properties of this kind, including properties that measure quantitative aspects of a system, can be expressed in the QATEX probabilistic temporal logic [4], and can be model checked using the VeStA tool [124]. See [2] for a substantial case study specifying a DoS-resistant TCP/IP protocol as a PMaude module, performing Monte Carlo simulations by means of its associated Maude module, and formally analyzing in VeStA its properties, expressed as QATEX specifications, according to the methodology just described. More recently, several object-based stochastic hybrid system case studies have been specified in an extension of both PMaude and Real-Time Maude called SHYMaude [99] and have been simulated in Maude. Relevant formal properties for each case study, expressed as QATEX specifications, have been statistically model checked in VeStA using Monte Carlo simulations performed in Maude [99].

4.3   Narrowing: Eqlog Revisited

Narrowing is a symbolic procedure like rewriting, except that rules, instead of being applied by matching a subterm, are applied by unifying the lefthand side with a nonvariable subterm. Traditionally, narrowing has been used as a method to solve equations in a confluent and terminating equational theory. In rewriting logic, Prasanna Thati and I have generalized narrowing to a procedure for *symbolic reachability analysis* [132]. That is, instead of solving equational goals $\exists x.\ t = t'$, we solve reachability goals $\exists x.\ t \longrightarrow t'$, stating that there is an instance of t from which we can reach by rewriting with rules R modulo equations E an instance of t'.

For arbitrary rewrite theories narrowing, though sound, is not a complete procedure [132]. However, for large classes of theories of interest, including theories specifying distributed object systems, narrowing is complete and provides a complete semidecision procedure for solving reachability problems [132]. Further recent work on narrowing with rewrite theories focuses on: (1) generalizing the procedure to so-called "back-and-forth narrowing," so as to ensure completeness under very general assumptions about the rewrite theory $\mathcal{R}$ [133]; and (2) efficient lazy strategies to restrict as much as possible the narrowing search space [45].

Narrowing with rewrite theories has important applications to the analysis of cryptographic protocols. A relevant point is that, since narrowing with a rewrite theory $\mathcal{R} = (\Sigma, E, R)$ is performed *modulo* the equations E, this allows more sophisticated analyses than those performed under the usual Dolev-Yao "perfect cryptography assumption." It is well-known that protocols that had been proved secure under this assumption can be broken if an attacker uses knowledge of the algebraic properties satisfied by the underlying cryptographic functions. In rewriting logic we can specify a cryptographic protocol with a type of rewrite theory $\mathcal{R} = (\Sigma, E, R)$ for which narrowing is complete, and can model those algebraic properties as equations in E. Very recent work in this direction by Escobar, Meadows and myself [44,43] uses rewriting logic and narrowing to give a precise rewriting semantics to the inference system of one of the most effective analysis tools for cryptographic protocols, namely the NRL Protocol Analyzer [84].

Equational narrowing is a special case of rewriting logic narrowing, namely the case where we solve reachability goals of the form $\exists \boldsymbol{x}.\ equal(t, t') \longrightarrow true$ using the equations E as rewrite rules and adding the extra rule $equal(x, x) \longrightarrow true$. Furthermore, Horn logic with equality can be conservatively embedded in rewriting logic [89,81]. Indeed, in this embedding narrowing with the resulting rewrite theory is complete and agrees with SLD resolution modulo the equations E. This means that we reencounter our old friend Eqlog within the broader perspective of rewriting logic narrowing.

4.4   The Open Calculus of Constructions

Rewriting logic is an expressive *logical framework*, in which many other logics can be naturally represented [81]. Furthermore, by exploiting its reflective features in conjunction with the inductive nature of initial models, it has also good properties as a *meta-logical framework*, so that we can not only represent logics, but can also reason within the framework about their meta-logical properties [5,6].

But how good and general is it anyway? For example, how does it compare with the higher-order type theory formalisms that have been proposed by different authors as logical frameworks? Mark-Oliver Stehr and I tried to give an answer to this question using transitivity of representation mappings. If we could show that a higher-order type theory can be easily and naturally represented in rewriting logic in a conservative way, then any representation of a logic into such a type theory would automatically yield one in rewriting logic by composition. This would not be the simplest representation of that logic that one could define directly in rewriting logic, but it would prove that anything one can represent in the higher-order framework can likewise be represented in rewriting logic. Even so, some people might still be skeptical. Maybe you did it for Martin-Löf type theory, but how do I *know* that you can also do it for the Calculus of Constructions? All this could be dragged *ad nauseam*. So, what Mark-Oliver and I did in [130] was to specify a single *parametric* map (using parameterization in Maude) faithfully representing *pure type systems* (PTS) [8] into rewriting logic. Since pure type systems encompass a large class of type theories with simple types, type parameters, and type families, including the lambda cube, our skeptical colleagues would now have to come up with more exotic type theories outside the PTS general fold. At the meta-logical framework level, a careful comparison with higher-order type theories used for that purpose was given by David Basin, Manuel Clavel and myself in [6].

In fact, Mark-Oliver and I defined in [130] several representation mappings for pure type systems at different levels of abstraction. The more abstract, textbook-like representation mapped isomorphically the textbook syntax of pure type systems. But in order to give a more computational representation that would take care automatically of all the binding and substitution paraphernalia, we also gave a more concrete representation using Mark-Oliver's CINNI calculus of explicit substitutions [126] and showed it equivalent to the textbook one. Similarly, typing inference systems were represented in Maude in a computational way by means of rewrite rules [130]. This more concrete representation map was used by Mark-Oliver in his thesis [127] to implement in Maude his Open Calculus of Constructions (OCC) [127,128,129]. Since the Coquand-Huet calculus of constructions (CC) [28] is one of the instances of pure type systems, one could of course obtain an implementation of CC in Maude that way. But Mark-Oliver went considerably further. One of the sore points with higher-order type theories is their very limited and awkward way of dealing with *equalities*: an equational reasoning system like Maude can perform millions of equational deduction steps automatically in a second; but to represent such deduction steps within a given constructive type theory one needs to justify each of those equality steps constructively. By generating proof objects for the deductions of an external tool, for example for membership equational logic deduction [121], one can partly get around the problem. But Mark-Oliver's solution was more radical. By dropping the constructive interpretation and allowing simple set-theoretic models for OCC, he solved this problem directly: equality steps are allowed inside OCC, even modulo axioms like associativity and commutativity. Furthermore, OCC distinguishes several notations for equality, making clear whether they can be handled automatically by equational simplification, or need to be performed by explicit deduction steps. Likewise, a notation for relations representing rewrite rules in the rewriting logic sense is also provided. All this means that OCC can be viewed as a natural conceptual unification of the Calculus of Constructions and of rewriting logic. In particular, Maude can be naturally regarded as a sublanguage of OCC. As shown in [127,128,129], all the nice reasoning capabilities of the Calculus of Constructions, including its extensions with inductive and co-inductive principles, can be represented in OCC, that can carry out highly nontrivial proof tasks [127,128,129].

4.5   BMaude

In some sense, Maude, and languages like CafeOBJ [46] and BOBJ [54] that support hidden logic and behavioral equivalence, push the envelope in different directions of the specification language design space. Yet, there is a natural question about how these languages are all related. For example, both Maude and those languages have equational logic sublanguages. CafeOBJ itself provided some answer to this question in the form of the CafeOBJ "cube" of institutions [46], in which equational logic, hidden equational logic, and rewriting logic are related and unified. But the unification of rewriting logic and hidden logic proposed in [29] and used in [46,31] has some limitations regarding its model theory, and the matter seems to deserve further research.

While leaving open the issue of whether a more satisfactory unification of hidden logic and rewriting logic can be found, what Grigore Roşu and I did in [96] was to develop a hidden/behavioral extension of membership equational logic called *behavioral membership equational logic*. We were interested in this extension because of theoretical and practical reasons. Theoretically, the greater generality and expressiveness of MEL over, say, order-sorted equational logic resulted in a more expressive behavioral logic. Practically, the reflective features of Maude make it easy to develop an extension of Maude called BMaude in which theories in behavioral membership equational logic can be specified as modules, and to support deduction in such modules by behavioral rewriting [120,122]. Work ahead in this direction includes passing from the present theoretical foundations and BMaude language design to a prototype implementation, and finding a more general behavioral extension of rewriting logic itself.

5   Conclusions

Science is a dialogue. This gets somewhat distorted by the unidirectional character of publications, including this one; and by the impossibility of making always explicit the many influences shaping our ideas. This festive occasion provides an opportunity for reflecting, with gratitude, on such influences; and for looking in hindsight at the road already traveled, and forward to the ways ahead. I have tried to do a little of all this from a limited and subjective perspective, but one that I am at least very familiar with: some of the ways in which the OBJ, Eqlog, and FOOPS ideas have influenced Maude. And some of the directions in which the current Maude ideas are expanding.

One way to wrap all this up is with a picture describing the relationship between the different languages I have been discussing. I call it a *language genealogy*. Solid lines describe language inclusions (or near inclusions). Dashed lines describe a weaker relationship, namely one of *influence* between different languages. Not all influences are reflected in the picture: to avoid too much cluttering, only those that I think are more *direct* are depicted. One point to bear in mind is that some of these languages are currently under construction, or even in their design phase. For example, only a first prototype of PMaude exists at present, and BMaude and SHYMaude are only language designs at this point.

Acknowledgments. In this paper I have reflected on *some* of the ways in which Joseph's ideas have influenced mine. But there are many others, both scientific and nonscientific: so much so that an actual enumeration would be both impossible and futile. It is with deep gratitude that I wish to thank Joseph, not only for his ideas and his example, but above all for his friendship. I have already mentioned by name all the colleagues who were involved in the OBJ1–3 collaborations. To all of them I also extend my sincere thanks.

Furthermore, although the references make all this clear, I want to point out that: (1) the work on Maude is joint work with all the members of the Maude team at SRI, UIUC, and the Universities of Madrid and Málaga; (2) the work on Maude tools is joint work with Manuel Clavel, Francisco Durán, Santiago Escobar, Joseph Hendrix, Salvador Lucas, Claude Marché, Hitoshi Ohsaki, Peter Ölveczky, Miguel Palomino, and Xavier Urbain; (3) the work on real-time rewrite

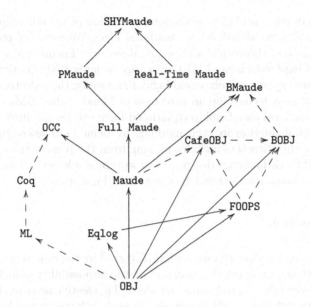

Fig. 1. A language genealogy ($\rightarrow$ inclusion; - - > influence)

theories is joint work with Peter Ölveczky at the University of Oslo; (4) the work on probabilistic rewrite theories and on stochastic hybrid systems is joint work with Gul Agha, Nirman Kumar, Koushik Sen, and Raman Sharykin at UIUC; (6) the work on OCC is entirely Mark-Oliver Stehr's; and (7) BMaude and its foundations are joint work with Grigore Roşu at UIUC. Several of these collaborators have also given me very useful comments to improve the final version of this paper.

This research has been supported by Grants ONR N00014-02-1-0715 and NSF CNS 05-24516, and by a bilateral CNRS-UIUC research project on "Rewriting calculi, logic and behavior."

References

1. G. Agha. *Actors.* MIT Press, 1986.
2. G. Agha, C. Gunter, M. Greenwald, S. Khanna, J. Meseguer, K. Sen, and P. Thati. Formal modeling and analysis of DoS using probabilistic rewrite theories. In *Proc. Workshop on Foundations of Computer Security (FCS'05) (Affiliated with LICS'05)*, 2005.
3. G. Agha and C. Hewitt. Concurrent programming using actors. In A. Yonezawa and M. Tokoro, editors, *Object-Oriented Concurrent Programming*, pages 37–53. MIT Press, 1988.
4. G. Agha, J. Meseguer, and K. Sen. PMaude: Rewrite-based specification language for probabilistic object systems. In *3rd Workshop on Quantitative Aspects of Programming Languages (QAPL'05)*, 2005.

5. D. Basin, M. Clavel, and J. Meseguer. Rewriting logic as a metalogical framework. In S. Kapoor and S. Prasad, editors, *FST TCS 2000*, pages 55–80. Springer LNCS, 2000.
6. D. Basin, M. Clavel, and J. Meseguer. Rewriting logic as a metalogical framework. *ACM Transactions on Computational Logic*, 5:528–576, 2004.
7. D. Basin and G. Denker. Maude versus Haskell: an experimental comparison in security protocol analysis. In K. Futatsugi, editor, *Proc. 3rd. Intl. Workshop on Rewriting Logic and its Applications*, volume 36. ENTCS, Elsevier, 2000.
8. S. Berardi. Towards a mathematical analysis of the Coquand-Huet calculus of constructions and other systems in barendregt's cube. Technical Report, Carnegie-Mellon University and Università di Torino, 1988.
9. G. Bernot, M. Bidoit, and T. Knapik. Observational specifications and the indistinguishability assumption. *Theoretical Comp. Science*, 139(1-2):275–314, 1995.
10. N. Berregeb, A. Bouhoula, and M. Rusinowitch. Observational proofs with critical contexts. In *Proceedings of FASE'98*, volume 1382 of *LNCS*. Springer, 1998.
11. M. Bidoit and R. Hennicker. Observer complete definitions are behaviourally coherent. In *OBJ/CafeOBJ/Maude at Formal Methods'99*, pages 83–94. Theta, 1999.
12. A. Bouhoula, J.-P. Jouannaud, and J. Meseguer. Specification and proof in membership equational logic. *Theoretical Computer Science*, 236:35–132, 2000.
13. R. Bruni and J. Meseguer. Generalized rewrite theories. In J. Baeten, J. Lenstra, J. Parrow, and G. Woeginger, editors, *Proceedings of ICALP 2003, 30th International Colloquium on Automata, Languages and Programming*, volume 2719 of *Springer LNCS*, pages 252–266, 2003.
14. R. Burstall and J. A. Goguen. The semantics of Clear, a specification language. In D. Bjorner, editor, *Proceedings of the 1979 Copenhagen Winter School on Abstract Software Specification*, pages 292–332. Springer LNCS 86, 1980.
15. M. Clavel. Reflection in general logics and in rewriting logic, with applications to the Maude language. Ph.D. Thesis, University of Navarre, 1998.
16. M. Clavel. *Reflection in Rewriting Logic: Metalogical Foundations and Metaprogramming Applications*. CSLI Publications, 2000.
17. M. Clavel, F. Durán, S. Eker, P. Lincoln, N. Martí-Oliet, and J. Meseguer. Metalevel computation in Maude. *Proc. 2nd Intl. Workshop on Rewriting Logic and its Applications*, ENTCS, Vol. 15, North Holland, 1998.
18. M. Clavel, F. Durán, S. Eker, P. Lincoln, N. Martí-Oliet, J. Meseguer, and J. Quesada. Maude: specification and programming in rewriting logic. *Theoretical Computer Science*, 285:187–243, 2002.
19. M. Clavel, F. Durán, S. Eker, P. Lincoln, N. Martí-Oliet, J. Meseguer, and C. Talcott. Maude Manual (Version 2.2). December 2005, http://maude.cs.uiuc.edu.
20. M. Clavel, F. Durán, S. Eker, and J. Meseguer. Building equational proving tools by reflection in rewriting logic. In *CAFE: An Industrial-Strength Algebraic Formal Method*. Elsevier, 2000. http://maude.cs.uiuc.edu.
21. M. Clavel, F. Durán, S. Eker, J. Meseguer, and M.-O. Stehr. Maude as a formal meta-tool. In J. Wing and J. Woodcock, editors, *FM'99 — Formal Methods*, volume 1709 of *Springer LNCS*, pages 1684–1703. Springer-Verlag, 1999.
22. M. Clavel, F. Durán, and N. Martí-Oliet. Polytypic programming in Maude. ENTCS 36, Elsevier, 2000. Proc. 3rd. Intl. Workshop on Rewriting Logic and its Applications.
23. M. Clavel and J. Meseguer. Axiomatizing reflective logics and languages. In G. Kiczales, editor, *Proceedings of Reflection'96, San Francisco, California, April 1996*, pages 263–288, 1996. http://jerry.cs.uiuc.edu/reflection/.

24. M. Clavel and J. Meseguer. Reflection and strategies in rewriting logic. In J. Meseguer, editor, *Proc. First Intl. Workshop on Rewriting Logic and its Applications*, volume 4 of *Electronic Notes in Theoretical Computer Science*. Elsevier, 1996.

25. M. Clavel and J. Meseguer. Reflection in conditional rewriting logic. *Theoretical Computer Science*, 285:245–288, 2002.

26. M. Clavel, J. Meseguer, and M. Palomino. Reflection in membership equational logic, many-sorted equational logic, Horn logic with equality, and rewriting logic. In F. Gadducci and U. Montanari, editors, *Proc. 4th. Intl. Workshop on Rewriting Logic and its Applications*. ENTCS, Elsevier, 2002.

27. M. Clavel and M. Palomino. The ITP tool's manual. Universidad Complutense, Madrid, April 2005, http://maude.sip.ucm.es/itp/.

28. T. Coquand and G. Huet. The calculus of constructions. *Information and Computation*, 76:95–120, 1988.

29. R. Diaconescu. Hidden sorted rewriting logic. In J. Meseguer, editor, *Proc. First Intl. Workshop on Rewriting Logic and its Applications*, volume 4 of *Electronic Notes in Theoretical Computer Science*. Elsevier, 1996.

30. R. Diaconescu and K. Futatsugi. Behavioral coherence in object-oriented algebraic specification. *Journal of Universal Computer Science*, 6(1):74–96, 2000.

31. R. Diaconescu and K. Futatsugi. Logical foundations of CafeOBJ. *Theoretical Computer Science*, 285:289–318, 2001.

32. F. Durán. A reflective module algebra with applications to the Maude language. Ph.D. Thesis, University of Málaga, 1999.

33. F. Durán. Coherence checker and completion tools for Maude specifications. Manuscript, Computer Science Laboratory, SRI International, http://maude.cs.uiuc.edu/papers, 2000.

34. F. Durán. Termination checker and Knuth-Bendix completion tools for Maude equational specifications. Manuscript, Computer Science Laboratory, SRI International, http://maude.cs.uiuc.edu/papers, 2000.

35. F. Durán, S. Eker, P. Lincoln, and J. Meseguer. Principles of Mobile Maude. In *Agent Systems, Mobile Agents, and Applications, ASA/MA 2000*, volume 1882 of *Springer LNCS*, pages 73–85. Springer-Verlag, 2000.

36. F. Durán, S. Lucas, J. Meseguer, C. Marché, and X. Urbain. Proving termination of membership equational programs. In P. Sestoft and N. Heintze, editors, *Proc. of ACM SIGPLAN 2004 Symposium on Partial Evaluation and Program Manipulation, PEPM'04*, pages 147–158. ACM Press, 2004.

37. F. Durán and J. Meseguer. An extensible module algebra for Maude. *Proc. 2nd Intl. Workshop on Rewriting Logic and its Applications*, ENTCS, Vol. 15, North Holland, 1998.

38. F. Durán and J. Meseguer. A Church-Rosser checker tool for Maude equational specifications. Manuscript, Computer Science Laboratory, SRI International, http://maude.cs.uiuc.edu/papers, 2000.

39. F. Durán and J. Meseguer. On parameterized theories and views in Full Maude 2.0. In K. Futatsugi, editor, *Proc. 3rd. Intl. Workshop on Rewriting Logic and its Applications*. ENTCS 36, Elsevier, 2000.

40. S. Eker. Term rewriting with operator evaluation strategy. *Proc. 2nd Intl. Workshop on Rewriting Logic and its Applications*, ENTCS, Vol. 15, North Holland, 1998.

41. S. Eker, J. Meseguer, and A. Sridharanarayanan. The Maude LTL model checker. In F. Gadducci and U. Montanari, editors, *Proc. 4th. Intl. Workshop on Rewriting Logic and its Applications*. ENTCS, Elsevier, 2002.

42. S. Eker, J. Meseguer, and A. Sridharanarayanan. The Maude LTL model checker and its implementation. In *Model Checking Software: Proc. 10th Intl. SPIN Workshop*, volume 2648, pages 230–234. Springer LNCS, 2003.

43. S. Escobar, C. Meadows, and J. Meseguer. A rewriting-based inference system for the NRL protocol analyzer and its meta-logical properties. Submitted for publication, 2005.

44. S. Escobar, C. Meadows, and J. Meseguer. A rewriting-based inference system for the NRL protocol analyzer: Grammar generation. In *Proc. FMSE'05, Formal Methods in Security Engineering (Alexandria, Virginia, Nov. 2005)*, pages 1–12. ACM Press, 2005.

45. S. Escobar, J. Meseguer, and P. Thati. Natural narrowing for general term rewriting systems. In *Rewriting Techniques and Applications, 16th Intl. Conference RTA 2005*, volume 3467, pages 279–293. Springer LNCS, 2005.

46. K. Futatsugi and R. Diaconescu. *CafeOBJ Report*. World Scientific, AMAST Series, 1998.

47. K. Futatsugi, J. Goguen, J.-P. Jouannaud, and J. Meseguer. Principles of OBJ2. In B. Reid, editor, *Proceedings of 12th ACM Symposium on Principles of Programming Languages*, pages 52–66. ACM, 1985.

48. P. Glynn. The role of generalized semi-Markov processes in simulation output analysis, 1983.

49. J. Goguen. Order sorted algebra. Technical Report Semantics and Theory of Computation Report 14, UCLA, 1978.

50. J. Goguen. Types as theories. In *Topology and Category Theory in Computer Science*, pages 357–390. Oxford, 1991.

51. J. Goguen and R. Diaconescu. Towards an algebraic semantics for the object paradigm. In *Proceedings of WADT*, volume 785 of *LNCS*. Springer, 1994.

52. J. Goguen, J.-P. Jouannaud, and J. Meseguer. Operational semantics of order-sorted algebra. In W. Brauer, editor, *Proceedings, 1985 International Conference on Automata, Languages and Programming*, volume 194 of *Springer LNCS*, pages 221–231. Springer-Verlag, 1985.

53. J. Goguen, C. Kirchner, H. Kirchner, A. Mégrelis, J. Meseguer, and T. Winkler. An introduction to OBJ3. In J.-P. Jouannaud and S. Kaplan, editors, *Proceedings, Conference on Conditional Term Rewriting, Orsay, France, July 8-10, 1987*, pages 258–263. Springer LNCS 308, 1988.

54. J. Goguen, K. Lin, and G. Roşu. Circular coinductive rewriting. In *Proceedings, 15th International Conference on Automated Software Engineering (ASE'00)*. Institute of Electrical and Electronics Engineers Computer Society, 2000. Grenoble, France, 11-15 September 2000.

55. J. Goguen and G. Malcolm. Hidden coinduction: Behavioral correctness proofs for objects. *Mathematical Structures in Computer Science*, 9(3):287–319, 1999.

56. J. Goguen and G. Malcolm. A hidden agenda. *J. of TCS*, 245(1):55–101, 2000.

57. J. Goguen and J. Meseguer. Correctness of recursive flow diagram programs. In *Proc. 6th Symp. Math. Found. Comp. Sci.*, pages 580–595. Springer LNCS 53, 1977.

58. J. Goguen and J. Meseguer. Universal realization, persistent interconnection and implementation of abstract modules. In M. Nielsen and E. M. Schmidt, editors, *Proceedings, 9th International Conference on Automata, Languages and Programming*, pages 265–281. Springer LNCS 140, 1982.

59. J. Goguen and J. Meseguer. Equality, types, modules and (why not?) generics for logic programming. *Journal of Logic Programming*, 1(2):179–210, 1984.

60. J. Goguen and J. Meseguer. Models and equality for logical programming. In H. Ehrig, G. Levi, R. Kowalski, and U. Montanari, editors, *Proceedings TAP-SOFT'87*, volume 250 of *Springer LNCS*, pages 1–22. Springer-Verlag, 1987.

61. J. Goguen and J. Meseguer. Unifying functional, object-oriented and relational programming with logical semantics. In B. Shriver and P. Wegner, editors, *Research Directions in Object-Oriented Programming*, pages 417–477. MIT Press, 1987.

62. J. Goguen and J. Meseguer. Order-sorted algebra I: Equational deduction for multiple inheritance, overloading, exceptions and partial operations. *Theoretical Computer Science*, 105:217–273, 1992.

63. J. Goguen, J. Meseguer, and D. Plaisted. Programming with parameterized abstract objects in OBJ. In D. Ferrari, M. Bolognani, and J. Goguen, editors, *Theory and Practice of Software Technology*, pages 163–193. North-Holland, 1983.

64. J. Goguen and G. Roşu. Hiding more of hidden algebra. In *Proceeding of FM'99*, volume 1709 of *LNCS*, pages 1704–1719. Springer, 1999.

65. J. Goguen, T. Winkler, J. Meseguer, K. Futatsugi, and J.-P. Jouannaud. Introducing OBJ. In *Software Engineering with OBJ: Algebraic Specification in Action*, pages 3–167. Kluwer, 2000.

66. J. A. Goguen and G. Malcolm, editors. *Software Engineering with OBJ: Algebraic Specification in Action*, volume 2 of *Advances in Formal Methods*. Kluwer Academic Publishers, Boston, 2000. ISBN 0-7923-7757-5.

67. J. Hendrix, M. Clavel, and J. Meseguer. A sufficient completeness reasoning tool for partial specifications. In *Rewriting Techniques and Applications, 16th Intl. Conference RTA 2005*, volume 3467, pages 165–174. Springer LNCS, 2005.

68. R. Hennicker and M. Bidoit. Observational logic. In *Proceedings of AMAST'98*, volume 1548 of *LNCS*, pages 263–277. Springer, 1999.

69. G. Holzmann. *The Spin Model Checker - Primer and Reference Manual.* Addison-Wesley, 2003.

70. C. Kirchner, H. Kirchner, and J. Meseguer. Operational semantics of OBJ3. In T. Lepistö and A. Salomaa, editors, *Proceedings, 15th Intl. Coll. on Automata, Languages and Programming, Tampere, Finland, July 11-15, 1988*, pages 287–301. Springer LNCS 317, 1988.

71. P. Kosiuczenko and M. Wirsing. Timed rewriting logic with application to object-oriented specification. Technical report, Institut für Informatik, Universität München, 1995.

72. N. Kumar, K. Sen, J. Meseguer, and G. Agha. Probabilistic rewrite theories: Unifying models, logics and tools. Technical Report UIUCDCS-R-2003-2347, CS Dept., University of Illinois at Urbana-Champaign, May 2003.

73. N. Kumar, K. Sen, J. Meseguer, and G. Agha. A rewriting based model of probabilistic distributed object systems. Proc. of Formal Methods for Open Object-Based Distributed Systems, FMOODS 2003, Springer LNCS Vol. 2884, 2003.

74. E. Lien. Formal modeling and analysis of the NORM multicast protocol in Real-Time Maude. Master's thesis, Dept. of Linguistics, University of Oslo, April 2004. http://wo.uio.no/as/WebObjects/theses.woa/wo/0.3.9.

75. S. Lucas. Context-sensitive computations in functional and functional logic programs. *J. Functl. and Log. Progr.*, 1(4):446–453, 1998.

76. S. Lucas. Termination of on-demand rewriting and termination of obj programs. In *Proc. PPDP'01*, pages 82–93. ACM, 2001.

77. S. Lucas. Termination of rewriting with strategy annotations. In *Proceedings of LPAR 2001*, volume 2250 of *LNAI*, pages 669–684. Springer-Verlag, 2001.

78. S. Lucas. Context-sensitive rewriting strategies. *Inf. Comput.*, 178(1):294–343, 2002.
79. S. Lucas, C. Marché, and J. Meseguer. Operational termination of conditional term rewriting systems. *Information Processing Letters*, 95(4):446–453, 2005.
80. E. Manes, editor. *Proceedings of the First International Symposium on Category Theory Applied to Computation and Control, San Francisco, California, February 25–26 1974*. Springer LNCS Vol. 25, 1975.
81. N. Martí-Oliet and J. Meseguer. Rewriting logic as a logical and semantic framework. In D. Gabbay and F. Guenthner, editors, *Handbook of Philosophical Logic, 2nd. Edition*, pages 1–87. Kluwer Academic Publishers, 2002. First published as SRI Tech. Report SRI-CSL-93-05, August 1993.
82. N. Martí-Oliet and J. Meseguer. Rewriting logic: roadmap and bibliography. *Theoretical Computer Science*, 285:121–154, 2002.
83. N. Martí-Oliet, J. Meseguer, and A. Verdejo. Towards a strategy language for Maude. In N. Martí-Oliet, editor, *Proc. 5th. Intl. Workshop on Rewriting Logic and its Applications*, pages 417–441. ENTCS, Vol. 117, Elsevier, 2004.
84. C. Meadows. The NRL protocol analyzer: An overview. *Journal of Logic Programming*, 26(2):113–131, 1996.
85. J. Meseguer. General logics. In H.-D. E. et al., editor, *Logic Colloquium'87*, pages 275–329. North-Holland, 1989.
86. J. Meseguer. A logical theory of concurrent objects. In *ECOOP-OOPSLA'90 Conference on Object-Oriented Programming, Ottawa, Canada, October 1990*, pages 101–115. ACM, 1990.
87. J. Meseguer. Rewriting as a unified model of concurrency. In *Proceedings of the Concur'90 Conference, Amsterdam, August 1990*, pages 384–400. Springer LNCS 458, 1990.
88. J. Meseguer. Conditional rewriting logic as a unified model of concurrency. *Theoretical Computer Science*, 96(1):73–155, 1992.
89. J. Meseguer. Multiparadigm logic programming. In H. Kirchner and G. Levi, editors, *Proc. 3rd Intl. Conf. on Algebraic and Logic Programming*, pages 158–200. Springer LNCS 632, 1992.
90. J. Meseguer. A logical theory of concurrent objects and its realization in the Maude language. In G. Agha, P. Wegner, and A. Yonezawa, editors, *Research Directions in Concurrent Object-Oriented Programming*, pages 314–390. MIT Press, 1993.
91. J. Meseguer. Solving the inheritance anomaly in concurrent object-oriented programming. In O. M. Nierstrasz, editor, *Proc. ECOOP'93*, pages 220–246. Springer LNCS 707, 1993.
92. J. Meseguer. Membership algebra as a logical framework for equational specification. In F. Parisi-Presicce, editor, *Proc. WADT'97*, pages 18–61. Springer LNCS 1376, 1998.
93. J. Meseguer. A rewriting logic sampler. In *Proc. International Colloquium on Theoretical Aspects of Computing ICTAC05 (Hanoi, Vietnam, October 2005)*, volume 3722 of *LNCS*, pages 1–28. Springer, 2005.
94. J. Meseguer and J. Goguen. Initiality, induction and computability. In M. Nivat and J. Reynolds, editors, *Algebraic Methods in Semantics*, pages 459–541. Cambridge University Press, 1985.
95. J. Meseguer and G. Roşu. A total approach to partial algebraic specification. In *Proc. ICALP'02*, pages 572–584. Springer LNCS 2380, 2002.
96. J. Meseguer and G. Roşu. Towards behavioral Maude: Behavioral membership equational logic. In *Proc. CMCS'02*. Elsevier ENTCS, 2002.

97. J. Meseguer and G. Roşu. Rewriting logic semantics: From language specifications to formal analysis tools. In *Proc. Intl. Joint Conf. on Automated Reasoning IJCAR'04, Cork, Ireland, July 2004*, pages 1–44. Springer LNAI 3097, 2004.

98. J. Meseguer and G. Roşu. The rewriting logic semantics project. In *Proc. SOS 2005*. Elsevier ENTCS, 2005.

99. J. Meseguer and R. Sharykin. Specification and analysis of distributed object-based stochastic hybrid systems. In *Hybrid Systems, HSCC 2006*, pages 460–475. Springer LNCS 3927, 2006.

100. J. Meseguer and C. Talcott. Semantic models for distributed object reflection. In *Proceedings of ECOOP'02, Málaga, Spain, June 2002*, pages 1–36. Springer LNCS 2374, 2002.

101. J. Moore, R. Krug, H. Liu, and G. Porter. Formal models of Java at the JVM level – a survey from the ACL2 perspective. In *Proc. Workshop on Formal Techniques for Java Programs, in association with ECOOP 2001*, 2002.

102. P. C. Ölveczky. *Specification and Analysis of Real-Time and Hybrid Systems in Rewriting Logic*. PhD thesis, University of Bergen, Norway, 2000. http://maude.csl.sri.com/papers.

103. P. C. Ölveczky and M. Caccamo. Formal simulation and analysis of the CASH scheduling algorithm in Real-Time Maude. In *Proc. FASE 2006, LNCS 3922*, pages 357–372. Springer, 2005.

104. P. C. Ölveczky, M. Keaton, J. Meseguer, C. Talcott, and S. Zabele. Specification and analysis of the AER/NCA active network protocol suite in Real-Time Maude. In *Proc. of FASE'01, 4th Intl. Conf. on Fundamental Approaches to Software Engineering*, volume 2029 of *Springer LNCS*, pages 333–348. Springer-Verlag, 2001.

105. P. C. Ölveczky, P. Kosiuczenko, and M. Wirsing. An object-oriented algebraic steam-boiler control specification. In J.-R. Abrial, E. Börger, and H. Langmaack, editors, *The Steam-Boiler Case Study Book*, pages 379–402. Springer-Verlag, 1996. Vol. 1165.

106. P. C. Ölveczky and J. Meseguer. Specifying real-time systems in rewriting logic. In J. Meseguer, editor, *Proc. First Intl. Workshop on Rewriting Logic and its Applications*, volume 4 of *Electronic Notes in Theoretical Computer Science*. Elsevier, 1996.

107. P. C. Ölveczky and J. Meseguer. Real-Time Maude: a tool for simulating and analyzing real-time and hybrid systems. volume 36. ENTCS, Elsevier, 2000. Proc. 3rd. Intl. Workshop on Rewriting Logic and its Applications.

108. P. C. Ölveczky and J. Meseguer. Specification of real-time and hybrid systems in rewriting logic. *Theoretical Computer Science*, 285:359–405, 2002.

109. P. C. Ölveczky and J. Meseguer. Real-Time Maude 2.1. In N. Martí-Oliet, editor, *Proc. 5th. Intl. Workshop on Rewriting Logic and its Applications*, pages 285–314. ENTCS, Vol. 117, Elsevier, 2004.

110. P. C. Ölveczky and J. Meseguer. Specification and analysis of real-time systems using Real-Time Maude. In T. Margaria and M. Wermelinger, editors, *Fundamental Approaches to Software Engineering (FASE 2004)*, volume 2984 of *Springer LNCS*, pages 354–358. Springer-Verlag, 2004.

111. P. C. Ölveczky and J. Meseguer. Abstraction and completeness for Real-Time Maude. In G. Denker and C. Talcott, editors, *Proc. 6th. Intl. Workshop on Rewriting Logic and its Applications*. ENTCS, Elsevier, 2006.

112. P. C. Ölveczky and J. Meseguer. Semantics and pragmatics of Real-Time Maude. *Higher-Order and Symbolic Computation*, 2006. To appear.

113. P. C. Ölveczky, J. Meseguer, and C. L. Talcott.   Specification and analysis of the AER/NCA active network protocol suite in Real-Time Maude. Technical Report UIUCDCS-R-2004-2467, Department of Computer Science, University of Illinois at Urbana-Champaign, 2004.   Available at http://www.ifi.uio.no/RealTimeMaude.

114. P. C. Ölveczky and S. Thorvaldsen. Formal modeling and analysis of wireless sensor network algorithms in Real-Time Maude. In *The 14th International Workshop on Parallel and Distributed Real-Time Systems (WPDRTS) 2006*. IEEE Computer Society Press, 2006.

115. P. Padawitz. Swinging data types: Syntax, semantics, and theory. In *Proceedings, WADT'95*, volume 1130 of *LNCS*, pages 409–435. Springer, 1996.

116. P. Padawitz. Towards the one-tiered design of data types and transition systems. In *Proceedings of WADT'97*, volume 1376 of *LNCS*, pages 365–380. Springer, 1998.

117. P. Padawitz. Swinging types = functions + relations + transition systems. *Theoretical Computer Science*, 243:93–165, 2000.

118. M. Palomino. A predicate abstraction tool for Maude. Manuscript, Universidad Complutense, 2005, http://maude.sip.ucm.es/~miguelpt/papers/pa-tool.pdf.

119. M. Puterman. *Markov Decision Processes: Discrete Stochastic Dynamic Programming*. John Wiley and Sons, 1994.

120. G. Roşu. *Hidden Logic*. PhD thesis, University of California at San Diego, 2000.

121. G. Roşu, S. Eker, P. Lincoln, and J. Meseguer. Certifying and synthesizing membership equational proofs. In *Proc. FM'03*, volume 2805, pages 359–380. Springer LNCS, 2003.

122. G. Roşu and J. Goguen. Hidden congruent deduction. In *Automated Deduction in Classical and Non-Classical Logics*, volume 1761 of *LNAI*. Springer, 2000.

123. R. Segala. *Modelling and Verification of Randomized Distributed Real Time Systems*. PhD thesis, Massachusetts Institute of Technology, 1995.

124. K. Sen, M. Viswanathan, and G. Agha. On statistical model checking of stochastic systems. In *17th conference on Computer Aided Verification (CAV'05)*, volume 3576 of *LNCS*, pages 266–280, Edinburgh, Scotland, 2005. Springer.

125. L. Steggles and P. Kosiuczenko. A timed rewriting logic semantics for SDL: a case study of the alternating bit protocol. *Proc. 2nd Intl. Workshop on Rewriting Logic and its Applications*, ENTCS, Vol. 15, North Holland, 1998.

126. M.-O. Stehr. CINNI - a generic calculus of explicit substitutions and its application to lambda-, sigma- and pi-calculi. ENTCS 36, Elsevier, 2000. Proc. 3rd. Intl. Workshop on Rewriting Logic and its Applications.

127. M.-O. Stehr. Programming, Specification, and Interactive Theorem Proving — Towards a Unified Language based on Equational Logic, Rewriting Logic, and Type Theory. Doctoral Thesis, Universität Hamburg, Fachbereich Informatik, Germany, 2002. http://www.sub.uni-hamburg.de/disse/810/.

128. M.-O. Stehr. The Open Calculus of Constructions: An equational type theory with dependent types for programming, specification, and interactive theorem proving (Part I). *Fundamenta Informaticae*, 68(1–2):131–174, 2005.

129. M.-O. Stehr. The Open Calculus of Constructions: An equational type theory with dependent types for programming, specification, and interactive theorem proving (Part II). *Fundamenta Informaticae*, 68(3):249–288, 2005.

130. M.-O. Stehr and J. Meseguer. Pure type systems in rewriting logic: Specifying typed higher-order languages in a first-order logical framework. In *Essays in Memory of Ole-Johan Dahl*, pages 334–375. Springer LNCS Vol. 2635, 2004.

131. W. J. Stewart. *Introduction to the Numerical Solution of Markov Chains*. Princeton, 1994.
132. P. Thati and J. Meseguer. Symbolic reachability analysis using narrowing and its application to the verification of cryptographic protocols. In N. Martí-Oliet, editor, *Proc. 5th. Intl. Workshop on Rewriting Logic and its Applications*, pages 153–182. ENTCS, Vol. 117, Elsevier, 2004.
133. P. Thati and J. Meseguer. Complete symbolic reachability analysis using back-and-forth narrowing. In *Proceedings of CALCO 2005*, volume 3629 of *LNCS*, pages 379–394. Springer, 2005.
134. S. Thorvaldsen and P. C. Ölveczky. Formal modeling and analysis of the OGDC wireless sensor network algorithm in Real-Time Maude. http://www.ifi.uio.no/RealTimeMaude/OGDC, 2005.
135. W. Tracz. Parametrized programming in LILEANNA. In *Proc. 1993 ACM/SIGAPP Symp. on Applied Computing (SAC '93)*, pages 77–86, 1993.
136. P. Viry. Equational rules for rewriting logic. *Theoretical Computer Science*, 285:487–517, 2002.

Constructive Action Semantics in OBJ

Peter D. Mosses

Department of Computer Science, Swansea University, Wales, UK
p.d.mosses@swan.ac.uk,
http://www.cs.swan.ac.uk/~cspdm/

Abstract. Goguen and Malcolm specify semantics of programming languages in OBJ. Here, we consider how the extensibility and reusability of their specifications could be improved. We propose using the notation and modular structure of the Constructive Action Semantics framework in OBJ, and give a simple illustration. The reader is assumed to be familiar with OBJ.

1 Introduction

Conventional semantic descriptions of programming languages suffer from poor modularity. In denotational semantics, for instance, descriptions are usually divided into three sections, defining (abstract) syntax, semantic entities, and semantic functions. The semantic functions map parts of programs compositionally to their denotations (which are themselves usually functions) and are defined inductively by so-called semantic equations. All the definitions have global visibility throughout the description of a particular language. Moreover, when developing a denotational semantics, adding a new construct to the syntax of the described language may require extensive reformulation of the definition of the semantic functions. The need for reformulation can be largely eliminated using monadic notation instead of pure λ-notation, but there are still no named modules that could be reused or extended in semantic descriptions of other languages. Similarly, conventional structural operational semantics lacks explicit modules, and adding a new construct may require reformulating all the previously-defined rules to take account of a new component of configurations (although the latter problem can be eliminated rather easily using MSOS [12]).

Goguen and Malcolm [3] specify semantics of programming languages using OBJ [4]. Their descriptions are a hybrid of denotational, operational, and algebraic semantics. Importantly, the OBJ system supports validation of the semantic description by running programs and proving properties about them. The introduction of named modules with restrictions on the visibility of their definitions helps to identify which parts could be affected when a definition is changed. However, just as with conventional denotational and operational semantics, adding a new construct to a language may still require extensive reformulation of the description of the original constructs. Their modules are also quite large, and unlikely to be reused directly in descriptions of different languages. These points will be discussed further and illustrated in Sect. 2.

K. Futatsugi et al. (Eds.): Goguen Festschrift, LNCS 4060, pp. 281–295, 2006.

We propose to improve the *extensibility* of semantics in OBJ by introducing *actions*. Actions are used as the denotations of programming constructs in Action Semantics [8,9,14], and expressed using *Action Notation* (AN). This notation plays the same role in action semantic descriptions as λ-notation does in conventional denotational semantics, but also provides primitives and combinators to specify control and data flow, scopes of bindings, effects on storage, and interactive processes. The design of AN is such that previous specifications of denotations never need reformulation when adding a new construct to the described language. For instance, the specification of the action semantics of an arithmetic expression can remain the same, regardless of whether the sub-expressions might have side-effects, raise exceptions, spawn processes, be non-deterministic, or diverge. Further details and illustrations will be given in Sect. 3.

Using AN dramatically improves the reusability of parts of semantic descriptions. For instance, the semantic equations defining the denotations of arithmetic expressions could now be reused verbatim in descriptions of different languages. However, specification reuse by copying has major disadvantages: it leaves no trace of the origin of the specification, and it is not apparent whether the copied specification has been subsequently edited. Verbatim reuse should be made explicit by *referring* directly to the original specification.

To maximize the possibility of verbatim reuse, we propose two further changes to the style of specifying semantics in OBJ. Both are rather radical, and form the basis for a novel approach to developing semantic descriptions called *Constructive Semantics* [13]. The first change concerns modular structure, where we intend to use a *separate module* for the description of each *individual* language construct. Such a module contains a single semantic equation, specifying the action semantics of the construct concerned using AN to combine the action semantics of its components. The second change is to map the concrete syntax of each language construct to the constructs of a *language-independent abstract syntax*. The semantics of each concrete construct is then derived by composing this map with the semantics of the abstract construct. Complex concrete constructs can be mapped to combinations of simpler abstract constructs; this is similar to reducing a language to its kernel constructs, except that we do not insist on the abstract constructs being themselves directly expressible in the concrete language, nor that the abstract constructs are themselves irreducible. Specification of constructive semantics in OBJ will be illustrated in Sect. 4.

The Constructive Action Semantics framework was originally developed in collaboration with Doh [1], and further enhanced in collaboration with van den Brand and Iversen [15]. A constructive action semantics for Core ML has been specified together with Iversen [6], and used for semantics-based compiler generation [5]. A constructive version of MSOS [12] has been used in connection with teaching operational semantics [10]. The general architecture of constructive semantics is advocated as a useful paradigm for the development of any kind of truly modular semantics [11,13].

The present paper is based on tentative experiments using OBJ. It includes excerpts from the full specifications developed by the author, which are available for downloading at http://www.cs.swan.ac.uk/~cspdm/Goguen-FS/. Please note that the author is not an expert user of OBJ; suggestions for improvements to the use of OBJ in the specifications are welcome.

2 Algebraic Semantics in OBJ

We shall start by recalling how Goguen and Malcolm specify the semantics of imperative programming languages algebraically in OBJ [3]. We shall then assess the extensibility and reusability of such semantic descriptions.

2.1 A Simple Example

Goguen and Malcolm introduce a small language called Simple, and specify both its syntax and semantics in OBJ. Some excerpts from their specification are given below (the full OBJ code is available at http://www.cs.ucsd.edu/users/goguen/sys/code1.html).

Goguen and Malcolm start by specifying the data types of Simple. For technical reasons (and to facilitate proofs), they specify a module ZZ which enriches the built-in module INT with the operation _is_ : Int Int -> Bool, and with various equations and conditional equations concerning integer operations and relations.

OBJ allows rather general mixfix operation symbols. Goguen and Malcolm exploit this to declare abstract syntax constructors that correspond to a grammar for concrete syntax, so that concrete programs can be parsed as terms by OBJ:

```
obj EXP is pr ZZ .
  dfn Var is QID .
  sorts  Exp Arvar Arcomp .
  subsorts  Var Int Arcomp < Exp .
  ops  a b c : -> Arvar .
  op  _+_  : Exp Exp -> Exp [prec 10] .
  op  _*_  : Exp Exp -> Exp [prec 8] .
  op  -_   : Exp -> Exp [prec 1] .
  op  _-_  : Exp Exp -> Exp [prec 10] .
  op  _[_] : Arvar Exp -> Arcomp [prec 1] .
endo
```

Goguen and Malcolm specify stores in terms of their relationship to expressions, Boolean-valued tests, and assignment statements:

```
th STORE is pr BPGM .
            pr ARRAY .
  sort Store .
  op initial : -> Store .
  op  _[[_]] : Store Exp -> Int [prec 65] .
  ...
```

```
op   _;_ : Store BPgm -> Store [prec 60] .
var  S : Store .
vars X1 X2 : Var .
...
eq  S [[E1 + E2]]  =  (S[[E1]]) + (S[[E2]]) .
...
eq  S ; X1 := E1 [[X1]]  =  S [[E1]] .
cq  S ; X1 := E1 [[X2]]  =  S [[X2]]   if  X1 =/= X2 .
eq  S ; X1 := E1 [[AV]]  =  S [[AV]] .
...
endth
```

Goguen and Malcolm's specification of the semantics of structured programs in Simple is formulated as equations between store terms, but the equations can easily be understood operationally:

```
obj SEM is pr PGM .
          pr STORE .
  sort EStore .
  subsort Store < EStore .
  op   _;_ : EStore Pgm -> EStore [prec 60] .
  var S : Store .
  var T : Tst .
  var P1 P2 : Pgm .
  eq  S ; skip  = S .
  eq  S ; (P1 ; P2)  =  (S ; P1) ; P2 .
  cq  S ; if T then P1 else P2 fi  =  S ; P1
    if  S[[T]] .
  cq  S ; if T then P1 else P2 fi  =  S ; P2
    if  not(S[[T]]) .
  cq  S ; while T do P1 od  =  (S ; P1) ; while T do P1 od
    if  S[[T]] .
  cq  S ; while T do P1 od  =  S
    if  not(S[[T]]) .
endo
```

(The supersort EStore is introduced because of the possibility of non-terminating while-programs.)

2.2 Extensibility

Goguen and Malcolm's semantics of Simple is algebraic (in the sense that it is specified as an initial algebra, using algebraic axioms). It also appears to be reasonably modular. But how *extensible* is it? Can we expect to be able to keep it largely unchanged when adding new constructs to the described language?

Inspection of the main modules STORE and SEM reveals that their formulation depends on two assumptions:

– expressions do not have side-effects, and
– the store is the only auxiliary information processed by expressions and
  structured programs.

If we were to extend Simple by adding side-effects to expressions, representing
local bindings by environments, allowing expressions to raise exceptions, or many
other language features, we would violate one or both of these assumptions.

The need to reformulate large parts of semantic descriptions when adding
new features to language constructs is familiar from conventional denotational
and operational semantics. In the next section, we shall see how the use of
the action notation provided by action semantics can avoid the need for such
reformulation, and ensure better extensibility.

2.3   Reusability

To what extent could parts of Goguen and Malcolm's semantics of Simple be
reused in descriptions of different languages? For instance, suppose we were to
describe the semantics of a corresponding sublanguage of C (with expressions
restricted to have no side-effects, etc.); which modules would we be able to reuse
verbatim?

Clearly, the modules specifying the data types of a programming language
are highly reusable. For example, the module ZZ that specifies (an enrichment of)
the usual integers would probably be appropriate in the semantic descriptions
of most programming languages; any further operations required could easily be
added after importing it.

However, several aspects of the other modules significantly hinder their reuse:

– The notation for expressions and structured programs is intended to reflect
  their *concrete* syntax. OBJ allows notation to be changed when importing
  a module, but when widespread changes would be needed (e.g., when going
  from the syntax for structured programs in Simple to that in C) it would
  surely be simpler and more perspicuous to copy and edit the original modules
  than to import them and specify the renaming of operations.
– The module *hierarchy* is relatively deep. If a module such as EXP in Goguen
  and Malcolm's example semantics were to be reused by importing and enriching it, any module that imports EXP would require a corresponding enrichment – unless it was copied and edited to refer to the module importing
  and enriching EXP.
– Particular sets of constructs are *bundled together* in the same module. OBJ
  does not allow operations to be hidden, so for instance when the module
  EXP is imported, the concrete syntax for array variables (Arvar) and array
  components (Arcomp) is included, whether one wants it or not. It appears
  that copying and editing is the only way of removing declared operations.

In Sect. 4 we shall see how to remove all the above hindrances to reuse.

3   Action Semantics in OBJ

Action Semantics is a hybrid of denotational and operational semantics. As usual in denotational semantics, semantic functions map programs and their components to denotations that represent their contribution to overall program behaviour. The semantic functions are compositional (i.e., the denotation of a construct depends only on the denotations of its components) and defined inductively by semantic equations. The main difference between action semantics and conventional denotational semantics is that denotations in action semantics are so-called actions, rather than higher-order functions on domains. The notation used to express actions, called simply Action Notation (AN), is itself defined operationally, in contrast to the λ-notation used in denotational semantics, which has a pure mathematical interpretation. When performed, actions may be given and compute data, refer to bindings, inspect and update storage; they may terminate normally, terminate exceptionally, fail, or never terminate. As we shall see below, AN is quite expressive, and provides primitives and combinators for specifying data and control flow, scopes of bindings, and effects on storage.[1]

3.1   Data

The items of data processed by actions consist of the usual data types of programming languages (numbers, arrays, etc.) together with identifiers, environments (representing bindings), storage cells (locations), and entities representing various kinds of procedural and data abstraction (such as packages and classes). Actions are given and may return arbitrary finite sequences of such data. The constructors for such sequences can be declared in OBJ as follows:

```
obj DATA is
sorts Data Datum .
subsorts Datum < Data .
op no-data : -> Data .
op _ , _ : Data Data -> Data [assoc id: no-data] .
endo
```

All other operations on Data are represented by constants of sort Op:

```
obj DATA/OP is
ex DATA .
sort Op .
op _ ! _ : Op Data -> Data . *** result of application
op _ ? _ : Op Data -> Bool . *** definedness of result
endo
```

Note that also subsorts of Data are represented by constants (written in lowercase, e.g. datum for the subsort Datum), and the corresponding retracts are represented by applying the operation 'the' to them (e.g. 'the datum').

---

[1] AN also provides primitives for asynchronous threads and message-passing, but these are omitted here.

3.2   Kernel AN

The primitives and combinators of the kernel of AN are declared below:

```
obj KERNEL-AN is
pr DATA DATA/BOOL DATA/INT DATA/SEQ DATA/BINDINGS DATA/STORE .
sorts Action Atomic-Action .
subsorts            Atomic-Action < Action .
op copy            :              -> Atomic-Action .
op result _        : Data         -> Atomic-Action .
op skip            :              -> Atomic-Action .
eq skip = result no-data .
op give _          : Op           -> Atomic-Action .
op _ then _        : Action Action -> Action [assoc] .
op _ and _         : Action Action -> Action [assoc] .
op _ and-then _    : Action Action -> Action [assoc] .
op indivisibly _   : Action       -> Action .
```

The difference between **then** and **and-then** is that in A1 **then** A2, the data computed by A1 is passed to A2, whereas in A1 **and-then** A2, the data computed by A1 is concatenated with that computed by A2. The difference between **and-then** and **and** is that the former insists on sequential execution, whereas the latter leaves the order unspecified, allowing interleaving.

```
op throw           :              -> Atomic-Action .
op thrown _        : Data         -> Atomic-Action .
op err             :              -> Atomic-Action .
eq err = thrown no-data .
op _ catch _       : Action Action -> Action [assoc] .
op _ and-catch _   : Action Action -> Action [assoc] .
```

The above notation is used for actions that can terminate exceptionally, throwing data. Note that when the given data is not in the domain of definition of an operation O, the outcome of **give** O is the same as that of **err**.

```
op fail            :              -> Atomic-Action .
op check _         : Op           -> Atomic-Action .
op _ else _        : Action Action -> Action [assoc] .
```

Explicit failure of an action is distinguished from throwing an exception, and **else** allows combination of alternative actions to recover from failure.

```
op unfold          :              -> Action .
op unfolding _     : Action       -> Action .
```

The above notation is used to express iteration.

```
op copy-bindings :                -> Atomic-Action .
op _ scope _       : Action Action -> Action [assoc] .
op recursively _ : Action         -> Action .
```

In `A1 scope A2` above, the bindings computed by `A1` are the current bindings
for `A2`. In `recursively A` the scope of the bindings computed by `A` includes `A`
itself.

```
op create     :               -> Atomic-Action .
op inspect    :               -> Atomic-Action .
op update     :               -> Atomic-Action .
```

The above notation is used for actions concerned with stored data.

```
op enact      :               -> Atomic-Action .
```

When the action `enact` is given an action as data, it performs that action. `Action`
is a subsort of `Datum`, and (constants corresponding to) action combinators are
included `Op`. Using the combinator `scope` allows the current bindings to be incorporated in actions before they are enacted, which supports both static and
dynamic bindings.

The following is an excerpt from the OBJ specification of the operational
semantics of kernel AN. It was developed primarily to support validation the
action semantics of Simple using OBJ. An action has to be supplied with data
and bindings, as well as access to the store, before it can be performed. The
outcome of the execution is computed data, thrown data, or failure, together
with the updated store.

```
op { _ } _ _ _   : Action Data Bindings Store   -> Action .
op { _ } _ _     : Action Data Bindings         -> Action .
op { _ } _       : Action Store                 -> Action .
op { _ } _       : Action Data                  -> Action .
op { _ } _       : Action Bindings              -> Action .
vars A A1 A2      : Action .
vars D* D1* D2*   : Data .
vars O            : Op .
vars BS BS1 BS2   : Bindings .
vars S            : Store .

eq {copy}D* BS S = {result D*}S .

eq {result D1*}D* BS S = {result D1*}S .

eq {give O}D* BS S =
   if O ? D* then {result (O ! D*)}S else {err}S fi .

eq {A1 then A2}D* BS S = {A1}D* BS S then {A2}BS .
eq {result D1*}S then {A2}BS = {A2}D1* BS S .
eq {thrown D1*}S then {A2}BS = {thrown D1*}S .
eq {fail}S then {A2}BS = {fail}S .
```

3.3 Full AN

The primitives and combinators of the full AN are declared below. Note that, in contrast to the original version of AN, so-called yielders are not part of the kernel.

```
obj AN is
pr KERNEL-AN .
sort Yielder .
subsorts Data Op < Yielder .
op _ , _  : Yielder Yielder        -> Yielder [assoc] .
op _ _    : Op Yielder             -> Yielder .
op _ _ _  : Yielder Op Yielder     -> Yielder .
op give _ : Yielder                -> Action .
op _ _    : Atomic-Action Yielder  -> Action .
```

Yielders allow compositions of data operations to be applied to the given data. When Y is a yielder, the action give Y gives the result of Y. The action AA Y makes the result of Y the data for the (atomic) action AA. For example, if the action update is used alone, it has to be given a cell and a storable value as data, whereas the action update(the cell, 0) is equivalent to (give the cell and result 0) then update, and is given only a cell.

```
op check _ : Yielder -> Action .
op maybe _ : Action  -> Action .
```

The action check Y merely tests whether the value is true (failing otherwise). When A terminates exceptionally, maybe A fails (so maybe give 0 fails is the given data is not in the domain of 0).

```
op furthermore _ : Action         -> Action .
op _ before _    : Action Action -> Action .
```

The action furthermore A lets the bindings computed by A override the current bindings, so that (furthermore A1) scope A2 corresponds to a block. The action A1 before A2 combines the bindings computed by A1 and A2, letting the scope of the former bindings include A2.

```
op bind              :          -> Atomic-Action .
op current-bindings  :          -> Yielder .
op closure _         : Yielder -> Yielder .
```

The action bind merely computes a binding from the identifier and bindable value given to it. The yielder closure A results in an action which, when enacted, performs A in the scope of the bindings that were current for the yielder.

```
op bound-to _  : Yielder -> Yielder .
op stored-at _ : Yielder -> Yielder .
```

The yielders bound-to Y and stored-at Y refer to components of the current bindings and of the store, respectively.

The following equations illustrate how the expansion of full AN to kernel AN is specified in OBJ:

```
eq give D* = result D* .
eq give (Y1, Y2) = give Y1 and give Y2 .
eq give (0 Y) = give Y then give 0 .
eq give (Y1 0 Y2) = (give Y1 and give Y2) then give 0 .
eq give current-bindings = copy-bindings .
eq give bound-to Y = (copy-bindings and give Y) then give bound .
eq give stored-at Y = give Y then inspect .

eq AA Y = give Y then AA .

eq maybe A = A catch fail .

eq furthermore A = copy-bindings then give overriding .

eq A1 before A2 =
        (copy-bindings and A1) then
        (give #2 and (give overriding scope A2)) then
        give overriding .

eq bind = give binding .

eq closure Y = (result current-bindings) scope Y .
```

3.4   Action Semantics

The action semantics of Simple's concrete expressions and structured programs could be specified as follows.

First, a semantic function is declared for each sort of concrete syntactic construct, e.g.:

```
op evaluate _ : Exp -> Action .
```

(It would be appropriate to specify what sorts of data the action denotations of the different sorts of construct may return, but the required notation for subsorts of Action has not yet been specified in OBJ.)

After importing AN and the relevant specifications of data types, the semantic functions are defined by semantic equations, e.g.:

```
eq evaluate (E1 + E2) =
    (evaluate E1 and evaluate E2) then give plus .
```

The action combinator A1 and A2 corresponds to so-called target-tupling of functions: it performs A1 and A2 (in an unspecified order) and if they both terminate normally, it concatenates the data that they computed. In contrast,

A1 then A2 corresponds to functional composition: it performs A1 first, and if that terminates normally, it gives any data computed by A1 to the performance of A2. There is also a combinator written A1 and-then A2, which is the same as A1 and A2 regarding data flow, but insists on sequential performance of the sub-actions. For any data operation O, the primitive action give O applies O to its data to compute a result (terminating exceptionally when the data is not in the domain of definition of the operation).

Apart from their use of common notation for data, actions, and semantic functions, the semantic equations for the various constructs are completely independent. For instance, the formulation of the semantic equations for arithmetic expressions does not depend at all on whether expression evaluation might have side-effects, raise exceptions, never terminate, etc. The only crucial feature of expression evaluation is that if it terminates normally, it returns a single data item.

Thanks to the independence provided by the use of AN, the semantic equations for different constructs never need reformulation when the constructs are combined in the same language. Thus the semantic equation for a particular concrete construct can be the same in different languages, which promises a high degree of reusability.

However, recall the hindrances to explicit reuse mentioned in Sect. 2.3: the dependence on concrete syntax, the relatively deep module hierarchy, and the bundling of constructs together in single modules. The next section shows how these hindrances can be removed in OBJ. In conjunction with the use of action semantics as described above, this leads to extreme reusabiity of parts of semantic descriptions in OBJ.

4 Constructive Semantics in OBJ

As outlined in the introduction, constructive semantics involves two main departures from the conventional style of semantic description:

– concrete language constructs are mapped to language-independent abstract constructs, and
– the semantics of each abstract construct is specified in a separate module.

Together, the above features allow the development of a repository containing semantic descriptions of individual abstract constructs, as well as the efficient reuse of these descriptions in connection with the semantics of concrete languages.

Below, we shall illustrate the ideas of constructive semantics by showing excerpts from a constructive action semantics of Simple, written in OBJ. See [13] for further details of the approach, [6] for a constructive action semantics of Core ML, and [15] for alternative tool support for constructive action semantics based on the ASF+SDF Meta-Environment [7].

4.1   Mapping Concrete Languages to Abstract Constructs

The concrete constructs of Simple can be found in some form or other in most high-level general-purpose languages. To avoid bias toward particular families of programming language, we eschew the use of mixfix notation and concrete symbols (such as reserved words or mathematical signs) when declaring abstract constructs: the operation symbols are generally abbreviated words[2] and ordinary prefix notation is used when writing applications. Here are some examples:

```
op assign : Var Exp -> Cmd .
op cond : Exp Cmd Cmd -> Cmd .
```

The abstract constructs are classified as variables (Var), expressions (Exp), commands (Cmd), etc., according to what kind of values they might compute.

A mapping from concrete Simple programs to abstract (language-independent) constructs is specified inductively in OBJ as illustrated in the excerpt below:

```
op [[ _ ]] : Exp.LANG/SIMPLE/EXP    -> Exp.CONS/EXP/SYN .
op [[ _ ]] : Tst.LANG/SIMPLE/TST    -> Exp.CONS/EXP/SYN .
op [[ _ ]] : BPgm.LANG/SIMPLE/BPGM  -> Cmd.CONS/CMD/SYN .
op [[ _ ]] : Pgm.LANG/SIMPLE/PGM    -> Cmd.CONS/CMD/SYN .
...
*** variables X :
eq [[X]]    = X .
...
*** expressions E:
eq [[I]]    = I .
eq [[E1 + E2]]  = app(plus, [[E1]], [[E2]]) .
...
*** tests T:
eq [[B]]    = B .
eq [[E1 < E2]]  = app(lt, [[E1]], [[E2]]) .
...
*** basic programs:
eq [[X := E]]   = assign(X, [[E]]) .
...
*** programs P:
eq [[skip]] = skip .
eq [[P1 ; P2]]  = seq([[P1]], [[P2]] ) .
eq [[if T then P1 else P2 fi]] = cond([[T]], [[P1]], [[P2]]) .
eq [[while T do P od]]   = cond-loop([[T]], [[P]]) .
```

Notice that if Simple were to be extended with an if-then structured program, it could be mapped to the obvious combination of previously introduced abstract constructs, thus avoiding the introduction of a further abstract construct:

```
eq [[if T then P fi]] = cond([[T]], [[P]], skip) .
```

---

[2] Unabbreviated words can be too long for use in lectures and exercise classes.

Readers who are already familiar with the notation and intended interpretation of the abstract constructs concerned may find that the specification of the mapping from concrete to abstract constructs is sufficient explanation of the semantics of the concrete constructs. Other readers should consult the action semantic descriptions of the abstract constructs involved in the mapping, e.g.:

```
eq evaluate app(O, E1, E2) =
    (evaluate E1 and evaluate E2) then give O .
```

4.2   Modular Structure of Constructive Action Semantics

The declaration of the action semantic function for each sort of abstract construct is a separate module, e.g.:

```
obj CONS/CMD/ACT is
 pr CONS/CMD/SYN AN .
  op execute _ : Cmd -> Action .
endo
```

Hierarchical module names such as CONS/CMD/ACT facilitate navigation among large collections of modules, and avoid accidental clashes between names. The imported module CONS/CMD/SYN merely declares the sort Cmd, and is therefore available for use in connection with alternative styles of constructive semantics. The module AN, in contrast, declares the full Action Notation, rather than just the sort Action. (Modularization of the OBJ specification of AN would be possible, but it is irrelevant to the main issues addressed here.)

The action semantic description of each individual abstract construct is also a separate module, e.g.:

```
obj CONS/CMD/ASSIGN/ACT is
 pr CONS/CMD/ASSIGN/SYN .
 us CONS/CMD/ACT .
 pr CONS/VAR/ACT CONS/EXP/ACT .
  var V : Var. var E : Exp .
  eq execute assign(V, E) =
        (locate V and evaluate E) then update .
endo
```

It needs to import (i.e., depends on) only the modules that declare the action semantic functions for the sorts mentioned in the signature of the constructor, as well as any data types directly involved in the semantics of the abstract construct. It *never* imports other modules concerned with individual abstract constructs. This discipline ensures a very flat modular structure, with no bundling of abstract constructs together. Notice that the declaration of each variable used in the semantic equation may have to be repeated in many different modules; giving these declarations in the modules that introduce the sorts of abstract constructs would allow their "importation" using the vars-of feature of OBJ, and help to maintain uniformity of names for such variables, but on balance it seems preferable to exhibit the sorts of variables locally.

Thanks to the systematic naming of modules, most of the modules that need to be imported for the action semantics of an individual construct are determined by the signature of the construct itself. It might be advantageous to generate the OBJ modules from more concise specifications where using a sort automatically imports *the* module that declares it (e.g., in OBJ files in the current directory, or in a specified search path), and similarly for semantic functions. Such "auto-loading" is familiar from Lisp, and has already been employed to considerable advantage in ASDF, the Action Semantic Description Formalism developed for use in connection with the Action Environment [5,6,15].

Finally, the specification of the mapping from a particular concrete language imports the concrete syntax of the language and all the modules declaring the abstract constructs used in the target of the mapping. The complete action semantics of the concrete language imports moreover the modules that specify the action semantics of the abstract constructs. The action semantics of a concrete program is obtained by mapping it to an abstract program and applying the appropriate semantic function, and the resulting action can then be performed. See http://www.cs.swan.ac.uk/~cspdm/Goguen-FS/ for the full details.

5   Conclusion

We have shown how constructive action semantics can be specified in OBJ, and given excerpts from such a description of Goguen and Malcolm's Simple illustrative language. Compared to the algebraic semantics of the same language given by Goguen and Malcolm, it would appear considerably easier to reuse entire modules of our specification when describing extensions or different languages (although full-scale case studies supporting this claim have yet to be carried out). However, we had to introduce a separate module for each construct, and all the explicit imports are somewhat tedious (both to write and to read).

Acknowledgement. This paper was written on the occasion of Joseph Goguen's 65th birthday. I would like to conclude it with some comments in acknowledge-ment of the influence that Joseph's work has had on my own research. I first met Joseph at MFCS in Gdańsk, Poland, in 1976. The invited paper that Joseph presented there on "Some Fundamentals of Order-Algebraic Semantics" [16] was particularly interesting to me in connection with my interest in denotational se-mantics – but what sparked my lasting interest in his work most of all was his impromptu evening session on the initial algebra approach to the specification of abstract data types and its relationship to the observability-based approach of Montanari and his colleagues [2]. Following Joseph's subsequent work, I became an enthusiastic user of order-sorted algebra. In 1985, I spent an intensive month with Joseph's group at SRI International, trying to specify some (half-baked and overambitious) ideas regarding action notation in a somewhat shaky new version of OBJ (about halfway between OBJ2 and OBJ3, I think). Despite not having used OBJ much at all since then, I was pleasantly surprised to find how easy it was to resume that project in connection with the preparation of this paper, and how robust the portable BOBJ implementation of OBJ has become (thanks to Kai Lin for promptly fixing the single bug that my specifications revealed). I hope that the approach presented in this paper will stimulate further interest in using OBJ for specifying semantics of programming languages.

The author is grateful to the anonymous referees for helpful suggestions.

References

1. K.-G. Doh and P. D. Mosses. Composing programming languages by combining action-semantics modules. *Sci. Comput. Programming*, 47(1):3–36, 2003.
2. V. Giarratana, F. Gimona, and U. Montanari. Observability concepts in abstract data type specification. In *MFCS'76, Proc. 5th. Symp. on Mathematical Foundations of Computer Science*, volume 45 of *LNCS*, pages 576–587. Springer, 1976.
3. J. A. Goguen and G. Malcolm. *Algebraic Semantics of Imperative Programs*. The MIT Press, 1996.
4. J. A. Goguen, T. Winkler, J. Meseguer, K. Futatsugi, and J.-P. Jouannaud. Introducing OBJ. In *Software Engineering with OBJ: Algebraic Specification in Action*. Kluwer, 2000.
5. J. Iversen. *Formalisms and Tools Supporting Constructive Action Semantics*. PhD thesis, University of Aarhus, 2005.
6. J. Iversen and P. D. Mosses. Constructive action semantics for Core ML. *IEE Proceedings-Software*, 152:79–98, 2005. Special issue on Language Definitions and Tool Generation.
7. The ASF+SDF Meta-Environment. http://www.cwi.nl/projects/MetaEnv/.
8. P. D. Mosses. *Action Semantics*, volume 26 of *Cambridge Tracts in Theoretical Computer Science*. Cambridge University Press, 1992.
9. P. D. Mosses. Theory and practice of Action Semantics. In *MFCS '96, Proc. 21st Int. Symp. on Mathematical Foundations of Computer Science, Cracow, Poland*, volume 1113 of *LNCS*, pages 37–61. Springer, 1996.
10. P. D. Mosses. Fundamental concepts and formal semantics of programming languages. Lecture Notes. Version 0.4, available from the author, 2004.
11. P. D. Mosses. Modular language descriptions. In *Generative Programming and Component Engineering: Third International Conference, GPCE 2004, Vancouver, Canada, Proceedings*, volume 3286 of *LNCS*, pages 489–490. Springer, 2004.
12. P. D. Mosses. Modular structural operational semantics. *J. Logic and Algebraic Programming*, 60–61:195–228, 2004. Special issue on SOS.
13. P. D. Mosses. A constructive approach to language definition. *Journal of Universal Computer Science*, 11(7):1117–1134, July 2005.
14. P. D. Mosses and D. A. Watt. The use of action semantics. In *Formal Description of Programming Concepts III, Proc. IFIP TC2 Working Conference, Gl. Avernæs, 1986*, pages 135–166. North-Holland, 1987.
15. M. G. J. van den Brand, J. Iversen, and P. D. Mosses. An action environment. In *Proceedings of the Fourth Workshop on Language Descriptions, Tools, and Applications (LDTA 2004)*, volume 110 of *ENTCS*, pages 149–168. Elsevier, 2004.
16. E. G. Wagner, J. B. Wright, J. A. Goguen, and J. W. Thatcher. Some fundamentals of order-algebraic semantics. In *MFCS'76, Proc. 5th. Symp. on Mathematical Foundations of Computer Science*, volume 45 of *LNCS*, pages 153–168. Springer, 1976.

Horizontal Composability Revisited*

Donald Sannella[1] and Andrzej Tarlecki[2,3]

[1] Laboratory for Foundations of Computer Science, University of Edinburgh
[2] Institute of Informatics, Warsaw University
[3] Institute of Computer Science, Polish Academy of Sciences

Abstract. We recall the contribution of Goguen and Burstall's 1980 CAT paper and its powerful influence on theories of specification implementation that were emerging at about the same time, via the introduction of the notions of *vertical* and *horizontal composition* of implementations. We then give a different view of implementation which we believe provides a more adequate reflection of the rather subtle interplay between implementation, specification structure and program structure.

1 Introduction

Goguen and Burstall's CAT paper [GB80] is surely the most influential paper in the algebraic specification literature never to be properly published in a workshop or conference proceedings or in a journal. The topic of the paper was the notion of specification implementation—also known as refinement—as a relation on specifications, used for the step-by-step development of a program from a specification of requirements. We write $SP \rightsquigarrow SP'$ to denote that SP' is an implementation of SP, with the informal meaning that SP' captures all the requirements expressed by SP but in a way that incorporates more design decisions and is thus closer to being a program. A hot question at the time was how to properly formalise this intuition. Earlier work that was relevant to this question was Hoare's work on data refinement [Hoa72] which had been incorporated into VDM [Jon80], and Milner's work on simulations [Mil71]; first approaches in the algebraic specification literature were [GTW78] and (early versions of) [Ehr82] and [EKMP82].

The main contribution of [GB80] was to sketch a compelling two-dimensional view of implementations, with implementations composing both vertically and horizontally. Composition along the vertical dimension corresponds to composition of consecutive implementations: if $SP \rightsquigarrow SP'$ and also $SP' \rightsquigarrow SP''$, then one would expect to have $SP \rightsquigarrow SP''$. This justifies the correctness of the principle of *stepwise refinement*. (This was called *vertical* composition because Goguen and Burstall drew their implementations vertically, with SP at the top; we draw them horizontally here, except in a few diagrams, to save

---

* This work was funded in part by the European IST FET programme under the IST-2005-015905 MOBIUS and IST-2005-016004 SENSORIA projects, and by the British–Polish Research Partnership Programme.

space.) Horizontal composition is about composing implementations of parts of a specification to give an implementation of the whole: if $SP_1 \rightsquigarrow SP_1'$ and $SP_2 \rightsquigarrow SP_2'$, then one would expect to have $SP_1 \oplus SP_2 \rightsquigarrow SP_1' \oplus SP_2'$ for any specification-building operation $\oplus$. In particular, this should hold for composition of parameterised specifications: if $P_1 \rightsquigarrow P_1'$ and $P_2 \rightsquigarrow P_2'$ then one would expect to have $P_1;P_2 \rightsquigarrow P_1';P_2'$. Finally, it was suggested that vertical and horizontal composition should satisfy the *double law*, which says that given a diagram of implementations admitting both vertical and horizontal composition of implementations, the result is the same whether vertical composition is done before or after horizontal composition.

In Section 2 we recall this work. A vertical composition theorem was the main result in many accounts of implementations that were emerging at about the same time, sometimes under more or less restrictive conditions on the specifications or implementations in question. Horizontal composition proved more elusive; in most cases it remained a topic for the "Future Research" section. Recent approaches go further. For instance, [GT00] (cf. [Gog96]) provides some algebraic laws that link vertical and horizontal structure, but with what seems to be a somewhat different understanding of the vertical dimension. Another example is [LF97] where horizontal composability is achieved for colimits of specification diagrams in the context of specifications for reactive systems. Still, to our knowledge, no theory of implementations ever entirely fulfilled the dream of CAT.

In Section 4 we give a different view of implementations which we believe properly reflects the subtle interplay between implementations, specification structure and program structure, and observe that it trivially satisfies a vertical composition theorem. In Section 5 we consider horizontal composition, and conclude that it does not hold in general but neither is it desirable. The problem with horizontal composition arises from the lack of correspondence in general between the structure of a specification and the structure of a program that implements it, and the difference between operations for combining specifications on one hand and operations for combining program components on the other.

2   CAT

The CAT paper [GB80] outlines a vision for a future interactive programming system to be used for the development and maintenance of programs from specifications, in which program components were to be equipped with specifications of their properties. The processes by which implementations are carried out were to be fully modularised and parameterised, and all concepts in CAT were to have a full semantic definition in order to support formal proofs of correctness. Complete system designs were to be obtained by composing a number of implementations, each one expressing an elementary design decision. Such a degree of formalization and modularization would be useful not just for achieving correctness but also for restricting the scope of re-checking required when the system is modified subsequently. Scherlis and Scott's Inferential Programming paper [SS83], which led to the Ergo project at CMU [LPRS88], contains some more detailed ideas along similar lines.

The important part of [GB80] is only a few pages long, sandwiched between a quick review of the features of the then brand-new CLEAR specification language [BG80] and a long OBJ definition that is only marginally relevant (see [GT79] for a presentation of OBJ as it was at the time). The key insight is the recognition of a distinction between so-called *vertical* and *horizontal* structure:

> "One basic intuition behind CAT is that the *process* of implementing a large program from its specification has a two-dimensional structure. One dimension of structure, the *horizontal*, corresponds to the structure of the specification. The second dimension, the *vertical*, corresponds to the sequence of successive refinements of the specification into actual code; the specification is at the *top*, and the code is at the *bottom*. ... A major purpose of the CAT project is to render this intuition much more precise."   J.A. Goguen and R.M. Burstall [GB80]

In elaborating this point, Goguen and Burstall make reference mainly to the structure of specifications arising from parameterised specifications, known as *theory procedures* in CLEAR, which provide a specification of requirements that any actual parameter needs to satisfy as well as a specification of the result. Implementation of one such procedure P by another one P' having the same "metasource" and "metatarget" specifications SP and SP' respectively (where any actual argument specification must extend SP and then the result will extend SP' in a corresponding way) would be represented by the following diagram:

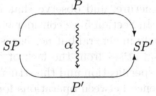

where α gives the relationship between P and P'.

Nowadays the authors would presumably agree with us (see e.g. [Gog96]) that the proper entities here are *specifications of parameterised programs*, see [SST92], that is, descriptions of functions mapping algebras to algebras, rather than CLEAR theory procedures which map specifications (descriptions of classes of algebras) to specifications. See Section 3.

Such implementations should compose both vertically and horizontally. Horizontal composition of implementations refers to composition of implementations of parts of a specification to give an implementation of the whole. Given the following diagram:

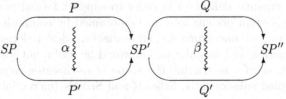

horizontal composition would give

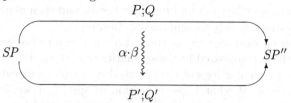

where "·" denotes horizontal composition of implementations and ";" stands for composition of specifications of parameterised programs. The same idea applies to other specification-building operations: given $\alpha : SP_1 \rightsquigarrow SP'_1$ and $\alpha' : SP_2 \rightsquigarrow SP'_2$, one would expect to have $\alpha \odot \alpha' : SP_1 \oplus SP_2 \rightsquigarrow SP'_1 \oplus SP'_2$ for any specification-building operation $\oplus$. This depends on having an operation $\odot$ for combining implementations that corresponds to each operation $\oplus$ for combining specifications. But according to [GB80]:

> "Questions remain about how the CLEAR operations can be extended from specifications to implementations."

Vertical composition of implementations corresponds to stepwise refinement:

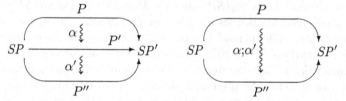

The composed implementation $\alpha;\alpha'$ combines the design decisions in α with those in α': for instance, if α shows how to implement graphs using sets, and α' shows how to implement sets using lists, then $\alpha;\alpha'$ shows how to implement graphs using lists.

Now, suppose we have a structured specification with consecutive implementations of its components, like so:

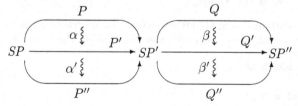

In this situation we may apply vertical composition to give implementations $\alpha;\alpha'$ and $\beta;\beta'$, and then apply horizontal composition to give an implementation $(\alpha;\alpha')\cdot(\beta;\beta') : P;Q \rightsquigarrow P'';Q''$. Alternatively, we may first apply horizontal composition to give implementations $\alpha\cdot\beta$ and $\alpha'\cdot\beta'$, and then apply vertical composition to give an implementation $(\alpha\cdot\beta);(\alpha'\cdot\beta') : P;Q \rightsquigarrow P'';Q''$. Goguen and Burstall conjecture that these two implementations should be *the same*: the order of composition should not matter. If this "double law" holds then implementations form a *two-dimensional category*, see [Mac71] (where the double law is called the "interchange law"). They speculate that the double law may not hold for some specification-building operations, and then extra care must be taken at such points during the implementation process.

All of this discussion is set in the context of an arbitrary institution [GB92]— a concept which first appeared in the semantics of CLEAR [BG80]—abstracting away from the particular logical system used to write specifications. There is no formal definition of what implementation of specifications means. Goguen and Burstall also suggest that the CAT framework would be appropriate for use with various different programming languages and programming paradigms. Although functional languages are the most obvious fit, they speculate that the use of imperative languages and assembly languages should not pose any insurmountable obstacles.

3   Specifications and Programs

The precise syntax of specifications is not very important in this paper. More significant is the way that the semantics of specifications is defined: for each specification SP, we define its signature $Sig(SP)$ and its class of models, $Mod(SP)$, where each SP-model is a $Sig(SP)$-algebra: $Mod(SP) \subseteq Alg(Sig(SP))$. The signature of a specification defines an interface giving names to the required program components, while its models represent programs that are considered to be its correct realizations. If $Sig(SP) = \Sigma$ we will say that SP is a Σ-specification.

The framework we are describing is independent of any particular institution [GB92]. It can therefore be used with different programming paradigms by selecting a notion of model that reflects the features of the paradigm in question. However, for the sake of concreteness and simplicity let us concentrate on standard many-sorted algebras over standard algebraic signatures, specified using axioms in first-order logic with equality. These capture a subset of Standard ML programs (so-called *structures*) over Standard ML signatures [MTHM97], comprising first-order non-polymorphic datatypes and first-order non-polymorphic properly-terminating functions.

Example 3.1. The following signature defines an interface for a program to sort lists of elements with respect to an order relation on the type of elements:

```
signature SORTELEM =
  sig
    type elem
    val ord : elem * elem -> bool
    type listelem
    val nil : listelem
    val cons : elem * listelem -> listelem
    val sort : listelem -> listelem
  end
```

A structure over this signature provides code for the required components, including such a sorting program:

```
structure SortElem : SORTELEM =
  struct
    type elem = int
    fun ord(x,y) = x >= y
    datatype listelem = nil | cons of elem * listelem
    fun sort l = ... (* code for sort *) ...
  end
```

The semantics of Standard ML [MTHM97] can be used to interpret the above code as a definition of an algebraic signature, call it [[SORTELEM]], and a particular algebra over this signature [[SortElem]] ∈ Alg([[SORTELEM]]).

Example 3.2. The following specification has the above program as a correct realization:

```
specification SORTELEMSPEC =
  spec
    type elem
    val ord : elem * elem -> bool
    axiom ... (* ord is transitive, reflexive and antisymmetric *) ...

    datatype listelem = nil | cons of elem * listelem
    val sort : listelem -> listelem
    axiom ... (* sort produces a permutation of its input *) ...
    axiom ... (* the output of sort is ordered according to ord *) ...
  end
```

Then Sig(SORTELEMSPEC) = [[SORTELEM]] and [[SortElem]] ∈ Mod(SORTELEMSPEC) ⊆ Alg([[SORTELEM]]).

For the sake of example, one often considers the following rudimentary ways of building specifications:

basic specifications: For any signature Σ and set Φ of Σ-sentences, the *basic specification* $\langle \Sigma, \Phi \rangle$ is a Σ-specification with $Mod(\langle \Sigma, \Phi \rangle) = \{M \in Alg(\Sigma) \mid M \models \Phi\}$. (SORTELEMSPEC above is a basic specification.)

union: For any Σ, given Σ-specifications SP_1 and SP_2, their *union* $SP_1 \cup SP_2$ is a Σ-specification with $Mod(SP_1 \cup SP_2) = Mod(SP_1) \cap Mod(SP_2)$.

translation: For any signature morphism $\sigma: \Sigma \to \Sigma'$ and Σ-specification SP, **translate** SP **by** σ is a Σ'-specification with $Mod(\textbf{translate } SP \textbf{ by } \sigma) = \{M' \in Alg(\Sigma') \mid M'|_\sigma \in Mod(SP)\}$.[1]

hiding: For any $\sigma: \Sigma \to \Sigma'$ and Σ'-specification SP', **derive from** SP' **by** σ is a Σ-specification with $Mod(\textbf{derive from } SP' \textbf{ by } \sigma) = \{M'|_\sigma \mid M' \in Mod(SP')\}$.[1]

---

[1] For any signature morphism $\sigma: \Sigma \to \Sigma'$ and algebra $M' \in Alg(\Sigma')$, $M'|_\sigma \in Alg(\Sigma)$ is the *reduct* of M' with respect to σ, see e.g. [ST99]. When σ is a signature inclusion, $M'|_\sigma$ may be written as $M'|_\Sigma$.

This follows ASL [SW83, ST88a] and is different from CLEAR, where specification expressions denoted *theories* which in turn have model classes, see [ST97] for a discussion of the difference. The operations are more primitive but are similarly expressive: for instance "+" in CLEAR corresponds to union of suitably translated specifications over different signatures, where the translations respect shared subspecifications.

This defines a number of so-called *specification-building operations* which map specifications to more complex specifications: we have constant specification-building operations (basic specifications), one binary specification-building operation (union) and two unary ones (translation and hiding). In fact, each of these may be viewed as a family of operations, indexed by signatures (union) and specification morphisms (translation and hiding). Once this "static" indexing is fixed, each specification-building operation semantically amounts to a function on appropriate classes of models.

One property of the above specification-building operations will prove crucial for further considerations: an n-ary specification-building operation op is *monotone* if it is monotone as a function on model classes. That is: for any specifications SP_1, SP_1', ..., SP_n, SP_n', such that $Sig(SP_i) = Sig(SP_i')$ and $Mod(SP_i) \subseteq Mod(SP_i')$ for $i = 1, \ldots, n$, we also have $Mod(op(SP_1, \ldots, SP_n)) \subseteq Mod(op(SP_1', \ldots, SP_n'))$.

All the above specification-building operations, and therefore any operation that may be defined using them, are monotone. In fact, nearly all specification-building operations one may find in the literature are monotone. The only exception we are aware of are operations that select initial or free models of specifications—one may argue though that such an operation should be viewed as simply imposing an additional *constraint* on the class of models of a specification, like an axiom, rather than as specification-building operations in their own right (see for instance *data constraints* in [GB92]).

Structured specifications in CASL [BM04, CoF04] are based on the operations above as well; somewhat more convenient notation is introduced there, which we will use in examples too. For instance, union (not limited to specifications with identical signatures) is written with **and**, translation along surjective signature morphisms is written with **with** (followed by the mapping of symbols), hiding is written with **reveal** or **hide** (followed by a list of symbols). Perhaps most useful is **then**, which is an obvious combination of a translation along a signature inclusion with union to build an *extension* of a specification by new sorts, operations and/or axioms.

Example 3.3. Here are some examples of structured specifications:

```
specification ELEMSPEC =
  spec
    type elem
    val ord : elem * elem -> bool
    axiom ... (* ord is transitive, reflexive and antisymmetric *) ...
  end
```

```
specification ELEMLISTSPEC =
  ELEMSPEC then
    datatype listelem = nil | cons of elem * listelem
  end
specification PERMELEMSPEC =
  ELEMLISTSPEC then
    val perm : listelem -> listelem
    axiom ... (* perm produces a permutation of its input *) ...
  end
specification ORDERELEMSPEC =
  ELEMLISTSPEC then
    val order : listelem -> listelem
    axiom ... (* the output of order is ordered w.r.t. ord *) ...
  end
specification STRUCTSORTELEMSPEC =
  {PERMELEMSPEC with perm |-> sort}
  and
  {ORDERELEMSPEC with order |-> sort}
```

Specifications SORTELEMSPEC of Example 3.2 and STRUCTSORTELEMSPEC above are equivalent: they have the same signature ([[SORTELEM]] in both cases, see Example 3.1) and the same class of models.

In common with all work on algebraic specification we have taken the view that algebras model programs. But in general we are interested in program *components* which define new sorts and operations in terms of some existing ones. These may be *generic* components, where the parameters are supplied explicitly, or components that explicitly import or implicitly build on other components. In each case, we need to model components as functions mapping algebras to algebras; in the case of explicit or implicit imports this reflects the way that the newly-defined sorts and operations depend on the imports.

Definition 3.4. *Let Σ and Σ' be signatures. A $(\Sigma \rightarrow \Sigma')$-constructor is a function[2] mapping Σ-algebras to Σ'-algebras.*

In the standard algebraic institution, constructors correspond most directly to Standard ML *functors* defining first-order non-polymorphic datatypes and first-order non-polymorphic properly-terminating functions, where the input and output signatures are explicit.

Example 3.5. Here is an example of a constructor in Standard ML:

```
signature ELEM =
sig
  type elem
```

---

[2] In general, we need to consider *partial* constructors, where the result may not be defined for every algebra over the parameter signature but only for those that satisfy additional constraints. See [ST89]. For simplicity, we restrict attention to total constructors here, with a few comments in footnotes concerning partial constructors.

```
    val ord : elem * elem -> bool
end

functor Sort(X: ELEM) : SORTELEM =
  struct
    open X
    datatype listelem = nil | cons of elem * listelem
    fun sort l = ... (* code for sort *) ...
  end
```

The semantics of Standard ML can be used to interpret the above code as defining a function mapping [[ELEM]]-algebras to [[SORTELEM]]-algebras, i.e. an ([[ELEM]] → [[SORTELEM]])-constructor. One important property of this function is that it is *persistent*: the argument structure is extended to the result structure, preserving the interpretation of parameter types and values.

Any $(\Sigma \to \Sigma')$-constructor κ determines a specification-building operation, written κ as well, that takes any Σ-specification SP to a Σ'-specification having the image of $Mod(SP)$ under κ as its models: $Mod(\kappa(SP)) = \{\kappa(M) \mid M \in Mod(SP)\}$. Hiding is one such specification-building operation, determined by reduct. The other specification-building operations discussed above do not arise in such a way, in general. Translation is determined by a total constructor only when it is with respect to a bijective renaming[3], and then it coincides with hiding with respect to the inverse of that renaming. CASL union is not determined by a total constructor unless there is no overlap ("sharing") between the signatures of the arguments.[4]

Constructors may themselves be specified. For the same reason as ordinary specifications describe classes of algebras, constructor specifications describe classes of constructors, that is, classes of functions mapping algebras to algebras [SST92].

Definition 3.6. *Given specifications SP and SP', the constructor specification $SP \to SP'$ specifies the class of $(Sig(SP) \to Sig(SP'))$-constructors that map models of SP to models of SP': $Mod(SP \to SP') = \{F: Alg(Sig(SP)) \to Alg(Sig(SP')) \mid$ for each $A \in Mod(SP), F(A) \in Mod(SP')\}$.*[5]

Moreover, when $Sig(SP)$ overlaps with $Sig(SP')$ then the specified constructors should preserve the interpretation of the overlapping sorts and operations. In particular, when $Sig(SP)$ is a subsignature of $Sig(SP')$, then as in CASL we require the functions in $Mod(SP \to SP')$ to be persistent: when $F: Alg(Sig(SP)) \to Alg(Sig(SP')) \in Mod(SP \to SP')$ then for every model $A \in Mod(SP), F(A) \in Mod(SP')$ is such that $F(A)|_{Sig(SP)} = A$.

---

[3] Translations along surjective signature morphisms are determined by partial constructors, in general.

[4] When there is overlap, CASL union is determined by a partial constructor which amalgamates models that coincide on the shared subsignature.

[5] If partial constructors are considered, an additional requirement here would be that their domain contains $Mod(SP)$.

Example 3.7. Recall Examples 3.1–3.3. Then ELEMSPEC → SORTELEMSPEC is a specification of (persistent) constructors $F\colon Alg(\llbracket\text{ELEM}\rrbracket) \to Alg(\llbracket\text{SORTELEM}\rrbracket)$ that when given a model $E \in Mod(\text{ELEMSPEC})$ extends it to a model $F(E) \in Mod(\text{SORTELEMSPEC})$. One example of such a constructor is the functor Sort $\in Mod(\text{ELEMSPEC} \to \text{SORTELEMSPEC})$, presented in Example 3.5. Constructor specifications correspond to functor specifications in Extended ML, see [KST97].

The generalisation to n-ary constructors and constructor specifications is straightforward.

4   Implementations and Vertical Composition

A very simple notion of specification implementation is the following:

Definition 4.1. *Let SP and SP' be specifications such that $Sig(SP) = Sig(SP')$. Then SP' is a* simple implementation *of SP, written $SP \rightsquigarrow SP'$, if $Mod(SP) \supseteq Mod(SP')$.*

This simply requires that all of the correct realizations of SP' are correct realizations of SP. That is, SP' incorporates all the requirements that are in SP, and perhaps other constraints that result from additional design decisions.

For simplicity, the definition of simple implementation requires the signatures of both specifications to be the same. The hiding operation may be used to adjust the signatures (for example, by removing auxiliary functions from the signature of the implementing specification) if this is not the case.

The fact that simple implementations vertically compose is an immediate consequence of the transitivity of the subset relation:

Proposition 4.2. *If $SP \rightsquigarrow SP'$ and $SP' \rightsquigarrow SP''$ then $SP \rightsquigarrow SP''$.*

The notion of simple implementation is powerful enough (in the context of a sufficiently rich specification language) to handle all concrete examples of interest. However, it is not very convenient. During the process of developing a program, the successive specifications incorporate more and more details arising from successive design decisions. Thereby, some parts become fully determined, and remain unchanged as a part of the specification until the final program is obtained. The following diagram is a visual representation of this situation, where $\kappa_1, \ldots, \kappa_n$ label the parts that become determined at consecutive steps.

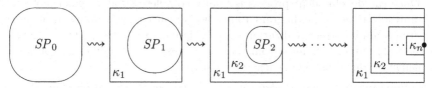

It is more convenient to avoid such clutter by separating the finished parts from the specification, putting them aside, and proceeding with the development of the unresolved parts only:

$$SP_0 \xrightarrow[\kappa_1]{\rightsquigarrow} SP_1 \xrightarrow[\kappa_2]{\rightsquigarrow} SP_2 \xrightarrow[\kappa_3]{\rightsquigarrow} \cdots \xrightarrow[\kappa_n]{\rightsquigarrow} \bullet SP_n = \text{EMPTY}$$

where EMPTY is a specification for which a standard implementation *empty* is available.

It is important for the finished parts $\kappa_1, \ldots, \kappa_n$ to be independent of the particular choice of realization for what is left: they should extend *any* realization of the unresolved part to a realization of what is being implemented. This is exactly what is required by the notion of a *constructor* defined in Sect. 3: κ_i is a *function* taking models of SP_i to models of SP_{i-1}. These considerations motivate a more elaborate version of the notion of implementation:

Definition 4.3 ([ST88b]). *Given specifications SP and SP' and constructor $\kappa : Alg(Sig(SP')) \to Alg(Sig(SP))$, we say that SP' is a constructor implementation of SP via κ, written $SP \xrightarrow[\kappa]{\rightsquigarrow} SP'$, if $\kappa \in Mod(SP' \to SP)$.*

Thus, in the development diagram above, $\kappa_i : Alg(Sig(SP_i)) \to Alg(Sig(SP_{i-1}))$ with $\kappa_i \in Mod(SP_i \to SP_{i-1})$ for $1 \leq i \leq n$; that is, each κ_i corresponds to a parameterised program with input interface SP_i and output interface SP_{i-1}. Given a model M of SP_i, κ_i may be applied to yield a model $\kappa_i(M)$ of SP_{i-1}.

Example 4.4. From Example 3.7, we have SORTELEMSPEC $\xrightarrow[\text{Sort}]{\rightsquigarrow}$ ELEMSPEC. That is, the task of implementing sorting of lists of elements with respect to a function ord is reduced by means of the constructor Sort to the task of implementing elem and ord.

The definition of constructor implementation generalises smoothly to implementations of constructor specifications. This requires *higher-order* constructors; for details see [ST97].

It is easy to see that constructor implementations compose vertically:

Proposition 4.5. *If $SP \xrightarrow[\kappa]{\rightsquigarrow} SP'$ and $SP' \xrightarrow[\kappa']{\rightsquigarrow} SP''$ then $SP \xrightarrow[\kappa';\kappa]{\rightsquigarrow} SP''$.*

So, a constructor implementation via $\kappa : Alg(Sig(SP')) \to Alg(Sig(SP))$ composed with a constructor implementation via $\kappa' : Alg(Sig(SP'')) \to Alg(Sig(SP'))$ yields a constructor implementation via $\kappa';\kappa : Alg(Sig(SP'')) \to Alg(Sig(SP))$, which is just the composition of the functions κ' and κ written in diagrammatical order.

Once the development process is finally complete (that is, when nothing is left unresolved, as in the diagram above) we can successively apply the constructors to obtain a correct realization of the original specification. The correctness of the final outcome follows from the correctness of the individual constructor implementation steps via vertical composition.

Proposition 4.6. *Given a chain of constructor implementation steps*

$$SP_0 \xrightarrow[\kappa_1]{\rightsquigarrow} SP_1 \xrightarrow[\kappa_2]{\rightsquigarrow} \cdots \xrightarrow[\kappa_n]{\rightsquigarrow} SP_n = \text{EMPTY}$$

we have $(\kappa_n; \cdots; \kappa_2; \kappa_1)(empty) \in Mod(SP_0)$.

Many approaches to implementation in the literature make use of a restrictive kind of constructor defined by a parameterised program having a particular rigid form: for example, the notion of implementation in [EKMP82] corresponds to the use of a constructor obtained by composing a free construction with a reduct, then a restriction to a subalgebra, and finally a quotient, in that order. Then the vertical composition of two implementations is required to yield an implementation of the same form, which is only possible under certain additional conditions on the specifications involved. This amounts to a requirement that the composition of parameterised programs be forced into some given normal form, which corresponds to requiring programs to be written in a rather restricted programming language.

5   Horizontal Composition

In Sect. 3 we have recalled a few basic specification-building operations, which form the backbone of many *specification languages*. Since the pioneering work on CLEAR [BG80], a number of such languages have been designed and used, with CASL [BM04, CoF04] as a prime recent example. They all aim at providing a convenient way to build specifications in a structured manner, where specification-building operations are used to gradually construct more and more complex specifications out of simpler component specifications. This *horizontal structure* of specifications (in the terminology of [GB80]) is indispensable for facilitating the understanding and use of any practical (hence: large and complex) specification. Typical ways in which the horizontal structure of specifications has been successfully exploited include the compositional semantics of complex specifications languages like CASL [BCH+04] and compositional proof systems for consequences of specifications, as introduced in [ST88a] and analyzed in [Bor02], even if for practical specification languages compositionality may sometimes be sacrified [MHAH04].

Under a mild assumption of monotonicity of the specification-building operations involved, the horizontal structure of specifications may also be exploited in the development process:

Proposition 5.1. *Suppose that op is a monotone n-ary specification-building operation. If $SP_1 \leadsto SP'_1$, ..., $SP_n \leadsto SP'_n$ then $op(SP_1, \ldots, SP_n) \leadsto op(SP'_1, \ldots, SP'_n)$.*

For simple implementations, Prop. 5.1 captures the essence of horizontal composition, as introduced in [GB80]. For constructor implementations this takes the following form:

Proposition 5.2. *Suppose that op is a monotone n-ary specification-building operation. If $SP_1 \underset{\kappa_1}{\leadsto} SP'_1$, ..., $SP_n \underset{\kappa_n}{\leadsto} SP'_n$ then $op(SP_1, \ldots, SP_n) \leadsto op(\kappa_1(SP'_1), \ldots, \kappa_n(SP'_n))$.*

Note that κ_1 in $\kappa_1(SP'_1)$ refers to the specification-building operation determined by the constructor κ_1—see Sect. 3—and similarly for the other constructors.

The strength and usefulness of Props. 5.1 and 5.2 are severely limited by two fundamental problems.

First, the consistency of specifications is not preserved under such refinement in general. In Prop 5.1, $op(SP_1, \ldots, SP_n)$ may be a perfectly implementable (consistent) specification, while $op(SP'_1, \ldots, SP'_n)$ is inconsistent, and hence cannot be implemented, even if implementation of each of the refined individual component specifications $SP'_1, \ldots, SP'_n$ is unproblematic.

Example 5.3. Consider the following trivial example:

```
specification EVEN =
    spec val a : int
        axiom exists k : int . a = 2 * k
    end
specification SMALL =
    spec val a : int
        axiom a > 0 andalso a < 10
    end
specification SMALL_EVEN = SMALL and EVEN
```

The last specification is formed as a union of two simpler specifications, and thus combines the requirements they impose. (Obviously, algebras in ⟦SMALL_EVEN⟧ have a $\in \{2, 4, 6, 8\}$.)

Since and is monotone, Prop. 5.1 allows one to refine SMALL_EVEN by refining its component specifications independently. Consider for instance:

```
specification VERY_EVEN =
    spec val a : int
        axiom exists k : int . a = 8 * k
    end
specification VERY_SMALL =
    spec val a : int
        axiom a > 0 andalso a < 5
    end
specification VERY_SMALL_VERY_EVEN = VERY_SMALL and VERY_EVEN
```

Clearly, we have then EVEN ⤳ VERY_EVEN and SMALL ⤳ VERY_SMALL, and so by Prop. 5.1,

$$\text{SMALL\_EVEN} \rightsquigarrow \text{VERY\_SMALL\_VERY\_EVEN}.$$

However, even though both VERY_SMALL and VERY_EVEN are consistent and separately can be easily implemented, the specification VERY_SMALL_VERY_EVEN is inconsistent, and so taking this implementation step cannot lead to a final realization of SMALL_EVEN.

The above problem with consistency of the refined specification may arise even with a unary specification-building operation op (for instance, consider translation along a non-injective signature morphism). However, it does not arise if the operation op is determined by a constructor.

The other problem with refinement based on horizontal composability is perhaps even more fundamental. Although the horizontal structure of a specification is crucial for its understanding and use, in general this structure may well be quite different from the modular structure of the final program that implements it. The aims of horizontal structure at the level of the original, high-level, abstract requirements specification are quite separate from the aims of modular structure in the final program. An interesting and convincing example is presented in [FJ90] in a somewhat different framework, but the case study and the general line of reasoning carry over here as well. The conclusion from this is that while horizontal composability (with respect to monotone specification-building operations) yields sound refinements and so may be used when appropriate, it cannot be the only way to implement structured specifications. We need separate means to explicitly mark design decisions that fix the final modular structure of the program under development, which requires the top-level specification-building operations to be determined by constructors. Once such a *design specification* [AG97] has been fixed, this top-level horizontal structure is to be preserved in programs resulting from the development process, and further development proceeds for each component specification separately. The final result is then obtained by applying the top-level constructors to the outcomes of these separate developments.

Consider for instance an n-ary constructor *op*. Abusing slightly the notation of *architectural specifications* [BST02] as provided by CASL [BM04, CoF04], a design specification that designates the top-level constructor *op* to be preserved and used at the top level of the modular structure of the final program may take the following form:

```
arch spec OP_DESIGN =
    units U_1 : SP_1
          ...
          U_n : SP_n
    result op(U_1,...,U_n)
```

This introduces names (U_1, ..., U_n) of *units* (or *modules*) to be further developed as realizations of their specifications (SP_1, ..., SP_n, respectively) and then put together using the constructor op to yield the overall realization of the system.[6] An architectural specification can be compared with ordinary specifications by defining its models to be all the possible result units that may be built in this way. Then one may consider refinements involving architectural specifications, like $SP \rightsquigarrow$ OP_DESIGN. This captures a design decision to implement the specification SP by a modular system, where the top-level modules U_1, ..., U_n, fulfilling specifications SP_1, ..., SP_n, respectively, are put together using the constructor op.

In particular, we always have: op(SP_1,..., SP_n) $\rightsquigarrow$ OP_DESIGN. Note that op refers here to the specification-building operation determined by the constructor op, see Sect. 3.

---

[6] If op is partial, it is necessary to ensure that no tuple of models which may potentially be given as an argument to op is outside its domain. See [BST02].

For unary constructors K, the constructor implementation $SP \underset{K}{\leadsto} SP'$ corresponds exactly to the refinement $SP \leadsto$ K_DESIGN, where

```
arch spec K_DESIGN = unit U : SP' result K(U)
```

An important twist in CASL architectural specifications is that the units used here may in fact be *generic* modules, that is, constructors with specifications taking the form discussed in Sect. 3. This allows one to delegate "coding" of constructors (as, say, Standard ML functors) to further development of the corresponding units, and to limit the vocabulary of the constructors in use in the result unit expression to a few basic constructs including the application of a generic unit to an argument.

Example 5.4. Recall the specifications in Examples 3.1–3.7. Note that the specification SORTELEMSPEC requires a sorting program sort for *some* realization for the type elem and ordering predicate ord chosen by the implementor. The following architectural specification decomposes this task by separating out on one hand the task to build such a realization for elem and ord, and on the other hand, the task of providing a sorting program sort that will work for *any* such realization. The overall result is then given by instantiating the outcome of the latter task to the outcome of the former one.

```
arch spec SORT_SPEC =
   units E : ELEMSPEC
         S : ELEMSPEC -> SORTELEMSPEC
   result S(E)
```

Then SORTELEMSPEC $\leadsto$ SORT_SPEC. We also have STRUCTSORTELEMSPEC $\leadsto$ SORT_SPEC even though the structure of SORT_SPEC does not match the structure of STRUCTSORTELEMSPEC.

The main point of architectural specifications as sketched above is that further developments of the specified units may proceed independently from each other, and the final results of these developments, which fulfill the unit specifications, may then be put together as prescribed by the result unit expression. Soundness of this procedure is guaranteed by the horizontal composability of implementations, Props. 5.1 and 5.2—however, with the additional effect that consistency of the result is ensured provided that each refined component specification remains consistent.

Note that horizontal composability follows from the following properties of implementation steps involving individual component specifications. Let op be a monotone n-ary specification-building operation.

– If $SP_1 \leadsto SP_1'$ then $op(SP_1, \ldots, SP_n) \leadsto op(SP_1', \ldots, SP_n)$.

  . . .

– If $SP_n \leadsto SP_n'$ then $op(SP_1, \ldots, SP_n) \leadsto op(SP_1, \ldots, SP_n')$.

Prop. 5.1 then follows by a simple application of vertical composability (Prop. 4.2).

Similarly, for constructor implementations we have:

- If $SP_1 \rightsquigarrow_{\kappa_1} SP'_1$ then $op(SP_1, \ldots, SP_n) \rightsquigarrow op(\kappa_1(SP'_1), \ldots, SP_n)$.
  ...
- If $SP_n \rightsquigarrow_{\kappa_n} SP'_n$ then $op(SP_1, \ldots, SP_n) \rightsquigarrow op(SP_1, \ldots, \kappa_n(SP'_n))$.

Prop. 5.2 now follows easily by Prop. 4.2.

The refinements of component specifications here are entirely independent from each other, and so may be taken in an arbitrary order. "Composition" of such independent refinements in any chosen order always yields the same result.

The key case here is when op is a constructor, and the specification considered is the architectural specification OP_DESIGN as above. In the notation of [MST04], refinements of individual unit specifications can be defined as follows:

```
refinement R_1 = U_1: SP_1 refined to arch spec
                              unit X_1 : SP'_1
                              result K_1(X_1)

...

refinement R_n = U_n: SP_n refined to arch spec
                              unit X_n : SP'_n
                              result K_n(X_n)
```

In [MST04], we have introduced the possibility of composing refinements, and indeed, according to the formal semantics given there, the above refinements can be composed in an arbitrary order, and each such composition yields the same result. For instance:

```
refinement R_1_to_n = R_1 then ... then R_n
refinement R_n_to_1 = R_n then ... then R_1
```

yields R_1_to_n = R_n_to_1. The fact that these refinements coincide in the case $n = 2$ captures the "double law" of [GB80], see Sect. 2.

In fact, [MST04] provides for the possibility of writing down the corresponding fragment of a development tree as follows:

```
arch spec DEVELOP =
    units U_1 : SP_1 refined to arch spec
                              unit X_1 : SP'_1
                              result K_1(X_1)

          ...

          U_n : SP_n refined to arch spec
                              unit X_n : SP'_n
                              result K_n(X_n)
    result op(U_1,...,U_n)
```

It should be clear (and this can be formally proved within the framework of [MST04]) that this is equivalent to the following architectural specification:

```
arch spec OP_DESIGN' =
    units X_1 : SP'_1
          ...
          X_n : SP'_n
    result op(K_1(X_1),...,K_n(X_n))
```

This explicitly captures the composition of the design decision to use op as the top-level constructor (captured by OP_DESIGN) with the constructor implementations for components in an arbitrary order. Note that this easily generalises to implementations of individual components that lead to further decomposition, again given by architectural specifications.

Example 5.5. Continuing Examples 5.4 and 3.1–3.7, consider the following additional specification:

```
specification INSERTELEMLISTSPEC =
  ELEMLISTSPEC then
    val insert : elem * listelem -> listelem
    axiom ... (* if l is ordered then insert(e,l) puts e into l
                      so that the result is ordered *) ...
```

Then the architectural specification SORT_SPEC may be refined as follows:

```
arch spec SORT_SPEC' =
  units E: ELEMSPEC
        S: ELEMSPEC -> SORTELEMSPEC
           refined to
             arch spec
               units L: ELEMSPEC -> ELEMLISTSPEC
                     I: ELEMLISTSPEC -> INSERTELEMLISTSPEC
                     IS: INSERTELEMLISTSPEC -> SORTELEMSPEC
               result lambda X: ELEMSPEC . IS(I(L(X)))
  result S(E)
```

We can also make the resulting overall design explicit as follows:

```
arch spec SORT_SPEC'' =
  units E: ELEMSPEC
        L: ELEMSPEC -> ELEMLISTSPEC
        I: ELEMLISTSPEC -> INSERTELEMLISTSPEC
        IS: INSERTELEMLISTSPEC -> SORTELEMSPEC
  result IS(I(L(E)))
```

Of course, we then have SORTELEMSPEC ⤳ SORT_SPEC''. Further development may involve for instance direct implementations of the generic units L, I and IS as Standard ML functors, entirely independent from each other.

The above example is misleadingly simple since there is no requirement for sharing between the units involved in the design. In general this need not be the case. Suppose that the task of implementing a specification SP_{big} is decomposed into the tasks of implementing specifications SP_1 and SP_2 where $[\![SP_1$ and $SP_2]\!] \subseteq [\![SP_{big}]\!]$ but the signatures of SP_1 and SP_2 overlap. If a realization of SP_{big} is to be obtained by combining realizations of SP_1 and SP_2, these two realizations need to share the realization of their common part. This is handled as in [Bur84]: we provide a specification SP of the common part and add its realization as a new task, and then use (persistent) generic units to separately extend the resulting unit to realizations of SP_1 and SP_2, thus ensuring that they share this common part and so can be put together.

Formalizing this: if $Sig(SP) \supseteq Sig(SP_1) \cap Sig(SP_2)$ and $[\![SP \text{ and } SP_1 \text{ and } SP_2]\!] \subseteq [\![SP_{big}]\!]$, then $SP_{big} \rightsquigarrow$ SHARING_SPEC where

```
arch spec SHARING_SPEC =
  units U: SP
       F1: SP->SP1
       F2: SP->SP2
  result F1(U) and F2(U)
```

Here, "and" is a partial binary constructor which amalgamates two models provided that they coincide on their common subsignature—see footnote 4 and note that the requirement mentioned in footnote 6 is satisfied. Note again that further refinements of the components may proceed independently from each other.

6 Conclusions

What emerged from [GB80] was a powerful and stimulating view of the process of systematic development of software from high-level formal specifications. What was insightful, new and perhaps ahead of its time then was the stress on *structure* as the only realistic means to master the size and complexity of practical software development projects.

The CAT paper identified formally two orthogonal aspects of structure in the process of software development: the vertical dimension, the structure of the development process as such; and the horizontal dimension, the structure of the specifications involved in development. Making this distinction was crucial to separating the two dimensions, for separate study, with vertical and horizontal composability as the key result to aim for. These separate lines of research resulted in a lot of interesting work, crucial for an adequate formalisation of the development process.

The vertical dimension proved easier for the theory: in spite of technical difficulties, in many frameworks the key vertical compositionality result has been established, with our composition of constructor implementations (further generalised to composition of *abstractor implementations*, not discussed here, see [ST88b, ST97]) covering the previous work as special cases—with the results recalled in Sect. 4.

The horizontal dimension attracted much work and research as well (including the pioneering work by Goguen and Burstall themselves on CLEAR [BG80]) with many specification languages designed that included various forms of horizontal structuring of specifications, and many key results on the use of this horizontal structure for proper understanding and use of large specifications. However, the interaction of the horizontal structure with development, formulated in [GB80] as horizontal composability, and the double law used to capture the interplay between the two dimensions, proved much tougher. In fact, there are hints in [GB80] which indicate that the authors viewed this idea as somewhat speculative, and foresaw potential obstacles in making it effective. We have already quoted their thought that the task to design implementation composition operations corresponding to all specification-building operations in CLEAR might be difficult. They also mention that the structure of a specification, with horizontal composition as the way to build its implementations, may constitute an "implementation bias", thus (perhaps unnecessarily) preventing implementations having a different structure. From our current perspective, it seems a bit unrealistic to claim that "this kind of bias seems to be actually desirable for large specifications, because it helps the implementer in his difficult task of structuring the overall program design." Indeed, this may well be the case sometimes, but it is certainly not always true.

As presented at length in Sect. 5, we are very far from the view that horizontal composability is unimportant. However, we believe that one should carefully distinguish and keep separate two conceptually different roles that the horizontal structure of a specification may play. One is the usual structuring of specifications, used to present the concepts of the problem space in a clear and perspicuous way. The horizontal structure obtained in this way is in principle irrelevant for vertical development, although it may be used when appropriate. The other role is the design of the modular structure of the system to be developed. This may be viewed as a very special kind of horizontal structure, which indeed is required to be preserved throughout development. Horizontal composability with respect to *this* structure is crucial, of course, and the double law is a natural and useful consequence. We proposed architectural specifications as a tool for capturing horizontal structure of this latter kind. We feel that the overall picture of vertical development and its interplay with this horizontal structure, as imposed by architectural specifications and sketched in Sects. 4 and 5, give a well-founded account of the ideas that were put forward in [GB80].

Acknowledgements: Hearty congratulations to Joseph on his 65th birthday and our thanks to him for the many novel ideas that over the years have stimulated much of our own work as well!

References

[AG97]   R. Allen and D. Garlan. A formal basis for architectural connection. *ACM Transactions on Software Engineering and Methodology*, 6(3):213–249, 1997.

[BCH⁺04] H. Baumeister, M. Cerioli, A. Haxthausen, T. Mossakowski, P.D. Mosses, D. Sannella, and A. Tarlecki. CASL semantics. [CoF04], part III, pages 115–273. D. Sannella and A. Tarlecki, editors.

[BG80]   R.M. Burstall and J.A. Goguen. The semantics of CLEAR, a specification language. In *Proceedings of the Abstract Software Specifications, 1979 Copenhagen Winter School*, Springer LNCS 86, pages 292–332, 1980.

[BM04]   M. Bidoit and P.D. Mosses. CASL *User Manual*. Springer LNCS 2900 (IFIP Series). 2004. With chapters by T. Mossakowski, D. Sannella, and A. Tarlecki.

[Bor02]   T. Borzyszkowski. Logical systems for structured specifications. *Theoretical Computer Science*, 286:197–245, 2002.

[BST02]   M. Bidoit, D. Sannella, and A. Tarlecki. Architectural specifications in
            CASL. *Formal Aspects of Computing*, 13:252–273, 2002.
[Bur84]   R.M. Burstall. Programming with modules as typed functional program-
            ming. In *Proc. Intl. Conference on Fifth Generation Computing Systems,
            Tokyo*, pages 103–112, 1984.
[CoF04]   CoFI (The Common Framework Initiative). CASL *Reference Manual*.
            Springer LNCS 2960 (IFIP Series). 2004.
[Ehr82]   H.-D. Ehrich. On the theory of specification, implementation and pa-
            rameterization of abstract data types. *Journal of the Association for
            Computing Machinery*, 29:206–227, 1982.
[EKMP82]   H. Ehrig, H.-J. Kreowski, B. Mahr, and P. Padawitz. Algebraic implemen-
            tation of abstract data types. *Theoretical Computer Science*, 20:209–263,
            1982.
[FJ90]   J. Fitzgerald and C.B. Jones. Modularizing the formal description of
            a database system. In *Proc. 3rd Intl. Symp. VDM Europe: VDM and
            Z, Formal Methods in Software Development*, Springer LNCS 428, pages
            189–210, 1990.
[GB80]   J.A. Goguen and R.M. Burstall. CAT, a system for the structured elabora-
            tion of correct programs from structured specifications. Technical Report
            CSL-118, Computer Science Laboratory, SRI International, 1980.
[GB92]   J.A. Goguen and R.M. Burstall. Institutions: Abstract model theory for
            specification and programming. *Journal of the Association for Computing
            Machinery*, 39(1):95–146, January 1992. An early version appeared under
            the title "Introducing Institutions" in *Logics of Programs*, Springer LNCS
            164, 221–256, 1984.
[Gog96]   J.A. Goguen. Parameterized programming and software architecture. In
            Proc. 4th Intl. IEEE Conf. on Software Reuse, pages 2–11, 1996.
[GT79]   J.A. Goguen and J. Tardo. An introduction to OBJ: A language for
            writing and testing software specifications. In M. K. Zelkowitz, editor,
            Specification of Reliable Software, pages 170–189. IEEE Press, Cambridge
            (MA, USA), 1979. Reprinted in *Software Specification Techniques*, N.
            Gehani and A. McGettrick, editors, Addison-Wesley, 1985, pages 391–
            420.
[GT00]   J.A. Goguen and W. Tracz. An implementation-oriented semantics for
            module composition. In *Foundations of Component-Based Systems*, pages
            231–263. Cambridge University Press, 2000. Edited by G. Leavens and
            M. Sitaraman.
[GTW78]   J.A. Goguen, J.W. Thatcher, and E.G. Wagner. An initial algebra ap-
            proach to the specification, correctness and implementation of abstract
            data types. In *Current Trends in Programming Methodology, Vol. 4: Data
            Structuring*, pages 80–149. 1978. Edited by R.T. Yeh.
[Hoa72]   C.A.R. Hoare. Correctness of data representations. *Acta Informatica*,
            1:271–281, 1972.
[Jon80]   C.B. Jones. *Software Development: A Rigorous Approach*. Prentice Hall,
            1980.
[KST97]   S. Kahrs, D. Sannella, and A. Tarlecki. The definition of Extended ML:
            A gentle introduction. *Theoretical Computer Science*, 173:445–484, 1997.
[LF97]   A. Lopes and J. Fiadeiro. Preservation and reflection in specification. In
            *Proc. 6th Intl. Conference on Algebraic Methodology and Software Tech-
            nology, AMAST 1997*, Springer LNCS 1349, pages 380–394, 1997.

[LPRS88]   P. Lee, F. Pfenning, G. Rollins, and W. Scherlis. The Ergo support system: An integrated set of tools for prototyping integrated environments. In *Proc. 3rd ACM SIGSOFT/SIGPLAN Software Engineering Symposium on Practical Software Development Environments*, pages 25–34, 1988.

[Mac71]   S. MacLane. *Categories for the Working Mathematician*. Springer, 1971.

[MHAH04]   T. Mossakowski, P. Hoffman, S. Autexier, and D. Hutter. CASL logic. [CoF04], part IV, pages 275–361. T. Mossakowski, editor.

[Mil71]   R. Milner. An algebraic definition of simulation between programs. In *Proc. 2nd Intl. Joint Conf. on Artificial Intelligence*, pages 481–489, 1971.

[MST04]   T. Mossakowski, D. Sannella, and A. Tarlecki. A simple refinement language for CASL. In *Recent Trends in Algebraic Development Techniques: Selected Papers from WADT 2004*, Springer LNCS 3423, pages 162–185, 2004.

[MTHM97]   R. Milner, M. Tofte, R. Harper, and D. MacQueen. *The Definition of Standard ML (Revised)*. MIT Press, 1997.

[SS83]   W. Scherlis and D. Scott. First steps towards inferential programming. In *IFIP Congress*, pages 199–212, 1983.

[SST92]   D. Sannella, S. Sokołowski, and A. Tarlecki. Toward formal development of programs from algebraic specifications: Parameterisation revisited. *Acta Informatica*, 29(8):689–736, 1992.

[ST88a]   D. Sannella and A. Tarlecki. Specifications in an arbitrary institution. *Information and Computation*, 76:165–210, 1988.

[ST88b]   D. Sannella and A. Tarlecki. Toward formal development of programs from algebraic specifications: Implementations revisited. *Acta Informatica*, 25:233–281, 1988.

[ST89]   D. Sannella and A. Tarlecki. Toward formal development of ML programs: Foundations and methodology. In *Proc. Colloq. on Current Issues in Programming Languages. Intl. Joint Conf. on Theory and Practice of Software Development (TAPSOFT'89)*, Springer LNCS 352, pages 375–389, 1989.

[ST97]   D. Sannella and A. Tarlecki. Essential concepts of algebraic specification and program development. *Formal Aspects of Computing*, 9:229–269, 1997.

[ST99]   D. Sannella and A. Tarlecki. Algebraic preliminaries. In E. Astesiano, H.-J. Kreowski, and B. Krieg-Brückner, editors, *Algebraic Foundations of Systems Specification*, chapter 2. Springer, 1999.

[SW83]   D. Sannella and M. Wirsing. A kernel language for algebraic specification and implementation. In *Proc. 1983 Intl. Conf. on Foundations of Computation Theory*, Springer LNCS 158, pages 413–427, 1983.

Composition by Colimit
and
Formal Software Development

Douglas R. Smith

Kestrel Institute, Palo Alto, California 94304 USA

Abstract. Goguen emphasized long ago that colimits are how to compose systems [7]. This paper corroborates and elaborates Goguen's vision
by presenting a variety of situations in which colimits can be mechanically applied to support software development by refinement. We illustrate the use of colimits to support automated datatype refinement,
algorithm design, aspect weaving, and security policy enforcement.

1 Introduction

Goguen emphasized long ago that colimits are how one composes systems [7].
In particular, Burstall and Goguen focused on specifications as presentations
of theories and the composition of specifications by colimit in the CLEAR and
CAT system proposals [3,11]. In a sense this paper serves to corroborate and
elaborate Goguen's insight through its applicability to software development by
refinement of specifications.

Kestrel's Specware system [29,12] is a descendant of CLEAR and CAT that
uses the cocomplete category of specifications over higher-order logic. Specware
is used to support the refinement of specifications into correct code in various
target programming languages, including CommonLisp, C, and Java. The role
of category theory is to organize the larger-scale structure of specifications and
the refinement process. The objects of the category are specifications, diagrams
represent structured specifications, and morphisms represent inclusions, parameters, and refinements. Specware uses a colimit algorithm to compose specifications and it uses pushouts to instantiate parameterized specifications (as in [8]).
Most of the detailed design work in software development is logical in nature and
is performed inside specifications (i.e. below the level of the category). No deep
results of category theory are used, but the structuring provided by the categorical framework has been conceptually useful and has guided the implementation
of Specware. The Specware system has been used for a variety of applications
involving both high assurance (e.g. [4]) properties and high performance (e.g.
[1]).

The most basic and straightforward use of colimits in a category of specifications is to build large specifications out of smaller specifications [2]. We briefly
review the technicalities of this usage, but the main focus of the paper is on

K. Futatsugi et al. (Eds.): Goguen Festschrift, LNCS 4060, pp. 317–332, 2006.

how to use composition by colimit to construct refinements. In particular, we discuss (1) how to represent design abstractions as specifications and specification morphisms and how to apply a design abstraction by colimit, and (2) how to express some kinds of policy requirements by automata and how to enforce such policies by a suitable colimit. The concepts are illustrated by examples from automated datatype refinement, algorithm design, aspect weaving, and security policy enforcement.

2   Preliminaries

We briefly review the category of specifications over classical higher-order logics, since all the examples and discussion build on it.

A *specification* (or *spec*) is the finite presentation of a theory. The signature of a specification provides the vocabulary for describing objects, operations, and properties in some domain of interest, and the axioms constrain the meaning of the symbols. For example, the following specification for partial orders is expressed in the MetaSlang specification language of Specware. It introduces a type symbol E and an infix binary predicate on E, called *le*, which is constrained by the usual axioms.

> spec *Partial-Order* is
>   type E
>   op $\_le\_ : E, E \rightarrow Boolean$
>   axiom *reflexivity* is $x\ le\ x$
>   axiom *transitivity* is $x\ le\ y\ \wedge\ y\ le\ z\ \Longrightarrow\ x\ le\ z$
>   axiom *antisymmetry* is $x\ le\ y\ \wedge\ y\ le\ x\ \Longrightarrow\ x = y$
> end-spec

A *specification morphism* translates the language of one specification into the language of another specification, preserving the property of provability, so that any theorem in the source specification remains a theorem under translation. In Specware, a specification morphism $m : T \rightarrow T'$ is given by a map from the type and operator symbols of the *domain* spec T to the symbols of the *codomain* spec T'. To be a specification morphism it is sufficient to show that every axiom of T translates to a theorem of T'. It then follows that a specification morphism translates theorems of the domain specification to theorems of the codomain.

For example, a specification morphism from *Partial-Order* to *Integer* can be presented by:

> morphism *Partial-Order-to-Integer* : *Partial-Order* $\rightarrow$ *Integer* is
> $\{E \mapsto Integer,\ le \mapsto\ \leq\}$

where *Integer* is a specification for the integers that includes the usual constants (such as 0), comparison relations (such as lesser-or-equal $\leq$), functions (such as addition), and so on.

Translation of an expression by a morphism is by straightforward application of the symbol map, so, for example, the *Partial-Order* axiom $\forall (x : E)\ x\ le\ x$

translates to $\forall(x : Integer)\ x \leq x$ With a reasonable axiomatization of the integers it is easy to verify that the three axioms of a partial order remain provable in *Integer* theory after translation.

Specification morphisms compose in a straightforward way as the composition of finite maps. It is easily checked that specifications and specification morphisms form a category SPEC. Colimits exist in SPEC and are easily computed. Suppose that we want to compute the colimit of $B \xleftarrow{\ i\ } A \xrightarrow{\ j\ } C$. First, form the disjoint union of all sort and operator symbols of A, B, and C, then define an equivalence relation on those symbols:

$$s \approx t \ \textit{iff}\ (i(s) = t\ \lor\ i(t) = s\ \lor\ j(s) = t\ \lor\ j(t) = s).$$

The signature of the colimit (also known as pushout in this case) is the collection of equivalence classes wrt $\approx$. The cocone morphisms take each symbol into its equivalence class. The axioms of the colimit are obtained by translating and collecting each axiom of A, B, and C. The colimit can be scalably computed in near-linear time.

For example, suppose that we want to build up the theory of partial orders by composing simpler theories.

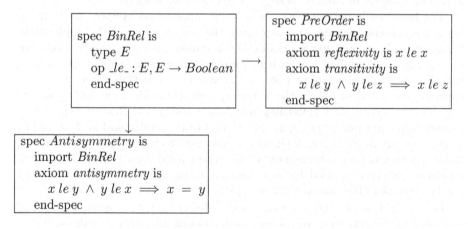

The pushout of *Antisymmetry* $\leftarrow$ *BinRel* $\rightarrow$ *PreOrder* is isomorphic to the specification for *Partial-Order* given above. In detail: the morphisms are $\{E \mapsto E,\ le \mapsto le\}$ from *BinRel* to both *PreOrder* and *Antisymmetry*. The equivalence classes are then $\{\{E, E, E\},\ \{le, le, le\}\}$, so the colimit spec has one type (which we rename E), and one operator (which we rename le). Furthermore, the axioms of *BinRel*, *Antisymmetry*, and *PreOrder* are each translated to become the axioms of the colimit. Thus we have *Partial-Order*.

The universal property of the colimit means that there exists a unique specification morphism from the constructed *Partial-Order* specification to any other specificaton that refines both *PreOrder* and *Antisymmetry*. Intuitively, *Partial-Order* is the simplest specification that composes the logical content of *PreOrder* and *Antisymmetry*.

Although the definitions above are given in higher-order logic, the concepts presented below essentially assume a cocomplete category of specifications over an institution [9].

For purposes of refinement, a loose semantics is natural. Semantics of a refinement morphism is given by a contravariant functor into CAT, the category of small categories. That is, each spec denotes a category of models, and each morphism denotes a functorial mapping that takes each codomain model into a domain model. Particular semantics are enforced by applying appropriate refinements and when performing the institution morphism from the spec language to a programming language.

3   Composing and Refining Specifications

Kestrel's work emphasizes automated tools for the refinement of specifications. There are several reasons for taking this approach to software development: (1) enhanced productivity through automated code generation, (2) enhanced assurance due to the correct-by-construction characteristic of refinement-based derivations, and (3) enhanced software quality and performance due to to automated application of codified best-practice design knowledge.

The first step in developing a new software application in Specware is building a domain specification and capturing the requirements of the application. Composition by colimit plays a major role in building domain specifications. An example from scheduling is shown in Figure 1. Generally, scheduling is about the allocation of resources to tasks so as to satisfy constraints on timeliness, capacity, cost, and so on. In the figure, specifications for *Time* and *Quantity* are shared between *Task* (modeling scheduling tasks) and *Resource* (modeling resources to carry out tasks). *Quantity* is used to model demand in *Task* and to model capacity in *Resource*. A pushout is also used to instantiate a spec SET of finite sets that is parameterized on a base type (called *1-Sort* here). The actual requirements are expressed by input/output constraints (pre/post-conditions) on the scheduler (for more details, see [28]).

In a refinement setting, a formal specification of system requirements is refined to code by incrementally adding design detail. Increments of implementation detail are expressed as morphisms between specifications (in an appropriate category). There is an active community of researchers and practitioners exploring the issues of building requirement specifications out of the (sometimes conflicting) agendas of various stakeholders. What has been missing in this picture is a focus on how to construct refinements – are they mostly ad-hoc, or can they be derived from reusable design abstractions? Most approaches to refinement in the literature (e.g. VDM, Z, RAISE, B) rely on manual invention of refinements, followed, if desired, by verification of the refinement conditions. Our approach, implemented in KIDS and Specware/Designware, has hypothesized that most code is derived from reusable design abstractions and that these can be codified and mechanically applied. A key component of our research has been collecting and formalizing principles of excellent design practice, as found in algorithm de-

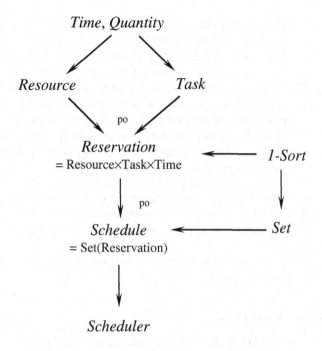

Fig. 1. Scheduling Domain Specification

sign textbooks and practice, system design patterns/architectures/frameworks, and so on.

The purpose of this paper is to highlight the ways in which colimits, in suitable categories, play a central role in composing these sources of information with the evolving design in order to mechanically generate refinements. Since the colimit is scalably computable in the categories of interest, it can play a central role in a refinement-oriented mechanized system development environment.

4   Design by Classification

Design knowledge typically has two essential components: its content and a characterization of situations in which the content applies. We represent these two components as the codomain and domain of a morphism, respectively. That is, abstract design knowledge about datatype refinement, algorithm design, software architectures, program optimization rules, visualization displays, and so on, can be expressed as refinements (i.e. morphisms). The codomain embodies a design constraint – the effect is a reduction in the set of possible implementations. The domain of one such refinement represents the abstract structure that is required in a user's specification in order to apply the embodied design knowledge. The codomain of the refinement contains new structures and definitions that are composed with the user's requirement specification.

The figure to the left shows the application of a library refinement $A \to B$ to a given specification $Spec_0$. First the library refinement is selected. The applicability of the refinement to $Spec_0$ is shown by constructing a *classification arrow* from A to $Spec_0$ which classifies $Spec_0$ as having A-structure by making explicit how $Spec_0$ has at least the structure of A. Finally the refinement is applied by computing the pushout. The colimit algorithm generates both the refined specification (the apex shown in the lower right) and the cocone morphisms, including the refinement morphism $Spec_0 \to Spec_1$. The creative work lies in constructing the classification arrow [21,22].

Furthermore we can organize the design theories into libraries with a taxonomic structure – more general theories refine to more specialized theories. Mechanisms for incrementally accessing and applying design theories from such a library as discussed in [22].

The next two subsections elaborate these notions in the context of datatype refinement and algorithm design respectively.

4.1 Datatype Refinement

Abstract data types (ADTs) allow us to think about data structures in terms of their essential operations and properties. To work effectively with ADTs we must add back in the implementation detail that is abstracted away in a, say, algebraic presentation of an ADT. Refinements serve this purpose. Specifically a morphism between an ADT theory and a (more concrete) datatype theory presents a way to implement the ADT (or dually, a way to view the implementing datatype as the ADT).

Some specific examples includes finite set theory mapping to lists or B-trees or splay trees. Another example: finite sets over a small finite type mapping to hash tables or bit vectors.

Each of these refinements/interpretations can be represented and stored in a library. To apply a datatype refinement we compute the following pushout:

$$
\begin{array}{ccc}
AbstractDT & \longrightarrow & Spec_0 \\
\downarrow & & \downarrow \\
ConcreteDT & \longrightarrow & Spec_1
\end{array}
$$

We sketch a simple example to illustrate the representation of abstract design knowledge as morphisms. In particular, we can refine finite sets to bit vectors as follows. Finite sets over the range 1..32 are partially specified by

```
spec FiniteSet is
    type FSet
    type Elt = 1..32                              % range from 1 to 32
    op {} : FSet                                  % empty set
    op _ with _ : FSet × Elt → FSet              % add an element
    op _ ∪ _ : FSet × FSet → FSet                % union
    op _ ∩ _ : FSet × FSet → FSet                % intersection
```

. . .

axiom *commutativity* is $((S \text{ with } a) \text{ with } b) = ((S \text{ with } b) \text{ with } a)$
axiom *idempotence* is $((S \text{ with } a) \text{ with } a) = (S \text{ with } a)$
. . .

end-spec

Bit vectors of length 32 are partially specified by

spec *BitVector32* is
    type *BV32*
    type *Index* = 1..32
    op *zero* : $BV32$ % the zero bit vector
    op *set* : $BV32 \times Index \to BV32$ % set index bit to 1
    op _ | _ : $BV32 \times BV32 \to BV32$ % bitwise OR
    op _ & _ : $BV32 \times BV32 \to BV32$ % bitwise AND
    op _ << _ : $BV32 \times Index \to BV32$ % left shift
    . . .

end-spec

A refinement of *FiniteSet* to *BitVector32* is presented by the morphism

morphism *FSet-to-BitVector32* is
  {   *FSet* $\mapsto$ *BV32*
      Elt $\mapsto$ Index
      {} $\mapsto$ zero
      with $\mapsto$ set
      $\cup$ $\mapsto$ |
      $\cap$ $\mapsto$ &
      . . .
  }

If we have a specification S that imports *FSet*, then taking a pushout of *FSet-to-BitVector32* with the import morphism *FSet* $\to$ *S* yields a refinement of S in which finite sets are implemented as bit vectors.

As another example, in Specware, the splay tree refinement is the default implementation given to sets due to its good performance profile. Programmers might tempted to avoid working with splay trees since their implementation is a little more complex than simpler representations of sets. A refinement setting allows developers to work with appropriate abstractions *and* obtain good performance.

4.2 Algorithm Design

Just as an algebraic presentation of a datatype aims to capture the abstract essence of the type, an algorithm theory aims to capture the abstract essence of a class of algorithms [27]. For example, consider the class of greedy algorithms (which work to build a solution by iteratively adding the best available component to the incremental solution, until no more components remain). The greedy algorithm can be abstractly represented by a program scheme, which is a definition in a theory that contains partially specified function symbols. A sufficient condition that the scheme generates an optimal solution is given by the matroid property (which is comprised of four conditions; see e.g. [15]). We can represent this package (sufficient structure plus program scheme) as a morphism, prove it once, and store it in a library.

A pushout can be used to apply such an algorithm refinement to a particular problem. For example, the problem of finding a minimum spanning tree can be solved by applying the greedy algorithm theory, yielding Kruskal's algorithm (or Prim's algorithm depending how on the classification arrow is constructed).

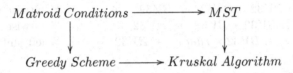

This approach to automated algorithm design was first implemented in the KIDS system [20] and more clearly in Specware/Designware [24,23]. A series of complex high-performance scheduling algorithms for Air Force applications were developed using this approach in KIDS [28] and a domain-specific variant of Designware called Planware [1].

5   Policy Enforcement

The previous two sections describe a means for applying abstract design knowledge to generate algorithms. When we turn to system design, there are issues to contend with that arise less obviously in algorithm design. In particular, cross-cutting concerns are one source of the extra complexity that arises in system design. A concern is cross-cutting if its manifestation cuts across the dominant hierarchical structure of a program. Cross-cutting concerns explain a significant fraction of the code volume and interdependencies of a system. The interdependencies complicate the understanding, development, and evolution of the system.

In this section, we illustrate two forms of cross-cutting concerns and how they can be expressed and mechanically enforced. We call these concerns *policies* to emphasize that (1) they are really requirements, and (2) they tend to reflect non-functional concerns, such as auditing, security, and so on.

The following colimit shows the intention of our approach: to use a colimit in a suitably defined category to enforce policies on a system design.

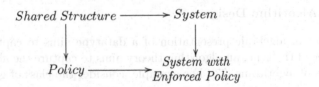

One issue that arises in this context is knowing where the policy applies. For example, a security policy must be applied pervasively in order to provide assurance. In our approach, static analysis [5,18] is used to find all occurrences and to set up the cospan (i.e. the *Shared Structure* specification above).

5.1   AOP as Invariant Maintenance

A simple example of a cross-cutting concern is an error logging policy – the requirement to log all errors in a system in a standard format. Error logging necessitates the addition of code that is distributed throughout the system code, even though the concept is easy to state in itself.

Aspect-oriented programming (AOP), as exemplified by AspectJ [13], provides a modular way to treat cross-cutting concerns. However, AspectJ aspects are expressed at a programming language level which obscures their intention. The reason for this, of course, is to lower the barriers to usage amongst the broad Java programming community. In [25] we proposed some techniques for specifying cross-cutting concerns as logical invariants to be maintained. For example, to express an error-logging policy as an invariant, we assert that the error-log data structure is equal to the list of all previous errors that have occurred during the course of the computation. To formalize this invariant, we need to reify the history of the computation, purely for specification purposes [25]. The counterpart to aspect weaving is (1) to use static analysis to find all code locations where the invariant might be violated, and (2) to specify and synthesize code to reestablish the invariant. For the error-logging example, static analysis would find all potential code locations where an error might be thrown, and the composition process would compose the throw with an update of the error-log data structure.

By expressing cross-cutting concerns as invariants, we capture their intention more clearly and we can use algorithmic means (static analysis) to determine the complete extent of their application, in contrast to the manual coding of join points in AspectJ.

Our point here is that one of the key mechanisms underlying the enforcement of an invariant is a suitable pushout. To see this most clearly, we switch to a category of abstract state machines over a suitable specification language; e.g. see *especs* in [16,17]. Here the objects are state machines and the morphisms/refinements represent the simulates relation between automata. An abstract state is given by a specification (for especs we use the higher-order specifications of Specware). For our purposes here, an abstract transition will be specified by a pre/post-condition pair: $A - \overset{[Pre,Post]}{- - - -} \blacktriangleright B$ (we use dashed arrows for transitions to distinguish them from morphisms in a diagram).

In a category of state machine, particularly especs, refinement means simulation and colimit serves (1) to compose the corresponding state specifications (the pushout of A and C is denoted $A \oplus C$) and (2) to superpose the actions on abstract transitions (the pushout of actions effectively conjoins their effects so the composite action achieves both simultaneously). The following diagram illustrates the composition of one step of the source system $A - - \blacktriangleright B$ with a step of the policy $C -- \succ D$. The policy asserts that I is to hold invariantly at states and the effect of composition is to add the invariance requirement to the system.

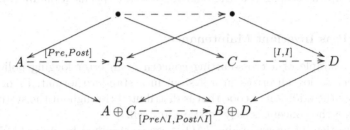

$$A \oplus C \underset{[Pre \wedge I, Post \wedge I]}{------\succ} B \oplus D$$

Static analysis is used to find the association of system steps and policy steps, then a pushout, as above, is used to compose the two. A further synthesis step is needed in order to synthesize an action that achieves the composed action specification $[Pre \wedge I, Post \wedge I]$. For a variety of detailed examples see [25].

5.2   Enforcing Automata-Based Security Policies

The previous section described a simple kind of policy, based on invariant state properties, and the composition and synthesis mechanisms that underlie enforcement. A more general kind of policy can be specified by means of automata or by temporal logic formulas.

As a concrete example, consider the following simple security policy which is adapted from Schneider [19]. Whenever a process reads from a particular file f, it is not allowed to send any messages. The policy states a particular kind of information flow constraint. The policy can be expressed as a policy automaton:

The transitions are labeled in the form *name* : *event/action*. The events are expressed as source-code patterns that either succeed (with bindings of pattern variables) or fail. If an action is omitted, then it is a no-op. This policy has only has one prescription of an action to take in a particular context – in policy state 1, if a *send* is attempted, then abort the program. The effect of enforcement will be to terminate any behaviors that do not implement the policy (a send following a read of file *f*). For examples of the enforcement of automata-based policies that prescribe behavior, see the error-handling policies in [26].

Colimits can be used to enforce policies specified by a policy automaton. However, there are interesting issues that arise. The foremost is that the effect of enforcing this policy is to sometimes cause the program to abort (terminate abnormally) when the system would otherwise continue normally. There are two problems here: (1) how to handle conflicting constraints on the system (here the system may satisfy constraints that conflict with the policy), and (2) how to define an appropriate notion of refinement (morphism) that allows termination of behaviors.

One approach to handling conflicting requirements is to treat system requirements as having a linear priority order. The idea is that a system satisfies a priority ordering of constraints if whenever the system fails to satisfy one constraint C, then it must satisfy some other higher-priority constraint. For example, it is often the case in system code that safety and security constraints dominate functional constraints. We make this approach more precise in the following.

Let $\langle R, \prec \rangle$ be a linearly ordered set of temporal formulas [14], and S a program. We say that a behavior b satisfies R if for each formula F in R, either S satisfies F or it satisfies some other formula $G \in R$ such that $F \prec G$. S satisfies R if every behavior of S satisfies R.

Technically, there is no extra expressive power in priority-ordered requirements. Consider the simplest situation in which there are just two requirements A and B together with the order $A \prec B$. An equivalent specification has the two requirements $A \vee B$ and B without an order. Clearly this notion of satisfaction is weak, since it admits programs that satisfy B but not A. While one could pursue this to obtain a stronger theoretical definition of satisfaction (e.g. by considering maximal satisfaction of dominated constraints), we take a pragmatic approach that addresses the problem via the design process. That is, our approach will be to perform design starting with the bottommost requirements of the order – typically these are the basic functional constraints. Then, we iteratively select dominating requirements in order and enforce them by colimit in the evolving design. In this way, whenever we enforce a requirement, the composition process will only override dominated constraints. The result is a design that will tend to satisfy the base functionality requirements as much as possible, but with some behaviors that accord with overriding policy constraints.

Mobile code provides a clear scenario in which this bottom-up design approach makes sense. Mobile code typically cannot be designed to anticipate all environments that it might run in. One host environment may have local policies that must be enforced, and it can do so by, say, composing the policies at the byte-code level at upload time. This way, the local environment's policies are maintained even if it means disallowing behaviors of the mobile code that might be acceptable in other environments.

Our point here is not to fully define a new approach to program satisfaction, nor a new design methodology, but simply to show another context in which composition by colimit provides basic support to system development.

The second problem mentioned above, a suitable notion of refinement that allows behavior termination, can be addressed as follows. In the category of especs [16,17], abstract states are given by specifications and abstract transitions are modeled by suitable morphisms – a state machine is then a diagram over a category of specs. Each abstract state naturally has the identity self-transition which is the identity morphism on specs. Semantically, the behaviors of such an espec includes arbitrary stuttering (no-op transitions that do not change state). In the literature, behaviors that stutter are often ruled out, although they play a crucial role in refinement. We propose to go farther and admit all such stuttering behaviors, including behaviors in which the machine stutters forever on some state. There are at least two reasons to adopt this rather loose semantics. First, it allows us to model failure in the underlying computation substrate. Most formal models of behavior assume a perfect computational model and ignore the unreliability of the hardware/software platform on which software executes. Second, it allows us to treat as refinement the notion of policy enforcement that works by terminating bad behaviors. In both cases, the idea is that for any behavior that successfully reaches a final/accepting state (or does so infinitely often), the semantics also includes all prefixes of that behavior. Each proper prefix corresponds to a computation that is terminated (due to failure of computational service, or to policing action, etc.). As a consequence, we obtain the conventional notion of trace-containment semantics for refinement. That is, every behavior of the codomain machine (including abnormally terminated behaviors) maps-to/simulates a behavior of the domain machine.

Enforcement of a policy automaton occurs in two stages. In the first stage, static analysis is used to simulate the automaton by matching the event patterns against the control-flow of the system source code. Recent progress has produced scalable low-order polynomial time algorithms for policy simulation [10,6]. These algorithms work by simulating the policy forward through the source code, recording the policy states and transitions in labels on the control-flow graph of the source code. When matching a policy transition labeled *event/action*, if the event pattern matches a source-code transition, then the policy transition (instantiated with the bindings from the match) is associated with the source transition. The algorithms terminate when a fixpoint is reached.

In effect, static analysis creates a refinement of the policy automaton that has the same essential shape as the source code, thus enabling automatic composition by colimit.

Consider for example the code

```
int c;
if c=0 then read f;
send m;
...
```

which is represented by a state machine in Figure 2. The figure also shows the results of policy simulation/analysis – each state of the code is labeled with the states of the policy automaton that it could possibly be in for some input, and each transition is labeled with the set of possible policy transitions that it simulates for some input.

Conceptually, the static analysis sets up a cospan in the category of especs [16,17]. Figure 3 shows both the cospan and the cocone. The static analysis allows us to set up a refinement of the policy automaton (shown on the right of the cospan) and the abstract shape that is common to the source code and the policy instance. In the example, the key feature is the policy ambiguity that results from the conditional: after the conditional, the system state is in either

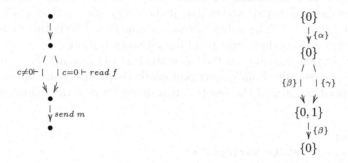

Fig. 2. Results of Static Analysis

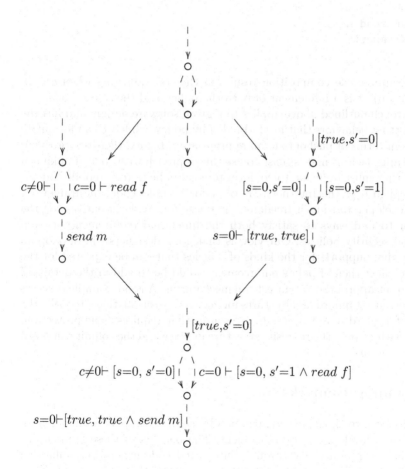

Fig. 3. Colimit to Enforce Policy

policy state 0 or 1 depending on which branch was taken. Crucially then, the **send** command is either (i) acceptable, if the policy state is 0, or (ii) forbidden if the policy state is 1. The policy instance automaton reflects this by recording the policy transitions that correspond to system transitions.

Computing the pushout has the essential effect of enforcing the security policy in the source code. Finally, program synthesis processes are applied to the pushout specification and the result is translated back to the following source-level code:

```
int c;
int s;  /* state variable */
s := 0;
if c=0
   then {read f || s := 1}
if s=0
   then send m
   else abort;
...
```

In the example, the composition results in the code aborting when $c = 0$. The pushout object is a refinement of both the policy and the source code.

The approach outlined above applies to a given software design, and has the effect of aborting behaviors that are forbidden by policy. while we can formulate this process in terms of pushouts in an appropriate category, there are pros and cons to this approach. It makes sense to use this approach with code of unknown provenance that must be made to conform to local policies (e.g. mobile code or services supplied over the Internet). However for bespoke code, the framework gives the developer too much freedom – it doesn't provide incentives for the programmer to find ways to satisfy both the functional requirements as well as safety and security policies. Our view is that good designers will develop an architecture that supports for the kinds of policies that can be expected for the system. The effect then of policy enforcement would be to add in the details of the policy to the appropriate architectural mechanisms. A good example is access control. There are standard architectures for access control [30] that prescribe the mediation of a guard in any access to a resource that requires some protection. The design pattern puts the requisite structure in place and the colimit composes in the policy details.

6   Concluding Remarks

Our goal has been to show how composition by colimit can play a fundamental role in software development by refinement. The benefits of these foundations include enhanced productivity through automated code generation, enhanced assurance due to the correct-by-construction characteristic of refinement-based derivations, and potentially enhanced software quality and performance due to to automated application of codified best-practice design knowledge.

Acknowledgments: This work was partially supported by the US Department of Defense and by the Office of Naval Research under Grant N00014-04-1-0727.

References

1. BECKER, M., GILHAM, L., AND SMITH, D. R. Planware II: Synthesis of schedulers for complex resource systems. Tech. rep., Kestrel Technology, 2003.
2. BURSTALL, R. M., AND GOGUEN, J. A. Putting theories together to make specifications. In *Proceedings of the Fifth International Joint Conference on Artificial Intelligence* (Cambridge, MA, August 22–25, 1977), IJCAI, pp. 1045–1058.
3. BURSTALL, R. M., AND GOGUEN, J. A. The semantics of CLEAR, a specification languge. In *Proceedings, 1979 Copenhagen Winter School on Abstract Software Specification*, D. Bjorner, Ed. Springer LNCS 86, 1980.
4. COGLIO, A. Toward automatic generation of provably correct Java Card applets. In *Proc. 5th ECOOP Workshop on Formal Techniques for Java-like Programs* (July 2003).
5. COUSOT, P., AND COUSOT, R. Abstract interpretation: a unified lattice model for static analysis of programs by construction or approximation of fixpoints. In *Conference Record of the Fourth Annual ACM SIGPLAN-SIGACT Symposium on Principles of Programming Languages* (1977), ACM, pp. 238–252.
6. DAS, M., LERNER, S., AND SEIGLE, M. ESP: Path-sensitive program verification in polynomial time. In *SIGPLAN 2002 Conference on Programming Language Design and Implementation (PLDI'02)* (2002).
7. GOGUEN, J. A. Categorical foundations for general systems theory. In *Advances in Cybernetics and Systems Research*, F. Pichler and R. Trappl, Eds. Transcripta Books, 1973, pp. 121–130.
8. GOGUEN, J. A. Parameterized programming. *IEEE Transactions on Software Engineering SE-10*, 5 (September 1984), 528–543.
9. GOGUEN, J. A., AND BURSTALL, R. M. Institutions: Abstract model theory for computer science. *Journal of the ACM 39*, 1 (1992), 95–146.
10. HALLEM, S., CHELF, B., XIE, Y., AND ENGLER, D. A system and language for building system-specific, static analyses. In *SIGPLAN 2002 Conference on Programming Language Design and Implementation (PLDI'02)* (2002).
11. J. GOGUEN AND R. BURSTALL. CAT: a system for the structured elaboration of correct programs from structured specifications. Tech. Rep. CSL-118, SRI International, 1988.
12. KESTREL INSTITUTE. *Specware System and documentation*, 2003. http://www.specware.org/.
13. KICZALES, G., AND ET AL. An Overview of AspectJ. In *Proc. ECOOP, LNCS 2072, Springer-Verlag* (2001), pp. 327–353.
14. MANNA, Z., AND PNUELI, A. *The Temporal Logic of Reactive and Concurrent Systems*. Springer-Verlag, New York, 1992.
15. PAPADIMITRIOU, C. H., AND STEIGLITZ, K. *Combinatorial Optimization: Algorithms and Complexity*. Prentice Hall, Englewood Cliffs, NJ, 1982.
16. PAVLOVIC, D., AND SMITH, D. R. Composition and refinement of behavioral specifications. In *Proceedings of Automated Software Engineering Conference* (2001), IEEE Computer Society Press, pp. 157–165.
17. PAVLOVIC, D., AND SMITH, D. R. Evolving specifications. Tech. rep., Kestrel Institute, 2004.

18. REPS, T., HORWITZ, S., AND SAGIV, M. Precise interprocedural dataflow analysis via graph reachability. In *Conference Record of the Twenty-Second ACM Symposium on Principles of Programming Languages* (1995), ACM, pp. 49–61.

19. SCHNEIDER, F. Enforceable security policies. *ACM Transactions on Information and System Security 3*, 1 (February 2000), 30–50.

20. SMITH, D. R. KIDS – a semi-automatic program development system. *IEEE Transactions on Software Engineering Special Issue on Formal Methods in Software Engineering 16*, 9 (1990), 1024–1043.

21. SMITH, D. R. Constructing specification morphisms. *Journal of Symbolic Computation, Special Issue on Automatic Programming 15*, 5-6 (May-June 1993), 571–606.

22. SMITH, D. R. Toward a classification approach to design. In *Proceedings of Algebraic Methodology and Software Technology (AMAST)* (1996), vol. LNCS 1101, Springer-Verlag, pp. 62–84.

23. SMITH, D. R. Designware: Software development by refinement. In *Proceedings of the Eighth International Conference on Category Theory and Computer Science* (1999), M. Hoffman, D. Pavlovic, and P. Rosolini, Eds., pp. 355–370.

24. SMITH, D. R. Mechanizing the development of software. In *Calculational System Design, Proceedings of the NATO Advanced Study Institute*, M. Broy and R. Steinbrueggen, Eds. IOS Press, Amsterdam, 1999, pp. 251–292.

25. SMITH, D. R. A generative approach to aspect-oriented programming. In *Proceedings of the Third International Conference on Generative Programming and Component Engineering* (2004), Springer-Verlag LNCS 3286, pp. 39–54.

26. SMITH, D. R., AND HAVELUND, K. Automatic enforcement of error-handling policies. Tech. rep., Kestrel Technology, September 2004.

27. SMITH, D. R., AND LOWRY, M. R. Algorithm theories and design tactics. *Science of Computer Programming 14*, 2-3 (October 1990), 305–321.

28. SMITH, D. R., PARRA, E. A., AND WESTFOLD, S. J. Synthesis of planning and scheduling software. In *Advanced Planning Technology* (1996), A. Tate, Ed., AAAI Press, Menlo Park, pp. 226–234.

29. SRINIVAS, Y. V., AND JÜLLIG, R. Specware: Formal support for composing software. In *Proceedings of the Conference on Mathematics of Program Construction*, B. Moeller, Ed. LNCS 947, Springer-Verlag, Berlin, 1995, pp. 399–422.

30. THE OPEN GROUP. Security design patterns. Tech. rep., http://www.opengroup.org/security/gsp.htm, 2004.

Proving Behavioral Refinements of COL-specifications[*]

Michel Bidoit[1] and Rolf Hennicker[2]

[1] Laboratoire Spécification et Vérification (LSV), CNRS & ENS de Cachan, France
[2] Institut für Informatik, Ludwig-Maximilians-Universität München, Germany

Abstract. The COL institution (constructor-based observational logic) has been introduced as a formal framework to specify both generation- and observation-oriented properties of software systems. In this paper we consider behavioral refinement relations between COL-specifications taking into account implementation constructions. We propose a general strategy for proving the correctness of such refinements by reduction to (standard) first-order theorem proving with induction. Technically our strategy relies on appropriate proof rules and on a lifting construction to encode the reachability and observability notions of the COL institution.

1 Introduction

Within the theory of algebraic specifications, behavioral (or observational) aspects of software systems have been considered since more than twenty years in many approaches in the literature. One of the first studies exposing the importance of a behavioral view for the formalization of implementation notions has been provided by Goguen and Meseguer in [11]. It is motivated by many examples which show that it is essential to abstract from internal implementation details and to rely only on the observable behavior of programs.

As discussed in [7], behavioral refinement concepts can be classified into two principal trends. The first one, pursued e.g. in [18, 19, 17, 3], uses an explicit behavioral abstraction operator to relax the standard model class semantics of the specification to be implemented. The second one uses specifications with built-in features to express behavioral properties. Examples are the hidden algebra institution developed by Goguen and his research group (see e.g. [12]), the CafeOBJ language [9] and the COL institution (constructor-based observational logic [4]). Each of these approaches is equipped with a notion of signature containing a distinguished set of observer operations (to build observable experiments) and with a notion of behavioral satisfaction such that the equality symbol is interpreted by the observational equality of elements (where two elements of an algebra are observationally equal if they cannot be distinguished by observable experiments). In the COL institution signatures contain additionally to the observers a distinguished set of constructor operations which specify those elements

---

[*] This work is partially supported by the GLOWA-Danube project (01LW0303A) sponsored by the German Federal Ministry of Education and Research.

K. Futatsugi et al. (Eds.): Goguen Festschrift, LNCS 4060, pp. 333–354, 2006.
© Springer-Verlag Berlin Heidelberg 2006

which are of interest from the user's point of view thus determining a subpart of an algebra (called the contructor-generated part). Hence a COL-signature $\Sigma_{\mathrm{COL}} = (\Sigma, \mathrm{OP}_{\mathrm{Cons}}, \mathrm{OP}_{\mathrm{Obs}})$ consists of a (standard) many-sorted signature $\Sigma = (S, \mathrm{OP})$ together with distinguished sets $\mathrm{OP}_{\mathrm{Cons}}$ of constructor operations and $\mathrm{OP}_{\mathrm{Obs}}$ of observer operations. The behavioral satisfaction of formulas is then further relaxed to the COL-satisfaction relation $\models_{\Sigma_{\mathrm{COL}}}$ which takes into account only constructor-generated values for the valuation of variables (thus abstracting from junk values).

A simple refinement relation between COL-specifications can be defined if the specification $\mathrm{SP}_{\mathrm{COL}}$ to be implemented and the implementing specification $\mathrm{SPI}_{\mathrm{COL}}$ have the same COL-signature Σ_{COL}. In this case $\mathrm{SPI}_{\mathrm{COL}}$ is a behavioral refinement of $\mathrm{SP}_{\mathrm{COL}}$ if its model class $\mathcal{M}od[\mathrm{SPI}_{\mathrm{COL}}]$ is included in the model class $\mathcal{M}od[\mathrm{SP}_{\mathrm{COL}}]$ of $\mathrm{SP}_{\mathrm{COL}}$. To prove the correctness of the refinement one has to show, assuming that $\mathrm{SP}_{\mathrm{COL}}$ is a (flat) specification of the form $\langle \Sigma_{\mathrm{COL}}, \mathrm{Ax} \rangle$, that $\mathcal{M}od[\mathrm{SPI}_{\mathrm{COL}}] \models_{\Sigma_{\mathrm{COL}}} \mathrm{Ax}$, i.e that all models of $\mathrm{SPI}_{\mathrm{COL}}$ behaviorally satisfy the axioms of the abstract specification $\mathrm{SP}_{\mathrm{COL}}$. For this purpose one can directly apply the proof techniques for behavioral consequences of COL-specifications developed in [4]. In general, however, the assumption that both specifications have the same signature is much too restrictive because an implementation usually involves some construction steps which led to the concept of a constructor implementation introduced in [19]. In the context of the COL institution this idea is formalized by the notion of a COL-implementation constructor κ_{COL} which can be applied to the models of the implementing specification $\mathrm{SPI}_{\mathrm{COL}}$ to produce models of the specification $\mathrm{SP}_{\mathrm{COL}}$ to be implemented. Hence, to prove the correctness of the refinement one has to show

$$(*) \quad \kappa_{\mathrm{COL}}(\mathcal{M}od[\mathrm{SPI}_{\mathrm{COL}}]) \models_{\Sigma_{\mathrm{COL}}} \mathrm{Ax}$$

with $\mathrm{SP}_{\mathrm{COL}} = \langle \Sigma_{\mathrm{COL}}, \mathrm{Ax} \rangle$ as above. Unfortunatley there is no obvious way for discharging this proof obligation since we cannot expect that κ_{COL} is compatible with COL-satisfaction, i.e. we cannot reduce the proof to $\mathcal{M}od[\mathrm{SPI}_{\mathrm{COL}}] \models_{\Sigma I_{\mathrm{COL}}} \mathrm{Ax}^*$ (with an appropriate syntactic adjustment Ax^* of Ax and ΣI_{COL} being the signature of $\mathrm{SPI}_{\mathrm{COL}}$). For instance if we consider an implementation of sets by lists, κ_{COL} would be the (standard) reduct functor along a (standard) signature morphism that would not preserve the usual observer operations where sets are observed by the membership test *isin* and lists are observed by *head* and *tail*. Hence, the reduct used for the implementation construction would not be compatible with the COL-satisfaction relations for sets and lists resp. These considerations are in accordance with Goguen's and Malcolm's study on the difference between vertical signature morphisms used for refinements and horizontal signature morphisms used for modular constructions of system specifications; see [15].

In this paper we propose a strategy to discharge the proof obligation (*) which consists of two major steps. In the first step (see Section 4) we show that instead of the COL-specification $\mathrm{SPI}_{\mathrm{COL}}$ used for the implementation it is sufficient to consider the (standard) first-order specification SPI obtained from $\mathrm{SPI}_{\mathrm{COL}}$ by forgetting the observer and constructor opertions. For the correctness of the corresponding proof rule it is essential that COL-implementation constructors must preserve observational equivalences between algebras (a property which is strongly related to Schoett's notion of stability; see [20]).

As a consequence of the first step it remains to show that the (standard) models of SPI behaviorally satisfy the axioms Ax of the specification SP$_{\mathrm{COL}}$ to be implemented. Therefore, in the next step (see Section 5), we investigate how the proof of behavioral consequences (w.r.t. $\models_{\Sigma_{\mathrm{COL}}}$) of an arbitrary class of Σ-algebras can be reduced to standard first-order reasoning (plus induction). Technically this is achieved by a "lifting" construction providing an appropriate axiomatization of observational equalities and generated parts. Our proof techniques are illustrated by an example (see Section 6) considering a behavioral refinement of sets by non-redundant lists.

2 Basic Concepts

In this section we summarize the basic concepts that are needed to study behavioral refinements of COL-specifications and corresponding proof techniques.

2.1 Algebraic Preliminaries

We assume that the reader is familiar with the basic notions of algebraic specifications (see, e.g., [22, 14, 1]), like the notions of (many-sorted) *signature* $\Sigma = (S, \mathrm{OP})$ (where S is a set of *sorts* and OP is a set of *operation symbols* $op : s_1, \ldots, s_n \to s$), *signature morphism* $\sigma : \Sigma \to \Sigma'$, *(total) Σ-algebra* $A = ((A_s)_{s \in S}, (op^A)_{op \in \mathrm{OP}})$, Σ-*term algebra* $T_\Sigma(X)$ over a family $X = (X_s)_{s \in S}$ of pairwise disjoint sets X_s of variables of sort s and *interpretation* $I_\alpha : T_\Sigma(X) \to A$ w.r.t. a *valuation* $\alpha : X \to A$. The class of all Σ-algebras is denoted by $\mathrm{Alg}(\Sigma)$. Together with Σ-morphisms this class forms a category which, for simplicity, is also denoted by $\mathrm{Alg}(\Sigma)$. For any signature morphism $\sigma : \Sigma \to \Sigma'$, the *reduct functor* $\_|_\sigma : \mathrm{Alg}(\Sigma') \to \mathrm{Alg}(\Sigma)$ is defined as usual and the reduct of a Σ'-algebra A w.r.t. σ is denoted by $A|_\sigma$. In particular, the reduct of A to a subsignature $\Sigma \subseteq \Sigma'$ is denoted by $A|_\Sigma$. In the following we assume that signatures are finite.

The notion of an institution was introduced by Goguen and Burstall [13] to formalize the general concept of a logical system from a model-theoretic point of view; see [21] for an overview. An important example is the institution FOLEq of many-sorted first-order logic with equality as detailed, e.g., in [2]. In FOLEq signatures are many-sorted signatures, models are Σ-algebras and sentences are arbitray first-order Σ-formulas. The satisfaction of a first-order Σ-formula φ by a Σ-algebra A, denoted by $A \models \varphi$, is defined as usual in the first-order predicate calculus with equality. The notation $A \models \varphi$ is extended in a straightforward way to classes of algebras and sets of formulas. The institution CFOLEq is an extension of the FOLEq institution where, in addition to first-order sentences, we consider as extra sentences *sort-generation constraints* of the form $\mathrm{SGC}(S_{\mathrm{Cons}}, \mathrm{OP}_{\mathrm{Cons}})$. A Σ-algebra A satisfies a sort-generation constraint $\mathrm{SGC}(S_{\mathrm{Cons}}, \mathrm{OP}_{\mathrm{Cons}})$ if it is reachable w.r.t. $\mathrm{OP}_{\mathrm{Cons}}$, i.e. if each element of a carrier set A_s with constrained sort $s \in S_{\mathrm{Cons}}$ can be constructed by the interpretations of the constructors $\mathrm{OP}_{\mathrm{Cons}}$ starting from constants and from arbitrary elements of non-constrained sorts, if any. It is well-known that a *free sort-generation constraint* is just an abbreviation for the corresponding sort-generation constraint plus a finite set of first-order sentences to state that all distinct constructor terms (up to variable renaming) denote distinct values. Therefore, in the following, we will also assume that the CFOLEq institution is equipped with free sort-generation constraints of the form $\mathrm{FSGC}(S_{\mathrm{Cons}}, \mathrm{OP}_{\mathrm{Cons}})$, with the meaning described above (see [16, pp. 152–153]).

Any institution provides a suitable framework to define specifications. The semantics of a specification SP is determined by its signature, denoted by $Sig[\mathrm{SP}]$, and by its class of models, denoted by $Mod[\mathrm{SP}]$. In this paper we will only consider basic specifications $\langle \Sigma, \mathrm{Ax} \rangle$ consisting of a signature Σ and a set Ax of Σ-sentences, also called the *axioms* of the specification, with semantics:

$$Sig[\langle \Sigma, \mathrm{Ax} \rangle] \stackrel{\mathrm{def}}{=} \Sigma$$
$$Mod[\langle \Sigma, \mathrm{Ax} \rangle] \stackrel{\mathrm{def}}{=} \{M \in \mathrm{Mod}(\Sigma) \mid M \models_\Sigma \mathrm{Ax}\}.$$

Notations. If SP is a specification and φ is a $Sig[\mathrm{SP}]$-sentence, we write $\mathrm{SP} \models \varphi$ for $Mod[SP] \models \varphi$ and similarly for sets of $Sig[\mathrm{SP}]$-sentences. In the context of the CFOLEq institution we will also consider the sum $\mathrm{SP}_1 + \mathrm{SP}_2$ of two specifications SP_1 and SP_2 with semantics:

$$Sig[\mathrm{SP}_1 + \mathrm{SP}_2] \stackrel{\mathrm{def}}{=} Sig[\mathrm{SP}_1] \cup Sig[\mathrm{SP}_2]$$
$$Mod[\mathrm{SP}_1 + \mathrm{SP}_2] \stackrel{\mathrm{def}}{=} \{A \in \mathrm{Alg}(Sig[\mathrm{SP}_1] \cup Sig[\mathrm{SP}_2]) \mid$$
$$A|_{Sig[\mathrm{SP}_1]} \in Mod[\mathrm{SP}_1] \text{ and } A|_{Sig[\mathrm{SP}_2]} \in Mod[\mathrm{SP}_2]\}.$$

By analogy, for any class C of Σ-algebras and any specification SP, we denote by $C + \mathrm{SP}$ the class of $\Sigma \cup Sig[\mathrm{SP}]$-algebras defined by:

$$C + \mathrm{SP} \stackrel{\mathrm{def}}{=} \{A \in \mathrm{Alg}(\Sigma \cup Sig[\mathrm{SP}]) \mid A|_\Sigma \in C \text{ and } A|_{Sig[\mathrm{SP}]} \in Mod[\mathrm{SP}]\}.$$

2.2 A Brief Introduction to the Constructor-Based Observational Logic COL

The COL institution has been introduced as a formal framework to capture the observational aspects of system specifications; see [4]. The basic idea is to consider distinguished sets of constructor and observer operations. Intuitively, the constructor operations determine those elements which are of interest from the user's point of view while the observer operations determine a set of observable experiments that a user can perform to examine hidden states. Thus we can abstract from junk elements and also from concrete state representations whereby two states are considered to be "observationally equal" if they cannot be distinguished by observable experiments.

Formally, a *constructor operation* is an operation symbol $cons : s_1, \ldots, s_n \rightarrow s$ with $n \geq 0$. The result sort s of $cons$ is called a *constrained sort*. An *observer operation* is a pair (obs, i) where obs is an operation symbol $obs : s_1, \ldots, s_n \to s$ with $n \geq 1$ and $1 \leq i \leq n$. The distinguished argument sort s_i of obs is called a *state sort* (or *hidden sort*). If $obs : s_1 \to s$ is a unary observer we simply write obs instead of $(obs, 1)$. A COL-*signature* $\Sigma_{\mathrm{COL}} = (\Sigma, \mathrm{OP}_{\mathrm{Cons}}, \mathrm{OP}_{\mathrm{Obs}})$ consists of a standard many-sorted signature $\Sigma = (S, \mathrm{OP})$ together with a distinguished set $\mathrm{OP}_{\mathrm{Cons}} \subseteq \mathrm{OP}$ of constructor operations and a distinguished set $\mathrm{OP}_{\mathrm{Obs}}$ of observer operations (obs, i) with $obs \in \mathrm{OP}$. We implicitly assume in the following that whenever we consider a COL-signature Σ_{COL} the underlying (standard) signature is Σ and similarly for Σ'_{COL} etc.

The set $S_{\mathrm{Cons}} \subseteq S$ of *constrained sorts* (w.r.t. $\mathrm{OP}_{\mathrm{Cons}}$) consists of all sorts s such that there exists at least one constructor in $\mathrm{OP}_{\mathrm{Cons}}$ with range s. The set $S_{\mathrm{Loose}} \subseteq S$ of *loose sorts* consists of all non-constrained sorts, i.e. $S_{\mathrm{Loose}} = S \backslash S_{\mathrm{Cons}}$. The set $S_{\mathrm{State}} \subseteq S$ of *state sorts* (or *hidden sorts*, w.r.t. $\mathrm{OP}_{\mathrm{Obs}}$) consists of all sorts s_i such that there exists at least one observer (obs, i) in $\mathrm{OP}_{\mathrm{Obs}}$, $obs :$ $s_1, \ldots, s_i, \ldots, s_n \to s$. The set $S_{\mathrm{Obs}} \subseteq S$ of *observable sorts* consists of all sorts which are not a state sort, i.e. $S_{\mathrm{Obs}} = S \setminus S_{\mathrm{State}}$. An observer $(obs, i) \in \mathrm{OP}_{\mathrm{Obs}}$, $obs : s_1, \ldots, s_i, \ldots, s_n \to s$ is called a *direct observer* if $s \in S_{\mathrm{Obs}}$, otherwise it is an *indirect observer*.

The set $\mathrm{OP}_{\mathrm{Cons}}$ of constructor operations (of a COL-signature Σ_{COL}) determines a set of *constructor terms*. A constructor term is a term t of a constrained sort $s \in S_{\mathrm{Cons}}$ which is built only from constructor operations of $\mathrm{OP}_{\mathrm{Cons}}$ and from variables of loose sorts. In particular, if all sorts are constrained, i.e., $S_{\mathrm{Cons}} = S$, the constructor terms are exactly the $(S, \mathrm{OP}_{\mathrm{Cons}})$-ground terms which are built by the constructor symbols. The set of constructor terms determines, for any Σ-algebra A, an S-sorted family of subsets of the carrier sets of A, called the *generated part* and denoted by $\mathrm{Gen}_{\Sigma_{\mathrm{COL}}}(A)$. For each constrained sort $s \in S_{\mathrm{Cons}}$, the corresponding subset $\mathrm{Gen}_{\Sigma_{\mathrm{COL}}}(A)_s \subseteq A_s$ consists of those elements that can be constructed by the interpretations of the given constructors (starting from constants and from arbitrary elements of loose sorts, if any). For each loose sort $s \in S_{\mathrm{Loose}}$, $\mathrm{Gen}_{\Sigma_{\mathrm{COL}}}(A)_s = A_s$. The Σ_{COL}-generated part represents those elements which are of interest from the user's point of view according to the given constructor operations. A Σ-algebra A is *reachable* (w.r.t. Σ_{COL}) if its carrier sets coincide with its Σ_{COL}-generated part.

The set $\mathrm{OP}_{\mathrm{Obs}}$ of observer operations (of a COL-signature Σ_{COL}) determines a set of *observable contexts* which represent the observable experiments that a user can perform. Observable contexts are defined in a coinductive style which will be reflected in the encoding of observable contexts in Section 5.

Definition 1 (Observable context). *Let Σ_{COL} be a COL-signature, let $X = (X_s)_{s \in S}$ be a family of pairwise disjoint, countably infinite sets X_s of variables of sort s and let $Z = (\{z_s\})_{s \in S_{\mathrm{State}}}$ be a disjoint family of singleton sets (one for each state sort). The sets $\mathcal{C}(\Sigma_{\mathrm{COL}})_{s \to s'}$ of observable Σ_{COL}-contexts with "application sort" s and "observable result sort" s', with $s \in S_{\mathrm{State}}$ and $s' \in S_{\mathrm{Obs}}$, are the least sets such that:*

1. For each direct observer (obs, i) with $obs : s_1, \ldots, s_i, \ldots, s_n \to s'$ and pair-wise disjoint variables $x_1{:}s_1, \ldots, x_n{:}s_n$,
   $obs(x_1, \ldots, x_{i-1}, z_{s_i}, x_{i+1}, \ldots, x_n) \in \mathcal{C}(\Sigma_{\mathrm{COL}})_{s_i \to s'}$.
2. For each observable context $c \in \mathcal{C}(\Sigma_{\mathrm{COL}})_{s \to s'}$, for each indirect observer (obs, i) with $obs : s_1, \ldots, s_i, \ldots, s_n \to s$, and pairwise disjoint variables $x_1{:}s_1, \ldots, x_n{:}s_n$ not occurring in c,
   $c[obs(x_1, \ldots, x_{i-1}, z_{s_i}, x_{i+1}, \ldots, x_n)/z_s] \in \mathcal{C}(\Sigma_{\mathrm{COL}})_{s_i \to s'}$
   where $c[obs(x_1, \ldots, x_{i-1}, z_{s_i}, x_{i+1}, \ldots, x_n)/z_s]$ denotes the term obtained from c by substituting the term $obs(x_1, \ldots, x_{i-1}, z_{s_i}, x_{i+1}, \ldots, x_n)$ for z_s .

We assume that for any state sort $s \in S_{\mathrm{State}}$ there exists an observable context with application sort s.

The set of observable contexts determines, for any Σ-algebra A, an indistin-guishability relation, called *observational equality*. The observational equality on A is an S-sorted binary relation $\approx_{\Sigma_{\mathrm{COL}}, A}$ such that for any two elements $a, b \in A$, $a \approx_{\Sigma_{\mathrm{COL}}, A} b$ holds if either $a = b$ and a, b are observable (i.e. belong to a carrier set of observable sort $s \in S_{\mathrm{Obs}}$) or if a and b are hidden (i.e. belong to a carrier set of a state sort $s \in S_{\mathrm{State}}$) but cannot be distinguished by the application of observable contexts. The application of observable contexts is defined in the usual way apart from the fact that for variables in X (occurring in an observable context c) we consider only valuations in the generated part $\mathrm{Gen}_{\Sigma_{\mathrm{COL}}}(A)$ (i.e. junk values are disregarded because they should not contribute to distinguish elements). A Σ-algebra A is *fully abstract* if the observational equality coincides (on all carrier sets) with the set-theoretic equality.

The constructor and the observer operations induce certain constraints on Σ-algebras. First, since the constructor operations determine the values of inter-est, we require that the non-constructor operations should (up to observational equality) respect the constructor-generated part of an algebra, i.e. by the ap-plication of non-constructor operations one should at most be able to obtain elements which are observationally equal to some element of the constructor-generated part $\mathrm{Gen}_{\Sigma_{\mathrm{COL}}}(A)$. Technically this means that for a given Σ-algebra A we first consider the smallest Σ-subalgebra $\langle \mathrm{Gen}_{\Sigma_{\mathrm{COL}}}(A) \rangle_\Sigma$ of A containing the Σ_{COL}-generated part because this subalgebra represents the only elements a user can compute (over the loose carrier sets) by invoking operations of Σ. Then we require that each element of $\langle \mathrm{Gen}_{\Sigma_{\mathrm{COL}}}(A) \rangle_\Sigma$ is observationally equal to some element of the Σ_{COL}-generated part $\mathrm{Gen}_{\Sigma_{\mathrm{COL}}}(A)$ of A. This condition is called *reachability constraint*.

Furthermore, since the declaration of observer operations determines a par-ticular observational equality on any Σ-algebra A, the (interpretations of the) non-observer operations should respect this observational equality, i.e. a non-observer operation should not contribute to distinguish non-observable elements. To ensure this we require that the observational equality is a Σ-congruence on the subalgebra $\langle \mathrm{Gen}_{\Sigma_{\mathrm{COL}}}(A) \rangle_\Sigma$. (It is sufficient to consider $\langle \mathrm{Gen}_{\Sigma_{\mathrm{COL}}}(A) \rangle_\Sigma$ in-stead of A because computations performed by a user can only lead to elements in the Σ-subalgebra $\langle \mathrm{Gen}_{\Sigma_{\mathrm{COL}}}(A) \rangle_\Sigma$.) This condition is called *observability con-straint*.

A Σ-algebra A which satisfies both the reachability and the observability constraints induced by a COL-signature Σ_{COL} is called a Σ_{COL}-algebra (or simply a COL-algebra). Obviously any Σ-algebra A which is reachable and fully abstract w.r.t. Σ_{COL} is a Σ_{COL}-algebra. The class of all Σ_{COL}-algebras is denoted by $\text{Alg}_{\text{COL}}(\Sigma_{\text{COL}})$. It can be extended to a category by an appropriate notion of Σ_{COL}-morphism which reflects behavioral relationships between Σ_{COL}-algebras (see [4] for details).

The satisfaction of the reachability and observability constraints allows us to construct for each Σ_{COL}-algebra A its *black box view* which is a reachable and fully abstract algebra representing the behavior of A from the user's point of view. The black box view is constructed in two steps. First, we *restrict* to the Σ_{COL}-generated subalgebra $\langle \text{Gen}_{\Sigma_{\text{COL}}}(A) \rangle_\Sigma$ of A thus forgetting junk values. Then, we *identify* all elements of $\langle \text{Gen}_{\Sigma_{\text{COL}}}(A) \rangle_\Sigma$ which are observationally equal. Hence the black box view of a Σ_{COL}-algebra A is given by the quotient algebra of $\langle \text{Gen}_{\Sigma_{\text{COL}}}(A) \rangle_\Sigma$ w.r.t. $\approx_{\Sigma_{\text{COL}},A}$ which, for simplicity, will be denoted by $A/\approx_{\Sigma_{\text{COL}},A}$. Two Σ_{COL}-algebras A and B are *observationally equivalent*, denoted by $A \equiv_{\Sigma_{\text{COL}}} B$, if their black box views $A/\approx_{\Sigma_{\text{COL}},A}$ and $B/\approx_{\Sigma_{\text{COL}},B}$ are isomorphic Σ-algebras. Observationally equivalent Σ_{COL}-algebras are isomorphic w.r.t. Σ_{COL}-morphisms (see [4]).

A crucial concept to obtain a built-in behavioral semantics for specifications is the COL-*satisfaction relation*, denoted by $\models_{\Sigma_{\text{COL}}}$, which generalizes the standard satisfaction relation of first-order logic by abstracting with respect to reachability and observability. First, from the reachability point of view, the valuations of variables are restricted to the elements of the Σ_{COL}-generated part only. From the observability point of view, the idea is to interpret the equality symbol "=" occurring in a first-order formula φ not by the set-theoretic equality but by the observational equality of elements.

Definition 2 (COL-satisfaction relation). *For any COL-signature Σ_{COL}, the COL-satisfaction relation between Σ-algebras and first-order Σ-formulas (with variables in X) is denoted by $\models_{\Sigma_{\text{COL}}}$ and defined as follows. Let $A \in \text{Alg}(\Sigma)$.*

1. *For any two terms $t, r \in T_\Sigma(X)_s$ of the same sort s and for any valuation $\alpha : X \to \text{Gen}_{\Sigma_{\text{COL}}}(A)$, $A, \alpha \models_{\Sigma_{\text{COL}}} t = r$ holds if $I_\alpha(t) \approx_{\Sigma_{\text{COL}},A} I_\alpha(r)$.*
2. *For any Σ-formula φ and for any valuation $\alpha : X \to \text{Gen}_{\Sigma_{\text{COL}}}(A)$, $A, \alpha \models_{\Sigma_{\text{COL}}} \varphi$ is defined by induction over the structure of the formula φ in the usual way. In particular, $A, \alpha \models_{\Sigma_{\text{COL}}} \forall x{:}s.\, \varphi$ if for all valuations $\beta : X \to \text{Gen}_{\Sigma_{\text{COL}}}(A)$ with $\beta(y) = \alpha(y)$ for all $y \neq x$, $A, \beta \models_{\Sigma_{\text{COL}}} \varphi$.*
3. *For any Σ-formula φ, $A \models_{\Sigma_{\text{COL}}} \varphi$ holds if for all valuations $\alpha : X \to \text{Gen}_{\Sigma_{\text{COL}}}(A)$, $A, \alpha \models_{\Sigma_{\text{COL}}} \varphi$ holds.*

The notation $A \models_{\Sigma_{\text{COL}}} \varphi$ is extended in the usual way to classes of algebras and sets of formulas. The COL-satisfaction relation is defined not only for Σ_{COL}-algebras but also for arbitrary Σ-algebras which will be important when we consider proof techniques for behavioral refinement relations.

Fact 1 *Let Σ_{COL} be a COL-signature, let φ be a Σ-formula and let A be a Σ_{COL}-algebra. Then:*

$$A \models_{\Sigma_{COL}} \varphi \text{ if and only if } A/\approx_{\Sigma_{COL},A} \models \varphi.$$

The above definitions provide the basic ingredients that lead to the COL institution. In particular, the COL-satisfaction relation satisfies the satisfaction condition of institutions w.r.t. COL-signature morphisms which are standard signature morphisms fulfilling additional properties related to the preservation of constructor and observer operations (see [4] for details). A basic COL specification $SP_{COL} = \langle \Sigma_{COL}, Ax \rangle$ consists of a COL-signature Σ_{COL} and a set Ax of Σ-sentences (the axioms of the specification). The semantics of SP_{COL} is given by its signature Σ_{COL} and by its class of models:

$$Mod[SP_{COL}] = \{A \in \mathrm{Alg}_{COL}(\Sigma_{COL}) \mid A \models_{\Sigma_{COL}} Ax\}.$$

3   Behavioral Refinements

Generally, specification refinement is a relation between an abstract specification to be implemented and a more concrete specification which satisfies the requirements of the given abstract specification. Taking into account the observable behavior described by COL-specifications, a COL-specification SPI_{COL} is considered as a behavioral refinement of a COL-specification SP_{COL} if SPI_{COL} respects the behavioral properties required by SP_{COL}. Formally, a simple behavioral refinement relation between two COL-specifications can be defined by requiring that both specifications have the same signature and that the model class of the implementing specification SPI_{COL} is included in the model class of SP_{COL}. Remember that for the sake of simplicity we restrict to basic specifications in the framework of this paper.

Definition 3 (Behavioral refinement: simple case).
Let $SP_{COL} = \langle \Sigma_{COL}, Ax \rangle$ and $SPI_{COL} = \langle \Sigma_{COL}, AxI \rangle$ be two COL-specifications with the same signature Σ_{COL}. SPI_{COL} is a behavioral refinement of SP_{COL}, denoted by $SP_{COL} \rightsquigarrow SPI_{COL}$, if

$$Mod[SPI_{COL}] \subseteq Mod[SP_{COL}].$$

To prove that $SP_{COL} \rightsquigarrow SPI_{COL}$ holds, one has to show that:

$$SPI_{COL} \models_{\Sigma_{COL}} \varphi \text{ for all axioms } \varphi \in Ax,$$

i.e. that the axioms of SP_{COL} are observable consequences of SPI_{COL}. For this purpose one can directly apply the proof techniques for COL-specifications studied in [4] (since Σ_{COL} is also the signature of SPI_{COL}).

In general, however, one has to take into account that an implementation involves some construction step, an idea which has been formalized by the notion of constructor implementation introduced in [19] (and similarly in other implementation concepts; see [17, 10] for an overview). According to [19] an implementation constructor is a function which maps algebras over the signature of the implementing specification to algebras over the signature of the abstract specification. Since it is sufficient if an implementation construction is defined on the models of the implementing specification implementation constructors are, in general, partial functions. We assume that implementation constructions are performed in a uniform way, i.e. preserve isomorphisms. It is obvious that the concept of an implementation constructor can be easily transferred to behavioral refinements of COL-specifications. In particular, the requirement that isomorphisms are preserved means in the context of the COL institution that a COL-implementation constructor preserves COL-isomorphisms, i.e. observational equivalences of COL-algebras (see Section 2.2).

Definition 4 (COL-implementation constructor). *Let* Σ_{COL}, ΣI_{COL} *be two COL-signatures. A COL-implementation constructor from* ΣI_{COL} *to* Σ_{COL} *is a partial function* κ_{COL} : $\mathrm{Alg}_{\mathrm{COL}}(\Sigma I_{\mathrm{COL}}) \rightarrow \mathrm{Alg}_{\mathrm{COL}}(\Sigma_{\mathrm{COL}})$ *which is COL-iso-preserving, i.e. for all* $AI, BI \in \mathrm{Alg}_{\mathrm{COL}}(\Sigma I_{\mathrm{COL}})$,

if $AI \equiv_{\Sigma I_{\mathrm{COL}}} BI$ *and* $\kappa_{\mathrm{COL}}(AI)$ *is defined*
then $\kappa_{\mathrm{COL}}(BI)$ *is defined and* $\kappa_{\mathrm{COL}}(AI) \equiv_{\Sigma_{\mathrm{COL}}} \kappa_{\mathrm{COL}}(BI)$.

The definition domain of κ_{COL} *is denoted by* $Dom(\kappa_{\mathrm{COL}})$.

Using the notion of a COL-implementation constructor we can generalize Definition 3 to the case where the abstract and implementing specifications have different signatures.

Definition 5 (Behavioral refinement w.r.t. an implementation constructor). *Let* $\mathrm{SP}_{\mathrm{COL}}$, $\mathrm{SPI}_{\mathrm{COL}}$ *be two COL-specifications with signatures* Σ_{COL}, ΣI_{COL} *resp. and let* κ_{COL} *be a COL-implementation constructor from* ΣI_{COL} *to* Σ_{COL}. $\mathrm{SPI}_{\mathrm{COL}}$ *is a* behavioral refinement *of* $\mathrm{SP}_{\mathrm{COL}}$ *w.r.t.* κ_{COL}, *denoted by* $\mathrm{SP}_{\mathrm{COL}} \leadsto^{\kappa_{\mathrm{COL}}} \mathrm{SPI}_{\mathrm{COL}}$, *if*

$$\mathcal{M}od[\mathrm{SPI}_{\mathrm{COL}}] \subseteq Dom(\kappa_{\mathrm{COL}}) \ and \ \kappa_{\mathrm{COL}}(\mathcal{M}od[\mathrm{SPI}_{\mathrm{COL}}]) \subseteq \mathcal{M}od[\mathrm{SP}_{\mathrm{COL}}].$$

As discussed in [7] an important question is, of course, which implementation constructors are appropriate for behavioral refinements. As a first approach one could simply consider COL-signature morphisms σ_{COL} : $\Sigma_{\mathrm{COL}} \rightarrow \Sigma I_{\mathrm{COL}}$. Since COL is an institution, the corresponding COL-reduct functor $\_\_|_{\sigma_{\mathrm{COL}}}$: $\mathrm{Alg}_{\mathrm{COL}}(\Sigma I_{\mathrm{COL}}) \rightarrow \mathrm{Alg}_{\mathrm{COL}}(\Sigma_{\mathrm{COL}})$ preserves COL-isomorphisms, i.e. is a COL-implementation constructor. Hence it is tempting to consider COL-refinements where the syntactic relationship between the specification $\mathrm{SP}_{\mathrm{COL}}$ to be implemented and the implementing specification $\mathrm{SPI}_{\mathrm{COL}}$ is established by a COL-signature morphism. This approach has, however, a serious drawback because the implementing specification $\mathrm{SPI}_{\mathrm{COL}}$ usually has constructor and observer operations $\mathrm{OPI}_{\mathrm{Cons}}$, $\mathrm{OPI}_{\mathrm{Obs}}$ which are unrelated to the constructor and observer operations OP_{Cons}, OP_{Obs} of the specification SP_{COL} to be implemented. As a simple example we consider in Section 6 the implementation of sets by lists where the observer for sets is the membership test *isin* while the observer operations for lists are, as usual, the *head* and *tail* operations. Hence the COL-specifications of sets and lists cannot be related by a COL-signature morphism which would require the preservation of constructor and observer operations. This is the reason why we want to consider standard signature morphisms and their reduct functors as implementation constructors for COL-specifications.

But before let us still point out that our viewpoint has been inspired by the following remarkable sentences by Goguen and Malcolm [15]: *"Signature morphisms perform two distinct roles. One role is to express the importation of one specification into another... referred to as* horizontal composition... *so that when a specification of a class of objects is imported into a larger specification, the properties of the imported object classes are preserved. The other role performed by signature morphisms is to compare two different specifications. This is referred to as* vertical composition, *and pertains to relationships between layers... In such a case we would not expect that signature morphisms encapsulate object class specifications, but rather expect that signature morphisms preserve the behaviour of object classes... "*.

Interpreting these considerations in the COL framework this means that it is indeed adequate not to stick to COL-signature morphisms when we construct implementations. COL-signature morphisms are the appropriate tool to ensure encapsulation of COL-specifications (formally expressed by the satisfaction condition of an institution) which is indeed important when we construct large design specifications in a modular way (i.e. by *horizontal composition*). But when we discuss refinements by relating abstract and concrete specifications (*vertical composition*) this is a totally different matter where it makes no sense to talk about encapsulation.

Let us now consider two COL-specifications SP_{COL}, SPI_{COL} with signatures Σ_{COL}, ΣI_{COL} resp. together with a (standard) signature morphism $\sigma : \Sigma \to \Sigma I$ (where Σ and ΣI are the underlying standard signatures of Σ_{COL} and ΣI_{COL} resp.). Moreover, let us consider the reduct functor $\_|_\sigma : \mathrm{Alg}(\Sigma I) \to \mathrm{Alg}(\Sigma)$ as a partial function $\_|_\sigma : \mathrm{Alg}_{COL}(\Sigma I_{COL}) \to \mathrm{Alg}_{COL}(\Sigma_{COL})$,[3] where:

$\_|_\sigma(AI) \stackrel{def}{=} AI|_\sigma$ if $AI|_\sigma$ is a Σ_{COL}-algebra,

$\_|_\sigma(AI)$ is undefined otherwise.

Then we have the following fact (see Lemma 1 in [7]).

Fact 2 $\_|_\sigma : \mathrm{Alg}_{COL}(\Sigma I_{COL}) \to \mathrm{Alg}_{COL}(\Sigma_{COL})$ *is a* COL-*implementation constructor if* $\sigma(S_{Obs}) \subseteq SI_{Obs}$ *and* $\sigma(S_{Loose}) \subseteq SI_{Loose}$ *where* S_{Obs}, SI_{Obs} *are the observable sorts and* S_{Loose}, SI_{Loose} *are the loose sorts induced by* Σ_{COL}, ΣI_{COL} *respectively (see Section 2.2).*

---

[3] By abuse of notation we use the same symbol $\_|_\sigma$ for the (total) reduct functor on $\mathrm{Alg}(\Sigma I)$ and for its induced partial reduct function on $\mathrm{Alg}_{COL}(\Sigma I_{COL})$.

Let us stress that vertical signature morphisms used for refinements in [15] satisfy the above conditions due to the fixed universe of visible data. Hence vertical signature morphisms in the sense of [15] are special cases of COL-implementation constructors.

4   Proof Rules for Behavioral Refinements: Part I

In the following we are interested in proof rules for proving behavioural refinement relations $SP_{COL} \leadsto^{\kappa_{COL}} SPI_{COL}$. Obviously, the following basic proof rule follows directly from Definition 5:

For any COL-specifications $SP_{COL} = \langle \Sigma_{COL}, Ax \rangle$, $SPI_{COL} = \langle \Sigma I_{COL}, AxI \rangle$, and COL-implementation constructor $\kappa_{COL} : Alg_{COL}(\Sigma I_{COL}) \to Alg_{COL}(\Sigma_{COL})$:

$$\text{(Basic)} \quad \frac{\begin{array}{ll} \text{(B1)} & Mod[SPI_{COL}] \subseteq Dom(\kappa_{COL}), \\ \text{(B2)} & \kappa_{COL}(Mod[SPI_{COL}]) \models_{\Sigma_{COL}} Ax \end{array}}{SP_{COL} \leadsto^{\kappa_{COL}} SPI_{COL}}$$

Note that in (B2) $\kappa_{COL}(Mod[SPI_{COL}])$ consists of Σ-algebras and that $\models_{\Sigma_{COL}}$ has been defined not only for COL-algebras but for arbitary Σ-algebras. Of course, the central question is how to prove (B1) and (B2)? For this purpose, we will follow a strategy which consists of two crucial steps. The idea of the first step, elaborated in this section, is to consider instead of the COL-specification SPI_{COL} a standard specification SPI (over the FOLEq institution) and instead of κ_{COL} an implementation constructor κ on standard algebras. This idea is related to the (behavioral) refinement notion in [20] and to the concept of an abstractor implementation in [19] where behavioral refinement is, by definition, a relation between the standard interpretation of the implementing specification and the behavioral interpretation of the specification to be implemented. The idea of the second step, elaborated in Section 5, is to reduce the proof of consequences w.r.t. the COL-satisfaction relation $\models_{\Sigma_{COL}}$ to proofs w.r.t. the standard satisfaction relation of first-order logic with equality.

Let us start by considering COL-implementation constructors which are induced by standard implementation constructors. Given two signatures Σ and ΣI a standard implementation constructor from ΣI to Σ is a function $\kappa : Alg(\Sigma I) \to Alg(\Sigma)$ which is iso-preserving. For simplicity, let us assume that κ is total. Since any COL-algebra is also a (standard) algebra it is obvious that any implementation constructor $\kappa : Alg(\Sigma I) \to Alg(\Sigma)$ gives rise to a (partial) function $\kappa_{COL} : Alg_{COL}(\Sigma I_{COL}) \to Alg_{COL}(\Sigma_{COL})$ where:

$\kappa_{COL}(AI) \stackrel{\text{def}}{=} \kappa(AI)$ if $\kappa(AI)$ is a Σ_{COL}-algebra,
$\kappa_{COL}(AI)$ is undefined otherwise.

344 Michel Bidoit and Rolf Hennicker

If this partial function is COL-iso-preserving then κ_{COL} is a COL-implementation constructor *induced* by κ.[4] For instance, Fact 2 provides a simple criterion when reduct functors along standard signature morphisms induce COL-implementation constructors.

To state our second proof rule we consider for any COL-specification $\mathrm{SPI}_{\mathrm{COL}}$ its associated standard specification SPI obtained by forgetting the constructor and observer operations declared in $\mathrm{SPI}_{\mathrm{COL}}$. Then we have for any specifications $\mathrm{SP}_{\mathrm{COL}} = \langle \Sigma_{\mathrm{COL}}, \mathrm{Ax} \rangle$, $\mathrm{SPI}_{\mathrm{COL}} = \langle \Sigma I_{\mathrm{COL}}, \mathrm{AxI} \rangle$, $\mathrm{SPI} = \langle \Sigma I, \mathrm{AxI} \rangle$, where ΣI is the underlying standard signature of ΣI_{COL}, and for any $\kappa : \mathrm{Alg}(\Sigma I) \to \mathrm{Alg}(\Sigma)$ and COL-implementation constructor $\kappa_{\mathrm{COL}} : \mathrm{Alg}_{\mathrm{COL}}(\Sigma I_{\mathrm{COL}}) \to \mathrm{Alg}_{\mathrm{COL}}(\Sigma_{\mathrm{COL}})$ induced by κ:

$$(\mathrm{Forget}_{\mathrm{COL}}) \quad \frac{\begin{array}{ll} (\mathrm{F1}) & \kappa(\mathcal{M}od[\mathrm{SPI}]) \subseteq \mathrm{Alg}_{\mathrm{COL}}(\Sigma_{\mathrm{COL}}), \\ (\mathrm{F2}) & \kappa(\mathcal{M}od[\mathrm{SPI}]) \models_{\Sigma_{\mathrm{COL}}} \mathrm{Ax} \end{array}}{\begin{array}{ll} (\mathrm{B1}) & \mathcal{M}od[\mathrm{SPI}_{\mathrm{COL}}] \subseteq \mathit{Dom}(\kappa_{\mathrm{COL}}), \\ (\mathrm{B2}) & \kappa_{\mathrm{COL}}(\mathcal{M}od[\mathrm{SPI}_{\mathrm{COL}}]) \models_{\Sigma_{\mathrm{COL}}} \mathrm{Ax} \end{array}}$$

Lemma 1. *The proof rule* $(\mathrm{Forget}_{\mathrm{COL}})$ *is correct.*

Proof. Assume (F1) and (F2). To prove (B1) and (B2) let $AI \in \mathcal{M}od[\mathrm{SPI}_{\mathrm{COL}}]$. Then $AI \models_{\Sigma I_{\mathrm{COL}}} \mathrm{AxI}$ and hence, by Fact 1, $AI/\!\approx_{\Sigma I_{\mathrm{COL}}, AI} \models \mathrm{AxI}$. Thus $AI/\!\approx_{\Sigma I_{\mathrm{COL}}, AI} \in \mathcal{M}od[\mathrm{SPI}]$. By assumption (F1), $\kappa(AI/\!\approx_{\Sigma I_{\mathrm{COL}}, AI})$ is a Σ_{COL}-algebra. Hence, since κ_{COL} is induced by κ, $\kappa_{\mathrm{COL}}(AI/\!\approx_{\Sigma I_{\mathrm{COL}}, AI})$ is defined. Since $AI \equiv_{\Sigma I_{\mathrm{COL}}} AI/\!\approx_{\Sigma I_{\mathrm{COL}}, AI}$ and κ_{COL} is a COL-implementation constructor, $\kappa_{\mathrm{COL}}(AI)$ is defined as well. Hence, $\mathcal{M}od[\mathrm{SPI}_{\mathrm{COL}}] \subseteq \mathit{Dom}(\kappa_{\mathrm{COL}})$, i.e. (B1) holds.

Moreover, since $AI/\!\approx_{\Sigma I_{\mathrm{COL}}, AI} \in \mathcal{M}od[\mathrm{SPI}]$, the assumption (F2) implies that $\kappa(AI/\!\approx_{\Sigma I_{\mathrm{COL}}, AI}) \models_{\Sigma_{\mathrm{COL}}} \mathrm{Ax}$. Then, since κ_{COL} is induced by κ, we have $\kappa_{\mathrm{COL}}(AI/\!\approx_{\Sigma I_{\mathrm{COL}}, AI}) \models_{\Sigma_{\mathrm{COL}}} \mathrm{Ax}$. Since κ_{COL} is a COL-implementation constructor and $AI \equiv_{\Sigma I_{\mathrm{COL}}} AI/\!\approx_{\Sigma I_{\mathrm{COL}}, AI}$, we conclude that $\kappa_{\mathrm{COL}}(AI) \equiv_{\Sigma_{\mathrm{COL}}} \kappa_{\mathrm{COL}}(AI/\!\approx_{\Sigma I_{\mathrm{COL}}, AI})$. But then $\kappa_{\mathrm{COL}}(AI) \models_{\Sigma_{\mathrm{COL}}} \mathrm{Ax}$ holds as well. Hence, $\kappa_{\mathrm{COL}}(\mathcal{M}od[\mathrm{SPI}_{\mathrm{COL}}]) \models_{\Sigma_{\mathrm{COL}}} \mathrm{Ax}$, i.e. (B2) holds. $\qquad\square$

The proof rule $(\mathrm{Forget}_{\mathrm{COL}})$ is also complete if $\mathcal{M}od[\mathrm{SPI}]$ is closed under behavioral quotients, i.e. if any ΣI-algebra $AI \in \mathcal{M}od[\mathrm{SPI}]$ is a ΣI_{COL}-algebra such that $AI/\!\approx_{\Sigma I_{\mathrm{COL}}, AI} \in \mathcal{M}od[\mathrm{SPI}]$. (The proof relies on the fact that, under this assumption, $\mathcal{M}od[\mathrm{SPI}] \subseteq \mathcal{M}od[\mathrm{SPI}_{\mathrm{COL}}]$.)

According to the given proof rules (Basic) and $(\mathrm{Forget}_{\mathrm{COL}})$, the remaining task to prove behavioral refinements is to prove (F1) and (F2). A possible approach to discharge (F1) will be explained with the example in Section 6. A general technique to discharge (F2) is studied in the next section.

---

[4] In particular this means that κ is compatible with observational equivalences between COL-algebras, a property which is related to the notion of stability introduced by Schoett [20].

5   Proof Rules for Behavioral Refinements: Part II

In this section we focus on how to handle the proof obligation (F2) arising from the proof rule (Forget$_{\text{COL}}$). Basically we have to show that a set of formulas is behaviorally satisfied by some class of arbitrary algebras. The difficulty here is that these algebras are not COL-algebras w.r.t. the same COL-signature as the one used for the behavioral satisfaction considered, hence we cannot reuse the ideas and proof techniques detailed in [4]. However, we can rely on another idea, similar to the one introduced in [5], where the proof of behavioral consequences is replaced by the proof of standard consequences using a so-called *"lifting encoding"*. The main difference to [5] is, first, that in COL we have distinguished sets of observer and constructor operations which lead to much less observable contexts and constructor terms than in the case of partial observational equalities considered in [5]. Hence, the ideas of [5], which were mainly of theoretical interest, now become practically relevant. Secondly, in contrast to [5], we follow a coinductive style for the encoding of observable contexts which is more appropriate for proving behavioral theorems.

Our lifting encoding relies on a syntactic counterpart of both the constructor terms and the observable contexts. Therefore we need a few preliminary definitions. Remember that given a COL-signature Σ_{COL}, for each state sort s and observable sort s', $\mathcal{C}(\Sigma_{\text{COL}})_{s \to s'}$ denotes the set of the observable Σ_{COL}-contexts with application sort s and result sort s'.

Definition 6 (Lifted signature $\mathcal{AL}(\Sigma_{\text{COL}})$ associated to a COL-signature Σ_{COL}). *Let $\Sigma_{\text{COL}} = (\Sigma, \text{OP}_{\text{Cons}}, \text{OP}_{\text{Obs}})$ be a* COL-*signature. The induced lifted signature $\mathcal{AL}(\Sigma_{\text{COL}})$ is defined as follows:*

$$\mathcal{AL}(\Sigma_{\text{COL}}) \stackrel{\text{def}}{=} \Sigma \cup \Delta(\text{OP}_{\text{Cons}}) \cup \Lambda(\text{OP}_{\text{Cons}}) \cup \Delta(\text{OP}_{\text{Obs}}) \cup \Lambda(\text{OP}_{\text{Obs}})$$

where $\Delta(\text{OP}_{\text{Cons}})$ *is the signature fragment containing:*
- *for each constrained sort $s \in S_{\text{Cons}}$, a new sort $c[s]$;*
- *for each constructor cons $: s_1, \ldots, s_n \to s \in \text{OP}_{\text{Cons}}$,*
  a new operation $cons^* : \overline{s_1}, \ldots, \overline{s_n} \to c[s]$
  where here and in the following, for any sort $r \in S$, $\overline{r} \stackrel{\text{def}}{=} r$ if $r \in S_{\text{Loose}}$ and $\overline{r} \stackrel{\text{def}}{=} c[r]$ if $r \in S_{\text{Cons}}$;

where $\Lambda(\text{OP}_{\text{Cons}})$ *is the signature fragment containing:*
- *for each constrained sort $s \in S_{\text{Cons}}$,*
  a new (overloaded) operation $inj : c[s] \to s$;
- *for each constrained sort $s \in S_{\text{Cons}}$,*
  a new (overloaded) unary predicate $G : s$ on the sort s;

where $\Delta(\text{OP}_{\text{Obs}})$ *is the signature fragment containing:*
- *for each state sort $s \in S_{\text{State}}$ and observable sort $s' \in S_{\text{Obs}}$, if $\mathcal{C}(\Sigma_{\text{COL}})_{s \to s'}$ is not empty, a new sort $Cont[s \to s']$;*[5]

---

[5] Otherwise, i.e. if $\mathcal{C}(\Sigma_{\text{COL}})_{s \to s'}$ is empty, no new sort is added to reduce the syntactic complexity of the encoding.

- *for each direct observer $(obs, i) \in \mathrm{OP_{Obs}}$ with $obs : s_1, \ldots, s_i, \ldots, s_n \to$
  s', a new operation $obs_i^* : \overline{s_1}, \ldots, \overline{s_{i-1}}, \overline{s_{i+1}}, \ldots, \overline{s_n} \to Cont[s_i{\to}s'];$[6]*
- *for each indirect observer $(obs, i) \in \mathrm{OP_{Obs}}$ with $obs : s_1, \ldots, s_i, \ldots, s_n \to$
  s, and for all observable sorts $s' \in S_{\mathrm{Obs}}$ such that $\mathcal{C}(\Sigma_{\mathrm{COL}})_{s \to s'}$ is not
  empty,[7] new (overloaded) operations
  $obs_i^* : Cont[s{\to}s'], \overline{s_1}, \ldots, \overline{s_{i-1}}, \overline{s_{i+1}}, \ldots, \overline{s_n} \to Cont[s_i{\to}s'];$*

and where $\Lambda(\mathrm{OP_{Obs}})$ *is the signature fragment containing:*
- *for each new sort $Cont[s{\to}s']$, a new (overloaded) operation
  $apply : Cont[s{\to}s'], s \to s';$*
- *for each state sort $s \in S_{\mathrm{State}}$, a new (overloaded) binary predicate
  $\sim : s, s.$*

**Definition 7 (Lifting axioms $\mathrm{Ax}(\Sigma_{\mathrm{COL}})$ associated to a COL-signature
Σ_{COL}).** *Let $\Sigma_{\mathrm{COL}} = (\Sigma, \mathrm{OP_{Cons}}, \mathrm{OP_{Obs}})$ be a COL-signature. The lifting axioms
$\mathrm{Ax}(\Sigma_{\mathrm{COL}})$ associated to the lifted signature $\mathcal{AL}(\Sigma_{\mathrm{COL}})$ introduced in Definition 6
are defined as follows:*

$$\mathrm{Ax}(\Sigma_{\mathrm{COL}}) \stackrel{\mathrm{def}}{=} \mathrm{SGC}(\Delta(\mathrm{OP_{Cons}})) \cup \mathrm{Ax}_{\Sigma_{\mathrm{COL}}}(inj) \cup \mathrm{Ax}_{\Sigma_{\mathrm{COL}}}(G) \cup$$
$$\mathrm{FSGC}(\Delta(\mathrm{OP_{Obs}})) \cup \mathrm{Ax}_{\Sigma_{\mathrm{COL}}}(apply) \cup \mathrm{Ax}_{\Sigma_{\mathrm{COL}}}(\sim)$$

where $\mathrm{SGC}(\Delta(\mathrm{OP_{Cons}}))$ *is the sort-generation constraint induced by the new
   sorts $c[s]$ and by the new operations $cons^*$;*
where $\mathrm{Ax}_{\Sigma_{\mathrm{COL}}}(inj)$ *states that the operations inj are injective and homomorphic w.r.t. $\mathrm{OP_{Cons}}$, i.e. $\mathrm{Ax}_{\Sigma_{\mathrm{COL}}}(inj)$ is the union of:*
- *for each constrained sort $s \in S_{\mathrm{Cons}}$, the conditional equation:
  $\forall x, y : c[s].\ inj(x) = inj(y) \Rightarrow x = y;$*
- *for each constrained sort $s \in S_{\mathrm{Cons}}$ and constructor $cons \in \mathrm{OP_{Cons}}$ with
  $cons : s_1, \ldots, s_n \to s$, the implicitly universally quantified equation:
  $inj(cons^*(x_1, \ldots, x_n)) = cons(\zeta x_1, \ldots, \zeta x_n)$
  where here and in the following, $\zeta x_i = x_i$ if the sort of x_i is in S_{Loose},
  and $\zeta x_i = inj(x_i)$ otherwise (i.e., if the sort of x_i is of the form $c[s_i]$
  with $s_i \in S_{\mathrm{Cons}}$);*
where $\mathrm{Ax}_{\Sigma_{\mathrm{COL}}}(G)$ *is the set of sentences:*
- *for each constrained sort $s \in S_{\mathrm{Cons}}$: $\forall x : s.\ G(s) \Leftrightarrow \exists y : c[s].\ x = inj(y);$*
where $\mathrm{FSGC}(\Delta(\mathrm{OP_{Obs}}))$ *is the free sort-generation constraint induced by the
   signature fragment $\Delta(\mathrm{OP_{Obs}})$, i.e., by the new sorts $Cont[s{\to}s']$ and the new
   operations obs_i^*;*
where $\mathrm{Ax}_{\Sigma_{\mathrm{COL}}}(apply)$ *is the set of equations:*
- *for each direct observer $(obs, i) \in \mathrm{OP_{Obs}}$ with $obs : s_1, \ldots, s_i, \ldots, s_n \to$
  s', the equation:
  $\forall x_1 : \overline{s_1}, \ldots, x_{i-1} : \overline{s_{i-1}}, x_{i+1} : \overline{s_{i+1}}, \ldots, x_n : \overline{s_n}.\ \forall x_i : s_i.$
  $apply(obs_i^*(x_1, \ldots, x_{i-1}, x_{i+1}, \ldots, x_n), x_i) =$
  $\qquad\qquad\qquad\qquad obs(\zeta x_1, \ldots, \zeta x_{i-1}, x_i, \zeta x_{i+1}, \ldots, \zeta x_n);$*

---

[6] The existence of the direct observer (obs, i) entails the non-emptiness of
$\mathcal{C}(\Sigma_{\mathrm{COL}})_{s_i \to s'}$, hence the existence of the new sort $Cont[s_i{\to}s']$.

[7] Hence, the new sort $Cont[s{\to}s']$ exists, and so does the new sort $Cont[s_i{\to}s']$.

- *for each indirect observer* $(obs, i) \in OP_{Obs}$ *with* $obs : s_1, \ldots, s_i, \ldots, s_n \to s$, *and for all observable sorts* $s' \in S_{Obs}$ *such that the new sort* $Cont[s \to s']$ *exists, the equations:*

$$\forall c: Cont[s \to s'], x_1:\overline{s_1}, \ldots, x_{i-1}:\overline{s_{i-1}}, x_{i+1}:\overline{s_{i+1}}, \ldots, x_n:\overline{s_n}. \ \forall x_i:s_i.$$
$$apply(obs_i^*(c, x_1, \ldots, x_{i-1}, x_{i+1}, \ldots, x_n), x_i) =$$
$$apply(c, obs(\zeta x_1, \ldots, \zeta x_{i-1}, x_i, \zeta x_{i+1}, \ldots, \zeta x_n));$$

where $Ax_{\Sigma_{COL}}(\sim)$ *is the set of sentences:*

- *for each state sort* $s \in S_{State}$*:*

$$\forall x, y:s. \left(\bigwedge_{Cont[s \to s']} \forall c: Cont[s \to s']. \ apply(c, x) = apply(c, y) \right) \Leftrightarrow x \sim y .^{8}$$

The main idea underlying the above definitions is that, to any Σ-algebra A (where Σ is the standard signature underlying the COL-signature Σ_{COL}), corresponds a unique (up to isomorphism) "lifted" $AL(\Sigma_{COL})$-algebra $AL(A)$ which extends A (i.e., $AL(A)|_{\Sigma} = A$) and satisfies the lifting axioms $Ax(\Sigma_{COL})$.[9] Moreover, this lifted algebra $AL(A)$ is defined in a way which ensures a one to one correspondence between:

- values in the Σ_{COL}-generated part of the Σ-algebra A and values in the carriers of the new sorts $c[s]$ in $AL(A)$;
- observable contexts, together with appropriate valuations of their variables (in the Σ_{COL}-generated part of A), and values in the (carriers of the) syntactic counterparts $Cont[s \to s']$ in $AL(A)$. Hence the new sorts $Cont[s \to s']$ reflect the observable contexts in $C(\Sigma_{COL})_{s \to s'}$ and they are generated by the constructors obs_i^*. Note that our definition of the constructors of the new sorts $Cont[s \to s']$ follows the coinductive definition of observable contexts given in Definition 1.

We still need a further definition to state the main result of this section.

Definition 8 (Lifted formula $\mathcal{L}(\varphi)$). *Let* $\Sigma_{COL} = (\Sigma, OP_{Cons}, OP_{Obs})$ *be a COL-signature and* φ *be an arbitrary* Σ*-formula. The lifted formula* $\mathcal{L}(\varphi)$ *is the* $AL(\Sigma_{COL})$*-formula defined by:* [10]

$$\mathcal{L}(\varphi) \overset{def}{=} \left(\bigwedge_{y:s \in FreeVar(\varphi) \text{ and } s \in S_{Cons}} G(y) \right) \Rightarrow \varphi^*$$

where $FreeVar(\varphi)$ *denotes the free variables of* φ*, if any, and where* φ^* *is defined by induction on the structure of* φ *as follows:*

---

[8] This sentence is finite, since for any state sort $s \in S_{State}$, there is only a finite number of sorts $Cont[s \to s']$, where $s' \in S_{Obs}$ is an observable sort.

[9] In other words, the lifting axioms $Ax(\Sigma_{COL})$ induce a strongly persistent free functor from Σ-algebras to $AL(\Sigma_{COL})$-algebras.

[10] Note however that the only extra (not in Σ) symbols used in $\mathcal{L}(\varphi)$ are the predicates G and $\sim$. Moreover, note the similarity with [5, Def. 4.1-(iv)].

1. *If φ is an equation $l = r$ between two terms of sort s:*

   if $s \in S_{\mathrm{Obs}}$ then $\varphi^* \stackrel{\mathrm{def}}{=} l = r$, otherwise $s \in S_{\mathrm{State}}$ and $\varphi^* \stackrel{\mathrm{def}}{=} l \sim r$;

2. $(\neg\varphi)^* \stackrel{\mathrm{def}}{=} \neg(\varphi^*)$, $(\varphi_1 \wedge \varphi_2)^* \stackrel{\mathrm{def}}{=} \varphi_1^* \wedge \varphi_2^*$, $(\varphi_1 \vee \varphi_2)^* \stackrel{\mathrm{def}}{=} \varphi_1^* \vee \varphi_2^*$;

3. *If $s \in S_{\mathrm{Loose}}$ then $(\forall x{:}s.\ \varphi)^* \stackrel{\mathrm{def}}{=} \forall x{:}s.\ \varphi^*$,*

   otherwise $s \in S_{\mathrm{Cons}}$ and $(\forall x{:}s.\ \varphi)^* \stackrel{\mathrm{def}}{=} \forall x{:}s.\ [G(x) \Rightarrow \varphi^*]$.

Obviously $\mathcal{L}(\varphi)$ coincides with φ^* if φ is a closed Σ-formula.

Theorem 3. *Let $\Sigma_{\mathrm{COL}} = (\Sigma, \mathrm{OP}_{\mathrm{Cons}}, \mathrm{OP}_{\mathrm{Obs}})$ be a COL-signature. For any class $C \subseteq \mathrm{Alg}(\Sigma)$ of Σ-algebras and any Σ-formula φ, we have:*

$$C \models_{\Sigma_{\mathrm{COL}}} \varphi \quad \text{if and only if} \quad C + \langle \mathcal{AL}(\Sigma_{\mathrm{COL}}), \mathrm{Ax}(\Sigma_{\mathrm{COL}}) \rangle \models \mathcal{L}(\varphi).$$

Proof. For lack of space we only detail here the main steps of the proof.

Step 1: In a first step we introduce a semantic lifting of Σ-algebras as follows. Let $\mathcal{L}(\Sigma)$ be the signature Σ enriched by the predicates G and $\sim$ (as they are introduced in Definition 6). Remember that $\mathcal{L}(\varphi)$ is indeed a $\mathcal{L}(\Sigma)$-formula, as pointed out in Definition 8. Now the semantic lifting $\mathcal{L}(A)$ of a Σ-algebra A is defined as being the unique $\mathcal{L}(\Sigma)$-algebra extension of A defined by:

1. $\mathcal{L}(A)|_{\Sigma} \stackrel{\mathrm{def}}{=} A$;
2. For any constrained sort $s \in S_{\mathrm{Cons}}$, and $a \in \mathcal{L}(A)_s = A_s$, $G^{\mathcal{L}(A)}(a)$ if and only if $a \in \mathrm{Gen}_{\Sigma_{\mathrm{COL}}}(A)_s$;
3. For any state sort $s \in S_{\mathrm{State}}$, and $a, b \in \mathcal{L}(A)_s = A_s$, $a \sim^{\mathcal{L}(A)} b$ if and only if $a \approx_{\Sigma_{\mathrm{COL}}, A, s} b$.

Now we have:

$$A \models_{\Sigma_{\mathrm{COL}}} \varphi \quad \text{if and only if} \quad \mathcal{L}(A) \models \mathcal{L}(\varphi).$$

The proof of this fact is similar to the proof of Theorem 4.2 in [5].

Step 2: In a second step we prove that, for any Σ-algebra A:

$$\mathcal{AL}(A)|_{\mathcal{L}(\Sigma)} = \mathcal{L}(A)$$

This indeed results directly from Definitions 6 and 7 (see the comments after the later definition), and from the definitions of $\mathrm{Gen}_{\Sigma_{\mathrm{COL}}}(A)$ and of $\approx_{\Sigma_{\mathrm{COL}}, A}$.

Step 3: From the above we conclude that:

$A \models_{\Sigma_{\mathrm{COL}}} \varphi$ if and only if, according to Step 1,

$\mathcal{L}(A) \models \mathcal{L}(\varphi)$ if and only if, according to Step 2,

$\mathcal{AL}(A)|_{\mathcal{L}(\Sigma)} \models \mathcal{L}(\varphi)$ if and only if, according to the satisfaction condition in CFOLEq, $\mathcal{AL}(A) \models \mathcal{L}(\varphi)$. This is enough to conclude the proof of the theorem, since $\{\,\mathcal{AL}(A) \mid A \in C\,\} = C + \langle \mathcal{AL}(\Sigma_{\mathrm{COL}}), \mathrm{Ax}(\Sigma_{\mathrm{COL}}) \rangle$.   □

As a direct consequence of Theorem 3 we obtain the following proof rule. For any COL-specification $\mathrm{SP}_{\mathrm{COL}} = \langle \Sigma_{\mathrm{COL}}, \mathrm{Ax} \rangle$, CFOLEq-specification $\mathrm{SPI} = \langle \Sigma I, \mathrm{AxI} \rangle$ and for any $\kappa : \mathrm{Alg}(\Sigma I) \to \mathrm{Alg}(\Sigma)$:

$$\text{(Lifting)} \quad \frac{\text{(L)} \quad \kappa(\mathcal{M}od[\mathrm{SPI}]) + \langle \mathcal{AL}(\Sigma_{\mathrm{COL}}), \mathrm{Ax}(\Sigma_{\mathrm{COL}}) \rangle \models \mathcal{L}(\mathrm{Ax})}{\text{(F2)} \quad \kappa(\mathcal{M}od[\mathrm{SPI}]) \models_{\Sigma_{\mathrm{COL}}} \mathrm{Ax}}$$

6 Example: Implementation of Sets by Non-redundant Lists

In this section we illustrate the use of our proof rules and proof techniques on a small but non-trivial example.

6.1 The Behavioral Refinement Relation

The following specification SET-COL specifies properties of sets over a loose domain of arbitrary elements. As constructors for sets we use the operations *empty* and *add* and as an observer for sets we use the membership test *isin*.[11]

spec SET-COL =
      sorts *bool, elem, set*
      ops   *true, false* : *bool*;
             empty : *set*;
             add : *elem* × *set* → *set*;
             remove : *elem* × *set* → *set*;
             isin : *elem* × *set* → *bool*;
      constructors *empty, add*
      observer (*isin, 2*)
      axioms
      $\forall x, y$: *elem*; *s* : *set*
      %% standard axioms for booleans, plus
- $isin(x, empty) = false$
- $isin(x, add(x, s)) = true$
- $x \neq y \Rightarrow isin(x, add(y, s)) = isin(x, s)$
- $isin(x, remove(x, s)) = false$
- $x \neq y \Rightarrow isin(x, remove(y, s)) = isin(x, s)$
- $add(x, add(x, s)) = add(x, s)$
- $add(x, add(y, s)) = add(y, add(x, s))$

end

As a refinement for sets we consider a classical implementation of sets by non-redundant lists where the set operation *add* is implemented in a such a way that it inserts an element *x* into a list only if *x* does not yet occur in the list and the set operation *remove* just removes the first occurrence of an element.

spec LIST-COL =
      sorts *bool, elem, list*
      ops   *true, false* : *bool*;
             empty : *list*;
             cons : *elem* × *list* → *list*;
             head : *list* → *elem*;
             tail : *list* → *list*;
             isin : *elem* × *list* → *bool*;

---

[11] All our examples are expressed using a syntactic sugar similar to the one of CASL [8].

$add : elem \times set \to set;$

$remove : elem \times set \to set;$

constructors $empty, cons$

observers $head, tail$

axioms

$\forall x, y : elem;\ l : list$

%% standard axioms for booleans, plus

- $head(cons(x, l)) = x$
- $tail(cons(x, l)) = l$
- $isin(x, empty) = false$
- $isin(x, cons(x, l)) = true$
- $x \neq y \Rightarrow isin(x, cons(y, l)) = isin(x, l)$
- $isin(x, l) = true \Rightarrow add(x, l) = l$
- $isin(x, l) = false \Rightarrow add(x, l) = cons(x, l)$
- $remove(x, empty) = empty$
- $remove(x, cons(x, l)) = l$
- $x \neq y \Rightarrow remove(x, cons(y, l)) = cons(y, remove(x, l))$

end

To state the refinement relation we still need an appropriate COL-implementation constructor. Since LIST-COL provides already all set operations the simple idea is to forget the list operations $cons, head$ and $tail$ and to perform an appropriate renaming to match the sorts set and $list$. For this purpose we consider the (standard) signature morphism $\sigma_{\text{SET}as\text{LIST}} : \Sigma_{\text{SET}} \to \Sigma_{\text{LIST}}$ where Σ_{SET} denotes the underlying (standard) signature of $Sig[\text{SET-COL}]$,[12] similarly, Σ_{LIST} denotes the underlying (standard) signature of $Sig[\text{LIST-COL}]$ and $\sigma_{\text{SET}as\text{LIST}}(set) \overset{\text{def}}{=} list$, $\sigma_{\text{SET}as\text{LIST}}(x) \overset{\text{def}}{=} x$ otherwise.

Since $bool$ and $elem$ are the observable sorts of both SET-COL and LIST-COL, Fact 2 implies that the reduct functor $\_\_|_{\sigma_{\text{SET}as\text{LIST}}} : \text{Alg}(\Sigma_{\text{SET}}) \to \text{Alg}(\Sigma_{\text{LIST}})$ on standard algebras induces a COL-implementation constructor:

$$\_\_|_{\sigma_{\text{SET}as\text{LIST}}} : \text{Alg}_{\text{COL}}(Sig[\text{LIST-COL}]) \to \text{Alg}_{\text{COL}}(Sig[\text{SET-COL}]).$$

Then, we claim that LIST-COL is indeed a behavioral refinement of SET-COL w.r.t. $\_\_|_{\sigma_{\text{SET}as\text{LIST}}}$, i.e. SET-COL $\rightsquigarrow^{\_\_|_{\sigma_{\text{SET}as\text{LIST}}}}$ LIST-COL.

6.2   Proof of the Refinement

For the proof of the above refinement relation the combination of the rules (Basic) and (Forget$_{\text{COL}}$) provided in Section 4 shows that it is enough to consider the standard specification LIST obtained from LIST-COL by omitting the declarations of the constructors and observers. Then we have the following two proof obligations:

---

[12] i.e. Σ_{SET} consists of all sorts and operations of the COL-signature $Sig[\text{SET-COL}]$ without any constructor or observer declaration.

(F1) $Mod[\text{LIST}]|_{\sigma_{\text{SETasLIST}}} \subseteq \text{Alg}_{\text{COL}}(Sig[\text{SET-COL}])$

(F2) $Mod[\text{LIST}]|_{\sigma_{\text{SETasLIST}}} \models_{Sig[\text{SET-COL}]} \varphi$ for all axioms φ of SET-COL

To prove (F1) one has to check that the reducts of all models of LIST satisfy the reachability and observability constraints induced by $Sig[\text{SET-COL}]$; see Section 2. To check the reachability constraint we consider the generated parts of sort *set* which are constructed by *empty* and *add*. Obviously, due to the implementation of *add*, those parts represent lists without duplicates. Moreover, from the axioms of LIST it follows that the only non-constructor operation *remove* does not introduce duplicates, i.e. the constructor-generated parts are already subalgebras and therefore the reachability constraint is trivially satisfied. For the proof of the observability constraint one has to show that both non-observer operations *add* and *insert* are congruent, i.e. are compatible with the observational equality for sets. For this purpose one can use the lifting encoding considered below and prove the congruence axioms for $\sim$. Another strategy would be first to verify (F2) and then to conclude that both *add* and *remove* are congruent operations since the axioms of SET-COL provide sufficiently complete definitions for *add* and for *remove*.[13]

For the proof of (F2) we will apply the rule (Lifting) of the previous section which says that it is sufficient to prove that for all axioms φ of SET-COL,

$$Mod[\text{LIST}]|_{\sigma_{\text{SETasLIST}}} + \langle \mathcal{AL}(Sig[\text{SET-COL}]), \text{Ax}(Sig[\text{SET-COL}]) \rangle \models \mathcal{L}(\varphi)$$

or equivalently, since the (standard) satisfaction relation is compatible with reducts of (standard) algebras,

$$\text{LIST}^* + \langle \mathcal{AL}(Sig[\text{SET-COL}]), \text{Ax}(Sig[\text{SET-COL}]) \rangle \models \mathcal{L}(\varphi),$$

where LIST^* is the same specification as LIST but with the sort *list* renamed into *set*. For this purpose, we first compute, according to Definition 6, the lifted signature:

$$\mathcal{AL}(Sig[\text{SET-COL}]) \overset{\text{def}}{=} \Sigma_{\text{SET}} \cup \Delta(\text{OP}_{\text{Cons}}) \cup \Lambda(\text{OP}_{\text{Cons}}) \cup \Delta(\text{OP}_{\text{Obs}}) \cup \Lambda(\text{OP}_{\text{Obs}})$$

where $\Delta(\text{OP}_{\text{Cons}})$ consists of
  − the new sort $c[set]$
  − the new operations $empty^* : c[set]$; $add^* : elem \times c[set] \rightarrow c[set]$;
where $\Lambda(\text{OP}_{\text{Cons}})$ consists of
  − the new operation $inj : c[set] \rightarrow set$;
  − the new unary predicate $G : set$;
where $\Delta(\text{OP}_{\text{Obs}})$ consists of
  − the new sort $Cont[set \rightarrow bool]$;
  − the new operation $isin^* : elem \rightarrow Cont[set \rightarrow bool]$;
and where $\Lambda(\text{OP}_{\text{Obs}})$ consists of
  − the new operation $apply : Cont[set \rightarrow bool] \times set \rightarrow bool$;
  − the new binary predicate $\sim : set \times set$.

---

[13] This idea follows a general result presented in [6] for observational logic and equational specifications which still has to be extended to COL and conditional equations.

In the next step, we compute, according to Definition 7, the lifted axioms:

$$\text{Ax}(Sig[\text{SET-COL}]) \stackrel{\text{def}}{=} \text{SGC}(\Delta(\text{OP}_{\text{Cons}})) \cup \text{Ax}_{Sig[\text{SET-COL}]}(inj) \cup \text{Ax}_{Sig[\text{SET-COL}]}(G) \cup$$
$$\text{FSGC}(\Delta(\text{OP}_{\text{Obs}})) \cup \text{Ax}_{Sig[\text{SET-COL}]}(apply) \cup \text{Ax}_{Sig[\text{SET-COL}]}(\sim)$$

where $\text{SGC}(\Delta(\text{OP}_{\text{Cons}}))$ is the sort-generation constraint
  generated type $c[set] ::= empty^* \mid add^*(elem; c[set])$;
where $\text{Ax}_{Sig[\text{SET-COL}]}(inj)$ consists of
  – the conditional equation:
    $\forall s, s':c[set].\ inj(s) = inj(s') \Rightarrow s = s'$;
  – the implicitly universally quantified equations:
    $inj(empty^*) = empty$
    $inj(add^*(x, s)) = add(x, inj(s))$
where $\text{Ax}_{Sig[\text{SET-COL}]}(G)$ consists of
  – $\forall s:set.\ G(s) \Leftrightarrow \exists s':c[set].\ s = inj(s')$;
where $\text{FSGC}(\Delta(\text{OP}_{\text{Obs}}))$ is the free sort-generation constraint
  free type $Cont[set \rightarrow bool] ::= isin^*(elem)$;
where $\text{Ax}_{Sig[\text{SET-COL}]}(apply)$ is the equation:
  – $\forall x:elem, s:set.\ apply(isin^*(x), s) = isin(x, s)$
where $\text{Ax}_{Sig[\text{SET-COL}]}(\sim)$ is the sentence:
  – $\forall s, s':set.$
    $(\forall c:Cont[set \rightarrow bool].\ apply(c, s) = apply(c, s')\ ) \Leftrightarrow s \sim s'$

According to the above axioms, the unary predicate G characterizes those lists (of type set because of the performed renaming) which are built with $empty$ and add. These lists are exactly the lists with no duplicates used for the representation of sets. On the other hand, the above axioms provide also a specification of the binary predicate $\sim$ which relates any two lists containing the same elements (independently of the order and the number of occurrences of these elements).

Let us now compute the lifting $\mathcal{L}(\varphi)$ of all axioms φ of SET-COL which leads, according to Definition 8, to the following set of sentences:

$\forall x, y : elem;\ s : set$
- $isin(x, empty) = false$
- $G(s) \Rightarrow isin(x, add(x, s)) = true$
- $G(s) \Rightarrow (x \neq y \Rightarrow isin(x, add(y, s)) = isin(x, s))$
- $G(s) \Rightarrow isin(x, remove(x, s)) = false$ \hfill (1)
- $G(s) \Rightarrow (x \neq y \Rightarrow isin(x, remove(y, s)) = isin(x, s))$
- $G(s) \Rightarrow add(x, add(x, s)) \sim add(x, s)$
- $G(s) \Rightarrow add(x, add(y, s)) \sim add(y, add(x, s))$ \hfill (2)

Of course, the remaining task is to show that the lifted axioms given above are consequences of LIST* + $\langle \mathcal{AL}(Sig[\text{SET-COL}]), \text{Ax}(Sig[\text{SET-COL}]) \rangle$. In most cases the proof is already a direct consequence of the axioms of the LIST specification without the need of the predicates G and $\sim$. The situation is, however, different for the sentences (1) and (2). In the case of (1) the relativization w.r.t. $G(s)$

is indeed crucial, because $isin(x, remove(x, s)) = false$ holds only for those list interpretations of s which have no duplicates, but these are exactly the lists characterized by the predicate symbol G. In the case of (2), the use of $\sim$ instead of "$=$" is crucial as well, since two lists (also two non-redundant lists) are different if they contain the same elements but in a different order. In this case they are, however, observationally equal which is axiomatized by $\sim$.[14]

7   Conclusion

We have provided proof techniques to verify behavioral refinements of COL-specifications based on a reduction to first-order specifications and (standard) inductive reasoning. Hence, any inductive theorem prover can be used to prove behavioral refinements. Let us stress that we do not use coinduction to prove the behavioral validity of equations but we use an encoding of observational equalities and generated parts which works for arbitrary first-order formulas. Typical proofs of consequences of the encoding are then performed by induction on the (coinductive) structure of observable contexts. Next steps are the extension of our approach to take into account structured specifications and the study of further examples of implementation constructors like, e.g., specification extension.

Acknowledgement. We are grateful to the anonymous referee of a previous version of this paper for valuable remarks.

References

1. E. Astesiano, H.-J. Kreowski, and B. Krieg-Brückner, editors. *Algebraic Foundations of Systems Specification.* Springer, 1999.
2. M. Bidoit, M.-V. Cengarle, and R. Hennicker. Proof systems for structured specifications and their refinements. In *[1]*, chapter 11, pages 385–433. Springer, 1999.
3. M. Bidoit and R. Hennicker. Modular correctness proofs of behavioural implementations. *Acta Informatica,* 35:951–1005, 1998.
4. M. Bidoit and R. Hennicker. Constructor-based observational logic. *Journal of Logic and Algebraic Programming,* 67 (1-2):3–51, 2006. Preliminary version available at www.lsv.ens-cachan.fr/Publis/PAPERS/PDF/BID-HEN-JLAP.pdf.
5. Michel Bidoit and Rolf Hennicker. Behavioural theories and the proof of behavioural properties. *Theoretical Computer Science,* 165(1):3–55, 1996.
6. Michel Bidoit and Rolf Hennicker. Observer complete definitions are behaviourally coherent. In *Proc. OBJ/CafeOBJ/Maude Workshop at Formal Methods'99, Toulouse, France, Sep. 1999,* pages 83–94. THETA, 1999.

---

[14] As a side remark the reader may note that in the lifted sentence requiring the idempotency of *add* the predicate symbol $\sim$ could indeed be replaced by "$=$" due to the relativization w.r.t. $G(s)$ and the implementation of *add*.

7. Michel Bidoit and Rolf Hennicker. Externalized and internalized notions of behavioral refinement. In Dang Van Hung and Martin Wirsing, editors, *Proceedings of the 2nd International Colloquium on Theoretical Aspects of Computing (ICTAC'05)*, volume 3722 of *Lecture Notes in Computer Science*, pages 334–350, Hanoi, Vietnam, October 2005. Springer.

8. Michel Bidoit and Peter D. Mosses. CASL *User Manual – Introduction to Using the Common Algebraic Specification Language*, volume 2900 of *Lecture Notes in Computer Science*. Springer, 2004.

9. R. Diaconescu and K. Futatsugi. *CafeOBJ Report: The Language, Proof Techniques, and Methodologies for Object-Oriented Algebraic Specification*, volume 6 of *AMAST Series in Computing*. World Scientific, 1998.

10. H. Ehrig and H.-J. Kreowski. Refinement and implementation. In *[1]*, chapter 7, pages 201–242. Springer, 1999.

11. J. Goguen and J.A. Meseguer. Universal realization, persistent interconnection and implementation of abstract modules. In *Proc. ICALP'82*, volume 140 of *Lecture Notes in Computer Science*, pages 265–281. Springer, 1982.

12. J. Goguen and G. Roşu. Hiding more of hidden algebra. In J.M. Wing, J. Woodcock, and J. Davies, editors, *Proc. Formal Methods (FM'99)*, volume 1709 of *Lecture Notes in Computer Science*, pages 1704–1719. Springer, 1999.

13. Joseph Goguen and Rod Burstall. Institutions: abstract model theory for specification and programming. *Journal of the ACM*, 39(1):95–146, 1992.

14. J. Loeckx, H.-D. Ehrich, and M. Wolf. *Specification of Abstract Data Types*. Wiley and Teubner, 1996.

15. G. Malcolm and J. Goguen. Proving correctness of refinement and implementation. Technical Report PRG-114, Oxford University Computing Laboratory, 1994.

16. Peter D. Mosses, editor. CASL *Reference Manual*, volume 2960 of *Lecture Notes in Computer Science*. Springer, 2004.

17. F. Orejas, M. Navarro, and A. Sanchez. Implementation and behavioural equivalence. In *Recent Trends in Data Type Specification*, volume 655 of *Lecture Notes in Computer Science*, pages 93–125. Springer, 1993.

18. D. Sannella and A. Tarlecki. On observational equivalence and algebraic specification. *Journal of Computer and System Sciences*, 34:150–178, 1987.

19. D.T. Sannella and A. Tarlecki. Toward formal development of programs from algebraic specifications: implementation revisited. *Acta Informatica*, 25:233–281, 1988.

20. O. Schoett. Data abstraction and correctness of modular programming. Technical Report CST-42-87, University of Edinburgh, 1987.

21. Andrzej Tarlecki. Institutions: An Abstract Framework for Formal Specification. In *[1]*, chapter 4, pages 105–130. Springer, 1999.

22. Martin Wirsing. Algebraic Specification. In J. van Leeuwen, editor, *Handbook of Theoretical Computer Science*, chapter 13, pages 676–788. Elsevier Science Publishers B.V., 1990.

The Reactive Engine for Modular Transducers

Gérard Huet and Benoît Razet

INRIA Rocquencourt,
BP 105, 78153 Le Chesnay Cedex, France

Abstract. This paper explains the design of the second release of the
Zen toolkit [5–7]. It presents a notion of reactive engine which simulates
finite-state machines represented as shared *aums* [8]. We show that it
yields a modular interpreter for finite state machines described as *local*
transducers. For instance, in the manner of Berry and Sethi, we define a
compiler of regular expressions into a scheduler for the reactive engine,
chaining through aums labeled with phases — associated with the letters
of the regular expression. This gives a modular composition scheme for
general finite-state machines.

Many variations of this basic idea may be put to use according to cir-
constances. The simplest one is when aums are reduced to dictionaries,
i.e. to (minimalized) acyclic deterministic automata recognizing finite
languages. Then one may proceed to adding supplementary structure
to the aum algebra, namely non-determinism, loops, and transduction.
Such additional choice points require fitting some additional control to
the reactive engine. Further parameters are required for some functionalities. For instance, the local word access stack is handy as an argument to
the output routine in the case of transducers. Internal virtual addresses
demand the full local state access stack for their interpretation.

A characteristic example is provided, it gives a complete analyser for
compound substantives. It is an abstraction from a modular version of
the Sanskrit segmenter presented in [9]. This improved segmenter uses
a regular relation condition relating the phases of morphology generation, and enforcing the correct geometry of morphemes. Thus we obtain
compound nouns from iic*.(noun+iic.ifc), where iic and ifc are the respectively prefix and suffix substantival forms for compound formation.

Dedicated to Joseph Goguen for his 65th birthday

1   Regular Morphology

We first consider the simplest framework for finite automata, where the state
transition graph is a dictionary structure (lexical tree or trie). Such structures
represent acyclic deterministic finite-state automata, with maximal sharing of
initial paths. Every state is accessible from the initial state, and we may also
assume that every state is on an accepting path. When we minimize the tree
as a dag, we obtain the corresponding minimal deterministic automaton. Such

K. Futatsugi et al. (Eds.): Goguen Festschrift, LNCS 4060, pp. 355–374, 2006.
© Springer-Verlag Berlin Heidelberg 2006

automata recognize finite languages. They are adequate for representing the lexicons of natural languages.

In a framework of generative morphology, we want to model the construction of lexemes from smaller chunks called morphemes: radical stems, prefixes and suffixes. It is convenient to sort the morphemes into categories, and to enforce structural conditions on these categories, restricting the geometry of lexemes. For instance, we may describe this geometry by a regular expression over the alphabet of lexical categories. The language generated by the regular expression, substituting each category by its corresponding morpheme lexicon, is recognized by a modular reactive engine, which chains the morphemes dictionary lookup with transitions corresponding to the regular expression recognizer. We shall use for this setup variants of the compiling algorithm of Berry and Sethi [2].

1.1   Automaton Interface

We use as algorithmic description language Pidgin ML, a core applicative subset of Objective Caml. Thus our algorithms may be read as rigorous higher-order inductive definitions, while being directly executable, in the spirit of *literate programming*.

We first recall the basic structures of the Zen toolkit [5].

We use as basic alphabet the natural numbers provided by the hardware processor:

module Word : **sig**

```
type letter = int
and word = list letter;
end;
```

Thus the basic morphology operations will rely on list processing, and *not* on string processing (and certainly not on encoding formats such as Unicode-UTF8, which are meant for data exchange portability and should not be used for core computation).

Here is the interface to our simplistic automata, reduced to deterministic transitions over a lexicon tree. Each state is labeled with a boolean (indicating whether or not it is an accepting state), and points to the list of its successor states, labeled with a letter.

module Auto : **sig**

```
type auto = [ State of ( bool × deter ) ]
and deter = list (Word. letter × auto );
end;
```

We assume that at most one transition issued from a given state is labeled with a given letter. The datatype *auto* is here isomorphic to lexical trees, or dictionary, also called *tries*. We may also assume that dead alleys, i.e. states which do not have an accepting node as a substructure, are ruled out. Thus the contraction of the tree as a dag, using for instance the corresponding instance of the *sharing functor* [5], yields the minimal automaton that recognizes the finite language stored in the dictionary.

1.2 Dispatching

We call *phases* the lexical categories, which constitute the alphabet of the regular expression defining the morphological geometry. We compile this regular expression using the Berry-Sethi method, which linearizes the expression, and computes the local automaton associated to this linearization [2, 3].

We recall that local automata (also called Glushkov automata) are finite automata such that all transitions labeled with a given letter lead to the same state, characteristic of this letter. States may thus be named with letters, here phases. It is this locality condition which is a key to modularity.

A local automaton is described by an *initial* phase, a set of *terminal* phases, here represented as a boolean function over phases, and a *dispatch* transition function, mapping each phase to a set of following phases, sequentialized here as a list. In the notations of [2], *initial* is called *1*, *dispatch* is called *follow*, and *terminal* is implicit from the use of an endmarker symbol. In the terminology of Eilenberg [4], the set of non-empty words recognized by a local automaton is a local set over phases.

In the Zen toolkit implementation, the *Dispatch* module is actually generated by meta-programming, i.e. it compiled from the regular expression, as we shall explain in section 4.

1.3 Scheduling

We are now ready to start the description of the reactive engine, as a functor taking a module *Dispatch* as parameter, and using its *dispatch* function as a local scheduler. Here is the corresponding specification of our *React* module. We assume the utility programming functions *fold_right* (list iterator), *assoc*, *length*, *mem*, etc. from the *List* standard library.

```
module React
   (Dispatch : sig
      type phase = α;
      value transducer : phase → auto;
      value initial : phase;
      value terminal : phase → bool;
      value dispatch : phase → list phase;
   end) = struct
type input = Word.word
and backtrack = [ Advance of phase and input ]
and resumption = list backtrack;
```

A *resumption* value stores as a datum what is necessary to resume our reactive engine as a coroutine.

The scheduler gets its phase transitions from *dispatch*. It respects the order of dispatching.

```
value schedule phase input cont =
  let add phase cont = [ Advance phase input :: cont ] in
  fold_right add (dispatch phase) cont;
```

1.4 React

The reactive engine originates from the Sanskrit segmenter described in [9], generalized to the framework of mixed automata defined in [8].

Here we have a much simpler framework, since we do not have transducer output, but we get a *modular* interpreter, driven by the phase scheduler.

In the following definition, *phase* is the current phase, *input* is the input tape represented as a *word*, *back* is the backtrack stack of type *resumption*, and *state* is the current state of type *auto*. We favor deterministic transitions within a phase to non-deterministic transitions to the next phase(s). Within a phase, we favor longer words over shorter ones. Phase transitions are effected in *dispatch* order. We have a mutual inductive definition between the reactive engine, reading forward, and the continuation manager, backtracking on failure.

```
exception Finished ;

value rec react phase input back state = match state with
  [ State (b,det) →
  let deter cont = match input with
          [ [] → continue cont
          | [ letter :: rest ] →
            try let state' = assoc letter det in
                react phase rest cont state'
            with [ Not_found → continue cont ]
          ] in
  if b (* accepting *) then
      if input =[] (* end of input *) then
          if terminal phase then back (* solution found *)
          else continue back
      else let cont = schedule phase input back in
          deter cont
  else deter back
  ]
and continue = fun
  [ [] → raise Finished
  | [ resume :: back ] → match resume with
    [ Advance phase input →
          react phase input back (transducer phase)
    ]
  ]
];
```

1.5   Usage

The initialization of the reactive engine consists in setting the backtrack stack to the single initial state given by *Dispatch.initial*, *input* being initialized to the full sentence:

value init_react sentence = [Advance initial sentence];

We may now recognize a string as belonging to the rational language described by the regular expression by calling the reactive continuation manager with this initial resumption:

value react1 sentence = continue (init_react sentence);

If the sentence belongs to the language, *react1* will return with a resumption value, otherwise it will throw the exception *Finished*. The resumption value is not of use in this simple model, where the interpreter is used as a mere recognizer. In more elaborate versions below, *react* may be used as a coroutine in order to compute a stream of transductions.

Note that classical formal languages theory abstracts a language as a set of words, or occasionally as a multiset (hiding structural idempotence) when multiplicities matter. Here we hide structural commutativity as well, obtaining streams of solutions, where computational details such as fairness, essential for completeness, may be revealed and discussed.

1.6   Correctness, Completeness

Let us be given a module *Dispatch* by its components *phase* (an ordered list of discrete phase values defining the alphabet), *initial* (the initial phase), and functions *transducer* : phase $\rightarrow$ auto, *terminal* : phase $\rightarrow$ bool and *dispatch* : phase $\rightarrow$ list phase.

Let $L(\phi)$ be the language recognized by the automaton *transducer*(ϕ), for a given phase ϕ. We assume that $L(initial)$ is the singleton $\{\epsilon\}$ where ϵ is the empty word [], that $L(\phi)$ does not contain ϵ for any other phase ϕ, and that for every phase ϕ the list *dispatch*(ϕ) does not contain *initial*. These invariants will be enforced by the Berry-Sethi compiler presented in section 4.

Let us say that a sequence $((\phi_1, w_1), \dots (\phi_n, w_n))$ is a valid *analysis* of a given word w whenever, taking $\phi_0 = initial$, we get $w = w_1 \cdot w_2 \cdot \dots w_n$ with $(0 < i \leq n)$ $w_i \in L(\phi_i)$, $(0 \leq i < n)$ $\phi_{i+1} \in dispatch(\phi_i)$, and $terminal(\phi_n) = True$. For $i > 0$, we know from the assumptions above that $L(\phi_i)$ does not contain the empty word, so there is a finite number of such analyses.

We define a total ordering on analyses by the lexicographical ordering generated by $(\phi, w) < (\phi', w')$ iff either ϕ precedes ϕ' in the common dispatch list where ϕ and ϕ' belong, or else $\phi = \phi'$ and w' is a strict initial prefix of w.

The correction and completeness of the *react* algorithm may be established by proving that it generates the set of valid analyses of an input word w in the sense that it implicitly builds a sequence of pairs of analyses of w and resumptions $((\alpha_1, r_1), \dots (\alpha_N, r_N))$ such that, taking $r_0 = init\_react\ w$, for each i $(0 \leq i < N)$

the evaluation of *continue* r_i terminates with value r_{i+1}, and the evaluation of *continue* r_N raises the exception *Finished*. Furthermore the list $(\alpha_1, ...\alpha_N)$ contains all the valid analyses of w, listed increasingly with respect to the above ordering.

We shall not give a formal proof of this rather fastidious property, which can be established by computational induction.

We remark that this argument makes explicit the fact that, within a given phase, we search for longer partial solutions before shorter ones. This is a rather arbitrary heuristic, which is convenient for the segmenting application.

2   Modular Aums

So far our automata have been mere recognizers for finite sets of words, i.e. dictionaries. Chaining them through phases, we may for instance model simple segmentation problems, where a sentence is defined as a list of words separated by blanks or punctuation signs, and words are defined as compounds of morphemes, according to prefix, suffix, or other finite-state regimes. Such a segmenter may be composed with a tagger, when the word dictionaries are decorated with morphological derivation annotations, using the structure of *revmaps* [5], which allows efficient sharing of morphological regularities.

We now allow more complex automata for the various phases. For instance, we may allow a notion of transition with virtual addresses, allowing both non-deterministic moves (including ϵ-transitions), and cycles.

Virtual addresses, as opposed to pointers and explicit cyclic structures, provide a declarative mechanism respecting sharing. In the original presentation of aums [8], two varieties of virtual addresses are proposed: absolute addresses, indexing a state by its absolute access path in the forest of deterministic skeletons, and relative addresses, indexing a state in the current covering trie by the shorthest path in the tree, encoded as a *differential word* pairing a natural number (how many levels in the tree you should go up) with a word (indexing the target state down from the closest common ancestor). These differential words are used for instance in the *revmap structure*, to store the reverse morphology.

In the next section, we shall ignore relative addresses, which necessitate a slightly more complex apparatus for their proper evaluation, since sharing makes ambiguous the inheritance relation, and thus access paths must be maintained in the automaton structure for proper interpretation. We shall present first simple absolute addresses. Furthermore, the role of the forest index will be played by the phase: to each phase corresponds a unique *auto* structure, covering all the states pertaining to this phase.

2.1   Mixed Automata with Virtual Absolute Addresses

A transition (w, v) recognizes word w on the input tape (the "guard" of the transition), and jumps to the state absolutely adressed by v in the next phase.

module Auto : **sig**

type transition = (Word. word × Word. word)
and choices = list transition ;

type auto = [State **of** (deter × choices)]
and deter = list (Word. letter × auto);
end ;

We take as convention that the state $State(d, c)$ is accepting iff c is not empty. We now define acceptance as the condition on external transitions (w, v) when the input is empty, the (next) phase is terminal, and the access parameter v verifies a *final* condition which we shall not precise further. Typically, v is final if it is empty or if it consists in a special end of sentence marker.

2.2 Service Routine

Our resumptions are now more complex, since we have non-deterministic choice points:

type backtrack =
  [Choose **of** phase **and** input **and** auto **and** choices
  | Advance **of** phase **and** input **and** word
  ]
and resumption = list backtrack ;

exception Finished ;

Here are two service routines to manage guard management.

exception Guard ;
value rec advance n w = **if** n = 0 **then** w **else** **match** w **with**
  [[] → **raise** Guard
  | [_ :: tl] → advance (n−1) tl
  ];

Thus $advance\ n\ [a_1;\ ...\ a_N] = [a_p;\ ...\ a_N]$, where $p = N - n$, whenever $n \leq N$; otherwise the exception *Guard* is raised.

(* [access : phase → word → auto] *)
value access phase = acc (transducer phase)
  where rec acc state = **fun**
    [[] → state
    | [c :: rest] → **match** state **with**
      [State (deter ,_) →
        acc (List. assoc c deter) rest
      ]
    ];

2.3   React for Aums

We use a similar *schedule* function as previously, it now stores the v access path for the next phase transition.

```
value schedule phase input v cont =
  let add phase cont = [ Advance phase input v :: cont ]
  in fold_right add (dispatch phase) cont;
```

We are now ready to present the reactive engine. It consists in three simultaneous inductions, the main one *react* managing the deterministic search, while stacking non-deterministic choice points, the second *choose* managing non-deterministic jumps, and the third *continue* backtracking in case of dead end. We favor deterministic transitions over non-deterministic ones.

```
(* phase is the parsing phase,
   input is the input tape represented as a word,
   back is the backtrack stack of type resumption,
   state is the current state of type auto *)
value rec react phase input back state =
  match state with
  [ State (det,choices) →
    (* we explore the deterministic space first *)
    let cont = if choices=[] then back else
        [ Choose phase input state choices :: back ]
    in match input with
        [ [] → continue cont
        | [ letter :: rest ] →
            try let next_state = assoc letter det in
                react phase rest cont next_state
            with [Not_found → continue cont]
        ]
  ]
and choose phase input back state = fun
  [ [] → continue back
  | [ (w,v) :: others ] →
    let cont = if others=[] then back else
        [ Choose phase input state others :: back ]
    in try let tape = advance (length w) input in
            if tape = [] (* input finished *) then
                if terminal phase && final v then cont
                else continue cont
            else continue (schedule phase tape v cont)
        with [Guard → continue cont]
  ]
and continue = fun
  [ [] → raise Finished
  | [ resume :: back ] → match resume with
```

```
[ Choose phase input state choices →
     choose phase input back state choices
| Advance phase input word →
     try let next_state = access phase word
         in react phase input back next_state
     with [Not_found → continue back]
]
];
```

Finally, here is the initialisation routine, building the initial resumption:

value init_react input = [Advance initial input []];

As previously, we may recognize a sentence using:

value react1 sentence = continue (init_react sentence);

2.4   Correctness, Completeness

Similarly to the previous section, we may prove the correctness and completeness of the construction, provided the guard w of each non-deterministic transition is non-empty. We may refine this condition as follows.

Definition: Guard condition. There is no cycle of transitions of an aum all of which have an empty guard: $(\epsilon, w_1); (\epsilon, w_2); ...(\epsilon, w_n)$. By cycle we mean that, for some access word w_0 in the current phase ϕ_0 leading in $transducer(\phi_0)$ to state σ_0, σ_0 has among its choices (ϵ, w_1), ϕ_1 in $dispatch(\phi_0)$ with w_1 leading in $transducer(\phi_1)$ to state σ_1, etc, until $\sigma_n = \sigma_0$.

   We claim that *react* terminates on an input word whenever the guard condition is verified. Note that this is a global condition on the family of aums, which requires the knowledge of the phase transition relation, but which may be checked in time linear in the cumulated size of the aum family.

3   Modular Aum Transducers

We now give the final refinement of our construction, with aums having both local and global virtual addresses.

3.1   Transducers

module Auto : **sig**

type continuation = (Word.word × Word.word)
and transition =
```
   [ External of (Word.word × continuation)
   | Internal of (Word.word × Word.delta)
   ];
```

```
type auto = [ State of (deter × choices) ]
and deter = list (Word.letter × auto)
and choices = list transition;
end;
```

An internal transition $Internal(w, d)$ recognizes w on the input tape and jumps to the state relatively addressed by d within the same phase. This uses the notion of *differential word* [5] from module Word:

```
type delta = (int × word); (* differential words *)
```

A differential word is a notation permitting to retrieve a word w from another word w' sharing a common prefix. It denotes the minimal path connecting the words in a trie, as a sequence of ups and downs: if $\delta = (n, u)$ we go up n times and then down along word u. In order to interpret the n part, we need to keep the stack of states leading locally to the current state. We keep along this *stack* the corresponding word path as well — this is useful as a parameter to the output computation.

An external transition $External(w, c)$ recognizes w on the input tape and executes the continuation c in a following phase. A continuation (u, v) returns words u as output parameter and v as access parameter in the next phase transducer.

As above we define acceptance as the condition on external transition when the input is empty, the phase is terminal, and the access parameter v verifies a *final* condition which we shall not precise further.

3.2 Modular Transducers

We now produce output, as words labeled by their phase.

```
type input = Word.word
and output = list (phase × Word.word);
```

The access stack has a letter component and a state component. The state component is necessary to interpret the part of the internal virtual address which concerns going up, whereas the letter component, i.e. the absolute name of the state in the current phase is useful for computing the transducer output.

```
type stack = list (Word.letter × auto);

type backtrack =
    [ Choose of phase and input and output
            and auto and stack and choices
    | Advance of phase and input and output and Word.word
    ]
and resumption = list backtrack;
```

Since the *Advance* resumption has now an output component and an access component (anticipating a prefix of the next phase component), we parameterize the scheduler accordingly:

```
value schedule phase input output access cont =
  let add phase cont =
    [ Advance phase input output access :: cont ]
  in fold_right add (dispatch phase) cont;
```

The service routine *access* manages the access stack, the functions *pop* and *push* are used to interpret internal jumps.

```
(* access : phase → word → ( auto × stack ) *)
value access phase = acc (transducer phase) []
  where rec acc state stack = fun
  [ [] → (state, stack)
  | [ c :: rest ] → match state with
    [ State (deter, _) →
      acc (assoc c deter) [ (c, state) :: stack ] rest
    ]
  ];

value rec pop n state stack =
  if n=0 then (state, stack)
  else match stack with
    [ [] → raise (Failure "Wrong_Internal_jump")
    | [ (_, st) :: rest ] → pop (n−1) st rest
    ]
and push w state stack = match w with
    [ [] → (state, stack)
    | [ c :: rest ] → match state with
      [ State (deter, _) →
        push rest (assoc c deter) [ (c, state) :: stack ]
      ]
    ];

value jump (n,w) state stack =
  let (state0, stack0) = pop n state stack
  in push w state0 stack0;
```

We provide the access stack as an output parameter via an extracting routine:

```
value extract stack (_,(u,_)) =
  fold_left unstack u stack
    where unstack acc (c, _) = [ c :: acc ];
```

3.3   Modular Reacting Transducers

We have a similar structure of three mutually recursive functions, but now *choose* has two cases, for the two transition constructors.

```
value rec react phase input output back stack state =
  match state with
  [ State (det,choices) →
    let cont = if choices=[] then back else
      [Choose phase input output state stack choices :: back]
    in match input with
      [ [] → continue cont
      | [ letter :: rest ] →
        try let state' = assoc letter det
          and stack' = [ (letter,state) :: stack ] in
            react phase rest output cont stack' state'
        with [ Not_found → continue cont ]
      ]
  ]
and choose phase input output back state stack = fun
  [ [] → continue back
  | [ External((w,(u,v)) as rule) :: others ] →
    let cont = if others=[] then back else
      [Choose phase input output state stack others :: back]
    in try let tape = advance (length w) input
        and out = [(phase,extract stack rule) :: output]
        in if tape = [] (* input finished *) then
            if terminal phase && final v then (out,cont)
            else continue cont
          else continue (schedule phase tape out v cont)
      with [ Guard → continue cont ]
  | [ Internal(w,delta) :: others ] →
    let cont = if others=[] then back else
      [Choose phase input output state stack others :: back]
    in try let tape = advance (length w) input
        and (state',stack') = jump delta state stack
        in react phase tape output cont stack' state'
      with [ Guard → continue cont ]
  ]
and continue = fun
  [ [] → raise Finished
  | [ resume :: back ] → match resume with
      [ Choose phase input output state stack choices →
          choose phase input output back state stack choices
      | Advance phase input output word →
        try let (state',stack') = access phase word
          in react phase input output back stack' state'
        with [ Not_found → continue back ]
      ]
  ];
```

3.4   Correctness, Completeness

The definitions of trace and analysis may be extended to the case of transducers, and the correctness and completeness of our engine may be formally proved in the sense that all transductions of the input word are properly generated, for a notion of left-to-right transduction. We omit here the full formal development.

In the case of non overlapping junction transductions, as defined in [9], the construction simplifies, since *Internal* transitions are not needed. The proofs of termination, correctness and completeness of the reactive engine are carried out in full in [9], for the simple case of one phase junction relations verifying a non-overlapping criterion. This criterion allows parallel computation of the relation along phases, without the need to cascade the transductions. Furthermore, such relations are invertible, and the reactive engine may thus be used to invert euphony and return segmentation solutions, even when the euphony relation is not length-preserving.

Other variations may be considered, since the presence or absence of output transitions is orthogonal to the structure of virtual addresses. We have considered virtual addresses of two kinds, internal and external. We may also imagine other encodings of jumps, potentially relevant for specific applications. For instance, specific encodings, relying on the fact that the underlying alphabet is boolean, may be used to represent boolean circuits, in the manner of BDD structures.

The general problem of compiling an arbitrary finite-state machine description into some variety of our *aum* structures is not addressed in the current paper. This problem has many degrees of freedom, since there is a choice between mapping state transitions into the deterministic skeleton, on one hand, and the non-deterministic choices sequences, on the other; in the latter case, there is a further choice between External and Internal jumps. Finally, the partition into phases may be more or less coarse, and extra encoding letters, disjoint from the input alphabet, may be used to attach orphan states. We should not expect one uniform best solution to this problem anyway, and compiling strategies may well depend on the application domain.

Remark.

In [9], section 8.1, the recursive call from *choose* calls *react* with *occ* parameter v, instead of *rev v* as effected above for *next_stack*. This is a local optimisation for the case of *sandhi*, where the junction rules are such that the length of component v is at most 1.

4   Dispatch Synthesis from Regular Expressions

We now explain how to synthesize the *dispatch* function from a regular expression representation of the phase language, using the Berry-Sethi algorithm [2]. The basic idea is that we compose a number of finite automata/transducers, each named with a *phase*. Phases are the letters of an alphabet, and we define the admissible joint behaviour of our automata as a rational language over the phase alphabet, specified by a regular expression.

4.1   Regular Expressions and Their Linearization

Here is the type of regular expressions. The type parameter α is used to abstract from the symbol representation.

```
type regexp α =
  [ One
  | Symb of α
  | Union of regexp α and regexp α
  | Conc of regexp α and regexp α
  | Star of regexp α
  | Epsilon of regexp α
  | Plus of regexp α
  ];
```

We use a specific constructor *Plus* rather than defining R^+ as the macro $R \cdot R^*$, because of the blow-up due to its non-linearity.

We mark symbols with an integer to linearize the regular expression.

```
type marked α = (α × int);
```

A symbol s is mapped to $(s, 0)$ if it occurs only one, and to $(s, 1)$, $(s, 2)$, etc. otherwise. Marked symbols are used as states of the recognizing automaton. The type *local* represents local automata, in the sense of Eilenberg, as a 4-tuple defining its initial state, the other states, the transitions, and the terminal states:

```
type local α =
  ( marked α × list (marked α)
  × list (marked α × list (marked α))
  × list (marked α)
  );
```

We skip the details of the linearization function *mark*, which is straightforward. The function *mark* takes as argument a *regexp* α, and returns a pair of type *regexp(marked* $\alpha) \times list(marked\ \alpha)$, consisting of the marked expression, and the list of marked symbols which will be used as states of the local automaton.

4.2   The Berry-Sethi Compiler

We basically follow the construction given in [2], with the addition of the *Plus* operation. We need an intermediate structure of *discriminating* regular expressions, which makes explicit whether the associated rational language contains the empty word ϵ or not.

```
type d_regexp α =
  [ DOne
  | DSymb of α
  | DUnion of bool and d_regexp α and d_regexp α
  | DConc of bool and d_regexp α and d_regexp α
  | DStar of d_regexp α
```

```
  | DEpsilon of d_regexp α
  | DPlus of bool and d_regexp α
  ];
```

We can tell in unit time this property with function *delta*, and translate in linear time a regexp in a discriminating regexp with function *discr*.

```
value delta = fun
  [ DOne → True
  | DSymb _ → False
  | DUnion b _ _ | DConc b _ _ → b
  | DStar _ | DEpsilon _ → True
  | DPlus b _ → b
  ];
```

```
(* discr : regexp α → d_regexp α *)
value rec discr = fun
  [ One → DOne
  | Symb s → DSymb s
  | Union e1 e2 →
      let de1 = discr e1 and de2 = discr e2 in
      DUnion (delta de1 || delta de2) de1 de2
  | Conc e1 e2 →
      let de1 = discr e1 and de2 = discr e2 in
      DConc (delta de1 && delta de2) de1 de2
  | Star e → DStar (discr e)
  | Epsilon e → DEpsilon (discr e)
  | Plus e →
      let de = discr e in
      DPlus (delta de) de
  ];
```

The core of the algorithm is the computation of sets *first*, *follow* and *last*.

```
(* first : list α → d_regexp α → list α *)
value rec first l = fun
  [ DOne → l
  | DSymb d → [ d :: l ]
  | DUnion _ e1 e2 → first (first l e2) e1
  | DConc _ e1 e2 →
      if delta e1 then first (first l e2) e1
      else first l e1
  | DStar e | DEpsilon e | DPlus _ e → first l e
  ];
```

```
(* follow : α → regexp α → list (α × list α) *)
value follow initial exp =
  let rec f1 exp l fol =
```

```
match exp with
[ DOne → fol
| DSymb d → [ (d,l) :: fol ]
| DUnion _ e1 e2 →
    let fol2 = f1 e2 l fol in f1 e1 l fol2
| DConc _ e1 e2 →
    let fol2 = f1 e2 l fol in
    let l1 = if delta e2 then first l e2
                else first [] e2 in
    f1 e1 l1 fol2
| DStar e | DPlus _ e →
    let l_res = first l e in
    f2 e l_res fol
| DEpsilon e → f1 e l fol
]
and f2 exp l fol = (* (first [] exp) already in l *)
  match exp with
  [ DOne → fol
  | DSymb d → [ (d,l) :: fol ]
  | DUnion _ e1 e2 →
      let fol2 = f2 e2 l fol in f2 e1 l fol2
  | DConc _ e1 e2 →
      let b1 = delta e1
      and b2 = delta e2 in
      if b1 (* l1 and l2 in l *)
      then if b2
              then f2 e1 l (f2 e2 l fol)
              else f1 e1 (first [] e2) (f2 e2 l fol)
      else if b2
              then f2 e1 (first l e2) (f1 e2 l fol)
              else f1 e1 (first [] e2) (f1 e2 l fol)
  | DStar e | DEpsilon e | DPlus _ e → f2 e l fol
  ] in
let fol_sets = f1 exp [] []
and initials = first [] exp in
[ (initial,initials) :: fol_sets ];
```

Functions *f1* and *f2* both compute the follow sets of Berry-Sethi but with different assertions on their arguments; precisely, a call *(f1 exp l fol)* is such that first elements of *exp* are not in *l*, and the contrary assertion obtains for *f2*. Thus we never attempt to add elements already present in *l*, which maintains a constant cost of adding an element in *l*.

```
(* last : α → regexp α → list α *)
value last initial e =
  let rec last_rec l = fun
```

```
    [ DOne → 1
    | DSymb d → [ d :: 1 ]
    | DUnion _ e1 e2 →
         last_rec (last_rec 1 e2) e1
    | DConc _ e1 e2 →
          if delta e2 then last_rec (last_rec 1 e2) e1
          else last_rec 1 e2
    | DStar e | DEpsilon e | DPlus _ e → last_rec 1 e
    ] in
  let 1 = last_rec [] e in
  if delta e then [ initial :: 1 ] else 1;
```

Now we have all the ingredients to compile a regular expression:

```
(* compile : marked α → regexp α → local α *)
value compile initial exp =
  let (exp_m, states) = mark exp in
  let exp_d = discr exp_m in
  let fol = follow initial exp_d
  and lasts = last initial exp_d in
  (initial , states , fol , lasts );
```

4.3   Parametric Regular Expressions

We now define systems of regular expressions over parametric alphabets whose symbols are associated to aums. Meta-variables allow sharing in such descriptions. We skip the details of the syntax, and present just an example of such a finite machine description, actually a subproblem of Sanskrit morphology, namely noun phrases representing compound substantives.

```
initial init epsilon_aum

alphabet noun ; iic ; ifc end

automaton Disp
  node SUBST = iic* . (noun | iic.ifc)
end
```

Here we specify that the initial phase is called init, that the user must provide a value epsilon_aum for the aum recognizing just the empty word, as well as aum values noun, iic and ifc for recognizing the corresponding languages. We are interested in the language iic* . (noun | iic.ifc). In the intended application, *SUBST* is the language of substantive forms, containing *noun* forms as well as compounds, formed with prefix *iic* forms which may be iterated, and suffix *ifc* forms.

We skip the details of the parsing of such a description. In the current syntax, we allow systems of regular expressions, allowing sharing, and the compiler unfolds the system into a flattened expression.

We use the meta-programming facilities provided by the Camlp4 preprocessor, which allows macro-generation of an Ocaml program at the level of abstract syntax. Skipping the details of this meta-programming, we obtain mechanically, for the above example, the following module text.

```
module Automata (Auto : sig type auto = 'a; end) =
  struct
    type auto_vect =
      { epsilon_aum : Auto.auto;
        noun : Auto.auto; iic : Auto.auto; ifc : Auto.auto };
    module Disp (Fsm : sig value autos : auto_vect; end) =
      struct
        type phase =
          [ Init | Iic1 | Noun | Iic2 | Ifc ];
        value transducer = fun
          [ Init  → Fsm.autos.epsilon_aum
          | Iic1  → Fsm.autos.iic
          | Noun  → Fsm.autos.noun
          | Iic2  → Fsm.autos.iic
          | Ifc   → Fsm.autos.ifc
          ];
        value dispatch =
          fun
          [ Init  → [ Iic1; Noun; Iic2 ]
          | Iic1  → [ Iic1; Noun; Iic2 ]
          | Noun  → []
          | Iic2  → [ Ifc ]
          | Ifc   → []
          ];
        value initial = Init;
        value terminal phase = List.mem phase [ Noun; Ifc ];
      end;
  end;
```

We now have all the components we wish to assemble, since the module instanciation (Automata Auto), for Auto one of the aum description modules given in the previous sections, creates a module Dispatch=(Disp Fsm) having the right functionality, with module Fsm holding the aum implementations. In this simple example these implementations are the various lexicons corresponding to the respective lexical categories. In the Sanskrit platform, these aums are decorated with non-deterministic transitions (using external addressing) corresponding to sandhi prediction.

Remarks. 1. During the Berry-Sethi compiling process, the candidate regular expression is linearized when a phase occurs more than once. However, the corresponding automata are shared via the *transducer* component, recovering the proper sharing.

2. Our Sanskrit platform[1] now uses this modular methodology, which enforces the right geometry for morphological *chunks*, taking care of preverb affixes, proper recognition of compound forms and periphrastic verbal constructions, and proper analysis of absolutive forms (with suffixes in *-tvā* for roots and *-ya* for verbs admitting preverbs).

3. As usual, we may augment our automata descriptions with weights reflecting (possibly conditional) probabilities in order to get stochastic automata whose behaviour reflects hidden Markov chains in the data. Note that the correctness criteria are invariant with the permutation of choices induced by priority selection according to these weights.

4.4   A Variant Using Antimirov's Compiling Algorithm

V. Antimirov proposed in [1] another algorithm for compiling regular expressions, using a notion of *partial derivative*. This algorithm produces automata that may be significantly smaller than the ones obtained by the Berry-Sethi algorithm. Such automata do not have the locality condition, and now the modularity of the construction obtains by a more complex mapping, since the transducer invocation does not simply depend on the states, but on the transitions. We shall not develop further this variant construction in this paper.

5   Conclusion

We have presented a methodology for constructing finite-state machines, such as finite automata and transducers, in a modular way. Regular expressions over an alphabet of phases express a composition of machines under a finite-state-controlled constraint. This corresponds to considering a regular expression not as the mere denotation of a rational language over the alphabet of its symbols seen as string generators, but rather as a rational polynomial over its symbols, abstracting themselves rational sets. The algebraic property of closure of rational sets over substitution (mapping symbols to rational sets), together with the local automaton representation of finite-state machines, provide the natural foundation for the modular composition of finite-state machines.

Our mechanism allows the controlled interaction of machines compiled as mixed automata (aums). This is useful for instance for shallow parsing in computational linguistics applications. For the Sanskrit platform built by the first author, this allows to build a tagger composing machines which invert phonology (sandhi analysis) and morphology, with separate machines for distinct lexical classes, constrained by the geometrical conditions defining admissible compounds, preverb management, and periphrastic constructions with auxiliary verbs.

Our design exploits and justifies our functional programming methodology as follows:

---

[1] http://sanskrit.inria.fr/

- Applicative programming leads to robust well-structured programs, amenable to formal proofs and to journal publication, in the spirit of literate programming — all our programs are rigorously expressed as inductive definitions over higher-order types.
- Functionality is essential to the concise expression of powerful control paradigms such as continuations, essential for the definition of coroutine interpreters for non-deterministic search.
- Modularity of the programming language is the essence of the parametricity underlying algebraic closure operations, and thus is an essential abstraction paradigm.
- Powerful macro-generation mechanisms lead to an effective meta-programming methodology, tailoring general algorithms to the specific needs of applications.
- Despite this very high-level view of software architecture, the resulting programs are efficient enough for their integration in real size applications, as witnessed by their use in computational linguistic platforms [9].

References

1. V. Antimirov. Partial derivatives of regular expressions and finite automaton constructions. *Theoretical Computer Science*, 155:291–319, 1996.
2. G. Berry and R. Sethi. From regular expressions to deterministic automata. *Theoretical Computer Science*, 48:117–126, 1986.
3. J. Berstel and J.-E. Pin. Local languages and the Berry-Sethi algorithm. *Theoretical Computer Science*, 155:439–446, 1996.
4. S. Eilenberg. *Automata, Languages, and Machines, volume A*. Academic Press, 1974.
5. G. Huet. The Zen computational linguistics toolkit. Technical report, ESSLLI Course Notes, 2002. !http://pauillac.inria.fr/ huet/ZEN/esslli.pdf!
6. G. Huet. The Zen computational linguistics toolkit: Lexicon structures and morphology computations using a modular functional programming language. In *Tutorial, Language Engineering Conference LEC'2002*, 2002.
7. G. Huet. Linear contexts and the sharing functor: Techniques for symbolic computation. In F. Kamareddine, editor, *Thirty Five Years of Automating Mathematics*. Kluwer, 2003. !http://pauillac.inria.fr/ huet/PUBLIC/DB.pdf!
8. G. Huet. Automata mista. In N. Dershowitz, editor, *Verification: Theory and Practice: Essays Dedicated to Zohar Manna on the Occasion of His 64th Birthday*, pages 359–372. Springer-Verlag LNCS vol. 2772, 2004. !http://pauillac.inria.fr/ huet/PUBLIC/zohar.pdf!
9. G. Huet. A functional toolkit for morphological and phonological processing, application to a Sanskrit tagger. *J. Functional Programming*, 15,4:573–614, 2005. +http://pauillac.inria.fr/ huet/PUBLIC/tagger.pdf+.
10. E. Roche and Y. Schabes. *Finite-State Language Processing*. MIT Press, 1997.
11. R. Sproat. *Morphology and Computation*. MIT Press, 1992.

A Bialgebraic Review of Deterministic Automata, Regular Expressions and Languages

Bart Jacobs

Institute for Computing and Information Sciences, Radboud University Nijmegen
P.O. Box 9010, 6500 GL Nijmegen, The Netherlands.
B.Jacobs@cs.ru.nl   http://www.cs.ru.nl/B.Jacobs

To Joseph Goguen on the occasion of his 65th birthday

Abstract. This papers reviews the classical theory of deterministic automata and regular languages from a categorical perspective. The basis is formed by Rutten's description of the Brzozowski automaton structure in a coalgebraic framework. We enlarge the framework to a so-called bialgebraic one, by including algebras together with suitable distributive laws connecting the algebraic and coalgebraic structure of regular expressions and languages. This culminates in a reformulated proof via finality of Kozen's completeness result. It yields a complete axiomatisation of observational equivalence (bisimilarity) on regular expressions. We suggest that this situation is paradigmatic for (theoretical) computer science as the study of "generated behaviour".

1   Introduction

In the early seventies Joseph Goguen described automata within a categorical perspective (see for instance [11,12,13]), together with colleagues Arbib and Manes [1]. This paper fits in that tradition, using a more modern, bialgebraic setting, where algebra meets coalgebra. A bialgebra is a combined algebra and coalgebra $F(X) \to X \to G(X)$ on a common carrier (or state space) X, satisfying a certain compatibility requirement wrt. a distributive law connecting the two functors F, G. These bialgebras found application within the abstract, combined description of operational and denotational semantics started explicitly by Turi and Plotkin [35,34]—and more implicitly by Rutten and Turi [32]. This is now an active line of work [26,20,5,18].

Goguen has always shown an interest in methodological and philosophical issues surrounding computing. The work in this paper also lends itself to such reflections. It is often claimed that data processing is the subject of the discipline of computer science. We think it is more to the point to describe the subject of computer science as *generated behaviour*. This is the behaviour that can be observed on the outside, for instance via a screen or printer. It arises in interaction with the environment, as a result of the computer executing instructions.

This behaviouristic approach allows us to understand the relation with natural sciences: biology is about "spontaneous" behaviour, and physics concentrates on lifeless natural phenomena, without autonomous behaviour. The generated behaviour that we claim to be the subject of computer science arises by a computer executing a program

K. Futatsugi et al. (Eds.): Goguen Festschrift, LNCS 4060, pp. 375–404, 2006.
© Springer-Verlag Berlin Heidelberg 2006

according to strict operational rules. The behaviour is typically observed via the computer's I/O. Abstractly, the program can be understood as an element in an inductively defined set P of terms. This set thus forms a suitable initial algebra $F(P) \to P$, where the functor F captures the signature of the operations for forming programs. The operational rules for the behaviour of programs are described by a coalgebra $P \to G(P)$, where the functor G captures the kind of behaviour that can be displayed—such as deterministic, or with exceptions. We see that in abstract form, generated computer behaviour amounts to the repeated evaluation of an (inductively defined) coalgebra structure on an algebra of terms. Hence the bialgebras that form the basic structures used in this paper are at the heart of computer science.

One of the big challenges of computer science is to develop techniques for effectively establishing properties of generated behaviour. Often such properties are formulated positively as wanted, functional behaviour. But these properties may also be negative, like in computer security, where unwanted behaviour must be excluded. However, an appropriate logical view about program properties within the combined algebraic/coalgebraic setting has not been fully elaborated yet.

A distributive law is a natural transformation $FG \Rightarrow GF$ that describes (in the current setting) the proper interaction of term-formation and computational behaviour. The basic observation of [35,34], further elaborated [5], is that such natural transformations correspond to specification formats for operational rules on (inductively defined) programs. A bialgebra is an algebra-coalgebra pair satisfying a compatibility requirement wrt. a given distributive law. These bialgebras, as already claimed, form very fundamental structures in computing, because they combine algebraic structure with the associated computational behaviour. The compatibility requirement entails elementary properties like: observational equivalence (*i.e.* bisimulation wrt. the coalgebra) is a congruence (wrt. the algebra).

This paper concentrates on deterministic automata, regular expressions and languages. They form the very basic structures in computer science (see for instance [28]) which are studied early on in standard curricula in computing. The main contribution of this paper is the demonstration that these classic structures fit perfectly in the bialgebraic framework. In fact, they may be considered as a paradigmatic example. The paper does not contain new results on regular expressions / automata / languages as such, but on the way they can (or should) be organised. The proper mathematical language for this organisation is categorical. The reader is assumed to be familiar with basic notions like functor, natural transformation, (co)monad and adjunction, such as can be found in any introductory text on category theory. Our investigations take place in the category **Sets** of ordinary sets and functions. We are well aware that many results generalise to other categories, but we do not always strive for the highest level of generality.

There is already a large body of algebraic work on regular expressions, automata and languages, for instance within the context of regular algebras [9]. The coalgebraic perspective on this topic was introduced by Rutten [30,33,31], who demonstrated its fruitfulness especially for proving equalities via coinduction (using bisimulations). Rutten's work exploits the automaton structure on regular expressions introduced by Brzozowski [8,9]. Here we go a step further by developing the bialgebraic (combined algebraic-coalgebraic) perspective. This involves a number of new technical results:

- a general mechanism for obtaining distributive laws and bialgebras for deterministic automata in Section 3;
- a description of the free algebra and Brzozowski coalgebra structure on regular expressions as a bialgebra wrt. a (categorical) GSOS law in Subsection 4.3;
- a new proof of Kozen's completeness result [23,24] for regular expressions and languages in Section 5, by describing the coalgebra of regular expressions modulo equations as a final object. This shows that Kozen's axioms and rules give a complete axiomatisation of observational equivalence (bisimilarity) on regular expressions.

Throughout the paper we heavily rely on previous work, notably [35,31,24].

We expect that the bialgebraic picture that is emerging constitutes a paradigm which also applies to many more computational models (as already suggested in [35]). After all, regular expressions are extremely elementary, and capture only a very limited form of computation. Hence the bialgebraic paradigm is still in need of further instantiation, confirmation, and elaboration.

2 Deterministic Automata as Coalgebras

This section collects some standard facts about deterministic automata, described as coalgebras, in order to determine the setting and fix the notation.

We use two arbitrary sets A and B, where the elements of A may be understood as letters of an alphabet, and the elements of B as outputs. A **deterministic automaton** with A as input and B as output set consists of two functions:

$$\delta: X \longrightarrow X^A \quad \text{for transition} \qquad \varepsilon: X \longrightarrow B \quad \text{for output}$$

acting on a state space X. The transition function δ maps a state $x \in X$ and an input letter $a \in A$ to a successor state $x' = \delta(x)(a) \in X$. In that case one may write $x \xrightarrow{a} x'$. The output function ε gives for a state $x \in X$ the associated observable output $\varepsilon(x) \in B$.

The one-step transition function δ can be extended to a multiple-step transition function $\delta^\star$. The latter takes a state $x \in X$ and a sequence $\sigma \in A^\star$ of inputs to a successor state obtained by consecutively executing the steps in σ.

$$X \xrightarrow{\ \delta^\star\ } X^{A^\star} \qquad \text{defined as} \qquad \begin{cases} \delta^\star(x)(\langle\rangle) = x \\ \delta^\star(x)(a \cdot \sigma) = \delta^\star(\delta(x)(a))(\sigma) \end{cases} \quad (1)$$

This extended transition function $\delta^\star$ gives rise to the multiple-step transition notation: $x \xrightarrow{\sigma}{}^\star x'$ stands for $x' = \delta^\star(x)(\sigma)$, and means that x' is the (non-immediate) successor state of x obtained by applying the inputs from the sequence $\sigma \in A^\star$, from left to right. The behaviour $\mathsf{beh}(x): A^\star \to B$ of a state $x \in X$ is then obtained as the function that maps a finite sequence $\sigma \in A^\star$ of inputs to the observable output

$$\mathsf{beh}(x)(\sigma) = \varepsilon(\delta^\star(x, \sigma)) \in B \quad (2)$$

The transition and output functions δ and ε of a deterministic automaton can be combined into a tuple $\langle \delta, \varepsilon \rangle \colon X \to X^A \times B$ forming a coalgebra of the functor $\mathcal{D} = \mathcal{D}_{A,B}$ given by $U \mapsto U^A \times B$. A coalgebra homomorphism from $(\langle \delta_1, \varepsilon_1 \rangle \colon X_1 \to X_1^A \times B)$ to $(\langle \delta_2, \varepsilon_2 \rangle \colon X_2 \to X_2^A \times B)$ consists of a function $h \colon X_1 \to X_2$ between the underlying state spaces satisfying:

$$\mathcal{D}(f) \circ \langle \delta_1, \varepsilon_1 \rangle = \langle \delta_2, \varepsilon_2 \rangle \circ f,$$

That is, $f^A \circ \delta_1 = \delta_2 \circ f$ and $\varepsilon_1 = \varepsilon_2 \circ f$. Or, more concretely, $f(\delta_1(x)(a)) = \delta_2(f(x))(a)$ and $\varepsilon_1(x) = \varepsilon_2(f(x))$, for all $x \in X$ and $a \in A$.

This describes morphisms in a category $\mathbf{CoAlg}(\mathcal{D})$. The following result, occurring for example in [2,29,16], is simple but often useful. It gives an explicit description of the final object in the category $\mathbf{CoAlg}(\mathcal{D})$. The proof is easy, and left to the reader.

Proposition 1. *The final coalgebra of the functor $\mathcal{D} = (-)^A \times B$ for deterministic automata is given by the set of behaviour functions B^{A^*}, with structure:*

$$B^{A^*} \xrightarrow{\ \langle D, E \rangle\ } \left(B^{A^*}\right)^A \times B$$

given by:

$$D(\varphi)(a) = \lambda \sigma \in A^*. \varphi(a \cdot \sigma) \qquad and \qquad E(\varphi) = \varphi(\langle\rangle). \qquad \square$$

As is well-known—after Lambek—the structure map of a final coalgebra is an isomorphism. The carrier B^{A^*} of the final coalgebra collects all possible behaviours of deterministic automata. Two special cases are worth mentioning explicitly.

Example 1. Consider the above final coalgebra $B^{A^*} \xrightarrow{\cong} \left(B^{A^*}\right)^A \times B$ of the deterministic automata functor $\mathcal{D} = (-)^A \times B$.

1. When A is a singleton set $1 = \{0\}$, so that $A^* = \mathbb{N}$, the resulting functor $\mathcal{D} = (-) \times B$ captures stream coalgebras $X \to X \times B$. Its final coalgebra is the set $B^{\mathbb{N}}$ of infinite sequences (streams) of elements of B, with (tail, head) structure,

$$B^{\mathbb{N}} \xrightarrow{\cong} B^{\mathbb{N}} \times B \quad \text{given by} \quad \varphi \mapsto (\lambda n \in \mathbb{N}. \varphi(n+1), \varphi(0))$$

2. When $B = 2 = \{0,1\}$ describing final (or accepting) states of the automaton, the final coalgebra B^{A^*} is the set $\mathcal{L}(A) = \mathcal{P}(A^*)$ of languages over the alphabet A, with structure:

$$\mathcal{L}(A) \xrightarrow{\cong} \mathcal{L}(A)^A \times 2 \quad \text{given by} \quad L \mapsto (\lambda a \in A. D(L)(a), E(L))$$

where $D(L)(a)$ is the so-called a-derivative, introduced by Brzozowski [8], and defined as:

$$D(L)(a) = \{\sigma \in A^* \mid a \cdot \sigma \in L\},$$

and where $E(L) = 1 \iff \langle\rangle \in L$.
Given an arbitrary automaton $X \to X^A \times \{0,1\}$ of this type, the resulting behaviour map $\mathsf{beh} \colon X \to \mathcal{P}(A^*)$ thus describes the language $\mathsf{beh}(x) \subseteq A^*$ **accepted** by this automaton with $x \in X$ considered as initial state.

Both these final coalgebras $B^{\mathbb{N}}$ and $\mathcal{L}(A) = \mathcal{P}(A^\star)$ are studied extensively by Rutten, see [30,33,31]. One of the things that he emphasises is the use of bisimulation as a reasoning principle. Here we only sketch the main points, for deterministic automata.

Definition 1. *Consider two coalgebras $\langle \delta_1, \varepsilon_1 \rangle \colon X_1 \to X_1^A \times B$ and $\langle \delta_2, \varepsilon_2 \rangle \colon X_2 \to X_2^A \times B$. A **bisimulation** between them is a relation $R \subseteq X_1 \times X_2$ on the underlying state spaces that satisifies for all $x_1 \in X_1, x_2 \in X_2$,*

$$R(x_1, x_2) \Longrightarrow \begin{cases} \varepsilon_1(x_1) = \varepsilon_2(x_2), \text{ and} \\ R(\delta_1(x_1)(a), \delta_2(x_2)(a)), \text{ for all } a \in A. \end{cases}$$

We write $y_1 \leftrightarrow y_2$ and call y_1, y_2 bisimilar if there is a bisimulation R with $R(y_1, y_2)$.

Bisimilarity expresses observational equality, that is, equality as far as one can observe with the available (coalgebraic) operations. This explains the following result.

Proposition 2. *In the situation of the previous definition one has: $y_1 \leftrightarrow y_2$ if and only if $beh_{\langle \delta_1, \varepsilon_1 \rangle}(y_1) = beh_{\langle \delta_2, \varepsilon_2 \rangle}(y_2)$.*

Proof. The implication ($\Rightarrow$) is easy, since if $y_1 \leftrightarrow y_2$, say via a bisimulation R with $R(y_1, y_2)$, then by induction, $R(\delta_1^\star(y_1)(\sigma), \delta_2^\star(y_2)(\sigma))$, for each $\sigma \in A^\star$. This yields $beh_{\langle \delta_1, \varepsilon_1 \rangle}(y_1) = \varepsilon_1(\delta_1^\star(y_1)(\sigma)) = \varepsilon_2(\delta_2^\star(y_2)(\sigma)) = beh_{\langle \delta_2, \varepsilon_2 \rangle}(y_2)$. For the reverse implication ($\Leftarrow$) one uses that the relation $\{(x_1, x_2) \mid beh_{\langle \delta_1, \varepsilon_1 \rangle}(x_1) = beh_{\langle \delta_2, \varepsilon_2 \rangle}(x_2)\}$ is a bisimulation. This follows directly because the beh maps are homomorphisms. $\square$

States are thus bisimilar if and only if they are equal when mapped to the final coalgebras. Bisimulations provide a means to prove equations via "single-step" arguments. This makes coinductive reasoning similar to ordinary inductive approaches. See [15] for an abstract account of the underlying dualities.

Here is a very simple example—already using the regular algebra structure on languages from Example 3 later on. For each letter $a \in A$ one has $(1 + a)^\star = a^\star$ in $\mathcal{L}(A)$. This can be proven via the bisimulation $R = \{\langle (1 + a)^\star, a^\star \rangle\} \cup \{\langle \emptyset, \emptyset \rangle\}$.

At some stage we shall need the modal "eventually" operator $\Diamond$. Let $\langle \delta, \varepsilon \rangle \colon X \to X^A \times B$ be an arbitrary coalgebra / automaton. For a predicate (or subset) $P \subseteq X$ we define $\Diamond(P) \subseteq X$ as the set of all states that are reachable from P:

$$\Diamond(P) = \{\delta^\star(x)(\sigma) \mid x \in P, \sigma \in A^\star\}.$$

For a single state we write $\Diamond(x)$ for $\Diamond(\{x\})$. Note that $\Diamond(P)$ is a subcoalgebra / subautomaton, because it is by construction closed under transitions. It may be described as the least invariant containing P, see [17]. The greatest invariant $\square(P)$ contained in P is the predicate $\{x \mid \forall \sigma \in A^\star. \delta^\star(x)(\sigma) \in P\}$.

3 Structured Output Sets and Distributive Laws

In [31, Section 9] the situation is studied where the output set B of a coalgebra $X \to X^A \times B$ is a semiring. This generalises the situations studied in [31] of final coalgebras of real-valued streams ($B = \mathbb{R}$) and languages ($B = 2$). It is shown that the sum and multiplication operations on B can be extended to the final coalgebras involved.

Here we go a step further and assume an algebra structure $\beta: T(B) \to B$, for a monad $T:$ **Sets** $\to$ **Sets** with unit $\eta:$ id $\Rightarrow T$ and multiplication $\mu: T^2 \Rightarrow T$. Semirings then form a special case, see Subsection 3.4. We show how this T-algebra structure on the output set B induces a distributive law $T\mathcal{D} \Rightarrow \mathcal{D}T$, and a strengthened form of coinduction using "T-automata", following the approach of [36,5]. We shall give several illustrations involving different types of automata, for various concrete monads. These investigations go a bit beyond what is strictly needed for deterministic automata and regular languages.

To start, we recall that for an arbitrary monad T and functor G acting on the same category, a **distributive law** $\lambda: TG \Rightarrow GT$ is a natural transformation that interacts appropriately with the monads unit η and multiplication μ. This means that the following two diagrams commute.

$$
\begin{array}{ccc}
GX \xrightarrow{\eta_{GX}} TGX & \qquad T^2GX \xrightarrow{T(\lambda_X)} TGTX \xrightarrow{\lambda_{TX}} GT^2X \\
\searrow{\scriptstyle G(\eta_X)} \quad \downarrow{\scriptstyle \lambda_X} & \mu_{GX} \downarrow \qquad\qquad\qquad\qquad\qquad \downarrow G(\mu_X) \\
\qquad\quad GTX & TGX \xrightarrow{\qquad\qquad \lambda_X \qquad\qquad} GTX
\end{array}
$$

Example 2. The next two illustrations will be used frequently. They both involve the so-called strength map.

1. For each functor T on the category **Sets** and for each set X there is a natural transformation st: $T(-)^X \Rightarrow (-)^X T$. It is usually called **strength**, and given as map $T(Y^X) \to (TY)^X$ by the formula:

$$
\mathsf{st}(u)(x) = T\big(\lambda h \in Y^X . h(x)\big)(u).
$$

In case T happens to carry a monad structure, the strength map becomes a distributive law. The above two diagrams then translate into:

$$
\begin{array}{ccc}
Y^X \xrightarrow{\eta_{Y^X}} T(Y^X) & \qquad T^2(Y^X) \xrightarrow{T(\mathsf{st})} T((TY)^X) \xrightarrow{\mathsf{st}} (T^2Y)^X \\
\searrow{\scriptstyle (\eta_Y)^X} \quad \downarrow{\scriptstyle \mathsf{st}} & \mu_{Y^X} \downarrow \qquad\qquad\qquad\qquad\qquad \downarrow (\mu_Y)^X \\
\qquad\quad (TY)^X & T(Y^X) \xrightarrow{\qquad\qquad \mathsf{st} \qquad\qquad} (TY)^X
\end{array}
$$

(The diligent reader may have noticed that strength is also natural in the functor, in the sense that for a natural transformation $\sigma: F \Rightarrow G$ one has $\mathsf{st}^G_{X,Y} \circ \sigma_{Y^X} = (\sigma_Y)^X \circ \mathsf{st}^F_{X,Y}$.)

One useful point about strength for monads is that it allows pointwise construction of algebras on function spaces: if $\alpha: T(Y) \to Y$ is an Eilenberg-Moore algebra, then so is $\alpha^X \circ \mathsf{st}: T(Y^X) \to (TY)^X \to Y^X$.

2. We have formulated the notion of a distributive law for a monad and a functor. There are several "obvious" variations, for instance for a functor and a comonad. The next example again involves strength, and is related to the final coalgebra construction in Proposition 1.

   To start, let $(M, \cdot, e)$ be an arbitrary monoid. It gives rise to a functor $(-)^M \colon \mathbf{Sets} \to \mathbf{Sets}$ that turns out to be a comonad. The counit $E_X \colon X^M \to X$ uses the monoids unit in $E_X(\varphi) = \varphi(e)$, and the comultiplication $C_X \colon X^M \to (X^M)^M$ works via the monoids multiplication in $C_X(\varphi) = \lambda a \in M. \, \lambda b \in M. \, \varphi(a \cdot b)$.

   We claim that for an arbitrary functor F, there is a distributive law $F(-)^M \Rightarrow (-)^M F$ over the comonad $(-)^M$. This law is again given by strength, and satisfies the following two "dual" properties.

$$
\begin{array}{ccc}
F(X^M) \xrightarrow{\ \mathsf{st}\ } (FX)^M & \qquad & F(X^M) \xrightarrow{\hspace{4cm}\mathsf{st}\hspace{4cm}} (FX)^M \\
\ \ \searrow_{\!\!F(E_X)} \quad \Big\downarrow{\scriptstyle E_{FX}} & & \Big\downarrow{\scriptstyle F(C_X)} \qquad\qquad\qquad\qquad\qquad\qquad \Big\downarrow{\scriptstyle C_{FX}} \\
F(X) & & F((X^M)^M) \xrightarrow[\mathsf{st}]{} (F(X^M))^M \xrightarrow[\mathsf{st}^M]{} ((FX)^M)^M
\end{array}
$$

Why is all this relevant? Well, the final coalgebra structure described in Proposition 1 arises in this manner via the (free) monoid $(A^\star, \cdot, \langle\rangle)$ of strings with concatenation: its observation map $E \colon B^{A^\star} \to B$ is precisely the above counit E_B, and its transition map $D \colon B^{A^\star} \to (B^{A^\star})^A$ arises from the comultiplication C_B, via restriction to singleton sequences: $D(\varphi)(a)(\sigma) = C(\varphi)(\langle a\rangle)(\sigma)$. The fact that strength forms a distributive law will be used in the proof of Proposition 4 below.

As stated in the beginning of this section, we assume an Eilenberg-Moore algebra $\beta \colon T(B) \to B$. By definition it satisfies the algebra laws $\beta \circ \eta = \mathrm{id}$ and $\beta \circ T(\beta) = \beta \circ \mu$. Then we can define a distributive law of the monad T over the automata functor $\mathcal{D} = (-)^A \times B$ from the previous section, namely:

$$
T\mathcal{D} \xRightarrow{\ \lambda\ } \mathcal{D}T \qquad \text{with components} \qquad T(X^A \times B) \xrightarrow{\ \lambda_X\ } (TX)^A \times B
$$

This law is obtained as composite:

$$
T(X^A \times B) \xrightarrow{\ \langle T(\pi_1), T(\pi_2)\rangle\ } T(X^A) \times TB \xrightarrow{\ \mathsf{st} \times \beta\ } (TX)^A \times B
$$

The next result summarises what we have found so far.

Proposition 3. *Each Eilenberg-Moore algebra $T(B) \to B$ induces a distributive law $\lambda \colon T\mathcal{D} \Rightarrow \mathcal{D}T$ for the deterministic automata functor $\mathcal{D} = (-)^A \times B$.* $\qquad\square$

When we have an arbitrary monad T, functor G, and a distributive law $\lambda \colon TG \Rightarrow GT$ the relevant associated notion is that of a λ-**bialgebra**: a pair of maps:

$$
TX \xrightarrow{\ a\ } X \xrightarrow{\ b\ } GX
$$

where:

- a is an Eilenberg-Moore algebra;
- a and b are compatible via λ, which means that the following diagram commutes.

$$
\begin{array}{ccc}
TX & \xrightarrow{\ a\ } X & \xrightarrow{\ b\ } GX \\
{\scriptstyle T(b)}\downarrow & & \uparrow{\scriptstyle G(a)} \\
TGX & \xrightarrow[\ \lambda_X\ ]{} & GTX
\end{array}
$$

A **map of λ-bialgebras**, from $(TX \xrightarrow{\ a\ } X \xrightarrow{\ b\ } GX)$ to $(TY \xrightarrow{\ c\ } Y \xrightarrow{\ d\ } GY)$ is a map $f\colon X \to Y$ that is both a map of algebras and of coalgebras: $f \circ a = c \circ T(f)$ and $d \circ f = G(f) \circ b$.

The next two results are standard, see for *e.g.* [5,18], and are given without proof.

Lemma 1. *Assume a distributive law $\lambda\colon TG \Rightarrow GT$, and let $\zeta\colon Z \xrightarrow{\cong} GZ$ be a final coalgebra. It carries an Eilenberg-Moore algebra obtained by finality in:*

$$
\begin{array}{ccc}
GTZ & \xdashrightarrow{\ \ G(\alpha)\ \ } & GZ \\
{\scriptstyle \lambda_Z}\uparrow & & \\
TGZ & & \cong{\scriptstyle\ \big|\ \zeta} \\
{\scriptstyle T(\zeta)}\uparrow{\scriptstyle\cong} & & \\
TZ & \xdashrightarrow[\ \ \alpha\ \ ]{} & Z
\end{array}
$$

The resulting pair $(TZ \xrightarrow{\ \alpha\ } Z \xrightarrow{\ \zeta\ } GZ)$ is then a final λ-bialgebra. $\Box$

Lemma 2. *In presence of a distributive law $\lambda\colon TG \Rightarrow GT$, there exists a bijective correspondence between GT-coalgebras $e\colon X \to GTX$ (also called equations) and λ-bialgebras $(T^2X \xrightarrow{\ \mu_X\ } TX \xrightarrow{\ d\ } GTX)$ with free algebra μ_X.*

Moreover, let $(TY \xrightarrow{\ a\ } Y \xrightarrow{\ b\ } GY)$ be a λ-bialgebra. Then there is a bijective correspondence between "solutions of e" $f\colon X \to Y$ in:

$$
\begin{array}{ccc}
GTX & \xrightarrow{\ \ GT(f)\ \ } & GTY \\
{\scriptstyle e}\uparrow & & \downarrow{\scriptstyle G(a)} \\
& & GY \\
& & \uparrow{\scriptstyle b} \\
X & \xrightarrow[\ \ f\ \ ]{} & Y
\end{array}
$$

and λ-bialgebra maps $g\colon TX \to Y$—for the associated equations and λ-bialgebras. $\Box$

Proposition 4. *The assumed algebra* $\beta\colon TB \to B$ *induces on the carrier* B^{A^*} *of the final* $\mathcal{D}$*-coalgebra from Proposition 1 another* T*-algebra via a pointwise construction, namely,*

$$\widehat{\beta} \stackrel{def}{=} \left(T(B^{A^*}) \xrightarrow{\;\mathsf{st}\;} (TB)^{A^*} \xrightarrow{\;\beta^{A^*}\;} B^{A^*} \right)$$

so that $E\colon B^{A^*} \to B$ *is a homomorphism of algebras. This* $\widehat{\beta}$ *is the unique coalgebra homomorphism from Lemma 1,*

$$
\begin{array}{ccc}
\mathcal{D}T(B^{A^*}) & \xdashrightarrow{\;\;\mathcal{D}(\widehat{\beta})\;\;} & \mathcal{D}(B^{A^*}) \\[4pt]
\lambda_{B^{A^*}} \uparrow & & \\
T\mathcal{D}(B^{A^*}) & & \cong \Big\uparrow \langle D, E \rangle \\
T(\langle D, E \rangle) \uparrow \cong & & \\
T(B^{A^*}) & \xdashrightarrow[\;\;\widehat{\beta}\;\;]{} & B^{A^*}
\end{array}
$$

using the distributive law from Proposition 3. Hence, this $\widehat{\beta}$ *together with the final coalgebra forms the final* λ*-bialgebra:* $T(B^{A^*}) \xrightarrow{\;\widehat{\beta}\;} B^{A^*} \xrightarrow{\;\langle D,E \rangle\;} \mathcal{D}(B^{A^*})$.

Proof. According to Lemma 1 it suffices to prove that $\widehat{\beta}$ is a homomorphism of coalgebras. Here we use that strength is a distributive law as described in Example 2.(2).

$$
\begin{aligned}
\mathcal{D}(\widehat{\beta}) &\circ \lambda \circ T(\langle D, E \rangle) \\
&= \mathcal{D}(\beta^{A^*} \circ \mathsf{st}) \circ \langle \mathsf{st} \circ T(\pi_1), \beta \circ T(\pi_2) \rangle \circ T(\langle D, E \rangle) \\
&= \langle (\beta^{A^*})^A \circ \mathsf{st}^A \circ \mathsf{st} \circ T(D), \beta \circ T(E) \rangle \\
&= \langle (\beta^{A^*})^A \circ D \circ \mathsf{st}, \beta \circ E \circ \mathsf{st} \rangle \\
&= \langle (\beta^{A^*})^A \circ D, \beta \circ E \rangle \circ \mathsf{st} \\
&= \langle D, E \rangle \circ \beta^{A^*} \circ \mathsf{st} \\
&= \langle D, E \rangle \circ \widehat{\beta}. \qquad\qquad \square
\end{aligned}
$$

The coinduction principle associated with a final λ-bialgebra is called λ-**coinduction** in [5]. In the current situation, with the functor $\mathcal{D}$ for deterministic automata, the principle yields a strengthened form of coinduction for "T-automata".

Theorem 1. *For each* T*-automaton* $\langle \delta, \varepsilon \rangle\colon X \to \mathcal{D}(TX) = (TX)^A \times B$—*where* B *carries a* T*-algebra* $\beta\colon TB \to B$—*there is a unique map* $\mathsf{beh}\colon X \to B^{A^*}$ *making the following diagram commute.*

$$
\begin{array}{ccc}
(TX)^A \times B = \mathcal{D}TX & \xrightarrow{\;\;\mathcal{D}T(\mathsf{beh})\;\;} & \mathcal{D}T(B^{A^*}) \\[4pt]
& & \downarrow \mathcal{D}(\widehat{\beta}) \\
\langle \delta, \varepsilon \rangle \uparrow & & \mathcal{D}(B^{A^*}) \\
& & \uparrow \langle D, E \rangle \\
X & \xrightarrow[\;\;\mathsf{beh}\;\;]{} & B^{A^*}
\end{array}
$$

Proof. This result is a direct consequence of Lemmas 1 and 2, but we like to give the concrete construction, as in the proof of Proposition 1. First we define an extension $\delta^\star \colon X \to (TX)^{A^*}$ of δ like in (1) by induction:

$$\delta^\star(x)(\langle\rangle) = \eta(x) \qquad \delta^\star(x)(a \cdot \sigma) = \mu\Big[\mathsf{st}\big(T(\delta^\star)(\delta(x)(a))\big)(\sigma)\Big].$$

Then we can define the required map as:

$$\mathsf{beh} = \Big(X \xrightarrow{\ \delta^\star\ } (TX)^{A^*} \xrightarrow{\ (T\varepsilon)^{A^*}\ } (TB)^{A^*} \xrightarrow{\ \beta^{A^*}\ } B^{A^*} \Big). \qquad \square$$

In the remainder of this section we shall investigate several instantiations of the monad T in the results above.

3.1   The Identity Monad and Deterministic Automata

If we take $T = \mathrm{id}$, with $\beta = \mathrm{id}$ as identity algebra we get $\lambda = \mathrm{id}$ and $\hat{\beta} = \mathrm{id}$, so that λ-coinduction is just the ordinary form of coinduction for deterministic automata.

3.2   The Powerset Monad and Non-deterministic Automata

In the above context we now consider the situation where the monad T is the powerset monad $\mathcal{P}$ and where the output set B is $2 = \{0, 1\}$. An Eilenberg-Moore algebra of $\mathcal{P}$ is a complete lattice (see *e.g.* [25, Chapter VI.2, Exerice 1]), *i.e.* a poset with joins (and hence also meets) of all subsets. Since $2 = \mathcal{P}(1)$, we have a free monad structure $\bigcup \colon \mathcal{P}(2) \to 2$ given by union. The strength map $\mathsf{st} \colon \mathcal{P}(Y^X) \to \mathcal{P}(Y)^X$ is $\mathsf{st}(u)(x) = \{f(x) \mid f \in u\}$. The resulting distributive law, say $\lambda^{\mathcal{P}} \colon \mathcal{P}\mathcal{D} \Rightarrow \mathcal{D}\mathcal{P}$, is given by:

$$\mathcal{P}(X^A \times 2) \xrightarrow{\quad\quad\quad \lambda^{\mathcal{P}}_X \quad\quad\quad} \mathcal{P}(X)^A \times 2$$
$$U \longmapsto \langle \lambda a \in A. \{f(a) \mid \exists b. (f, b) \in U\},\ \exists f. (f, 1) \in U \rangle$$

The final coalgebra is in this case the set $2^{A^*} = \mathcal{P}(A^*) = \mathcal{L}(A)$ of languages over the "alphabet" A, see Example 1 (ii). The induced algebra structure $\mathcal{P}(\mathcal{L}(A)) \to \mathcal{L}(A)$ is simply union $\bigcup$.

The $\lambda^{\mathcal{P}}$-coinduction principle from Theorem 1 tells how a state x of a non-deterministic automaton is mapped to the associated language (that is accepted starting from x as initial state):

$$
\begin{array}{ccc}
\mathcal{P}(X)^A \times 2 = \mathcal{D}\mathcal{P}(X) & \dashrightarrow & \mathcal{D}\mathcal{P}\mathcal{L}(A) \\
\uparrow & & \downarrow \mathcal{D}(\bigcup) \\
\big| & & \mathcal{D}\mathcal{L}(A) = \mathcal{L}(A)^A \times 2 \\
\big| & & \uparrow \cong \\
X & \dashrightarrow & \mathcal{L}(A)
\end{array}
$$

This was first noted in [5, Corollary 4.4.6].

3.3 The Multiset Monad and Weighted Automata

It is well-known that the Kleene-star or list monad $X \mapsto X^\star$ has monoids as Eilenberg-Moore algebras. The monad $\mathcal{M}$ for commutative monoids is given by multisets:

$$\mathcal{M}(X) = \{\varphi \in \mathbb{N}^X \mid \varphi \text{ has finite support}\},$$

where the support of φ is the set $\mathsf{supp}(\varphi) = \{x \in X \mid \varphi(x) \neq 0\}$. Such a φ can thus be represented as finite sum $n_1 x_1 + \cdots + n_k x_k$ of elements $x_i \in X$ with "multiplicities" $n_i = \varphi(x_i) \in \mathbb{N}$. The action $\mathcal{M}(f)$ on such a representation is then simply $n_1 f(x_1) + \cdots + n_k f(x_k)$. The unit of this monad is $x \mapsto 1x$ and multiplication is $n_1 \varphi_1 + \cdots + n_k \varphi_k \mapsto \lambda x \in X . n_1 \varphi_1(x) + \cdots + n_k \varphi_k(x)$.

An $\mathcal{M}$-automaton $\langle \delta, \varepsilon \rangle \colon X \to \mathcal{M}(X)^A \times 2$ is then a so-called weighted automaton. For a state $x \in X$ and letter $a \in A$ there may then be several result states x_i in the outcome $\delta(x)(a) = n_1 x_1 + \cdots + n_k x_k$, each with a particular "weight" n_i.

The set 2 forms a commutative monoid via finite disjunctions $\top, \vee$—and also via conjunctions. The disjunctions induce a commutative monoid structure on $\mathcal{L}(A)$ given by union of languages. Since this is an idempotent monoid, the structure of multiplicities is ignored when a state is mapped to the associated language.

3.4 The Semiring Monad

A basic observation is that there is a distributive law of monads $\pi \colon (-)^\star \circ \mathcal{M} \Rightarrow \mathcal{M} \circ (-)^\star$ between the list and multiset monads. It is given by multiplication in $\mathbb{N}$:

$$\mathcal{M}(X)^\star \xrightarrow{\quad \pi_X \quad} \mathcal{M}(X^\star)$$

$$\langle \varphi_1, \ldots, \varphi_n \rangle \longmapsto \sum \{\varphi_1(x_1) \cdots \varphi_n(x_n)\langle x_1, \ldots, x_n \rangle \mid x_i \in \mathsf{supp}(\varphi_i)\}$$

$$= \lambda \langle y_1, \ldots, y_m \rangle \in X^\star . \begin{cases} 0 & \text{if } m \neq n \\ \varphi_1(y_1) \cdots \varphi_n(y_n) & \text{otherwise.} \end{cases}$$

With some perseverance one can prove that π is a natural transformation that commutes appropriately with the monad structures.

It is a standard result that in presence of a distributive law like $\pi \colon (-)^\star \circ \mathcal{M} \Rightarrow \mathcal{M} \circ (-)^\star$ the composite $\mathcal{M} \circ (-)^\star$ is again a monad, see for instance [6,19,4]. Moreover, the multiset monad $\mathcal{M}$ can be lifted to a monad $\overline{\mathcal{M}}$ on the category of $(-)^\star$-algebra (monoids), such that the algebras of the composite monad $\mathcal{M} \circ (-)^\star$ are the same as $\overline{\mathcal{M}}$-algebras. This functor $\overline{\mathcal{M}}$ maps a monoid $(X, \cdot, 1)$ to $(\mathcal{M}(X), \bullet, \eta(1))$ with multiplication $\bullet$ given by:

$$\varphi \bullet \psi = \sum \{\varphi(x)\psi(y)(x \cdot y) \mid x \in \mathsf{supp}(\varphi), y \in \mathsf{supp}(\psi)\}.$$

An Eilenberg-Moore algebra $(\mathcal{M}(X), \bullet, \eta(1)) \to (X, \cdot, 1)$ for the monad $\overline{\mathcal{M}}$ consists of a commutative monoid $m \colon \mathcal{M}(X) \to X$ whose structure map m preserves the monoid structure. Such an algebra of the composite monad is thus a **semiring**. Therefore we call the monad the semiring monad, and write it as $\mathcal{S}(X) = \mathcal{M}(X^\star)$.

Rutten [31, Section 9] explicitly considers deterministic automata $X \to X^A \times B$ where the set B is a semiring, *i.e.* carries an Eilenberg-Moore algebra $\mathcal{S}(B) \to B$. This includes his main examples $B = \mathbb{R}$ and $B = 2$. In those cases the final coalgebra B^{A^*} is also a semiring, via pointwise construction. Theorem 1 yields for a "semiring" automaton $X \to \mathcal{S}(X)^A \times 2$ a mapping $X \to \mathcal{L}(A)$ to languages over A.

3.5   The Language Monad

The language monad $\mathcal{L}(X) = \mathcal{P}(X^*)$ can be constructed similarly to the semiring monad $\mathcal{S}(X) = \mathcal{M}(X^*)$, namely via a distributive law. The algebras of the language monad are Kleene algebras with arbitrary joins, also known as unital quantales, see [18] for more information. Theorem 1 then yields behaviours for states of "language automata" $X \to \mathcal{L}(X)^A \times B$. They resemble alternating automata [27].

4   Regular Expressions

As is well-known, regular expressions are built up from constants $0, 1$, letters $a \in A$ from a given alphabet A, sum $s + t$, composition $s \cdot t$ and Kleene-star s^*. These operations form an algebra of the functor:

$$\mathcal{R}(X) = 1 + 1 + (X \times X) + (X \times X) + X$$

where we ignore the alphabet for a moment—because it will turn up in the associated monad below. The initial algebra of this functor $\mathcal{R}$ is not so interesting: it consists of the (closed) terms that can be obtained from $0, 1$ via $+, \cdot, (-)^*$. Notice that at this stage there are no equations involved. They will appear in the next section.

Example 3.   For an arbitrary set U, the set of languages $\mathcal{L}(U) = \mathcal{P}(U^*)$ over U carries an $\mathcal{R}$-algebra structure $\mathcal{R}(\mathcal{L}(U)) \to \mathcal{L}(U)$. It is given by the familiar definitions

$$\begin{cases} \text{zero term case:} & 0 \longmapsto \emptyset \\ \text{one term case:} & 1 \longmapsto \{\langle\rangle\} \\ \text{sum case:} (L_1, L_2) & \longmapsto L_1 \cup L_2 \\ \text{product case:} (L_1, L_2) & \longmapsto \{\sigma_1 \cdot \sigma_2 \mid \sigma_i \in L_i\} \\ \text{star case:} & L \longmapsto \bigcup_{n \in \mathbb{N}} L^n. \end{cases}$$

since a single (algebra) map $\mathcal{R}(\mathcal{L}(U)) \to \mathcal{L}(U)$ jointly describes five maps of the form $1 \to \mathcal{L}(U)$, $1 \to \mathcal{L}(U)$, $\mathcal{L}(U) \times \mathcal{L}(U) \to \mathcal{L}(U)$, $\mathcal{L}(U) \times \mathcal{L}(U) \to \mathcal{L}(U)$ and $\mathcal{L}(U) \to \mathcal{L}(U)$, giving the individual operations of regular algebra.

For the special case where $U = \emptyset$ we get an algebra structure on $\mathcal{L}(\emptyset) = \mathcal{P}(\emptyset^*) = \mathcal{P}(1) = 2$. This structure $\mathcal{R}(2) \to 2$ uses $0, \vee$ and $1, \wedge$ as additive and multiplicative monoids, and the constant map $x \mapsto 1$ as star operation.

Usually one considers regular expressions over an alphabet A. It means that the letters $a \in A$ are used as atoms to build up regular expressions. This can be done via the free monad $\mathcal{R}^*$ generated by $\mathcal{R}$. It is defined on a set A as the initial algebra of the functor $X \mapsto A + \mathcal{R}(X)$. We shall sometimes write Re_A for the carrier $\mathcal{R}^*(A)$ of regular expressions over A, or simply Re if the alphabet A is clear from the context. This set Re is built up inductively from $0, 1, a \in A$ using the regular operations $+, \cdot, (-)^*$.

We thus have an initiality isomorphism $[\eta_A, \tau_A] \colon A + \mathcal{R}(\mathsf{Re}) \xrightarrow{\cong} \mathsf{Re}$, where the map $\tau_A \colon \mathcal{R}(\mathcal{R}^*(A)) \to \mathcal{R}^*(A)$ is the free $\mathcal{R}$-algebra on A. The extension map $\sigma \colon \mathcal{R} \Rightarrow \mathcal{R}^*$ is then given by $\sigma = \tau \circ \mathcal{R}(\eta)$.

The next result collects the basics about this situation.

Lemma 3. *In the situation described above:*

1. *The functor $A \mapsto \mathcal{R}^*(A)$ is a monad, whose category of Eilenberg-Moore algebras is isomorphic to the category of $\mathcal{R}$-algebras. The multiplication of this monad is defined by initiality in:*

$$
\begin{array}{ccc}
\mathcal{R}^*A + \mathcal{R}(\mathcal{R}^*\mathcal{R}^*A) & \xrightarrow{\;id + \mathcal{R}(\mu_A)\;} & \mathcal{R}^*A + \mathcal{R}(\mathcal{R}^*A) \\
{\scriptstyle [\eta,\tau]}\Big\downarrow{\scriptstyle\cong} & & \Big\downarrow{\scriptstyle [id,\tau]} \\
\mathcal{R}^*\mathcal{R}^*A & \xrightarrow[\;\;\;\;\;\;\mu_A\;\;\;\;\;\;]{} & \mathcal{R}^*A
\end{array}
$$

2. *The $\mathcal{R}$-algebra on 2 from Example 3 yields a distributive law $\lambda \colon \mathcal{R}^*\mathcal{D} \Rightarrow \mathcal{D}\mathcal{R}^*$ for the deterministic automaton functor $\mathcal{D} = (-)^A \times 2$.*

Proof. The first point is standard, and the second is a special case of Proposition 3. □

With this result, an $\mathcal{R}$-algebra from Example 3, say $r \colon \mathcal{R}(\mathcal{L}(U)) \to \mathcal{L}(U)$ corresponds to a unique Eilenberg-Moore algebra $\bar{r} \colon \mathcal{R}^*(\mathcal{L}(U)) \to \mathcal{L}(U)$ with $\bar{r} \circ \sigma = r$. Especially for $U = \emptyset$ this yields an algebra $\mathcal{R}^*(2) \to 2$ that will be used in (4) below. The multiplication μ maps a term $s(t_1, \ldots, t_n)$ built up from other terms $t_1, \ldots, t_n$ as atoms, to the term $s[t_1, \ldots, t_n]$ obtained by substituting these t_i into s.

Example 4. The standard interpretation of the set Re_A regular expressions over an alphabet A in the set $\mathcal{L}(A)$ of languages over A may be understood as the unique homomorphism of algebras:

$$
\begin{array}{ccc}
\mathcal{R}^*\mathcal{R}^*A & \xrightarrow{\;\mathcal{R}^*([\![-]\!])\;} & \mathcal{R}^*(\mathcal{L}(A)) \\
{\scriptstyle \mu_A}\Big\downarrow & & \Big\downarrow \\
\mathsf{Re}_A = \mathcal{R}^*A & \xrightarrow[\;\;\;\;[\![-]\!]\;\;\;\;]{} & \mathcal{L}(A)
\end{array}
\qquad \text{with} \quad [\![\,\eta(a)\,]\!] = \{\langle a \rangle\}.
$$

The Eilenberg-Moore algebra on $\mathcal{L}(A)$ arises from the $\mathcal{R}$-algebra from Example 3. Freeness of μ_A and the inclusion $\{\langle - \rangle\} \colon A \to \mathcal{L}(A)$ does the rest.

Usually one does not make a clear distinction between an expression like $s = 1 + a^*ba^* \in \mathsf{Re}_A$ and its interpretation $[\![\,s\,]\!] = 1 \cup a^*ba^* \in \mathcal{L}(A)$. Here however, we like to keep the two apart, and use an explicit interpretation function $[\![\,-\,]\!]$.

4.1 Two Questions

Given this basic set-up, we ask ourselves the following two questions.

1. Is there a coalgebra/automaton structure $\langle D, E \rangle$ on regular expressions such that the above interpretation $[\![-]\!]$ is also a homomorphism of coalgebras, as in:

$$
\begin{array}{ccc}
\mathcal{R}^* \mathcal{R}^* A & \xrightarrow{\ \mathcal{R}^*([\![-]\!])\ } & \mathcal{R}^*(\mathcal{L}(A)) \\
\mu_A \downarrow & & \downarrow \\
\mathsf{Re}_A = \mathcal{R}^* A & \xrightarrow{\ [\![-]\!]\ } & \mathcal{L}(A) \\
\langle D, E \rangle \downarrow \ ?? & & \cong \downarrow \langle \delta, \varepsilon \rangle \\
(\mathcal{R}^* A)^A \times 2 & \xrightarrow{\ [\![-]\!]^A \times 2\ } & \mathcal{L}(A)^A \times 2
\end{array}
\tag{3}
$$

2. Is this diagram a map between two κ-bialgebras, for a suitable distributive law κ.

We address this matter in the next two subsections. The first question can be answered positively, and involves Brzozowski's "derivative" and "non-empty word" operations on regular expressions from [8,9]. The second question will be solved by a special kind of distributive law, following the so-called GSOS format. It puts the concrete construction of Brzozowski in the general framework developed in [35].

4.2 Regular Expressions as Coalgebras

From a coalgebraic perspective the most interesting part of regular expressions is that they form a deterministic automaton $\langle D, E \rangle \colon \mathsf{Re} \to \mathsf{Re}^A \times 2 = \mathcal{D}(\mathcal{R}^*(A))$.

The output operation $E \colon \mathsf{Re} \to 2$ is obtained by freeness as the unique map in

$$
\begin{array}{ccc}
\mathcal{R}^*(\mathsf{Re}) & \dashrightarrow^{\ \mathcal{R}^*(E)\ } & \mathcal{R}^*(2) \\
\mu \downarrow & & \downarrow \\
\mathsf{Re} & \dashrightarrow_{\ E\ } & 2
\end{array}
\qquad \text{with} \quad E(\eta(a)) = 0
\tag{4}
$$

where the algebra structure $\mathcal{R}^*(2) \to 2$ is as described before Example 4. Commutation of the diagram (4) yields the equations $E(0) = 0$, $E(1) = 1$, $E(s+t) = E(s) \vee E(t)$, $E(s \cdot t) = E(s) \wedge E(t)$ and $E(s^*) = 1$. This operation E describes what is sometimes called the empty word property.

Since the values of $E(s) \in 2$ are either 0 or 1, we shall often treat $E(s)$ as a term in Re.

By induction on the structure of a term $s \in \mathsf{Re}$ one checks the first bi-implication:

$$
E(s) = 1 \Longleftrightarrow \langle \rangle \in [\![s]\!] \Longleftrightarrow (\varepsilon \circ [\![-]\!])(s) = 1
$$

Hence $\varepsilon \circ [\![-]\!] = E$, which is one part of the lower square in (3).

The "derivative" operation $D: \mathsf{Re} \to \mathsf{Re}^A$ is more complicated. It is due to Brzozowski [8], see also [9]. We shall use the common notation $D_a(s)$ for the successor term $D(s)(a)$. The derivative is defined by the following clauses (or rules).

$$D_a(0) = 0 \qquad\qquad D_a(s+t) = D_a(s) + D_a(t)$$
$$D_a(1) = 0 \qquad\qquad D_a(s \cdot t) = D_a(s) \cdot t + E(s) \cdot D_a(t)$$
$$D_a(b) = \begin{cases} 1 \text{ if } b = a \\ 0 \text{ otherwise.} \end{cases} \qquad D_a(s^*) = D_a(s) \cdot s^*. \tag{5}$$

Is this a proper inductive definition? The problem is in the clause for composition, where the term t is used in the subterm $D_a(s) \cdot t$ in original form. Similarly for s in the star case. Hence we cannot use an inductive/freeness definition like for E in (4). We have to use recursion to deal with the additional parameter. The remainder of this subsection elaborates the required formulation of recursion.

A categorical analysis of strengthened induction principles for a functor F is given in [36] in terms of distributive laws between F and a comonad—dual to the approach underlying Theorem 1. We shall use this approach in the current situation where F is the functor $A + \mathcal{R}(-)$ for regular expressions described in the beginning of this section and the comonad is simply $(-) \times D$ for a set D, with coalgebra $\Delta = \langle \mathrm{id}, \mathrm{id} \rangle$. We concentrate on the result, and refer to [36] for the distributive law involved.

Theorem 2 (Recursion following [36]). *An initial algebra $\alpha: F(D) \xrightarrow{\cong} D$ satisfies the following strengthened induction property: for each map $f: F(X \times D) \to X$ there is a unique map $h: D \to X$ making the following diagram commute.*

$$
\begin{array}{ccc}
F(D \times D) & \xrightarrow{\ F(h \times D)\ } & F(X \times D) \\
F(\Delta) \uparrow & & \\
F(D) & & \downarrow f \\
\alpha \downarrow \cong & & \\
D & \xrightarrow{\quad h \quad} & X
\end{array}
$$

Proof. We shall give a direct proof, ignoring the distributivity properties involved. Let $f: F(X \times D) \to X$ therefore be given. Write $f' = \langle f, \alpha \circ F(\pi_2) \rangle: F(X \times D) \to X \times D$. It gives by initiality rise to a unique map $k: D \to X \times D$ with $k \circ \alpha = f' \circ F(k)$. Then $\pi_2 \circ k = \mathrm{id}$ by uniqueness of algebra maps $\alpha \to \alpha$. Hence we take $h = \pi_1 \circ k$. ☐

With this theorem the derivative operation $D: \mathsf{Re} \to \mathsf{Re}^A$ can be obtained by recursion from a map $[f_1, f_2]: A + \mathcal{R}(\mathsf{Re}^A \times \mathsf{Re}) \to \mathsf{Re}^A$ in:

$$
\begin{array}{ccc}
A + \mathcal{R}(\mathsf{Re}) & \xrightarrow{\ \mathrm{id} + \mathcal{R}(\langle D, \mathrm{id}\rangle)\ } & A + \mathcal{R}(\mathsf{Re}^A \times \mathsf{Re}) \\
[\eta, \tau] \downarrow \cong & & \downarrow [f_1, f_2] \\
\mathsf{Re} & \xrightarrow{\qquad D \qquad} & \mathsf{Re}^A
\end{array} \tag{6}
$$

The map $f_1: A \to \mathsf{Re}^A$ is defined as $f_1(a) = \lambda b \in A.$ if $b = a$ then 1 else 0. And $f_2: \mathcal{R}(\mathsf{Re}^A \times \mathsf{Re}) \to \mathsf{Re}^A$ is given by the following cases.

$$
\left\{
\begin{array}{rl}
\text{zero term case:} & 0 \longmapsto \lambda a \in A.\, 0 \\
\text{one term case:} & 1 \longmapsto \lambda a \in A.\, 0 \\
\text{sum case: } (\langle \varphi_1, s_1 \rangle, \langle \varphi_2, s_2 \rangle) & \longmapsto \lambda a \in A.\, \varphi_1(a) + \varphi_2(a) \\
\text{product case: } (\langle \varphi_1, s_1 \rangle, \langle \varphi_2, s_2 \rangle) & \longmapsto \lambda a \in A.\, \varphi_1(a) \cdot s_2 + E(s_1) \cdot \varphi_2(a) \\
\text{star case: } (\varphi, s) & \longmapsto \lambda a \in A.\, \varphi(a) \cdot s^*.
\end{array}
\right.
$$

Commutation of the diagram (6) now yields the appropriate clauses (5) for the derivative function. Further, by induction on $s \in \mathsf{Re}$ one proves:

$$
\begin{aligned}
[\![D(s)(a)]\!] = D([\![s]\!])(a) & \quad \text{as in Example 1 (ii)} \\
& = \{ \sigma \in A^* \mid a \cdot \sigma \in [\![s]\!] \}.
\end{aligned}
$$

This means that $[\![-]\!]$ is a homomorphism of both algebras and coalgebras in (3). This settles our first question from Subsection 4.1. In particular, the operational semantics ($[\![-]\!]$ as coalgebra homomorphism) is compositional (*i.e.* is an algebra homomorphism).

We now turn to the second question from Subsection 4.1.

4.3   Regular Expressions as Bialgebras

Since the derivative operation $D: \mathsf{Re} \to \mathsf{Re}^A$ is defined by recursion (instead of induction), the distributive laws and bialgebras described in Section 3 do not work in this situation. Interestingly, the so-called GSOS format does work. It has been developed in syntactic form for process calculi [7,14], and formulated categorically in [35]. We follow the latter approach—see also [5]. The main point is that these GSOS laws have an extra parameter—like in recursion.

Definition 2. *For a monad T and functor G, a **GSOS law** is a distributive law of the form $\lambda: T(G \times id) \Rightarrow (G \times id)T$ with $\pi_2 \circ \lambda = T(\pi_2)$.*

*A λ-model, or **GSOS model**, for such a GSOS law λ, consists of an Eilenberg-Moore algebra $a: TX \to X$ and a coalgebra $b: X \to GX$ on the same state space, such that the pair $TX \xrightarrow{a} X \xrightarrow{\langle b, id \rangle} GX \times X$ is a λ-bialgebra; equivalently, such that the following diagram commutes.*

$$
\begin{array}{ccccc}
TX & \xrightarrow{\ a\ } & X & \xrightarrow{\ b\ } & GX \\
{\scriptstyle T(\langle b, id \rangle)}\big\downarrow & & & & \big\uparrow{\scriptstyle Ga} \\
T(GX \times X) & & \xrightarrow[\ \pi_1 \circ \lambda\ ]{} & & GTX
\end{array}
$$

The formulation of GSOS law that we use is not quite the same as in [35]. The latter handles the special case where the monad T is free, *i.e.* of the form F^*. The next result shows that this special case of our definition is equivalent to the "natural transformation" formulation used in [35].

Proposition 5. *Let F be an arbitrary endofunctor with associated free monad F^*. There is a bijective correspondence between:*

$$\frac{\text{GSOS laws}\ \ F^*(G \times id) \xRightarrow{\ \lambda\ } (G \times id)F^*}{\text{natural transformations}\ \ F(G \times id) \underset{\rho}{\Longrightarrow} GF^*}$$

We use an overline-notation $\lambda \mapsto \overline{\lambda}, \rho \mapsto \overline{\rho}$ for this correspondence, in both directions.

Correspondingly, $F^*X \xrightarrow{a} X \xrightarrow{b} GX$ is a λ-model (as in Definition 2) if and only if the following diagram commutes.

$$
\begin{array}{ccccc}
FX & \xrightarrow{\ a \circ \sigma\ } & X & \xrightarrow{\ b\ } & GX \\
{\scriptstyle F(\langle b, id \rangle)}\downarrow & & & & \uparrow {\scriptstyle Ga} \\
F(GX \times X) & & \xrightarrow[\ \overline{\lambda}\ ]{} & & GF^*X
\end{array}
$$

In view of this result, we shall often also call a natural transformation $F(G \times id) \Rightarrow GF^*$ a GSOS law.

Proof. We only describe the constructions, and leave the details to the interested readers. For the correspondence between GSOS laws and natural transformations, first assume a GSOS law $\lambda: F^*(G \times id) \Rightarrow (G \times id)F^*$. It gives rise to a natural transformation:

$$\overline{\lambda}_X = \left(F(GX \times X) \xrightarrow{\ \sigma\ } F^*(GX \times X) \xrightarrow{\ \lambda_X\ } GF^*X \times F^*X \xrightarrow{\ \pi_1\ } GF^*X \right)$$

Conversely, for $\rho: F(G \times id) \Rightarrow GF^*$ we define a distributive law $\overline{\rho} = \langle \rho_1, \rho_2 \rangle: F^*(G \times id) \Rightarrow (G \times id)F^*$ where $\rho_2 = F^*(\pi_2)$ and ρ_1 is defined by recursion (following Theorem 2) in:

$$
\begin{array}{ccc}
\left(\begin{array}{c} (GX \times X) + \\ F\big(F^*(GX \times X)\big) \end{array}\right) & \xdashrightarrow{\ id + F(\langle \rho_1, id \rangle)\ } & \left(\begin{array}{c} (GX \times X) + \\ F\big(GF^*X \times F^*(GX \times X)\big) \end{array}\right) \\
{\scriptstyle [\eta, \tau]}\downarrow {\scriptstyle \cong} & & \downarrow {\scriptstyle [G\eta \circ \pi_1,\, G\mu \circ \rho \circ F(id \times F^*(\pi_2))]} \\
F^*(GX \times X) & \xdashrightarrow[\ \ \ \ \ \ \ \ \ \rho_1\ \ \ \ \ \ \ \ \ ]{} & GF^*X
\end{array}
$$

The equivalence with respect to models amounts for $F^*X \xrightarrow{a} X \xrightarrow{b} GX$ to:

$$
\begin{array}{ccc}
F^*X \xrightarrow{\;a\;} X \xrightarrow{\langle b,\,\mathrm{id}\rangle} (G \times \mathrm{id})X & & FX \xrightarrow{a \circ \sigma} X \xrightarrow{\;b\;} GX \\
\Big\downarrow{\scriptstyle F^*(\langle b,\mathrm{id}\rangle)} \qquad \Big\uparrow{\scriptstyle Ga \times a} & \mathrm{iff} & \Big\downarrow{\scriptstyle F(\langle b,\mathrm{id}\rangle)} \qquad \Big\uparrow{\scriptstyle Ga} \\
F^*(GX \times X) \xrightarrow[\;\lambda\;]{} (G \times \mathrm{id})F^*X & & F(GX \times X) \xrightarrow[\;\lambda\;]{} GF^*X
\end{array}
$$

The direction from left to right is straightforward, and the reverse direction requires the use of uniqueness in recursion.   □

Example 5. The regular expression functor $R(X) = 1 + 1 + (X \times X) + (X \times X) + X$ and the deterministic automaton functor $D(X) = X^A \times 2$ are connected via a GSOS law:

$$
\begin{array}{rcl}
R(X^A \times 2 \times X) & \xrightarrow{\quad\quad\quad \rho X \quad\quad\quad} & R^*(X)^A \times 2 \\[4pt]
0 & \xmapsto{\quad\quad \text{zero} \quad\quad} & (\lambda a \in A.\, 0, 0) \\[4pt]
1 & \xmapsto{\quad\quad \text{one} \quad\quad} & (\lambda a \in A.\, 0, 1) \\[4pt]
\langle(\varphi_1, b_1, x_1), (\varphi_2, b_2, x_2)\rangle & \xmapsto{\quad\quad \text{plus} \quad\quad} & (\lambda a \in A.\, \varphi_1(a) + \varphi_2(a), b_1 \vee b_2) \\[4pt]
\langle(\varphi_1, b_1, x_1), (\varphi_2, b_2, x_2)\rangle & \xmapsto{\quad \text{product} \quad} & (\lambda a \in A.\, \varphi_1(a) \cdot x_2 + b_1 \cdot \varphi_2(a), b_1 \wedge b_2) \\[4pt]
(\varphi, b, x) & \xmapsto{\quad\quad \text{star} \quad\quad} & (\lambda a \in A.\, \varphi(a) \cdot x^*, 1).
\end{array}
$$

One recognises the clauses/rules for D and E as described in the previous subsection. Their format can thus be expressed via a GSOS law; see [5] for more information about such correspondences. We shall illustrate that this law is fundamental, in the sense that it induces familiar structure (and associated results) on regular expressions.

There are a number of general results about GSOS laws that put our running example in perspective. We shall concentrate on these results first, and return to the example of regular expressions at the end of this subsection. The next two results are the analogues for GSOS laws of Lemmas 1 and 2. The proof of the second one uses a form of recursion for Eilenberg-Moore algebras.

Lemma 4. *If we have a GSOS law* $\lambda: T(G \times \mathrm{id}) \Rightarrow (G \times \mathrm{id})T$, *then a final coalgebra* $\zeta: Z \xrightarrow{\;\cong\;} GZ$ *induces a final λ-model with algebra* $\alpha: TZ \to Z$ *defined by coinduction:*

$$
\begin{array}{ccc}
GTZ & \xdashrightarrow{\quad G\alpha \quad} & GZ \\[4pt]
{\scriptstyle \pi_1 \circ \lambda}\Big\uparrow & & \Big\uparrow{\scriptstyle \cong}{\scriptstyle \zeta} \\[4pt]
T(GZ \times Z) & & \\[4pt]
{\scriptstyle T(\langle\zeta,\mathrm{id}\rangle)}\Big\uparrow & & \\[4pt]
TZ & \xdashrightarrow[\quad \alpha \quad]{} & Z
\end{array}
$$

Proof. By uniqueness one obtains that α is an Eilenberg-Moore algebra. By construction the pair (α, ζ) is a λ-model. It is final because for an arbitrary λ-model $TX \overset{a}{\rightarrow}$ $X \overset{b}{\rightarrow} GX$ the induced coalgebra map $X \rightarrow Z$ is also an algebra map—again proven by uniqueness.   □

Lemma 5. *Given a GSOS law* $\lambda: T(G \times id) \Rightarrow (G \times id)T$ *there is a bijective correspondence between GT-coalgebras and λ-models with free algebra:*

$$\frac{\text{``equations''} \quad X \overset{e}{\longrightarrow} GTX}{\lambda\text{-models} \quad TTX \xrightarrow[\mu]{} TX \xrightarrow[d]{} GTX}$$

and also between corresponding solutions and bialgebra maps.

Proof. The proof relies on the following "recursion" version of freeness for Eilenberg-Moore algebras: for each $f: X \rightarrow Y$ and $a: T(Y \times TX) \rightarrow Y$ there is a unique map g in:

$$
\begin{array}{ccc}
T^2X & \overset{T(\langle g, \mathrm{id}\rangle)}{\dashrightarrow} & T(Y \times TX) \\
{\scriptstyle\mu}\downarrow & & \downarrow{\scriptstyle a} \\
TX & \underset{g}{\dashrightarrow} & Y
\end{array}
\qquad \text{with} \quad g \circ \eta = f \qquad (7)
$$

provided that a satisfies $a \circ \eta = \pi_1$ and $a \circ \mu = a \circ T(\langle a, \mu \circ T(\pi_2)\rangle)$. The proof of this property is much like the proof of Theorem 2 and left to the reader.

We only describe the correspondence between equations and GSOS models, and leave the rest to the interested reader. Given $e: X \rightarrow GTX$ define $\bar{e}$ via (7) in:

$$
\begin{array}{ccc}
T^2X & \overset{T(\langle \bar{e}, \mathrm{id}\rangle)}{\dashrightarrow} & T(GTX \times TX) \\
{\scriptstyle\mu}\downarrow & & \downarrow{\scriptstyle G\mu \,\circ\, \pi_1 \,\circ\, \lambda} \\
TX & \underset{\bar{e}}{\dashrightarrow} & GTX
\end{array}
\qquad \text{with} \quad \bar{e} \circ \eta = e
$$

By construction this forms a λ-model. In the reverse direction, given $d: TX \rightarrow GTX$ one takes $\bar{d} = d \circ \eta: X \rightarrow GTX$. Then $\bar{\bar{e}} = \bar{e} \circ \eta = e$. And $\bar{\bar{d}} = d$ follows by uniqueness, using that (μ, d) is a GSOS model: $G\mu \circ \pi_1 \circ \lambda \circ T(\langle d, \mathrm{id}\rangle) = d \circ \mu$.   □

Remark 1.   1. If we apply the construction of the previous lemma starting from a law $\rho: F(G \times id) \Rightarrow GF^*$ like in Proposition 5, then the GSOS model $F^*F^*X \overset{\mu}{\longrightarrow}$ $F^*X \overset{d}{\longrightarrow} GF^*X$ associated with an equation $e: X \rightarrow GF^*X$ can be described via recursion (like in Theorem 2) as:

$$
\begin{array}{ccc}
X + F(F^*X) & \xrightarrow{\ \ \mathrm{id} + F(\langle d, \mathrm{id}\rangle)\ \ } & X + F(GF^*X \times F^*X) \\[2mm]
{\scriptstyle [\eta, \tau]}\Big\downarrow \cong & & \Big\downarrow {\scriptstyle [e, G\mu \,\circ\, \rho]} \\[2mm]
F^*X & \xrightarrow{\hspace{4cm} d \hspace{0.5cm}} & GF^*X
\end{array}
$$

This will be used later.

2. In [35, Proposition 5.1] it is shown that a (GSOS) law $\rho\colon F(G\times\mathrm{id}) \Rightarrow GF^*$ induces a lifting of the free monad F^* to the category $\mathbf{CoAlg}(G)$. The construction uses the previous point: it takes a coalgebra $b\colon X \to GX$ to the coalgebra-part of the bialgebra corresponding to the equation $G(\eta) \circ b\colon X \to GF^*X$.

With all these general GSOS results in place we are finally in a position to analyse the situation of regular expressions and languages, using the GSOS law from Example 5.

Theorem 3. *1. The "equation" $A \to \mathcal{D}(\mathcal{R}^*(A))$ that is given by the two maps*

$$
\begin{array}{lcr}
A \xrightarrow{\hspace{3cm}} \mathcal{R}^*(A)^A & \qquad & A \xrightarrow{\hspace{1cm}} 2 \\[2mm]
a \longmapsto \lambda b \in A.\ \textit{if } b = a \textbf{ then } 1 \textbf{ else } 0 & & a \longmapsto 0
\end{array}
$$

corresponds by Lemma 5 to the free algebra and Brzozowski automaton structure on the set $\mathsf{Re} = \mathcal{R}^*(A)$ of regular expressions:

$$
\mathcal{R}^*(\mathsf{Re}) \xrightarrow{\ \ \mu\ \ } \mathsf{Re} \xrightarrow{\ \langle D, E\rangle\ } \mathsf{Re}^A \times 2
$$

2. *The final $\mathcal{D}$-coalgebra $\mathcal{L}(A) \xrightarrow{\cong} \mathcal{L}(A)^A \times 2$ of languages yields by Lemma 4 the final bialgebra:*

$$
\mathcal{R}^*(\mathcal{L}(A)) \xrightarrow{\hspace{3cm}} \mathcal{L}(A) \xrightarrow{\ \cong\ } \mathcal{L}(A)^A \times 2
$$

with the standard algebra of regular expressions.

3. *The interpretation $[\![-]\!]\colon \mathsf{Re} \to \mathcal{L}(A)$ introduced via freeness in (3) can also be obtained as $\mathsf{beh}\colon \mathsf{Re} \to \mathcal{L}(A)$ by finality using the previous two points.*

4. *Bisimilarity between regular expressions is a congruence: $s \leftrightarrow s'$ and $t \leftrightarrow t'$ implies $s + t \leftrightarrow s' + t'$, $s \cdot t \leftrightarrow s' \cdot t'$ and $s^* \leftrightarrow s'^*$.*

Proof. 1. Let's write $e\colon A \to \mathsf{Re}^A \times 2$ for the equation. We need to check that the Brzozowski structure $\langle D, E\rangle$ from Subsection 4.2 fits in the description in Remark 1.(1), *i.e.* that the following diagram commutes,

$$
\begin{array}{ccc}
A + \mathcal{R}(\mathsf{Re}) & \xrightarrow{\ \ \mathrm{id} + \mathcal{R}(\langle\langle D, E\rangle, \mathrm{id}\rangle)\ \ } & A + \mathcal{R}(\mathsf{Re}^A \times 2 \times \mathsf{Re}) \\[2mm]
{\scriptstyle [\eta, \tau]}\Big\downarrow \cong & & \Big\downarrow {\scriptstyle [e, (\mu^A \times \mathrm{id}) \,\circ\, \rho]} \\[2mm]
\mathsf{Re} & \xrightarrow{\hspace{3cm} \langle D, E\rangle \hspace{0.5cm}} & \mathsf{Re}^A \times 2
\end{array}
$$

where ρ is as described in Example 5. This diagram commutes because the Brzo-zowski structure $\langle D, E \rangle$ precisely follows the GSOS law ρ.

2. Similarly we need to show that the standard interpretation $\alpha \colon \mathcal{R}(\mathcal{L}(A)) \to \mathcal{L}(A)$ yields a commuting diagram in Lemma 4. This means that $\langle \delta, \varepsilon \rangle \circ \alpha = (\alpha^A \times \mathrm{id}) \circ \rho \circ \mathcal{R}(\langle\langle \delta, \varepsilon \rangle, \mathrm{id}\rangle)$, which can be checked easily—where $\langle \delta, \varepsilon \rangle$ is the final coalgebra structure on $\mathcal{L}(A)$.

3. Obvious, since $[\![-]\!]$ is also a map of coalgebras.

4. The bisimilarity relation $\underline{\leftrightarrow} \rightarrowtail \mathsf{Re} \times \mathsf{Re}$ is the equaliser e at the bottom row below, because of Proposition 2 and because $[\![-]\!] = \mathsf{beh}$ by the previous point.

$$
\begin{array}{ccccc}
\mathcal{R}^*(\underline{\leftrightarrow}) & \xrightarrow{\;\;d\;\;} & \mathcal{R}^*(\mathsf{Re}) \times \mathcal{R}^*(\mathsf{Re}) & \underset{\mathcal{R}^*([\![-]\!]) \circ \pi_2}{\overset{\mathcal{R}^*([\![-]\!]) \circ \pi_1}{\rightrightarrows}} & \mathcal{R}^*(\mathcal{L}(A)) \\[2ex]
\Big\downarrow & & \Big\downarrow {\scriptstyle \mu \times \mu} & & \Big\downarrow \\[2ex]
\underline{\leftrightarrow} & \underset{e}{\rightarrowtail} & \mathsf{Re} \times \mathsf{Re} & \underset{[\![-]\!] \circ \pi_2}{\overset{[\![-]\!] \circ \pi_1}{\rightrightarrows}} & \mathcal{L}(A)
\end{array}
$$

The map $d = \langle \mathcal{R}^*(\pi_1 \circ e), \mathcal{R}^*(\pi_2 \circ e) \rangle$ induces an algebra structure on the relation $\underline{\leftrightarrow}$, as indicated. This makes $\underline{\leftrightarrow}$ a congruence. $\qquad\square$

The map $[\![-]\!] \colon \mathsf{Re} \to \mathcal{L}(A)$ defined by initiality is by construction "compositional", in the sense that it preserves the operations. This map describes what may be called the denotational semantics of regular expressions. In contrast, the map $\mathsf{beh} \colon \mathsf{Re} \to \mathcal{L}(A)$ obtained by finality describes the operational semantics, because it is induced by the dynamical (coalgebra) structure on regular expressions. The equality of denotational $[\![-]\!]$ and operational beh semantics in point 3 of the previous theorem says in particular that the operational semantics is compositional, so that for instance the behaviour of a sum expression is the sum of the behaviours of the two summands. Many coincidences of operational and denotational semantics are described in more concrete form in [3].

5 Regular Expressions with Equations

An equational logic for regular expressions is formulated by Kozen in [23], for which a completeness theorem is proved. An alternative proof of completenes (again by Kozen) is given in [24]. Here we shall give a coalgebraic review of the situation, which leads to a third completeness proof. It is similar, but shorter, than the proof in [24].

Throughout this section we fix a *finite* alphabet A. We shall indicate where we need this finiteness (in Definition 4).

The definition of **Kleene algebra** from [23] involves a particular formulation of the rules for the star operation. It requires for an algebra $[0, 1, +, \cdot, (-)^*] \colon \mathcal{R}(Y) \to Y$ that $(Y, 0, 1, +, \cdot)$ is an idempotent semiring in which the star axioms and rules in point 2 below hold.

One can also turn the set Re of regular expressions into a Kleene algebra via a suitable quotient. For clarity we shall use a special symbol $\doteq\, \subseteq \mathsf{Re} \times \mathsf{Re}$ for the least relation satisfying the next three points.

1. $(\text{Re}, +, 0, \cdot, 1)$ is an idempotent semiring, *i.e.*

   – $(\text{Re}, +, 0)$ is an idempotent commutative monoid, in which one defines a partial order by $s \leq t \iff s + t \doteq t$.
   – $(\text{Re}, \cdot, 1)$ is a monoid, where $\cdot$ preserves the additive monoid structure $+, 0$ in both arguments: $s \cdot (t + r) \doteq (s \cdot t) + (s \cdot r)$ and $(t + r) \cdot s \doteq (t \cdot s) + (r \cdot s)$, and also $s \cdot 0 \doteq 0$ and $0 \cdot s \doteq 0$.

2. The star inequalities and rules:

$$1 + s \cdot s^* \leq s^* \qquad 1 + s^* \cdot s \leq s^* \qquad \frac{s + t \cdot x \leq x}{t^* s \leq x} \qquad \frac{s + x \cdot t \leq x}{s \cdot t^* \leq x}$$

3. Axioms and rules making $\doteq$ a congruence, *i.e.* an equivalence relation preserved by the operations: $s \doteq s'$ and $t \doteq t'$ implies $s + t \doteq s' + t'$, $s \cdot t \doteq s' \cdot t'$ and $s^* \doteq s'^*$.

We shall write $\text{Re}/\doteq$ for the set of regular expressions modulo $\doteq$. By construction it forms a Kleene algebra. As usual, we often simply write s for the equivalence class $[s] = \{t \in \text{Re} \mid t \doteq s\} \in \text{Re}/\doteq$.

Of the many results that can be derived in Kleene algebras we shall need the following ones.

Lemma 6. *In an arbitrary Kleene algebra one has:*

1. $1 + s \cdot s^* = s^*$;
2. $s \cdot x = x \cdot t$ implies $s^* \cdot x = x \cdot t^*$.

And each term $s \in \text{Re}$ satisfies $s \geq \sum_{a \in A} a \cdot D_a(s) + E(s)$.

Proof. The inequality $1 + s \cdot s^* \leq s^*$ is one of the star axioms. And $s^* \leq 1 + s \cdot s^*$ is obtained by applying a star rule to the inequality $1 + s \cdot x \leq x$ for $x = 1 + s \cdot s^*$.

For the second point it suffices to show: if $s \cdot x \leq x \cdot t$ then $s^* \cdot x \leq x \cdot t^*$. The latter can be obtained via a star rule from $x + s \cdot (x \cdot t^*) \leq x \cdot t^*$, which follows from the assumption $s \cdot x \leq x \cdot t$.

The final inequality $s \geq \sum_{a \in A} a \cdot D_a(s) + E(s)$ is obtained by induction on the structure of $s \in \text{Re}$. $\qquad\square$

The following two standard lemmas (see *e.g.* [9,24,31]) must be made explicit first.

Lemma 7. *1. The derivative operation on regular expressions preserves equality, i.e. satisfies $s \doteq t \implies D_a(s) \doteq D_a(t)$, for each letter a. Similarly, $s \doteq t \implies E(s) = E(t)$.*

The Brzozowski coalgebra structure $\langle D, E \rangle \colon \text{Re} \to \text{Re}^A \times 2$ thus restricts to $\langle D, E \rangle \colon (\text{Re}/\doteq) \to (\text{Re}/\doteq)^A \times 2$, making the quotient map $[-] \colon \text{Re} \twoheadrightarrow \text{Re}/\doteq$ a homomorphism of coalgebras.

2. *If $s \doteq t$ then $[\![\, s \,]\!] = [\![\, t \,]\!]$, i.e. s, t yield the same languages. Hence the diagram (3) of bialgebras can be further refined by taking images:*

$$
\begin{array}{ccccccc}
\mathcal{R}(Re) & \longrightarrow & \mathcal{R}(Re/\!\doteq) & \longrightarrow & \mathcal{R}(\mathcal{L}_r(A)) & \longrightarrow & \mathcal{R}(\mathcal{L}(A)) \\
{\scriptstyle \mu}\downarrow & & \downarrow & & \downarrow & & \downarrow \\
Re & \longrightarrow & Re/\!\doteq & \xrightarrow{\;[\![-]\!]\;} & \mathcal{L}_r(A) \rightarrowtail & \longrightarrow & \mathcal{L}(A) \\
{\scriptstyle \langle D, E\rangle}\downarrow & & {\scriptstyle \langle D, E\rangle}\downarrow & & \downarrow & & {\scriptstyle \cong}\downarrow{\scriptstyle \langle \delta, \varepsilon\rangle} \\
Re^A \times 2 & \longrightarrow & (Re/\!\doteq)^A \times 2 & \longrightarrow & \mathcal{L}_r(A)^A \times 2 & \longrightarrow & \mathcal{L}(A)^A \times 2
\end{array}
\tag{8}
$$

where $\mathcal{L}_r(A)$ is the subset of regular (also called rational) languages obtained as interpretation $[\![\, s \,]\!]$ of a regular expression s.

The completeness result of [24] states that the (restricted) homomorphism $[\![-]\!]$ in the middle of (8) is an isomorphism, see Theorem 4 below.

Proof. By induction on the length of derivations of $\doteq$. □

The derivative operation $D \colon Re \to Re^A$ yields a multiple derivative $D^\star \colon Re \to Re^{A^\star}$ like in (1). Similarly we get $D^\star \colon Re/\!\doteq \to (Re/\!\doteq)^{A^\star}$ for expressions modulo equations. We shall also use the subscript notation in these situations (and drop the star), so that $D_\sigma(s) = D^\star(s)(\sigma)$ with cases $D_{\langle\rangle}(s) = s$ and $D_{a \cdot \sigma}(s) = D_\sigma(D_a(s))$.

Lemma 8. *Expressions modulo equations have only finitely many successors: for each term/state $s \in Re$ the set $\Diamond(s) = \{D_\sigma(s) \mid \sigma \in A^\star\} \subseteq Re/\!\doteq$ of successors of s in the coalgebra $Re/\!\doteq \to (Re/\!\doteq)^A \times 2$ is finite.*

Proof. The basic terms are easy, since $\Diamond(0) = \{D_\sigma(0) \mid \sigma \in A^\star\} = \{0\}$, $\Diamond(1) = \{1, 0\}$ and $\Diamond(a) = \{a, 1, 0\}$. For the compound terms one first proves the following equations.

$$
\begin{aligned}
D_\sigma(s + t) &\doteq D_\sigma(s) + D_\sigma(t) \\
D_\sigma(s \cdot t) &\doteq D_\sigma(s) \cdot t + \sum_{\tau \cdot \rho = \sigma; \rho \neq \langle\rangle} E(D_\tau(s)) \cdot D_\rho(t) \\
D_\sigma(s^*) &\doteq D_\sigma(1) + D_\sigma(s) \cdot s^* + \sum_{\tau \cdot \rho = \sigma; \tau, \rho \neq \langle\rangle} E(D_\tau(s)) \cdot D_\rho(s^*).
\end{aligned}
$$

These equations are obtained by induction on the length of $\sigma \in A^\star$.

If we now write $\# \Diamond(s) \in \mathbb{N}$ for the number of elements of $\Diamond(s)$, then:

$$
\begin{array}{ll}
\# \Diamond(0) = 1 & \# \Diamond(s + t) \leq \# \Diamond(s) \cdot \# \Diamond(t) \\
\# \Diamond(1) = 1 & \# \Diamond(s \cdot t) \leq \# \Diamond(s) \cdot 2^{\# \Diamond(t)} \\
\# \Diamond(a) = 3 & \# \Diamond(s^*) \leq \# \Diamond(s) \cdot 2^{\# \Diamond(s)}.
\end{array}
$$

Hence we can conclude that each subset $\Diamond(s) \subseteq Re/\!\doteq$ is finite. □

Next we shall define a category in which the Brzozowski automaton on $\mathsf{Re}/\doteq$ lives.

Definition 3. *We write $\mathbf{DetAut}_{fb}$ for the category of deterministic automata with finite behaviour. Objects are coalgebras $\langle \delta, \varepsilon \rangle \colon X \to X^A \times 2$ such that for each state $x \in X$ the set of successors $\Diamond(x) = \{ \delta^*(x)(\sigma) \mid \sigma \in A^* \} \subseteq X$ is finite. Maps in $\mathbf{DetAut}_{fb}$ are the usual homomorphisms of coalgebras.*

(Notice that we leave the set A of inputs implicit in the notation.)

It is not hard to see that if $\left(X \to X^A \times 2 \right) \overset{f}{\twoheadrightarrow} \left(Y \to Y^A \times 2 \right)$ is a surjective coalgebra homomorphism where $X \to X^A \times 2$ is in $\mathbf{DetAut}_{fb}$, then so is $Y \to Y^A \times 2$. The reason is that $f(\delta^*(x)(\sigma)) = \delta^*(f(x))(\sigma)$, and so $\Diamond(f(x)) \subseteq f[\Diamond(x)]$. Hence the automaton structure $\mathcal{L}_r(A) \to \mathcal{L}_r(A)^A \times 2$ from (8) is also in the category $\mathbf{DetAut}_{fb}$, via the surjection $[\![-]\!] \colon \mathsf{Re}/\doteq \twoheadrightarrow \mathcal{L}_r(A)$.

A basic property of Kleene algebras is that an inequality $x \geq s \cdot x + t$ has a least solution $s^* t$, via the star rule and via $s^* \cdot t \geq s \cdot (s^* \cdot t) + t$. Even stronger, the latter is actually an equality, since $s \cdot (s^* \cdot t) + t \doteq (s \cdot s^* + 1) \cdot t \doteq s^* \cdot t$.

This can be generalised to equations in multiple variables, using the standard fact that square matrices in Kleene algebras form again Kleene algebras, and can be used to solve equations, see [23, Section 3]. A system of n equations:

$$x_i = s_{i1} x_1 + \cdots + s_{in} x_n + t_i$$

has a least solution that can be described as vector $S^* \cdot T$ where

$$S = \begin{pmatrix} s_{11} & \cdots & s_{1n} \\ & \vdots & \\ s_{n1} & \cdots & s_{nn} \end{pmatrix} \quad \text{and} \quad T = \begin{pmatrix} t_1 \\ \vdots \\ t_n \end{pmatrix}$$

describe the equation as $\vec{x} = S \cdot \vec{x} + T$ and the star operation S^* is in the Kleene algebra of $n \times n$ matrices.

Definition 4. *Let $\langle \delta, \varepsilon \rangle \colon X \to X^A \times 2$ be an arbitrary coalgebra with finite behaviour (i.e. an object of $\mathbf{DetAut}_{fb}$). With each state $x \in X$ we associate a term $\ulcorner x \urcorner \in \mathsf{Re}/\doteq$ in the following way.*

By assumption $\Diamond(x)$ is finite, say $\Diamond(x) = \{ x_1, x_2, \ldots, x_n \}$ where $x_1 = x$. An $n \times n$ transition matrix $S_x = (s_{ij})$ and an output vector $T_x = (t_i)$ over $\mathsf{Re}/\doteq$ are constructed with elements

$$s_{ij} = \sum \{ a \in A \mid \delta(x_i)(a) = x_j \} \quad \text{and} \quad t_i = \varepsilon(x_i).$$

We then take $\ulcorner x \urcorner \in \mathsf{Re}/\doteq$ to be the first element of the least solution $S_x^* \cdot T_x$ of the associated equations. More formally, as vector product, $\ulcorner x \urcorner = (1\, 0 \ldots 0) \cdot S_x^* \cdot T_x$.

The sum $\sum$ in this definition exists because we have assumed that the alphabet A is finite. The sum over an empty set is 0, as usual. Notice that the ordering of the elements in $\Diamond(x)$ is not relevant.

One can understand S as a big square matrix $X \times X \to \mathsf{Re}/\doteq$ defined by $(x, x') \mapsto \sum \{ a \mid \delta(a)(x) = x' \}$ like in [24]. The matrix S_x in the definition is then the restriction of S to $\{ x_1, \ldots, x_n \} \subseteq X$.

Lemma 9. *The mapping* $x \mapsto \ulcorner x \urcorner$ *is a homomorphism of coalgebras.*

Proof. Consider $x = x_1 \in X$ as in Definition 4. We need to show:

$$E(\ulcorner x \urcorner) = \varepsilon(x) \quad \text{and} \quad D(\ulcorner x \urcorner)(a) = \ulcorner \delta(x)(a) \urcorner.$$

We notice that the vector of solutions in $\mathsf{Re}/\doteq$ can be described as $\overrightarrow{\ulcorner x_i \urcorner}$. Hence

$$\ulcorner x_1 \urcorner \doteq s_{11} \cdot \ulcorner x_1 \urcorner + \cdots + s_{1n} \cdot \ulcorner x_n \urcorner + \varepsilon(x_1),$$

where each s_{ij} is a sum of atoms/letters from A. Thus:

$$
\begin{aligned}
E(\ulcorner x \urcorner) &= E(s_{11} \cdot \ulcorner x_1 \urcorner + \cdots + s_{1n} \cdot \ulcorner x_n \urcorner + \varepsilon(x_1)) \\
&= \big(E(s_{11}) \wedge E(\ulcorner x_1 \urcorner)\big) \vee \cdots \vee \big(E(s_{1n}) \wedge E(\ulcorner x_n \urcorner)\big) \vee E(\varepsilon(x_1)) \\
&= \big(0 \wedge E(\ulcorner x_1 \urcorner)\big) \vee \cdots \vee \big(0 \wedge E(\ulcorner x_n \urcorner)\big) \vee \varepsilon(x_1) \\
&= \varepsilon(x_1) \\
D(\ulcorner x \urcorner)(a) &= D(s_{11} \cdot \ulcorner x_1 \urcorner + \cdots + s_{1n} \cdot \ulcorner x_n \urcorner + \varepsilon(x_1))(a) \\
&= D(s_{11} \cdot \ulcorner x_1 \urcorner)(a) + \cdots + D(s_{1n} \cdot \ulcorner x_n \urcorner)(a) \\
&= D(s_{11})(a) \cdot \ulcorner x_1 \urcorner + E(s_{11}) \cdot D(\ulcorner x_1 \urcorner)(a) + \cdots + \\
&\qquad D(s_{1n})(a) \cdot \ulcorner x_n \urcorner + E(s_{1n}) \cdot D(\ulcorner x_n \urcorner)(a) \\
&= D(s_{11})(a) \cdot \ulcorner x_1 \urcorner + \cdots + D(s_{1n})(a) \cdot \ulcorner x_n \urcorner \\
&= \ulcorner x_i \urcorner \text{ if } \delta(x)(a) = x_i \\
&= \ulcorner \delta(x)(a) \urcorner.
\end{aligned}
$$
$\qquad\square$

By finality this homomorphism $\ulcorner - \urcorner$ yields a commuting diagram:

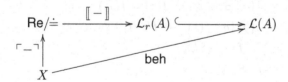

In particular, when $X = \mathcal{L}_r(A)$, we see that $\ulcorner - \urcorner$ is a section of $[\![-]\!]$.

Corollary 1. *The coalgebra* $\mathcal{L}_r(A) \to \mathcal{L}_r(A)^A \times 2$ *is final in the category* **DetAut**$_{fb}$.

Proof. Given a coalgebra $X \to X^A \times 2$ in **DetAut**$_{fb}$ there is a composition of homomorphisms $[\![-]\!] \circ \ulcorner - \urcorner \colon X \to \mathsf{Re}/\doteq \to \mathcal{L}_r(A)$. If we have two homomorphisms $f, g \colon X \to \mathcal{L}_r(A)$, then by postcomposition with the inclusion $\mathcal{L}_r(A) \hookrightarrow \mathcal{L}(A)$ we get two homomorphisms to the final $(-)^A \times 2$ coalgebra—which must thus be equal. Hence also $f = g$. $\qquad\square$

At this stage we can obtain Kleene's theorem [21], as point 2 below. Point 1 is [31, Theorem 10.1].

Corollary 2. *1. A language $L \in \mathcal{L}(A)$ is regular—i.e. belongs to $\mathcal{L}_r(A) \hookrightarrow \mathcal{L}(A)$—if and only if the set of derivatives $\Diamond(L)$ is finite.*
  2. A language $L \in \mathcal{L}(A)$ is regular if and only if it is accepted by a finite automaton (i.e. an automaton with a finite state space).

Proof. 1. If $L \in \mathcal{L}_r(A)$, then $\Diamond(L)$ is finite because $\mathcal{L}_r(A)$ is in **DetAut**$_{fb}$. Conversely, if $\Diamond(L)$ is finite, then $\Diamond(L)$ can be considered as a subcoalgebra $\Diamond(L) \hookrightarrow \mathcal{L}(A)$ that belongs to **DetAut**$_{fb}$. Hence it factors as $\Diamond(L) \hookrightarrow \mathcal{L}_r(A)$.
  2. If L is regular, then $\Diamond(L)$ is itself a finite automaton (by 1) with initial state $L \in \Diamond(L)$ whose behaviour $\mathsf{beh}(L) \in \mathcal{L}_r(A)$ is L itself. Conversely, if $L \in \mathcal{L}(A)$ is $\mathsf{beh}(x)$ for an initial state $x \in X$ of a finite automaton, then $\Diamond(x)$ is finite, so $L = \mathsf{beh}(x) \in \mathcal{L}_r(A)$ because $\mathcal{L}_r(A)$ is final in **DetAut**$_{fb}$. □

The next two lemmas and their proofs are reformulations of results in [24].

Lemma 10. *If $f\colon X \to Y$ is a homomorphism in **DetAut**$_{fb}$, then $\ulcorner f(x) \urcorner = \ulcorner x \urcorner$.*

Proof. If $\Diamond(x) = \{x_1, \ldots, x_n\}$ where $x_1 = x$, then $\Diamond(f(x)) = \{f(x_1), \ldots, f(x_n)\}$. The latter set may be smaller than the former. We shall consider the following three square matrices $S, \widehat{f}, S^f \colon \{1, \ldots, n\}^2 \to \mathsf{Re}/\doteq$.

$$S_{ij} = \sum \{a \mid x_i \xrightarrow{a} x_j\} \qquad (\widehat{f})_{ij} = \begin{cases} 1 \text{ if } f(x_i) = f(x_j) \\ 0 \text{ otherwise.} \end{cases}$$
$$(S^f)_{ij} = \sum \{a \mid f(x_i) \xrightarrow{a} f(x_j)\}$$

Then there is an equality of matrix products:

$$\begin{aligned}
(S \cdot \widehat{f})_{ij} &= \sum_k S_{ik} \cdot (\widehat{f})_{kj} \\
&= \sum \{\sum \{a \mid x_i \xrightarrow{a} z\} \mid z \in \Diamond(x) \wedge f(z) = f(x_j)\} \\
&= \sum \{a \mid \exists z \in \Diamond(x). x_i \xrightarrow{a} z \wedge f(z) = f(x_j)\} \\
&= \sum \{a \mid f(x_i) \xrightarrow{a} f(x_j)\} \\
&= \sum \{a \mid \exists z \in \Diamond(x). f(z) = f(x_i) \wedge f(z) \xrightarrow{a} f(x_j)\} \\
&= \sum \{\sum \{a \mid f(z) \xrightarrow{a} f(x_j)\} \mid z \in \Diamond(x) \wedge f(z) = f(x_i)\} \\
&= \sum_k (\widehat{f})_{ik} \cdot (S^f)_{kj} \\
&= (\widehat{f} \cdot S^f)_{ij}.
\end{aligned}$$

Lemma 6 (2) now yields $S^* \cdot \widehat{f} = \widehat{f} \cdot (S^f)^*$. If we write T for the vector of elements $\varepsilon(x_i) = \varepsilon(f(x_i))$, then $\widehat{f} \cdot T = T$, since

$$\begin{aligned}
(\widehat{f} \cdot T)_i &= \sum_k (\widehat{f})_{ik} \cdot T_k = \sum \{\varepsilon(x_k) \mid f(x_k) = f(x_i)\} \\
&= \sum \{\varepsilon(x_i) \mid f(x_k) = f(x_i)\} = \varepsilon(x_i) = T_i.
\end{aligned}$$

Hence:

$$\begin{aligned}
\ulcorner x \urcorner &= (1\,0 \ldots 0) \cdot S^* \cdot T = (1\,0 \ldots 0) \cdot S^* \cdot \widehat{f} \cdot T \\
&= (1\,0 \ldots 0) \cdot \widehat{f} \cdot (S^f)^* \cdot T \\
&= (\widehat{f}_{11}, \ldots, \widehat{f}_{1n}) \cdot (S^f)^* \cdot T \\
&= \sum \{\ulcorner f(x_i) \urcorner \mid f(x_i) = f(x_1)\} = \ulcorner f(x) \urcorner.
\end{aligned}$$

The last equation holds even though S^f may be "too big" a matrix, describing too many equations. These additional equations however are repeated equations, which do not influence the least solution.   □

Lemma 11. *The homomorphism* $\ulcorner - \urcorner \colon Re/\doteq \to Re/\doteq$ *is the identity.*

Proof. We first establish the following points.
1. $\ulcorner s \urcorner \le s$, for $s \in Re/\doteq$;
2. $\ulcorner 1 \urcorner \doteq 1$ and $\ulcorner 0 \urcorner \doteq 0$;
3. $s \le t$ implies $\ulcorner s \urcorner \le \ulcorner t \urcorner$;
4. $s \le \ulcorner s \urcorner$.

The first and fourth point then yield the required result.

As to the first point, for $s \in Re/\doteq$ we obtain $\ulcorner s \urcorner$ via the recipe in Definition 4, namely by considering the successor states/derivatives $\Diamond(s) = \{s_1, \ldots, s_n\}$ and the associated transition matrix. By Lemma 6 these terms $s_1, \ldots, s_n$ satisfy the defining inequality for $\ulcorner s_i \urcorner$, so that $\ulcorner s_i \urcorner \le s_i$, since $\ulcorner s_i \urcorner$ is the least solution.

The term 1 has one successor, namely 0. The associated single equation, following Definition 4, is $x = 1$, which has as (least) solution $\ulcorner 1 \urcorner \doteq 1$. Similarly $\ulcorner 0 \urcorner \doteq 0$.

For the third point we consider the product $X = Re/\doteq \times Re/\doteq$ as state space with two coalgebra structures $\langle D, E_1 \rangle, \langle D, E_2 \rangle \colon X \to X^A \times 2$, where

$$D(s,t)(a) = (D_a(s), D_a(t)) \qquad E_1(s,t) = E(s) \qquad E_2(s,t) = E(t).$$

The projections $\pi_i \colon X \to Re/\doteq$ are then homomorphisms from $\langle D, E_i \rangle$ to $\langle D, E \rangle$. Hence Lemma 10 applies. Given elements $s, t \in Re/\doteq$, let $S = S_{(s,t)}$ be the transition matrix associated with $(s,t) \in X$, and T_1, T_2 be the associated output vectors determined by the output functions E_1, E_2 respectively. Thus, if $s \le t$, then $E_1(s,t) \le E_2(s,t)$ and similarly for all successors of (s,t)—because D and E are order preserving. Hence $T_1 \le T_2$, and thus:

$$
\begin{aligned}
\ulcorner s \urcorner = \ulcorner \pi_1(s,t) \urcorner &= \ulcorner (s,t) \urcorner \quad \text{wrt. } \langle D, E_1 \rangle \\
&= (1\,0 \ldots 0) \cdot S^* \cdot T_1 \\
&\le (1\,0 \ldots 0) \cdot S^* \cdot T_2 \\
&= \ulcorner (s,t) \urcorner \quad \text{wrt. } \langle D, E_2 \rangle \\
&= \ulcorner \pi_2(s,t) \urcorner \\
&= \ulcorner t \urcorner.
\end{aligned}
$$

For the fourth point we proceed like in [24] and prove the stronger statement $\forall t \in Re.\ s \cdot \ulcorner t \urcorner \le \ulcorner s \cdot t \urcorner$ by induction on s. We are then done by taking $t = 1$, using point 2.

- $0 \cdot \ulcorner t \urcorner \doteq 0 \doteq \ulcorner 0 \urcorner \doteq \ulcorner 0 \cdot t \urcorner$.
- $1 \cdot \ulcorner t \urcorner \doteq \ulcorner t \urcorner \doteq \ulcorner 1 \cdot t \urcorner$.
- $\ulcorner b \cdot t \urcorner \ge \sum_{a \in A} a \cdot D_a(\ulcorner b \cdot t \urcorner) + E(\ulcorner b \cdot t \urcorner)$ by Lemma 6
  $\doteq \sum_{a \in A} a \cdot \ulcorner D_a(b \cdot t) \urcorner + E(b \cdot t)$ by Lemma 9
  $\doteq b \cdot \ulcorner t \urcorner$ by point 2.

$-\ (s_1 + s_2) \cdot \ulcorner t \urcorner \doteq s_1 \cdot \ulcorner t \urcorner + s_2 \cdot \ulcorner t \urcorner$

$\qquad \le \ulcorner s_1 \cdot t \urcorner + \ulcorner s_2 \cdot t \urcorner$ by induction hypothesis

$\qquad \le \ulcorner s_1 \cdot t + s_2 \cdot t \urcorner \quad$ by point 3.

$\qquad \doteq \ulcorner (s_1 + s_2) \cdot t \urcorner.$

$-\ (s_1 \cdot s_2) \cdot \ulcorner t \urcorner \doteq s_1 \cdot (s_2 \cdot \ulcorner t \urcorner)$

$\qquad \le s_1 \cdot \ulcorner s_2 \cdot t \urcorner \quad$ by induction hypothesis

$\qquad \le \ulcorner s_1 \cdot (s_2 \cdot t) \urcorner$ by induction hypothesis

$\qquad \doteq \ulcorner (s_1 \cdot s_2) \cdot t \urcorner.$

– Finally, $s^* \cdot \ulcorner t \urcorner \le \ulcorner s^* \cdot t \urcorner$ is obtained by applying the star rule to:

$$\ulcorner t \urcorner + s \cdot \ulcorner s^* \cdot t \urcorner \le \ulcorner t \urcorner + \ulcorner s \cdot (s^* \cdot t) \urcorner \text{ by induction hypothesis}$$

$$\le \ulcorner t + s \cdot (s^* \cdot t) \urcorner \quad \text{by point 3.}$$

$$\doteq \ulcorner (1 + s \cdot s^*) \cdot t \urcorner$$

$$\doteq \ulcorner s^* \cdot t \urcorner \qquad\qquad \text{by Lemma 6.} \qquad \square$$

Theorem 4 (Completeness [23,24]). *The Brzozowski coalgebra* $Re/\doteq\ \rightarrow\ Re/\doteq^A \times 2$ *is final in* **DetAut**$_{fb}$. *Hence the (bialgebra) homomorphism* $[\![-]\!]\colon Re/\doteq\ \rightarrow \mathcal{L}_r(A)$ *is an isomorphism.*

Proof. Each object $X \rightarrow X^A \times 2$ in **DetAut**$_{fb}$ yields a homomorphism $\ulcorner - \urcorner\colon X \rightarrow Re/\doteq$ by Lemma 9. Suppose we have two homomorphisms $f, g\colon X \rightarrow Re/\doteq$, then by Lemmas 10 and 11 we have:

$$f = \text{id}_{Re/\doteq} \circ f \overset{(11)}{=} \ulcorner - \urcorner \circ f \overset{(10)}{=} \ulcorner - \urcorner \overset{(10)}{=} \ulcorner - \urcorner \circ g \overset{(11)}{=} \text{id}_{Re/\doteq} \circ g = g.$$

Final object are unique up-to-isomorphism, so the coalgebra homomorphism $[\![-]\!] =$ beh: $Re/\doteq\ \rightarrow \mathcal{L}_r(A)$ is an isomorphism by Corollary 1. $\qquad \square$

Another way to formulate this result is: Kozen's axioms and rules give a complete axiomatisation of bisimilarity for regular expressions. Indeed, for $s, t \in Re$,

$$s \leftrightarrow t \iff \text{beh}(s) = \text{beh}(t) \text{ by Proposition 2}$$

$$\iff [\![s]\!] = [\![t]\!] \quad\quad \text{by Theorem 3.(3)}$$

$$\iff [s] = [t] \quad\quad\quad \text{by Theorem 4, where } [-]\colon Re \twoheadrightarrow Re/\doteq$$

$$\iff s \doteq t.$$

This gives a perfect bialgebraic match, where the equational logic on the algebra-side completely captures the observational equivalence on the coalgebra-side. Similar such results occur for instance within a line of work [10] in process algebra.

6 Conclusions

We have illustrated the effectiveness of the bialgebraic approach introduced by Turi and Plotkin [35] by showing how it neatly connects the elementary and classic structures of computer science, namely regular expressions, automata and languages. It thus forms a framework for what we consider to be the essence of computing: generated behaviour via matching algebra-coalgebra pairs. This framework may even guide developments in settings which are more complicated and possibly less well-developed, like extended regular expressions [22], or timed and probabilistic automata and their languages.

Acknowledgements

Thanks are due to Ichiro Hasuo for helpful discussions and for his valuable comments on the first draft of this paper.

References

1. M.A. Arbib and E.G. Manes. Foundations of system theory: Decomposable systems. *Automatica*, 10:285–302, 1974.
2. M.A. Arbib and E.G. Manes. *Algebraic Approaches to Program Semantics*. Texts and Monogr. in Comp. Sci.,. Springer, Berlin, 1986.
3. J.W. de Bakker and E. Vink. *Control Flow Semantics*. MIT Press, Cambridge, MA, 1996.
4. M. Barr and Ch. Wells. *Toposes, Triples and Theories*. Springer, Berlin, 1985. Revised and corrected version available from URL:
   www.cwru.edu/artsci/math/wells/pub/ttt.html.
5. F. Bartels. *On generalised coinduction and probabilistic specification formats. Distributive laws in coalgebraic modelling*. PhD thesis, Free Univ. Amsterdam, 2004.
6. J. Beck. Distributive laws. In B. Eckman, editor, *Seminar on Triples and Categorical Homolgy Theory*, number 80 in Lect. Notes Math., pages 119–140. Springer, Berlin, 1969.
7. B. Bloom, S. Istrail, and A.R. Meyer. Bisimulation can't be traced. *Journ. ACM*, 42(1):232–268, 1988.
8. J.A. Brzozowski. Derivatives of regular expressions. *Journ. ACM*, 11(4):481–494, 1964.
9. J.H. Conway. *Regular Algebra and Finite Machines*. Chapman and Hall, 1971.
10. W. Fokkink. On the completeness of the equations for the Kleene star in bisimulation. In M. Wirsing and M. Nivat, editors, *Algebraic Methodology and Software Technology*, number 1101 in Lect. Notes Comp. Sci., pages 180–194. Springer, Berlin, 1996.
11. J.A. Goguen. Minimal realization of machines in closed categories. *Bull. Amer. Math. Soc.*, 78(5):777–783, 1972.
12. J.A. Goguen. Realization is universal. *Math. Syst. Theor.*, 6(4):359–374, 1973.
13. J.A. Goguen. Discrete-time machines in closed monoidal categories. I. *Journ. Comp. Syst. Sci*, 10:1–43, 1975.
14. J.F. Groote and F. Vaandrager. Structured operational semantics and bisimulation as a congruence. *Inf. & Comp.*, 100(2):202–260, 1992.
15. C. Hermida and B. Jacobs. Structural induction and coinduction in a fibrational setting. *Inf. & Comp.*, 145:107–152, 1998.
16. B. Jacobs. Objects and classes, co-algebraically. In B. Freitag, C.B. Jones, C. Lengauer, and H.-J. Schek, editors, *Object-Orientation with Parallelism and Persistence*, pages 83–103. Kluwer Acad. Publ., 1996.

17. B. Jacobs. Exercises in coalgebraic specification. In R. Crole R. Backhouse and J. Gibbons, editors, *Algebraic and Coalgebraic Methods in the Mathematics of Program Construction*, number 2297 in Lect. Notes Comp. Sci., pages 237–280. Springer, Berlin, 2002.

18. B. Jacobs. Distributive laws for the coinductive solution of recursive equations. *Inf. & Comp.*, 2006, to appear. Earlier version in number 106 in Elect. Notes in Theor. Comp. Sci.

19. P.T. Johnstone. Adjoint lifting theorems for categories of algebras. *Bull. London Math. Soc.*, 7:294–297, 1975.

20. M. Kick. Bialgebraic modelling of timed processes. In P. Widmayer *et al.*, editor, *International Colloquium on Automata, Languages and Programming*, number 2380 in Lect. Notes Comp. Sci., pages 525–536. Springer, Berlin, 2002.

21. S.C. Kleene. Representation of events in nerve nets and finite automata. In C. E. Shannon and J. McCarthy, editors, *Automata Studies*, number 34 in Annals of Mathematics Studies, pages 3–41. Princeton University Press, 1956.

22. S. Koushik and G. Rosu. Generating optimal monitors for extended regular expressions. In *Runtime Verification (RV'03)*, number 89(2) in Elect. Notes in Theor. Comp. Sci. Elsevier, Amsterdam, 2003.

23. D. Kozen. A completeness theorem for Kleene algebras and the algebra of regular events. *Inf. & Comp.*, 110(2):366–390, 1994.

24. D. Kozen. Myhill-nerode relations on automatic systems and the completeness of Kleene algebra. In A. Ferreira and H. Reichel, editors, *Symposium on Theoretical Aspects of Computer Science*, number 2010 in Lect. Notes Comp. Sci., pages 27–38. Springer, Berlin, 2001.

25. S. Mac Lane. *Categories for the Working Mathematician*. Springer, Berlin, 1971.

26. M. Lenisa, J. Power, and H. Watanabe. Distributivity for endofunctors, pointed and co-pointed endofunctors, monads and comonads. In H. Reichel, editor, *Coalgebraic Methods in Computer Science*, number 33 in Elect. Notes in Theor. Comp. Sci. Elsevier, Amsterdam, 2000.

27. D.E. Muller and P.E. Schupp. Alternating automata on infinite trees. *Theor. Comp. Sci.*, 54(2/3):267–276, 1987.

28. D. Perrin. Finite automata. In J. van Leeuwen, editor, *Handbook of Theoretical Computer Science*, volume B, pages 1–55. Elsevier/MIT Press, 1990.

29. H. Reichel. An approach to object semantics based on terminal co-algebras. *Math. Struct. in Comp. Sci.*, 5:129–152, 1995.

30. J. Rutten. Automata and coinduction (an exercise in coalgebra). In D. Sangiorigi and R. de Simone, editors, *Concur'98: Concurrency Theory*, number 1466 in Lect. Notes Comp. Sci., pages 194–218. Springer, Berlin, 1998.

31. J. Rutten. Behavioural differential equations: a coinductive calculus of streams, automata, and power series. *Theor. Comp. Sci.*, 308:1–53, 2003.

32. J. Rutten and D. Turi. Initial algebra and final coalgebra semantics for concurrency. In J.W. de Bakker, W.P. de Roever, and G. Rozenberg, editors, *A Decade of Concurrency*, number 803 in Lect. Notes Comp. Sci., pages 530–582. Springer, Berlin, 1994.

33. J.J.M.M. Rutten. Automata, power series, and coinduction: Taking input derivatives seriously (extended abstract). In J. Wiedermann, P. van Emde Boas, and M. Nielsen, editors, *International Colloquium on Automata, Languages and Programming*, number 1644 in Lect. Notes Comp. Sci., pages 645–654. Springer, Berlin, 1999.

34. D. Turi. *Functorial operational semantics and its denotational dual*. PhD thesis, Free Univ. Amsterdam, 1996.

35. D. Turi and G. Plotkin. Towards a mathematical operational semantics. In *Logic in Computer Science*, pages 280–291. IEEE, Computer Science Press, 1997.

36. T. Uustalu, V. Vene, and A. Pardo. Recursion schemes from comonads. *Nordic Journ. Comput.*, 8(3):366–390, 2001.

Sheaves and Structures of Transition Systems

Grant Malcolm

Department of Computer Science
University of Liverpool, Liverpool L69 7ZF, UK

Abstract. We present a way of viewing labelled transition systems as sheaves: these can be thought of as systems of observations over a topology, with the property that consistent local observations can be pasted together into global observations. We show how this approach extends to hierarchical structures of labelled transition systems, where behaviour is taken as a limit construction. Our examples show that this is particularly effective when transition systems have structured states.

1   Introduction

Despite many advances in developing calculi and formal models for concurrent processes, it is still difficult to reason effectively about large systems, which may comprise many subcomponents related in intricate ways, and have a correspondingly large state space. It is therefore important to find compositional methods of specifying, analysing, and reasoning about hierarchical, distributed and concurrent processes based on coherent notions of observation and behaviour.

Goguen [11] has proposed sheaf theory as a semantic foundation for the study of concurrent and distributed systems. Sheaf theory is concerned with the transition from local to global properties, and a sheaf can be thought of as a system of observations made at various locations in a topology, with the key property that consistent local observations can be uniquely pasted together to provide a global observation. Thus, the semantics of a distributed system could be couched in terms of the topology of the system and the local observations that could be made of its various parts, and the overall, global behaviour of the system then emerges from the behaviour of its parts. Goguen's paper builds upon earlier work on Categorical General Systems Theory [7,8], and together these papers provide a rich variety of different kinds of systems, including musical pieces [9,12], that do indeed give rise very naturally to sheaves. The approach has been used to give semantics to Petri nets by Lilius [16], and to object-oriented languages, originally by Wolfram and Goguen [25], and also by Ehrich, Goguen and Sernadas [6], and by Cîrstea [4].

Many of the examples of Goguen's sheaf-theoretic approach use discrete time as a topology: here, behaviour is observed locally at particular intervals of time, and the global behaviour is the behaviour over the union of these intervals. Cîrstea's work also provides a relationship between sheaves on discrete time and transition systems, and strengthens the arguments for a sheaf-theoretic approach by showing that transition systems give rise to sheaves: a transition system has

K. Futatsugi et al. (Eds.): Goguen Festschrift, LNCS 4060, pp. 405–419, 2006.
© Springer-Verlag Berlin Heidelberg 2006

an 'underlying' sheaf. In this paper we further explore the possibility of using sheaf theory to provide a semantic foundation for distributed concurrent systems, by exploring their relationship to labelled transition systems. We present an adjunction that provides translations between labelled transition systems and sheaves on a topology of traces, i.e., prefix-closed sets of words over the alphabet of labels. However, our main interest is in systems built from subcomponents. We show that the adjunction extends to hierarchically structured transition systems by using a principle from Goguen's Categorical Systems Theory: behaviour is limit. Although colimits are also used to model ways of combining concurrent processes (see, e.g., [22]), we would suggest that for processes with a structured notion of state, limits provide the most useful ways of combining systems. Indeed, the limit constructions we consider provide ways of structuring states. The following section contains some examples; see also [17]. As a consequence of the emphasis that we place on states, we are less interested in notions such as bisimulation. We show that the adjunction between transition systems and sheaves extends to hierarchical systems, and that the translation from transition systems to sheaves preserves limits and, hence, behaviour.

We assume familiarity with basic notions from category theory: functor, natural transformation, limit and adjunction (see [10,1] for introductions).

This paper is dedicated with great affection to Joseph Goguen on his sixty fifth birthday. I had the privelege and pleasure of working as a research assistant with Joseph for several years; I can think of no better apprenticeship for a computer scientist. His wealth of ideas and breadth of vision were stimulating and inspirational, and I am delighted to dedicate this to him as inspiration, teacher, and friend.

2 Transition Systems and Sheaves

The following subsections review labelled transition systems and sheaves, and presents an adjunction between them. We begin by recalling some basic definitions concerning labelled transition systems, which we generalise to allow transitions to take labels in an arbitrary monoid.

2.1 Transition Systems

Definition 1. *A **labelled transition system** over L is a pair $(T, \longmapsto)$, where T is a set of **states**, and $\longmapsto \subseteq T \times L \times T$ is the **transition relation**. We write $t \overset{l}{\longmapsto} t'$ for $(t, l, t') \in \longmapsto$, and we will usually refer to a transition system $(T, \longmapsto)$ simply as T. A **pointed transition system** is a transition system T with a distinguished **initial state** $t_0 \in T$.*

A morphism of transition systems over L $(T_1, \longmapsto_1) \to (T_2, \longmapsto_2)$ is a function $f : T_1 \to T_2$ such that if $t \overset{l}{\longmapsto}_1 t'$ then $f(t) \overset{l}{\longmapsto}_2 f(t')$; a morphism of pointed transition systems in addition maps the initial state to the initial state.

When there is no risk of confusion, we will drop subscripts and decorations on the arrow '$\longmapsto$', and simply write, for example, 'if $t \xrightarrow{l} t'$ then $f(t) \xrightarrow{l} f(t')$.'

Note that there is another common definition of morphism in the literature (see, e.g., [23,22]), whereby morphisms 'lift' transitions, in the sense that for a morphism $f : T_1 \to T_2$, if $f(t) \xrightarrow{l} t'$ in T_2, then there is some $t_1 \in T_1$ such that $t \xrightarrow{l} t_1$ in T_1 and $f(t_1) = t'$. This definition is particularly useful in studying bisimulation, but since that is beyond the scope of the present paper, we use the simpler definition above.

Example 1. Any subset $S \subseteq L^*$ of lists over L gives rise to a transition system $(S, \longmapsto)$, where $w \xrightarrow{l} w'$ iff $w' = wl$. For example, take $S = \{\varepsilon, a, aa, ab\} \subseteq \{a, b\}^*$, where ε is the empty list. Then we have $\varepsilon \xrightarrow{a} a$, $a \xrightarrow{a} aa$, and $a \xrightarrow{b} ab$, describing a simple transition system with a 'fork' at state a. Any morphism $f : (S, \longmapsto) \to (T, \longmapsto)$ describes a similarly forking path (or *run*) in T: both

$$f(\varepsilon) \xrightarrow{a} f(a) \xrightarrow{a} f(aa) \text{ and}$$
$$f(\varepsilon) \xrightarrow{a} f(a) \xrightarrow{b} f(ab)$$

The following is an example that we will refer to later on in Section 3, where we consider hierarchical structures of transitions systems.

Example 2. Consider a coffee dispenser that dispenses coffee only after payment has been received. Later on, we will see an example concerning a coin slot that accepts payment; for the moment we simply assume a boolean value that says whether or not payment has been received. The states of the coffee dispenser are pairs consisting of a boolean value and a number between 0 and 20, indicating the level of coffee available; that is, the state set is $Bool \times \{0..20\}$.

There are three labels for transitions: d for dispensing coffee, n for notification that payment has been received, and r for refilling. Transitions are described exhaustively as follows:

- $(true, N) \xrightarrow{\text{d}} (false, N - 1)$ for all $0 < N \leq 20$,
- $(false, N) \xrightarrow{\text{n}} (true, N)$ for all $0 \leq N \leq 20$, and
- $(B, N) \xrightarrow{\text{r}} (B, 20)$ for all $B \in Bool$ and all $0 \leq N \leq 20$.

Any transition system over L extends to a transition system over L^* with

$$t \xrightarrow{\varepsilon} t' \text{ iff } t = t'$$
$$t \xrightarrow{wl} t' \text{ iff } t \xrightarrow{w} t'' \text{ and } t'' \xrightarrow{l} t' \text{ for some } t'' \ .$$

That is, transitions can be freely extended to paths of transitions. We can use this to provide a slightly more general notion of transition system that takes labels in a monoid.

Definition 2. *Let* $\mathcal{M} = (M, \cdot, \varepsilon)$ *be a monoid (we generally write mm' in place of $m \cdot m'$); we say m is a* **prefix** *of m', and write $m \leq m'$, iff $m' = mn$ for some $n \in M$, and we say that a subset $X \subseteq M$ is* **prefix-closed** *iff $x \leq y$ and $y \in X$ implies $x \in X$. We write $\Omega(\mathcal{M})$ for the set of all prefix-closed subsets of M (including M itself), and for $m \in M$, we write $m{\downarrow}$ for the set of all prefixes of m (including m itself)*

A **labelled transition system over** $\mathcal{M}$ *is a pair $(T, \longmapsto)$, with $\longmapsto \subseteq T \times M \times T$ such that*

$$t \xmapsto{\;\varepsilon\;} t' \; iff \; t = t'$$
$$t \xmapsto{\;mn\;} t'' \; iff \; t \xmapsto{\;m\;} t' \; and \; t' \xmapsto{\;n\;} t'' \; for \; some \; t' \in T \;.$$

A morphism $f : (T, \longmapsto) \to (T', \longmapsto)$ of transition systems over $\mathcal{M}$ is a function $f : T \to T'$ such that $t \xmapsto{\;m\;} t'$ implies $f(t) \xmapsto{\;m\;} f(t')$. This gives a category $\mathsf{LTS}_\mathcal{M}$ of transition systems over $\mathcal{M}$.

We usually refer to a labelled transition system $(T, \longmapsto)$ simply as T.

In the sequel, we will be interested in transition systems that are built from other transition systems by taking limits. For the present, we note

Proposition 1. *The category $\mathsf{LTS}_\mathcal{M}$ is complete.*

Limits are constructed from limits of the underlying state sets; we will see some examples in Section 3.

2.2   Sheaves

Sheaf theory is used in many branches of mathematics, the underlying theme in its various applications being the passage from local to global properties [13]. It provides a formal notion of coherent systems of observations: a number of consistent observations of various aspects of an object can be uniquely pasted together to give an observation over all of those aspects. The passage from local to global properties, and the pasting together of local observations of behaviour allow sheaf theory to be usefully applied in computer science, to give models for concurrent processes [19,5,16] and objects [11,6,25,4,17]. We give a basic definition of 'sheaf' below; fuller accounts can be found in [21,15].

We may consider a sheaf as giving a set of observations of an object's behaviour from a variety of 'locations'. The notion of location is formalised by the following

Definition 3. *A* **complete Heyting algebra** *is a partially ordered set $(C, \leq)$ such that:*

- *for all $c, d \in C$, there is a greatest lower bound $c \wedge d$*
- *for all subsets $\{c_i \mid i \in I\}$ of C, there is a least upper bound $\bigvee_{i \in I} c_i$*
- *greatest lower bounds distribute through least upper bounds:*

$$\left(\bigvee_{i \in I} c_i\right) \wedge d \; = \; \bigvee_{i \in I} (c_i \wedge d) \;.$$

For example, any topological space with the inclusion ordering between open sets is a complete Heyting algebra; also, any complete lattice is a complete Heyting algebra. In particular, the set of prefix-closed subsets of a monoid, $\Omega(\mathcal{M})$, is a complete Heyting algebra.

Like any preorder, a complete Heyting algebra C can be seen as a category; in particular, the opposite category C^{op} has the elements of the set as objects, and a unique arrow from c' to c precisely when $c \leq c'$.

Definition 4. *Let C be a complete Heyting algebra; a* **presheaf** *F on C is a functor from C^{op} to* Set. *That is, for each $c \in C$ there is a set $F(c)$, and for $c, d \in C$ such that $c \leq d$, there is a* **restriction function** *$F_{c \leq d} : F(d) \to F(c)$, subject to the following conditions:*

- *$F_{c \leq c} = \mathrm{id}_{F(c)}$, the identity on the set $F(c)$; and*
- *if $c \leq d \leq e$, then $F_{d \leq e}; F_{c \leq d} = F_{c \leq e}$.*

Notation 1 *For a presheaf F on C, if $c \leq d$ in C and $x \in F(d)$, we often write $x|_c$ for $F_{c \leq d}(x)$.*

A sheaf is a presheaf which allows families of consistent local observations to be pasted together to give a global observation.

Definition 5. *A presheaf F is a* **sheaf** *iff it satisfies the following* **pasting condition***:*

- *if $c = \bigvee_{i \in I} c_i$ and $x_i \in F(c_i)$ is a family of elements for $i \in I$ such that $x_i|_{c_i \wedge c_j} = x_j|_{c_i \wedge c_j}$ for all $i, j \in I$, then there is a unique $x \in F(c)$ such that $x|_{c_i} = x_i$ for all $i \in I$.*

A morphism of sheaves $\theta : F \to G$ is just a natural transformation from F to G viewed as presheaves.

We write $\mathsf{Sh}_{\mathcal{M}}$ for the category of sheaves over $\Omega(\mathcal{M})$, where $\mathcal{M}$ is a monoid.

Given $\theta : F \to G$, naturality of θ says that θ respects restrictions: given $Y \leq X$, and $e \in F(X)$,

$$\theta_Y(e|_Y) = \theta_X(e)|_Y \ .$$

Example 3. For $S \subseteq M$, if we write $\Omega(S)$ for the prefix-closed subsets of S, then Ω is a sheaf over $\Omega(\mathcal{M})$; given an inclusion $X \subseteq Y$ of prefix closed sets, then $\Omega_{X \subseteq Y}$ takes $V \subseteq Y$ (i.e., $V \in \Omega(Y)$) to $V \cap X \subseteq X$ (i.e., $V \cap X \in \Omega(X)$).

Example 4 (Eventually ϕ). Let T be a pointed transition system on L^* with a distinguished initial state $t_0 \in T$, and a subset of states $\phi \subseteq T$. The functor $\diamond \phi : \Omega(L^*) \to$ Set defined by

$$\diamond \phi(X) = \prod_{w \in X} \{p : w{\downarrow} \to T \mid p(\varepsilon) = t_0 \wedge$$
$$(\exists w' \in L^*, p' : w'{\downarrow} \to T) \, w \leq w' \wedge p'|_{w{\downarrow}} = p$$
$$\wedge \, p'(w') \in \phi\}$$

is a sheaf.

We will look in more detail at limits in Section 3; again we note

Proposition 2. *The category* $\mathsf{Sh}_\mathcal{M}$ *is complete.*

Limits are constructed pointwise from limits in Set. For example, given sheaves F and G, their product $F \times G$ is defined by $F \times G(X) = F(X) \times G(X)$.

2.3   An Adjunction Between Transition Systems and Sheaves

We present a translation from transition systems to sheaves, and a translation from sheaves to transition systems. The main result of this section is that these translations form an adjunction.

Given a transition system T, we construct a sheaf from T by considering sets of paths in T, as in Example 1. Recall that a (forking) path in T was just a transition system morphism $f : X \to T$ for some $X \in \Omega(\mathcal{M})$. If we have $f_1 : X_1 \to T$ and $f_2 : X_2 \to T$ such that $f_1|_{X_1 \cap X_2} = f_1|_{X_1 \cap X_2}$, then clearly these functions can be uniquely pasted to give a morphism, or path, $X_1 \cup X_2 \to T$.

Definition 6. *The functor* $\mathbf{Sh}_\mathcal{M} : \mathsf{LTS}_\mathcal{M} \to \mathsf{Sh}_\mathcal{M}$ *is defined by, for a prefix-closed subset* $X \in \Omega(\mathcal{M})$

$$\mathbf{Sh}_\mathcal{M}(T)(X) = \mathsf{LTS}_\mathcal{M}(X, T) \ .$$

Note that, because every $X \in \Omega(\mathcal{M})$ *is a transition system,* $\mathsf{LTS}_\mathcal{M}(\_, T)$ *is a functor* $\Omega(\mathcal{M})^{op} \to \mathsf{Set}$, *and so this definition also applies for morphisms (i.e., inclusions) in* $\Omega(\mathcal{M})$. *That is, restriction in* $\mathbf{Sh}_\mathcal{M}(T)$ *is restriction of paths.*

For $f : T \to U$ *in* $\mathsf{LTS}_\mathcal{M}$, *the natural transformation*

$$\mathbf{Sh}_\mathcal{M}(f) : \mathbf{Sh}_\mathcal{M}(T) \to \mathbf{Sh}_\mathcal{M}(U)$$

is defined by saying that for each $X \in \Omega(\mathcal{M})$, *the component* $\mathbf{Sh}_\mathcal{M}(f)_X$ *takes a* T-*path* $h : X \to T$ *to the* U-*path* $h; f : X \to U$. *This is in fact the action of the functor* $\mathsf{LTS}_\mathcal{M}(X, \_)$ *on* f, *and naturality of* $\mathbf{Sh}_\mathcal{M}(f)$ *is a consequence of this fact.*

Going the other way, we represent a sheaf by its set of 'elements' (m, e), where $m \in M$ and $e \in F(m\!\downarrow)$. transitions on these states are given by the restriction actions of F.

Definition 7. *The functor* $\mathbf{Tr} : \mathsf{Sh}_\mathcal{M} \to \mathsf{LTS}_\mathcal{M}$ *is defined by*

$$\mathbf{Tr}_\mathcal{M}(F) = \sum_{m \in M} F(m\!\downarrow) \ .$$

Transitions in this system are defined by $(m, e) \overset{n}{\longmapsto} (m', e')$ *iff* $m' = mn$ *and* $e'|_{m\downarrow} = e$. *For natural transformations* $\theta : E \to F$ *in* $\mathsf{Sh}_\mathcal{M}$,

$$\mathbf{Tr}_\mathcal{M}(\theta) : \mathbf{Tr}_\mathcal{M}(E) \to \mathbf{Tr}_\mathcal{M}(F)$$

takes $(m, e) \in \mathbf{Tr}_\mathcal{M}(E)$ *to* $(m, \theta_{m\downarrow}(e)) \in \mathbf{Tr}_\mathcal{M}(F)$.

Our main result of this section is that $\mathbf{Sh}_{\mathcal{M}}$ gives the 'underlying' sheaf of a transition system.

Theorem 1. $\mathbf{Tr}_{\mathcal{M}}$ *is left adjoint to* $\mathbf{Sh}_{\mathcal{M}}$.

Proof. The unit of the adjunction is given by $\eta_F : F \to \mathbf{Sh}_{\mathcal{M}}(\mathbf{Tr}_{\mathcal{M}}(F))$, which, for $X \in \Omega(\mathcal{M})$, takes $e \in F(X)$ to the path mapping $x \in X$ to $(x, e|_{x\downarrow})$. For any transition system T, a morphism $h : F \to \mathbf{Sh}_{\mathcal{M}}(T)$ uniquely extends to $h^{\sharp} : \mathbf{Tr}_{\mathcal{M}}(F) \to T$, which takes $(m, e) \in \mathbf{Tr}_{\mathcal{M}}(F)$ to $h_{m\downarrow}(e)(m)$.

This generalises a result of Winskel that gives an adjunction between standard transition systems and presheaves [3]. The adjunction applies more to presheaves than to sheaves. A further twist can be given by considering pointed transition systems. Let F be a sheaf, then $\mathbf{Tr}_{\mathcal{M}}(F)$ can be made a pointed transition system by designating $(\varepsilon, *)$ as initial state, where $*$ is the unique element of $F(\{\varepsilon\})$. Since morphisms of pointed transition systems preserve initial states, and if we take $\mathbf{Sh}_{\mathcal{M}}(\mathbf{Tr}_{\mathcal{M}}(F))(X)$ to be the set of pointed transition system morphisms from X (with initial state ε) to $\mathbf{Tr}_{\mathcal{M}}(F)$, then any such morphism corresponds uniquely to a consistent family of elements $e_{m\downarrow} \in F(m\downarrow)$ for $m \in X$, which, since F is a sheaf, corresponds uniquely to an element $e \in F(X)$. Thus, if we specialise the above adjunction to pointed transition systems, the unit of the adjunction is an isomorphism.

In the next section, we look at how behaviour of composite systems arises through limit constructions. Since $\mathbf{Sh}_{\mathcal{M}}$ is a right adjoint, our translation from transition systems to underlying sheaves preserves limits, and therefore, behaviour:

Corollary 1. $\mathbf{Sh}_{\mathcal{M}}$ *preserves limits.*

3   Hierarchical Systems

In this section we explore the notion of behaviour as limit for transition systems built from component parts. We start by allowing transition systems to vary over the monoids of their labels, and extend the completeness results of the previous section to this setting. Correspondingly, we also introduce morphisms between the 'trace' topologies of sheaves, and extend the adjunction of Theorem 1 to hierarchically structured transition systems.

We give an example based on the coffee dispenser of Example 2, which shows that the appropriateness of limits as giving behaviour of composite systems depends, to some extent, on our 'state-based' approach to transition systems.

Finally, we use a generalisation of the notion of sheaf to show that the notion of behaviour as limit is, in itself, sheaf-theoretical.

We begin by noting that the category Mon of monoids and monoid homomorphisms is complete. Limits of monoid homomorphisms capture the notion of synchronisation on actions.

Example 5. The pullback of $f_i : \mathcal{M}_i \to \mathcal{M}$ $(i = 1, 2)$ is the monoid with underlying set $\{(x_1, x_2) \in M_1 \times M_2 \mid f_1(x_1) = f_2(x_2)\}$, unit $\varepsilon = (\varepsilon, \varepsilon)$ and composition defined by $(x_1, x_2)(y_1, y_2) = (x_1 y_1, x_2 y_2)$, together with first and second projections to $\mathcal{M}_1$ and $\mathcal{M}_2$ respectively.

As a particular example, let $\mathcal{M} = \{c\}^*$, $\mathcal{M}_1 = \{a, b, c\}^*$ and $\mathcal{M}_2 = \{c, d\}^*$, with $f_1(a) = f_1(b) = \varepsilon = f_2(d)$, and $f_1(c) = c = f_2(c)$. Then the pullback contains all pairs (x, y) in $\{a, b, c\}^* \times \{c, d\}^*$ where x and y contain the same number of c's.

We can think of the monoid homomorphisms as taking a sequence of actions in some system $\mathcal{M}_i$ and 'restricting' them to a sequence of actions in a subsystem $\mathcal{M}$. In the particular pullback described above, the common subsystem $\mathcal{M}$ has only one action, c. We can think of these 'words' as expressing sequences of actions where the action of c is synchronised in $\mathcal{M}_1$ and $\mathcal{M}_2$.

Also note that

$$(a\,b\,c,\ d\,c\,d) = (\varepsilon, d)(a\,b,\ \varepsilon)(c, c)(\varepsilon, d) = (a\,b,\ \varepsilon)(\varepsilon, d)(c, c)(\varepsilon, d)$$

so unsynchronised actions from different components can occur in any order.

3.1   Behaviour as Limit

We begin by considering morphisms between transition systems over different label monoids.

Definition 8. *The category* LTS *has objects* $(\mathcal{M}, T)$, *where $\mathcal{M}$ is a monoid, and T is a labelled transition over $\mathcal{M}$. A morphism $\phi : (\mathcal{M}, T) \to (\mathcal{M}', T')$ is a pair $\phi = (f, g)$, with $f : \mathcal{M} \to \mathcal{M}'$ a monoid homomorphism, and $g : T \to T'$ such that if $t \overset{m}{\longmapsto} t'$, then $g(t) \overset{f(m)}{\longmapsto} g(t')$.*

Again, these morphisms can be thought of as expressing a restriction to a subsystem.

Example 6. Recall the coffee dispenser of Example 2 as a transition system over $\mathcal{M} = \{d, n, e\}^*$. We give an example of a morphism from the coffee dispenser to a simple coin slot that can accept coins. The state set of the coin-slot transition system is *Bool*, indicating whether a coin has been inserted. There are two labels for transitions: c for a coin being inserted, and e for ending a transaction (the coffee is dispensed and the coin chinks into the money box). The transitions are defined exhaustively by:

- $false \overset{c}{\longmapsto} true$
- $true \overset{e}{\longmapsto} false$

We describe a morphism from the coffee dispenser to the coin slot. The monoid homomorphism on labels is defined by

$$d \mapsto e$$
$$n \mapsto c$$
$$r \mapsto \varepsilon$$

and the mapping from the state of the coffee dispenser ($Bool \times \{0..20\}$) to that of the coin slot ($Bool$) is just the first projection. It is straightforward to check that these maps preserve transitions; for example, in the coffee dispenser,

$$(true, N) \xmapsto{\ \mathsf{d}\ } (false, N-1)$$

for all $0 < N \leq 20$, which translates to $true \xmapsto{\ \mathsf{e}\ } false$ in the coin slot.

In this example, we think of the coffee dispenser as actually comprising a coin slot as a subsystem. This is perhaps somewhat unnatural; a more realistic description might have both the coffee dispenser and the coin slot sharing a common subcomponent (essentially just the boolean value of the example above). We hope that the familiarity of this example to readers will make them more disposed to indulge such simplifications.

The case of two systems sharing a common subcomponent is treated in the next example, which illustrates

Proposition 3. *The category* LTS *is complete.*

Limits are taken componentwise, consisting of a limit of monoids, together with a limit of the associated transition systems.

We saw in Example 6 how a morphism from a coffee dispenser to a coin slot expressed the idea that the coin slot was a subcomponent of the coffee dispenser, so that coffee was only dispensed after a coin had been put in the slot. We first present another morphism from a money box to the coin dispenser (so that coins put in the slot eventually end up in the money box), and then show how the limit of these two morphisms behaves.

Example 7. A money box with a coin slot as a subcomponent can be specified as having states that are pairs whose first component is a boolean value, specifying whether there is a coin in the slot, and whose second component is a natural number specifying how many coins are in the money box.

Transitions are labelled by c for a coin entering the slot, t for the coin being taken from the slot to the money box, and m for all the money being taken out of the box. Transitions are defined exhaustively by

- $(false, N) \xmapsto{\ \mathsf{c}\ } (true, N)$ for all $N \geq 0$,
- $(true, N) \xmapsto{\ \mathsf{t}\ } (false, N+1)$ for all $N \geq 0$, and
- $(B, N) \xmapsto{\ \mathsf{m}\ } (B, 0)$ for all $N \geq 0$ and $B \in Bool$.

The monoid homomorphism to the coin slot is given by

$$\mathsf{c} \mapsto \mathsf{c}$$
$$\mathsf{t} \mapsto \mathsf{e}$$
$$\mathsf{m} \mapsto \varepsilon$$

and the first projection $(B, N) \mapsto B$ gives the mapping on states.

We now have two mappings to the coin slot, indicating that this is a subobject of both the coffee dispenser and the money box:

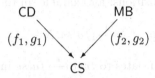

The limit of the monoid morphisms gives a label monoid of

$$\{(u, v) \in \{d, n, r\}^* \times \{c, t, m\}^* \mid f_1(u) = f_2(v)\} \qquad (1)$$

Since the coffee dispenser and money box synchronise on coins entering and leaving the coin slot, this requires every d (dispense coffee) to be paired with a t (take the coin), and every n (notify there's a coin in the slot) to be paired with a c (coin in the slot). Thus, the pullback monoid of labels is effectively the same as lists over $\{dt, nc, r, m\}$, but where occurrences of r and m (the unsynchronised actions) commute (dt represents the synchronized event of coffee being dispensed and the coin being taken from the slot, i.e., dt = (d, t), while nc represents the synchronised events of a coin being put in the slot and the coffee dispenser notified of this) i.e., nc = (n, c); for example:

$$nc\ r\ m\ dt\ m = nc\ m\ r\ dt\ m$$

are equal because it does not matter what order unsynchronised events occur in — if you like, they occur in separate frames of reference with no notion of simultaneity applying. Both sequences represent a coin being put into the slot, then the machine being refilled and its money box emptied (in either order, or even 'at the same time'), then coffee being dispensed, and then the money box being emptied once again.

The state set of the limiting transition system is

$$\{(b, x, b', y) \in Bool \times \{0..20\} \times Bool \times Nat \mid g_1(b, x) = g_2(b', y)\} \ .$$

Since in this case both g_1 and g_2 are the first projection, the requirement is simply that $b = b'$. In other words, the state of the common coin-slot subcomponent is shared by the coffee dispenser and the money box. Any changes in the coin slot's state must occur in both the coffee dispenser and the money box.

This synchronisation on a shared subcomponent is again reflected in the transitions of the limiting transition system: essentially, a label (u, v) as in (1) represents a u-transition on the coffee-dispenser part together with a v-transition on the money box part. Formally,

$$(b, x, b, y) \xrightarrow{(u,v)} (b', x', b', y') \quad \text{iff} \quad (b, x) \xrightarrow{u} (b', x') \quad \text{and} \quad (b, y) \xrightarrow{v} (b', y')$$

For example,

$$(true, x, true, y) \xrightarrow{dt} (false, x - 1, false, y + 1)$$

for $0 < x \leq 20$ and $y \geq 0$, represents coffee being dispensed and the coin being taken from the coin slot to the money box.

3.2   Sheaves of Hierarchical Systems

We extend the relationship described in Section 2 between transition systems and sheaves to the hierarchical systems described in the previous subsection. This gives a much more interesting notion of topology that arises from the way a hierarchical system is composed from its subcomponents. By 'more interesting', we mean that such a topology gives a more realistic notion of location at which to observe a system. We conclude by observing that the behaviour-as-limit approach to hierarchical systems naturally gives rise to sheaves on such a topology.

We begin by showing that monoid homomorphisms allow a translation between topologies $\Omega(\mathcal{M})$.

Definition 9. *A monoid homomorphism* $f : \mathcal{M} \to \mathcal{M}'$ *extends to a mapping* $\Omega(f) : \Omega(\mathcal{M}) \to \Omega(\mathcal{M}')$ *defined by*

$$\Omega(f)(X) = \{y \in M \mid y \leq f(x) \text{ for some } x \in X\} \ .$$

That is, $\Omega(f)(X)$ *is the prefix-closure of the image* $f(X)$.

The morphism also extends, contravariantly, to a functor $\mathsf{Sh}_{\mathcal{M}'} \to \mathsf{Sh}_{\mathcal{M}}$, *which we also denote* $\Omega(f)$, *defined by*

$$\Omega(f)(G)(X) = G(\Omega(f)(X))$$

for a sheaf G *on* $\Omega(\mathcal{M}')$.

Now we can define morphisms between sheaves on different trace topologies. Corresponding to the category LTS, we have a category Sh, where, intuitively, morphisms correlate to restrictions to subsystems.

Definition 10. *The category* Sh *has objects* $(\mathcal{M}, F)$, *where* $\mathcal{M}$ *is a monoid, and* F *is a sheaf on* $\Omega(\mathcal{M})$. *A morphism* $(\mathcal{M}, F) \to (\mathcal{M}', F')$ *is a pair* (f, θ), *where* $f : \mathcal{M} \to \mathcal{M}'$ *is a monoid homomorphism, and* $\theta : F \to \Omega(f)(F')$.

Given $(f, \theta) : (\mathcal{M}_1, F_1) \to (\mathcal{M}_2, F_2)$ *and* $(g, \kappa) : (\mathcal{M}_2, F_2) \to (\mathcal{M}_3, F_3)$, *the composite* $(f, \theta); (g, \kappa)$ *is* $(f; g, \ \theta; \kappa_{\Omega(f)}) : (\mathcal{M}_1, F_1) \to (\mathcal{M}_3, F_3)$. *To make sense of the second component, note that for* $X \in \Omega(\mathcal{M}_1)$,

$$\theta_X : F_1(X) \to \Omega(f)(F_2)(X) = F_2(\Omega(f)(X))$$

and

$$\kappa_{\Omega(f)(X)} : F_2(\Omega(f)(X)) \to \Omega(g)(F_3)(\Omega(f)(X)) = F_3(\Omega(f; g)(X))$$

This is an example of a Grothendieck category, with the following consequence (see, e.g., [20]):

Theorem 2. *The functor* $\mathbf{Sh}_{\mathcal{M}}$ *extends to a functor* $\mathbf{Sh} : \mathsf{LTS} \to \mathsf{Sh}$, *taking* $(\mathcal{M}, T)$ *to* $(\mathcal{M}, \mathbf{Sh}_{\mathcal{M}}(T))$; *also,* $\mathbf{Tr}_{\mathcal{M}}$ *extends to* $\mathbf{Tr} : \mathsf{Sh} \to \mathsf{LTS}$ *taking* $(\mathcal{M}, F)$ *to* $(\mathcal{M}, \mathbf{Tr}_{\mathcal{M}}(F))$. *Moreover,* $\mathbf{Tr}$ *is left adjoint to* $\mathbf{Sh}$.

As a corollary, since $\mathbf{Sh}$ is a right adjoint, it preserves limits, and therefore preserves the behaviour of a composite system constructed by taking limits, as in Example 7. We might use this to verify a property such as the 'eventually-ϕ' property of Example 4 by constructing an element of $\diamond\phi(X)$ by showing that every word $w \in X$ can be extended to a path ending in a state where ϕ holds. However, this $\Omega(\mathcal{M})$ topology really only says that the branching behaviour of a transition system arises by pasting together linear paths $w{\downarrow} \to T$. A more powerful approach to capturing global properties through local properties would be to construct sheaves on a topology representing the hierarchical structure of a composite system. Such topologies can arise through downwards-closure.

Let X be a preorder category, with a unique morphism $x \to y$ whenever $x \geq y$. For example, X might have objects 0, 1 and 2, with $0 \leq 1$ and $0 \leq 2$, and a functor $\delta : X \to \mathsf{LTS}$, for example, would then represent two transition systems with a shared subcomponent. The completion $\Omega(X)$ of downward-closed subsets of X, in this example, looks like

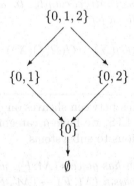

which represents all of the parts of the system: downwards-closure means that subcomponents are always included in a 'part'. Note that the example shows that moving from X to $\Omega(X)$ is very like moving from the basis of a pullback diagram to a pullback diagram. The top element that is added, $\{0, 1, 2\}$, corresponds to the limit, i.e., to the behaviour of the entire system, while $\{0, 1\}$ corresponds to the system on the left, together with its 'component', 0. In Example 7, this latter would be the coffee dispenser with its component coin slot.

Rather than extend the machinery of the previous sections to sheaves on topologies $\Omega(X)$, we conclude by showing that the notion of behaviour as limit is itself sheaf-like. For this, we need a generalisation of the notion of sheaf that seems to be due to Gray (cf. [14], Chapter 18), and allows for sheaves that take values in categories other than Set:

Definition 11. *A sheaf with values in a category* L *is a functor* F *from a complete Heyting algebra to* L *such that if* $X = \bigvee_{i \in I} X_i$, *then*

$$F(X) \longrightarrow \prod_{i \in I} F(X_i) \rightrightarrows \prod_{i,j \in I} F(X_i \wedge X_j)$$

is an equaliser diagram (where all the arrows arise from the obvious restrictions by the universal property of the target product).

The universal property of the equaliser diagram expresses that consistent families of 'elements' can be uniquely pasted together: a unique arrow to $F(X)$ arises from an arrow to $\prod_{i \in I} F(X_i)$ that equalises the parallel arrows, that is, 'elements' of each $F(X_i)$ that agree on overlaps $X_i \wedge X_j$.

The following follows directly from general properties of limits:

Proposition 4. *Let* X *be a preorder category, and let* $\delta : \mathsf{X} \to \mathsf{LTS}$. *Define* $\delta^* : \Omega(\mathsf{X}) \to \mathsf{LTS}$ *by* $\delta^*(X) = \lim(\delta\!\upharpoonright_X)$; *then* δ^* *is a sheaf of transition systems.*

This states that the behaviour of a composite system arises by pasting together the behaviours (i.e., limits) of its components. We could combine this with Theorem 2 to obtain that $\delta^*; \mathbf{Sh}$ is a sheaf of sheaves: the behaviour of a composite system arises by pasting together consistent paths through its component parts. We might thus, for example, verify that particular paths are possible globally by verifying that their restrictions to subcomponents are possible locally.

We can also show that the approach goes beyond an interleaving model of concurrency. One attempt to capture asynchronous or 'true concurrency' is Winskel and Nielsen's notion [24] of *transition systems with independence*; these are transition systems with a relation, $|$, on transitions, which specifies independence between transitions (e.g., they can occur truly concurrently). Such an independence relation is required to satisfy:

$$t \xmapsto{m} t_1 \sim t \xmapsto{m} t_2 \Rightarrow t_1 = t_2$$
$$t \xmapsto{m} t_1 \mid t \xmapsto{n} t_2 \Rightarrow (\exists u)\, t \xmapsto{m} t_1 \mid t_1 \xmapsto{n} u \wedge t \xmapsto{n} t_2 \mid t_2 \xmapsto{m} u$$
$$t \xmapsto{m} t_1 \mid t_1 \xmapsto{n} u \Rightarrow (\exists t_2)\, t \xmapsto{m} t_1 \mid s \xmapsto{n} t_2 \wedge t \xmapsto{n} t_2 \mid t_2 \xmapsto{m} u$$
$$t \xmapsto{m} t_1 \sim t_2 \xmapsto{m} u \mid w \xmapsto{n} w' \Rightarrow t \xmapsto{m} t_1 \mid w \xmapsto{n} w'$$

where $\sim$ is the equivalence relation freely generated by $\prec$, which is defined by $t \xmapsto{m} t_1 \prec t_2 \xmapsto{m} u$ iff there is an n with $t \xmapsto{m} t_1 \mid t \xmapsto{n} t_2$, $t \xmapsto{n} t_1 \mid t_1 \xmapsto{m} u$ and $t \xmapsto{n} t_2 \mid t_2 \xmapsto{n} u$. We can give a simpler characterisation of independence for sheaves of transition systems. Suppose F is a sheaf of transition systems on $\Omega(\mathsf{X})$, as in the example above, then transitions $t_1 \xmapsto{m} t_1'$ and $t_2 \xmapsto{n} t_2'$ in the limit are *independent* iff $m\!\upharpoonright_{\{0,1\}} = \varepsilon$ and $n\!\upharpoonright_{\{0,2\}} = \varepsilon$ (or vice-versa w.r.t. 1 and 2). That is, transitions are independent iff they are local to separate parts of the system. More generally, given $C = C_1 \cup C_2$, transitions at $F(C)$ are independent iff one restricts to ε at C_1 and the other restricts to ε at C_2.

Proposition 5. *Independence of transitions for sheaves gives transition systems with independence in the sense of Winskel and Nielsen [24].*

4 Conclusion

We have presented an adjunction between transition systems and sheaves on a topology of traces. The functor from transition systems to sheaves is right adjoint, and therefore preserves limits, which we consider to be the behaviour of hierarchical systems of transition systems.

The eventual aim of this work (from which we are still a long way off) is to provide semantic foundations for reasoning about hierarchical, distributed concurrent systems. Of paramount importance in such an endeavour are coherent notions of behaviour and observation. In this paper we have adopted the principle that the behaviour of composite, hierarchical, systems is given by a limit, and our results add to the argument [19,11] that viewing sheaves as systems of observations is coherent with the notion of behaviour as limit. As the examples in this paper illustrate, the notion of behaviour as limit leads to a more 'state-based' view of transition systems. Often, states in transition systems are viewed as little more than 'place holders' between transitions; there are advantages to such an approach, indeed it is almost necessary for process calculi and the study of bisimulation, and one area for future work is to relate the state-based approach to these established and successful fields.

Labelled transition systems are one of the fundamental structures in concurrency, and this paper establishes some relationships between transition systems and sheaves. It seems quite possible to further develop this, and relate sheaves usefully to other fundamental structures. To some extent, this has already been done, for example Monteiro and Pereira [19] consider event systems, Ehrich et al [6] and Goguen [11] apply sheaf-theoretic machinery to concurrent object systems, while Monteiro [18] applies related concepts to coalgebra. These all seem to be quite separate threads of development, and it would be instructive to find some means of drawing them together. One possibility lies in Lawvere's notion of 'control category', which determines a structure on observations, and is used by Bunge and Fiore [2] to give a general framework for considering concurrent processes.

References

1. Michael Barr and Charles Wells. *Category Theory for Computing Science*. Prentice Hall, 1990.
2. Marta Bunge and Marcelo Fiore. Unique factorization lifting functors. *Journal of Pure and Applied Algebra*, 2002.
3. Gian Luca Cattani and Glynn Winskel. Presheaf models for concurrency. In *Computer Science Logic: Tenth international Workshop, CSL'96, Annual Conference of the EACSL. Selected Papers*, number 1258 in Lecture Notes in Computer Science, pages 58–75. Springer-Verlag, 1997.
4. Corina Cîrstea. A distributed semantics for FOOPS. Technical Report PRG-TR-20-95, Programming Research Group, University of Oxford, 1995.
5. Rakesh Dubey. On a general definition of safety and liveness. Master's thesis, School of Electrical Engineering and Comp. Sci., Washington State Univ., 1991.

6. Hans-Dieter Ehrich, Joseph A. Goguen, and Amílcar Sernadas. A categorial theory of objects as observed processes. In J.W. de Bakker, Willem de Roever, and Gregorz Rozenberg, editors, *Foundations of Object Oriented Languages*. Springer-Verlag Lecture Notes in Computer Science 489, 1991.

7. Joseph A. Goguen. Mathematical representation of hierarchically organised systems. In E. O. Attinger, editor, *Global Systems Dynamics*, pages 111–129. S. Karger, 1970.

8. Joseph A. Goguen. Objects. *International Journal of General Systems*, 1:237–243, 1975.

9. Joseph A. Goguen. Complexity of hierarchically organized systems and the structure of musical experiences. *Int. Journal of General Systems*, 3:233–251, 1977.

10. Joseph A. Goguen. A categorical manifesto. *Mathematical Structures in Computer Science*, 1(1):49–67, 1991.

11. Joseph A. Goguen. Sheaf semantics for concurrent interacting objects. *Mathematical Structures in Computer Science*, 11:159–191, 1992.

12. Joseph A. Goguen. Musical qualia, context, time, and emotion. *Journal of Consciousness Studies*, 11:117–147, 2004.

13. John Gray. Fragments of the history of sheaf theory. In M.P. Fourman, C.J. Mulvey, and D.S. Scott, editors, *Applications of Sheaves*. Springer-Verlag Lecture Notes in Mathematics 753, 1980.

14. Joachim Lambek and Philip J. Scott. *Introduction to Higher Order Categorical Logic*. Cambridge University Press, 1986. Cambridge Studies in Advanced Mathematics, Volume 7.

15. Saunders Mac Lane and Ieke Moerdijk. *Sheaves in Geometry and Logic*. Springer-Verlag, 1992.

16. Johan Lilius. A sheaf semantics for Petri nets. Technical Report A23, Dept. of Computer Science, Helsinki University of Technology, 1993.

17. Grant Malcolm. Interconnection of object specifications. In Stephen Goldsack and Stuart Kent, editors, *Formal Methods and Object Technology*. Springer Workshops in Computing, 1996.

18. Luís Monteiro. Observation systems. *Electronic Notes in Theoretical Computer Science*, 33, 2000.

19. Luís Monteiro and Fernando Pereira. A sheaf-theoretic model of concurrency. In *Proc. Logic in Computer Science (LICS '86)*. IEEE Press, 1986.

20. Andrzej Tarlecki, Rod Burstall, and Joseph Goguen. Some fundamental algebraic tools for the semantics of computation, part 3: Indexed categories. *Theoretical Computer Science*, 91:239–264, 1991.

21. B.R. Tennison. *Sheaf Theory*, volume 20 of *London Mathematical Society Lecture Notes*. Cambridge University Press, 1975.

22. G. Winskel and W. Nielsen. Models for concurrency. Technical Report DAIMI PB – 463, Computer Science Department, Aarhus University, 1993.

23. Glynn Winskel. A compositional proof system on a category of labelled transition systems. *Information and Computation*, 87:2–57, 1990.

24. Glynn Winskel and Mogens Nielsen. Models for concurrency. In Samson Abramsky, editor, *Handbook of Logic and the Foundations of Computer Science*, volume 4. Oxford University Press, 1995.

25. David A. Wolfram and Joseph A. Goguen. A sheaf semantics for FOOPS expressions (extended abstract). In M. Tokoro, O. Nierstrasz, P. Wegner, and A. Yonezawa, editors, *Proceedings of the ECOOP'91 Workshop on Object-Based Concurrent Computing*, pages 81–98. Springer-Verlag Lecture Notes in Computer Science 612, 1992.

Uniform Functors on Sets

Lawrence S. Moss

Mathematics Department
Indiana University
Bloomington IN 47405 USA

Dedicated to Joseph Goguen on his 65th birthday

Abstract. This paper studies uniformity conditions for endofunctors on sets following Aczel [1], Turi [21], and others. The "usual" functors on sets are uniform in our sense, and assuming the Anti-Foundation Axiom *AFA*, a uniform functor H has the property that its greatest fixed point H^* is a final coalgebra whose structure is the identity map. We propose a notion of uniformity whose definition involves notions from recent work in coalgebraic recursion theory: completely iterative monads and completely iterative algebras (cias). Among our new results is one which states that for a uniform H, the entire set-theoretic universe V is a cia: the structure is the inclusion of HV into the universe V itself.

1 Introduction

I have considered Joseph Goguen to be one of my main teachers for many years. My first encounter with him was in an undergraduate course in the theory of computation given at UCLA around 1979. What I remember most is that serious students had to both write research papers and take an oral final exam, and looking back I see it as both a didactic move and a way to take seriously the thoughts of students. After hearing of my interests in mathematics and linguistics, he suggested that I write on representing inexact concepts in Montague grammar, thereby mixing topics that he considered interesting: formal semantics and fuzzy logic. Later, I took a graduate seminar that he and Charlotte Linde taught on natural language processing. I remember their strong opposition to generative grammar and advocacy of views that there was no "real world." Both of these were a real surprise. I also remember Joseph's sense of humor as well as his more serious side.

A few years later, I was a post-doc at Stanford's Center for the Study of Language and Information. Joseph had moved to SRI a few years earlier and was also at CSLI. I don't know how we started, but he and José Meseguer started meeting to decide on something to work on together. They pointed me to a conjecture of theirs on abstract data type computability which I settled and wrote up in a paper with the two of them. One of them told me I was getting "on-the-job training" in category theory, and this very paper is also on-the-job training. I also remember Joseph's delightful influence all over CSLI during that

K. Futatsugi et al. (Eds.): Goguen Festschrift, LNCS 4060, pp. 420–448, 2006.

time. I mainly lost touch with him after that, though at some point after he moved to UCSD we met again through my oldest friend, Martin Schapiro.

For me, one of the most long-lasting influences were various pointers to currents in the social sciences, along with indications that these deserved to be taken seriously in computer science and artificial intelligence. Another was the mingling and mixing of ideas from Western science and Eastern religions; despite being raised in Los Angeles and with Alan Watts' lectures regularly on my radio, I scarcely had met anyone who *lived* the ideas. I am still inspired by his wide-ranging concerns and penetrating insights into many subjects. In all of these, I am reminded of a character in the story of his namesake the Joseph of Genesis, the "man" in 37:15-17, someone who points people to important places and ideas. It is a pleasure to thank Joseph for his many years of direct inspiration to me and wish him many more years.

1.1   Whatever Happened to the Study of Recursive Program Schemes?

The title of our volume is "Algebra, Meaning, and Computation," and so I want to make the case that my contribution is related to all three points. As in the distantly-related areas of the semantics of natural language, the semantic project in computer science is to give some sort of mathematical model of meaning. Today the semantic project attracts less attention and prestige than the study of algorithmic complexity. (This is especially true in the USA.) Still, the area is important because if one wants to be sure that computer programs 'do what they are supposed to', then one quickly needs formally specified and tractable notions of meaning. I think it is fair to say that the centerpiece of the semantic project concerning computation is the treatment of *recursion*, the main mechanism of 'looping' for computer programs and algorithms. The work reported here is an offshoot of *coalgebraic recursion theory*, an application of ideas from coalgebra and closely related fields to circular phenomena and more recently to recursive program schemes. Many of the mathematical tools that are now common in semantics were first introduced for the study of recursive program schemes. I would like to think that some of the notions in coalgebraic recursion theory also will enter the mainstream of semantic research. I also think that some of this work allows one to approach the semantics of computation from an even more algebraic perspective than previous studies. For example, one of Joseph Goguen's early papers mentions the use of initial continuous algebras in connection with recursion. As a result of very recent work, it turns out that one can dispense with the domain-theoretic underpinnings of continuous algebras (or more precisely, one has a clearer understanding of the principles that make continuous algebras work in the first place). But none of this is a main point in this paper, however, and so I will only touch on these matters in passing. For a longer discussion, one could see [15] or [16].

Given the importance of recursion and recursive program schemes, one has to ask why the subject is not pursued so intensively these days. Here are two possible answers: (a) the work that had been done was mathematically challenging, requiring expertise in both algebra and domain theory; and at the same time it was not clear that the results coming out of it justified one's mastery of the field. And (b), there were many easier things to do, and many of them had a closer connection to computer science practice.

This paper is actually about a different subject, but with a connection: our concern here is the notion of *uniformity* for functors on sets that goes back to Peter Aczel's book [1] on non-wellfounded sets. The book contains a short discussion of what it called the *Special Final Coalgebra Theorem*, a result that gives a sufficient condition for a functor on the category of sets (actually the category of classes) to have a final coalgebra whose carrier is the greatest fixed point of the functor and whose structure map is the identity. This is a natural matter to investigate from the point of view of the book. However, the particular condition given was difficult to understand and work with, and so it fell to other researchers to clarify the matter. This has been the subject of a number of other papers such as [19,18,22,21]. This paper is another in the same line. It revisits the discussion in the light of concepts introduced by Adámek and his coworkers: mainly completely iterative monads and algebras. This paper formulates a new notion of uniformity for functors and studies it under *AFA*, and it also obtains some new results.

To read this paper, one should be conversant with the basics of category theory. At the same time, readers with this background only (that is, readers who have not worked with set theory or non-wellfounded sets) are likely to find the whole issue in this paper uninteresting. The reason for this is that we are interested in properties of functors which are not preserved under natural isomorphism. These properties are defined in terms of inclusion morphisms and greatest fixed points, and neither of these are preserved in this way. (However, the referee of this paper points out to me that the topic of the paper could be more interesting to those who have seen *inclusion systems*). Returning to the intended audience, it would also help to have seen the background notions that we expound in Section 2, but we have attempted to be concise and as (only as) complete as necessary.

2   Background

In this section, we present the background that we need in two parts: background from coalgebra, and background from set theory. In both cases, the background will be unusual. From coalgebra, we need a set of definitions from a handful of recent papers. I doubt that most people who glance at this paper will have heard of any of these notions, and I also know that the spare presentation here will not really help one to get a feeling for the substantial work in the area. On the set theoretic side, most of the background concerns non-standard subjects such as non-wellfounded sets and functors on the category of classes.

The next two sections may be read in either order.

2.1   Background from Coalgebra

Let $\mathcal{A}$ be category with a fixed finite coproduct operation $\oplus$.[1] An endofunctor $H : \mathcal{A} \to \mathcal{A}$ is *iteratable* [sic]: for each object A, the functor $H(\_) \oplus A$ has a final coalgebra. This condition of iteratabiltiy is satisfied by many functors of interest; it is perhaps most pertinent to this paper that results of Aczel and Mendler [4] later strengthened by Adámek et al [5] show that every endofunctor on the category of classes is iteratable. In the setting of this paper, it will be important to remember that the power set functor is iteratable on the category of classes but not on the category of sets. On the other hand the subfunctors $\mathcal{P}_\kappa$ are iteratable on the category of sets; here $\mathcal{P}_\kappa(X)$ is the set of subsets of X whose cardinality is at most κ.

If H is iteratable, then for each object A we have a final $H(\_) \oplus A$-coalgebra (TA, α_A). As the notation indicates, T extends to a functor and α to a natural transformation. T has many properties, but only a few are explicitly needed in this paper. For example, we need at one point that in the endofunctor category $[\mathcal{A}, \mathcal{A}]$, the functor $G \mapsto (H \cdot G) \oplus Id$ has (T, α) as a final coalgebra. Moreover, H brings not only T but also a free *completely iterative monad*. For our purposes, a completely iterative monad based on H is a monad $T = (T, \mu, \eta)$ together with natural transformations $\alpha : T \to HT \oplus Id$ and $\tau : HT \to T$ such that

1. For all objects A of $\mathcal{A}$, (TA, α_A) is a final coalgebra of $H(\_) \oplus A$.
2. $[\tau, \eta] : HT \oplus Id \to T$ is the pointwise inverse of α.
3. Every suitably guarded equation morphism has a unique solution.

Actually, the first point is the key; the other are consequences and/or strengthenings of it. We shall not need the precise formulation of the last point, so we omit it.

We shall always write κ for $\tau \cdot H\eta$. In general, an *ideal natural transformation into* T is one that factors through τ; so κ, for example, is ideal.

Proposition 1. *The diagrams below commute:*

$$
\begin{array}{ccc}
HTT & \xrightarrow{\ \tau T\ } & TT \\
\scriptstyle H\mu \downarrow & & \downarrow \scriptstyle \mu \\
HT & \xrightarrow[\ \tau\ ]{} & T
\end{array}
\qquad\qquad
\begin{array}{ccc}
HT & \xrightarrow{\ \tau\ } & T \\
\scriptstyle \kappa T \downarrow & \nearrow \scriptstyle \mu & \\
TT & &
\end{array}
$$

Completely iterative algebras The following notion is studied in Milius [14] and other papers. Let $H : \mathcal{A} \to \mathcal{A}$ be an endofunctor. By a *flat equation morphism* in an object A (*of parameters*) we mean a morphism of the form

$$ e : X \to HX \oplus A . $$

---

[1] We are using the symbol $\oplus$ in this section rather than the more usual symbol $+$ for coproducts. In this paper, $+$ will denote the specific coproduct on sets or classes given by the Kuratowski pairing operation (see Section 2.2). We use the different notations to help the reader with this distinction.

Let $(A, a : HA \to A)$ be an H-algebra. We say that $s : X \to A$ is a *solution* of e in (A, a) if the square

$$
\begin{array}{ccc}
X & \xrightarrow{\ e\ } & HX \oplus A \\
{\scriptstyle s}\downarrow & & \downarrow{\scriptstyle Hs \oplus A} \\
A & \xleftarrow[{[a,A]}]{} & HA \oplus A
\end{array}
$$

commutes. (Note that we often use the name of an object (such as A) as a name of the identity morphism on it.) And we call (A, a) a *completely iterative algebra* (or *cia*) for H if every flat equation morphism in A has a unique solution in it.

Example 1 We present a suggestive example, not so much for this paper but for other discussions. It is based on ideas of Peter Freyd [12] which figure in his presentation of the unit interval as the carrier of a final coalgebra structure on a certain category. Let $\mathcal{A}$ be the category of sets, let $HX = X + X$ in the usual way, and and let I be the unit interval $[0, 1]$. Consider the following algebra $a : I + I \to I$: $a(\mathsf{inl}(x)) = x/2$, and $a(\mathsf{inr}(x)) = (x+1)/2$. It turns out that (I, a) is a cia for H. In elementary terms, this means that every system of equations such as the one below has a unique solution in I:

$$
\begin{array}{rclcrcl}
x_1 & = & \tfrac{1}{2}x_2 + \tfrac{1}{2} & \quad & x_4 & = & \tfrac{1}{2}x_5 \\
x_2 & = & \tfrac{1}{2}x_3 & & x_5 & = & \tfrac{1}{2}x_6 + \tfrac{1}{2} \\
x_3 & = & \tfrac{1}{2}x_4 & & x_6 & = & .347
\end{array}
$$

On the right we can have one of three things: a variable multiplied by $1/2$, the sum of $1/2$ and a variable multiplied by $1/2$, or a constant from $[0, 1]$. The cia property says that every such system, even one with an infinite or uncountable set of variables, has a unique solution in $[0, 1]$. Incidentally, the easiest way to establish the cia property is to argue via more general results on complete metric spaces, eventually using the Banach Fixed Point Theorem.

In the statement below and wherever we refer to inverses of various maps, recall that final coalgebra morphisms are always categorical isomorphisms.

Proposition 2 (AMV [6], Milius [14]). *Concerning cias for H and completely iterative monads:*

1. *If (A, a^{-1}) is a final coalgebra for H, then (A, a) is a cia for H.*
2. *For every object A, (TA, τ_A) is a cia for H.*
3. *For every cia (A, a) for H, the solution to the flat equation morphism α_A is an Eilenberg-Moore algebra of the monad T. We write this solution morphism as $\widetilde{a} : TA \to A$.*
4. *Moreover, for every cia (A, a) for H, the triangle*

$$
\begin{array}{ccc}
HA & \xrightarrow{\ a\ } & A \\
{\scriptstyle \kappa_A}\downarrow & \nearrow{\scriptstyle \widetilde{a}} & \\
TA & &
\end{array}
$$

commutes.

A solution principle Given a final coalgebra (A, a^{-1}) for an endofunctor H, we can map any H-coalgebra uniquely into it. In this paper, we often will want to map other kinds of morphisms into it. This matter is related to the "flattening" constructions that one finds in the theory of non-wellfounded sets. For example if we have $f : X \to TX$ we shall want to define something like a coalgebra morphism $f^\dagger : X \to A$ and say what its properties should be. We shall use the notions mentioned in this section. Recall that (A, a) gives a cia for H, and we then have an Eilenberg–Moore algebra structure $\tilde{a} : TA \to A$. Now using $\tilde{a}$ and our $f : X \to TX$, we shall define the map $f^\dagger$ so that the triangle on the right commutes

$$\begin{array}{ccc} X & \xrightarrow{\ f\ } & TX \\ {\scriptstyle f^\dagger}\downarrow & {\scriptstyle [\![f^\dagger]\!]}\swarrow & \downarrow{\scriptstyle Tf^\dagger} \\ A & \xleftarrow[\tilde{a}]{} & TA \end{array} \qquad (1)$$

So there are two tasks. First, we must introduce the $[\![\ ]\!]$ operation on various morphisms and then spell out its relevant properties. Then after this we need to use this notation in principles of definition.

Definition 1. *Let H be iteratable, let (A, a^{-1}) be a final H-coalgebra, and let T be the associated monad. Recall that $\tilde{a} : TA \to A$ is the Eilenberg-Moore algebra structure associated to A; it is the solution to the flat equation morphism $\alpha_A : TA \to HTA \oplus A$ with parameters in A. For any morphism of the form $f : B \to A$, we let $[\![f]\!] : TB \to A$ be given by*

$$[\![f]\!] \ = \ \tilde{a} \cdot Tf.$$

Lemma 1. *Once again, let (A, a^{-1}) be a final H-coalgebra. Here are some properties of the morphisms $[\![f]\!]$, where $f : B \to A$.*

1. *$[\![f]\!] \cdot \eta_B = f$.*
2. *$[\![f]\!] \cdot \tau_B = a \cdot H[\![f]\!]$.*
3. *$[\![id_A]\!] \cdot T[\![f]\!] = [\![[\![f]\!]]\!] = [\![f]\!] \cdot \mu_B$, where μ is the multiplication of the monad T.*
4. *$[\![f]\!] = [\![id_A]\!] \cdot Tf$.*

The proofs are routine calculations using naturality and the definition of an Eilenberg-Moore algebra of a monad.

We also need what would be considered a folkloric result.

Lemma 2. *Let (A, a^{-1}) be a final H-coalgebra, so that (A, a) is a cia for H. Let $f : X \to TX \oplus A$ factor as on the left below.*

$$\begin{array}{ccc} X & \xrightarrow{\ f_0\ } & HX \oplus A \\ {\scriptstyle f}\searrow & & \downarrow{\scriptstyle \kappa_X \oplus A} \\ & & TX \oplus A \end{array} \qquad\qquad \begin{array}{ccc} X & \xrightarrow{\ f\ } & TX \oplus A \\ {\scriptstyle f^\dagger}\downarrow & & \downarrow{\scriptstyle Tf^\dagger \oplus A} \\ A & \xleftarrow[{[\tilde{a}, A]}]{} & TA \oplus A \end{array}$$

Then there is a unique $f^\dagger : X \to A$ such that $f^\dagger = [\tilde{a}, A] \cdot (Tf^\dagger \oplus A) \cdot f$.

426   Lawrence S. Moss

Proof   We have already mentioned the result in Milius [14] to the effect that (A, a) is a cia for H. Thus there is a unique morphism $f_0{}^\dagger$ making the square in the upper left below commute:

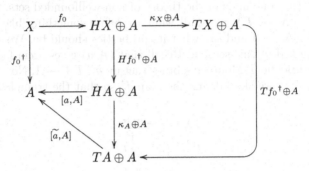

The triangle commutes using Proposition 1, and the square on the right by naturality of κ. So the outside of the figure commutes, showing that $f_0{}^\dagger$ is a morphism with the properties requested in our result. And if $g : X \to A$ is any morphism making the outside of the figure commute, then the square at the upper left commutes. Thus we have $g = f_0{}^\dagger$. This establishes the uniqueness of solutions.   ⊣

As this section comes to a close, we look back at the diagram in (1). We now have the promised result that this diagram defines $f^\dagger$ uniquely from f, assuming the relevant guardedness condition.

Lemma 3. *Let (A, a^{-1}) be a final H-coalgebra, and let $f : X \to TX$ factor through $\kappa_X : HX \to TX$. Then there is a unique $f^\dagger : X \to A$ such that*

$$f^\dagger \;=\; [\![f^\dagger]\!] \cdot f.$$

Proof   Apply Lemma 2 to $\mathsf{inl} \cdot f : X \to TX \oplus A$. There is a unique $g : X \to A$ such that

$$g \;=\; [\widetilde{a}, A] \cdot (Tg \oplus A) \cdot \mathsf{inl} \cdot f \;=\; \widetilde{a} \cdot Tg \cdot f \;=\; [\![g]\!] \cdot f.$$

We take g for the needed morphism $f^\dagger$. For the uniqueness, if $g = [\![g]\!] \cdot f$, then the same calculations as above show that $g = [\widetilde{a}, A] \cdot (Tg \oplus A) \cdot \mathsf{inl} \cdot f$; hence we are done by Lemma 2.   ⊣

We emphasize that the background in this section only contains a hint of a more extensive subject that is currently an active area. Not only have I omitted many motivational points connected to recursive program schemes, first- and second-order substitution, the very interesting notion of an *Elgot algebra*, and the like. I also have not even mentioned all of the results that this paper will call upon. the results that we are going to use directly. Two places to read about all of this and more is Stefan Milius' dissertation [15] and the paper on recursive program schemes and coalgebra [16].

2.2   Background from Set Theory

We remind the reader of the basic facts of set theory which will be relevant in this paper.

The Kuratowski ordered pair (a, b) of two sets a and b is $\{\{a\}, \{a, b\}\}$. In terms of this one defines and studies relations, functions, and the like. One also defines versions of the natural numbers by: $0 = \emptyset$, $1 = \{\emptyset\}$, etc. Finally, we shall fix a coproduct operation $+$ on sets by

$$
\begin{aligned}
a + b &= (\{0\} \times a) \cup (\{1\} \times b) \\
&= \{(0, x) : x \in a\} \cup \{(1, y) : y \in b\}
\end{aligned}
$$

For sets a and b, the coproduct injections $\mathsf{inl} : a \to a + b$ and $\mathsf{inr} : b \to a + b$ are then given by

$$
\begin{aligned}
\mathsf{inl}(x) &= (0, x) \\
\mathsf{inr}(y) &= (1, y)
\end{aligned}
$$

Henceforth in this paper, the symbol $+$ is used for this operation on sets (extended in the natural way to classes).

For any set a, $\bigcup a$ is the set of elements of elements of a. A set a is transitive if $\bigcup a \subseteq a$. The *transitive closure* of a is

$$
tc(a) \quad = \quad a \cup \bigcup a \cup \bigcup\bigcup a \cup \cdots.
$$

This is a set, and it is the smallest transitive set (under the inclusion ordering) which includes a.

If $a \subseteq b$, we write $i_{a,b}$ for the inclusion map of a into b. If $b = V$, then we generally drop it from the notation. So if $a \subseteq b$, we have $i_a = i_b \cdot i_{a,b}$.

Note also that if a is transitive, then $a \subseteq \mathcal{P}a$. Further $i_{a,\mathcal{P}a} : a \to \mathcal{P}a$ is a $\mathcal{P}$-coalgebra, and $i_a : a \to V$ is a $\mathcal{P}$-coalgebra morphism from it to $(V, i_{\mathcal{P}V}^{-1} = id_V)$.

The axioms of set theory are not about sets as much as they are about the *universe of sets*. One of the intuitive principles of the theory is that arbitrary collections of mathematical objects "should be" sets. Due to paradoxes, this intuitive principle is not directly formalized in standard set theories. In a sense, the axioms one does have are intended to give enough sets to constitute a mathematical universe while not having so many as to risk inconsistency. But it is natural in this connection to consider some collections of objects which are demonstrably not sets. These are called *proper classes*. The term *class* informally refers to a collection of mathematical objects. Classes are usually not first-class objects in set theory (certainly they are not in the most standard set theory, *ZFC*). Instead, a statement about classes is regarded as a paraphrase for some other (more complicated and usually less intuitive) statement about sets. This is probably not a good place to discuss the details of the formalization; one clear source is Chapter 1 of Azriel Levy's book [13] on set theory. For our purposes, classes may be taken as definable subcollections of sets. For example, if a is any set, then the class of all sets which do not contain a as an element is $\{x : a \notin x\}$. The class V of all sets is $\{x : x = x\}$. The definability here is in the first-order logic with just a symbol $\in$ for membership, and the quantifiers range over sets (not classes). If C is a class, the *power class of C*,

$$\mathcal{P}(C) \quad = \quad \{x : x \text{ is a set, and } (\forall y)(y \in x \to \varphi_C(y))\},$$

where φ_C is the formula that defines the class C.

We are interested in functors H on sets and classes which are *monotone* in the sense of preserving inclusions among objects: if $a \subseteq b$, then $Ha \subseteq Hb$.

Each set-based[2] monotone operation H on classes has a least fixed point H_* and a greatest fixed point H^*. For the least fixed point, we first define classes H_α by transfinite recursion: $H_0 = \emptyset$, $H_{\alpha+1} = H(H_\alpha)$ and for limit λ, $H_\lambda = \bigcup_{\beta<\lambda} H_\beta$. Then the class H_* is defined by $x \in H_*$ iff $(\exists \alpha) x \in H_\alpha$. The assumption that H be set-based, together with the Replacement Axiom, implies that H_* is a fixed point of H, and it is easy to see by induction that each H_α is a subset of any fixed point of H. So H_* is the least fixed point. In categorical terms, (H_*, id) is an initial H-algebra on the category of classes. We are especially concerned with the dual concept, greatest fixed points. As shown in Aczel [1],

$$H^* \quad = \quad \bigcup \{b : b \text{ is a set and } b \subseteq Hb\}.$$

H^* might well be a proper class.

For example, by Cantor's Theorem there are no sets which are fixed points of the power set functor, but on classes, the least fixed point exists and indeed is the class WF of wellfounded sets. Another fixed point is the class V of all sets. Saying that $\mathcal{P}V = V$ just means that every set of sets is a set, and that every set is a set of sets. (So this would contradict any axiom of *urelements*, and indeed usually set theories implicitly do not allow for urelements.) Note that i_V, $i_{\mathcal{P}V}$, $\mathcal{P}i_V$, and $\mathcal{P}i_{\mathcal{P}V}$ all denote the same operation, the identity on the universe.

Here are some further examples to orient the reader. The identity functor has the universe V as its greatest fixed point on the category of classes. The identity has not greatest fixed point on sets. But even on classes, the greatest fixed point is not the carrier of a final coalgebra structure, since that would be a mere singleton set. But consider the variant functor $H(a) = 1 \times a$. Here there are some differences, even though H is naturally isomorphic to the identity. Whether H has any fixed points besides $\emptyset$ is a question that is sensitive to the underlying set theory. Under the Foundation Axiom, the empty set $\emptyset$ is the only fixed point of H. Under the Anti-Foundation Axiom (formulated shortly), H has one additional fixed point (which therefore is the greatest fixed point): there is a unique set a such that $a = \{(0, a)\}$ (this uses AFA). And so $b = \{a\}$ satisfies $b = \{0\} \times \{a\} = 1 \times b$. Moreover, b is the only set with this property except for $\emptyset$.

In any case, the overall point is that properties of the greatest fixed points of various operations are sensitive to the underlying set theory. The topics of this paper are certain classes which form either final coalgebras or cias for various functors. Again, such classes do not exist in the usual set theory ZFC, due mainly

---

[2] The condition of set-based-ness introduced in Aczel [1] turned out to be unnecessary for functors on classes: see [5,9,10]. As a result, we suppress mention of this condition.

to the Foundation Axiom. In this connection, and in connection with other coalgebraic notions, it is more natural to work in the set theory ZFA obtained from ZFC by replacing the Foundation Axiom with a 'dual' statement, the Anti-Foundation Axiom first formulated by Forti and Honsell and then popularized in Peter Aczel's book [1].

The Anti-Foundation Axiom The *Anti-Foundation Axiom (AFA)* is the assertion that for every set b and every $e : b \to \mathcal{P}b$, there exists a unique $s : b \to V$ such that $s = \mathcal{P}s \cdot e$:

$$
\begin{array}{ccc}
b & \xrightarrow{\ e\ } & \mathcal{P}b \\
{\scriptstyle s}\big\downarrow & & \big\downarrow{\scriptstyle \mathcal{P}s} \\
V & =\!\!=\!\!= & \mathcal{P}V
\end{array}
\tag{2}
$$

The map s is called the *solution* to the *system* e.

To see how this is used, we mentioned above that under AFA, there is a unique set $a = \{(0, a)\}$. To see this, we let $b = \{v, w, x, y, z\}$ and consider $e : b \to \mathcal{P}b$ given by

$$
\begin{array}{rclcrcl}
e(v) & = & \{w\} & \qquad & e(y) & = & \{v, z\} \\
e(w) & = & \{x, y\} & & e(z) & = & \emptyset \\
e(x) & = & \{z\} & &&&
\end{array}
$$

Then if s is as in the statement of AFA, we have $s(v) = \{s(w)\}$, $s(w) = \{s(x), s(y)\}$, ..., $s(z) = \emptyset$. So $s(x) = \{0\}$, $s(y) = \{s(v), 0\}$, and

$$
s(w) \quad = \quad \{\{0\}, \{s(v), 0\}\} \quad = \quad (0, s(v)).
$$

Finally, $s(v) = \{(0, s(v))\}$. Thus $s(v)$ is a set which solves $a = \{(0, a)\}$. It is not hard to check that it is the only solution, because any solution to this equation gives a solution to the "flat system" e by unraveling a bit.

Lemma 4 (Turi [21], see also [18]). *AFA is equivalent to the assertion that* $(V, i_V) = (V, id_V)$ *is a final $\mathcal{P}$-coalgebra.*

Our overall setting in this paper is ZFA. (Actually, many of the results do not actually use AFA, especially those before Section 3.1. But the main results of the paper do use it.)

(By the way, the formulation of AFA in (2) above does not include any specific morphism between V and $\mathcal{P}V$. This is basically the way AFA is presented in Aczel's book [1], for example, and also my book with Jon Barwise [8]. The disadvantage of this kind of formalization is that it hides the fact that there are two different possible assertions:

$$
\begin{array}{ccc}
b & \xrightarrow{\ e\ } & \mathcal{P}b \\
{\scriptstyle s}\big\downarrow & & \big\downarrow{\scriptstyle \mathcal{P}s} \\
V & \xrightarrow[(i_{\mathcal{P}V})^{-1}]{} & \mathcal{P}V
\end{array}
\qquad \text{vs.} \qquad
\begin{array}{ccc}
b & \xrightarrow{\ e\ } & \mathcal{P}b \\
{\scriptstyle s}\big\downarrow & & \big\downarrow{\scriptstyle \mathcal{P}s} \\
V & \xleftarrow[i_{\mathcal{P}V}]{} & \mathcal{P}V
\end{array}
$$

When one reworks our statement of *AFA* using the first formulation, one can sense the connection to final coalgebras and Lemma 4. The second formulation would be closer to what we find in Lemma 6.)

The main problem for this papers and all previous ones on "uniformity" for functors is to propose a condition guaranteeing that the greatest fixed point of a monotone H be a final coalgebra together with the identity. This paper proposes and studies one such condition.

3   Standard Functors and Monads

At this point, we have all of the background we need to begin our study The first concept we need is that of a standard functor on sets or classes. An endofunctor H is *standard* if H preserves inclusion maps in the sense that $Hi_{a,b} = i_{Ha,Hb}$. This notion was introduced in a slightly stronger form in Adámek and Trnková's book [7]; Theorem 3.4.5 of that book shows that every functor on sets is naturally isomorphic to a standard functor in their sense.

Proposition 3. *The coproduct $+$ derived from the Kuratowski pair has the property that for all classes c, the endofunctor $\_ + c$ is standard.*

The proof is an easy calculation. Of course, the functors $c + \_$ are also standard. Here is a consequence of these: Let $x \subseteq x'$ and $y \subseteq y'$. Then the diagram

$$
\begin{array}{ccc}
x & \xrightarrow{\;\mathsf{inl}\;} & x + y \\
{\scriptstyle i_{x,x'}} \downarrow & & \downarrow {\scriptstyle i_{x,x'}+i_{y,y'}} \\
x' & \xrightarrow[\;\mathsf{inl}\;]{} & x' + y'
\end{array}
\tag{3}
$$

commutes.

Definition 2. *Let T be the free completely iterative monad on H. T is* standard *if for each a, $Ta = HTa + a$, and moreover $\alpha_a = id_{Ta}$.*

Lemma 5. *Let T be the free completely iterative monad on a standard functor H. If T is a standard monad, then T is a standard functor.*

Proof Let $a \subseteq b$, and write i for $i_{a,b}$. We know that Ti is the unique map such that $Ti \cdot \tau_a = \tau_b \cdot HTi$. (This follows from the Substitution Theorem of [3,2,18] applied to $\eta_b \cdot i$.) But if we take Ti to be $i_{Ta,Tb}$, then the equation is satisfied:

$$
i_{Ta,Tb} \cdot \tau_a \quad = \quad \tau_b \cdot i_{HTa,HTb} \quad = \quad \tau_b \cdot Hi_{Ta,Tb}.
$$

In this we are using the fact that τ is inl for a standard functor, and also equation (3). So by uniqueness, $Ti = i_{Ta,Tb}$.   ⊣

3.1 $\mathcal{P}$ Generates a Standard Iterative Monad

We check here that under *AFA*, $\mathcal{P}$ generates a standard iterative monad. The general idea of our work is to use this fact to show that many other functors also generate standard iterative monads. In fact, our definition of uniformity effects such a reduction.

Lemma 6 (See [18]). *Let H be standard. The following are equivalent:*

1. (H^*, id) *is a final H-coalgebra.*
2. (V, i_{HV}) *is a coalgebra-final H-algebra: for every class b and every $e : b \to Hb$, there exists a unique solution $s : b \to V$, a morphism such that $s = i_{HV} \cdot Hs \cdot e$:*

Proof We show first that (2) implies (1). Consider $e : b \to Hb$ and its solution s. Let $c = s[b]$ be the image of b under s. Then $Hs[Hb] \subseteq H(s[b]) = Hc$ (see, e.g., Proposition 5.1.2 of [18]). Condition (2) in our lemma implies that $c \subseteq Hs[Hb]$, and so we see that $c \subseteq Hc$. Let $t : b \to c$ be such that $i_c \cdot t = s$. Then all parts of the diagram on the left below commute, save for the top square.

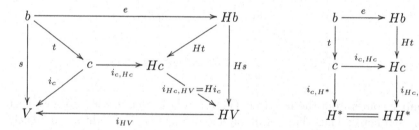

Thus that part also commutes. This is the top square on the right, and so it commutes. By the monotonicity of H, we have $c \subseteq H^*$. Thus the bottom square on the right commutes, and we see that $i_{c,H^*} \cdot t$ is a coalgebra morphism from (b, e) to (H^*, id).

Next, we argue the uniqueness of this morphism $i_{c,H^*} \cdot t$. Suppose that $f : b \to H^*$ is any coalgebra morphism. Let $c' = f[b]$, let $t' : b \to c'$, and write f as $i_{c',H^*} \cdot t'$. So we have a diagram similar to the one on the right above, but with t replaced by t', and c by c'. The overall outside commutes. And since i_{Hc',HH^*} is an inclusion and hence monic, we see that the top square commutes: $i_{c',Hc'} \cdot t' = Ht' \cdot e$. This means that the top square on the left commutes, mutatis mutandis. We then take s' to be $i'_c \cdot t'$ so that the two triangles on the left commute. By our statement (1), we have uniqueness of solutions; thus $s' = s$. It follows that $t' = t$ and $c' = c$. We conclude that $f = i_{c',H^*} \cdot t' = i_{c,H^*} \cdot t$, as desired.

Now we prove that (1) implies (2). Let (H^*, id) be final, We check that (2) indeed holds. Let $e : b \to Hb$. We have a final H-coalgebra morphism $e^* : b \to H^*$, and we consider $i_{H^*} \cdot e^*$. We see that

$$
\begin{aligned}
i_{HV} \cdot H(i_{H^*} \cdot e^*) \cdot e &= i_{HV} \cdot H i_{H^*} \cdot (He^* \cdot e) \\
&= i_{HV} \cdot i_{HH^*, HV} \cdot i_{H^*, HH^*} \cdot e^* \\
&= i_{H^*} \cdot e^*
\end{aligned}
$$

This shows that $i_{H^*} \cdot e^*$ is a solution to e in the sense of point (2) above. For the uniqueness, if s is a solution to e, then write $s = i_c \cdot t$ as in the work we did above in showing that (2)$\Rightarrow$(1). By the finality of H^*, we have $e^* = i_{c,H^*} \cdot t$. But now $s = i_c \cdot t = i_{H^*} \cdot i_{c,H^*} \cdot t = i_{H^*} \cdot e^*$. $\dashv$

In the next proposition, and in the rest of this paper, we let G_w be the constant functor with value w.

Proposition 4. *For every set w, $((\mathcal{P}+G_w)^*, id)$ is a final coalgebra for $\mathcal{P}+G_w$.*

Proof We apply Lemma 6. Let $e : b \to \mathcal{P}b + w$. Consider the diagram below:

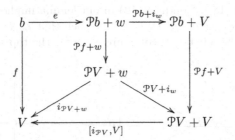

The map f comes from the fact that $(V, (i_{\mathcal{P}V})^{-1})$ is a cia for $\mathcal{P}$. (So note that *AFA* is used here.) Thus the overall outside commutes. The right square easily commutes. For the triangle, we use the general fact that $i_{a+b} = [i_a, i_b]$. (In fact, for classes a, b, and c such that $a \subseteq c$, and $b \subseteq c$, and $a + b \subseteq c$, we have $i_{a+b,c} = [i_{a,c}, i_{b,c}]$.) We conclude that the left square above commutes. This is the existence of the needed f in Lemma 6, and the uniqueness comes from the cia structure. $\dashv$

It follows from Proposition 4 that $\mathcal{P}$ generates a standard iterative monad on the category of classes.

4   The Class TV and the Map χ

As we now know, the power set functor determines a free completely iterative monad

$$
T = (T^{\mathcal{P}}, \mu^{\mathcal{P}}, \eta^{\mathcal{P}}).
$$

This monad is indeed standard. It also comes with additional natural transformations $\alpha^{\mathcal{P}}$ and $\tau^{\mathcal{P}}$. Because this is the most common monad in the rest of the paper, we drop the superscripts on all of this data related to it.

By AFA the inverse of inclusion gives a final coalgebra $(i_{\mathcal{P}V})^{-1} : V \to \mathcal{P}V$. Because so much of the rest of this paper uses the map $[\![i_{\mathcal{P}V}]\!]$, we shorten the notation to write

$$\chi \;=\; [\![i_{\mathcal{P}V}]\!] : TV \to V.$$

For a mnemonic on this, think of χ for χrunch. As we shall see, it takes elements of TV and collapses them back to sets. Those familiar with the *Mostowski collapse* in set theory might think of χ as a kind of non-wellfounded version of that map.

It is worthwhile to get a feeling for the class TV. To understand it better, we use Proposition 4, taking V for w. So TV is the greatest fixed point of the functor which takes a class X to

$$\mathcal{P}(X) + V \;=\; (\{0\} \times \mathcal{P}(X)) \cup (\{1\} \times V).$$

Hence TV is the largest collection C of sets with the property that each member of C is of one of the following forms:

1. $(0, x)$ for some subset $x \subseteq C$.
2. $(1, x)$ for some set x.

Note as well that $\eta : Id \to T$ is defined by $\eta_X(a) = (1, a)$ for all classes X and all sets $a \in X$. As for τ, standardness implies that its components are all inclusions.

We now turn to χ. The elements of TV *code* sets as follows:

1. $(0, x)$ codes the set of sets coded by the elements of x.
2. $(1, x)$ codes x itself.

The map χ is the *decoding* map.

Example 2 Here are some examples of χ at work:

1. For all sets a, $\chi(1, a) = a$, and thus $\chi(0, \{(1, a)\}) = \{a\}$.
2. $\chi(0, \emptyset) = \emptyset$.
3. $\chi(0, \{(0, \emptyset)\}) = \{\chi(0, \emptyset)\} = \{\emptyset\}$.
4. $\chi(0, \{(0, \emptyset), (1, x)\}) = \{\chi(0, \emptyset), \chi(1, x)\} = \{\emptyset, x\}$.
5. For all sets a and b,

$$(0, \{(0, \{(1, a)\}), (0, \{(1, a), (1, b)\})\})$$

belongs to TV, and χ applied to it is the ordered pair (a, b).

In all of these, we omit mention of α since it is the identity.

We record the following application of Lemma 1:

Proposition 5. *Concerning* $\chi : TV \to V$:

1. $\chi \cdot \eta_V = id_V$.
2. $\chi \cdot \tau_V = i_{\mathcal{P}V} \cdot \mathcal{P}\chi$.
3. $\chi \cdot T\chi = [\![\chi]\!] = \chi \cdot \mu_V$.

5   Uniformity

As our title indicates, this paper is about notions of uniformity for functors on sets and classes. We propose a new definition in Section 5.1 below. Before that, we want to mention the previous notions of uniformity in the literature, and the motivation for them.

The first place where some notion of "uniform functor" may be found is Aczel's book [1] on non-wellfounded sets. His definition is in terms of the "expanded universe ... [which] has an atom x_i for each pure set i." In our terminology, this is exactly $\mathcal{P}T$. (Recall that we are dropping the superscript, writing T for $T^{\mathcal{P}}$.) Were his definition to be translated into our notation, it would look similar to ours. It would involve for each class A a map $\pi_A : HA \to TA$ with some properties. However, the resulting π is not required to be a natural transformation (and indeed, it was not realized until several years later that T was even a functor, etc.). As a compensation, the definition requires another property on π. Incidentally, I have not worked extensively with Aczel's definition, but it seems to be hard to check that the uniform functors in his sense are closed under composition.

We also emphasize that the first motivation for uniformity is to provide a sufficient condition on a monotone functor H that its greatest fixed point H^* be a final H-coalgebra along with the identity as a structure map.

The first work to formulate uniformity in terms of natural transformations is that of Turi [21] (also presented in Turi and Rutten [22]). Our definition is similar to theirs, and to distinguish the two we call theirs *TR-uniformity*. Its definition is in terms of a different monad on sets, the monad W given by WX is the *least fixed point* of $X \mapsto \mathcal{P}X + X$. In addition, there is a unique morphism $\epsilon_V : WV \to V$ such that the composition

$$WV \longrightarrow \mathcal{P}WV + V \xrightarrow{[\mathcal{P}\epsilon_V, id_V]} \mathcal{P}V + V \xrightarrow{[i_{\mathcal{P}}V, id_V]} V$$

is ϵ_V. They require of a functor H that there be a natural transformation $\rho : H \to \mathcal{P}W$ such that

$$\begin{array}{ccc} HV & \xrightarrow{\rho_V} & \mathcal{P}WV \\ {\scriptstyle i_{HV}}\downarrow & & \downarrow{\scriptstyle \mathcal{P}\epsilon_V} \\ V & \xleftarrow{i_{\mathcal{P}}V} & \mathcal{P}V \end{array}$$

(The use of $\mathcal{P}WV$ corresponds to our requirement that that natural transformations involved in uniformity be *ideal*.) The main difference is that we use the monad T, a larger monad than W; hence more functors are uniform in our sense. (For example, the constant K_a functors whose value a are non-wellfounded sets are uniform in our sense but not in Turi and Rutten's sense. Furthermore, functors built from K_a in the expected ways will also turn out to be uniform in our sense; see Theorem 7.)

Once again, it is worthwhile mentioning that their motivation for uniformity again is the same as Aczel's. However, they recognize that there is also a different intuition, one related to substitution:

> Intuitively, an endofunctor on SET is uniform on maps [their terminology, following Aczel] if it is completely determined by is action on objects (i.e., classes). Most endofunctors are thus uniform on maps. For instance, consider the endofunctor $X \mapsto A \times X$ mapping a class X to its product with a fixed class A. Given a function $f : X \to Y$, the value of $A \times f$ at an element (a, x) of $A \times X$ is the pair $(a, f(x)) \in A \times Y$ which is obtained by applying f to the $x \in X$ in $A \times X$. This suggests that the class X should be regarded as a class of variables and that, in general, the action of a functor F uniform on maps on a function f should simply be the substitution of the variables x occurring in FX by $f(x)$. (Turi [21] p. 211; also Turi and Rutten [22], Sec. 5.5.)

For other approaches, see Devlin [11] and also Moss and Danner [19].

The upshot is that there are two intuitions at work in the definition of uniformity, or at least two different goals. One is to search for condition on functors F which guarantees that the greatest fixed point F^* of F be a final coalgebra with the identity as the structure map. I would like to emphasize, especially for readers with a background in category theory, that this kind of question is not "preserved under natural isomorphisms of functors". The identity functor will never be uniform under any reasonable definition, but functors like $1 \times x$ will turn out uniform under AFA.

A second intuition is mentioned in the quoted paragraph above. We could say that this has to do with the class TV and way that set theory is used to represent natural mathematical operations, and also with the matter of coding sets by elements of TV. The overall thrust of set theory as a foundational study is that natural mathematical operations are representable in a first-order way in the universe of sets. It is not always easy to spell out what this means, and most textbooks never get around to it. What we are doing in the definition of uniformity is to spell out the representability of natural mathematical operations, but not in terms of first-order logic but in terms of the iterative monad of the power set.

5.1 Our Definition

We now come to the main definition in this paper. We continue to write T for the monad determined by the power set functor, omitting the superscript $\mathcal{P}$ in most places. We also remind the reader that an *ideal* natural transformation is one which factors through τ.

Definition 3. *A functor H is* uniform *if there is an ideal natural transformation $\pi : H \to T$ such that for all classes a,*

$$[\![i_a]\!] \cdot \pi_a \;=\; i_{Ha}.$$

We call π a uniformity *for H.*

Uniformity is equivalent to standardness plus the identity $\chi \cdot \pi_V = i_{HV}$. This says that if we encode HV as a subclass of TV and then collapse back to V via χ, we have an inclusion. The reason why we want to do any encoding has to do with co-recursion: given $e : a \rightarrow Ha$, we want to use get a solution satisfying an appropriate recursion principle. There is no evident way to do this without extra maps. We use π to get a related map $e' : a \rightarrow T(a)$. Having this, we use Lemma 3 to get a map $a \rightarrow V$.

Lemma 7. *Let $\pi : H \rightarrow T$ be an ideal natural transformation. The following are equivalent:*

1. *H is uniform.*
2. *H is standard, and $\chi \cdot \pi_V = i_{HV}$.*

Proof   First, assume that π is a uniformity for H. Then in particular, $\chi \cdot \pi_V = i_{HV}$. The interesting point is to check that H is standard. Let $a \subseteq b$. In the diagram below,

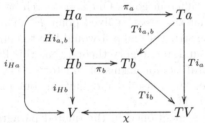

everything commutes except the region on the left: the top uses naturality of π; the triangle on the right is by applying T to the fact that $i_a = i_b \cdot i_{a,b}$; and uniformity is used in the overall outside and in the bottom square. So we see that $i_{Ha} = i_{Hb} \cdot Hi_{a,b}$. But now we notice a general fact: if x and y are any sets, and $f : x \rightarrow y$ is such that $i_x = i_y \cdot f$, then $x \subseteq y$ and $f = i_{x,y}$. It now follows that $Ha \subseteq Hb$ and that $Hi_{a,b} = i_{Ha,Hb}$, as desired.

Going the other way, suppose H is standard, and $\chi \cdot \pi_V = i_{HV}$. Let a be any class. Return to the diagram above, and replace b by V. Then our assumption that $\chi \cdot \pi_V = i_{HV}$ implies that the bottom square commutes, and the region on the left is by standardness. It follows that we have the desired uniformity equation $[\![i_a]\!] \cdot \pi_a = i_{Ha}$.   ⊣

The second formulation is often easier to check, since standardness is usually immediate for functors. We use Lemma 7 without further mention.

Our main results   The main results of this paper are as follows: the uniform functors contain the power set functor and the constants, and they are closed under a number of natural operations including composition and iteration. A uniform H has the property that H^* together with the identity is a final H-coalgebra, and V together with the inclusion of HV into it is a cia for H. The same generally holds for λ-uniform functors, a notion we introduce in Section 7 except that the only constant functors which are λ-uniform are those for sets in H_λ. If H is λ-uniform, then H^* is a subset of H_λ.

The rest of this paper is devoted to proofs of these assertions, and some additional discussion.

5.2   Examples and Closure Properties

Example 3 We establish the uniformity of the power set functor $\mathcal{P}$. This functor is easily standard. Let $\pi : \mathcal{P} \to T$ be $\tau \cdot \mathcal{P}\eta$ from the iterative monad determined by $\mathcal{P}$. Note that π is ideal. Furthermore,

$$
\begin{aligned}
\chi \cdot \tau_V \cdot \mathcal{P}\eta_V &= i_{\mathcal{P}V} \cdot \mathcal{P}\chi \cdot \mathcal{P}\eta_V \\
&= i_{\mathcal{P}V} \cdot \mathcal{P}id_V \\
&= i_{\mathcal{P}V}
\end{aligned}
$$

We used Proposition 5.

Example 4 Let w be a set; we show that the constant functor G_w with value w is uniform. Let $\overline{w}$ be the transitive closure of w. Since $\overline{w} \subseteq \mathcal{P}(\overline{w})$, we have an inclusion $i_{\overline{w}, \mathcal{P}(\overline{w})}$. To shorten our notation, we abbreviate this as i in this example. We regard i as a natural transformation between constant functors. We also have a natural transformation $G_{\overline{w}} \to \mathcal{P}G_{\overline{w}} \to \mathcal{P}G_{\overline{w}} + Id$. By a finality result concerning T in the functor category, we have a natural transformation $\pi_0 : G_{\overline{w}} \to T$ such that $\pi_0 = \tau^{\mathcal{P}} \cdot \mathcal{P}\pi_0 \cdot i$. It follows easily from this that $\chi \cdot \pi_0(V)$ is the inclusion $i_{G_{\overline{w}}V} = i_{\overline{w}}$. And the desired ideal natural transformation is $\pi_0 \cdot j$, where j is the inclusion $i_{w,\overline{w}}$ considered as a natural transformation.

Example 5 The identity functor I is *not* uniform. Here are two ways to see this. First, we argue directly, by contradiction. Suppose we had an ideal $\pi : I \to T$ such that $\chi \cdot \pi = id_V$. Then for all sets a, $id_a = \chi \cdot i_{T(a),T(V)} \cdot \pi_a$. In short, for all $x \in a$, $x = \chi(\pi_a x)$. Let $a = \{0, 1\}$. Then $\pi_a(0)$ must be $(0, \emptyset)$, as π is ideal, and $\chi^{-1}[\emptyset] = \{(0, \emptyset), (1, \emptyset)\}$. Let $f : a \to a$ be the transposition $f(0) = 1$ and $f(1) = 0$. By naturality, $\pi_a \cdot f = Tf \cdot \pi_a$. Applying this to 0, we see that $\pi_a(1) = Tf(0, \emptyset) = (0, \emptyset)$. But then we would have $1 = \chi \cdot \pi_a(1) = \chi(0, \emptyset) = \emptyset$; this is a contradiction.

A less elementary way to establish the non-uniformity is to use a result from later that for uniform H, the greatest fixed point H^* gives a final coalgebra with the identity map. For 1, we have $I^* = V$. But the final coalgebras of I are the singleton sets. So for this reason, I is not uniform.

Example 6 In contrast to this, the functor $H(a) = a + 0$ *is* uniform; this is the same as

$$
H(a) \quad = \quad 1 \times a \quad = \quad \{(0, x) : x \in a\}.
$$

The natural transformation $\pi : H \to T$ is given by

$$\pi_a(0,x) \quad = \quad (0,\{(0,\{(0,0)\}),(0,\{(0,0),(1,x)\})\}).$$

Similar to what we have seen in Example 2, part 5, for all sets x, $\chi(\pi_V(0,x)) = (0,x)$:

$$
\begin{aligned}
&\chi((0,\{(0,\{(0,0)\}),(0,\{(0,0),(1,x)\})\})) \\
=\ &\{\chi(0,\{(0,0)\}),\chi(0,\{(0,0),(1,x)\})\} \\
=\ &\{\{0\},\{0,x\}\} \qquad\qquad\qquad\qquad (*) \\
=\ &(0,x)
\end{aligned}
$$

The calculations in the line marked $(*)$ were performed in Example 2. Moreover, π is an ideal natural transformation because each $\pi_a(0,x)$ is an ordered pair beginning with 0; more formally, consider $\pi^* : H \to \mathcal{P}T$ given by

$$\pi_a^*(0,x) \quad = \quad \{(0,\{(0,0)\}),(0,\{(0,0),(1,x)\})\}.$$

Then $\pi = \tau \cdot \pi^*$. The most tedious part of the verification has to do with the naturality of π^*. Let $f : a \to b$. Note that $Tf(0,0) = (0,0)$, and for all $x \in a$, $Tf(1,x) = (1,fx)$. It follows that

$$Tf(0,\{(0,0),(1,x)\}) \ = \ \{(0,\mathcal{P}Tf\{(0,0),(1,x)\}\} \ = \ \{(0,\{(0,0),(1,fx)\}\}.$$

Therefore

$$
\begin{aligned}
\pi_b^*(1,fx) \quad &= \quad \{(0,\{(0,0)\}),(0,\{(0,0),(1,fx)\})\} \\
&= \quad \{Tf(0,\{(0,0)\})\},Tf(0,\{(0,0),(1,x)\})\} \\
&= \quad \mathcal{P}Tf\{(0,\{(0,0)\}),(0,\{(0,0),(1,x)\}\}\} \\
&= \quad \mathcal{P}Tf(\pi_a^*(0,x))
\end{aligned}
$$

The point is that we can "implement" the pairing machinery in a way which is recoverable by χ. In a similar fashion, we also have the following result:

Theorem 7. *If F and G are uniform, and a is any set, then the following functors are also uniform: $F + G$, $F \times G$, $1 + F$, F^a.*

Proof See [18] for many similar calculations involving the coding machinery.
⊣

Finally, we have the following proposition which shows that our notion of uniformity indeed generalizes TR-uniformity as defined in Section 5. This result is not needed for the rest of this paper, and the reader may omit it. We shall need a property of the monad W. The monad W also carries some extra structure. First of all, there is a natural transformation $[\gamma, \delta] : \mathcal{P}W + Id \to W$. $\mathcal{P}$ may be regarded as an endofunctor on the endofunctor category on classes; viz. $F \mapsto \mathcal{P} \cdot F$. We also get a related functor $\mathcal{P} \cdot \_ + Id$. For the functor W, the value of this functor at W is $\mathcal{P}W + Id$. So the natural transformation $[\gamma, \delta] : \mathcal{P}W + Id \to W$ may be regarded as a $(\mathcal{P} \cdot \_ + Id)$-algebra structure for $\mathcal{P}$. (For that matter, $[\tau, \eta] : \mathcal{P}T + Id \to T$ is another algebra structure.) Moreover, $(W, [\gamma, \delta])$ is an initial algebra of this functor $\mathcal{P} \cdot \_ + Id$. By initiality, there is a unique natural transformation $\beta : W \to T$ giving an algebra morphism from $[\gamma, \delta]$ to $[\tau, \eta]$.

Proposition 6. *Every standard TR-uniform functor H is uniform in our sense.*

Proof   Let ρ establish the TR-uniformity of H. Let $\pi = \tau \cdot \mathcal{P}\beta \cdot \rho$. Then π is ideal. We are going to use Lemma 7. We must check that the outside of the figure below commutes:

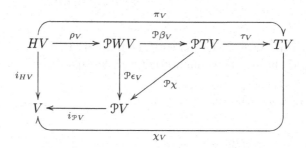

The square on the left commutes by definition of TR-uniformity, the region at the top is the definition of π, and the region at the right and bottom is by Proposition 5, part 2. For the triangle, we show that $\chi \cdot \beta_V = \epsilon_V$. To do this, consider the diagram below:

$$\mathcal{P}WV + V \xrightarrow{\mathcal{P}\beta_V + id_V} \mathcal{P}TV + V \xrightarrow{\mathcal{P}\chi + id_V} \mathcal{P}V + V$$

$$\downarrow [\gamma_V, \delta_V] \qquad\qquad \downarrow [\tau_V, \eta_V] \qquad\qquad \downarrow [i_{\mathcal{P}V}, id_V]$$

$$WV \xrightarrow{\quad\beta_V\quad} TV \xrightarrow{\quad\chi\quad} V$$

The square on the left commutes by the definition of β as an algebra morphism in the endofunctor category. The square on the right commutes by Proposition 5. Taken together, the two squares show that $\chi \cdot \beta_V$ satisfies the equation which uniquely defines ϵ_V. Hence $\chi \cdot \beta_V = \epsilon_V$, as desired.   ⊣

This result, together with our earlier remark about constant functors for non-wellfounded sets, shows that the standard TR-uniform functors are a proper subcollection of the uniform functors.

5.3   Closure Under Composition and Iteration

Theorem 8. *If F and G are uniform, then FG is also uniform.*

Proof   Let F be uniform via π, and G uniform via ρ. To see that $F \cdot G$ is uniform, let $\pi * \rho = \pi T \cdot F\rho$, and consider the natural transformation

$$\mu \cdot (\pi * \rho).$$

The following diagram shows π to be an ideal natural transformation:

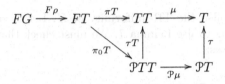

Here π_0 is a natural transformation with the property that $\pi = \tau \cdot \pi_0$; this gives the the triangle above. The commutativity of the square is a part of Proposition 1.

Consider next the following diagram:

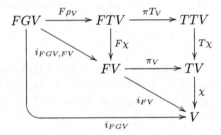

The upper triangle is obtained by applying F to the uniformity equation for G and using the standardness of F as well. Everything else commutes easily. We now use Proposition 5 to calculate:

$$\chi \cdot \mu_V \cdot \pi T_V \cdot F\rho_V \quad = \quad \chi \cdot T\chi \cdot \pi T_V \cdot F\rho_V \quad = \quad i_{FGV}.$$

This completes the proof that FG is uniform.   ⊣

Theorem 9. *Let (T^H, μ^H, η^H) be a standard iterative monad which is free on a uniform functor H. Then T^H is also uniform.*

Proof In this proof we have the monad of H and also the monad of $\mathcal{P}$. As in our statement, we write the data coming from the first free completely iterative monad with the superscript H, and we continue our practice of dropping the superscripts on the free completely iterative monad of $\mathcal{P}$.

We know that T^H is standard by Lemma 5. Let $\pi : H \to T$ be a uniformity for H. Let $\hat{\pi} : H \to \mathcal{P}T$ be such that $\pi = \tau \cdot \hat{\pi}$. By the fundamental freeness theorem of Aczel et al. [2], there is a unique ideal monad morphism $\pi^* : T^H \to T$ such that $\pi = \pi^* \cdot \kappa$. We check that $\chi \cdot \pi_V^* = i_{T^H V}$.

Consider the following diagram:

$$
\begin{array}{c}
T^H V \xrightleftharpoons[{[\tau_V^H, \eta_V^H]}]{\alpha_V^H} HT^H V + V \xrightarrow{\hat{\pi}_{T^H V} + V} \mathcal{P} T T^H V + V \xrightarrow{\tau_{T^H V} + V} T T^H V + V
\end{array}
$$

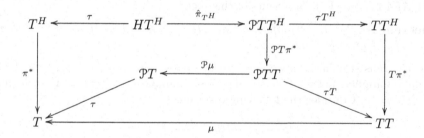

(4)

We claim that both halves commute. The bottom uses Proposition 5. For the top, it is best to begin at $HT^H V + V$ and argue separately for the two components. The right component commutes due to the fact that π^* is a monad morphism; specifically, $\pi^* \cdot \eta^H = \eta$. The left component is more involved. We drop V and consider the following diagram in the endofunctor category:

For the hexagonal region in the upper left, we appeal to Lemma 6.10 of [16]. The region on the right commutes by naturality of τ. The bottom square is an instance of Proposition 1.

At this point we know that (4) commutes. We conclude that $g = \chi \cdot \pi_V^*$ satisfies

$$
[\chi, V] \cdot Tg \cdot (\pi_V + V) \cdot \alpha_V^H = g.
$$

By our Solution Lemma 2, there is exactly one g which satisfies this equation. We check that $i_{T^H V}$ also satisfies it. We note that the diagram below commutes:

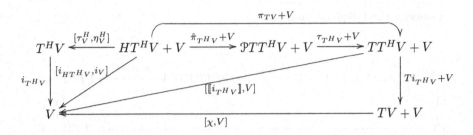

In the topmost region, we have used the fact that $\pi = \tau \cdot \hat{\pi}$. To see that the triangle on the left commutes, recall that $\alpha_V^H = [\tau_V^H, \eta_V^H]^{-1}$ is the identity and that $i_{HT^HV+V} = [i_{HT^HV}, i_V]$. We have used the fact that π is uniform in the middle region, and on the right we have the definition of $[\![i_{T^HV}]\!]$.

This concludes the proof that $\chi \cdot \pi_V^* = i_{T^HV}$. $\dashv$

6   Consequences of Uniformity

Theorem 10 below is an adaptation of the analogous result from [18], and ultimately the ideas come from Turi [21], following Aczel [1]. We remind the reader that *AFA* is needed in the results of this section.

Theorem 10. *Let H be uniform. Then (H^*, id) is a final H-coalgebra, where H^* is the greatest fixed point of H.*

Proof H is standard, so we may use Lemma 6. Since we are assuming *AFA*, Lemma 3 applies. Let $e : b \to Hb$. There is a unique $s = f^\dagger : b \to V$ such that $s = [\![s]\!] \cdot \pi_b \cdot e$. Consider the following diagram.

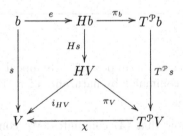

All the parts clearly commute except the left, and thus this does commute. This part shows that $s = i_{HV} \cdot Hs \cdot e$. For the uniqueness, note that s with our desired property determines a solution to $\pi_b \cdot e$. $\dashv$

Corollary 1. *If H is uniform, then H generates a standard iterative monad by taking for each class a, $Ta = (H + G_a)^*$, the greatest fixed point of $H + G_a$.*

Proof We know that for all sets a, $H + G_a$ is uniform and standard. So the result follows from Theorem 10. $\dashv$

As shown in Milius [14], if H is any iteratable functor (on any category with $+$) and T and τ are from its free completely iterative monad, then $(TA, \tau_A : HTA \to TA)$ is always a cia for H. The next fact does not follow from Milius' result.

Theorem 11. *If H is uniform, then (V, i_{HV}) is a cia for H.*

Proof   The proof is virtually the same as that of Theorem 10, so we merely indicate the idea and exhibit the diagram. Let $e : X \to HX + V$. Consider the diagram below:

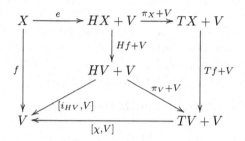

The map f comes from Lemma 2 applied to $(\pi_X + V) \cdot e$. The rest of the proof is the same. $\dashv$

7   A Variation: λ-Uniformity

For each cardinal λ, consider the functor $\mathcal{P}_\lambda$ giving the set of subsets of size less than λ: by

$$\mathcal{P}_\lambda(s) \quad = \quad \{t \subseteq s : |t| < \lambda\}.$$

We have a natural transformation $n_\lambda : \mathcal{P}_\lambda \to \mathcal{P}$ whose components are the evident inclusions.

Proposition 7. *$(V, i_{\mathcal{P}_\lambda V})$ is a cia for $\mathcal{P}_\lambda$. Specifically, given a flat equation morphism $e : X \to \mathcal{P}_\lambda X + V$, we have a flat equation morphism for $\mathcal{P}$: $((n_\lambda)_X + V) \cdot e$. The solution to these two are the same morphism.*

The natural transformation $\nu_\lambda : \mathcal{P}_\lambda \to T$ shows $\mathcal{P}_\lambda$ to be uniform, where $\nu_\lambda = \tau \cdot \mathcal{P}\eta \cdot n_\lambda$. That is, it is ideal, and

$$\chi \cdot (\tau \cdot \mathcal{P}\eta \cdot n_\lambda)_V \quad = \quad (\chi \cdot (\tau \cdot \mathcal{P}\eta)_V) \cdot (n_\lambda)_V \quad = \quad i_{\mathcal{P}V} \cdot i_{\mathcal{P}_\lambda V, \mathcal{P}V} \quad = \quad i_{\mathcal{P}_\lambda V}.$$

We have used the calculation in Example 3.

As a result, these functors $\mathcal{P}_\lambda$ generate standard iterative monads on the category of classes by taking greatest fixed points. Moreover, these functors have a property that $\mathcal{P}$ does not have: as functors on Set, they have final coalgebras. Indeed, the greatest fixed point of $\mathcal{P}_\lambda$ is the set H_λ of sets x such that $|tc(x)| < \lambda$, where $tc(x)$ is the transitive closure of x.[3] For λ an infinite regular cardinal, this is the same thing as saying that $|x| < \lambda$, and every $y \in tc(x)$ also has cardinality $< \lambda$. The properties of this collection H_λ are sensitive to the underlying set theory. But assuming either the Foundation or Anti-Foundation Axioms, it is a set and not just a class. As a result, the functors $\mathcal{P}_\lambda$ determine standard iterative monads T_λ on Set. We use a subscript λ to indicate the data from this monad.

Proposition 8. *The inclusion $i_{H_\lambda} : H_\lambda \to V$ is a morphism of cias for $\mathcal{P}_\lambda$.*

Using the freeness theorem of Aczel et al. [2], there is a unique ideal monad morphism $\overline{\nu_\lambda} : T_\lambda \to T$ such that $\nu_\lambda = \overline{\nu_\lambda} \cdot \tau_\lambda \cdot \mathcal{P}\eta_\lambda$. All of the components of $\overline{\nu_\lambda}$ are inclusions.

We rework the results of Sections 4 and 5 by replacing $\mathcal{P}$ by $\mathcal{P}_\lambda$ throughout. The first step is to comment on the morphisms $[\![f]\!]_\lambda$ associated to morphisms $f : B \to H_\lambda$. We define $[\![f]\!]_\lambda : T_\lambda B \to H_\lambda$ by $a^\# \cdot T_\lambda f$, where $a^\# : T_\lambda H_\lambda \to H_\lambda$ is the solution to $\alpha_A : T_\lambda \to H_\lambda T_\lambda + H_\lambda$ in the cia (H_λ, id).

Proposition 9. *For all sets B and all functions $f : B \to H_\lambda$, the diagram below commutes:*

$$
\begin{array}{ccc}
T_\lambda B & \xrightarrow{\;[\![f]\!]_\lambda\;} & H_\lambda \\
{\scriptstyle (\overline{\nu_\lambda})_B}\big\downarrow & & \big\downarrow{\scriptstyle i_{H_\lambda}} \\
TB & \xrightarrow[\;[\![i_{H_\lambda} \cdot f]\!]\;]{} & V
\end{array}
$$

Proof   We consider the following diagram:

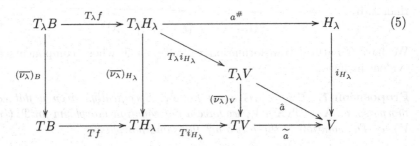

$$(5)$$

The leftmost two squares commutes by naturality. The morphism $\hat{a} : T_\lambda V \to V$ is the solution of the flat equation morphism $\alpha_V : T_\lambda V \to \mathcal{P}_\lambda T_\lambda V + V$. The square on the right takes an argument. Let G be the functor $a \mapsto \mathcal{P}_\lambda a + H_\lambda$. Then the greatest fixed point G^* gives a final coalgebra with the identity as structure map. Recall Lemma 6 for G and $\alpha_V : T_\lambda V \to G(T_\lambda V)$. We check that both $i_{H_\lambda} \cdot a^\#$ and $\hat{a} \cdot T_\lambda i_{H_\lambda}$ are solutions of α_V. The verifications are easy and we omit them.

---

[3] We do apologize for any notational confusion that could result from our use of H for a functor and to designate an operation from cardinals to sets.

The commutativity of the triangle also takes an argument. We show that $\tilde{a} \cdot (\overline{\nu_\lambda})_V$ has the property which uniquely defines $\hat{a}$; that is, that it is a solution to α_V. By Proposition 7, we only need to show that $\tilde{a} \cdot (\overline{\nu_\lambda})_V$ is a solution to $T_\lambda V \to \mathcal{P}T_\lambda V + V$. But this follows from the *compositionality identity* (see, e.g., [14]) and the fact that $(\overline{\nu_\lambda})_V$ reorganizes the flat morphism $T_\lambda V \to \mathcal{P}T_\lambda V + V$ to $TV \to \mathcal{P}TV + V$.

The commutativity of the outside of (5) implies the result of this lemma, in view of the definitions of $[\![f]\!]_\lambda$ and $[\![i_{H_\lambda} \cdot f]\!]$.   ⊣

In the definition and results below, we recall that a standard functor $H :$ Set $\to$ Set extends to a standard endofunctor on classes. We identify the two functors.

Definition 4. H *is* λ-*uniform if* H *is uniform via some* $\pi : H \to \mathcal{P}$ *such that for all sets* a,

$$[\![i_{a,H_\lambda}]\!]_\lambda \cdot \pi_a \quad = \quad i_{Ha,H_\lambda}.$$

Proposition 10. H *is* λ-*uniform iff there is some uniformity* $\rho : H \to T$ *which factors through* $\overline{\nu_\lambda}$.

Proof   First, suppose that H is λ uniform via $\pi : H \to T_\lambda$. Let $\rho = \overline{\nu_\lambda} \cdot \pi$. Then the diagram below shows that for all sets a, $[\![i_a]\!] \cdot \rho_a = i_{Ha}$.

We used Proposition 9. In the other direction, suppose that ρ is a uniformity for H which factors as $\rho = \overline{\nu_\lambda} \cdot \pi$. Then the lower passage above is an inclusion. So since i_{H_λ} is an inclusion, so is $[\![i_{a,H_\lambda}]\!]_\lambda \cdot \pi_a$.   ⊣

This shows that if H is λ-uniform, then H is uniform. And it is also easy to check that if $\lambda < \kappa$ and H is λ-uniform, then H is κ-uniform.

As we mentioned above, the results of this paper which we established for our notion of uniformity may be reworked for the refined versions of λ-uniformity. For example, the version of Theorem 10 gives the following result.

Proposition 11. *If* H *is* λ-*uniform, then* $H^* \subseteq H_\lambda$. *In particular, there is a final coalgebra for* H *which is a subset of the set of sets of hereditary cardinality* $< \lambda$.

Turning to the closure properties of the collection of λ-uniform functors, the main point is that then the constant functor with value w is λ-uniform iff $w \in H_\lambda$. And we see that any functor built from constants $w \in H_\lambda$, $\mathcal{P}_\lambda$, product, and coproduct has a final coalgebra which is a set and moreover is a subset of H_λ. This final result gives an application of our work to the topic of bounded functors on **Set**.

8 Concluding Remarks

The main point of this paper has been to rework the theory of uniformity using some of the machinery introduced in coalgebraic recursion theory in past years, including the notions of a completely iterative monad and a completely iterative algebra. As we have seen, there are two different intuitions at work, two different goals for the study. In a sense, one wants to find functors with the nice property that their greatest fixed points are final coalgebras, and then the technical details lead one to propose definitions that are about functors working by a general form of substitution.

As it happens, the notions of uniformity that attempt to get at the intuition that a functor is determined "by substitution" in some sense single out a smaller class than those which give final coalgebras by considering greatest fixed points. The referee to this paper mentions the functor which maps each set into the set of all its finite multisets as an example. I shall work instead with the distribution functor $\mathcal{D}$ on sets given by $\mathcal{D}(X)$ is the set of all finite partial functions μ from X to $(0,1]$ such that $\Sigma_{x \in X}\, \mu(x) = 1$. (Equivalently, one may work with total function which whose value is 0 at all by finitely many points. However, this alternative would not define a monotone functor.) On morphisms, $\mathcal{D}$ works by marginalization (summing). The details are technical but it seems intuitively clear that $\mathcal{D}$ is not uniform in our sense (or under any definition stated in terms of natural transformations and maps like χ). At the same time, it is the case that the greatest fixed point of $\mathcal{D}$ is a final coalgebra with the identity, and the universe is a cia for it with the inclusion. For $\mathcal{D}$ itself, this is easy to see as $\mathcal{D}^*$ is a singleton $x = \{\{(x,1)\}\}$. Things are more interesting for variants such as $H(x) = \mathcal{D}(x) + A$ for a fixed set A. We show by example that (H^*, id) is a final coalgebra, invoking Lemma 6. Let $b = \{w, x, y, z\}$, let $a \in A$, and consider $e : b \to Hb$ given on the left below:

$$
\begin{array}{llllll}
e(w) & = & \text{inl } \{(x,1/3),(y,1/3),(z,1/3)\} & f(w) & = & (0, \{(x,2/3),(z,1/3))\} \\
e(x) & = & \text{inl } \{(x,1)\} & f(x) & = & (0, \{(x,1)\}) \\
e(y) & = & \text{inl } \{(y,1)\} & & & \\
e(z) & = & \text{inr } a & f(z) & = & (1, a)
\end{array}
$$

To get the desired $s : b \to V$, we must identify x and y (since they are bisimilar in e); this is the reason why uniformity in the sense of this paper fails). We do this in the system f. This system has a unique solution s^*, by standard techniques. We then extend s^* to the desired s by $s(y) = s^*(x)$.

The results here extend to show that every functor built from $\mathcal{D}$ and the polynomial-forming operations (except of course for the identity functor) has the properties of interest in this paper. One can even imagine re-working the definition of uniformity in this paper to allow $\mathcal{D}$ and related functors to be uniform. However, doing this in an ad hoc manner gives no insight to help with a search for the most general uniformity notion.

Acknowledgments

I thank Jiří Adámek and Stefan Milius for several conversations on this topic. I also thank them and anonymous referee for corrections of errors, and for many suggestions which improved this paper.

References

1. Peter Aczel, *Non-Well-Founded Sets*. CSLI Lecture Notes Number 14, CSLI Publications, Stanford, 1988.
2. Peter Aczel, Jiří Adámek, Stefan Milius, Jiri Velebil, "Infinite trees and completely iterative Theories: a coalgebraic view." *Theoretical Computer Science*, 300 (2003), 1-45.
3. Peter Aczel, Jiří Adámek, and Jiří Velebil, "A coalgebraic view of infinite trees and iteration," Electronic Notes in Theoretical Computer Science 44.1 (2001).
4. Peter Aczel and Nax Mendler, "A final coalgebra theorem", in D. H. Pitt et al (eds.) *Category Theory and Computer Science*, Springer-Verlag, Heidelberg, 1989, 357–365.
5. Jiří Adámek, Stefan Milius, and Jiří Velebil, "On coalgebra based on classes," *Theoretical Computer Science* 316 (2004), no. 1-3, 3–23.
6. Jiří Adámek, Stefan Milius, and Jiří Velebil, "Elgot algebras," preprint, 2005.
7. Jiří Adámek and Věra Trnková, *Automata and Algebras in Categories*. Kluwer Academic Publishers Group, Dordrecht, 1990.
8. Jon Barwise and Lawrence Moss, *Vicious Circles*. CSLI Lecture Notes Number 60, CSLI Publications, Stanford, 1996.
9. Daniela Cancila, Ph.D. Dissertation, University of Udine Computer Science Department, 2003.
10. Daniela Cancila, Furio Honsell, and Marina Lenisa, "Properties of set functors," in F. Honsell et al. eds., *Proceedings of COMETA'03*, ENTCS, 104 , 2004, pp. 61-80.
11. Keith Devlin, *The Joy of Sets*, second edition. Springer-Verlag, 1993.
12. Peter Freyd, "Real coalgebra", post on categories mailing list, 22 December 1999, available via www.mta.ca/~cat-dist.
13. Azriel Levy, *Basic Set Theory*, Springer-Verlag 1979.
14. Stefan Milius, "Completely iterative algebras and completely iterative monads," *Inform. and Comput.* 196 (2005), 1–41.
15. Stefan Milius, Ph.D. Dissertation, Institute of Theoretical Computer Science, Technical University of Braunschweig, 2005.
16. Stefan Milius and Lawrence S. Moss, "The category theoretic solution of recursive program schemes," in J. L. Fiadero et al (eds.) the *Proceedings of CALCO 2005*, Springer LNCS 3629, 2005, 293–312.
17. Lawrence S. Moss, "Coalgebraic logic," *Annals of Pure and Applied Logic* 96 (1999), no. 1-3, 277–317.
18. Lawrence S. Moss, "Parametric corecursion," *Theoretical Computer Science* 260 (1–2), 2001, 139–163.
19. Lawrence S. Moss and Norman Danner, "On the foundations of corecursion," *Logic Journal of the IGPL* Vol. 5, No. 2 (1997) pp. 231–257.
20. J.J.M.M. Rutten, "Universal coalgebra: a theory of systems," *Theoretical Computer Science* 249(1), 2000, pp. 3-80.

21. Daniele Turi, *Functorial Operational Semantics and its Denotational Dual*, (Ph.D. thesis, CWI, Amsterdam, 1996).
22. Daniele Turi and J.J.M.M. Rutten, "On the foundations of final semantics: non-standard sets, metric spaces, partial orders," *Mathematical Structures in Computer Science* 8 (1998), no. 5, 481–540.

An Algebraic Approach to Regular Sets

Horst Reichel

Institut für Theoretische Informatik
Technische Universität Dresden
D–01062 Dresden, Germany
reichel@tcs.inf.tu-dresden.de

Abstract. In recent years an increasing interest in regular sets for different kinds of elements could be observed. The introduction of XML has led to investigations of regular sets of both ranked and unranked trees and also of attributed unranked trees.

The aim of this short note is to introduce a uniform notion of regularity. If instantiated for strings, ranked trees and unranked trees it will coincide with the existing concepts and it can easily be extended to arbitrary data types. This leads to a natural notion of regularity for different kinds of attributed unranked trees and also to regular sets of structured elements which have not yet been investigated. The approach takes advantage from freeness constraints and parametric abstract data types as offered by the algebraic specification language CASL

1 Introduction

It is well known that strings and ranked trees can be interpreted as ground terms of suitable ranked signatures (alphabets).

In the case of strings over a finite alphabet Σ the corresponding signature Ω_{String} consists of a constant ε and a unary operation for each letter $x \in \Sigma$. For each Ω_{String}–algebra $\mathbb{A}$ there exists a unique homomorphism $f_{\mathbb{A}} : \mathbb{T}(\Omega_{String}) \to \mathbb{A}$ where $\mathbb{T}(\Omega_{String})$ denotes the free term algebra for the signature Ω_{String}. For each term $t = \varepsilon x_1 \ldots x_n$ the homomorphism $f_{\mathbb{A}}$ maps t to the evaluation of t in $\mathbb{A}$. Now it is folklore that a subset $L \subseteq T(\Omega_{String})$ is regular if and only if there is a finite Ω_{String}–algebra $\mathbb{A}$ with a distinguished subset $A_0 \subseteq A$ such that

$$L = f_{\mathbb{A}}^{-1}(A_0).$$

This means that L is the homomorphic inverse image of an accepting set A_0 of states of the finite automaton $\mathbb{A}$.

It has been shown by Thatcher at that a corresponding characterization holds for regular sets of ranked trees.

It turns out that for the intended generalization it will be more convenient to work with *partial algebras* and *weak homomorphisms* between partial algebras.

In order to define unranked trees and other types of structured data, we will work with more general algebraic structures as usually used in Universal Algebra. The generalizations concern the domains of fundamental operations. In

K. Futatsugi et al. (Eds.): Goguen Festschrift, LNCS 4060, pp. 449–458, 2006.
© Springer-Verlag Berlin Heidelberg 2006

a many–sorted framework the domains of fundamental operations are assumed
to be products of sorts and the codomain to be one of the given sorts. We will
allow that the domain of a fundamental operation can be an abstract data type
on the given sorts. Such structures arise naturally if one works with many–sorted
algebras and freeness constraints which are basic in the algebraic specification
language CASL .

One example for this more general concept is given by *list–algebras*. For list–
algebras the domain of a fundamental operation may be the set of all finite lists
with elements out of the carrier set of the list–algebra.

Let us have a first look at the simplest case of list–algebras. We consider the
signature with just one sort s and one operation symbol of type $c : s\_lists \rightarrow$
s. What about the ground terms generated by that signature? Since there
is the empty list, denoted by [], there is the ground term $c([])$. This ground
term can be used to build for instance the list $[c([]), c([]), c([])]$ consisting of
three copies of the previously constructed ground term. This list yields a new
ground term $c([c([]), c([]), c([])])$. Similarly one could construct $c([c([]), c([])])$ and
$c([c([]), c([c([]), c([])]), c([])])$ and so on. Evidently, the ground terms represent the
construction of list of list of . . . list of the empty set which may also be be seen as
finite unranked ordered trees. This corresponds to the well known specification
of finite unranked ordered trees as an inductively defined data type.

It is well known that regular sets of finite unranked ordered trees can also
be characterized as inverse homomorphic images. But, taking only finite list–
algebras would lead to a more general concept. For the characterization of regular
sets of finite unranked trees one has to use an additional property which leads to
so–called *regular list–algebras*, where a list–algebra is called regular, if the set of
lists (sequences) of elements mapped by the fundamental operation to one and
the same element is always a regular set of lists (sequences). A finite regular list–
algebra is a finite deterministic bottom–up tree automaton in the terminology
of [1]. Therefore, in the following we will also speak of states if we talk about
elements of finite regular algebras.

This encourages us to call a subset of an inductively defined data type regular
if it is the inverse homomorphic image of an accepting subset of a finite regular
algebra (of a corresponding generalized type). In this way, the definition and
investigation of regular expressions can very generally be based on operations
on classes of finite regular algebras. This leads to a uniform view of regular
expressions for different types of structured data.

It is worth mentioning that finite regular algebras coincide with finite algebras
in the case of traditional algebras for ranked alphabets.

This short note may be seen as a straightforward generalization of J.W.
Thatcher's work on tree automata. This generalization does not include regularity for sets of infinite data structures like streams or infinite lists.

2   Algebraic Operations with Structured Domains

Traditional algebraic structures use on the meta level only the type constructors of Cartesian products. Algebraic structures in the framework of category theory use for typing arbitrary endofunctors $T : Set \to Set$ and define an algebra as a pair $(A, \alpha : T(A) \to A)$. In the case of many–sorted algebras one has to use endofunctors $T : Set^S \to Set^S$ with a finite set S of sort names.

In this note we will work with endofunctors that can be built up by finite Cartesian products and by generic free data types in the sense of the algebraic specification language CASL [4]. This requirement rules out for instance the powerset functor, but it allows the powerset functor $\mathcal{P}_\omega(\_)$ of finite subsets. This means that we use ideas and concepts which came up very early within the theory of abstract data type, see for instance [2], [8], [5], [6] and [7].

The following are some examples of *extended signatures* for these more general algebras:

```
SIG ListAlgebras IS
   SORTS  s
   OPS    c : s_lists ---> s
END

SIG Attributed1ListAlgebras IS
   SORTS  s1, s2
   OPS    c : s2 x s1_lists ---> s1
END

SIG SetAlgebras IS
   SORT   s
   OPS    c : s_sets ---> s
END

SIG Attributed2ListAlgebras IS
   SORTS  s1, s2
   OPS    c : s1_lists x s2_lists ---> s1
END
```

An algebra $\mathbb{A} = (A_{s1}, A_{s2}; c_{\mathbb{A}})$ for the signature Attributed2ListAlgebra is then given by an arbitrary set A_{s1}, the interpretation of the sort name s1, a second set A_{s2}, the interpretation of the sort name s2, and a mapping $c_{\mathbb{A}} : A_{s2}^* \times A_{s1}^* \to A_{s1}$. Accordingly an algebra $\mathbb{A} = (A_s; c_{\mathbb{A}} : \mathcal{P}_\omega(A_s) \to A_s)$ for the extended signature SetAlgebras is given by the interpretation of the sort name s and the interpretation of the operation symbol c which assigns to each finite subset of the carrier set an element of the carrier set.

Since we want to define regular subsets of intial algebras of extended signatures, we will first have a look at the intial algebras of the given extended signatures.

As described in the introduction the initial algebra of the extended signature `ListAlgebras` represents finite ordered unranked trees. However, the initial algebra of the signature `Attributed1ListAlgebras` is given by the empty set for both sort names. But, the intended meaning of the signature is the set of finite ordered unranked trees whose nodes are labelled with elements out of the interpretation of `s2`. By the same reason the initial algebra of the signature `Attributed2ListAlgebras` differs from the intended meaning. In that case the nodes should be labelled with lists of elements out of the interpretation of the sort name `s2`. In both cases the difference is caused by the fact that the initial algebras interpret the sort name `s2` by the empty set.

Finally, we see that the initial algebra of the signature `SetAlgebras` represents finite unordered and repetition free trees.

The problems described above can be solved by the use of *parameteric extended signatures*. If one uses the sort name `s2` as a parameter then each interpretation of this sort name produces an instantiated extended signature. Now, for each nonempty interpretation of the sort parameter the initial algebras of the instantiated extended signatures represent the intended meaning.

Parameterized extended signatures are just syntactic sugar for the representation of families of extended signatures. If one interprets the sort parameter `s` by a set M then the instantiated extended signature results by adding `s` as a sort symbol each element of M as a constant operation symbol `m : s`, and one has to fix the interpretation of the sort name `s` to the set M.

In the following we will work with one instance of the parametric version of `Attributed2ListAlgebra` where the parameter sort `s2` is instantiated by the alphabet $\{a, b, A, B\}$. The resulting extended signature is

```
SIG ALAlg IS
    SORTS  s,
    OPS    c : {a,b,A,B}* x s_lists ---> s
END
```

3   Regular Subsets of Initial Extended Algebras

In the case of strings and finite ranked trees regular subsets can be characterized as inverse homomorphic images of subsets of finite algebras. The example of finite ordered unranked trees shows that in general finite algebras are not sufficient to characterize regular subsets by inverse homomorphic images. More specific finite algebras are needed.

Definition 3.1. Let `Sig` be an extended signature such that for all type constructors used in definitions of the types of domains of the operation symbols the notion of regular subsets is known. Let $\mathbb{A}$ be a finite algebra for the given signature. The finite algebra is called *regular* if for each operation and each element of a carrier the inverse image of that element is a regular subset of the domain.

If the used type constructors of a signature preserve finite sets, then evidently each finite algebra is regular. Therefor, the concept of regularity of finite algebras is not needed in case of regular strings (lists) or finite ranked trees.

Definition 3.2. Let $\mathtt{Sig}$ be an extended signature such that for all type constructors used in definitions of the types of domains of the operation symbols the notion of regular subsets is known. A subset of the initial $\mathtt{Sig}$–algebra $\mathbb{T}(\mathtt{Sig})$ is called *regular* if it is the inverse homomorphic image of a subset of a finite regular $\mathtt{Sig}$–algebra.

The extended signatures above use only products and lists as type constructors. Therefore, this definition can be used to define regular sets for the defined types of trees.

In a next step trees could be used as typ constructors in order to define other structured data types. Definition 3.2 provides then notions of regular sets for the resulting structured data types. Since each interesting data type can be specified by freeness constraints using only finitely many auxiliary data types, also defined by freeness constraints, Definition 3.2 allows to define the concept of regular sets for all interesting data type, using a suitable hierarchy of type definitions by freeness constraints.

Since $\mathtt{Attributed1ListAlgebras}$–algebras are deterministic bottom–up automata as introduced in [1], the notion of regular sets for the extended signature $\mathtt{Attributed1ListAlgebras}$ coincides with the notion of tree regular languages according Definition 2.14 in [1].

Let us apply Definition 3.2 to the extended signature $\mathtt{ALAlg}$. A finite regular $\mathtt{ALAlg}$–algebra $\mathbb{A} = (A_s; c_\mathbb{A})$ assigns to pair of a finite list of elements out of A_s and a finite list of elements out of $\{a, b, A, B\}$ an element in A_s.

The unique homomorphism from the initial $\mathtt{ALAlg}$–algebra to a specific finite regular $\mathtt{ALAlg}$–algebra

$$\mathbb{A} = (A_s, ; c_\mathbb{A})$$

defines a *classification of the trees* where each class is given by the inverse homomorphic image of an element in A_s. The basic operation $c_\mathbb{A}$ assigns to each class a regular set over $\{a, b, A, B\}$ whose elements can be used as attributes for the root of trees out of the corresponding class.

We will illustrate this by an example:

Example 3.1: Let $\mathbb{E} = (E_s; c_\mathbb{E})$ be given by:

$$E_s = \{s_0, s_1, s_2, s_3, s_d\}$$
$$\text{for } (w, l) \in (\{a, b, A, B\}^* \times A_s^*) :$$

$$c_\mathbb{E}(w, l) = \begin{cases} s_0 & \text{if } w \in L(a^*) & l \in \{nil\} \\ s_1 & \text{if } w \in L(b^*) & l \in \{s_0 s_0, s_0 s_0 s_0\} \\ s_2 & \text{if } w \in L((A + B)a^*) & l \in \{s_1 s_1, s_0 s_1\} \\ s_3 & \text{if } w \in L((A + B)b^*) & l \in \{s_2 s_2 s_2, s_1 s_2 s_3\} \\ s_d & \text{else} & \text{else} \end{cases}$$

and let us assume that $\{s_3\} \subseteq E_s$ is the set of accepting states.

Then we have four interesting classes of attributed trees, defined by the inverse homomorphic images of s_0, s_1, s_2, s_3 respectively, and the complement of the union of this classes, represented by s_d.

s_0 represents the class of all trees with exactly one node labelled with a string $w \in L(a^*)$.

s_1 represents the class of trees with two or three sons, classified by s_0 and the root is labelled with a string $w \in L(b^*)$.

s_2 represents the class of trees with two sons such that the first is classified by s_1 or s_0 and the second son is classified by s_1. Finally the root is labelled with a string $L((A + B)a^*)$.

Finally, the accepting state s_3 represents the class of trees with exactly three sons, where either all of them are classified by s_2 or the first son is classified by s_1, the second by s_2 and the third again by s_3. The root is labelled by a string $w \in L((A + B)b^*)$.

The example shows that the explicit definition of the basic operation of a finite regular algebra for the extended signature ALAlg has great similarity with a tree grammar, where the elements act as meta variables.

The example shows also another aspect, it is basically a partial ALAlg–algebra with a one–point completion given by the state s_d. In terms of automata this completion point is a trap. If the computation once reaches the trap it can never leave it.

4   Regular Expressions Based on Colimits

The investigation of regular expressions can now be based on colimits in suitable categories. The well known interpretation of the operations of regular expression as operations on finite automata can now be extended to operations on finite regular extended algebras.

We will illustrate this by means of the category of finite partial regular ALAlg–algebras as objects and weak homomorphisms as morphisms.

Weak homomorphisms preserve the applicability of the basic operations but do not necessarily reflect this property. To be more precise, let $\mathbb{A}, \mathbb{B}$ be partial algebras. Then $c_{\mathbb{A}} : \{a, b, A, B\}^* \times A_s^* \to? A_s$ and $c_{\mathbb{B}} : \{a, b, A, B\}^* \times B_s^* \to? B_s$ are partial mappings. A total mapping $f : A_s \to B_s$ is a weak homomorphism if for each $(x, y) \in dom(c_{\mathbb{A}})$ the pair $(x, f^*(y)) \in dom(c_{\mathbb{B}})$ and $f(c_{\mathbb{A}}(x, y)) = c_{\mathbb{B}}(x, f^*(y))$, where $f^* : A_s^* \to B_s^*$ is the canonic extension to lists.

From category theory it is known that arbitrary finite colimits exist if sum and coequalizer exist [3]. Therefore, the most interesting constructions on algebras are *summation* and *quotient construction*.

Before we study this construction in detail we introduce a notation. For a given finite partial regular ALAlg–algebra $\mathbb{A}$ and a given subset $X \subseteq A_s$ of accepting states $L(\mathbb{A}, X)$ denotes the regular set of those ground terms for the extended signature ALAlg which can be evaluated in $\mathbb{A}$ to a value out of X. To be more formal, if $\mathbb{A}^t$ denotes the one–point completion of $\mathbb{A}$ and

$$f_{\mathbb{A}^t} : \mathbb{T}(\text{ALAlg}) \to \mathbb{A}^t$$

the unique homomorphism then we define

$$eval(\mathbb{A}) = \{t \in T(\texttt{ALAlg})|\quad f_{\mathbb{A}^t}(t) \in A_s\},$$
$$L(\mathbb{A}, X) = \{t \in T(\texttt{ALAlg})|\quad f_{\mathbb{A}^t}(t) \in X \subseteq A_s\}.$$

With this notation it is easy to see that the empty set of ground terms and the set of all ground terms are both regular. For the empty set one takes an algebra $\mathbb{A} = (A_s; c_\mathbb{A})$ where there is no $w \in \{$a,b,A,B$\}*$ with $(w, nil) \in dom(c_\mathbb{A})$ and for the second case one takes the total one–element algebra where the only element is also an accepting one.

For a finite partial regular algebra $\mathbb{A} = (A_s; c_\mathbb{A})$ we call an element $c_\mathbb{A}(w, nil)$, if it exists, an *initial state* of $\mathbb{A}$.

In the following we describe the sum of two algebras $\mathbb{A}, \mathbb{B}$. First we take the disjoint unions $A_s + B_s$. Let $in_A : A_s \to A_s + B_s, in_B : B_s \to A_s + B_s$ be the injections. $A_s + B_s$ becomes the carrier of $\mathbb{A} + \mathbb{B}$. The basic operation $c_{\mathbb{A}+\mathbb{B}}$ is defined as follows

$$c_{\mathbb{A}+\mathbb{B}}(w, l) = \begin{cases} c_\mathbb{A}(w, l) & \text{if } l = in_A^*(l') \text{ and } (w, l') \in dom(c_\mathbb{A}) \\ c_\mathbb{B}(w, l) & \text{if } l = in_B^*(l') \text{ and } (w, l') \in dom(c_\mathbb{B}) \\ \text{undefined} & \text{else} \end{cases}$$

It is a matter of routine to show that this construction gives a sum in the category of finite partial regular $\texttt{ALAlg}$–algebras with weak homomorphisms.

Summation can be used to show that the union of two regular subsets is again regular. If $X \subseteq A_s, Y \subseteq B_s$ are given sets of accepting states in $\mathbb{A}, \mathbb{B}$ respectively and $X \uplus Y$ denotes the *union* of their embeddings in $\mathbb{A} + \mathbb{B}$ then

$$L(\mathbb{A}, X) \cup L(\mathbb{B}, Y) = L(\mathbb{A} + \mathbb{B}, X \uplus Y).$$

That regular subsets in $T(\texttt{ALAlg})$ are closed under *intersection* can easily be seen by means of the Cartesian product of finite partial regular algebras:

$$L(\mathbb{A}, X) \cap L(\mathbb{B}, Y) = L(\mathbb{A} \times \mathbb{B}, X \times Y).$$

It is even simpler to see that the *complement* of a regular subset is regular. One takes just the complement of the accepting subset on states in a finite (total) regular algebra which represents the given regular subset and one gets the wanted algebra for the complement.

Above we have seen that the sum of algebras represents the union of regular sets and the sum of regular expressions. What about the composition and the star–operation of regular expressions. The semantics of these operations can be reduced to quotient construction of finite partial regular algebras. It is sufficient to define *how states can be fused together*.

Let $\mathbb{A} = (A_s; c_\mathbb{A})$ be a given partial algebra and $x, y \in A_s$ two elements. We define the quotient algebra $\mathbb{A}^{x \rightsquigarrow y}$ which results from $\mathbb{A}$ by *fusing x with y* as follows.

Definition 4.1: An equivalence relation $R \subseteq A_s \times A_s$ in the carrier set of a partial algebra $\mathbb{A}$ is called a *congruence* if for all $w \in \{a, b, A, B\}^*$

and all $(a_1, a'_1) \in R, \ldots, (a_n, a'_n) \in R, n \geq 0$ if $c_\mathbb{A}(w, [a_1, \ldots, a_n]) = a$ and $c_\mathbb{A}(w, [a'_1, \ldots, a'_n]) = a'$ then $(a, a') \in R$.

Definition 4.2: For a given partial **ALAlg**–algebra $\mathbb{A}$ and a congruence R in $\mathbb{A}$ the *quotient* $\mathbb{A}/R$ has the quotient set A_s/R as carrier and

$$c_{\mathbb{A}/R}(w, [\langle a_1 \rangle_R, \ldots, \langle a_n \rangle_R]) = \langle a \rangle_R$$

if there are representatives $a'_i \in \langle a_i \rangle_R$ for $i \in \{1, \ldots, n\}$ with

$$c_\mathbb{A}(w, [a'_1, \ldots, a'_n]) = a$$

where $\langle x \rangle_R$ denotes the congruence class containing x.

Since congruences in the sense of Definition 4.1 are closed under intersections there exists for each set X of pairs the smallest congruence containing X which is denoted by R_X. R_X is also called the *congruence generated by* X.

The algebra $\mathbb{A}^{x \leftrightsquigarrow y}$ can now be defined by

$$\mathbb{A}^{x \leftrightsquigarrow y} = \mathbb{A}/R_{\{(x,y)\}}.$$

Iterated application of this construction leads to the identification of a finite set $\{x_1, \ldots, x_n\}$ of states with the state y or of identifying x_1 with $y_1 \ldots x_n$ with y_n. The resulting quotient algebra will be denoted by

$$\mathbb{A}^{\{x_1, \ldots, x_n\} \leftrightsquigarrow y} \quad \text{and} \quad \mathbb{A}^{\{x_1 \leftrightsquigarrow y_1, \ldots, x_n \leftrightsquigarrow y_n\}}$$

respectively.

For a given **ALAlg**–algebra $\mathbb{A}$, a set $X \subseteq A_s$ of accepting states and a congruence relation R the set of accepting states in $\mathbb{A}/R$ is given by $\{\langle x \rangle_R \mid x \in X\}$.

By means of the introduced quotient construction on partial **ALAlg**–algebras one can define a construction on algebras which corresponds to the *–operation* of regular expressions. The corresponding algebra $\mathbb{A}^*$ can be constructed as quotient of the congruence relation which is generated by fusing each initial state with each accepting state.

With respect to Example 3.1 the algebra $\mathbb{E}$ has one initial state s_0 and one accepting state s_3. This implies

$$\mathbb{E}^* = \mathbb{E}^{s_3 \leftrightsquigarrow s_0}.$$

The quotient construction together with the sum can be used to define an operation on algebras which corresponds to the product of regular expressions or the sequential composition of automata. The basic idea for the construction of $\mathbb{A} \cdot \mathbb{B}$ is to define first a sum $\mathbb{A} + (\mathbb{B} + \ldots \mathbb{B})$ which contains as many copies of $\mathbb{B}$ as $\mathbb{A}$ has accepting states and fuse each accepting state with the initial states of the copy of $\mathbb{B}$ which corresponds to the accepting state.

There is one problem left. Which algebras correspond to the atomic regular expressions? The corresponding concept of an atomic partial algebra depends on the given extended signature. We will discuss the case of *atomic partial* **ALAlg**–algebras.

Let be given a finite set $\{a_0, a_1, \ldots, a_n\}, 0 \leq n$, and regular sets $L_1 \subseteq \{a, b, A, B\}^*, L_2 \subseteq \{a_1, \ldots, a_n\}^*$. Then we call a partial ALAlg–algebra $\mathbb{A}$ with the carrier set $\{a_0, a_1, \ldots, a_n\}$ atomic if $dom(c_\mathbb{A}) = L_1 \times L_2$ and $c_\mathbb{A}(w, l) = a_0$ for all $w \in L_1, l \in L_2$.

Finally we demonstrate that the partial ALAlg–algebra which results from Example 3.1 by removing the trap s_d can be constructed out of atomic partial ALAlg–algebras using summation and quotient construction.

We start with the following atomic algebras:

1. $\mathbb{A}$ with $A = \{a_0\}, L_1(\mathbb{A}) = L(a^*), L_2(\mathbb{A}) = \{nil\}$;
2. $\mathbb{B}$ with $B = \{b_0, b_1\}, L_1(\mathbb{B}) = L(b^*), L_2(\mathbb{A}) = \{b_1 b_1, b_1 b_1 b_1\}$;
3. $\mathbb{C}$ with $C = \{c_0, c_1, c_2\}, L_1(\mathbb{C}) = L((A + B)a^*), L_2(\mathbb{C}) = \{c_2 c_2, c_1 c_2\}$;
4. $\mathbb{D}$ with $D = \{d_0, d_1, d_2, d_3\}, L_1(\mathbb{D}) = L((A+B)b^*), L_2(\mathbb{D}) = \{d_2 d_2 d_2, d_1 d_2 d_3\}$;

Then

$$(\mathbb{A} + \mathbb{B} + \mathbb{C} + \mathbb{D}^{d_3 \leadsto d_0})^{\{a_0 \leadsto b_1, b_0 \leadsto c_2, a_0 \leadsto c_1, c_0 \leadsto d_2, b_0 \leadsto d_1\}}$$

is a representation (up to isomorphism) of the partial ALAlg–algebra from Example 3.1.

Based on this example it is not hard to see that each finite partial regular ALAlg–algebra can be constructed out of atomic partial ALAlg–algebras using summations and quotient constructions.

5   Conclusions

We have introduced a uniform notion of regular sets which is applicable to arbitrary structured data types and which coincides with the existing notions of regular sets of words, ranked and unranked trees. The introduced notion is based on the observation that regular sets can be seen as subsets of free data types which are inverse homomorphic images of finite regular algebras (for suitable signatures).

In this paper we have sketched by a representative example a purely algebraic approach to the notion of regular sets for arbitrary data type. By a suitable extension of the signatures of algebras the approach of Thatcher [14] to ranked trees could be generalized to arbitrary structured data types. This approach can be seen as an addition to approaches based on logics, see for instance [11] and [12].

Algebras without rank have earlier also been used by Indermark [9]. But Indermak uses unranked algebras in order to avoid many–sortedness.

The most closely related work is that of K. Hashiguchi, Y. Wada and S. Jimbo [10] which use classical algebraic structures. They introduce binoids which have two associative binary operations and an identity to each operation. One operation is used to represent the depth and the other to represent the width of trees. The formal framework of this approach becomes rather complicated, since a two–sorted structure has to be encrypted in a one–sorted structure and the unrestricted width and depth of unranked trees in two binary operations. This approach does not offer a framework which can easily be extended to arbitrary structured data types.

It remains for future work to give a complete formal presentation of the sketched approach. Additionally it would be interesting to know if the coalgebraic approach to regular expressions developed by J.J.M.M. Rutten [13] can be extended to the more general situation of regular sets of arbitrary structured data.

References

1. M. Murata A. Brüggemann-Klein and D. Wood. Regular tree and regular hedge languages over unranked alphabets: Version 1. Technical Report HKUST-TCSC-2001-0, The Hongkong University of Science and Technology, April 2001.
2. R.M. Burstall A. Tarlecki and J.A. Goguen. Some fundamental algebraic tools for the semantics of computation - part III: Indexed categories. *TCS*, (91):239–264, 1991.
3. M. Barr and Ch. Wells. *Category Theory for Computing Science*. International Series in Computer Science. Prentice–Hall, second edition edition, 195.
4. M. Bidoit and P.D. Mosses. *CASL - User Manual*, volume 2900 of *LNCS*. Springer, 2004.
5. Rod Burstall and Joseph Goguen. The semantics of CLEAR, a specification language. *Springer LNCS*, 86:292–332, 1979.
6. H. Ehrig and B. Mahr. *Fundamentals of Algebraic Specification 1: Equations and Initial Semantics*, volume 6. Springer–Verlag, eatcs monographs on theoretical computer science edition, 1985.
7. H. Ehrig and B. Mahr. *Fundamentals of Algebraic Specification 2: Module Specifications and Constraints*, volume 21. Springer–Verlag, eatcs monographs on theoretical computer science edition, 1990.
8. Joseph A. Goguen. A categorical manifesto. *Math. Struct. in Comp. Sci.*, 1(1):49–67, 1991.
9. K. Indermark. On rational definitions in complete algebras without rank. *Theoretical Computer Science*, 21:281–313, 1982.
10. Y. Wada K. Hashiguchi and S. Jimbo. Regular binoid expressions and regular binoid languages. *Theoretical Computer Science*, 304:291–313, 2003.
11. F. Neven. Automata, logic, and XML. In *Proceedings CSL 2002*, volume 2471 of *LNCS*, pages 2–26. Springer, 2002.
12. F. Neven. Automata theory for XML researchers. *SIGMOD Record*, 31(3), 2002.
13. J.J.M.M. Rutten. Behavioural differential equations: a coinductive calculus of streams, automata, and power series. *Theoretical Computer Science*, 308(1-3):1–53, 2003.
14. J.W. Thatcher and J.B. Wright. Generlized finite automata theory with an application to a decision problem of second–order logic. *Mathematical Systems Theory*, 2(1):57–81, 1968.

Elementary Algebraic Specifications of the Rational Complex Numbers

Jan A. Bergstra[1] and John V. Tucker[2]

[1] University of Amsterdam,
Informatics Institute,
Kruislaan 403,
1098 SJ Amsterdam,
The Netherlands.
janb@science.uva.nl
[2] Department of Computer Science,
University of Wales Swansea,
Singleton Park,
Swansea, SA2 8PP,
United Kingdom.
j.v.tucker@swansea.ac.uk

Abstract. From the range of techniques available for algebraic specifications we select a core set of features which we define to be the *elementary algebraic specifications*. These include equational specifications with hidden functions and sorts and initial algebra semantics. We give an elementary equational specification of the field operations and conjugation operator on the rational complex numbers $\mathbb{Q}(i)$ and discuss some open problems.

For Joseph Goguen

1 Introduction

Joseph Goguen has a vision for the theory of computation. It is algebraic, it is comprehensive, and it is focussed on the world's work. He uses a set of mathematical tools from category theory and universal algebra to explore a vast landscape of fundamental concepts, system architectures, emerging technologies, and contemporary practices in software development. He is a great explorer. His achievement is a fine example of just how much intellectual ground can be covered in the life time of a brilliant computer scientist with energy, curiosity, technical insight and a personal scientific agenda. He has reflected on his work on computing up to 1999 in Goguen [15]. There is so much to think about in this oeuvre.

One early line of thought is the role of initial algebras in semantic modelling and specification, expounded in [19]. Our own work in algebraic specifications from 1979 onwards owes a great debt to Joseph Goguen and his colleagues Jim Thatcher and Eric Wagner who, writing as the ADJ Group, provided a *perfect*

K. Futatsugi et al. (Eds.): Goguen Festschrift, LNCS 4060, pp. 459–475, 2006.

mathematical basis for modelling and specifying abstract data types, starting in [20]. The ADJ Group established most of the basic theory by combining the technical ideas of many sorted algebras, equations, conditional equations, hidden functions and sorts, term rewriting and initial algebras. We continue to use many of their notations and techniques. Thanks to Sam Kamin [22], a rapidly growing literature was organised and problems identified and clearly stated, such as when were hidden functions and sorts necessary? Some of the more difficult problems of partiality, errors and parameterization also showed themselves in their early writing, problems which, after many long papers and books, are still not quite under control. Joseph Goguen and Eric Wagner have reflected on the ADJ Group in [13] and [38], respectively.

Our general theory of the algebraic specification of computable data types analysed the relationship between computability of abstract data types and equational specifications. Between 1979 and 1995 we published a series of papers that classified the computable, semicomputable and cosemicomputable data types using algebraic specifications (see, for example, Bergstra and Tucker [2,3,4,5]). We proved several theorems that show that all computable data types have specifications that are very simple and small, or have good term rewriting properties, but always with hidden functions. Work has continued on this subject, refining notions such as finality (e.g., including Meseguer and Goguen [16] and Moss, Meseguer and Goguen, [31]), and on open questions (e.g., by Marongiu and Tulipani [26] and, most recently, by Khoussainov [23,24]).

We have returned to the foundations of the subject in [7], tackling the specification of basic data types such as the rational numbers, and we continue here. First, we will select a core set of features which we define to be the *elementary algebraic specifications*. These are close to the basic techniques of the ADJ Group of the 1970s.

The set $\mathbb{Q}$ of rational numbers is a number system designed to denote measurements. They are used to define the real and complex numbers via approximation. The rationals are the numbers with which we make finite computations. Algebras made by equipping $\mathbb{Q}$ with some constants and operations we call *rational arithmetics*. We usually calculate with the algebra $(\mathbb{Q}|0, 1, +, -, \cdot, ^{-1})$ which is called the *field* of rational numbers when the operations satisfy certain standard axioms.

In addition to rational arithmetics, of particular interest are field extensions of the rational number field. Through Galois Theory, field extensions play a fundamental role in our understanding of the algebra of numbers, including the theory of equations ([12,34]). One important field extension is the field of rational complex numbers, based on the set

$$\mathbb{Q}(i) = \{p + i \cdot q | p, q \in \mathbb{Q}\}.$$

This has special operations such as complex conjugation $cc(p + i \cdot q) = p - i \cdot q$.

The algebras of rational numbers, such as the field and its extensions by real and complex numbers, are among the truly fundamental data types. Despite the fact they have been known and used for over two millennia, they are neglected in the modern theory of data types. After over 30 years of data type theory, many questions about rational arithmetics and their extensions are open.

Now the common rational arithmetics and field extensions are all computable algebras. Indeed, in the *theory of computable rings and fields* there is a wealth of constructions of computable algebras that start with the rationals and the finite fields: see the introduction and survey Stoltenberg-Hansen and Tucker [36]. Therefore, according to our general theory of algebraic specifications of computable data types they have various equational specifications under initial and final algebra semantics. Computable algebras even have equational specifications that are also complete term rewriting systems ([5]). However, these general specification theorems for computable data types involve hidden functions and are based on equationally definable enumerations of data.

Recently, in Moss [30], algebraic specifications of the rationals were considered. Among several interesting observations, Moss showed that there exists an equational specification of the ring of rationals (i.e., without division) with just *one* unary hidden function. He used a special enumeration technique based on a remarkable enumeration theorem for the rationals in Calkin and Wilf [10]. He also gave specifications of other rational arithmetics and asked if hidden functions were necessary. In [7] we proved that there exists a finite equational specification under initial algebra semantics, *without* hidden functions, of the field of rational numbers. The pursuit of this result leads to a thorough axiomatic examination of the divisibility operator, in which some interesting new axioms and models were discovered, and related results on fields and other rational arithmetics.

In particular, here we prove:

Theorem 1. *There exists a finite equational specification under initial algebra semantics, and without hidden functions, of the algebra*

$$(\mathbb{Q}(i)|0, 1, i, +, -, \cdot, ^{-1}, cc)$$

of rational complex numbers with field and conjugate operations that are all total.

The structure of the paper is this. In Section 2, we discuss the basics of specification theory and define the elementary algebraic specifications. In Section 3, we describe the algebras and the axioms we will use to specify them. In Section 4 we prove the main theorem. Finally, in Section 5 we discuss some open problems.

This paper, and our [7,8,9], can be read independently but they are better viewed as a sequel to Bergstra and Tucker [4,5], which contain many complementary results.

2   Elementary Algebraic Specifications

Since the first examples of algebraic specifications of data types in the 1970s, there has been a steady growth in the features that one may add to the basic techniques to be found in the early ADJ papers such as [20]. The new techniques have been introduced for a number of obvious reasons: they have been found to be natural, or useful, or necessary to solve problems, or they have been used to extend or explore simpler techniques. The development of languages and tools (such as OBJ, ASF-SDF, Maude, CASL, etc.) for algebraic specification has increased the number and complexity of features in use.

So just what are the basic elements of this subject?

2.1   What Are the Elementary Algebraic Specifications?

Algebraic specification starts with the idea of modelling - e.g., data, processes, syntax, hardware, etc. - using sets and functions. Wherever there are sets and functions there are algebras! For example, the sets X, Y and function $f : X \rightarrow Y$ are combined to form the many sorted algebra $(X, Y | f)$. A particular algebra A is a mathematical model of a specific concrete representation of the system equipped with concrete operations. The need to understand the system, its representations and the extent to which they are unique leads to the concepts of (i) axiomatic theories for the chosen operators, and (ii) homomorphisms and isomorphisms for the comparison of algebras. The simplest axioms are equations. The simplest deductions are are those of equational logic based on the rewriting of terms. Any system can be modelled in this way. Therefore, we define the basic elements as follows.

Definition 1. *An algebraic specification (Σ', E') of a Σ algebra A is elementary if it involves only*

1. *A many sorted signature Σ' that is non-void. A signature is non-void if there is a closed term of every sort.*
2. *A set E' of equations or conditional equations.*
3. *An initial algebra semantics such that $I(\Sigma', E')|_{\Sigma} \cong A$.*

In particular, the elementary specifications *require* total functions, *allow* hidden functions and sorts, and may or may not be complete term rewriting systems. Clearly, there are plenty of restrictions in force: see 2.2.

A standard way of proving an elementary specification is to check these properties:

Definition 2. *An algebraic specification (Σ', E') of a Σ algebra A satisfies Goguen's conditions if it the following are true:*

No Junk or Minimality *The algebra A is Σ-minimal.*
No Confusion or Completeness *For all closed Σ terms t, t', we have*

$$A \models t = t' \text{ if, and only if, } E' \vdash t = t'.$$

In particular, the Goguen conditions imply that

$$I(\Sigma', E')|_{\Sigma} \cong A.$$

What makes these features *elementary*?

The purpose of developing a specification is to model, analyse and understand. In simple terms, these algebraic tools are fundamental for any modelling using sets and functions: they are used to abstract and analyse the properties of an idea, component, or system. One chooses a set of operators and postulates a set of laws they satisfy; the laws are expressed as equations or conditional equations. The terms express all possible operations that can be derived by combining operations, and the equational identities express the consequent facts about the model. The term rewriting is a completely basic mechanism for both abstract reasoning and computation. This view suggests the elementary character of the equations and that we cannot make do with less. There is also an argument that they need extension in special circumstances.

Now, the whole modelling and specification process for elementary specifications is mathematically *robust* in the sense that the syntax and semantics have virtually no special conditions, neither subtle or obvious.

In modelling using an elementary algebraic specification one simply starts playing with operators, equations and rewrites. There are no side conditions, side effects, and semantic errors to beware. The elementary algebraic specifications work simply in all cases. The only mistakes possible are mistakes in understanding what one is trying to model.

In our algebraic theory of computable data types, there are many results that show that if a many sorted algebra can be implemented on a computer then it possesses a range of *elementary equational specifications* with remarkable properties. Technically, all computable algebras can be specified with hidden functions, and all semicomputable algebras can be specified with hidden sorts and functions. In general this is the best possible. One theorem provides complete term rewriting systems ([5], Terese [37]). We need not worry about their power because:

The elementary algebraic specifications can specify everything that can be implemented on a computer in principle.

2.2   What Are the Non-elementary Algebraic Specifications?

What features have we excluded from the Definition 1 and hence have "declared" to be *not* elementary, and why?

We have excluded *final algebra semantics* because final algebras of equational specifications do not always exist and there are different interpretations possible (see Moss, Meseguer and Goguen [31]).

We have excluded *loose semantics* because we are focussed on specifying algebras up to isomorphism rather than classes of possible models.

We have excluded the following, too:

- **Generalisations of equations** One can use first order formulae that are "close" to equations such as Horn formulae. Since we exclude relations the Horn clauses are excluded. The *multi-equations* studied by Adamek et. al. [1] are also not simple enough.

- **Partial functions** Partiality is an essential aspect of computation. However, their logic is awfully complicated. Total functions are not without problems when specifying the stack, as we have seen in our [6]. However, as we showed in [7], it is not a problem to use total functions to specify the inverse on the rationals.
- **Errors and exceptions** The addition of new types of data, such as error flags, to familiar old friends, such as the natural numbers equipped with the predecessor $n-1$, leads to difficult specifications and semantic complications.
- **Subsorts and order sorted algebras** Subsorts occur naturally and help with modelling subtyping, errors, etc. However, there are different theories none of which are simple: see, for example, Mosses on unified algebras [32] or the survey [18].
- **Higher order** The higher-order theory is complicated from the start though it does possess a nice generalisation of the standard theory (see Meinke [27]).
- **Empty sorts** Empty sorts are tricky: see Goguen and Meseguer [17].
- **Priorities** Priorities for the equational rules are technically natural in developing software tools for algebraic specifications. However, they lead to complications since their term algebra representations do not satisfy the equations in general and must be considered pre-initial in some sense.
- **Modularity** Our elementary specifications are flat and do not have imports. Even the most simple notion of import introduces involved operations for flattening, see Rees et al [33].
- **Parameterization** There are many alternate treatments of parameterization, none of them simple.

Many of the features and techniques above that we have declared to be *not* elementary we certainly consider important. For example, features such as partiality and higher order equations are semantically fascinating and challenging to study, and are necessary to meet desires for certain kinds of specifications. However, they are not elementary.

Some of the features we have chosen for exclusion, such as subsorts, may *seem* less complicated to the user: they are not. For example, consider distinguishing the set $\mathbb{Q}_{\neq 0}$ of non-zero elements of $\mathbb{Q}$ using a subsort of a signature for the field of rationals: let $nzrat$ be the subsort of the sort rat. What is the type of the rational function $1/(1+x.x)$? Is it $nzrat$ and, if so, why? This kind of typing problem is complicated for if it were decidable then the diophantine problem over $\mathbb{Q}$ would be decidable - this remains an important open problem in computability theory. The types of open terms are problematic, and so are types of equations. Is the equation $(1 + x.x)/(1 + x.x) = 1$ usable as an axiom? If so, what does that imply about its type, or should that be given explicitly. But what can the type be: taking type $nzrat$ represents the axiom that this denominator is never 0 which to prove may require this very axiom, taking as type rat may be a type error.

We believe that none of the features in our list are elementary for users, and that combining them leads to significant complications.

2.3   Technical Preliminaries on Algebraic Specifications

We assume the reader is familiar with using equations and conditional equations and initial algebra semantics to specify data types. Some accounts of this are: ADJ [20], Meseguer and Goguen [29], or Wirsing [40].

The theory of algebraic specifications is based on theories of universal algebras (e.g., Wechler [39], Meinke and Tucker [28]), computable and semicomputable algebras (Stoltenberg-Hansen and Tucker [35]), and term rewriting (Klop [25], Terese [37]).

We use standard notations: typically, we let Σ be a many sorted signature and A a total Σ algebra. The class of all total Σ algebras is $Alg(\Sigma)$ and the class of all total Σ-algebras satisfying all the axioms in a theory T is $Alg(\Sigma, T)$. The word 'algebra' will mean total algebra.

3   Specifications for Rational Complex Numbers

3.1   Algebraic Specifications of the Rationals

We will build our specifications in stages. The primary signature Σ is simply that of the *field* of rational numbers:

> **signature** Σ
> **sorts** *field*
> **operations**
> $0: \rightarrow field;$
> $1: \rightarrow field;$
> $+: field \times field \rightarrow field;$
> $-: field \rightarrow field;$
> $\cdot: field \times field \rightarrow field;$
> $^{-1}: field \rightarrow field$
> **end**

The first set of axioms is that of a *commutative ring with* 1, which establishes the standard properties of $+$, $-$, and $\cdot$.

> **equations** CR

$$(x + y) + z = x + (y + z) \tag{1}$$
$$x + y = y + x \tag{2}$$
$$x + 0 = x \tag{3}$$
$$x + (-x) = 0 \tag{4}$$
$$(x \cdot y) \cdot z = x \cdot (y \cdot z) \tag{5}$$
$$x \cdot y = y \cdot x \tag{6}$$
$$x \cdot 1 = x \tag{7}$$
$$x \cdot (y + z) = x \cdot y + x \cdot z \tag{8}$$

> **end**

Our first set SIP of axioms for $^{-1}$ contain the following, which we call the *strong inverse properties*. They are "strong" because they are equations in involving $^{-1}$ *without any guards*, such as $x \neq 0$:

equations SIP

$$(-x)^{-1} = -(x^{-1}) \tag{9}$$
$$(x \cdot y)^{-1} = x^{-1} \cdot y^{-1} \tag{10}$$
$$(x^{-1})^{-1} = x \tag{11}$$

end

Our specification $CR \cup SIP$ draws attention to division by zero:

Lemma 1. *The following equation is provable from $CR \cup SIP$:*

$$0^{-1} = 0.$$

In particular, in our [7] (Theorem 3.5) we add a single axiom L to prove:

Theorem 2. *There exists a finite elementary equational specification $(\Sigma, CR \cup SIP \cup L)$, without hidden functions and under initial algebra semantics, of the rational numbers with field operations that are all total.*

In [7] we also add to $CR \cup SIP$ the *restricted inverse law* (Ril),

equations Ril

$$x \cdot (x \cdot x^{-1}) = x \tag{12}$$

end

which, using commutativity and associativity, expresses that $x \cdot x^{-1}$ is 1 in the presence of x.

Whilst the initial algebra of CR is the ring of integers, we find that

Lemma 2. *The initial algebra of $CR + SIP + Ril$ is a computable algebra but it is not an integral domain.*

The models of $CR + SIP + Ril$ are algebras with nice properties, in spite of not being fields nor even integral domains.

Definition 3. *A model of $CR + SIP + Ril$ is called a* meadow.

All fields are clearly meadows but not conversely (as the initial algebra is not a field).

Theorem 3. *For any closed terms $t, t' \in T(\Sigma)$, the following are equivalent*

1. *$t = t'$ is true in all totalised fields.*
2. *$t = t'$ is true in all totalised meadows.*

3.2 Algebraic Specifications of the Rational Complex Numbers

We add to the field signature Σ the complex conjugate operation $cc\colon field \rightarrow field$ to form the signature Σ_{cc}. Also to this signature we add the constant $i\colon \rightarrow field$ to form the signature $\Sigma_{cc,i}$. Consider these equations over the signature $\Sigma_{cc,i}$:

equations CC

$$i \cdot i = -1 \tag{13}$$
$$cc(1) = 1 \tag{14}$$
$$cc(\mathtt{i}) = -\mathtt{i} \tag{15}$$
$$cc(x + y) = cc(x) + cc(y) \tag{16}$$
$$cc(x \cdot y) = cc(x) \cdot cc(y) \tag{17}$$
$$cc(-x) = -cc(x) \tag{18}$$
$$cc(x^{-1}) = (cc(x))^{-1} \tag{19}$$
$$cc(x \cdot x^{-1}) = x \cdot x^{-1} \tag{20}$$

end

3.3 Totalised Fields and Algebras Satisfying the Specifications

The axioms of a field simply add to CR the following: the *general inverse law* (*Gil*)

$$x \neq 0 \implies x \cdot x^{-1} = 1$$

and the *axiom of separation* (*Sep*)

$$0 \neq 1.$$

Thus, let (Σ, T_{field}) be the axiomatic specification of fields, where

$$T_{field} = CR \cup Gil \cup Sep.$$

Clearly, this specification is not elementary as it contains negations; and, as it is commonly applied, allows partial functions in its models.

However, by definition, the class $Alg(\Sigma, T_{field})$ is the class of *total* algebras satisfying the axioms in T_{field}. For emphasis, we refer to these algebras as *totalised fields*.

For all totalised fields $A \in Alg(\Sigma, T_{field})$ and all $x \in A$, the inverse x^{-1} is defined. In particular, 0_A^{-1} is defined. The actual value $0_A^{-1} = a$ can be anything.

However, it is convenient to set $0^{-1} = 0$ (see [7], and compare, e.g., Hodges [21], p. 695). We use the specification $CR \cup SIP$ which forces $0^{-1} = 0$ (Lemma 1). A field with $0^{-1} = 0$ we call a *0-totalised field*.

The main Σ-algebras we are interested in are these: first,

$$Q_0 = (\mathbb{Q}|0, 1, +, -, \cdot, ^{-1})$$

where the inverse is total

$$x^{-1} = 1/x \qquad\qquad \text{if } x \neq 0;$$
$$= 0 \qquad\qquad \text{if } x = 0$$

This total algebra satisfies the axioms of a field T_{field} and is a 0-totalised field of rationals. Next, we are interested in the 0-totalised field extension $Q_0(\mathtt{i})$ and its expansion by conjugation $Q_0(\mathtt{i}, cc)$.

4   Proof of Main Theorem

Theorem 4. *There exists a finite elementary equational specification $(\Sigma_{cc,i}, E)$, without hidden functions, of the algebra $Q_0(cc, i)$ of rational complex numbers with field and conjugate operations that are all total, under initial algebra semantics. That is,*

$$I(\Sigma_{cc,i}, E) = \mathbb{Q}_{cc,i}$$

Proof. Let (Σ, E) be any elementary equational specification without hidden functions of the 0-totalised field of rationals $Q_0 = (\mathbb{Q}|0, 1, +, -, \cdot,^{-1})$ so $I(\Sigma, E) \cong Q_0$. By Theorem 2, there is such an elementary specification. The strategy is to build a specification of $Q_0(cc, i)$ using real and imaginary parts, which are rationals.

Let Σ_{cc} be the field signature Σ extended by the complex conjugation operator

$$cc\colon field \rightarrow field.$$

First, we look at conjugation on $\mathbb{Q}$.

Conjugation on $\mathbb{Q}$. Conjugation on $\mathbb{Q} \subset \mathbb{C}$ is the identity function, $cc(r) = r$ for $r \in \mathbb{Q}$. Let $Q_0(cc)$ be the 0-totalised field of rational numbers extended by conjugation cc. Let E^+ be the result of applying the following transformation of the equations in E: for each variable x in each equation of E substitute

$$\tfrac{1}{2}(x + cc(x)),$$

where $\frac{1}{2} = (1+1)^{-1}$. When applied to a complex number, the formula calculates its real part so when applied to a rational complex number from $Q_0(cc, i)$ it returns a rational number that would satisfy the equations of E.

Now define

$$E_{cc}^+ = E^+ \cup \{cc(x) = x\} \cup \{\tfrac{1}{2}(x + x) = x\},$$

Lemma 3. $I(\Sigma_{cc}, E_{cc}^+) \cong Q_0(cc)$

Proof. We use Goguen's conditions.

No Junk The algebra $Q_0(cc)$ is clearly Σ_{cc} minimal since Q_0 is Σ minimal.

No Confusion By inspection, $Q_0(cc) \models E_{cc}^+$. We have to show completeness.
First, we make the observation that

$$E_{cc}^+ \vdash E$$

since we can derive the equations of E by substituting back using the two axioms added as follows:

$$E_{cc}^+ \vdash \tfrac{1}{2}(x + cc(x)) = \tfrac{1}{2}(x + x) = x.$$

Now suppose that $Q_0(cc) \models t_1 = t_2$ for any closed terms. By using axiom $cc(x) = x$ in E_{cc}^+, we can delete cc in the terms $t_1, t_2 \in T(\Sigma_{cc})$ to form $t_1', t_2' \in T(\Sigma)$ such that

$$E_{cc}^+ \vdash t_1 = t_1' \text{ and } E_{cc}^+ \vdash t_2 = t_2'.$$

We know that

$$Q_0 \models t_1' = t_2'$$

and since (Σ, E) is an initial algebra specification, we have

$$E \vdash t_1' = t_2'.$$

Hence, by the above observation,

$$E_{cc}^+ \vdash t_1' = t_2'.$$

and by applying $cc(x) = x$ as often as t_1', t_2' contain occurrences of cc

$$E_{cc}^+ \vdash t_1 = t_2.$$

This completes the argument.

Let us replace the equation $cc(x) = x$ in E_{cc}^+ by the set

$$CCid = \{cc(t) = t \,|\, t \in T(\Sigma)\}$$

of all its closed Σ-term instances. We define:

$$E_{cc}^{++} = E^+ \cup CCid \cup \{\tfrac{1}{2}(x + x) = x\}.$$

Lemma 4. $I(\Sigma_{cc}, E_{cc}^{++}) \cong Q_0(cc)$

Proof. Replacing an equation by the set of all its closed instances does not change the initial algebra. In this case the cc's can be removed anyway.

Now we consider the complex numbers

Conjugation on $\mathbb{Q}(i)$. Now consider the signature $\Sigma_{cc,i} = \Sigma_{cc} \cup \{i: \ \to field\}$, and the algebra $Q_0(cc, i)$ of rational complex numbers. We define the set

$$T = CR \cup SIP \cup Ril \cup CC \cup \{2 \cdot 2^{-1} = 1\}$$

Theorem 5. $I(\Sigma_{cc,i}, E^+ \cup T) \cong Q_0(cc, i)$

Proof. We verify the Goguen conditions.
No Junk Clearly, $Q_0(cc, i)$ is $\Sigma_{cc,i}$ minimal.
No Confusion By inspection,

$$Q_0(cc, i) \models E^+ \cup T.$$

For this we use the fact that the substitution of $\frac{1}{2}(x + cc(x))$ for each x in E guarantees that E is restricted to rational values which are the real parts of x when evaluated in $Q_0(cc, i)$.

To complete the argument we need some lemmas.

Lemma 5. *For each $t \in T(\Sigma_{cc})$, we have $T \vdash cc(t) = t$.*

Proof. This is an easy induction on t.

Some useful consequences of this lemma are as follows. First, $T \vdash CCid$. Furthermore, we may deduce

$$T \vdash \tfrac{1}{2}(x + x) = \tfrac{1}{2} \cdot 2x = \tfrac{2}{2} \cdot x = 1 \cdot x = x$$

using CR and the axiom $2 \cdot 2^{-1} = 1$ in T. So we also have

$$E^+ \cup T \vdash E_{cc}^{++}.$$

Now suppose that $Q_0(cc, i) \models t_1 = t_2$ with $t_1, t_2 \in T(\Sigma_{cc,i})$. To show completeness we have to show that $E^+ \cup T \vdash t_1 = t_2$.

Lemma 6. *For any closed term $t \in T(\Sigma_{cc,i})$ there are terms $p, q \in T(\Sigma)$ such that*

$$T \vdash t = p + i \cdot q.$$

Proof. We prove this by induction on terms.

Basis By the ring axioms of CR, the constants are as follows:

$$0 = 0 + i \cdot 0$$
$$1 = 1 + i \cdot 0$$
$$i = 0 + i \cdot 1.$$

Induction Step There are five cases, one for each operation. The cases of $+, -, \cdot$ are easy - here is one:

Let $t = t_1 \cdot t_2$ and suppose as induction hypothesis:

$$T \vdash t_1 = p_1 + i \cdot q_1 \text{ and } T \vdash t_2 = p_2 + i \cdot q_2.$$

Then, substituting, we calculate:

$$T \vdash t = t_1 \cdot t_2 \qquad\qquad \text{by assumption}$$

$$\vdash t = (p_1 + i \cdot q_1) \cdot (p_2 + i \cdot q_2) \qquad \text{by induction hypothesis}$$

$$\vdash t = (p_1 \cdot p_2 - q_1 \cdot q_2) + i(q_1 \cdot p_2 + q_2 \cdot p_1) \qquad \text{by axioms in } CR \text{ and}$$

$$i \cdot i = -1$$

The other cases are more interesting.

Let $t = cc(t_0)$ and suppose as induction hypothesis:

$$T \vdash t_0 = p_0 + i \cdot q_0.$$

Then, substituting, we calculate:

$$T \vdash t = cc(t_0) \qquad\qquad \text{by assumption}$$

$$\vdash t = cc(p_0 + i \cdot q_0) \qquad \text{by induction hypothesis}$$

$$\vdash t = cc(p_0) - i \cdot cc(q_0) \qquad \text{by axioms of } CC \text{ in } T$$

$$\vdash t = p_0 + i \cdot q_0 \qquad \text{by Lemma 5.}$$

Let $t = r^{-1}$ and suppose as induction hypothesis:

$$T \vdash r = p + i \cdot q.$$

Then, substituting, we calculate:

$$T \vdash t = r^{-1} \qquad\qquad\qquad\qquad \text{by assumption}$$

$$\vdash t = \frac{1}{p + i \cdot q} \qquad\qquad\qquad \text{by induction hypothesis}$$

$$\vdash t = \frac{1}{p + i \cdot q} \cdot \frac{1}{p + i \cdot q} \cdot \left(\frac{1}{p + i \cdot q}\right)^{-1} \qquad \text{by Ril in } T$$

$$\vdash t = \frac{1}{p + i \cdot q} \cdot \frac{1}{p + i \cdot q} \cdot (p + i \cdot q) \qquad \text{by SIP}$$

$$\vdash t = \frac{1}{p + i \cdot q} \cdot \frac{p + i \cdot q}{p + i \cdot q} \qquad \text{by axioms of } CR$$

$$\vdash t = \frac{1}{p + i \cdot q} \cdot cc\left(\frac{p + i \cdot q}{p + i \cdot q}\right) \qquad \text{by axiom of } cc(x \cdot x^{-1}) = x \cdot x^{-1}$$

$$\vdash t = \frac{1}{p + i \cdot q} \cdot \frac{cc(p + i \cdot q)}{cc(p + i \cdot q)} \qquad \text{by axiom of CC}$$

$$\vdash t = \frac{1}{p + i \cdot q} \cdot \frac{cc(p) - i \cdot cc(q)}{cc(p) - i \cdot cc(q)} \qquad \text{by axioms of CC}$$

$$\vdash t = \frac{p - i \cdot q}{(p + i \cdot q) \cdot (p - i \cdot q)} \qquad \text{by Lemma 5}$$

$$\vdash t = \frac{p}{p^2 + q^2} - i \cdot \frac{q}{p^2 + q^2} \qquad \text{by axioms of T}$$

whch has the required form.

Finally, to finish the completeness, suppose $Q_0(cc, i) \models t_1 = t_2$ with $t_1, t_2 \in T(\Sigma_{cc,i})$. By Lemma 6, we suppose that

$$T \vdash t_1 = p_1 + i \cdot q_1 \text{ and } T \vdash t_2 = p_2 + i \cdot q_2.$$

where $p_1, p_2, q_1, q_1 \in T(\Sigma)$. Hence,

$$Q_0 \models p_1 = p_2 \text{ and } Q_0 \models q_1 = q_2.$$

By Lemma 4, $I(\Sigma_{cc}, E_{cc}^{++}) \cong Q_0(cc)$ and so by completeness

$$E_{cc}^{++} \vdash p_1 = p_2 \text{ and } E_{cc}^{++} \vdash q_1 = q_2.$$

Next, thanks to Lemma 5, a consequence is

$$E^+ \cup T \vdash E_{cc}^{++}.$$

Therefore,

$$E^+ \cup T \vdash p_1 = p_2 \text{ and } E^+ \cup T \vdash q_1 = q_2.$$

and so we are done with

$$E^+ \cup T \vdash p_1 + i \cdot q_1 = p_2 + i \cdot q_2.$$

This completes the proof of Theorem 5.

And hence the main theorem.

5 Concluding Remarks

There are open questions left over from the study of the rationals. For example, the following problem is quite basic:

Problem 1. Is there a finite elementary equational specification of the 0-totalised field Q_0, without hidden functions and under initial algebra semantics, which constitutes a complete term rewriting system?

We know from our [5] that there exists such a specification with hidden functions.

However, questions proliferate as one reflects on the number of algebras using the rational numbers ([36]). For example, we do not know the answer to these simple questions.

Problem 2. Is there a finite elementary equational specification of the field $Q_0(i)$ of rational complex numbers, without any hidden functions?

Problem 3. Is there a finite elementary equational specification of the algebra $Q_0(i, cc)$, (without further hidden functions), which constitutes a complete term rewriting system?

The rational numbers constitute the data type for measurement with a finite system of units and subunits. The real and complex numbers are constructed as completions of the rationals, using the idea of the approximation of measurements with unlimited accuracy. The real and complex numbers are the basis for vast range of data types used to model physical systems by means of measurement and equations (e.g., algebras of sequences, streams and signals, scalar and vector fields, continuous functions, probability distributions, and their abstractions). In general terms the data in these algebras are continuous data and they are built by some completion process from subalgebras containing discrete data, as the reals are made from the rationals.

Problem 4. To create a comprehensive theory of computing, specifying and reasoning with systems based on continuous data. Ideally, the theory should integrate discrete and continuous data.

At present this is a huge and complicated task as computation, specification and verification on continuous data are all active research areas. In fact, the task is a challenge in the special case of real numbers, see [7] for a discussion.

References

1. J. ADAMEK, M. HEBERT AND J. ROSICKY On abstract data types presented by multiequations *Theoretical Computer Science* 275 (2002) 427 - 462
2. J A BERGSTRA AND J V TUCKER, The completeness of the algebraic specification methods for data types, *Information and Control*, 54 (1982) 186-200
3. J A BERGSTRA AND J V TUCKER, Initial and final algebra semantics for data type specifications: two characterisation theorems, *SIAM Journal on Computing*, 12 (1983) 366-387.
4. J A BERGSTRA AND J V TUCKER, Algebraic specifications of computable and semicomputable data types, *Theoretical Computer Science*, 50 (1987) 137-181.
5. J A BERGSTRA AND J V TUCKER, Equational specifications, complete term rewriting systems, and computable and semicomputable algebras, *Journal of ACM*, 42 (1995) 1194-1230.
6. J A BERGSTRA AND J V TUCKER, The data type variety of stack algebras, *Annals of Pure and Applied Logic*, 73 (1995) 11-36.
7. J A BERGSTRA AND J V TUCKER, The rational numbers as an abstract data type, submitted.
8. J A BERGSTRA AND J V TUCKER, On fields and meadows of finite characteristic, submitted.
9. J A BERGSTRA, Elementary algebraic specifications of the rational function field, CIE 2006, Springer Lecture Notes in Computer Science, accepted for publication.
10. N CALKIN AND H S WILF, Recounting the rationals, *American Mathematical Monthly*, 107 (2000) 360-363.
11. E CONTEJEAN, C MARCHE AND L RABEHASAINA, Rewrite systems for natural, integral, and rational arithmetic, in *Rewriting Techniques and Applications 1997*, Springer Lecture Notes in Computer Science 1232, 98-112, Springer, Berlin,1997.
12. H EDWARDS, *Galois theory*, Springer, 1984.
13. J A GOGUEN, Memories of ADJ, *Bulletin of the European Association for Theoretical Computer Science*, 36 (October 1989), pp 96-102.

14. J A GOGUEN, A categorical manifesto, *Mathematical Structures in Computer Science*, 1 (1991), pp 49-67.

15. J A GOGUEN, Tossing algebraic flowers down the great divide, in C S Calude (ed.), *People and ideas in theoretical computer science*, Springer, Singapore, 1999, pp 93-129.

16. J MESEGUER AND J A GOGUEN, Initiality, induction, and computability, In M Nivat (editors) *Algebraic methods in semantics*, Cambridge University Press,1986 pp 459 - 541

17. J MESEGUER AND J A GOGUEN, Remarks on remarks on many-sorted algebras with possibly emtpy carrier sets, *Bulletin of the EATCS*, 30 (1986) 66-73.

18. J A GOGUEN AND R DIACONESCU An Oxford Survey of Order Sorted Algebra *Mathematical Structures in Computer Science* 4 (1994) 363-392.

19. J A GOGUEN, J W THATCHER, E G WAGNER AND J B WRIGHT, Initial algebra semantics and continuous algebras, *Journal of ACM*, 24 (1977), 68-95.

20. J A GOGUEN, J W THATCHER AND E G WAGNER, An initial algebra approach to the specification, correctness and implementation of abstract data types, in R.T Yeh (ed.) *Current trends in programming methodology. IV. Data structuring*, Prentice-Hall, Engelwood Cliffs, New Jersey, 1978, pp 80-149.

21. W HODGES, *Model Theory*, Cambridge University Press, Cambridge, 1993.

22. S KAMIN, Some definitions for algebraic data type specifications, SIGLAN Notices 14 (3) (1979), 28.

23. B KHOUSSAINOV, Randomness, computability, and algebraic specifications, *Annals of Pure and Applied Logic*, (1998) 1-15

24. B KHOUSSAINOV, On algebraic specifications of abstract data types, in *Computer Science Logic: 17th International Workshop*, Lecture Notes in Computer Science, Volume 2803, 299-313, 2003

25. J W KLOP, Term rewriting systems, in S. Abramsky, D. Gabbay and T Maibaum (eds.) *Handbook of Logic in Computer Science. Volume 2: Mathematical Structures*, Oxford University Press, 1992, pp.1-116.

26. G MARONGIU AND S TULIPANI, On a conjecture of Bergstra and Tucker, *Theoretical Computer Science*, 67 (1989), 87-97.

27. K MEINKE, Universal algebra in higher types, *Theoretical Computer Science*, 100 (1992) 385-417.

28. K MEINKE AND J V TUCKER, Universal algebra, in S. Abramsky, D. Gabbay and T Maibaum (eds.) *Handbook of Logic in Computer Science. Volume I: Mathematical Structures*, Oxford University Press, 1992, pp.189-411.

29. J MESEGUER AND J A GOGUEN, Initiality, induction and computability, in M Nivat and J Reynolds (eds.), *Algebraic methods in semantics*, Cambridge University Press, Cambridge, 1985, pp.459-541.

30. L MOSS, Simple equational specifications of rational arithmetic, *Discrete Mathematics and Theoretical Computer Science*, 4 (2001) 291-300.

31. L MOSS, J MESEGUER AND J A GOGUEN, Final algebras, cosemicomputable algebras, and degrees of unsolvability, *Theoretical Computer Science*, 100 (1992) 267-302.

32. P MOSSES, Unified algebras and institutions, *Proceedings 4th Logic in Computer Science*, IEEE Press, 1989, 304-312.

33. D REES, K STEPHENSON AND J V TUCKER, The algebraic structure of interfaces, *Science of Computer Programming*, 49 (2003), 47-88.

34. I STEWART, *Galois theory*, Chapman and Hall, 1973.

35. V STOLTENBERG-HANSEN AND J V TUCKER, Effective algebras, in S Abramsky, D Gabbay and T Maibaum (eds.) *Handbook of Logic in Computer Science. Volume IV: Semantic Modelling*, Oxford University Press, 1995, pp.357-526.

36. V STOLTENBERG-HANSEN AND J V TUCKER, Computable rings and fields, in E Griffor (ed.), *Handbook of Computability Theory*, Elsevier, 1999, pp.363-447.

37. TERESE, *Term Rewriting Systems*, Cambridge Tracts in Theoretical Computer Science 55, Cambridge University Press, Cambridge, 2003.

38. E WAGNER, Algebraic specifications: some old history and new thoughts, *Nordic Journal of Computing*, 9 (2002), 373 - 404.

39. W WECHLER, *Universal algebra for computer scientists*, EATCS Monographs in Computer Science, Springer, 1992.

40. M WIRSING, Algebraic specifications, in J van Leeuwen (ed.), *Handbook of Theoretical Computer Science. Volume B: Formal models and semantics*, North-Holland, 1990, pp 675-788.

From Chaos to Undefinedness

A Story About Recursion as Well as Termination, Underspecification, Nondeterminism, Fixpoints, Metric Treatment, and Logical Models

Dedicated to Joseph Goguen

Manfred Broy

Institut für Informatik, Technische Universität München
D-80290 München Germany, broy@in.tum.de
http://wwwbroy.informatik.tu-muenchen.de

Abstract. The semantic and logical treatment of recursion and of recursive definitions in computer science, in particular in requirements specification, in programming languages and related formalisms such as λ-calculus or recursively defined functions is one of the key issues of the semantic theory of programming and programming languages. As it has been recognised already in the early days of the theory of programming there are several options to formalise and give a theory of the semantics of recursive function declarations. In different branches of computer science, logics, and mathematics various techniques for dealing with the semantics of recursion have been developed and established. We outline, compare, and shortly discuss advantages and disadvantages of these different possibilities, illustrate them by a simple running example, and relate these approaches.

1 Introduction

In informatics, recursion appears - explicitly or implicitly - everywhere. Throughout this paper we study the following problem pattern. We assume that we are given a *heterogeneous algebra*, also called a *computation structure* in the foundation of *abstract data types*, consisting of a family of carrier sets (corresponding to data types) and a family of functions/operations over them. We furthermore assume that we are given a logical theory for which the algebra is a model. This theory need not necessarily be logically complete. In this case, there might exist many further essentially different models for the given theory.

We study the introduction of an additional function identified by a fresh function symbol f into this algebra and its logical theory, in particular. We carry out this extension by first fixing the functionality of the introduced function. The sorts and the associated carrier sets that form its domain and range determine it. After fixing the functionality we define the values (the "graph") of the function by an explicit, possibly recursive equation f(x) = E.

K. Futatsugi et al. (Eds.): Goguen Festschrift, LNCS 4060, pp. 476-496, 2006.
© Springer-Verlag Berlin Heidelberg 2006

A signature, a set of axioms, and a set of inference rules provide a *logical theory*. The signature provides symbols for constants, functions, logical variables, and, in the case of heterogeneous theories, of sorts (also called types) as well. Based on the signature we form terms and formulas. The set of axioms Γ and inference rules induce a deduction relation $\vdash$. We express by $\Gamma \vdash \theta$ the proposition that the formula θ follows logically from the axioms in Γ by the logical inference rules of the logical theory.

With logical theories we associate *models*. A model is an algebra. We work only with total algebras in the following. These are algebras where all functions are total. In the case of heterogeneous theories it is a heterogeneous algebra, which contains a "carrier" set for each sort of the signature, a data value for each constant and a function for each function symbol. This family of sets and functions allows us to interpret terms and formulas. Terms are interpreted by mapping them onto elements of the algebra's carrier sets, which represent their values. Formulas are interpreted by truth values.

For a model we require that all formulas θ for which the proposition $\Gamma \vdash \theta$ holds be mapped by the interpretation to the truth-value true. For each model we call the set of elements of the carrier sets for each of which a term exists whose interpretation yields this element the *standard elements* (also called the *term generated* elements). The other terms are called *non-standard elements*. A model that contains only standard elements is called a *standard* or a *term-generated model*. An equation of the form $f(x) = E$ is called *recursive* (for the function symbol f) if the function symbol f occurs in the term E.

The introduction of a new function symbol by recursion can be studied either in the *model-theoretic* or in the *logical* setting:

- In a logical approach, we extend the signature of the given logical theory by adding the fresh function symbol f to it. Then we add the specifying equation $f(x) = E$ as an axiom. By this approach, we transform the given theory into an extended one. Of course, we want to be sure that the step of adding the function symbol f and the equation provides a conservative extension of the logical theory (meaning that we do not introduce any additional properties to the given logical theory) and that the defining equation[1] characterises the function f uniquely.

- In a model-theoretic approach, we extend the algebra by a function called f that is required to fulfil the equation $f(x) = E$. To justify that step we want to be sure that such a function actually exists (in other words that the definition of f by the equation actually makes sense) and that it is uniquely determined by the equation.

Of course, both approaches are closely related. In the context of the logical approach we may consider the set of models of the theory. Then the idea of a conservative extension can be used, from which it follows that each of the models of the logical theory can be extended in a unique way adding a function called f that fulfils the defining equation.

---

[1] More precisely, for every ground term t of appropriate sort we would like to be able to reduce the term $f(t)$ to a ground term that does not contain the function symbol f.

Recursion is a funny concept. To "define" a function recursively seems like a missuse of the principle of definition, which requires that in a definition a new concept be defined exclusively in terms of known concepts. A recursive equation is circular by nature while it is a fundamental principle that definitions are required to be noncircular. This is in contrast to explicit, noncircular, defining equations where we define a function f by an explicit equation

(*) $f(x) = E$ where the function symbol f does not occur free in the term E

In the nonrecursive case E is an arbitrary term that may contain the identifier x but must not contain the function symbol f, however. Of course, by the nonrecursive equation (*) the function f is uniquely determined. Moreover, it is obvious that such a function f always exists. In other words, adding the equation (*) to define the meaning of a fresh function symbol f within a logical system does never introduce any contradiction. The extended theory is always a *conservative* extension by construction.

The function application $f(x)$ is then only an abbreviation for the term E. Similarly, for any term G the term $f(G)$ is just an abbreviation for the term $E[G/x]$. Here by $E[G/x]$ we denote the term formed by substituting the term G for the identifier x in the term E. It is obtained from E by replacing all (free) occurrences of the identifier x in the term E by the term G.

This simple situation of explicit (nonrecursive) equational definitions changes crucially if we allow for recursive equations. By recursion we declare a function f by the recursive equation

(**) $f(x) = E$

where the term E may contain arbitrary many applications of the function (symbol) f. In the case of such recursive definitions of functions the semantic and logical treatment gets way more complicated. We observe:

- There need not exist at all a function f that fulfils the equation (**); in other words, adding the equation (**) as the defining axiom for the fresh function symbol f to an axiomatic theory may introduce a logical *inconsistency* and allows the deduction of a *contradiction*.

- There are cases where there exist many distinct functions f that fulfil the defining equation (**). So adding the equation (**) as an axiom may lead to an extension of a complete theory into one that is incomplete.

- There are cases where the term $f(t)$ cannot be reduced with the help of the axioms and the equation (**) to a term that does not contain the function symbol f. Then new "nonstandard" elements that were not representable by the terms available so far may be the result of function calls of f; in other words, f may be chosen such that it yields results that are not elements in the original algebra. As a consequence there may be standard models of the extended logical theory that are not standard models for the original logical theory. In fact, there are even cases where there does not exist a standard model for the original logical theory the carrier sets of which form a standard model for the extended theory, if in the standard model a total function f that fulfils the equation does not exist.

Recursion has both a *logical* and an *operational* flavour. The function f is characterised by an equation $f(x) = E$ and this equation defines a rewrite rule $f(x) \rightarrow E$. This observation is the bridge between a descriptive and an algorithmic interpretation of the syntax of a programming and a specification language. In the following, we show a number of technical options to treat the semantics of recursion by logical and mathematical means.

One popular way to deal with the semantics of recursion is to give an operational semantics for recursively defined functions. Term rewriting can do this. This means that we introduce a rewriting relation $\rightarrow$ on terms defined by rewriting rules. In the case of a recursive definition, given a ground term t that contains the recursively defined function symbol f we assume that the reduction sequence

$$t \rightarrow t_1 \rightarrow ... \rightarrow t_n$$

for the term t either terminates leading to a term t_n that cannot be reduced anymore (meaning there does not exist a term t_{n+1} such that $t_n \rightarrow t_{n+1}$ holds) such that t_n is a term in normal form, which in our case, in particular, means that t_n does not contain applications of the function symbol f any longer (otherwise we could use the reduction $f(x) \rightarrow E$) or that the reduction can be continued forever resulting in an infinite reduction sequence.

This operational interpretation provides a strong guideline for the logical and denotational treatment of recursion. Given an operational interpretation (be it by an interpreter or by a term rewriting system) we have a clear reference for the logical treatment of recursion: the logical interpretation should match with the operational one, or, formulated in a more demanding way, it should reflect exactly the abstract behaviour induced by the operational semantics in terms of rewriting.

Before we go deeper into the semantic treatment of recursion let us be more precise on the used syntax. A recursive equation

$$f(x) = E \qquad\qquad (*)$$

is an equation where the expression E contains an arbitrary number of applications of the function f. Since the term E is finite it certainly contains only a finite number, say $n \in \mathbb{N}$, applications of the function f, say $f(G_1)$, ..., $f(G_n)$. In the following we sometimes want to identify the instances of the individual applications. This can be done by replacing the expression E by an expression E^* such that in E^* each of the fresh function symbols f_1, ..., f_n occurs exactly once (we assume that the identifiers f_1, ..., f_n do not occur in the term E) such that the following equation holds:

$$E = E^*[f/f_1, ..., f/f_n]$$

This way each application is marked individually by a function symbol f_i that occurs exactly once. In a model-theoretic approach we associate with the expression E^* a function (by using individual function symbols in each application)

(1) $\lambda f_1, ..., \lambda f_n: \lambda x: E^*$

or a function (using the function symbol f for each of the applications):

(2) $\lambda f: \lambda x: E$

The function given by (2) is called the *functional* associated with the recursive equation (∗). The function given by (1) is called the *multicall functional* associated with the equation (∗). If the expression E contains exactly one recursive application both coincide.

In general, the expression E contains conditional expressions to formulate case distinctions. These can be eliminated and transformed (under the assumption that C is a Boolean term) into conditional equations by replacing each equation of the form

$f(x) = $ **if** C **then** E_1 **else** E_2 **fi**

into the following two conditional equations:

$C = \text{true} \Rightarrow f(x) = E_1$

$C = \text{false} \Rightarrow f(x) = E_2$

If we assume then C is two valued ("tertium non datur") this translation is an equivalence relation. Moreover we also use rules like: replace the term

$f(\textbf{if } C \textbf{ then } E_1 \textbf{ else } E_2 \textbf{ fi})$

by the semantically equivalent term

if C **then** $f(E_1)$ **else** $f(E_2)$ **fi**.

By furthermore breaking up the terms in E in the equation $f(x) = E$ that way we can eliminate all conditional expressions by implications. This way recursive equations are transformed schematically into a set of conditional equations and vice versa, as long as all conditional equations have the form shown above.

All problems with recursive declarations disappear if we manage to choose our defining equation for f such that it defines the total function associated with the function symbol f uniquely. Special cases, where this applies, are definitions that can be interpreted inductively. This means that we can find a Noetherian partial order ≤ on the domain of the function f such that the following holds: in the recursive equation

$P \Rightarrow f(x) = E'$

where the expression E' contains only the recursive applications $f(G1), ... f(Gk)$ of the function f we can prove that the values of the terms $G1, ..., Gn$ are always elements that are in that ordering strictly below the original argument x1). In other words, we can prove

$P \Rightarrow G_k < x$

for all k. Then the recursive definition can be seen as an infinite set of explicit nonrecursive equations for the values of the function associated with the identifier f.

In the following, we shortly recapitulate and relate the various techniques to give a denotational or an axiomatic semantics to recursion. The main goal of this work is to

---

[1] In fact, working with conditional expressions the situation gets slightly more complicated. The recursive applications are guarded by conditions. Only if the conditions evaluate to true the modified parameters G_k have to be strictly below x.

integrate and justify a number of methodological decisions when working with a theory of program construction.

This work is motivated to a large extend by discussions in the IFIP Working Group 2.3 by discussions with Michel Sintzoff and the work of Tony Hoare, presented at the Marktoberdorf Summer School 1996, towards a integrating framework for the semantics and different semantical techniques to treat recursion in specification and programming languages.

2   Simple Running Example: Division

We demonstrate the various approaches to deal with the semantics of recursion by a very simple running example. This example has to be simple enough to keep the treatment short and concise but it should include and envisage the typical problems that arise when dealing with recursion. With this in mind we choose arithmetic division on the naturals as a running example.

Let $\mathbb{N}$ denote the set of natural numbers. Division on the naturals can be represented by a partial or by a total function on the naturals:

$$\text{div: } \mathbb{N} \times \mathbb{N} \to \mathbb{N}$$

or (note that functions are a special case of a relation) by a relation

$$\text{Div} \subseteq \mathbb{N} \times \mathbb{N} \times \mathbb{N}$$

or (isomorphic to a relation) by a set-valued function

$$\text{DIV: } \mathbb{N} \times \mathbb{N} \to \wp(\mathbb{N})$$

or (again isomorphic to a relation) by a predicate

$$\text{isdiv: } \mathbb{N} \times \mathbb{N} \times \mathbb{N} \to \mathbb{B}$$

or by a predicate that characterises a set of (partial or total) functions

$$\text{ISDIV: } (\mathbb{N} \times \mathbb{N} \to \mathbb{N}) \to \mathbb{B}$$

Strictly speaking according to the foundations of mathematics the function div represents also a relation which is a subset of product set $\mathbb{N} \times \mathbb{N} \times \mathbb{N}$. Since div is supposed to be a function we require that it contains for given numbers x, y at most one triple (x, y, z). This requirement for a relation to be called a function is known as the *Leibniz principle*. In other words, a function is a relation that fulfils the Leibniz principle. A function with two arguments is called *total*, if it contains for every pair of arguments x and y a triple (x, y, z); otherwise it is called partial. We are free to associate with a recursive definition a function or a general relation (allowing us to deal also with "nondeterminism" in our model).

Of course, the critical question is how to specify the result of division in the case where its second argument is 0. In the algebra of partial functions there is a simple answer. Working with partial functions, we easily express that the result of a function application is "not defined" or more precisely "does not exist". But for partial functions we pay the price that we now can write expressions that "do not of a value". For total functions, on the other hand, this simple solution is not available. For them for each pair of arguments a result has to be given, which, however, in the case of nontermination might be chosen quite arbitrarily. However, for partial functions the logical theory of equations is certainly less standard than that for total function.

In the case of set-valued functions, relations, or predicates, the idea of partial functions is easily incorporated. We may return the empty set for DIV(x, 0), define that every triple of the form (x, 0, t) is not in the relation DIV or that the predicate isdiv(x, 0, t) yields always false.

We may also represent undefined by "chaos" and return the set of all naturals for DIV(x, 0), including all triples (x, 0, z) into the relation Div and analogously define isdiv(x, 0, z) to be always true. In fact, we can choose many constructions between these two extremes. If we work with a predicate ISDIV that characterises a set of functions we can select any function that behaves like division in the case of the second argument being distinct to 0 and arbitrary otherwise. Or we may be more restrictive in the case the second argument is 0. All these options are discussed in more detail in the following.

3   Inductive Definitions, Total Functions

One simple way to cope with recursive definitions of functions is to follow the ideas of primitive recursion and their generalisations to inductive definitions. To do that we make sure that the recursive equation that defines the function is based on some kind of Noetherian ordering and therefore defines a function uniquely. The advantage of this approach is that it allows us to keep the logics classical and simple. For instance, we may restrict our model and our logics to total functions. Then all terms that can be formulated over the given signature denote well-defined elements.

Unfortunately by this technique, which is often used in type theory and in a number of verifications support systems like PVS (see [PVS 92]), recursive definitions need more work for their justification, since we have to prove termination. For that an appropriate Noetherian order has always to be introduced explicitly by which has to be proven that the definition is inductive. Only in simple cases this proof can be carried out by schematic or even automatic proof techniques. More remarkable, however, is that certain recursive definitions of practical importance cannot be treated at all that way or at least not in a straightforward manner. Famous examples are functions with nonrecursively enumerable codomains such as interpreters of programming languages of universal computability (an example is typed λ-calculus with μ-recursion). For these examples a constructive definition of the inductive ordering does not exist.

We use our running example of division on natural numbers to demonstrate the idea of inductive definitions. We work with the function symbol div that has two parameters. The critical question is, of course, which result the function div should return if the second parameter is 0.

A recursive definition of the total function associated with the symbol div is given by the following conditional equations (let n, m be of sort Nat):

$$\text{div}(m, n) = \qquad\qquad \Leftarrow n > m$$

$$\text{div}(m, n) = 1 + \text{div}(m-n, n) \quad \Leftarrow m \geq n \wedge n > 0$$

In the second equation - read from left to right as a rewrite rule - the first argument is decreased provided n > 0. Note that the condition n > 0 for the second equation is crucial according to the assumption that div is a total function. Leaving it away would lead to the equation

$$d = 1+d$$

for the natural number d = div(m, n) for the case n = 0, which introduces a contradiction into the theory of natural numbers.

By the conditional equations the result of applying div to the parameters (m, 0) is obviously not specified. Therefore we characterise this approach to the treatment of recursion by the sketch word *total functions with underspecification*.

Underspecification means, of course, that the axiomatisation of the introduced function is incomplete. As a consequence, there are several functions that fulfil the axioms. A simple, but not very elegant trick to specify the function div uniquely after all would be to give an arbitrary specification for the case that the second parameter is 0, for instance, by specifying:

$$div(m, 0) = 0.$$

However, this trick is by no means very elegant and even, in general, not always possible. The idea of restricting the equations by predicates that characterise the parameters for which the function is defined is not always so simple to achieve. In the case of functions with nonrecursive domains (more precisely, not recursively enumerable codomains), in fact, we cannot even formulate the domain restriction by a decidable (computable) condition.

The two equations given above provide, of course, not a classical inductive definition. In a classical inductive definition we work at the left-hand side of the equation with patterns of the form div(0, n) and div(m+1, n). For division, such a version of a specification is neither very efficient nor very elegant nor intuitive. Nevertheless, it can be rather easily formulated as follows:

$$div(0, n) \quad = \quad 0$$

$$div(m+1, n) \quad = \quad \textbf{if } m+1-div(m, n)*n \geq n \quad \textbf{then } div(m, n)+1$$
$$\textbf{else } div(m, n)$$
$$\textbf{fi}$$

This is in fact a classical inductive definition, working with the standard Noetherian ordering on the natural numbers. In the case n = 0 we easily deduce the equation

$$div(m, n) \quad = \quad m$$

by the two defining equations which is certainly a possible, but of course arbitrary and therefore somewhat artificial choice for the value of div(m, 0). We might call such an implicit specification of the result of a function by some rather arbitrarily chosen value an *overspecification*. It constraints the function div in a way not justified by its underlying theory of arithmetic.

4   Recursive Equations in CPOs and Metric Spaces

Another option that avoids - in contrast to inductive principles - the requirements of the existence of an inductive order is to introduce either a partial ordering or a metric distance into the function space of the recursively defined function. This construction is done such that these sets are turned into complete partially ordered sets, or into complete metric spaces. Then it has to be shown that the defining functional is monotonic, or, in the sense of the metric, strongly contracting. From this, by general theorems, the existence of a (least) fixpoint can be concluded, which is a particular solution for the recursive equations.

4.1   Least Fixpoints and CPOs of Partial Functions

In the partial order approach even recursive equations are treated which do not define fixpoints uniquely. A specific fixpoint ("least fixpoint") is associated with the recursive equation (or more precisely with the function associated with the right-hand side of the defining equation) by selecting the least function in that ordering that fulfils the equations. We may even treat recursive equations that way which, added in a naive approach for total functions, would introduce a contradiction. This is achieved by introducing elements representing "undefined" with specific logical properties[1]. This way partial functions are represented by total functions.

In the general case, however, there may exist many solutions (meaning several functions that fulfil the equations) for the recursive equation. The classical approach of fixpoint theory is then to choose the least solution, the so-called *least fixpoint* of the functional, associated with the defining equations.

As it is well-known we can turn the set of the natural numbers by a simple extension into a complete partially ordered set (cpo). We introduce a pseudo element $\perp$ that serves as a dummy for the result of function applications that does not have a well-defined result. Along these lines, we define the following "natural" extension of the set of natural numbers

$$\mathbb{N}^{\perp} = \mathbb{N} \cup \{\perp\}$$

We define the function

$$\text{div: } \mathbb{N}^{\perp} \times \mathbb{N}^{\perp} \rightarrow \mathbb{N}^{\perp}$$

on the set $\mathbb{N}$ extended by $\perp$. We specify the function div on this extended set by ($\forall$ m, n $\in \mathbb{N}$) the equations:

$$\text{div}(m, n) = 0 \qquad\qquad \Leftarrow n > m,$$
$$\text{div}(m, n) = \text{div}(m-n, n) + 1 \qquad \Leftarrow n \leq m.$$

Here we do not have to give any restrictions for the value of the argument n in the second equation. The reason is that for the case n = 0, although we now get the equation

$$\text{div}(m, 0) = \text{div}(m, 0) + 1$$

which does not lead to a contradiction since we may choose (and in this case can even deduce this equation from the defining equations since $\perp$ is the only element that fulfils the equation)

$$div(m, 0) = \perp$$

One might ask, why introducing the dummy $\perp$ not only into the range of the function div but also into its domain. The answer is simple. If we want to freely form expressions of nested function applications such as in the term

$$div(div(1, 0), 2)$$

we have to allow for the case that also the values of the arguments of the function div might be $\perp$. However, then we have also to specify the result of the function div in cases where one of its arguments is $\perp$. For that case we choose a simple solution. We assume that div and all other arithmetic functions are *strict* functions. This means that the result of a function is $\perp$ whenever by strictness one of its arguments is $\perp$. For div we get the equations:

$$div(x, \perp) = div(\perp, x) = \perp$$

This idea of a "strict" extension of total or partial functions to total functions on domains and ranges that are extended by the element $\perp$ can be used also for all the other functions schematically such as the arithmetic functions. This extension is required, anyhow, to be able to cope properly with terms like $div(n, 0) + 1$.

The set of strict functions on domains and ranges that are extended by the dummy $\perp$ is isomorphic to the set of partial functions on sets the same domains and ranges without the element $\perp$. It is not difficult to reformulate (see [Broy 86] all constructs for strict functions for the set of partial functions and vice versa. However, when interested in nonstrict functions the concept of partial functions is no longer powerful enough.

We introduce a partial order $\sqsubseteq$ on the set $\mathbb{N}^\perp$ as follows:

$$\forall\, m, n \in \mathbb{N}^\perp\colon m \sqsubseteq n \Leftrightarrow (m = \perp \vee n = m)$$

It extends to functions by pointwise application

$$f \sqsubseteq f \Leftrightarrow \forall\, m, n \in \mathbb{N}^\perp\colon f(m, n) \sqsubseteq f(m, n)$$

A related approach to domains for recursive equation are metric spaces. In the metric space approach the treatment of partiality and underspecification is not so simple. The functions for which we want to find a fixpoint are required to be strongly contracting, in general. If they are, they have a unique fixpoint[2].

We do not give a metric version of the treatment of the recursive equation for div since the classical approaches work only for functions that are total such that the defining functions are contracting and have unique fixpoints (see [de Bakker, Zucker 84]). Therefore they do not apply immediately to our example. We come back to the metric space approach in section 4.3 when working with sets of functions.

---

[2] We may work with set-valued functions instead of functions producing single elements as results. Then we may drop the requirements of strong contractivity and replace it by weak contractivity.

4.2   Complete Lattices of Predicates

A technique that is very similar to the cpo-based least fixpoint approach is the complete lattice of predicates. As it is well known, predicates are partially ordered by logical implication and form a complete lattice. We explain the idea with the help of our example. Let us consider the following functional

$$\tau: (\mathbb{N}^\perp \times \mathbb{N}^\perp \to \mathbb{N}^\perp) \to (\mathbb{N}^\perp \times \mathbb{N}^\perp \to \mathbb{N}^\perp)$$

that maps functions onto functions and that is specified by the term in the defining equation for division as follows:

$$\tau[f](m, n) = \textbf{if } m < n \textbf{ then } 0 \textbf{ else } f(m-n, n) \textbf{ fi}$$

The functional τ is induced by the recursive equation that is specifying div. The fixpoint equation then reads follows:

$$div = \tau[div].$$

We may replace the functional τ associated with the recursive equation by a predicate transformer that operates on predicates over functions:

$$T: ((\mathbb{N}^\perp \times \mathbb{N}^\perp \to \mathbb{N}^\perp) \to \textbf{B}) \to ((\mathbb{N}^\perp \times \mathbb{N}^\perp \to \mathbb{N}^\perp) \to \textbf{B})$$

It is specified by (note the similarity to the functional τ introduced above) the recursive equation

$$T[Q].f \equiv \ \forall m, n \in \mathbb{N}: \ m < n \Rightarrow f(m, n) = 0$$
$$\wedge \quad m \geq n \Rightarrow \exists f: Q[f] \wedge f(m, n) = 1 + f(m-n, n)$$

or expressed with the help of the function τ:

$$T[Q].f = \exists f: f = \tau[f] \wedge Q[f]$$

Obviously, the function T is an inclusion monotonic function on predicates[1]. Therefore it has (recall that the set of predicates forms a complete lattice) as well known from μ-calculus a weakest and a strongest fixpoint. It is not difficult to show that the strongest fixpoint of the predicate transformer T is the predicate λ f: false and the weakest fixpoint is the predicate

$$\lambda f: \forall m, n \in \mathbb{N}: \ (m < n \Rightarrow f(m, n) = 0)$$
$$\wedge (m \geq n \wedge n > 0 \Rightarrow f(m, n) = 1 + f(m-n, n))$$

Hence the weakest fixpoint is the predicate on functions that characterises all functions that fulfil the defining equations for div but produce arbitrary results in the case the second parameter is 0.

In general, we may treat recursive equations along the lines explained above as follows. Introducing a function

$$f: D \to R$$

---

[1] We call T also a *predicate transformer*.

specified by the recursive equation

$$f(x) = E$$

we may translate the recursive equation into a predicate

$$Q: (D \to R) \to \mathbf{B}$$

specified by the equivalence

$$Q[f] \equiv \quad \forall \, x: \exists \, f_1, ..., f_n: Q[f_1] \wedge ... \wedge Q[f_n] \wedge f(x) = E^*$$

where the expression E^* is defined as in the introduction where we defined

$$E = E^*[f/f_1, ..., f/f_n]$$

This definition of the predicate Q does never introduce a contradiction since the right-hand side is inclusion monotonic (or, in other terms, implication monotonic) in the predicate Q. According to μ-calculus a strongest and a weakest solution exist. We can easily show, moreover, that the weakest solution is never the strongest predicate λ f: false provided the equation $f(x) = E$ specifying the function f has a solution. We only have to choose that solution for all the function $f_1, ..., f_n$ to obtain one solution. But there are many other solutions, in general. More precisely there may be functions f that fulfil the proposition Q[f] where Q is the weakest solution. These functions need not by fixpoints of τ. In the case of the function div as defined by the cpo approach these functions f all have the property div $\sqsubseteq$ f.

We may replace the definition of Q by the more liberal equation

$$Q.f = \exists \, f': \tau[f'] \sqsubseteq f' \wedge f' \sqsubseteq f$$

We prove that the predicate

$$Q.f \equiv div \sqsubseteq f$$

is a fixpoint of T where

$$T[Q].f = \exists \, f' : Q.f' \wedge \tau[f'] \sqsubseteq f$$

as follows:

$$T[Q].f$$

$$\equiv \exists \, f' : \tau[f'] \sqsubseteq f \wedge Q[f']$$

$$\equiv \exists \, f' : \tau[f'] \sqsubseteq f \wedge div \sqsubseteq f'$$

Monotonicity of τ shows

div	{fixpoint property}
$= \tau[div]$	{monotonicity of τ, div $\sqsubseteq$ f'}
$\sqsubseteq \tau[f']$	
$\sqsubseteq f$	

Thus $T[Q].f \Rightarrow div \sqsubseteq f$.

Now assume div $\sqsubseteq$ f. Then (div is a fixpoint)

$$\tau[\text{div}] \sqsubseteq f$$

Thus we have

$$\exists\ f' : \tau[f'] \sqsubseteq f \wedge Q.f$$

This shows that div $\sqsubseteq$ f $\Rightarrow$ T[Q].f.

4.3   Metric Spaces

Similar as above where we work with predicates on functions we may work with metric spaces over the set of functions or relations. We introduce a metric distance on the function space $(D \rightarrow R)$:

$$d: (D \rightarrow R) \times (D \rightarrow R) \rightarrow \mathbf{R}$$

such that $(D \rightarrow R)$ is a complete metric space. A function

$$\Phi: (D \rightarrow R) \rightarrow (D \rightarrow R)$$

is called *weakly contracting* if for all sets of functions $g_1, g_2 \in (D \rightarrow R)$:

$$d(g_1, g_2) \geq d(\Phi(g_1), \Phi(g_2))$$

We call the function Φ *strongly contracting* if there exists a real number $\varepsilon \in \mathbf{R}$ with $0 < \varepsilon < 1$ such that

$$\varepsilon\ d(g_1, g_2) \geq d(\Phi(g_1), \Phi(g_2))$$

The idea of metric spaces for proving the existence of fixpoints is well known. Given a complete metric space (X, d) and a strongly contracting function

$$\tau: X \rightarrow X$$

there exists a unique fixpoint of τ.

The critical issue is to select an appropriate metric distance on functions. In our example of functions in $\mathbf{N} \times \mathbf{N} \rightarrow \mathbf{N}$ we may define a metric distance as follows (where ε is a number with $0 < \varepsilon < 1$):

$$d(f_1, f_2) = \max\ \{\varepsilon^{nm}: f_1(n, m) \neq f_2(n, m)\}$$

This metric induces a metric distance on sets of functions and turns the function space into a complete metric space. In fact, with this metric distance every inductive definition leads to a contractive functional. Note, that in our running example the functional τ is not strongly contracting. The technique of metric spaces as we introduced it works only for recursive equations that define least fixpoints that total functions.

5   Lattice of Predicate Logics

When following logic oriented approaches to give semantics to recursion we work with inclusion-monotonic functions on predicates called *predicate transformers* where the specified functions are represented as relations or predicates. Recursive equations can be mapped onto predicate transformers and for predicate transformers we can apply constructions from μ-calculus. This way we can associate strongest or weakest predicates with recursive equations specifying functions. These predicates characterise functions or relations.

Note that both the choice of the weakest as well as the choice of the strongest solution of recursive equations leads to interesting interpretations of recursive equations. An example for such a treatment is given already at the end of the previous section.

We may define the ternary predicate and the naturals Div (representing a relation) also directly recursively by the following logical equivalence:

$$\mathrm{Div}(m, n, r) \equiv \textbf{if} \quad m < n \quad \textbf{then} \quad r = 0$$
$$\textbf{elif} \quad r = 0 \quad \textbf{then} \quad \text{false}$$
$$\textbf{else} \quad \mathrm{Div}(m\text{-}n, n, r\text{-}1)$$
$$\textbf{fi}$$

which reads in a more logical style (translating the **if-then-else-fi** into classical logical connectors)

$$\mathrm{Div}(m, n, r) \equiv (m < n \wedge r = 0) \vee (r > 0 \wedge m \geq n \wedge \mathrm{Div}(m\text{-}n, n, r\text{-}1))$$

It is not difficult to show that from these equivalences we obtain directly the following conditional equivalences

$$m < n \Rightarrow \mathrm{Div}(m, n, r) \equiv (r = 0)$$

$$m \geq n \wedge r = 0 \Rightarrow \mathrm{Div}(m, n, r) \equiv \text{false}$$

$$m \geq n \wedge r > 0 \Rightarrow \mathrm{Div}(m, n, r) \equiv \mathrm{Div}(m\text{-}n, n, r\text{-}1)$$

From these formulas we can deduce (via an easy proof by induction on the naturals) the following logical consequence:

$$n > 0 \Rightarrow \mathrm{Div}(m, n, r) \equiv (0 \leq m\text{-}n*r < n)$$

In the case n = 0 we obtain the equivalences

$$r = 0 \Rightarrow \mathrm{Div}(m, 0, r) \equiv \text{false}$$

$$r > 0 \Rightarrow \mathrm{Div}(m, 0, r) \equiv \mathrm{Div}(m, 0, r\text{-}1)$$

So by straightforward induction the only choice for the logical value of Div(m, 0, r) is therefore

$$\forall\, m, r \in \mathbf{N}: \mathrm{Div}(m, 0, r) \equiv \text{false}$$

In contrast to functions modelling division in propositions Div(m, n, r) and in their isomorphic representation by ternary relations we do not have any indication which of the three arguments are considered as input and which as output for the operation to be defined. It is therefore more explicit to work instead with a set-valued function

Manfred Broy

DIV: $\mathbb{N} \times \mathbb{N} \rightarrow \wp(\mathbb{N})$

that is specified by the following equation

DIV(m, n) ≡   **if** m < n   **then** {0}
                       else {y+1: y ∈ DIV(m-n, n)}
    fi

Recall, that the powerset is a complete lattice ordered by set inclusion. Moreover, consider the functional Θ, defined by the equation

Θ[F](m, n) ≡   **if** m < n   **then** {0}
                   else {y+1: y ∈ F(m-n, n)}
            fi

We have DIV = Θ(DIV). Θ is monotonic with respect to the set inclusion ordering. More precisely, it is monotonic with respect to the ordering on set-valued functions induced by pointwise application of the inclusion ordering on sets. Therefore, since the sets form a complete lattice there exists an inclusion least and an inclusion greatest fixpoint according to Knaster-Tarski. The least fixpoint is described by the set-valued function

$$\lambda \ m, n: \{r \in \mathbb{N}: 0 \leq m - n * r < n\}$$

The greatest fixpoint is given by the same set-valued function. The set-valued functions are isomorphic to relations. They stress, however, which of the elements of the tuples in a relation are input and which are considered as output.

Of course, we may also work with sets of natural numbers extended by the element ⊥ as a dummy for undefined when associating relations or set-valued functions with recursive definitions. This way we obtain combinations of the partial order approach and the lattice of sets approach.

6   Sets of Models

When dealing with algebraic equations for the specification of functions which can also be seen as recursive equations for functions it is common by now to work with so called *loose semantics* approaches (see [Broy, Wirsing 82]). This means that we associate not only exactly one model with a set of axioms, such as for instance an initial model, but, in general, a set of models in terms of heterogeneous algebras with an algebraic specification (which is a logical theory). If we consider a purely equational specification and restricted forms of axioms we can identify extreme models in the class of models such as initial or terminal algebras. This works even in the case of conditional equations. These initial or terminal algebras are closely related to strongest and weakest solutions in a form of predicates that are associated with the logical treatment of recursion (see [Broy, Wirsing 80]).

Let the following specification of natural numbers be given (we follow closely the syntax and concepts of Larch, see [Larch 93]):

SPEC NAT =

{   based_on BOOL

    sort Nat

0 : Nat,
succ, pred :   Nat → Nat,
iszero :   Nat → Bool,
+, *,- :   Nat, Nat → Nat,   *Infix*

Nat **generated_by** 0, succ,

iszero(0) = true,
iszero(succ(x)) = false,

pred(succ(x)) = x,

0+y = y,
succ(x)+y = succ(x+y),

x-0 = x,

0-y = y,

succ(x)-succ(y) = x-y,

0*y = 0,

succ(x)*y = y+(x*y)   }

It defines the natural numbers by the help of an induction principle and some basic operations (for details see [Larch 93]). Recall that functions in Larch are assumed to be total. If we add the following specification fragment (extending the specification NAT)

div: Nat, Nat → Nat

$\text{div}(m, n) = 0$ $\Leftarrow m < n$ $(*)$

$\text{div}(m, n) = \text{div}(m-n, n)+1$ $\Leftarrow m \geq n$ $(**)$

to the specification NAT above, we get a contradiction (all functions are assumed to be total), since with the help of the generation principles which gives the basis for induction proofs we may deduce the proposition: $\forall\ n \in$ Nat: $n \neq n+1$. As demonstrated before the equation

$$\text{div}(m, 0) = \text{div}(m, 0)+1$$

leads to a contradiction. If we drop the term generation principle (this is the principle to consider standard models only) then induction is no longer available as a proof principle and the contradiction can no longer be deduced since div(m, 0) may be a non-standard-value.

However, giving up induction would hurt. The other option to avoid the inconsistency while maintaining the principle of induction is to use the following conditional defining equation:

$$\text{div}(m, n) = \text{div}(m\text{-}n, n)+1 \Leftarrow m \geq n \wedge n > 0$$

instead. Adding only this equation and the equation (∗) there exist many models for the enriched specification. Each of these models contains a function div with arbitrary choices for the results of function application div(m, 0). This approach corresponds again exactly to the idea of underspecification.

7 The Herbrand Universe

Another area where recursive declarations are used is logic programming. Here we work with the concept of term models called *Herbrand models*. These so called Herbrand models are used to interpret "recursive" Horn clauses.

When dealing with ideas from logic programming we do not represent operations like division by functions but by relations or by predicates. Along these lines we may describe division by the predicate

$$\text{Div}: \mathbb{N} \times \mathbb{N} \times \mathbb{N} \to \mathbb{B}$$

specified by the following Horn-clauses:

$$\text{Div}(m, n, 0) \qquad\qquad \Leftarrow m < n$$

$$\text{Div}(m\text{+}n, n, r\text{+}1) \qquad \Leftarrow \text{Div}(m, n, r)$$

Of course, there are many predicates Div that fulfil these axioms. The weakest of these predicates is true (more precisely the predicate λ x, y, r: true). The strongest predicate corresponds to the so called *closed world assumption*, which leads to the strongest predicate that fulfils the Horn-clauses. It is specified by the equivalence

$$\text{Div}(m, n, r) \equiv (0 \leq m\text{-}n*r < n)$$

This relation Div directly represents the partial function div discussed extensively above. The closed world assumption simply assumes that all facts that are not explicitly stated as being true (more precisely that cannot be deduced logically from the axioms) are false. This is exactly mirrored by the possible computations (which may be seen as logical deductions) and also by the strongest fixpoint. This is the standard semantics used in logic programming.

Of course, there are many other solutions (other predicates that fulfil the equations). For every natural number $k \in \mathbb{N}$ we get a predicate Div_k specified by the equation

$$\text{Div}_k(m, n, r) \equiv (0 \leq m\text{-}n*r < n) \vee k \leq r.$$

These relations Div_k and the strongest fixpoint are examples for fixpoints (solutions) of the defining Horn-clauses for Div.

8 Conclusion

Recursion is a fundamental concept in computer science. Recursion is used both explicitly, such as in recursive data type declarations, recursive function declarations, or in formal languages, and implicitly, such as in loops, everywhere. There are many options to treat the semantics of recursive equations. They all have serious impacts on the logical theories and mathematical models of programming.

Let us finally survey the considered options for the treatment of recursion shortly once more: to treat recursive declarations we have to observe the following facts:

Models with total functions without any extension to "undefined" cannot be extended by recursive equations without running into contradictions for certain recursive equations. We have to be careful to add conditions to those equations to avoid contradictions. However, in the worst case, such conditions are not recursive and thus not computable.

We can extend the logic of total functions to partial functions, functions on cpos, relations, predicates, set-valued functions of even sets of functions. All such extensions can be used to treat recursion. In fact, such an extension makes the logic more sophisticated, in general.

As we have demonstrated the different possibilities may be combined. For instance, we can work with sets of total functions or with sets of partial functions. Each of these approaches has its advantages and disadvantages. Working with total functions allows us to keep the logics simple, for the price that not all computable functions can be described by computable expressions and that contradictions and incompleteness may be introduced.

A second disadvantage, from a practical point of view perhaps a more serious one, of the concept of total functions with underspecification is the fact that the arguments for which, operationally speaking, the recursion does not terminate are not distinguished logically from the cases where the values of the function are well-defined by a terminating recursion. This is awful and unacceptable from the point of view of software engineering since reasoning about exceptions, termination, and definedness is an important part of the specification and analysis of reliability and verification of programs. We want to be able to distinguish bad and unacceptable arguments from good acceptable ones! Therefore we are in favour of explicit representations of undefined (see the discussion in [Hehner 74]).

Appendix

As we have shown, a purely equational treatment and characterisation of solutions of recursive equations is difficult. However, there is one logical "trick" that allows us to work with total functions even in cases of recursive definitions that lead within a logical setting to least fixpoints that are function with non-recursively enumerable codomains. For each function

$$f: D \to R$$

that is specified by the recursive equation

(*) $f(x) = E$

we introduce together with the function symbol f a corresponding predicate

dom$_f$: $D \to \mathbb{B}$

that characterises the subset of the domain D for which the function f has to have a well-specified result and replace therefore the recursive equation (*) by the weaker implication

dom$_f(x) \Rightarrow f(x) = E$

Of course, this implication alone is not strong enough to characterize the function associated with the symbol f and certainly not at all to characterize the predicate dom$_f$ the way we want it. A trivial choice to fulfil the conditional equation would be to choose dom$_f(x)$ = false. Then the conditional equation is trivially fulfilled. Therefore we need additional axioms for specifying the domain restriction predicate dom$_f$.

We assume for simplicity that all given function and operation symbols g: $D \to R$ in our signature have an associated domain predicate

dom$_g$: $D \to \mathbb{B}$

We introduce a syntactic rewrite procedure that produces for every term E syntactically a logical formula DEF[E] that characterizes the proposition the value of the expression "E is well defined". It is specified as follows:

DEF[x] $\equiv$ true   for
identifiers x

DEF[h(E_1,...,E_n)] $\equiv$ dom$_h(E_1, ..., E_n) \wedge$ DEF[E_1] $\wedge$ ... $\wedge$ DEF[E_n]

DEF[**if** C **then** E_1 **else** E_2 **fi**] $\equiv$

DEF[C] $\wedge$ (C $\Rightarrow$ DEF[E_1]) $\wedge$ ($\neg$C $\Rightarrow$ DEF[E_2])

For total functions g we simply choose dom$_g(x) \equiv$ true for all x.

Based on these definitions we replace the recursive equation

$f(x) = E$

by the following two axioms

(**)   dom$_f(x) \Rightarrow f(x) = E$

DEF[E] $\Rightarrow$ dom$_f(x)$

If all function symbols g occurring in function calls in the expression E are totally defined in the sense of dom$_g(x)$ = true for all except the function f we can simplify this treatment. The formula DEF[E] then only refers to the definedness of the recursive calls in the term E.

By this simple encoding of the domain predicate we work with underspecification both for the function f and the domain predicate dom$_f$. Formally, dom$_f$ is a predicate and has nothing to do with the function f. However, the predicate is used as a guard for the recursive defining equation for f. So the defining equation is not required to be valid if $\text{dom}_f(x)$ is false. This avoids contradictions, since in the case the function application $f(x)$ does not terminate we cannot derive $\text{dom}_f(x) = \text{true}$ and thus always may choose $\text{dom}_f(x) = \text{false}$.

We demonstrate how our idea works in the case of our simple example. In the case of division we get the following defining equations for div:

$$\text{dom}_{\text{div}}(m, n) \wedge n > m \Rightarrow \text{div}(m, n) = 0$$

$$\text{dom}_{\text{div}}(m, n) \wedge n \leq m \Rightarrow \text{div}(m, n) = 1 + \text{div}(m-n, n)$$

and the following axioms for dom_{div}:

$$n > m \Rightarrow \text{dom}_{\text{div}}(m, n)$$

$$\text{dom}_{\text{div}}(m-n, n) \Rightarrow \text{dom}_{\text{div}}(m, n)$$

By this we can prove the definedness of the function div for all its arguments (m, n) with $n > 0$ introduced by a recursive equation. This way we exactly mimic the way computations are executed.

However, the introduction of a domain predicate is only a logical trick which encodes the classical idea of fixpoint theory for partial functions into logics of total functions. In the case of our example we can prove

$$\text{dom}_{\text{div}}(m, n) \Leftarrow n \neq 0$$

Moreover, by contradiction, (assuming div is a total function) we can prove $\neg\text{dom}_{\text{div}}(m, 0)$ since $\text{dom}_{\text{div}}(m, 0)$ would lead to a contradiction since by this we could deduce

$$\text{div}(m, 0) = 1 + \text{div}(m, 0)$$

Hence a proof by contradiction yields $\neg\text{dom}_{\text{div}}(m, 0)$.

Note that, in general, however, the predicate, dom_f is not uniquely specified by the axioms (**). If several fixpoints exist for the recursive equation $f(x) = E$ then also several solutions for domain predicate dom_f exist. If we choose the strongest predicate for dom_f this reflects the idea of the least defined fixpoint.

References

[Broy 86]
  M. Broy: Partial interpretations of higher order algebraic types. (Invited lecture) In: J. Gruska (ed): Mathematical Foundations of Computer Science-13th Symposium, Lecture Notes in Computer Science 233, Berlin-Heidelberg-New York-Tokyo: Springer 1986, 29-43
[Broy, Wirsing 80]
  M. Wirsing, M. Broy: Abstract data types as lattices of finitely generated models. In: P. Dembinski (ed.): Mathemarical Foundations of Computer Science - 9th Symposium. Rydzyna 1980, Lecture Notes in Computer Science 88, Berlin-Heidelberg-New York: Springer 1980, 673-685
[Broy, Wirsing 82]
  M. Broy, M. Wirsing: Initial versus terminal algebra semantics for partially defined abstract types. Technische Universität München, Institut für Informatik, TUM-I8018, December 1981. Revidierte Fassung: Partial Abstract Types, Acta Informatica 18, 1982, 47-64

[Broy, Pepper, Wirsing 87]
   M. Broy, M. Pepper, M. Wirsing: On the algebraic definition of programming languages. Technische Universität München, Institut für Informatik, TUM-I8204, 1982. Revised version in TOPLAS 9:1 (1987) 54-99
[de Bakker, Zucker 84]
   J. W. de Bakker and J. I. Zucker. Processes and the denotational semantics of concurrency. Information and Control, 54(1/2):70-120
[Hehner 84]
   E.C.R. Hehner: Predicative Programming. Comm. ACM 27:2, 1984, 134-151
[Knaster-Tarski]
   A. Tarski: A lattice-theoretical fixpoint theorem and its application. Pacific Journal of Mathematics Vol. 5, 1955, 285-309
[Larch 93]
   John V. Guttag and James J. Horning, with S.J. Garland, K.D. Jones, A. Modet, and J.M. Wing: Larch: Languages and Tools for Formal Specification, Springer-Verlag Texts and Monographs in Computer Science, 1993
[λ-calculus 81]
   H.P. Barendregt: The Lambda Calculus: Its Syntax and Semantics. North-Holland 1981
[μ-calculus 73]
   P. Hitchcock, D. Park: Induction rules and termination proofs. M. Nivat (ed.): Proc. Ist ICALP. North Holland 73
[π-calculus 99]
   R. Milner: Communication and mobile systems: the π-calculus. Cambridge University Press 1999
[Prolog/Herbrand Universe 87]
   J. Lloyd. Foundations of Logic Programming: 2nd Edition. Springer-Verlag, 1987
[PVS 92]
   S. Owre, J. M. Rushby, N. Shankar: PVS: A Prototype Verification System. In: Deepak Kapur (ed.):11th Conference on Automated Deduction, Saratoga, NY, Jun, 1992
[Schieder, Broy 99]
   B. Schieder, M. Broy: Adapting Calculational Logic to the Undefined. The Computer Journal, Vol. 42, No. 2, 1999
[Sintzoff 87]
   M. Sintzoff: Expressing program developments in a design calculus. M. Broy (ed.): Logic of programming and calculi of discrete design. Springer NATO ASI Series, Series F: Computer and System Sciences, Vol. 36, 1987, 343-365
[Scott 81]
   D. Scott: Lectures on a mathematical theory of computation. In: Theoretical Foundations of Programming Methodology, edited by M. Broy and G. Schmidt. D. Reidel Publishing Company, 1982, pp. 145 - 292

Completion Is an Instance of Abstract Canonical System Inference

Guillaume Burel[1] and Claude Kirchner[2]

[1] Ecole Normale Supérieure de Lyon & LORIA*
[2] INRIA & LORIA*

Abstract. Abstract canonical systems and inference (ACSI) were introduced to formalize the intuitive notions of good proof and good inference appearing typically in first-order logic or in Knuth-Bendix like completion procedures.

Since this abstract framework is intended to be generic, it is of fundamental interest to show its adequacy to represent the main systems of interest. This has been done for ground completion (where all equational axioms are ground) but was still an open question for the general completion process.

By showing that the standard completion is an instance of the ACSI framework we close the question. For this purpose, two proof representations, proof terms and proofs by replacement, are compared to built a proof ordering that provides an instantiation adapted to the abstract canonical system framework.

Classification: Logic in computer science, rewriting and deduction, completion, good proof, proof representation, canonicity.

1 Introduction

The notion of good proof is central in mathematics and crucial when mechanizing deduction, in particular for defining useful and efficient tactics in proof assistant and theorem provers. Motivated on one hand by this quest for *good proof* theory and on the other by the profound similarities between many proof search approaches, N. Dershowitz and C. Kirchner proposed in [17, 18] a general framework based on ordering the set of proofs. In this context the best proofs are simply the minimal one. Once one has defined what the best proofs are by the mean of a proof ordering, the next step is to obtain the best presentation of a theory, i.e. the set of axioms necessary for obtaining the best proofs for all the theory, but not containing anything useless.

To formalize this, the notion of *good inference* was introduced by M.P. Bonacina and N. Dershowitz [6]. Given a theory, its canonical presentation is defined as the set of the axioms needed to obtain the minimal proofs. It is general enough to produce all best proofs, leading to a notion of *saturation*, but

---

* UMR 7503 CNRS-INPL-INRIA-Nancy2-UHP

K. Futatsugi et al. (Eds.): Goguen Festschrift, LNCS 4060, pp. 497–520, 2006.

it does not contain any redundant informations, hence the notion of *contraction*. Presentations, i.e. sets of axioms, are then transformed using appropriate deduction mechanisms to produce this canonical presentation.

This leaded to the Abstract Canonical Systems and Inference (ACSI) generic framework presented in [18, 6].

The ACSI framework got its sources of inspiration from three related points. First, the early works on *Proof orderings* as introduced in [3] and [4] to prove the completeness of completion procedures *a la* Knuth-Bendix. Second, the developments about redundancy [24, 5] to focus on the important axioms to perform further inferences. Last but not least, by the completion procedure [31], central in most theorem proving tools where an equality predicate is used. This procedure has been refined, mainly for two purposes: to have a more specific and thus more efficient algorithm when dealing with particular cases, or to increase the efficiency although remaining general. For the first case, a revue of specific completion procedures for specific algebraic structures can be found in [33]. For the second case, completion has been extended to equational completion [25, 36, 28]; inductionless induction, initiated by J.A. Goguen [21] and D. Musser [35]; and ordered completion [32, 24, 4], to mention only a few. One important application of the completion procedure is rewrite based programming, either based on matching or on unification. The seminal work of J.A. Goguen on OBJ and its various incarnations [22] plays a preeminent role in this class of algebraic languages and has directly inspired CafeOBJ [20], ELAN [8] or Maude [14]. When the operational semantics of the language is based on unification, we find logic programming languages of the Prolog family, where EQLOG [23] is also a preeminent figure. Good syntheses about completion based rewrite programs can be found in [15, 7].

Several works intend to uniform this different completion procedures, and to make it a special case of a more general process. The notion of critical-pair completion procedure was introduced by [10] and covers not only standard completion, but also Buchberger algorithm for Gröbner basis [9, 42] and resolution [37]. Indeed, R. Bündgen shown that Buchberger's algorithm can be simulated by standard completion [11]. This concept of critical-pair completion was categorically formalized by K. Stokkermans [40]. Other generalizations can be found in works of M. Schorlemmer [39], M. Aiguier and D. Bahrami [1] or in the PhD of G. Struth [41], where standard completion, Buchberger's algorithm and resolution are shown to be special instantiation of a non-symmetric completion procedure.

But, even if initially motivated by these three points, the ACSI framework has been developed as a full stand alone theory. This theory provide important abstract results based on basic hypothesis on proofs and a few postulates.

Therefore, a main question remains: is this framework indeed useful? Does this theory allows to *uniformly* understand and prove the main properties of a proof system, centered around the appropriate ordering on proofs?

At the price of a slight generalization of two postulates, it is shown in [12], that good proofs in natural deduction are indeed the cut free proofs as soon as proofs are compared using the ordering induced by beta reduction over the simply typed lambda-terms. For ground completion, the adequacy of the framework has been shown in [16], leaving the more general question of standard completion open.

This paper proves the adequacy to the framework for the standard completion procedure, generalizing in a non trivial way the result of [16] and showing the usefulness of abstract canonical systems. This brings serious hopes that the ACSI framework is indeed well adapted and useful to uniformly understand and work with other algorithms, in particular all the ones based on critical-pair completion.

The next section will summarize the framework of abstract canonical systems, as defined in [18, 6], and briefly recall the standard completion. Section 3 deals with two representations of proofs in equational logic, namely as proof terms in the rewriting logic [34], and as proof by replacement [3]. We will show how to combine them to keep the tree structure of the first one, and the ordering associated with the second one, which is well adapted to prove the completeness of the standard completion. Finally, in Section 4, we will apply the abstract canonical systems framework to this proof representation to show the completeness of the standard completion. The proofs details are given in the Appendix.

2 Presentation

2.1 Abstract Canonical Systems

The results in this section are extracted from [18, 6], which should be consulted for motivations, details and proofs.

Let $\mathbb{A}$ be the set of all formulæ over some fixed vocabulary. Let $\mathbb{P}$ be the set of all proofs. These sets are linked by two functions: $[\cdot]^{Pm} : \mathbb{P} \to 2^{\mathbb{A}}$ gives the *premises* in a proof, and $[\cdot]_{Cl} : \mathbb{P} \to \mathbb{A}$ gives its *conclusion*. Both are extended to sets of proofs in the usual fashion. The set of proofs built using assumptions in $A \subseteq \mathbb{A}$ is noted by[3]

$$Pf(A) \overset{!}{=} \left\{ p \in \mathbb{P} : [p]^{Pm} \subseteq A \right\} .$$

The framework proposed here is predicated on two *well-founded* partial orderings over $\mathbb{P}$: a *proof ordering* $>$ and a *subproof relation* $\rhd$. They are related by a monotonicity requirement (postulate E). We assume for convenience that the proof ordering only compares proofs with the same conclusion $(p > q \Rightarrow [p]_{Cl} = [q]_{Cl})$, rather than mention this condition each time we have cause to compare proofs.

We will use the term *presentation* to mean a set of formulæ, and *justification* to mean a set of proofs. We reserve the term *theory* for deductively closed presentations:

$$Th\,A \overset{!}{=} [Pf(A)]_{Cl} = \{[p]_{Cl} : p \in \mathbb{P},\ [p]^{Pm} \subseteq A\} .$$

---

[3] $\overset{!}{=}$ is used for definitions.

Theories are monotonic:

Proposition 1 (Monotonicity). *For all presentations A and B:*

$$A \subseteq B \Rightarrow Th\,A \subseteq Th\,B$$

Presentations A and B are *equivalent* $(A \equiv B)$ if their theories are identical: $Th\,A = Th\,B$. In addition to this, we assume the two following postulates:

Postulate A (Reflexivity). *For all presentations A:*

$$A \subseteq Th\,A$$

Postulate B (Closure). *For all presentations A:*

$$Th\,Th\,A \subseteq Th\,A$$

We call a proof *trivial* when it proves only its unique assumption and has no subproofs other than itself, that is, if $[p]^{Pm} = \{[p]_{Cl}\}$ and $p \trianglerighteq q \Rightarrow p = q$, where $\trianglerighteq$ is the reflexive closure of the subproof ordering $\triangleright$. We denote by $\widehat{a}$ such a trivial proof of $a \in \mathbb{A}$ and by $\widehat{A}$ the set of trivial proofs of each $a \in A$.

We assume that proofs use their assumptions (postulate C), that subproofs don't use non-existent assumptions (postulate D), and that proof orderings are monotonic with respect to subproofs (postulate E):

Postulate C (Trivia). *For all proofs p and formulæ a:*

$$a \in [p]^{Pm} \Rightarrow p \trianglerighteq \widehat{a}$$

Postulate D (Subproofs Premises Monotonicity). *For all proofs p and q:*

$$p \trianglerighteq q \Rightarrow [p]^{Pm} \supseteq [q]^{Pm}$$

Postulate E (Replacement). *For all proofs p, q and r:*

$$p \triangleright q > r \Rightarrow \exists v \in Pf([p]^{Pm} \cup [r]^{Pm}).\ p > v \triangleright r$$

We make no other assumptions regarding proofs or their structure. As remarked in [6], the subproof relation essentially defines a tree structure over proof: a "leaf" is a proof with no subproofs but itself, and direct subproofs, i.e. subproofs that are not subproofs of another subproof, can be considered as "subtrees". These trees can be infinitely branching, but their height is finite because of the wellfoundedness of $\triangleright$.

The proof ordering $>$ is lifted to an ordering $\succsim$ over presentations:

$$A \succsim B \text{ if } A \equiv B \text{ and } \forall p \in Pf(A)\ \exists q \in Pf(B).\ p \geq q\ .$$

We define what a *normal-form proof* is, i.e. one of the minimal proofs of $Pf(Th\,A)$:

$$Nf(A) \;\overset{!}{=}\; \mu Pf(Th\,A) \;\overset{!}{=}\; \{p \in Pf(Th\,A) \;:\; \neg\exists q \in Pf(Th\,A).\, p > q\} \;.$$

The *canonical presentation* contains those formulæ that appear as assumptions of normal-form proofs:

$$A^\sharp \;\overset{!}{=}\; [Nf(A)]^{Pm} \;.$$

So, we will say that A is *canonical* if $A = A^\sharp$.

A presentation A is *saturated* if it supports all possible normal form proofs:

$$Pf(A) \supseteq Nf(A) \;.$$

The set of all *redundant formulæ* of a given presentation A will be denoted as follows:

$$Red\,A \;\overset{!}{=}\; \{r \in A \colon A \succsim A \setminus \{r\}\} \;.$$

and a presentation A is *contracted* if

$$Red\,A = \emptyset \;.$$

The following main result can then be derived [17]:

Theorem 1. *A presentation is canonical iff it is saturated and contracted.*

We now consider inference and deduction mechanisms. A *deduction mechanism* $\rightsquigarrow$ is a function from presentations to presentations and we call the relation $A \rightsquigarrow B$ a *deduction step*. A sequence of presentations $A_0 \rightsquigarrow A_1 \rightsquigarrow \cdots$ is called a *derivation*. The *result* of the derivation is, as usual, its *persisting* formulæ:

$$A_\infty \;\overset{!}{=}\; \liminf_{j \to \infty} A_j \;=\; \bigcup_{j > 0} \bigcap_{i > j} A_i \;.$$

A deduction mechanism $\rightsquigarrow$ is *sound* if $A \rightsquigarrow B$ implies $Th\,B \subseteq Th\,A$. It is *adequate* if $A \rightsquigarrow B$ implies $Th\,A \subseteq Th\,B$. It is *good* if proofs only get better:

$$\rightsquigarrow \;\subseteq\; \succsim \;.$$

A derivation $A_0 \rightsquigarrow A_1 \rightsquigarrow \cdots$ is *good* if $A_i \succsim A_{i+1}$ for all i.

We now extend the notion of saturation and contraction to derivation:

- A derivation $\{A_i\}_i$ is *saturating* if A_∞ is saturated.
- It is *contracting* if A_∞ is contracted.
- It is *canonical* if both saturating and contracting.

A canonical derivation can be used to build the canonical presentation of the initial presentation:

Theorem 2. *A good derivation is canonical if and only if*

$$A_\infty = A_0^\sharp \;.$$

2.2   The Standard Completion

The standard completion algorithm was first introduced by Knuth and Bendix in [31], hence the name it is often called. Its correctness was first shown by Huet in [26], using a fairness hypothesis. We use here a presentation of this algorithm as inference rules (see Fig. 1), as can be found in [3]. For basics on rewritings and completions, we refer to [2, 29].

The Knuth-Bendix algorithm consists of 6 rules which apply to a couple E, R of a set of equational axioms and a set of rewriting rules. It takes a reduction ordering $\gg$ over terms as argument. The rules are presented in Fig. 1.

Deduce: If (s, t) is a critical pair of R

$$E, R \quad \leadsto \quad E \cup \{s = t\}, R$$

Orient: If $s \gg t$

$$E \cup \{s = t\}, R \quad \leadsto \quad E, R \cup \{s \to t\}$$

Delete:

$$E \cup \{s = s\}, R \quad \leadsto \quad E, R$$

Simplify: If $s \xrightarrow{R} u$

$$E \cup \{s = t\}, R \quad \leadsto \quad E \cup \{u = t\}, R$$

Compose: If $t \xrightarrow{R} u$

$$E, R \cup \{s \to t\} \quad \leadsto \quad E, R \cup \{s \to u\}$$

Collapse^a: If $s \xrightarrow{v \to w \in R} u$, and $s \blacktriangleright v$,

$$E, R \cup \{s \to t\} \quad \leadsto \quad E \cup \{u = t\}, R$$

Fig. 1. Standard Completion Inference Rules.

---
^{a} $\blacktriangleright$ designate the encompassment ordering, $s \blacktriangleright t$ if a subterm of s in an instance of t but not vice versa.

Since [26], standard completion is associated with a fairness assumption (see [3, Lemma 2.8]): at the limit, all equations are oriented ($E_\infty = \emptyset$) and all persistent critical pairs coming from R_∞ are treated by **Deduce** at least once.

Because we work with terms with variables, the reduction ordering $\gg$ cannot be total, so that **Orient** may fail. Therefore, the standard completion algorithm may either:

- terminate with success and yield a terminating, confluent set of rules;
- terminate with failure; or
- not terminate.

Here, the completeness of the standard completion will only be shown using the ACSI framework for the first case.

3   Proof Representations

Our goal is now to use the ACSI framework to directly show that standard completion inference rules are correct and complete. We have therefore first to find the right order on proofs. We have two main choices that we are now defining and relating.

3.1   Proof Terms

Let us first consider the proof representation coming from the one used in rewriting logic (introduced by Meseguer [34], see also [30]). Consider a signature Σ, and a set of variable V. The set of terms built upon these signature and variables is noted $\mathcal{T}(\Sigma, V)$. Consider also a set of equational axioms E and a set of rewrite rules R based on this signature. To simplify the notations of proof terms, equational axioms and rewrite rules are represented by labels not appearing in the signature Σ. An equational axiom or a rewrite rule $(l, r) \in E \cup R$ will be also noted $(l(x_1, \ldots, x_n), r(x_1, \ldots, x_n))$ where $x_1, \ldots, x_n$ are the free variables of both sides. We consider the rules of the equational logic given in the Fig. 2. These inference rules define the *proof term* associated with a proof. The notation $\pi : t \longrightarrow t'$ means that π is a proof term—that could also be seen as a trace—showing that the term t can be rewritten to the term t'.

By definition, $\mathcal{T}(\Sigma, V)$ is plunged into the proof terms when they are formed with the rules **Reflexivity** and **Congruence**. Also, **Reflexivity** for $t \longrightarrow t$ is not essential because it can be replaced by a tree of **Congruence** isomorph to t. The proof terms associated are furthermore the same in both case: t. Notice that these proof terms are a restricted form of rho-terms [13].

Example 1. Consider the rewrite rules and equational axiom

$$\ell_1 : g(x) \longrightarrow d(x), \quad \ell_2 : s = t, \quad \ell_3 : l \longrightarrow r,$$

- r is a proof term of $r = r$,
- $f(\ell_1(\ell_2), (\ell_3; r)^{-1})$ is a proof term of $f(g(s), r) = f(d(t), l)$.

Some proof terms defined here are "essentially the same". For instance, the transitivity operator should be considered as associative, so that the proofs $(\pi_1; \pi_2); \pi_3$ and $\pi_1; (\pi_2; \pi_3)$ are equal. This can be done by quotienting the proof terms algebra by the congruence rules of Fig. 3. In particular, in proof terms, parallel rewriting can be combined in one term without transitivity. The **Parallel Moves Lemma** equivalence corresponds to the fact that this parallel rewriting can be decomposed by applying first the outermost rule, then the innermost, or conversely. (About the Parallel Moves Lemma, see for instance [27].)

504   Guillaume Burel and Claude Kirchner

Reflexivity:
$$t : t \longrightarrow t$$

Congruence:
$$\pi_1 : t_1 \longrightarrow t'_1 \quad \ldots \quad \pi_n : t_n \longrightarrow t'_n$$
$$\overline{f(\pi_1, \ldots, \pi_n) : f(t_1, \ldots, t_n) \longrightarrow f(t'_1, \ldots, t'_n)}$$

Replacement: For all rules or equational axioms
$\ell = (g(x_1, \ldots, x_n), d(x_1, \ldots, x_n)) \in E \cup R$,
$$\pi_1 : t_1 \longrightarrow t'_1 \quad \ldots \quad \pi_n : t_n \longrightarrow t'_n$$
$$\overline{\ell(\pi_1, \ldots, \pi_n) : g(t_1, \ldots, t_n) \longrightarrow d(t'_1, \ldots, t'_n)}$$

Transitivity:
$$\pi_1 : t_1 \longrightarrow t_2 \quad \pi_2 : t_2 \longrightarrow t_3$$
$$\overline{\pi_1; \pi_2 : t_1 \longrightarrow t_3}$$

Symmetry:
$$\pi : t_1 \longrightarrow t_2$$
$$\overline{\pi^{-1} : t_2 \longrightarrow t_1}$$

Fig. 2. Inference Rules for Equational Logic

Example 2. From the rules **Associativity**, **Identities** and **Inverse** we can deduce that the proofs $(\pi_1; \pi_2)^{-1}$ and $\pi_2^{-1}; \pi_1^{-1}$ are equivalent:

$$
\begin{aligned}
(\pi_1; \pi_2)^{-1} &\equiv (\pi_1; \pi_2)^{-1}; t \\
&\equiv (\pi_1; \pi_2)^{-1}; \pi_1; \pi_1^{-1} \\
&\equiv (\pi_1; \pi_2)^{-1}; \pi_1; t'; \pi_1^{-1} \\
&\equiv (\pi_1; \pi_2)^{-1}; \pi_1; \pi_2; \pi_2^{-1}; \pi_1^{-1} \\
&\equiv t''; \pi_2^{-1}; \pi_1^{-1} \\
&\equiv \pi_2^{-1}; \pi_1^{-1} .
\end{aligned}
$$

We similarly have $f(\pi_1, \ldots, \pi_n)^{-1}$ equivalent to $f(\pi_1^{-1}, \ldots, \pi_n^{-1})$, because

$$
\begin{aligned}
f(\pi_1^{-1}, \ldots, \pi_n^{-1}) &\equiv f(\pi_1^{-1}, \ldots, \pi_n^{-1}); f(t_1, \ldots, t_n) \\
&\equiv f(\pi_1^{-1}, \ldots, \pi_n^{-1}); (f(\pi_1, \ldots, \pi_n); f(\pi_1, \ldots, \pi_n)^{-1}) \\
&\equiv (f(\pi_1^{-1}, \ldots, \pi_n^{-1}); f(\pi_1, \ldots, \pi_n)); f(\pi_1, \ldots, \pi_n)^{-1} \\
&\equiv f(\pi_1^{-1}; \pi_1, \ldots, \pi_n^{-1}; \pi_n); f(\pi_1, \ldots, \pi_n)^{-1} \\
&\equiv f(t'_1, \ldots, t'_n); f(\pi_1, \ldots, \pi_n)^{-1} \\
&\equiv f(\pi_1, \ldots, \pi_n)^{-1} .
\end{aligned}
$$

3.2   Proofs by Replacement of Equal by Equal

This proof representation was introduced by [3] to prove the completeness of the Knuth-Bendix completion algorithm, using an ordering over such proofs that decreases for every completion step.

Associativity: For all proof terms π_1, π_2, π_3,

$$\pi_1; (\pi_2; \pi_3) \equiv (\pi_1; \pi_2); \pi_3$$

Identities: For all proof terms $\pi : t \longrightarrow t'$,

$$\pi; t' \equiv t; \pi \equiv \pi$$

Preservation of Composition: For all proof terms $\pi_1, \ldots, \pi_n, \pi_1', \ldots, \pi_n'$, for all function symbols f,

$$f(\pi_1; \pi_1', \ldots, \pi_n; \pi_n') \equiv f(\pi_1, \ldots, \pi_n); f(\pi_1', \ldots, \pi_n')$$

Parallel Moves Lemma: For all rewrite rules or equational axiom $\ell = (g(x_1, \ldots, x_n), d(x_1, \ldots, x_n)) \in E \cup R$, for all proof terms $\pi_1 : t_1 \longrightarrow t_1', \ldots, \pi_n : t_n \longrightarrow t_n'$,

$$\ell(\pi_1, \ldots, \pi_n) \equiv \ell(t_1, \ldots, t_n); d(\pi_1, \ldots, \pi_n)$$
$$\equiv g(\pi_1, \ldots, \pi_n); \ell(t_1', \ldots, t_n')$$

Inverse: For all proof terms $\pi : t \longrightarrow t'$,

$$\pi; \pi^{-1} \equiv t$$
$$\pi^{-1}; \pi \equiv t'$$

Fig. 3. Equivalence of Proof Terms

An *equational proof step* is an expression $s \xleftrightarrow{p}_{e} t$ where s and t are terms, e is an equational axiom $u = v$, and p is a position of s such that $s_{|p} = \sigma(u)$ and $t = s[\sigma(v)]_p$ for some substitution σ.

An *equational proof* of $s_0 = t_n$ is any finite sequence of equational proof steps $\left(s_i \xleftrightarrow{p_i}_{e_i} t_i \right)_{i \in \{0, \ldots, n\}}$ such that $t_i = s_{i+1}$ for all $i \in \{0, \ldots, n-1\}$. It is noted:

$$s_0 \xleftrightarrow{p_0}_{e_0} s_1 \xleftrightarrow{p_1}_{e_1} s_2 \cdots s_n \xleftrightarrow{p_n}_{e_n} t_n .$$

A *rewrite proof step* is an expression $s \xrightarrow{p}_{\ell} t$ or $t \xleftarrow{p}_{\ell} s$ where s and t are terms, ℓ is a rewrite rule $u \to v$, and p is a position of s such that $s_{|p} = \sigma(u)$ and $t = s[\sigma(v)]_p$ for some substitution σ.

An *proof by replacement (of equal by equal)* of $s_0 = t_n$ is any finite sequence of equational proof steps and rewrite proof step $\left(s_i \stackrel{p_i}{\underset{\ell_i}{\leftrightharpoons}}_i t_i \right)_{i \in \{0, \ldots, n\}}$ where $\leftrightharpoons_i \in \{\longleftrightarrow, \longrightarrow, \longleftarrow\}$ for $i \in \{0, \ldots, n\}$ and such that $t_i = s_{i+1}$ for all $i \in \{0, \ldots, n-1\}$. It is noted:

$$s_0 \stackrel{p_0}{\underset{\ell_0}{\leftrightharpoons}}_0 s_1 \stackrel{p_1}{\underset{\ell_1}{\leftrightharpoons}}_1 s_2 \cdots s_n \stackrel{p_n}{\underset{\ell_n}{\leftrightharpoons}}_n t_n .$$

Example 3. Consider the rewrite rules and equational axiom:

$$\ell_1 : g(x) \longrightarrow d(x), \quad \ell_2 : s = t, \quad \ell_3 : l \longrightarrow r,$$

- r is a proof by replacement of $r = r$ (empty sequence),
- $f(g(s), r) \xrightarrow[\ell_1]{1} f(d(s), r) \xleftarrow[\ell_2]{11} f(d(t), r) \xleftarrow[\ell_3]{2} f(d(t), l)$ is a proof by replacement of $f(g(s), r) = f(d(t), l)$.

3.3 From Proof Terms to Proofs by Replacement

In order to have a one to one correspondence between proof representations, we use the equivalence of proof terms defined in Fig. 3. We can refine them to the proof term rewrite system $\leadsto$ given in Fig. 4, in which $\pi, \pi', \pi_1, \ldots$ range over proof terms, $t, t', t_1, \ldots$ over Σ-terms, f, g, d over function symbols, ℓ over rules and equational axioms labels and i and k over $\{1, \ldots, n\}$.

Delete Useless Identities:
$$\left. \begin{array}{c} \pi; t' \\ t; \pi \end{array} \right\} \leadsto \pi$$

Sequentialization: If $\pi_k : t_k \longrightarrow t'_k$ and there exists $i \neq j \in \{1, \ldots, n\}$ such that $\pi_i \neq t_i$ and $\pi_j \neq t_j$,

$$f(\pi_1, \ldots, \pi_n) \leadsto f(\pi_1, t_2, \ldots, t_n); f(t'_1, \pi_2, \ldots, t_n); \ldots; f(t'_1, t'_2, \ldots, \pi_n)$$

Composition Shallowing: If $\pi_i : t_i \longrightarrow t'_i$ and $\pi'_i : t'_i \longrightarrow t''_i$,

$$f(t_1, \ldots, \pi_i; \pi'_i, \ldots, t_n) \leadsto f(t_1, \ldots, \pi_i, \ldots, t_n); f(t_1, \ldots, \pi'_i, \ldots, t_n)$$

Parallel Moves: If $\ell = (g(x_1, \ldots, x_n), d(x_1, \ldots, x_n))$, $\pi_1 : t_1 \longrightarrow t'_1, \ldots, \pi_n : t_n \longrightarrow t'_n$, and if there exists $i \in \{1, \ldots, n\}$ such that $\pi_i \neq t_i$,

$$\ell(\pi_1, \ldots, \pi_n) \leadsto \ell(t_1, \ldots, t_n); d(\pi_1, \ldots, \pi_n)$$

Delete Useless Inverses:
$$t^{-1} \leadsto t$$

Inverse Congruence: If $\pi_i : t_i \longrightarrow t'_i$,

$$f(t_1, \ldots, \pi_i^{-1}, \ldots, t_n) \leadsto f(t_1, \ldots, \pi_i, \ldots, t_n)^{-1}$$

Inverse Composition:

$$(\pi_1; \pi_2)^{-1} \leadsto \pi_2^{-1}; \pi_1^{-1}$$

Fig. 4. Rewrite System for Proof Terms

The associativity is still considered in the congruence, so that all proof terms rewrite rules must be considered modulo the associativity of ; which will be noted $\sim$. The class rewrite system that we consider will be therefore noted $\rightsquigarrow / \sim$. As it is linear, we can use the framework and results from [25].

We first prove that this rewrite system is included in the equivalence relation of Fig. 3.

Proposition 2 (Correctness). *For all proof terms* π_1, π_2, *if* $\pi_1 \rightsquigarrow \pi_2$ *then* $\pi_1 \equiv \pi_2$.

The converse is false: for instance $f(\ell_1, \ell_2) \equiv f(t_1, \ell_2); f(\ell_1, t_2')$ but we do not have $f(\ell_1, \ell_2) \overset{*}{\underset{\rightsquigarrow}{\leftrightarrow}} f(t_1, \ell_2); f(\ell_1, t_2')$.

Proposition 3 (Termination and Confluence). *The proof term rewrite system* $\rightsquigarrow$ *modulo* $\sim$ *is terminating and confluent modulo* $\sim$.

The proof terms rewrite system $\rightsquigarrow$ allow us to give a correspondence between proof terms and proofs by replacement of equal by equal: normal forms of proof terms correspond exactly to proofs by replacement. This fact is expressed in the following theorem, which is indeed a generalization of Lemma 3.6 in [34] for equational logic. We also have operationalized the way to construct the chain of "one-step sequential rewrites".

Theorem 3 (Correspondence between Proof Representations). *The normal form of a proof term* π *for the rewrite system* $\rightsquigarrow$, *noted* $\mathrm{nf}(\pi)$, *has the following form: For some* $n \in \mathbb{N}$, *some contexts* $w_1[], \ldots, w_n[]$, *some indices* $i_1, \ldots, i_n \in \{-1, 1\}$, *some rule labels* $\ell_1, \ldots, \ell_n$ *and some terms* $t_1^1, \ldots, t_{m_1}^1, \ldots, t_1^n, \ldots, t_{m_n}^n$:

$$\mathrm{nf}(\pi) = (w_1[\ell_1(t_1^1, \ldots, t_{m_1}^1)])^{i_1}; \ldots; (w_n[\ell_n(t_1^n, \ldots, t_{m_n}^n)])^{i_n}$$

where for all proof terms ν, ν^1 *is a notation for* ν.

Such a proof term correspond with the following proof by replacement of equal by equal:

$$w_1[g_1(t_1^1, \ldots, t_{m_1}^1)] \overset{p_1}{\underset{\ell_1}{\leftrightharpoons}}_1 w_1[d_1(t_1^1, \ldots, t_{m_1}^1)] \overset{p_2}{\underset{\ell_2}{\leftrightharpoons}}_2 \cdots \overset{p_n}{\underset{\ell_n}{\leftrightharpoons}}_n w_n[d_n(t_1^n, \ldots, t_{m_n}^n)]$$

where for all $j \in \{1, \ldots, n\}$ *we have:*

- $\ell_j = (g_j, d_j)$,
- p_j *is the position of* $[]$ *in* $w_j[]$,
- $\leftrightharpoons_j \; = \; \longrightarrow$ *if* $i_j = 1$ *and* $\ell_j \in R$,

  $\longleftarrow$ *if* $i_j = -1$ *and* $\ell_j \in R$,

  $\longleftrightarrow$ *if* $\ell_j \in E$.
- *if* $j \neq n$, $w_j[d_j(t_1^j, \ldots, t_{m_j}^j)] = w_{j+1}[g_{j+1}(t_1^{j+1}, \ldots, t_{m_{j+1}}^{j+1})]$.

Example 4. Consider $\pi = f(\ell_1(\ell_2), (\ell_3; r)^{-1})$ where $\ell_1 : g(x) \longrightarrow d(x)$, $\ell_2 : s = t$, $\ell_3 : l \longrightarrow r$, we have:

$$\pi \underset{\rightsquigarrow}{\longrightarrow} f(\ell_1(s); d(\ell_2), (\ell_3; r)^{-1}) \qquad \text{(\textbf{Parallel Moves})}$$
$$\underset{\rightsquigarrow}{\longrightarrow} f(\ell_1(s); d(\ell_2), r); f(d(t), (\ell_3; r)^{-1}) \qquad \text{(\textbf{Sequentialization})}$$
$$\underset{\rightsquigarrow}{\longrightarrow} f(\ell_1(s); d(\ell_2), r); f(d(t), r^{-1}; \ell_3^{-1}) \qquad \text{(\textbf{Inverse Composition})}$$
$$\underset{\rightsquigarrow}{\longrightarrow} f(\ell_1(s); d(\ell_2), r); f(d(t), r; \ell_3^{-1}) \qquad \text{(\textbf{Delete Useless Inverses})}$$
$$\underset{\rightsquigarrow}{\longrightarrow} f(\ell_1(s); d(\ell_2), r); f(d(t), \ell_3^{-1}) \qquad \text{(\textbf{Delete Useless Identities})}$$
$$\underset{\rightsquigarrow}{\longrightarrow} f(\ell_1(s), r); f(d(\ell_2), r); f(d(t), \ell_3^{-1}) \qquad \text{(\textbf{Composition Shallowing})}$$
$$\underset{\rightsquigarrow}{\longrightarrow} f(\ell_1(s), r); f(d(\ell_2), r); f(d(t), \ell_3)^{-1} \qquad \text{(\textbf{Inverse Congruence})}$$

This last term is the normal form proof term, and it is equivalent to the proof by replacement $f(g(s), r) \xrightarrow[\ell_1]{1} f(d(s), r) \xleftarrow[\ell_2]{11} f(d(t), r) \xleftarrow[\ell_3]{2} f(d(t), l)$.

Due to this theorem, normal forms of proof terms can be considered in the following indifferently as proof terms or as proofs by replacement.

3.4   Proofs Ordering

The representation of Bachmair by the mean of proof by replacement was defined to introduce an order on proofs [3]: given a reduction ordering $\gg$, to each single proof steps $s \overset{p}{\underset{\ell}{\leftrightarrow}} t$ is associated a *cost*. The cost of an equational proof step $s \overset{p}{\underset{u=v}{\longleftrightarrow}} t$ is the triple $(\{s, t\}, u, t)$. The cost of a rewrite proof step $s \overset{p}{\underset{u \rightarrow v}{\longrightarrow}} t$ is $(\{s\}, u, t)$. Proof steps are compared with each other according to their cost, using the lexicographic combination of the multiset $\gg_{mult}$ extension of the reduction ordering over terms in the first component, the encompassment ordering $\blacktriangleright$ on the second component, and the reduction ordering $\gg$ on the last component. Proofs are compared as multisets of their proof steps. For two proofs by replacement p, q, we will write $p >_{rep} q$ if p is greater than q for such an ordering.

Using theorem 3, we can translate Bachmair's proof ordering to proof terms:

Definition 1 (Bachmair's Ordering on Proof Terms).
For all proof terms π_1, π_2, we say that $\pi_1 >_B \pi_2$ iff

$$\text{nf}(\pi_1) >_{rep} \text{nf}(\pi_2) \ .$$

Example 5. Suppose we have $\Sigma = \{f^1, a^0, b^0, c^0\}$ where the exponents of function symbols denote their arity, and a precedence $f > a > b > c$.

Consider $\pi_1 = f(\ell_1^{-1}; \ell_2)$ and $\pi_2 = f(\ell_3)$ where $\ell_1 = a \longrightarrow b$, $\ell_2 = a \longrightarrow c$ and $\ell_3 = b = c$, and suppose $a > b > c$.

We have $\text{nf}(\pi_1) = f(b) \xleftarrow[\ell_1]{1} f(a) \xrightarrow[\ell_2]{1} f(c)$ and $\text{nf}(\pi_2) = f(b) \xrightarrow[\ell_3]{1} f(c)$. The cost of $\text{nf}(\pi_1)$ is $\{(\{f(a)\}, a, f(b)), (\{f(a)\}, a, f(c))\}$, the cost of $\text{nf}(\pi_2)$ is $\{((\{f(b), f(c)\}, b, f(c))\}$, so $\text{nf}(\pi_1) >_{rep} \text{nf}(\pi_2)$ and $\pi_1 >_B \pi_2$.

As we can see, the way we define the ordering over proofs is not trivial. The question remains if we could have defined it more directly, without using the representation as proof by replacement. The following statement give a beginning of answer: we cannot hope to extend an RPO on Σ-terms to a RPO[4] $>_{rpo}$ on proof terms so that $>_B$ and $>_{rpo}$ coincide for the normal forms of proof terms:

Counter-example 6. With the same hypothesis as in Example 5, let $\ell_f = f(a) \longrightarrow c$ and $\ell_b = b \longrightarrow c$.

We now want to extend the precedence to ℓ_f and ℓ_b in order to extend the RPO to proof terms. If we have $\ell_f < \ell_b$, $f(a) \xrightarrow[\ell_f]{\epsilon} c >_{rep} b \xrightarrow[\ell_b]{\epsilon} c$ but $\ell_f <_{rpo} \ell_b$.

If we suppose $f > \ell_f > \ell_b$ we have $f(a) \xrightarrow[\ell_f]{\epsilon} c >_{rep} f(b) \xrightarrow[\ell_b]{1} f(c)$ but $\ell_f <_{rpo} f(\ell_b)$.

If we suppose $\ell_f > \ell_b$ and $\ell_f > f$, then $f(f(b)) \xrightarrow[\ell_b]{11} f(f(c)) >_{rep} f(a) \xrightarrow[\ell_f]{\epsilon} c$ but $f(f(\ell_b)) <_{rpo} \ell_f$.

Such an extension is therefore impossible, there is no extension of $>_{rpo}$ on proof terms such that for all proof terms π_1, π_2, we have $\mathrm{nf}(\pi_1) >_{rpo} \mathrm{nf}(\pi_2)$ if and only if $\mathrm{nf}(\pi_1) >_B \mathrm{nf}(\pi_2)$.

In other words, the ordering we defined above can *not* be defined as a RPO over proof terms.

In the following, proofs will be represented by proof terms, the proof ordering $>$ between them will be the ordering $>_B$ restricted to proofs with the same conclusion, and the subproof relation $\rhd$ will be the subterm relation.

4 Standard Completion Is an Instance of Abstract Canonical System

4.1 Adequacy to the Postulates

Adequacy to postulates A, B, C and D comes from the tree structure of the proof terms representation.

Postulate E is not trivially verified, because of the definition of the ordering as translation of an ordering over proof by replacement. Nevertheless:

Theorem 4 (Postulate E for Equational Proofs). *For all contexts $w[]$, for all proof terms q, r:*

$$q > r \text{ implies } w[q] > w[r] \ .$$

The deduction mechanism $\rightsquigarrow$ used here will be of course the standard completion. We now show that it has the required properties.

---

[4] Or better an ordering compatible with associativity, such as the AC-RPO [38].

4.2 Standard Completion Is Sound and Adequate

This is shown in [3, Lemma 2.1]: if $E, R \rightsquigarrow E', R'$, then $\xleftrightarrow{*}_{E \cup R}$ and $\xleftrightarrow{*}_{E' \cup R'}$ are the same. To prove this, one has simply to verify it for each inference rule of standard completion.

4.3 Standard Completion Is Good

This is shown in [3, Lemma 2.5, 2.6]: if $E, R \rightsquigarrow E', R'$, then proofs in E, R can be transformed to proofs in E', R' using following rules:

$$s \xleftrightarrow{E} t \;\twoheadrightarrow\; s \xrightarrow{R'} t \qquad \textbf{(Orient)}$$

$$s \xleftrightarrow{E} t \;\twoheadrightarrow\; s \xrightarrow{R'} u \xleftrightarrow{E'} t \qquad \textbf{(Simplify)}$$

$$s \xleftrightarrow{E} s \;\twoheadrightarrow\; s \qquad \textbf{(Delete)}$$

$$s \xleftarrow{R} u \xrightarrow{R} t \;\twoheadrightarrow\; s \xleftrightarrow{E'} t \qquad \textbf{(Deduce)}$$

$$s \xleftarrow{R} u \xrightarrow{R} t \;\twoheadrightarrow\; s \xrightarrow{R'}^{*} v \xleftarrow{R'}^{*} t$$

$$s \xrightarrow{R} t \;\twoheadrightarrow\; s \xrightarrow{R'} v \xleftarrow{R'} t \qquad \textbf{(Compose)}$$

$$s \xrightarrow{R} t \;\twoheadrightarrow\; s \xrightarrow{R'} v \xleftrightarrow{E'} t \qquad \textbf{(Collapse)}$$

We have $\xrightarrow{}_{\twoheadrightarrow} \;\subseteq\; >$, so these proofs become indeed better.

4.4 Standard Completion Is Canonical

We can now show the following theorem:

Theorem 5 (Completeness of Standard Completion). *Standard completion results—at the limit, when it terminates without failure—in the canonical, Church-Rosser basis.*

Proof. We can show $R_\infty = E_0^\sharp$, and because standard completion is good we can use Theorem 2.

Remark 1. When standard completion does not terminate, we can show that $E_0^\sharp = R_\infty^\sharp \subseteq R_\infty$. Consequently, the resulting set R_∞ is then *saturated*, but it is not necessarily *contracted*.

This shows that the standard completion is an instance of the framework of the abstract canonical systems, when we choose the convenient proof representation.

5   Conclusion

We presented a proof that standard completion can be seen as an instance of the abstract canonical systems and inference framework. This led us to make precise the relation between different equational proof representations. The first one, proof terms as presented in [34], is convenient to consider proofs as terms, with a subterm relation and substitutions. The other one, initiated in [3], is well adapted to the study of the completeness of the standard completion procedure. We presented a way to pass from one representation to another by the mean of the proof term rewrite rules presented in Fig. 4. Thanks to this, we extended the ordering introduced with the proof by replacement to the proof terms and thus combine the advantages of both representations. This therefore positively answer to the question whether the abstract canonical systems, centered in a quite general way around the notion of proof ordering, are indeed the right framework to uniformly prove the completeness of completion.

We plan now to understand how the results we have presented here can be extended to other completion procedures. Bachmair introduced another proof ordering to prove the completeness of the completion modulo [3], so that the generalization seems rather natural. We plan also to look at other kinds of deduction mechanisms, such as Buchberger's algorithm or resolution. For this, we may show that Struth's non-symmetric completion [41], which subsumes both procedures, is also an instance of the ACSI framework.

Furthermore, proof terms as presented by [34, 30] are specific terms of the rewriting calculus [13] [http://rho.loria.fr]. The link between the completion procedure and the sequent systems mentioned above can probably be found here and be related to Dowek's work proving that confluent rewrite rules can be linked with **Cut**-free proofs of some sequent systems [19].

Acknowledgments This paper benefited greatly from suggestions, discussions and the enthusiasm of Nachum Dershowitz. We thank also Georg Struth for his useful comments and the anonymous referees for their careful reading and constructive suggestions.

References

[1] M. Aiguier and D. Bahrami. Structures for abstract rewriting. *Journal of Automated Reasoning*, 2006. To appear.

[2] F. Baader and T. Nipkow. *Term Rewriting and all That*. Cambridge University Press, 1998.

[3] L. Bachmair. *Proof methods for equational theories*. PhD thesis, University of Illinois, Urbana-Champaign, (Ill., USA), 1987. Revised version, August 1988.

[4] L. Bachmair and N. Dershowitz. Equational inference, canonical proofs, and proof orderings. *Journal of Association for Computing Machinery*, 41(2):236–276, 1994.

[5] L. Bachmair and H. Ganzinger. Resolution theorem proving. In A. Robinson and A. Voronkov, editors, *Handbook of Automated Reasoning*, volume I, chapter 2, pages 19–99. Elsevier Science, 2001.

[6] M. P. Bonacina and N. Dershowitz. Abstract Canonical Inference. *ACM Transactions on Computational Logic*, 2006. To appear.

[7] M. P. Bonacina and J. Hsiang. On rewrite programs: semantics and relationship with Prolog. *Journal of Logic Programming*, 14(1 & 2):155–180, October 1992.

[8] P. Borovansky, C. Kirchner, H. Kirchner, and P.-E. Moreau. ELAN from a rewriting logic point of view. *Theoretical Computer Science*, 2(285):155–185, July 2002.

[9] B. Buchberger. *An algorithm for finding a basis for the residue class ring of a zero-dimensional polynomial ideal.* PhD thesis, University of Inssbruck (Austria), 1965. (in German).

[10] B. Buchberger. A critical-pair/completion algorithm for finitely generated ideals in rings. In E. Börger, G. Hasenjaeger, and D. Rödding, editors, *Proceedings of Logic and Machines: Decision problems and Complexity*, volume 171 of *Lecture Notes in Computer Science*, pages 137–161. Springer-Verlag, 1983.

[11] R. Bündgen. Simulating Buchberger's algorithm by Knuth-Bendix completion. In R. V. Book, editor, *Rewriting Techniques and Applications: Proc.of the 4th International Conference RTA-91*, pages 386–397. Springer, Berlin, Heidelberg, 1991.

[12] G. Burel. Systèmes Canoniques Abstraits : Application à la Déduction Naturelle et à la Complétion. Master's thesis, Université Denis Diderot – Paris 7, 2005.

[13] H. Cirstea and C. Kirchner. The rewriting calculus — Part I *and* II. *Logic Journal of the Interest Group in Pure and Applied Logics*, 9(3):427–498, May 2001.

[14] M. Clavel, S. Eker, P. Lincoln, and J. Meseguer. Principles of Maude. In J. Meseguer, editor, *Proceedings of the first international workshop on rewriting logic*, volume 4, Asilomar (California), September 1996. Electronic Notes in Theoretical Computer Science.

[15] N. Dershowitz. Computing with rewrite systems. *Information and Control*, 65(2/3):122–157, 1985.

[16] N. Dershowitz. Canonicity. *Electronic Notes in Theoretical Computer Science*, 86(1), June 2003.

[17] N. Dershowitz and C. Kirchner. Abstract saturation-based inference. In P. Kolaitis, editor, *Proceedings of LICS 2003*, Ottawa, Ontario, June 2003. ieee.

[18] N. Dershowitz and C. Kirchner. Abstract Canonical Presentations. *Theorical Computer Science*, To appear, 2006.

[19] G. Dowek. Confluence as a cut elimination property. In R. Nieuwenhuis, editor, *RTA*, volume 2706 of *Lecture Notes in Computer Science*, pages 2–13. Springer, 2003.

[20] K. Futatsugi and A. Nakagawa. An overview of CAFE specification environment – an algebraic approach for creating, verifying, and maintaining formal specifications over networks. In *Proceedings of the 1st IEEE Int. Conference on Formal Engineering Methods*, 1997.

[21] J. A. Goguen. How to prove algebraic inductive hypotheses without induction, with applications to the correctness of data type implementation. In W. Bibel and R. Kowalski, editors, *Proceedings 5th International Conference on Automated Deduction, Les Arcs (France)*, volume 87 of *Lecture Notes in Computer Science*, pages 356–373. Springer-Verlag, 1980.

[22] J. A. Goguen and G. Malcolm, editors. *Software Engineering with OBJ: algebraic specification in practice*, volume 2 of *Advances in Formal Methods*. Kluwer Academic Publishers, Boston, 2000.

[23] J. A. Goguen and J. Meseguer. EQLOG: Equality, types, and generic modules for logic programming. In D. DeGroot and G. Lindstrom, editors, *Logic Programming: Functions, Relations, and Equations*, pages 295–363. Prentice-Hall, Englewood Cliffs, NJ, 1986.

[24] J. Hsiang and M. Rusinowitch. Proving refutational completeness of theorem proving strategies: The transfinite semantic tree method. *Journal of the ACM*, 38(3):559–587, July 1991.

[25] G. Huet. Confluent reductions: Abstract properties and applications to term rewriting systems. *Journal of the ACM*, 27(4):797–821, 1980.

[26] G. Huet. A complete proof of correctness of the Knuth–Bendix completion algorithm. *Journal of Computer and System Sciences*, 23(1):11–21, August 1981.

[27] G. Huet and J.-J. Lévy. Computations in orthogonal rewriting systems, I. In J.-L. Lassez and G. Plotkin, editors, *Computational Logic*, chapter 11, pages 395–414. The MIT press, 1991.

[28] J.-P. Jouannaud and H. Kirchner. Completion of a set of rules modulo a set of equations. *SIAM Journal of Computing*, 15(4):1155–1194, 1986.

[29] C. Kirchner and H. Kirchner. Rewriting, solving, proving. A preliminary version of a book available at www.loria.fr/~ckirchne/rsp.ps.gz, 1999.

[30] C. Kirchner, H. Kirchner, and M. Vittek. Designing constraint logic programming languages using computational systems. In P. Van Hentenryck and V. Saraswat, editors, *Principles and Practice of Constraint Programming. The Newport Papers.*, chapter 8, pages 131–158. The MIT press, 1995.

[31] D. E. Knuth and P. B. Bendix. Simple word problems in universal algebras. In J. Leech, editor, *Computational Problems in Abstract Algebra*, pages 263–297. Pergamon Press, Oxford, 1970.

[32] D. Lankford. Canonical inference. Technical report, Louisiana Tech. University, 1975.

[33] P. Le Chenadec. *Canonical Forms in Finitely Presented Algebras*. John Wiley & Sons, 1986.

[34] J. Meseguer. Conditional rewriting logic as a unified model of concurrency. *Theoretical Computer Science*, 96(1):73–155, 1992.

[35] D. Musser. On proving inductive properties of abstract data types. In *Proceedings, Symposium on Principles of Programming Languages*, volume 7. Association for Computing Machinery, 1980.

[36] G. Peterson and M. Stickel. Complete sets of reductions for some equational theories. *Journal of the ACM*, 28:233–264, 1981.

[37] J. A. Robinson. A machine-oriented logic based on the resolution principle. *Journal of the ACM*, 12:23–41, 1965.

[38] A. Rubio and R. Nieuwenhuis. A total AC-compatible ordering based on RPO. *Theoretical Computer Science*, 142(2):209–227, 1995.

[39] W. M. Schorlemmer. Rewriting logic as a logic of special relations. *Electr. Notes Theor. Comput. Sci.*, 15, 1998.

[40] K. Stokkermans. A categorical formulation for critical-pair/completion procedures. In M. Rusinowitch and J.-L. Remy, editors, *CTRS*, volume 656 of *Lecture Notes in Computer Science*, pages 328–342. Springer, 1992.

[41] G. Struth. *Canonical Transformations in Algebra, Universal Algebra and Logic*. Dissertation, Institut für Informatik, Universität des Saarlandes, Saarbrücken, Germany, June 1998.

[42] F. Winkler. Knuth-Bendix procedure and Buchberger algorithm - A synthesis. In *Proceedings of the ACM-SIGSAM 1989 International Symposium on Symbolic and Algebraic Computation*, pages 55–67, Portland (Oregon, USA), 1989. ACM Press.

A   Proofs for Section 3 and 4

A.1   From Proof Terms to Proof by Replacement

To prove the termination of $\leadsto / \sim$, we need a reduction ordering compatible with associativity. We consider only associativity here, although most of the existing works use associativity and commutativity. Therefore, we need the following lemma.

Lemma 1. *If $A \subseteq B$ then $>$ is B-compatible implies $>$ is A-compatible.*

Proof. Just notice that $s' \overset{*}{\underset{A}{\longleftrightarrow}} s > t \overset{*}{\underset{A}{\longleftrightarrow}} t'$ implies $s' \overset{*}{\underset{B}{\longleftrightarrow}} s > t \overset{*}{\underset{B}{\longleftrightarrow}} t'$.

We can therefore use the AC-RPO ordering: a total AC-compatible simplification ordering on ground terms is defined in [38], as an extension of the RPO. To compare terms, they are interpreted using flattening and interpretation rules. As we consider here that the associative commutative symbols have the lowest precedence, we do not need the interpretation rules, and we will only present the flattening rules: terms are reduced using a set of rules

$$f(x_1, \ldots, x_n, f(y_1, \ldots, y_r), z_1, \ldots, z_m) \to f(x_1, \ldots, x_n, y_1, \ldots, y_r, z_1, \ldots, z_m) \tag{1}$$

for all AC-symbols f with $n + m \geq 1$ and $r \geq 2$. Such a rewrite system is terminating as shown in [38].

For all terms t, let $snf(t)$ denote the *set of normal forms* of t using rules (1).

Given a precedence $>$ on function symbols, let $>_{rpo}$ denote the recursive path ordering with precedence $>$ where AC function symbols have multiset status and other symbols have lexicographic status.

If $f(s_1, \ldots, s_n)$ is the normal form of a term s rewriting by (1) only at topmost position, then $tf(s) \overset{!}{=} (s_1, \ldots, s_n)$.

Definition 2 (AC-RPO). *For all terms s, t, $s >_{AC-rpo} t$ if:*

- $\forall t' \in snf(t)\ \exists s' \in snf(s),\ s' >_{AC-rpo} t'$ *or*
- $\forall t' \in snf(t)\ \exists s' \in snf(s),\ s' \geq_{rpo} t'$ *and* $tf(s) = f(s_1, \ldots, s_m)$ *and* $tf(t) = (t_1, \ldots, t_n)$ *and*
  - *if the head of s is AC then* $\{s_1, \ldots, s_m\} >_{AC-rpo_{mult}} \{t_1, \ldots, t_n\}$ *or*
  - *if the head of s is not AC then* $(s_1, \ldots, s_m) >_{AC-rpo_{lex}} (t_1, \ldots, t_n)$.

Proposition 4 ([38]). *The AC-RPO is an AC-compatible simplification ordering which is total for non AC-equivalent ground terms.*

We define a precedence $>$ such that for all function symbols f and for all rule labels ℓ we have $\ell > f > \cdot^{-1} > ;\ .$ The AC-RPO built with this precedence will be noted $\succ$.

To show termination, we also need the following lemma:

Lemma 2. *For all proof terms* $\pi : t \longrightarrow t'$, *we have* $\pi \succeq t$ *and* $\pi \succeq t'$.

Proof. By induction on the structure of the proof term π.

For **Reflexivity**, $\pi = t = t'$.

For **Congruence**, $\pi = f(\pi_1, \ldots, \pi_n)$, $t = f(t_1, \ldots, t_n)$ and $t' = f(t'_1, \ldots, t'_n)$. By induction hypothesis, for all $i \in \{1, \ldots, n\}$, we have $\pi_i \succeq t_i, t'_i$. Furthermore, π is not reducible on the top position using rules (1), so that $snf(\pi) = \{f(\pi'_1, \ldots, \pi'_n) : \forall i, \pi'_i \in snf(\pi_i)\}$, whereas t and t' are not reducible. Consequently, by definition of an AC-RPO, $\pi \succeq t, t'$.

For **Replacement**, $\pi = \ell(\pi_1, \ldots, \pi_n)$, $t = g(t_1, \ldots, t_n)$ and $t' = d(t'_1, \ldots, t'_n)$ where $\ell = (g, d) \in E \cup R$. With the same arguments than for **Congruence**, we can conclude that $\pi \succeq t, t'$ (recall that $\ell > g, d$).

For **Transitivity**, $\pi = \pi_1; \pi_2$ where $\pi_1 : t \longrightarrow t''$ and $\pi_2 : t'' \longrightarrow t'$. By induction hypothesis, $\pi_1 \succeq t$ and $\pi_2 \succeq t'$. As $\succ$ is a simplification ordering, $\pi \succ \pi_1, \pi_2 \succeq t, t'$.

For **Symmetry**, $\pi = \pi'^{-1}$ where $\pi' : t' \longrightarrow t$. By induction hypothesis and because $\succ$ is a simplification ordering, $\pi \succ \pi' \succeq t', t$.

Proposition 5 (Termination). *The rewrite system* $\rightsquigarrow$ *of Fig. 4 modulo* $\sim$ *is terminating for ground proof terms.*

Proof. We can show that $\rightsquigarrow \subseteq \succ$, thus proving the termination of $\rightsquigarrow / \sim$:

For **Delete Useless Identities**, it comes from the fact that $\succ$ is a simplification ordering.

For **Sequentialization**, rules (1) are not applicable on the left side whereas they lead on the right side to $; (f(\pi_1, t_2, \ldots, t_n), f(t'_1, \pi_2, \ldots, t_n), \ldots, f(t'_1, t'_2, \ldots, \pi_n))$. We have $f >;$, thus by definition of a RPO, we must then prove that for all $i \in \{1, \ldots, n\}$ we have $f(\pi_1, \ldots, \pi_n) \succ_{RPO} f(t'_1, \ldots, t'_{i-1}, \pi_i, t_{i+1}, \ldots, t_n)$, i.e. $(\pi_1, \ldots, \pi_n) \succ^{lex}_{RPO} (t'_1, \ldots, t'_{i-1}, \pi_i, t_{i+1}, \ldots, t_n)$. By hypothesis there exists at least a $j \in \{1, \ldots, n\} \setminus \{i\}$ such that $\pi_j \neq t_j$, so we can conclude with the preceding lemma.

For **Composition Shallowing**, both sides are not reducible using rules (1). We have $f >;$, thus we have to show: $f(t_1, \ldots, \pi_i; \pi'_i, \ldots, t_n) \succ_{RPO} f(t_1, \ldots, \pi_i, \ldots, t_n)$ and $f(t_1, \ldots, \pi_i; \pi'_i, \ldots, t_n) \succ_{RPO} f(t_1, \ldots, \pi'_i, \ldots, t_n)$. Both comparisons hold by definition of a RPO.

For **Parallel Moves**, both sides are not reducible using rules (1). We have $\ell >;$, thus we have to prove that $\ell(\pi_1, \ldots, \pi_n) \succ_{RPO} \ell(t_1, \ldots, t_n)$ and $\ell(\pi_1, \ldots, \pi_n) \succ_{RPO} d(\pi_1, \ldots, \pi_n)$. The first comparison holds because of the lemma and because there exists a $i \in \{1, \ldots, n\}$ such that $\pi_i \neq t_i$; the second one holds because $\ell > d$.

For **Delete Useless Inverses**, this comes from the fact that $\succ$ is a simplification ordering.

516   Guillaume Burel and Claude Kirchner

For **Inverse Congruence**, both sides are not reducible using rules (1), therefore this is a consequence of $f > \cdot^{-1}$.

For **Inverse Composition**, both sides are not reducible using rules (1), therefore this is a consequence of $\cdot^{-1} >;$.

We can also prove confluence:

Proposition 6 (Confluence). *The rewrite system $\leadsto$ is confluent modulo $\sim$ on ground proof terms.*

Proof. The class rewrite system is linear and terminating, so we just have to check that the critical pairs are confluent [25].

For $\underset{R}{\longleftarrow} \circ \underset{R}{\longrightarrow}$, it is easy to check for most of the critical pairs that they are confluent. We only detail the most problematic one. For two possible applications of **Sequentialization**, we have for instance $f(g(\nu_1,\ldots,\nu_m),\pi_1,\ldots,\pi_n)$ that can be rewritten to $f(g(\nu_1,\ldots,\nu_m),t_1,\ldots,t_n); f(g(s_1,\ldots,s_m),\pi_1,\ldots,t_n);\ldots;$ $f(g(s_1,\ldots,s_m),t_1',\ldots,\pi_n)$ and to $f(g(\nu_1,\ldots,s_m);\ldots;g(s_1',\ldots,\nu_m),\pi_1,\ldots,\pi_n)$. Both of them reduce to $f(g(s_1,\ldots,s_m);\ldots;g(s_1',\ldots,\nu_m),t_1,\ldots,t_n);$ $f(g(s_1,\ldots,s_m),\pi_1,\ldots,t_n);\ldots;f(g(s_1,\ldots,s_m),t_1',\ldots,\pi_n)$.

For $\underset{R}{\longleftarrow} \circ \underset{A}{\longleftrightarrow}$, the only rules that can interfere with $\sim$ are **Delete Useless Identities**, **Composition Shallowing** and **Inverse Composition**. We can check that all critical pairs are confluent.

Theorem 6 (Correspondence between Proof Representations). *The normal form of a proof term π for the rewrite system $\leadsto$, noted $\mathrm{nf}(\pi)$, has the following form: For some $n \in \mathbb{N}$, some contexts $w_1[],\ldots,w_n[]$, some indices $i_1,\ldots,i_n \in \{-1,1\}$, some rule labels $\ell_1,\ldots,\ell_n$ and some terms $t_1^1,\ldots,t_{m_1}^1,\ldots,t_1^n,\ldots,t_{m_n}^n$:*

$$\mathrm{nf}(\pi) = (w_1[\ell_1(t_1^1,\ldots,t_{m_1}^1)])^{i_1};\ldots;(w_n[\ell_n(t_1^n,\ldots,t_{m_n}^n)])^{i_n}$$

where ν^1 is a notation for ν.

We will denote by $\mathrm{nf}(\pi)$ the normal form of a proof term π.

Such a proof term correspond with the following proof by replacement of equal by equal:

$$w_1[g_1(t_1^1,\ldots,t_{m_1}^1)] \overset{p_1}{\underset{\ell_1}{\leftrightarrows}}_1 w_1[d_1(t_1^1,\ldots,t_{m_1}^1)] \overset{p_2}{\underset{\ell_2}{\leftrightarrows}}_2 \cdots \overset{p_n}{\underset{\ell_n}{\leftrightarrows}}_n w_n[d_n(t_1^n,\ldots,t_{m_n}^n)]$$

where for all $j \in \{1,\ldots,n\}$ we have:

- $\ell_j = (g_j,d_j)$,
- p_j is the position of $[]$ in $w_j[]$,
   $\longrightarrow$ *if $i_j = 1$ and $\ell_j \in R$,*
- $\leftrightarrows_j = \longleftarrow$ *if $i_j = -1$ and $\ell_j \in R$,*
   $\longleftrightarrow$ *if $\ell_j \in E$.*
- *if $j \neq n$, $w_j[d_j(t_1^j,\ldots,t_{m_j}^j)] = w_{j+1}[g_{j+1}(t_1^{j+1},\ldots,t_{m_{j+1}}^{j+1})]$.*

Proof. We first have to check that proof terms in that form are indeed irreducible by $\rightsquigarrow$, what is left to the reader.

Then, suppose that we have an irreducible proof term. Because **Sequentialization** cannot be applied, there is at most one ; under all function symbols. Because **Composition Shallowing** cannot be applied, there are no ; under all function symbols. Because **Inverse Congruence** and **Inverse Composition** cannot be applied, $\cdot^{-1}$ is applied between ; and function symbols. Irreducible proof term are therefore application of ; over eventually $\cdot^{-1}$ over base terms composed of function symbols and rule labels.

Because **Delete Useless Identities** and **Delete Useless Inverse** cannot be applied, there is a least one non-trivial proof (i.e a proof with a label in it) in each of these base terms. Because **Sequentialization** cannot be applied, there is at most one non-trivial proof in each of them. Because **Parallel Moves** cannot be applied, the subterms of the labels are Σ-terms. Consequently, each base term contains one and only one rule label, applied to Σ-terms.

A.2   Adequacy to the Postulates

Postulate A: The proof of $(u, v) \in E \cup R$ labeled by ℓ is $\ell(x_1, \ldots, x_n)$ where $x_1, \ldots, x_n$ are the free variables of (u, v).

Postulate B: We can replace the assumption $\ell(\pi_1, \ldots, \pi_n)$ of something proved by its proof where the free variables are replaced by the proofs $\pi_1, \ldots, \pi_n$.

Postulate C and D: These postulates hold because of the tree structure of proofs.

Postulate E: This one does not trivially hold. We first show the following lemma:

Lemma 3. *For all function symbols f of arity $n + 1$, for all proof terms $\pi_1, \ldots, \pi_n$, q and r:*

$$q > r \text{ implies } f(\pi_1, \ldots, q, \ldots, \pi_n) > f(\pi_1, \ldots, r, \ldots, \pi_n) \ .$$

Proof. Suppose $q > r$, thus by definition $\mathrm{nf}(q) >_{rep} \mathrm{nf}(r)$. To compare $f(\pi_1, \ldots, q, \ldots, \pi_n)$ and $f(\pi_1, \ldots, r, \ldots, \pi_n)$, we have to transform them to proof by replacement. As $\xrightarrow[\rightsquigarrow/\sim]{}$ is Church-Rosser, the way it is applied does not matter.

We have

$$f(\pi_1, \ldots, q, \ldots, \pi_n)$$
$$\xrightarrow[\rightsquigarrow]{}^* f(\pi_1, t_2, \ldots, t_n); \ldots; f(t'_1, \ldots, q, \ldots, t_n); \ldots; f(t'_1, \ldots, \pi_n)$$
$$\xrightarrow[\rightsquigarrow]{}^* f(\pi_1, t_2, \ldots, t_n); \ldots; \underline{f(t'_1, \ldots, \mathrm{nf}(q), \ldots, t_n)}; \ldots; f(t'_1, \ldots, \pi_n)$$

Then, if $\mathrm{nf}(q)$ contains ; the underlined term will be split by **Composition Shallowing**. If it contains $^{-1}$ the rule **Inverse Congruence** will be applied.

Some terms outside the underline corresponding to identity will be removed by **Delete Useless Identities**, and the normal form will look like:

$$f(\pi_1, t_2, \ldots, t_n); \ldots; \underline{f(t'_1, \ldots, q_1, \ldots, t_n)^{i_1}}; \ldots; \underline{f(t'_1, \ldots, q_m, \ldots, t_n)^{i_m}}; \ldots;$$
$$f(t'_1, \ldots, \pi_n)$$

with $\mathrm{nf}(q) = q_1^{i_1}; \ldots; q_m^{i_m}$.

The same will apply with r, and therefore, to compare the initial proofs, we just have to compare the costs of the underlined terms.

The cost of $\mathrm{nf}(q)$ will look like $\{(\{s_1\}, u_1, h_1), \ldots, (\{s_m\}, u_m, h_m)\}$. Then the cost of $f(t'_1, \ldots, q_1, \ldots, t_n)^{i_1}; \ldots; f(t'_1, \ldots, q_m, \ldots, t_n)^{i_m}$ will be:

$$\left\{ \begin{array}{l} (\{f(t'_1, \ldots, s_1, \ldots, t_n)\}, u_1, f(t'_1, \ldots, h_1, \ldots, t_n)), \ldots, \\ (\{f(t'_1, \ldots, s_m, \ldots, t_n)\}, u_m, f(t'_1, \ldots, h, m, \ldots, t_n)) \end{array} \right\}.$$

For $\mathrm{nf}(r)$ they will be respectively $\{(\{g_1\}, v_1, d_1), \ldots, (\{g_p\}, v_p, d_p)\}$ and:

$$\left\{ \begin{array}{l} (\{f(t'_1, \ldots, g_1, \ldots, t_n)\}, v_1, f(t'_1, \ldots, d_1, \ldots, t_n)), \ldots, \\ (\{f(t'_1, \ldots, g_p, \ldots, t_n)\}, v_p, f(t'_1, \ldots, d_p, \ldots, t_n)) \end{array} \right\}.$$

$\gg$, which is used to compare the first and the third components of each part of the cost, is a reduction ordering, so that $\mathrm{nf}(q) >_{rep} \mathrm{nf}(r)$ implies for instance $f(t'_1, \ldots, q_1, \ldots, t_n)^{i_1}; \ldots; f(t'_1, \ldots, q_m, \ldots, t_n)^{i_m} >_{rep} f(t'_1, \ldots, r_1, \ldots, t_n)^{i_1}; \ldots; f(t'_1, \ldots, r_p, \ldots, t_n)^{i_p}$.

The same is true for labels:

Lemma 4. *For all rule labels ℓ, for all proof terms $\pi_1, \ldots, \pi_n$, q and r:*

$$q > r \text{ implies } \ell(\pi_1, \ldots, q, \ldots, \pi_n) > \ell(\pi_1, \ldots, r, \ldots, \pi_n)$$

Proof. $\ell(\pi_1, \ldots, q, \ldots, \pi_n)$ and $\ell(\pi_1, \ldots, r, \ldots, \pi_n)$ can be reduced by **Parallel Moves** to $\ell(t_1, \ldots, t_n); d(\pi_1, \ldots, q, \ldots, \pi_n)$ and $\ell(t_1, \ldots, t_n); d(\pi_1, \ldots, r, \ldots, \pi_n)$. We can therefore conclude using the preceding lemma.

This allows us to show

Theorem 7 (Postulate E for Equational Proofs). *For all proof terms p, r, for all position i of p:*

$$p_{|i} > r \text{ implies } p > p[r]_i.$$

Proof. This is proved by induction on i. For $i = \epsilon$ this is trivial. For $i \neq \epsilon$, by induction hypothesis, the result holds for the subproofs of p. For the head of p:

- for **Symmetry**, it is trivial;
- for **Transitivity**, it comes from the fact that equational proofs are compared as the multiset of their equational proof steps;
- for **Congruence**, it comes from lemma 3;
- for **Replacement**, it comes from lemma 4.

A.3   Standard Completion Is Canonical

Remember that by fairness assumption, $E_\infty = \emptyset$.

Lemma 5. *For all standard completion derivations* $(E_i, R_i)_i$:

$$E_0^\sharp \subseteq R_\infty .$$

Proof. By contradiction, suppose there is $(a, b) \in E_0^\sharp \setminus R_\infty$, labeled ℓ. Because completion is adequate, there exists $p \in \mu Pf(R_\infty)$ proving $a = b$. Because $a = b \in E_0^\sharp$, $\ell(x_1, \ldots, x_n) \in Nf(E_0) = Nf(R_\infty)$ where $(x_i)_i$ are the free variables of ℓ, so that

$$p > \ell(x_1, \ldots, x_n)$$

- If there are no peak in nf(p), then nf(p) is a valley proof, and it is easy to show that it is smaller than $\ell(x_1, \ldots, x_n)$, which is a contradiction with the preceding comparison.
- If there is a parallel peak, for instance $s[c, e] \xleftarrow[\ell_1]{i} s[d, e] \xrightarrow[\ell_2]{j} s[d, f]$, then the proof by replacement where this peak is replaced by $s[c, e] \xrightarrow[\ell_2]{j} s[c, f] \xleftarrow[\ell_1]{i} s[d, f]$ is smaller, thus leading to a contradiction with the minimality of p in $Pf(R_\infty)$.
- If there is a critical peak, then by fairness assumption there is some step k where this critical peak is treated by **Deduce**. The proof of the conclusion of the critical peak at the step $k + 1$ is therefore smaller. Because standard completion is good, it can only go smaller, so that at the limit we can find by replacement of the critical peak by this proof a smaller proof of $a = b$, thus leading to a contradiction with the minimality of p in $Pf(R_\infty)$.

Lemma 6. *For all standard completion derivations* $(E_i, R_i)_i$ *which* terminate without failure:

$$R_\infty \subseteq E_0^\sharp .$$

Proof. By contradiction, suppose there is $(a, b) \in R_\infty \setminus E_0^\sharp$, labeled by ℓ. Then there exists a proof $p \in \mu Pf(E_0^\sharp)$ such that $\ell(x_1, \ldots, x_n) > p$ where $x_1, \ldots, x_n$ are the free variables of ℓ.

Rules comes from orientation of equational axioms through **Orient**, so that $a \gg b$. The cost of $\ell(x_1, \ldots, x_n)$ is then $\{(\{a\}, a, b)\}$. Consider the leftmost step of nf(p). It is of the form $a \xleftrightarrow[(c,d)]{i} a[d]_i$ where $c = a_{|i}$. If it is $a \xrightarrow[d \to c]{i} a[d]_i$ then the cost of this proof step would be $\{(\{a[d]_i\}, d, a)\}$, which is then greater than $\{(\{a\}, a, b)\}$, thus leading to a contradiction with the fact that $\ell(x_1, \ldots, x_n) > p$. If $a \xleftrightarrow[c=d]{i} a[d]_i$ then the cost of this proof step would be $\{(\{a, a[d]_i\}, c, a[d]_i)\}$, which is then greater than $\{(\{a\}, a, b)\}$, thus leading to a contradiction with the fact that $\ell(x_1, \ldots, x_n) > p$. If it is $a \xrightarrow[c \to d]{i} a[d]_i$ then there is a critical pair $(b, a[d]_i)$

in R_∞ (we just proved that $E_0^\sharp \subseteq R_\infty$). The fairness assumption will therefore apply, and therefore **Deduce** will produce the equational axiom $b = a[d]_i$, which will be oriented, and $a \longrightarrow b \in R_\infty$ will be simplified through **Compose** or **Collapse**. Because $a \longrightarrow b$ is persisting, it must be generated once again, thus contradicting the termination of the completion.

Theorem 8 (Completeness of Standard Completion). *Standard completion results — at the limit, when it terminates without failure — in the canonical, Church-Rosser basis.*

Proof. There is nothing more to prove, because we have $R_\infty = E_0^\sharp$, and standard completion is good so we can use Theorem 2.

Eliminating Dependent Pattern Matching

Healfdene Goguen[1], Conor McBride[2], and James McKinna[3]

[1] Google, New York, New York
[2] School of Computer Science and Information Technology, University of Nottingham
[3] School of Computer Science, University of St Andrews

Abstract. This paper gives a reduction-preserving translation from Coquand's *dependent pattern matching* [4] into a traditional type theory [11] with universes, inductive types and relations and the axiom **K** [22]. This translation serves as a proof of termination for structurally recursive pattern matching programs, provides an implementable compilation technique in the style of functional programming languages, and demonstrates the equivalence with a more easily understood type theory.

Dedicated to Professor Joseph Goguen on the occasion of his 65th birthday.

1   Introduction

Pattern matching is a long-established notation in functional programming [3,19], combining discrimination on constructors and selection of their arguments safely, compactly and efficiently. Extended to dependent types by Coquand [4], pattern matching becomes still more powerful, managing more complexity as we move from simple inductive datatypes, like Nat defined as follows,

$$\mathsf{Nat} : \star \;=\; \mathsf{zero} : \mathsf{Nat} \;\mid\; \mathsf{suc}\,(n\!:\!\mathsf{Nat}) : \mathsf{Nat}$$

to work with inductive families of datatypes [6] like Fin, which is indexed over Nat (Fin n is an n element enumeration), or Fin's ordering relation, $\leq$, indexed over indexed data.[4]

$$
\begin{aligned}
\mathsf{Fin}\,(n\!:\!\mathsf{Nat}) : \star \quad &= \quad \mathsf{fz}_n & &: \mathsf{Fin}\,(\mathsf{suc}\,n) \\
&\mid\; \mathsf{fs}_n(i\!:\!\mathsf{Fin}\,n) & &: \mathsf{Fin}\,(\mathsf{suc}\,n) \\
(i\!:\!\mathsf{Fin}\,n) \leq_n (j\!:\!\mathsf{Fin}\,n) : \star \quad &= \quad \mathsf{leqz}_{n;j} & &: \mathsf{fz}_n \leq_{(\mathsf{suc}\,n)} j \\
&\mid\; \mathsf{leqs}_{n;i;j}\,(p\!:\!i \leq_n j) : \mathsf{fs}_n\,i \leq_{(\mathsf{suc}\,n)} \mathsf{fs}_n\,j
\end{aligned}
$$

Pattern matching can make programs and proofs defined over such structures just as simple as for their simply-typed analogues. For example, the proof of transitivity for $\leq$ works just the same for Fin as for Nat:

$$
\begin{aligned}
\mathbf{trans}\,(p\!:\!i \leq j;\, q\!:\!j \leq k) &: i \leq k \\
\mathbf{trans}\quad \mathsf{leqz}_{n;j} \qquad\qquad q \qquad &\mapsto\; \mathsf{leqz}_{n;k} \\
\mathbf{trans}\,(\mathsf{leqs}_{n;i';j'}\,p')\,(\mathsf{leqs}_{n;j';k'}\,q') &\mapsto\; \mathsf{leqs}_{n;i';k'}\,(\mathbf{trans}\,p'\,q')
\end{aligned}
$$

---

[4] Here we write as subscripts arguments which are usually inferrable; informally, and in practice, we omit them entirely.

K. Futatsugi et al. (Eds.): Goguen Festschrift, LNCS 4060, pp. 521–540, 2006.
© Springer-Verlag Berlin Heidelberg 2006

There is no such luxury in a traditional type theory [14,20], where a datatype is equipped only with an *elimination constant* whose type expresses its induction principle and whose operational behaviour is primitive recursion. This paper provides a translation from dependent pattern matching in Coquand's sense to such a type theory—Luo's UTT [11], extended with the Altenkirch-Streicher **K** axiom 'uniqueness of identity proofs' [22]. Coquand observed that his rules admit **K**; Hofmann and Streicher have shown that **K** does not follow from the usual induction principle for the identity relation [9]. We show that (a variant of) **K** is *sufficient* to bridge the gap: it lets us encode the constructor-based unification which Coquand built directly into his rules.

Our translation here deploys similar techniques to those in [18], but we now ensure both that the translated pattern matching equations hold as *reductions* in the target theory and that the equations can be given a conventional operational semantics [1] directly, preserving termination and confluence. By doing so, we justify pattern matching as a language construct, in the style of ALF [13], without compromising the rôle of the elimination constant in characterising the meaning of data.

An early approximant of our translation was added to the LEGO system [12] and demonstrated at 'Types 1998'. To date, McBride's thesis [15] is the only account of it, but there the treatment of the empty program is unsatisfying, the computational behaviour is verified only up to conversion, and the issue of unmatched but trusted terms in pattern matching rules is barely considered.

Our recent work describes the key equipment. The account of elimination in [16] uses a *heterogeneous* equality to express unification constraints over dependently typed data. Hence where Coquand's pattern matching invokes an *external* notion of unification and of structural recursion, we have built the tools we need *within* type theory [17]. Now, finally, we assemble these components to perform dependent pattern matching by elimination.

Overview   The rest of the paper is organised as follows. Section 2 examines pattern matching with dependent types, and develops basic definitions, including that of *specialisation* in patterns, as well as the programs which will eventually be translatable to type theory. The key technical definition here is that of *splitting tree*; novel here is the recording of explicit evidence for impossible case branches. Section 3 describes the target type theory. This is extended by function symbols with defining equations which determine reduction rules, subject to certain conditions. The allowable such function definitions arise from the existence of *valid* splitting trees. Finally, Section 4 shows how such function definitions may be eliminated in favour of closed terms in the type theory with the same reduction behaviour; the valid splitting trees precisely correspond to the terms built from constructor case analysis and structural recursion on inductive families, modulo the heterogeneous equality Eq.

2 Dependent Pattern Matching

Let us first take a look at what dependent pattern matching is, and why it is a more subtle notion than its simply typed counterpart. Inductive families gain their precision from the way their constructors have *specialised* return types. For example, the constructors of Fin can only make elements of sets whose 'size' is non-zero. Consider writing some function $\mathbf{p}\ (i : \mathsf{Nat}; x : \mathsf{Fin}\ i)\ : \ \cdots$. Trying to match on x without instantiating i is an error. Rather, one *must* take account of the fact that i is sure to be a suc, if $\mathbf{p}$ is to typecheck:

$$\cdots \nvdash \ \mathbf{p}\ i\ \mathsf{fz} \quad\quad : \mathsf{Nat} \qquad\qquad \mathbf{p}\ (\mathsf{suc}\ j)\ \mathsf{fz} \quad \mapsto \cdots$$
$$\cdots \nvdash \ \mathbf{p}\ i\ (\mathsf{fs}\ y)\ : \mathsf{Nat} \qquad\qquad \mathbf{p}\ (\mathsf{suc}\ j)\ (\mathsf{fs}\ y) \mapsto \cdots$$

Of course, there need not be any actual check at run time whether these $(\mathsf{suc}\ j)$ patterns match—the type system guarantees that they *must* if the patterns for x do. This is not merely a convenient optimisation, it is a new and necessary phenomenon to consider. For example, we may define the property of 'being in the image of $\mathbf{f}$' for some fixed $\mathbf{f} : S \to T$, then equip $\mathbf{f}$ with an 'inverse':

$$\mathsf{Imf}\ (t{:}T)\ :\ \star = \mathsf{imf}\ (s{:}S)\ :\ \mathsf{Imf}\ (\mathbf{f}\ s) \qquad\qquad \begin{aligned} &\mathsf{inv}\ (t{:}T; p{:}\mathsf{Imf}\ t)\ :\ S \\ &\mathsf{inv}\ (\mathbf{f}\ s)\ (\mathsf{imf}\ s) \mapsto s \end{aligned}$$

The typing rules force us to write $(\mathbf{f}\ s)$ for t, but there is no way in general that we can compute s from t by inverting $\mathbf{f}$. Of course, we actually get s from the constructor pattern $(\mathsf{imf}\ s)$ for p, together with a *guarantee* that t is $(\mathbf{f}\ s)$.

We have lost the ability to consider patterns for each argument independently. Moreover, we have lost the distinction of patterns as the sub-language of terms consisting only of the *linear constructor forms*, and with this, the interpretation of defining equations as rewriting rules is insufficient. It is not enough just to assign dependent types to conventional programs: specialised patterns change what programs can be.

Let us adapt to these new circumstances, and gain from specialisation, exploiting the information it delivers 'for free'. For example, in a fully decorated version of the step case of the above definition of the trans function,

$$\mathbf{trans}_{(\mathsf{suc}\ n);(\mathsf{fs}_n\ i);(\mathsf{fs}_n\ j);(\mathsf{fs}_n\ k)}\ (\mathsf{leqs}_{n;i;j}\ p')\ (\mathsf{leqs}_{n;j;k}\ q')\ \mapsto$$
$$\mathsf{leqs}_{n;i;k}\ (\mathbf{trans}_{n;i;j;k}\ p'\ q')$$

it is precisely specialisation that ensures the p' and q' are not arbitrary $\leq$ proofs, but rather appropriate ones, which justify the recursive call to **trans**. Meanwhile, we need not analyse the case

$$\cdots \nvdash\ \mathbf{trans}_{(\mathsf{suc}\ n);(\mathsf{fs}_n\ i);?;k}\ (\mathsf{leqs}_{n;i;j}\ p')\ \mathsf{leqz}_{n;k}\ :\ \mathsf{fs}\ i \leq_{(\mathsf{suc}\ n)} k$$

because the two proof patterns demand incompatible specialisations of the middle value upon which they must agree. In general, specialisation is given by the *most general unifier* for the type of the value being analysed and the type of the pattern used to match it. Later, we shall be precise about how this works, but let us first sketch how we address its consequences.

2.1   Patterns with Inaccessible Terms

The key to recovering an operational interpretation for these defining equations is to find the distinction between those parts which *require* constructor matching, and those which merely *report* specialisation. We shall show how to translate the *terms* on the left-hand sides of definitional equations written by the programmer into *patterns* which, following Brady [2], augment the usual linear constructor forms with a representation for the arbitrary terms reported by specialisation and *presupposed* to match.

Definition 1 (Patterns)

$$
\begin{array}{llll}
pat := & x & \lceil x \rceil \Longrightarrow x & \text{AV}(x) \Longrightarrow \{x\} \\
 & \mid \text{c } pat^* & \lceil \text{c } \vec{p} \rceil \Longrightarrow \text{c } \lceil \vec{p} \rceil & \text{AV}(\text{c } \vec{p}) \Longrightarrow \text{AV}(\vec{p}) \\
 & \mid \underline{term} & \lceil \underline{t} \rceil \Longrightarrow t & \text{AV}(\underline{t}) \Longrightarrow \emptyset \\
lhs := & \text{f } pat^* & \lceil \text{f } \vec{p} \rceil \Longrightarrow \text{f } \lceil \vec{p} \rceil & \text{AV}(\text{f } \vec{p}) \Longrightarrow \text{AV}(\vec{p})
\end{array}
$$

We say the terms marked $\underline{t}$ are *inaccessible* to the matcher and may not bind variables. The partial map $\text{AV}(-)$ computes the set of *accessible variables*, where $\text{AV}(\vec{p})$ is the *disjoint* union, $\biguplus_i \text{AV}(p_i)$, hence $\text{AV}(-)$ is defined only for linear patterns. The map $\lceil - \rceil$ takes patterns back to terms.

We can now make sense of our inv function: its left-hand side becomes

$$\text{inv } \underline{(\text{f } s)} \text{ (im } s)$$

Matching for these patterns is quite normal, with inaccessible terms behaving like 'don't care' patterns, although our typing rules will always ensure that there is actually no choice! We define MATCH to be a partial operation yielding a matching substitution, throwing a CONFLICT exception[5], or failing to make progress only in the case of non-canonical values in a nonempty context.

Definition 2 (Matching) *Matching is given as follows:*

$$
\begin{array}{lll}
\text{MATCH}(x, & t) & \Longrightarrow [x \mapsto t] \\
\text{MATCH}(\text{chalk } \vec{p}, \text{chalk } \vec{t}) & \Longrightarrow & \text{MATCHES}(\vec{p}, \vec{t}) \\
\text{MATCH}(\text{chalk } \vec{p}, \text{cheese } \vec{t}) & \Uparrow & \text{CONFLICT} \\
\text{MATCH}(\underline{u}, & t) & \Longrightarrow \varepsilon
\end{array}
$$

$$
\begin{array}{lll}
\text{MATCHES}(\varepsilon, & \varepsilon) & \Longrightarrow \varepsilon \\
\text{MATCHES}(p; \vec{p}, t; \vec{t}) & \Longrightarrow & \text{MATCH}(p, t); \text{MATCHES}(\vec{p}, \vec{t})
\end{array}
$$

So, although definitional equations are not admissible as rewriting rules just as they stand, we can still equip them with an operational model which relies only on constructor discrimination. This much, at least, remains as ever it was.

Before we move on, let us establish a little equipment for working with patterns. In our discussion, we write $p[x]$ to stand for p with an accessible x abstracted. We may thus form the instantiation $p[p']$ if p' is a pattern with variables

---

[5] We take chalk and cheese to stand for an arbitrary pair of distinct constructors.

disjoint from those free in $p[-]$, pasting p' for the accessible occurrence of x and $\lceil p' \rceil$ for the inaccessible copies. In particular, $p[c\ \vec{y}]$ is a pattern, given fresh $\vec{y}$. Meanwhile, we shall need to apply specialising substitutions to patterns:

Definition 3 (Pattern Specialisation) *If σ is a substitution from variables Δ to terms over Δ' with $\mathrm{AV}(p) = \Delta \uplus \Delta'$ (making σ idempotent), we define the specialisation σp, lifting σ to patterns recursively as follows:*

$$\sigma x \implies \underline{\sigma x} \ \ if\ x \in \Delta \qquad\qquad \sigma(c\ \vec{p}) \implies c\ \sigma\vec{p} \qquad\qquad \sigma\underline{t} \implies \underline{\sigma t}$$
$$\sigma x \implies x \ \ \ if\ x \in \Delta'$$

Observe that $\mathrm{AV}(\sigma p) = \Delta'$.

Specialisations, being computed by unification, naturally turn out to be idempotent. Their effect on a pattern variable is thus either to retain its accessibility or to eliminate it entirely, replacing it with an inaccessible term. Crucially, specialisation preserves the availability of a matching semantics despite apparently introducing nonlinearity and non-constructor forms.

2.2 Program Recognition

The problem we address in this paper is to recognize programs as total functions in UTT+K. Naturally, we cannot hope to decide whether it is possible to construct a functional value exhaustively specified by a set of arbitrary equations. What we can do is fix a recognizable and total fragment of those programs whose case analysis can be expressed as a *splitting tree* of constructor discriminations and whose recursive calls are on *structurally decreasing* arguments.

The idea is to start with a candidate left-hand side whose patterns are just variables and to grow a partition by analysing a succession of pattern variables into constructor cases. This not only gives us an efficient compilation in the style of Augustsson [1], it will also structure our translation, with each node mapping to the invocation of an eliminator. Informally, for trans, we build the tree

$$\text{trans } p\ q$$
$$p : i \leq j \begin{cases} \text{trans leqz } q\ \mapsto\ \text{leqz} \\ \text{trans (leqs } p')\ q \\ \quad q : \text{fs } j \leq k \begin{cases} \text{\sout{trans (leqs } p'\text{) leqz}} \\ \text{trans (leqs } p')\ \text{(leqs } q')\ \mapsto\ \text{leqs (trans } p'\ q') \end{cases} \end{cases}$$

The program just gives the leaves of this tree: finding the whole tree guarantees that it partitions the possible input. The recursion reduces the size of one argument (both, in fact, but one is enough), so the function is total.

However, if we take a 'program' just to be a set of definitional equations, even this recognition problem is well known to be undecidable [4,15,21]. The difficulty for the recognizer is the advantage for the programmer: specialisation can prune the tree! Above, we can see that q must be split to account for (leqs q'), and having split q, we can confirm that no leqz case is possible. But consider the signature empty $(i:\text{Fin zero}) : X$. We have the splitting tree:

$$\text{empty } i$$

$$i : \text{Fin zero} \quad \begin{cases} \text{~~empty fz~~} \\ \text{~~empty (fs i')~~} \end{cases}$$

If we record only the leaves of the tree for which we return values, we shall not give the recognizer much to work from! More generally, it is possible to have arbitrarily large splitting trees with no surviving leaves—it is the need to recover these trees from thin air that makes the recognition of equation sets undecidable. Equations are insufficient to define dependently typed functions, so we had better allow our programs to consist of something more. We extend the usual notion of program to allow clauses $\text{f } \vec{t} \ \text{⋔} \ x$ which *refute* a pattern variable, requiring that splitting it leaves no children. For example, we write

$$\text{empty } (i:\text{Fin zero})$$
$$\text{empty } i \ \text{⋔} \ i$$

We now give the syntax for programs and splitting trees.

Definition 4 (Program, Splitting Tree)

$$
\begin{aligned}
program &:= \text{f}\,(context) : term & splitting &:= compRule \\
&\quad\ clause^{+} & &\mid\ [context]\ lhs \\
clause &:= \text{f}\ term^{*}\ rhs & &\quad\ x\ \{splitting^{+} \\
rhs &:= \mapsto term & compRule &:= [context]\ lhs\ rhs \\
&\mid\ \text{⋔}\ x
\end{aligned}
$$

We say that a splitting tree solves the programming problem $[\Delta]\,\text{f}\,\vec{p}$, *if these are the context and left-hand side at its root node. Every such programming problem must satisfy* $\text{AV}(\vec{p}) = \Delta$, *ensuring that every variable is accessible.*

To recognize a program with clauses $\{\text{f }\vec{t_i}\ r_i \mid 0 \le i \le n\}$ is to find a valid splitting tree with computation rules $\{[\Delta_i]\ \text{f }\vec{p_i}\ r_i \mid 0 \le i \le n\}$ such that $\lceil\text{f }\vec{p_i}\rceil = \text{f }\vec{t_i}$ and to check the guardedness of the recursion. We defer the precise notion of 'valid' until we have introduced the type system formally, but it will certainly be the case that if an internal node has left-hand side $\text{f }\vec{p}[x]$, then its children (numbering at least one) have left-hand sides $\text{f }\sigma\vec{p}[\text{c }\vec{y}]$ where c is a constructor and σ is the specialising substitution which unifies the datatype indices of x and $\text{c }\vec{y}$.

We fix unification to be *first-order* with datatype constructors as the rigid symbols [10]—we have systematically shown constructors to be injective and disjoint, and that inductive families do not admit cyclic terms [17]. Accordingly, we have a terminating unification procedure for two vectors of terms which will either *succeed positively* (yielding a specialising substitution), *succeed negatively* (establishing a constructor conflict or cyclic equation), or *fail* because the problem is too hard. Success is guaranteed if the indices are in constructor form.

We can thus determine if a given left-hand side may be split at a given pattern variable—we require all the index unifications to succeed—and generate specialised children for those which succeed positively. We now have:

Lemma 5 (Decidable Coverage) *Given* $f(\vec{x}:\vec{S}) : T$; $\{f\,\vec{t_i}\,r_i \mid 0 \le i \le n\}$, *it is decidable whether there exists a splitting tree, with root* $[\vec{x}:\vec{S}]\,f\,\vec{x} : T$ *and computation rules* $\{[\Delta_i]\,f\,\vec{p_i}\,r_i \mid 0 \le i \le n\}$ *such that* $\lceil f\,\vec{p_i}\rceil = f\,\vec{t_i}$.

Proof   The total number of constructor symbols in the subproblems of a splitting node strictly exceeds those in the node's problem. We may thus generate all candidate splitting trees whose leaves bear at most the number of constructors in the program clauses and test if any yield the program.   □

Coquand's specification of a *covering* set of patterns requires the construction of a splitting tree: if we can find a covering for a given set of equations, we may read off one of our programs by turning the childless nodes into refutations. As far as recursion checking is concerned, we may give a criterion a little more generous than Coquand's original [4].

Definition 6 (Guardedness, Structural Recursion) *We define the binary relation* $\prec$, 'is guarded by', *inductively on the syntax of terms:*

$$\frac{}{t_i \prec c\,t_1\,\dots\,t_n}\,1 \le i \le n \qquad \frac{f \prec t}{f\,s \prec t} \qquad \frac{r \prec s \quad s \prec t}{r \prec t}$$

We say that a program $f(\vec{x}:\vec{S}) : T$; $\{f\vec{t_i}r_i \mid 0 \le i \le n\}$ *is* structurally recursive *if, for some argument position* j, *we have that every recursive call* $f\,\vec{s}$ *which is a subterm of some* r_i *satisfies* $s_j \prec t_{ij}$.

It is clearly decidable whether a program is structurally recursive in this sense. Unlike Coquand, we do permit one recursive call within the argument of another, although this distinction is merely one of convenience. We could readily extend this criterion to cover lexicographic descent on a number of arguments, but this too is cosmetic. Working in a higher-order setting, we can express the likes of Ackermann's function, which stand beyond *first-order* primitive recursion. Of course, the interpreter for our own language is beyond it.

3   Type Theory and Pattern Matching

We start from a predicative subsystem of Luo's UTT [11], with rules of inference given in Figure 1. UTT's dependent types and inductive types and families are the foundation for dependent pattern matching. Programs with pattern matching are written over types in the base type universe $\star_0$, which we call *small* types. Eliminations over types to solve unification are written in $\star_1$, and the Logical-Framework-level universe $\square$ is used to define a convenient presentation of equality from the traditional I, **J** and **K**. Our construction readily extends to

validity $\boxed{context \vdash valid}$ $\dfrac{}{\mathcal{E} \vdash \underline{valid}}$ $\dfrac{\Gamma \vdash S : \alpha}{\Gamma; x : S \vdash \underline{valid}}$ $\alpha \in \{\star_0, \star_1, \Box\}$

typing $\boxed{context \vdash term : term}$

$$\dfrac{\Gamma \vdash valid}{\Gamma \vdash x : S} \; x : S \in \Gamma \qquad \dfrac{\Gamma \vdash valid}{\Gamma \vdash \star_0 : \star_1} \qquad \dfrac{\Gamma \vdash valid}{\Gamma \vdash \star_1 : \Box}$$

$$\dfrac{\Gamma \vdash S : \alpha \quad \Gamma; x : S \vdash T : \alpha}{\Gamma \vdash \Pi x : S. T : \alpha} \quad \alpha \in \{\star_0, \star_1, \Box\}$$

$$\dfrac{\Gamma; x : S \vdash t : T}{\Gamma \vdash \lambda x : S. t : \Pi x : S. T} \qquad \dfrac{\Gamma \vdash f : \Pi x : S. T \quad \Gamma \vdash s : S}{\Gamma \vdash f s : [x \mapsto s] T}$$

$$\dfrac{\Gamma \vdash t : S \quad S \preceq T \quad \Gamma \vdash T : \sigma}{\Gamma \vdash t : T}$$

reduction $\boxed{term \leadsto term}$ $\dfrac{}{(\lambda x : S. t) \, s \leadsto [x \mapsto s] t}$ plus contextual closure

conversion $\boxed{term \cong term}$ equivalence closure of $\leadsto$

cumulativity $\boxed{term \preceq term}$

$$\dfrac{}{\star_0 \preceq \star_1} \qquad \dfrac{}{\star_1 \preceq \Box} \qquad \dfrac{S_1 \cong S_2 \quad T_1 \preceq T_2}{\Pi x : S_1. T_1 \preceq \Pi x : S_2. T_2} \qquad \dfrac{S \cong T}{S \preceq T} \qquad \dfrac{R \preceq S \quad S \preceq T}{R \preceq T}$$

Fig. 1. Luo's UTT (functional core)

the additional hierarchy of universes of full UTT. The impredicative universe of propositions in UTT is not relevant to explaining pattern matching through the primitive constructs of type theory, and so we omit it.

We identify terms that are equivalent up to the renaming of bound variables, and we write $[x \mapsto s]t$ for the usual capture-free substitution of s for the free variable x in t.

UTT is presented through the Logical Framework, a meta-language with typed arities for introducing the constants and equalities that define a type theory. While the Logical Framework is essential to the foundational understanding of UTT, it is notationally cumbersome, and we shall hide it as much as possible. We shall not distinguish notationally between framework Π kinds and object-level Π types, nor between the framework and object representations of types. We justify this by observing that $\Box$ represents the types in the underlying framework, and that $\star_0$ and $\star_1$ are universes with names of specific types within $\Box$. However, informality with respect to universes may lead to size issues if we are not careful, and we shall explicitly mention the cases where it is important to distinguish between the framework and object levels.

There is no proof of the standard metatheoretic properties for the theory UTT plus **K** that we take as our target language. Goguen's thesis [8] establishes the metatheory for a sub-calculus of UTT with the Logical Framework, a single universe and higher-order inductive types but not inductive families or the **K** combinator. Walukiewicz-Chrzaszcz [23] shows that certain higher-order rewrite rules are terminating in the Calculus of Constructions, including inductive families and the **K** combinator, but the rewrite rules do not include higher-order inductive types, and the language is not formulated in the Logical Framework.

However, our primary interest is in justifying dependent pattern matching by translation to a traditional presentation of type theory, and UTT plus **K** serves this role very well. Furthermore, the extensions of additional universes, inductive relations and the **K** combinator to the language used in Goguen's thesis would complicate the structure of the existing proof of strong normalization but do not seem to represent a likely source of non-termination.

3.1   Telescope Notation

We shall be describing general constructions over dependent datatypes, so we need some notational conveniences. We make considerable use of de Bruijn's *telescopes* [5]—dependent sequences of types—sharing the syntax of contexts. We also use Greek capitals to stand for them. We may check telescopes (and constrain the universe level α of the types they contain) with the following judgment:

$$\frac{\Gamma \vdash \text{valid}}{\Gamma \vdash \mathcal{E} \ \underline{\text{tele}}(\alpha)} \qquad \frac{\Gamma \vdash S : \alpha \quad \Gamma; x{:}S \vdash \Delta \ \underline{\text{tele}}(\alpha)}{\Gamma \vdash x : S; \Delta \ \underline{\text{tele}}(\alpha)}$$

We use vector notation $\vec{t}$ to stand for sequences of terms, $t_1; \ldots; t_n$. We identify the application $f\ t_1; \ldots; t_n$ with $f\ t_1\ \ldots\ t_n$. Simultaneous substitutions from a telescope to a sequence are written $[\Theta \mapsto \vec{t}]$, or $[\vec{t}]$ if the domain is clear. Substituting through a telescope textually yields a sequence of typings $t_1{:}T_1; \ldots; t_n{:}T_n$ which we may check by iterating the typing judgment. We write $\vec{t} : \Theta$ for the sequence of typings $[\vec{t}]\Theta$, asserting that the t's may instantiate Θ. We also let $\Gamma \vdash \sigma\Delta$ assert that σ is a type-correct substitution from Δ to Γ-terms.

We write $\Pi\Delta.\ T$ to iterate the Π-type over a sequence of arguments, or $\Delta{\to}T$ if T does not depend on Δ. The corresponding abstraction is $\lambda\Delta.\ t$. We also let telescopes stand as the sequence of their variables, so if $f\ :\ \Pi\Delta.\ T$, then $\Delta \vdash f\ \Delta : T$. The empty telescope is $\mathcal{E}$, the empty sequence, ε.

3.2   Global Declarations and Definitions

A development in our type theory consists of a *global context* Γ containing declarations of datatype families and their constructors, and definitions of function symbols. To ease our translation, we declare global identifiers g with a telescope of arguments and we demand that they are applied to a suitable sequence wherever they are used. Each function $\mathsf{f}(\Delta)\ :\ T$ has a nonempty set of computation rules. We extend the typing and reduction rules (now contextualised) accordingly:

$$\frac{g(\Theta){:}T \in \Gamma \quad \Gamma; \Delta \vdash \vec{t} : \Theta}{\Gamma; \Delta \vdash g\ \vec{t} : [\vec{t}]T} \qquad \begin{array}{l} \mathsf{f}\ \vec{t} \leadsto_\Gamma \theta e \ \text{ if } [\Delta']\ \mathsf{f}\ \vec{p} \mapsto e \in \Gamma \\ \qquad\quad \text{MATCHES}(\vec{p}, \vec{t}) \implies \theta \end{array}$$

We take the following at least to be basic requirements for defined functions.

Definition 7 (Function Criteria) *To extend Γ with $f(\Delta) : T$ with computation rules $\{[\Delta_i]\, f\, \vec{p}_i\, r_i \mid 0 \leq i \leq n\}$, we require that:*

- $\Gamma; \Delta \vdash T : \square$,
- *the computation rules arise as the leaves of a splitting tree solving $[\Delta]\, f\, \Delta$,*
- *the corresponding program is structurally recursive,*
- *if r_i is $\mapsto e_i$, then $\Gamma; \Delta_i \vdash e_i : P_i$.*

We shall check basic properties of pattern matching computation shortly, but we first give our notion of data (and hence splitting) a firm basis.

Definition 8 (Inductive Families) *Inductive families with $n \geq 1$ constructors are checked for strict positivity and introduced globally as shown in figure 2. We write $\overline{D}$ for the telescope $\Xi; z : D\, \Xi$.*

$$\frac{\Gamma \vdash \Xi\ \underline{\text{tele}(\star_0)} \qquad \{\Gamma | D(\Xi) : \star_0 \vdash \Delta_i\ \underline{\text{con}(\vec{u}_i)} \mid i \leq n\};}{\begin{array}{l} \Gamma;\ D(\Xi) : \star_0;\quad \{c_i(\Delta_i) : D\ \vec{u}_i \mid i \leq n\}; \\ E_D(\{M_i : \Pi\Delta_i.\ \text{HYPS}(\Delta_i, \text{BIG}) \to \star_1 \mid i \leq n\}; \overline{D}) : \star_1 \\ \{[\vec{M}; \Delta_i]\ E_D\ \vec{M}\ \vec{u}_i\ (c_i\, \Delta_i) \mapsto M_i\, \Delta_i\ \text{RECS}(\Delta_i, E_D\, \vec{M}) \mid i \leq n\}; \\ e_D(P : \overline{D} \to \star_1; \{m_i : \Pi\Delta_i.\ \text{HYPS}(\Delta_i, \text{LITTLE}(P)) \to P\, \vec{u}_i\, (c\, \Delta_i) \mid i \leq n\}; \overline{D}) : P\, \overline{D} \\ \{[P; \vec{m}; \Delta_i]\ e_D\ P\ \vec{M}\ \vec{u}_i\ (c_i\, \Delta_i) \mapsto m_i\, \Delta_i\ \text{RECS}(\Delta_i, e_D\, P\, \vec{m}) \mid i \leq n\} \\ \vdash \underline{\text{valid}} \end{array}}$$

where $\qquad$ $\text{BIG}(\_, \_) \implies \star_1 \qquad \text{LITTLE}(P)(\vec{v}, x) \implies P\, \vec{v}\, x$

$$\frac{\Gamma; \Theta \vdash \vec{u} : \Xi}{\Gamma | D(\Xi) : \star_0; \Theta \vdash \varepsilon\ \underline{\text{con}(\vec{u})}} \qquad \frac{\Gamma; \Theta \vdash A : \star_0 \quad \Gamma | D(\Xi) : \star_0; \Theta; a : A \vdash \Delta\ \underline{\text{con}(\vec{u})}}{\Gamma | D(\Xi) : \star_0; \Theta \vdash a : A; \Delta\ \underline{\text{con}(\vec{u})}}$$

$$\text{HYPS}(\varepsilon, H) \implies \varepsilon \qquad\qquad \text{HYPS}(a : A; \Delta, H) \implies \text{HYPS}(\Delta, H)$$
$$\text{RECS}(\varepsilon, f) \implies \varepsilon \qquad\qquad \text{RECS}(a : A; \Delta, f) \implies \text{RECS}(\Delta, f)$$

$$\frac{\Gamma; \Theta \vdash \Phi\ \underline{\text{tele}(\star_0)} \quad \Gamma; \Theta; \Phi \vdash \vec{v} : \Xi \quad \Gamma | D(\Xi) : \star_0; \Theta \vdash \Delta\ \underline{\text{con}(\vec{u})}}{\Gamma | D(\Xi) : \star_0; \Theta \vdash r : \Pi\Phi.\ D\ \vec{v}; \Delta\ \underline{\text{con}(\vec{u})}}$$

$$\text{HYPS}(r : \Pi\Phi.\ D\ \vec{v}; \Delta, H) \implies r' : \Pi\Phi.\ H(\vec{v}, r\, \Phi); \text{HYPS}(\Delta, H)$$
$$\text{RECS}(r : \Pi\Phi.\ D\ \vec{v}; \Delta, f) \implies (\lambda\Phi.\ f\, \vec{v}\, (r\, \Phi)); \text{RECS}(\Delta, f)$$

Fig. 2. Declaring inductive types with constructors

In Luo's presentation [11], each inductive datatype is an inhabitant of $\square$; it is then given a *name* in the universe $\star_0$. There is a single framework-level eliminator whose kind is much too large for a UTT type. Our presentation is implemented on top: D really computes Luo's *name* for the type; our UTT eliminators are readily simulated by the framework-level eliminator. This definition behaves as usual: for Nat, we obtain

$$\mathsf{Nat} : \star_0; \quad \mathsf{zero} : \mathsf{Nat}; \quad \mathsf{suc}(n{:}\mathsf{Nat}){:}\mathsf{Nat};$$
$$\mathbf{E}_{\mathsf{Nat}}(Z{:}\star_1; S{:}\mathsf{Nat} \to \star_1 \to \star_1; n{:}\mathsf{Nat}){:}\star_1;$$
$$[Z;S] \; \mathbf{E}_{\mathsf{Nat}} \; Z \; S \; \mathsf{zero} \quad \mapsto Z$$
$$[Z;S;n] \; \mathbf{E}_{\mathsf{Nat}} \; Z \; S \; (\mathsf{suc} \; n) \mapsto S \; n \; (\mathbf{E}_{\mathsf{Nat}} \; Z \; S \; n)$$
$$\mathsf{e}_{\mathsf{Nat}}(P{:}\mathsf{Nat} \to \star_1; z{:}P \; \mathsf{zero}; s{:}\Pi n{:}\mathsf{Nat}. \; P \; n \to P \; (\mathsf{suc} \; n); n{:}\mathsf{Nat}){:}P \; n;$$
$$[P;z;s] \; \mathsf{e}_{\mathsf{Nat}} \; P \; z \; s \; \mathsf{zero} \quad \mapsto z$$
$$[P;z;s;n] \; \mathsf{e}_{\mathsf{Nat}} \; P \; z. \; s \; (\mathsf{suc} \; n) \mapsto s \; n \; (\mathsf{e}_{\mathsf{Nat}} \; P \; z \; s \; n)$$

Given this, the Fin declaration yields the following (we suppress $\mathbf{E}_{\mathsf{Fin}}$):

$$\mathsf{Fin}(n{:}\mathsf{Nat}){:}\star_0; \quad \mathsf{fz}(n{:}\mathsf{Nat}){:}\mathsf{Fin} \; (\mathsf{suc} \; n); \quad \mathsf{fs}(n{:}\mathsf{Nat}; i{:}\mathsf{Fin} \; n){:}\mathsf{Fin} \; (\mathsf{suc} \; n);$$
$$\mathbf{E}_{\mathsf{Fin}} \cdots;$$
$$\mathsf{e}_{\mathsf{Fin}}(P : \Pi n{:}\mathsf{Nat}. \; \mathsf{Fin} \; n \to \star_1;$$
$$\quad z : \Pi n{:}\mathsf{Nat}. \; P_{(\mathsf{suc} \; n)} \; (\mathsf{fz}_n); \quad s : \Pi n{:}\mathsf{Nat}; i{:}\mathsf{Fin} \; n. \; P_n \; i \to P_{(\mathsf{suc} \; n)} \; (\mathsf{fs}_n \; i);$$
$$\quad n{:}\mathsf{Nat}; i{:}\mathsf{Fin} \; n) \; : \; P_n \; i$$
$$[P;z;s;n] \; \mathsf{e}_{\mathsf{Fin}} \; P \; z \; s \; \underline{(\mathsf{suc} \; n)} \; (\mathsf{fz}_n) \quad \mapsto z \; n$$
$$[P;z;s;n;i] \; \mathsf{e}_{\mathsf{Fin}} \; P \; z \; s \; \underline{(\mathsf{suc} \; n)} \; (\mathsf{fs}_n \; i) \mapsto s \; n \; i \; (\mathsf{e}_{\mathsf{Fin}} \; P \; z \; s \; n \; i)$$

All of our eliminators will satisfy the function criteria: each has just one split, resulting in specialised, inaccessible patterns for the indices. As the indices may be arbitrary terms, this is not merely convenient but essential. Rewriting with the standard equational laws which accompany the eliminators of inductive families is necessarily confluent.

Meanwhile, empty families have eliminators which refute their input.

$$\frac{\Gamma \vdash \Xi \; \underline{\mathsf{tele}(\star_0)}}{\begin{array}{l} \Gamma; \; \mathsf{D}(\Xi){:}\star_0; \; \mathbf{E}_{\mathsf{D}}(\overline{\mathsf{D}}){:}\star_1; \qquad\qquad [\Xi;x] \; \mathbf{E}_{\mathsf{D}} \quad \Xi \; x \; \pitchfork \; x; \\ \qquad\quad \mathsf{e}_{\mathsf{D}}(P : \overline{\mathsf{D}} \to \star_1; \overline{\mathsf{D}}){:}P \; \overline{\mathsf{D}}; \quad [P;\Xi;x] \; \mathsf{e}_{\mathsf{D}} \; P \; \Xi \; x \; \pitchfork \; x \\ \vdash \underline{\mathsf{valid}} \end{array}}$$

We have constructed families over elements of sets, but this does not yield 'polymorphic' datatypes, parametric in sets themselves. As Luo does, so we may also parametrise a type constructor, its data constructors and eliminators uniformly over a fixed initial telescope of UTT *types*, including $\star_0$.

3.3   Valid Splitting Trees and Their Properties

In this section, we deliver the promised notion of 'valid splitting tree' and show it fit for purpose. This definition is very close to Coquand's original construction of 'coverings' from 'elementary coverings' [4]. Our contribution is to separate the empty splits (with explicit refutations) from the nonempty splits (with nonempty subtrees), and to maintain our explicit construction of patterns in linear constructor form with inaccessible terms resulting from specialisation.

Definition 9 (Valid Splitting Tree) *A* valid splitting tree *for* $f(\Delta) : T$ *has root problem* $[\Delta] f \Delta$. *At each node,*

- *either we have* $\Delta' \vdash e : [\![\vec{p}]\!]T$ *and computation rule*

$$[\Delta'] f \vec{p} \mapsto e$$

- *or we have problem* $[\Delta^x; x : D\vec{v}; {}^x\!\Delta] f \vec{p}[x]$ *and for each constructor* $c(\Delta^c) : D\vec{u}$, *unification succeeds for* $\vec{u}$ *and* $\vec{v}$, *in which case*
  - *either all succeed negatively, and the node is the computation rule*

$$[\Delta^x; x : D\,\vec{v}; {}^x\!\Delta] f \vec{p}[x] \pitchfork x$$

  - *or at least one succeeds positively, and the node is a split of form*

$$[\Delta^x; x : D\,\vec{v}; {}^x\!\Delta] f \vec{p}[x]$$
$$x \, \{S$$

  Each positive success yields a pair (Δ', σ) *where* σ *is a most general idempotent unifier for* $\vec{u}$ *and* $\vec{v}$ *satisfying* $\Delta' \vdash \sigma\Delta^c; \sigma\Delta^x$ *and* $\mathrm{DOM}(\sigma) \uplus \Delta' = \Delta^c \uplus \Delta^x$, *and contributes a subtree to* S *with root*

$$[\Delta'; \sigma[x \mapsto c\,\Delta^c]{}^x\!\Delta] f \sigma\vec{p}[c\,\Delta^c]$$

We shall certainly need to rely on the fact that matching well typed terms yields type-correct substitutions. We must also keep our promise to use inaccessible terms in patterns only where there is no choice.

Definition 10 (Respectful Patterns) *For a function* $f(\Delta) : T$, *we say that a programming problem* $[\Delta'] f \vec{p}$ *has* respectful patterns *provided*

- $\Delta' \vdash \lceil \vec{p} \rceil : \Delta$
- *if* $\Theta \vdash \vec{a} : \Delta$ *and* $\mathrm{MATCHES}(\vec{p}, \vec{a}) \implies \theta$, *then* $\Theta \vdash \theta\Delta'$ *and* $\theta\lceil \vec{p} \rceil \cong \vec{a}$.

Let us check that valid splitting trees maintain the invariant.

Lemma 11 (Functions have respectful patterns) *If* $f(\Delta) : T$ *with computation rules* $\{[\Delta_i] f \vec{p}_i\, r_i \mid 0 \le i \le n\}$ *satisfies the function criteria, then* $[\Delta_i] f \vec{p}_i$ *has respectful patterns.*

Proof   The root problem $[\Delta] f \Delta : T$ readily satisfies these properties. We must show that splitting *preserves* them. Given a typical split as above, taking $[\Delta^x; x : D\vec{v}; {}^x\!\Delta] f \vec{p}[x]$ to some $[\Delta'; {}^x\!\Delta] f \sigma\vec{p}[c\Delta^c]$. Let us show the latter is respectful.

We have $\Delta^x; x : D\,\vec{v}; {}^x\!\Delta \vdash \lceil \vec{p}[x] \rceil : \Delta$, hence idempotence of σ yields $\Delta'; x : D\,\sigma\vec{v}; \sigma^x\!\Delta \vdash \lceil \sigma\vec{p}[x] \rceil : \Delta$. But $c\,\sigma\Delta^c : D\,\sigma\vec{u} \cong D\,\sigma\vec{v}$, hence $\Delta'; {}^x\!\Delta \vdash \lceil \sigma\vec{p}[c\,\Delta^c] \rceil : \Delta$.

Now suppose $\mathrm{MATCHES}(\sigma\vec{p}[c\,\Delta^c], \vec{a}) \implies \phi$ for $\Phi \vdash \vec{a} : \Delta$. For some $\vec{b} : \Delta^c$, we must have $\mathrm{MATCHES}(\vec{p}[x], \vec{a}) \implies \theta; [x \mapsto c\,\vec{b}]$. By assumption, the $\vec{p}[x]$ are respectful, so $\Phi \vdash (\theta; [x \mapsto c\,\vec{b}])(\Delta^x; x : D\,\vec{v}; {}^x\!\Delta)$, hence $c\,\vec{b} : D\,\theta\vec{v} = D\,[\Delta^c \mapsto \vec{b}]\vec{u}$, and $\theta; [x \mapsto c\,\vec{b}]\lceil \vec{p}[x] \rceil \cong \vec{a}$. Rearranging, we get $\theta; [\Delta^c \mapsto \vec{b}]\lceil \vec{p}[c\,\Delta^c] \rceil \cong \vec{a}$.

But $\theta; [\Delta^c \mapsto \vec{b}]y$ unifies $\vec{u}$ and $\vec{v}$ and thus factors as $\theta' \cdot \sigma$ as σ is the most general unifier. By idempotence of σ, θ' and $\theta; [\Delta^c \mapsto \vec{b}]y$ coincide on Δ'. But ϕ coincides with $\theta; [\Delta^c \mapsto \vec{b}]$ on Δ' because they match the same subterms of the $\vec{a}$, so $\theta; [\Delta^c \mapsto \vec{b}] = \phi \cdot \sigma$, hence $\phi[\sigma \vec{p}[c\ \Delta^c]] \cong \vec{a}$. Moreover, we now have $\Phi \vdash (\phi \cdot \sigma)\Delta^c$ and $\Phi \vdash (\phi \cdot \sigma)(\Delta^x; x : D\ \vec{v}; {}^x\!\Delta)$, but idempotence makes Δ' a subcontext of $\sigma(\Delta^c; \Delta^x)$, so $\Phi \vdash \phi(\Delta'; \sigma^x\!\Delta)$ as required. □

Lemma 12 (Matching Reduction Preserves Type) *If* $\Theta \vdash f\ \vec{a} : A$ *and* $f(\Delta) : T$ *has a computation rule* $[\Delta']\ f\ \vec{p} \mapsto e$ *for which* MATCHES$(\vec{p}, \vec{a}) \implies \theta$, *then* $\Theta \vdash \theta e : A$.

Proof By inversion of the typing rules, we must have $[\vec{a}]T \preceq A$. By respectfulness, we have $\Theta \vdash \theta\Delta'$ and $\vec{a} \cong \theta[\vec{p}]$. By construction, $\Delta' \vdash e : [\![\vec{p}]\!]T$, hence $\Theta \vdash \theta e : [\theta[\vec{p}]]T \cong [\vec{a}]T \preceq A$. □

Lemma 13 (Coverage) *If a function* $f(\Delta) : T$ *is given by computation rules* $\{[\Delta_i]\ f\ \vec{p_i}\ r_i : P_i \mid 0 \le i \le n\}$, *then for any* $\Theta \vdash \vec{t} : \Delta$, *it is not the case that for each* i, MATCHES$(\vec{p_i}, \vec{t}) \Uparrow$ CONFLICT.

Proof An induction on splitting trees shows that if we have root problem $f\ \vec{p}$ and MATCHES$(\vec{p}, \vec{t}) \implies \theta$ for well typed arguments $\vec{t}$, matching cannot yield CONFLICT at all the leaf patterns. Either the root is the leaf and the result is trivial, or the root has a split at some $x : D\vec{v}$. In the latter case, we either have θx not in constructor form and matching gets stuck, or $\theta x = c\ \vec{b}$ where $\vec{c}(\Delta^c) : D\ \vec{v}$, hence unifying $\vec{u}$ and $\vec{v}$ must have succeeded positively yielding some σ for which we have a subtree whose root patterns, $\sigma \vec{p}[c\ \Delta^c]$ also match $\vec{t}$. Inductively, not all of this subtree's leaf patterns yield CONFLICT. □

It may seem a little odd to present coverage as 'not CONFLICT in all cases', rather than guaranteed progress for *closed* terms. But our result also treats the case of *open* terms, guaranteeing that progress can only be blocked by the presence of non-constructor forms.

Lemma 14 (Canonicity) *For global context* Γ, *if* $\Gamma \vdash t : D\vec{v}$, *with* t *in normal form, then* t *is* $c\ \vec{b}$ *for some* $\vec{b}$.

Proof Select a minimal counterexample. This is necessarily a 'stuck function', $f\ \vec{a}$. By the above reasoning, we must have some internal node in f's splitting tree $[\Delta^x; x : D\ \vec{v}; {}^x\!\Delta]\ f\ \vec{p}[x]$ with $\theta[\vec{p}[x]] \cong \vec{a}$ but $\Gamma \vdash \theta x : D\ \theta\vec{v}$ a non-constructor form. But θx is a proper subterm of $f\ \vec{a}$, hence a smaller counterexample. □

Lemma 15 (Confluence) *If every function defined in* Γ *satisfies the function criteria, then* $\leadsto_\Gamma$ *is confluent.*

Proof Function symbols and constructor symbols are disjoint. By construction, splitting trees yield left-hand sides which match disjoint sets of terms. Hence there are no critical pairs. □

4   Translating Pattern Matching

In this section, we shall give a complete translation from functions satisfying the function criteria and inhabiting small types to terms in a suitable extension of UTT, via the primitive elimination operators for inductive datatypes. We do this by showing how to construct terms corresponding to the splitting trees which give rise to the functions: we show how to represent programming problems as types for which splitting trees deliver inhabitants, and we explain how each step of problem reduction may be realised by a term.

4.1   Heterogeneous Equality

We must first collect the necessary equipment. The unification which we take for granted in splitting trees becomes explicit equational reasoning, step by step. We represent problems using McBride's *heterogeneous* equality [16]:

$$\mathsf{Eq}(S,T{:}\star_0; s{:}S; t{:}T){:}\star_1; \quad \mathsf{refl}(R{:}\star_0; r{:}R){:}\mathsf{Eq}_{R\,R}\ r\ r;$$
$$\mathbf{subst}(R{:}\star_0; s,t{:}R; q{:}\mathsf{Eq}_{R\,R}\ s\ t; P{:}R{\to}\star_1; p{:}P\ s){:}P\ t;$$
$$[R; r; P; p]\ \mathbf{subst}_{R;r;r}\ (\mathsf{refl}_R\ r)\ P\ p\ \mapsto p$$

Eq is not a standard inductive definition: it permits the expression of heterogeneous equations, but its eliminator **subst** gives the Leibniz property only for *homogeneous* equations. This is just a convenient repackaging of the traditional homogeneous identity type family I. The full construction can be found in [15].

It is to enable this construction that we keep equations in $\star_1$. We shall be careful to form equations over data sets, but not equality sets. We are unsure whether it is safe to allow equality sets in $\star_0$, even though this would not yield an independent copy of $\star_0$ in $\star_0$. At any rate, it is sufficient that we can form equations over data and eliminate data over equations.

We shall write $s \simeq t$ for $\mathsf{Eq}_{S\,T}\ s\ t$ when the types S, T are clear. Furthermore Eq precisely allows us to express equations between *sequences* of data in the same telescope: the constraints which require the specialisation of datatype indices take exactly this form. Note we always have $\overline{\mathsf{D}}\ \underline{\mathrm{tele}}(\star_0)$, hence if $\vec{s}, \vec{t} : \overline{\mathsf{D}}$, we may form the telescope of equations $\quad q_1{:}s_1 \simeq t_1; \ \dots\ ; q_n{:}s_n \simeq t_n\ \underline{\mathrm{tele}}(\star_1)\quad$ which we naturally abbreviate as $\vec{s} \simeq \vec{t}$. Correspondingly, we write refl $\vec{t} : \vec{t} \simeq \vec{t}$.

4.2   Standard Equipment for Inductive Datatypes

In [17], we show how to equip every datatype with some useful tools, derived from its eliminator, which we shall need in the constructions to come. To summarise,

case$_\mathsf{D}$ is just e$_\mathsf{D}$ weakened by dropping the inductive hypotheses.

Below$_\mathsf{D}(P : \overline{\mathsf{D}} \to \star_1; \overline{\mathsf{D}}) : \star_1$ is the 'course of values', defined inductively by Giménez [7]; simulated via $\mathbf{E}_\mathsf{D}$, Below$_\mathsf{D}\ P\ \Xi\ z$ computes an iterated tuple type asserting P for every value structurally smaller than z. For Nat we get

$$\begin{aligned}\mathsf{Below}_\mathsf{Nat}\ P\ \mathsf{zero}\ &\mapsto 1 \\ \mathsf{Below}_\mathsf{Nat}\ P\ (\mathsf{suc}\ n)\ &\mapsto \mathsf{Below}_\mathsf{Nat}\ P\ n \times P\ n\end{aligned}$$

$\text{below}_D(P:\overline{D} \to \star_1;\ p:\Pi\overline{D}.\ \text{Below}_D\ P\ \overline{D} \to P\ \overline{D};\ \overline{D}):\text{Below}_D\ P\ \overline{D}$ constructs the tuple, given a 'step' function, and is simulated via e_D:

$\text{below}_{\mathsf{Nat}}\ P\ p\ \text{zero} \quad\mapsto ()$
$\text{below}_{\mathsf{Nat}}\ P\ p\ (\text{suc}\ n) \mapsto (\lambda b:\text{Below}_{\mathsf{Nat}}\ P\ n.\ (b, p\ n\ b))\ (\text{below}_{\mathsf{Nat}}\ P\ p\ n)$

$\text{rec}_D(P:\overline{D} \to \star_1;\ p:\Pi\overline{D}.\ \text{Below}_D\ P\ \overline{D} \to P\ \overline{D};\ \overline{D}):P\ \overline{D}$ is the structural recursion operator for D, given by $\text{rec}_D\ P\ p\ \overline{D} \mapsto p\ \overline{D}\ (\text{below}_D\ P\ p\ \overline{D})$

We use case_D for splitting and rec_D for recursion. For unification, we need:

noConfusion_D is the proof that D's constructors are injective and disjoint— also a two-level construction, again by example:

$\text{NoConfusion}_{\mathsf{Nat}}(P : \star_1; x, y:\mathsf{Nat}):\star_1$
$\text{NoConfusion}_{\mathsf{Nat}}\ P \quad \text{zero} \quad\quad \text{zero} \ \mapsto P \to P$
$\text{NoConfusion}_{\mathsf{Nat}}\ P \quad \text{zero} \quad (\text{suc}\ y) \mapsto P$
$\text{NoConfusion}_{\mathsf{Nat}}\ P\ (\text{suc}\ x) \quad \text{zero} \ \mapsto P$
$\text{NoConfusion}_{\mathsf{Nat}}\ P\ (\text{suc}\ x)\ (\text{suc}\ y) \mapsto (x \simeq y \to P) \to P$

$\text{noConfusion}_{\mathsf{Nat}}(P : \star_1; x, y:\mathsf{Nat}; q:x \simeq y):\text{NoConfusion}_{\mathsf{Nat}}\ P\ x\ y$
$\text{noConfusion}_{\mathsf{Nat}}\ P\ \underline{\text{zero}} \quad \underline{\text{zero}} \quad (\text{refl zero}) \ \mapsto \lambda p:P.\ p$
$\text{noConfusion}_{\mathsf{Nat}}\ P\ \underline{(\text{suc}\ x)}\ \underline{(\text{suc}\ x)}\ (\text{refl}\ (\text{suc}\ n)) \mapsto \lambda p:x \simeq x \to P.\ p\ (\text{refl}\ x)$

NoConfusion_D is simulated by two appeals to $\mathbf{E}_D$; noConfusion_D uses subst once, then case_D to work down the 'diagonal'.
noCycle_D disproves any cyclic equation in D—details may be found in [17].

Lemma 16 (Unification Transitions) *The following (and their symmetric images) are derivable:*

$$\begin{aligned}
\textit{deletion}\quad &m:\Pi\Delta.\ P\\
&\vdash \lambda\Delta;\ q.\ m\ \Delta\\
&\quad :\Pi\Delta.\ t \simeq t \to P\\[4pt]
\textit{solution}\quad &m:\Pi\Delta^0.\ [x \mapsto t]\Pi\Delta^1.\ P\\
&\vdash \lambda\Delta;\ q.\ \text{subst}\ T\ t\ x\ q\ (\lambda x.\ \Pi\Delta^0;\Delta^1.\ P)\ m\ \Delta^0\ \Delta^1\\
&\quad :\Pi\Delta.\ t \simeq x \to P\\
&\quad \underline{\text{if}}\ \ \Delta \sim \Delta^0; x{:}T;\Delta^1\ \ \underline{\text{and}}\ \ \Delta^0 \vdash t : T\\[4pt]
\textit{injectivity}\quad &m:\Pi\Delta.\ \vec{s} \simeq \vec{t} \to P\\
&\vdash \lambda\Delta;\ q.\ \text{noConfusion}\ P\ (\mathsf{c}\ \vec{s})\ (\mathsf{c}\ \vec{t})\ q\ (m\ \Delta)\\
&\quad :\Pi\Delta.\ \mathsf{c}\ \vec{s} \simeq \mathsf{c}\ \vec{t} \to P\\[4pt]
\textit{conflict}\quad &\vdash \lambda\Delta;\ q.\ \text{noConfusion}\ P\ (\text{chalk}\ \vec{s})\ (\text{cheese}\ \vec{t})\ q\\
&\quad :\Pi\Delta.\ \text{chalk}\ \vec{s} \simeq \text{cheese}\ \vec{t} \to P\\[4pt]
\textit{cycle}\quad &\vdash \lambda\Delta;\ q.\ \text{noCycle}\ P\ \dots\ q\dots\\
&\quad :\Pi\Delta.\ x \simeq \mathsf{c}\ \lceil\vec{p}[x]\rceil \to P
\end{aligned}$$

Proof By construction. □

4.3   Elimination with Unification

In [16], McBride gives a general technique for deploying operators whose types resemble elimination rules. We shall use this technique repeatedly in our constructions, hence we recapitulate the basic idea here. Extending the previous account, we shall be careful to ensure that the terms we construct not only have the types we expect but also deliver the computational behaviour required to simulate the pattern matching semantics.

Definition 17 (Elimination operator) *For any telescope* $\Gamma \vdash \Xi \, \underline{\text{tele}}(\star_0)$*, we define a* Ξ*-elimination operator to be any*

$$e : \Pi P {:} \Pi \Xi. \star_1 . (\Pi \Delta_1. P \, \vec{s}_1) \to \cdots \to (\Pi \Delta_n. P \, \vec{s}_n) \to \Pi \Xi. P \, \Xi$$

Note that e_D is a $\overline{\mathsf{D}}$-elimination operator; case_D and rec_D are also. We refer to the Ξ as the *targets* of the operator as they indicate what is to be eliminated; we say P is the *motive* as it indicates why; the remaining arguments we call *methods* as they explain how to proceed in each case which may arise. Now let us show how to adapt such an operator to any *specific* sequence of targets.

Definition 18 (Basic analysis) *If* e *is a* Ξ*-elimination operator (as above),* $\Delta \, \underline{\text{tele}}(\star_0)$ *and* $\Delta \vdash T : \star_1$*, then for any* $\Delta \vdash \vec{t} : \Xi$*, the basic e-analysis of* $\Pi \Delta. T$ *at* $\vec{t}$ *is the (clearly derivable) judgment*

$$m_1 {:} \Pi \Delta_1; \Delta. \, \Xi \simeq \vec{t} \to T; \ldots; m_n {:} \Pi \Delta_n; \Delta. \, \Xi \simeq \vec{t} \to T$$
$$\vdash \lambda \Delta. \, e \, (\lambda \Xi. \, \Pi \Delta. \, \Xi \simeq \vec{t} \to T) \, m_1 \, \ldots \, m_n \, \vec{t} \, \Delta \, (\mathsf{refl} \, \vec{t}) \, : \, \Pi \Delta. \, T$$

Notice that when e is case_D and the targets are some $\vec{v}; x$ where $x : \mathsf{D} \, \vec{v} \in \Delta$, then for each constructor $\mathsf{c} \, (\Delta^\mathsf{c}) : \mathsf{D} \, \vec{u}$, we get a method

$$m_\mathsf{c} : \Pi \Delta^\mathsf{c}; \Delta. \, \vec{u} \simeq \vec{v} \to \mathsf{c} \, \Delta^\mathsf{c} \simeq x \to T$$

Observe that the equations on the indices are exactly those we must unify to allow the instantiation of x with $\mathsf{c} \, \Delta^\mathsf{c}$. Moreover, if we have such an instance for x, i.e. if θ unifies $\vec{u}$ and $\vec{v}$, and takes $x \mapsto \mathsf{c} \, \theta \Delta^\mathsf{c}$, then the analysis actually reduces to the relevant method:

$$\mathsf{case}_\mathsf{D} \, (\lambda \overline{\mathsf{D}}. \, \Pi \Delta. \, \Xi \simeq \vec{t} \to T) \, \vec{m} \, \theta \vec{v} \, (\mathsf{c} \, \theta \Delta^\mathsf{c}) \, \theta \Delta \, (\mathsf{refl} \, \theta \vec{v}) \, (\mathsf{refl} \, (\mathsf{c} \, \theta \Delta^\mathsf{c}))$$
$$\rightsquigarrow m_\mathsf{c} \, \theta \Delta^\mathsf{c} \, \theta \Delta \, (\mathsf{refl} \, \theta \vec{v}) \, (\mathsf{refl} \, (\mathsf{c} \, \theta \Delta^\mathsf{c}))$$

We may now simplify the equations in the method types.

Definition 19 (Specialisation by Unification) *Given any type of the form* $\Pi \Delta. \, \vec{u} \simeq \vec{v} \to T : \star_1$*, we may seek to construct an inhabitant—a* specialiser*—by exhaustively iterating the unification transitions from lemma 16 as applicable. This terminates by the usual argument [10], with three possible outcomes:*

negative success *a specialiser is found, either by* **conflict** *or* **cycle**;

positive success *a specialiser is found, given some* $m : \Pi\Delta'.\, \sigma T$ *for* σ *a most general idempotent unifer of* $\vec{u}$ *and* $\vec{v}$, *or*

failure *at some stage, an equation is reached for which no transition applies.*

Lemma 20 (Specialiser Reduction) *If specialisation by unification delivers*

$$m : \Pi\Delta'.\, \sigma T \vdash s \,:\, \Pi\Delta.\, \vec{u} \simeq \vec{v} \to T$$

then for any $\Theta \vdash \theta\Delta$ *unifying* $\vec{u}$ *and* $\vec{v}$ *we have* $s\,\theta\Delta\,(\mathsf{refl}\,\theta\vec{u}) \leadsto^* m\,\theta\Delta'$.

Proof   By induction on the transition sequence. The **deletion, solution** and **injectivity** steps each preserve this property by construction. $\qquad\square$

We can now give a construction which captures our notion of *splitting*.

Lemma 21 (Splitting Construction) *Suppose* $\Delta \vdash T : \star_1$, *with* $\Delta\,\underline{\mathsf{tele}}(\star_0)$, $\Delta^x;\, x\!:\!\mathsf{D}\vec{v};\,{}^x\!\Delta\underline{\mathsf{tele}}(\star_0)$ *and* $\Delta^x;\, x\!:\!\mathsf{D}\vec{v};\,{}^x\!\Delta \vdash \lceil\vec{p}[x]\rceil : \Delta$. *Suppose further that for each* $\mathsf{c}\,(\Delta^{\mathsf{c}}) : \mathsf{D}\,\vec{u}$, *unifying* $\vec{u}$ *with* $\vec{v}$ *succeeds. Then we may construct an inhabitant* $f : \Pi\Delta^x;\, x\!:\!\mathsf{D}\,\vec{v};\,{}^x\!\Delta.\, [\![\lceil\vec{p}[x]\rceil]\!]T$ *over a context comprising, for each* c *with* positive *success,*

$$m_{\mathsf{c}} : \Pi\Delta';\, \sigma[x \mapsto \mathsf{c}\,\Delta^{\mathsf{c}}]^x\!\Delta.\, [\![\sigma\vec{p}\lceil\mathsf{c}\,\Delta^{\mathsf{c}}\rceil]\!]T$$

for some most general idempotent unifier $\Delta' \vdash \sigma(\Delta^{\mathsf{c}};\Delta^x)$. *In each such case,*

$$f\,\sigma\Delta^x\,(\mathsf{c}\,\sigma\Delta^{\mathsf{c}})\,{}^x\!\Delta \leadsto^* m_{\mathsf{c}}\,\Delta'\,{}^x\!\Delta$$

Proof   The construction is by basic case$_{\mathsf{D}}$-analysis of $\Pi\Delta^x;\, x\!:\!\mathsf{D}\,\vec{v};\,{}^x\!\Delta.\, [\![\lceil\vec{p}[x]\rceil]\!]T$ at $\vec{v};\, x$, then specialisation by unification for each method. The required reduction behaviour follows from lemma 20. $\qquad\square$

4.4   Translating Structural Recursion

We are very nearly ready to translate whole functions. For the sake of clarity, we introduce one last piece of equipment:

Definition 22 (Computation Types) *When implementing a function* $\mathsf{f}\,(\Delta) : T$, *we introduce the family of* f-*computation types as follows:*

$$\mathsf{Comp\text{-}f}(\Delta) : \star_0; \quad \mathsf{return\text{-}f}(\Delta;\, t\!:\!T) : \mathsf{Comp\text{-}f}\,\Delta$$
$$\mathsf{call\text{-}f}(\overline{\mathsf{Comp\text{-}f}}) : T$$
$$\mathsf{call\text{-}f}\,\underline{\Delta}\,(\mathsf{return\text{-}f}\,\Delta\,t) \mapsto t$$

where $\mathsf{call\text{-}f}$ *is clearly definable from* $e_{\mathsf{Comp\text{-}f}}$.

Comp-f book-keeps the connection between f's high-level program and the low-level term which delivers its semantics. We translate each f-application to the corresponding call-f of an f-computation; the latter will compute to a return-f value exactly in correspondence with the pattern matching reduction. The translation takes the following form:

Definition 23 (Application Translation) *If* $f(\Delta) : T$ *is globally defined, but* $\Delta \vdash f : \mathsf{Comp\text{-}f}\,\Delta$ *for some* f *not containing* f, *the translation* $\{-\}_f^f$ *takes*

$$\{f\,\vec{t}\}_f^f \implies \mathsf{call\text{-}f}\,\{\vec{t}\}_f^f\,([\{\vec{t}\}_f^f]f)$$

and proceeds structurally otherwise. Recalling that we require global functions to be applied with at least their declared arity, this translation removes f *entirely.*

Theorem 24 *If* $f(\Delta) : T$ *has a small type and computation rules* $[\Delta_i]\,f\,\vec{p}_i\,r_i$ *satisfying the function criteria, then there exists an* f *such that*

$$\Delta \vdash f : \mathsf{Comp\text{-}f}\,\Delta \quad \text{and} \quad s \leadsto_{\Gamma;f} t \quad \text{implies} \quad \{s\}_f^f \leadsto_\Gamma^+ \{t\}_f^f$$

Proof It suffices to ensure that the pattern matching reduction schemes are faithfully translated. For each i such that r_i returns a value $\mapsto e_i$, we shall have

$$\{f\,\lceil\vec{p}_i\rceil\}_f^f = \mathsf{call\text{-}f}\,\lceil\vec{p}_i\rceil\,\llbracket\vec{p}_i\rrbracket f \leadsto_\Gamma^* \mathsf{call\text{-}f}\,\lceil\vec{p}_i\rceil\,(\mathsf{return\text{-}f}\,\lceil\vec{p}_i\rceil\,\{e_i\}_f^f) \leadsto_\Gamma \{e_i\}_f^f$$

Without loss of generality, let f be structurally recursive on some $x:D\,\vec{v}$, jth in Δ. The basic $\mathsf{rec_D}$-analysis of $\Pi\Delta.\,\mathsf{Comp\text{-}f}\,\Delta$ at $\vec{v}; x$ requires a term of type

$$\Pi\overline{D}.\,\mathsf{Below_D}\,P\,\overline{D} \to \Pi\Delta.\,\overline{D} \simeq \vec{v}; x \to \mathsf{Comp\text{-}f}\,\Delta$$

where $P = \lambda\overline{D}.\,\Pi\Delta.\,\overline{D} \simeq \vec{v}; x \to \mathsf{Comp\text{-}f}\,\Delta$. Specialisation substitutes $\vec{v}; x$ for $\overline{D}$, yielding a specialiser $[m]s$ of the required type, with

$$m : \Pi\Delta.\,\mathsf{Below_D}\,P\,\vec{v}\,x \to \mathsf{Comp\text{-}f}\,\Delta; \Delta \vdash \mathsf{rec_D}\,P\,[m]s\,\vec{v}\,x\,\Delta\,(\mathsf{refl}\,\vec{v}; x)$$
$$\leadsto_\Gamma^* m\,\Delta\,(\mathsf{below_D}\,P\,[m]s\,\vec{v}\,x) : \mathsf{Comp\text{-}f}\,\Delta$$

by definition of $\mathsf{rec_D}$ and specialisation reduction. We shall take the latter to be our f, once we have suitably instantiated m. To do so, we follow f's splitting tree: lemma 21 justifies the splitting construction at each internal node and at each $\Uparrow y$ leaf. Each programming problem $[\Delta']\,f\,\vec{p}$ in the tree corresponds to the task of instantiating some $m' : \Pi\Delta'.\,\mathsf{Below_D}\,P\,(\llbracket\vec{p}\rrbracket(\vec{v}; x)) \to \mathsf{Comp\text{-}f}\,\lceil\vec{p}\rceil$ where, again by lemma 21, $m\,\lceil\vec{p}\rceil \leadsto_\Gamma^* m'\,\Delta'$.

The splitting finished, it remains to instantiate the m_i corresponding to each $[\Delta_i]\,f\,\vec{p}_i \mapsto e_i$. Now, $[\Delta \mapsto \lceil\vec{p}_i\rceil]$ takes $x : D\,\vec{v}$ to some $\lceil p_{ij}\rceil : D\,\vec{u}$, so we may take

$$m_i \mapsto \lambda\Delta_i; H : \mathsf{Below_D}\,P\,\vec{u}\,\lceil p_{ij}\rceil.\,\mathsf{return\text{-}f}\,\lceil\vec{p}_i\rceil\,e_i^\dagger$$

where $e_i^\dagger$ is constructed by replacing each call $f\,\vec{r}$ in e_i by an appropriate appeal to H. As f is well typed and structurally recursive, so $[\Delta \mapsto \vec{r}]$ maps $x : D\,\vec{v}$ to $r_j : D\,\vec{w}$ where $r_j \prec \lceil p_{ij}\rceil$. By construction, $\mathsf{Below_D}\,P\,\vec{u}\,\lceil p_{ij}\rceil$ reduces to a tuple of the computations for subobjects of $\lceil p_{ij}\rceil$. Hence we have a projection g such that $g\,H : \Pi\Delta.\,\vec{w}; r_j \simeq \vec{v}; x \to \mathsf{Comp\text{-}f}\,\Delta$ and hence we take $\mathsf{call\text{-}f}\,\vec{r}\,(g\,H\,\vec{r}\,(\mathsf{refl}\,\vec{w}; r_j))$ to replace $f\,\vec{r}$, where by construction of $\mathsf{below_D}$,

$$\mathsf{call\text{-}f}\,\vec{r}\,(g\,(\mathsf{below_D}\,P\,[m]s\,\vec{u}\,\lceil p_{ij}\rceil)\,\vec{r}\,(\mathsf{refl}\,\vec{w}; r_j))$$
$$\leadsto_\Gamma^* \mathsf{call\text{-}f}\,\vec{r}\,([m]s\,\vec{w}\,r_j\,\vec{r}\,(\mathsf{refl}\,\vec{w}; r_j))$$
$$\leadsto_\Gamma^* \mathsf{call\text{-}f}\,\vec{r}\,(m\,\vec{r}\,(\mathsf{below_D}\,P\,[m]s\,\vec{w}\,r_j))$$
$$= \{f\,\vec{r}\}_f^f$$

So, finally, we arrive at

$$\{f \ulcorner \vec{p_i} \urcorner\}_f^f = \text{call-f } \ulcorner \vec{p_i} \urcorner (m \ulcorner \vec{p_i} \urcorner (\text{below}_D P [m]s \vec{u} \ulcorner p_{ij} \urcorner))$$
$$\leadsto_\Gamma^* \text{call-f } \ulcorner \vec{p_i} \urcorner (m_i \Delta_i (\text{below}_D P [m]s \vec{u} \ulcorner p_{ij} \urcorner))$$
$$\leadsto_\Gamma^* \text{call-f } \ulcorner \vec{p_i} \urcorner (\text{return-f } \ulcorner \vec{p_i} \urcorner [H \mapsto \text{below}_D P [m]s \vec{u} \ulcorner p_{ij} \urcorner] e_i^\dagger)$$
$$\leadsto_\Gamma^* \text{call-f } \ulcorner \vec{p_i} \urcorner (\text{return-f } \ulcorner \vec{p_i} \urcorner \{e_i\}_f^f)$$
$$= \{e_i\}_f^f$$

as required.   □

5   Conclusions

We have shown that dependent pattern matching can be translated into a powerful though notationally minimal target language. This constitutes the first proof that dependent pattern matching is equivalent to type theory with inductive types extended with the **K** axiom, at the same time reducing the problem of the termination of pattern matching as a first-class syntax for structurally recursive programs and proofs to the problem of termination of UTT plus **K**.

Two of the authors have extended the raw notion of pattern matching that we study here with additional language constructs for more concise, expressive programming with dependent types [18]. One of the insights from that work is that the technology for explaining pattern matching and other programming language constructs is as important as the language constructs themselves, since the technology can be used to motivate and explain increasingly powerful language constructs.

References

1. Lennart Augustsson. Compiling Pattern Matching. In Jean-Pierre Jouannaud, editor, *Functional Programming Languages and Computer Architecture*, volume 201 of *LNCS*, pages 368–381. Springer-Verlag, 1985.
2. Edwin Brady, Conor McBride, and James McKinna. Inductive families need not store their indices. In Stefano Berardi, Mario Coppo, and Ferrucio Damiani, editors, *Types for Proofs and Programs, Torino, 2003*, volume 3085 of *LNCS*, pages 115–129. Springer-Verlag, 2004.
3. Rod Burstall. Proving properties of programs by structural induction. *Computer Journal*, 12(1):41–48, 1969.
4. Thierry Coquand. Pattern Matching with Dependent Types. In Bengt Nordström, Kent Petersson, and Gordon Plotkin, editors, *Electronic Proceedings of the Third Annual BRA Workshop on Logical Frameworks (Båstad, Sweden)*, 1992.
5. Nicolas G. de Bruijn. Telescopic Mappings in Typed Lambda-Calculus. *Information and Computation*, 91:189–204, 1991.
6. Peter Dybjer. Inductive Sets and Families in Martin-Löf's Type Theory. In Gérard Huet and Gordon Plotkin, editors, *Logical Frameworks*. CUP, 1991.
7. Eduardo Giménez. Codifying guarded definitions with recursive schemes. In Peter Dybjer, Bengt Nordström, and Jan Smith, editors, *Types for Proofs and Programs, '94*, volume 996 of *LNCS*, pages 39–59. Springer-Verlag, 1994.

8. Healfdene Goguen. *A Typed Operational Semantics for Type Theory.* PhD thesis, Laboratory for Foundations of Computer Science, University of Edinburgh, 1994. Available from
   http://www.lfcs.informatics.ed.ac.uk/reports/94/ECS-LFCS-94-304/.
9. Martin Hofmann and Thomas Streicher. A groupoid model refutes uniqueness of identity proofs. In *Proc. Ninth Annual Symposium on Logic in Computer Science (LICS) (Paris, France)*, pages 208–212. IEEE Computer Society Press, 1994.
10. Jean-Pierre Jouannaud and Claude Kirchner. Solving equations in abstract algebras: A rule-based survey of unification. In Jean-Louis Lassez and Gordon Plotkin, editors, *Computational Logic: Essays in Honor of Alan Robinson*, pages 257–321. MIT Press, 1991.
11. Zhaohui Luo. *Computation and Reasoning: A Type Theory for Computer Science.* Oxford University Press, 1994.
12. Zhaohui Luo and Robert Pollack. LEGO Proof Development System: User's Manual. Technical Report ECS-LFCS-92-211, Laboratory for Foundations of Computer Science, University of Edinburgh, 1992.
13. Lena Magnusson and Bengt Nordström. The ALF proof editor and its proof engine. In Henk Barendregt and Tobias Nipkow, editors, *Types for Proofs and Programs*, LNCS 806. Springer-Verlag, 1994. Selected papers from the Int. Workshop TYPES '93, Nijmegen, May 1993.
14. Per Martin-Löf. A theory of types. Manuscript, 1971.
15. Conor McBride. *Dependently Typed Functional Programs and their Proofs.* PhD thesis, University of Edinburgh, 1999. Available from
   http://www.lfcs.informatics.ed.ac.uk/reports/00/ECS-LFCS-00-419/.
16. Conor McBride. Elimination with a Motive. In Paul Callaghan, Zhaohui Luo, James McKinna, and Robert Pollack, editors, *Types for Proofs and Programs (Proceedings of the International Workshop, TYPES'00)*, volume 2277 of *LNCS*. Springer-Verlag, 2002.
17. Conor McBride, Healfdene Goguen, and James McKinna. A few constructions on constructors. In Jean-Christophe Filliâtre, Christine Paulin, and Benjamin Werner, editors, *Types for Proofs and Programs, Paris, 2004*, LNCS. Springer-Verlag, 2004.
18. Conor McBride and James McKinna. The view from the left. *Journal of Functional Programming*, 14(1), 2004.
19. Fred McBride. *Computer Aided Manipulation of Symbols.* PhD thesis, Queen's University of Belfast, 1970.
20. Bengt Nordström, Kent Petersson, and Jan Smith. *Programming in Martin-Löf's type theory: an introduction.* Oxford University Press, 1990.
21. Carsten Schürmann and Frank Pfenning. A coverage checking algorithm for LF. In D. Basin and B. Wolff, editors, *Proceedings of the Theorem Proving in Higher Order Logics 16th International Conference*, volume 2758 of *LNCS*, pages 120–135, Rome, Italy, September 2003. Springer.
22. Thomas Streicher. Investigations into intensional type theory. Habilitation Thesis, Ludwig Maximilian Universität, 1993.
23. Daria Walukiewicz-Chrzaszcz. Termination of rewriting in the calculus of constructions. *J. Funct. Program.*, 13(2):339–414, 2003.

Iterative Lexicographic Path Orders

Jan Willem Klop[1,2,3], Vincent van Oostrom[4], and Roel de Vrijer[1]

[1] Vrije Universiteit, Department of Theoretical Computer Science,
De Boelelaan 1081a, 1081 HV Amsterdam, The Netherlands
jwk@cs.vu.nl, rdv@cs.vu.nl
[2] Radboud Universiteit Nijmegen, Department of Computer Science,
Toernooiveld 1, 6525 ED Nijmegen, The Netherlands
[3] CWI, P.O. Box 94079, 1090 GB Amsterdam, The Netherlands
[4] Universiteit Utrecht, Department of Philosophy,
Heidelberglaan 8, 3584 CS Utrecht, The Netherlands
Vincent.vanOostrom@phil.uu.nl

Abstract. We relate Kamin and Lévy's original presentation of *lexicographic path orders* (LPO), using an inductive definition, to a presentation, which we will refer to as *iterative lexicographic path orders* (ILPO), based on Bergstra and Klop's definition of recursive path orders by way of an auxiliary term rewriting sytem.

Dedicated to Joseph Goguen, in celebration of his 65th birthday.

1 Introduction

In his seminal paper [1], Dershowitz introduced the recursive path order (RPO) method to prove termination of a first-order term rewrite system (TRS) $\mathcal{T}$. The method is based on lifting a well-quasi-order $\preceq$ on the signature of a TRS to a well-quasi-order $\preceq_{rpo}$ on the set of terms over the signature [2]. Termination of the TRS follows if $l \succ_{rpo} r$ holds for every rule $l \to r$ of $\mathcal{T}$.

In Bergstra and Klop [3] an alternative definition of RPO is put forward, which we call the *iterative* path order (IPO), the name stressing the way it is generated—see also Bergstra, Klop and Middeldorp [4]. It is operational in the sense that it is itself defined by means of an (auxiliary) term rewrite system $\mathcal{L}$ex, the rules of which depend (only) on the given well-quasi-order $\preceq$.

What has been lacking until now is an understanding of the exact relationship between the recursive and iterative approaches to path orders. This will be the main subject of our investigation here. We show that both approaches coincide in the case of *transitive* relations (orders). Moreover, we provide a direct proof of termination for the iterative path order starting from an *arbitrary* terminating relation on the signature, employing a proof technique due to Buchholz [5]. Both proofs essentially rely on a natural-number-labelled variant $\mathcal{L}$ex$^{\omega}$ of the auxiliary TRS $\mathcal{L}$ex, introduced here for the first time.

For the sake of exposition we focus on the restriction of RPO due to Kamin and Lévy [6] known as the lexicographic path order (LPO)—see also Baader and

K. Futatsugi et al. (Eds.): Goguen Festschrift, LNCS 4060, pp. 541–554, 2006.

Nipkow [7]. Restricting the iterative approach accordingly gives rise to what we call the iterative lexicographic path order (ILPO), as formulated for the first time in the PhD thesis of Geser [8] and, in a slightly restricted form, by Klop [9]. As far as we know also in this case the correspondence between both has not been investigated in the literature.

The proofs that the iterative lexicographic path order is terminating and that LPO and ILPO coincide will constitute the body of the paper. In the conclusions, we put forward some ideas on the robustness of this correspondence, i.e. whether variations on LPO can be matched by corresponding variations on ILPO restoring their coincidence.

Acknowledgement We thank Alfons Geser for useful remarks.

2   The Iterative Lexicographic Path Order

The iterative lexicographic path order (ILPO) is a method to prove termination of a term rewrite system (TRS). Here a TRS $\mathcal{T}$ is *terminating* if its rewrite relation $\to_{\mathcal{T}}$ is so, i.e. if it does not allow an infinite reduction $t_0 \to_{\mathcal{T}} t_1 \to_{\mathcal{T}} t_2 \to_{\mathcal{T}} \cdots$. The method is based on iteratively lifting a terminating relation R on the signature to a terminating relation R_{ilpo} on the terms over the signature. The lifting is iterative in the sense that R_{ilpo} is defined via the *iteration* of reduction steps in an *atomic decomposition* TRS $\mathcal{L}$ex (depending on R), instead of *recursively* as in Dershowitz's recursive path order method [1]. By its definition via the atomic decomposition TRS, transitivity and closure under contexts and substitutions of R_{ilpo} are automatic, which combined with termination yields that R_{ilpo} is a so-called *reduction order* [10, Definition 6.1.2]. Therefore, for the TRS $\mathcal{T}$ to be terminating it suffices that $l\ R_{ilpo}\ r$ holds for each rule $l \to r$ in $\mathcal{T}$.

As a running example to illustrate ILPO, we take the terminating relation R given by M R A and A R S on the signature of the TRS $\mathcal{D}$ed of addition and multiplication on natural numbers with the rewrite rules of Table 1, going back to at least Dedekind [11].[5]

$$
\begin{aligned}
\mathsf{A}(x,0) &\to x \\
\mathsf{A}(x,\mathsf{S}(y)) &\to \mathsf{S}(\mathsf{A}(x,y)) \\
\mathsf{M}(x,0) &\to 0 \\
\mathsf{M}(x,\mathsf{S}(y)) &\to \mathsf{A}(x,\mathsf{M}(x,y))
\end{aligned}
$$

Table 1. Dedekind's rules for addition and multiplication

Clearly, the relation R is terminating and ILPO will lift it to a terminating relation R_{ilpo} such that $l\ R_{ilpo}\ r$ holds for each rule $l \to r$ in $\mathcal{D}$ed, implying termination of $\mathcal{D}$ed.

---

[5] Dedekind took 1 instead of 0 for the base case.

For a given relation R over the signature, the definition of R_{ilpo} proceeds in two steps: We first define the atomic decomposition TRS $\mathcal{L}$ex depending on R, over the signature extended with *control* symbols. Next, the relation R_{ilpo} is defined by restricting iteration of $\mathcal{L}$ex-reduction steps, i.e. of $\rightarrow^+_{\mathcal{L}ex}$, to terms over the original signature.

Definition 1. *Let R be a relation on a signature Σ, and let V be a signature of nullary symbols disjoint from Σ (called variables). The* atomic decomposition *TRS $\mathcal{L}$ex is $\langle \Sigma \uplus \Sigma^* \uplus V, \mathcal{R} \rangle$, where:*

1. *The signature Σ^* of* control *symbols is a copy of Σ, i.e. for each function symbol $f \in \Sigma$, Σ^* contains a fresh symbol f^* having the same arity f has.*
2. *The rules $\mathcal{R}$ are given in Table 2, for arbitrary function symbols f, g in Σ, with $\boldsymbol{x}$, $\boldsymbol{y}$, $\boldsymbol{z}$ disjoint vectors of pairwise disjoint variables of appropriate lengths.*

$$
\begin{aligned}
f(\boldsymbol{x}) &\rightarrow_{\text{put}} f^*(\boldsymbol{x}) \\
f^*(\boldsymbol{x}) &\rightarrow_{\text{select}} x_i \quad (1 \le i \le |\boldsymbol{x}|) \\
f^*(\boldsymbol{x}) &\rightarrow_{\text{copy}} g(f^*(\boldsymbol{x}), \ldots, f^*(\boldsymbol{x})) \quad (f \; R \; g) \\
f^*(\boldsymbol{x}, g(\boldsymbol{y}), \boldsymbol{z}) &\rightarrow_{\text{lex}} f(\boldsymbol{x}, g^*(\boldsymbol{y}), l, \ldots, l) \quad (l = f^*(\boldsymbol{x}, g(\boldsymbol{y}), \boldsymbol{z}))
\end{aligned}
$$

Table 2. The rules of the atomic decomposition TRS $\mathcal{L}$ex

The idea of the atomic decomposition $\mathcal{L}$ex is that marking the head symbol of a term, by means of the put-rule, corresponds to the obligation to make that term *smaller*, whereas the other rules correspond to *atomic* ways in which this can be brought about:

1. The select-rule expresses that selecting one of the arguments of a term makes it smaller.
2. The copy-rule expresses that a term t can be made smaller by putting copies of terms smaller than t below a head symbol g which is less heavy than the head symbol f of t.
3. The lex-rule expresses that a term t can be made smaller by making one of its subterms smaller. At the same time one may replace all the subterms to the *right* (whence the name lex) of this subterm by arbitrary terms that are smaller than the whole term t.

For our running example $\mathcal{D}$ed, the reduction $\text{A}(x, 0) \rightarrow_{\text{put}} \text{A}^*(x, 0) \rightarrow_{\text{select}} x$ in $\mathcal{L}$ex, is a decomposition of the first rule into atomic $\mathcal{L}$ex-steps. This also holds for the other rules of $\mathcal{D}$ed. E.g. the case of the fourth rule is displayed in Figure 1.

Remark 1. The atomic decomposition TRS $\mathcal{L}$ex is not minimal; in general it does not yield unique atomic decompositions of rules. For instance, assuming for the moment that M R 0 would hold, the third rule $\text{M}(x, 0) \rightarrow 0$ of $\mathcal{D}$ed

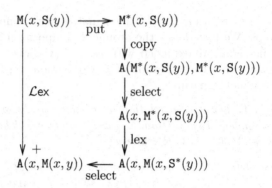

Fig. 1. Atomic $\mathcal{L}$ex-decomposition of the fourth Dedekind rule

could be atomically decomposed into both $M(x,0) \rightarrow_{\text{put}} M^*(x,0) \rightarrow_{\text{select}} 0$ and $M(x,0) \rightarrow_{\text{put}} M^*(x,0) \rightarrow_{\text{copy}} 0$; the term $M^*(x,0)$ is copied to all (zero!) arguments of the symbol 0.

Definition 2.

1. *The* iterative lexicographic path order R_{ilpo} *of a relation R on a signature Σ is the restriction of $\rightarrow^+_{\mathcal{L}\text{ex}}$ to $T(\Sigma \uplus V)$.*
2. *A TRS is* ILPO-terminating *if its rules are contained in R_{ilpo} for some terminating relation R.*

For the TRS $\mathcal{D}$ed we already saw that $l \rightarrow^+_{\mathcal{L}\text{ex}} r$ holds for each rule $l \rightarrow r$. Hence $\mathcal{D}$ed is ILPO-terminating.

An observation that plays a crucial role in many termination methods is that a TRS is terminating if and only if it admits a *reduction order*, i.e. iff its rules are contained in a terminating (order) relation which is closed under contexts and substitutions. (See e.g. [10, Prop. 6.1.3].) Note that transitivity and closure under contexts and substitutions of R_{ilpo} are 'built in' into its definition via the atomic decomposition TRS. Therefore, in order to show that R_{ilpo} is a reduction order, it only remains to be shown that it is terminating. This will be proved in Section 4. From that result we can then conclude that ILPO-termination implies termination.

But first, in Section 3, we present some further examples of ILPO-terminating TRSs.

Remark 2. Although by definition the iterative lexicographic path order R_{ilpo} is transitive, even in cases when R isn't, we do not put stress on this. In particular, transitivity is not used in the proof that termination lifts from R to R_{ilpo}.

Remark 3. The iterative lexicographic path order as presented here is a strengthening of the version of the iterative path order in [9] (which is there still called recursive path order). The difference is that in [9] instead of the lex-rule (cf. Table 2) the *down*-rule is employed:

$$f^*(\boldsymbol{x}, g(\boldsymbol{y}), \boldsymbol{z}) \rightarrow_{\text{down}} f(\boldsymbol{x}, g^*(\boldsymbol{y}), \boldsymbol{z})$$

It expresses that a term may be made smaller by making one of its arguments smaller. The down-rule is a derived rule in our system:

$$f^*(\boldsymbol{x}, g(\boldsymbol{y}), \boldsymbol{z}) \rightarrow_{\text{lex}} f(\boldsymbol{x}, g^*(\boldsymbol{y}), l) \twoheadrightarrow_{\text{select}} f(\boldsymbol{x}, g^*(\boldsymbol{y}), \boldsymbol{z})$$

where the ith select-step applies to the ith occurrence of $l = f^*(\boldsymbol{x}, g(\boldsymbol{y}), \boldsymbol{z})$, and then selects z_i. The implication in the other direction does not hold as witnessed by the one-rule TRS $f(a,b) \rightarrow f(b,a)$ which cannot be proven terminating by the method presented in [9], but which is ILPO-terminating for $a\ R\ b$:

$$f(a,b) \rightarrow_{\text{put}} f^*(a,b) \rightarrow_{\text{lex}} f(a^*, f^*(a,b)) \rightarrow_{\text{copy}} f(b, f^*(a,b)) \twoheadrightarrow_{\text{select}} f(b,a)$$

The *simplify-left-argument*-rule, introduced in the exercises in [9] in order to prove termination of the Ackermann TRS $\mathcal{A}$ck (see below), is also easily derived in our system: it simply is the lex-rule with $\boldsymbol{x}$ taken to be the empty vector, i.e. the leftmost argument must be made smaller. It is easy to see that also that version is strictly weaker than ILPO-termination.

3   Examples of ILPO-terminating TRSs

In this section the iterative lexicographic path order method is illustrated by applying it to some well-known TRSs.

The example of Dedekind's rules for addition and multiplication only employs trivial applications of the lex-rule, where $\boldsymbol{z}$ is empty. Proving ILPO-termination of the Ackermann function requires non-trivial applications of the lex-rule.

Example 1 (Ackermann's function). The TRS $\mathcal{A}$ck has a signature consisting of the nullary symbol 0, the unary symbol S, and the binary symbol Ack, with rules as in Table 3.

$$\begin{array}{|rcl|}
\hline
\mathsf{Ack}(0,y) & \to & \mathsf{S}(y) \\
\mathsf{Ack}(\mathsf{S}(x),0) & \to & \mathsf{Ack}(x,\mathsf{S}(0)) \\
\mathsf{Ack}(\mathsf{S}(x),\mathsf{S}(y)) & \to & \mathsf{Ack}(x,\mathsf{Ack}(\mathsf{S}(x),y)) \\
\hline
\end{array}$$

Table 3. Ackermann's function

For the relation R defined by Ack R S, the TRS $\mathcal{A}$ck is ILPO-terminating as witnessed by the following atomic decompositions of each of its rules:

- $\mathsf{Ack}(0,y) \rightarrow_{\text{put}} \mathsf{Ack}^*(0,y) \rightarrow_{\text{copy}} \mathsf{S}(\mathsf{Ack}^*(0,y)) \twoheadrightarrow_{\text{select}} \mathsf{S}(y)$.
- $\mathsf{Ack}(\mathsf{S}(x),0) \rightarrow_{\text{put}} \mathsf{Ack}^*(\mathsf{S}(x),0) \rightarrow_{\text{lex}}$
  $\mathsf{Ack}(\mathsf{S}^*(x), \mathsf{Ack}^*(\mathsf{S}(x),0)) \twoheadrightarrow_{\text{select}} \mathsf{Ack}(x, \mathsf{Ack}^*(\mathsf{S}(x),0)) \rightarrow_{\text{copy}}$
  $\mathsf{Ack}(x, \mathsf{S}(\mathsf{Ack}^*(\mathsf{S}(x),0))) \twoheadrightarrow_{\text{select}} \mathsf{Ack}(x,\mathsf{S}(0))$.

- $\mathrm{Ack}(\mathrm{S}(x),\mathrm{S}(y)) \to_{\mathrm{put}} \mathrm{Ack}^*(\mathrm{S}(x),\mathrm{S}(y)) \to_{\mathrm{lex}}$
  $\mathrm{Ack}(\mathrm{S}^*(x),\mathrm{Ack}^*(\mathrm{S}(x),\mathrm{S}(y))) \to_{\mathrm{select}} \mathrm{Ack}(x,\mathrm{Ack}^*(\mathrm{S}(x),\mathrm{S}(y))) \to_{\mathrm{lex}}$
  $\mathrm{Ack}(x,\mathrm{Ack}(\mathrm{S}(x),\mathrm{S}^*(y))) \to_{\mathrm{select}} \mathrm{Ack}(x,\mathrm{Ack}(\mathrm{S}(x),y)).$

Example 2 (Dershowitz and Jouannaud [12]). Consider the string rewrite system $\mathcal{DJ}$ given by the four rules in Table 4.

10	→ 0001
01	→ 1
11	→ 0000
00	→ 0

Table 4. String rewrite system on 0,1-words

So we have e.g. the reduction

$1101 \to 100011 \to 10011 \to 0001011 \to 001011 \to 00100000 \to 0000010000 \to \dots$

To capture this string rewrite system as a term rewrite system, the symbols 0 and 1 are perceived as unary function symbols and the rules are read accordingly; e.g. $10 \to 0001$ is the term rewrite rule

$$1(0(x)) \to 0(0(0(1(x))))$$

To show ILPO-termination for $\mathcal{DJ}$, we set 1 R 0 and check $l\ R_{ilpo}\ r$ for every rule $l \to r$. For the displayed rule the corresponding atomic decomposition is shown in Figure 2 (after dropping all parentheses).

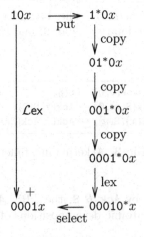

Fig. 2. Atomic $\mathcal{L}$ex-decomposition of the rule $10x \to 0001x$

Example 3 (Primitive recursion). Let a TRS on the natural numbers be given, having a unary function symbol g and a ternary function symbol h. Suppose we adjoin a binary symbol f to the signature, with defining rules

$$f(0, x) \rightarrow g(x)$$
$$f(S(x), y) \rightarrow h(f(x, y), x, y)$$

If the original TRS is ILPO-terminating, say for the terminating relation R on its signature, then the resulting system is ILPO-terminating again as can be easily seen after adjoining $f \ R \ g$ and $f \ R \ h$ to R (this yields a terminating relation again).

4   ILPO-termination Implies Termination

The remaining task is now to show that if a relation R is terminating, then R_{ilpo} is terminating as well. As explained in Section 2, R_{ilpo} is then a reduction order, and it follows that ILPO-termination implies termination.

Since the rewrite rules of R_{ilpo} are given by the restriction of $\rightarrow^+_{\mathcal{L}ex}$ to $T(\Sigma \uplus V)$, termination of the atomic decomposition TRS $\mathcal{L}$ex would be sufficient for termination of R_{ilpo}. However, $\mathcal{L}$ex is in general *not* terminating. For instance, in case of the running example we have, despite R being terminating:

$$\mathsf{A}(x, y) \rightarrow_{\text{put}} \mathsf{A}^*(x, y) \rightarrow_{\text{copy}} \mathsf{S}(\mathsf{A}^*(x, y)) \rightarrow_{\text{copy}} \mathsf{S}(\mathsf{S}(\mathsf{A}^*(x, y))) \rightarrow_{\text{copy}} \cdots$$

We even have cycles

$$\mathsf{A}^*(x, y) \rightarrow_{\text{copy}} \mathsf{S}(\mathsf{A}^*(x, y)) \rightarrow_{\text{put}} \mathsf{S}^*(\mathsf{A}^*(x, y)) \rightarrow_{\text{select}} \mathsf{A}^*(x, y)$$

In either case, non-termination is 'caused' by the left-hand side of the copy-rule being a subterm of its right-hand side; *a priori* such an iteration is not bounded. Similar examples can be given with the lex-rule, which is also self-embedding.

However, observe that in both of the infinite reductions the control symbol A^* is 'used' infinitely often — by the copy rule. We will show that this is necessary in *any* infinite reduction. More precisely, that if for each control symbol a bound is given in advance on how often it can be used in the copy- and lex-rules, this will yield a terminating TRS $\mathcal{L}ex^\omega$. Since in any given atomic decomposition $l \rightarrow^+_{\mathcal{L}ex} r$ of a rule $l \rightarrow r$, any control symbol is only used finitely often (certainly not more often than the length of the decomposition), the relations $\rightarrow^+_{\mathcal{L}ex}$ and $\rightarrow^+_{\mathcal{L}ex^\omega}$ coincide on the unmarked terms. We will exploit this fact by proving termination of R_{ilpo} via termination of $\mathcal{L}ex^\omega$.

Definition 3. *Let R be a relation on a signature Σ, and let V be a signature of nullary symbols disjoint from Σ. The TRS $\mathcal{L}ex^\omega$ is $\langle \Sigma \uplus \Sigma^\omega \uplus V, \mathcal{R}^\omega \rangle$:*

1. *The signature Σ^ω of ω-control symbols consists of ω copies of Σ, i.e. for each symbol $f \in \Sigma$ and natural number n, Σ^ω contains a fresh symbol f^n having the arity f has.*

$$
\begin{aligned}
f(\boldsymbol{x}) &\to_{\text{put}} f^n(\boldsymbol{x}) \\
f^n(\boldsymbol{x}) &\to_{\text{select}} x_i \quad (1 \le i \le |\boldsymbol{x}|) \\
f^{n+1}(\boldsymbol{x}) &\to_{\text{copy}} g(f^n(\boldsymbol{x}),\dots,f^n(\boldsymbol{x})) \quad (f \mathrel{R} g) \\
f^{n+1}(\boldsymbol{x},g(\boldsymbol{y}),\boldsymbol{z}) &\to_{\text{lex}} f(\boldsymbol{x},g^n(\boldsymbol{y}),l,\dots,l) \quad (l = f^n(\boldsymbol{x},g(\boldsymbol{y}),\boldsymbol{z}))
\end{aligned}
$$

Table 5. The rules of $\mathcal{L}ex^\omega$

2. *The rules $\mathcal{R}^\omega$ are given in the table, for arbitrary symbols f, g in Σ and natural number n, with $\boldsymbol{x}$, $\boldsymbol{y}$, $\boldsymbol{z}$ disjoint vectors of pairwise disjoint variables of appropriate lengths.*

The TRS $\mathcal{L}ex$ (Definition 1) is seen to be a homomorphic image of $\mathcal{L}ex^\omega$ by mapping f^n to f^*, for any natural number n. Vice versa, reductions in $\mathcal{L}ex$ can be 'lifted' to $\mathcal{L}ex^\omega$.

Lemma 1. $\to^+_{\mathcal{L}ex}$ *and* $\to^+_{\mathcal{L}ex^\omega}$ *coincide as relations restricted to* $T(\Sigma \uplus V)$.

Proof.

($\subseteq$) One shows that $\to^+_{\mathcal{L}ex}$ is included in $\to^+_{\mathcal{L}ex^\omega}$ by formalizing the reasoning already indicated above. In particular, one can translate (*lift*) a finite reduction of length n in $\mathcal{L}ex$ to $\mathcal{L}ex^\omega$ by replacing all marks ($*$) in the begin term by a natural number greater than n and, along the reduction, likewise each $*$ introduced by an application of the put-rule. Numerical values for the other marks then follow automatically by applying the $\mathcal{L}ex^\omega$-rules that correspond to the original $\mathcal{L}ex$-rules.

The result then follows, because, if $t \to^+_{\mathcal{L}ex} s$ and the terms t, s are in $T(\Sigma \uplus V)$, i.e. do not contain marks, then the transformation of a $\to_{\mathcal{L}ex}$-reduction from t to s to $\mathcal{L}ex^\omega$ leaves the begin and end terms t and s untouched.

($\supseteq$) Every $\to^+_{\mathcal{L}ex^\omega}$ step is a $\to^+_{\mathcal{L}ex}$ step, by the homomorphism. □

Example 4. The atomic decomposition displayed in Figure 1 can be lifted to:

- $\mathsf{M}(x,\mathsf{S}(y)) \to_{\text{put}} \mathsf{M}^2(x,\mathsf{S}(y)) \to_{\text{copy}} \mathsf{A}(\mathsf{M}^1(x,\mathsf{S}(y)),\mathsf{M}^1(x,\mathsf{S}(y))) \to_{\text{select}}$ $\mathsf{A}(x,\mathsf{M}^1(x,\mathsf{S}(y))) \to_{\text{lex}} \mathsf{A}(x,\mathsf{M}(x,\mathsf{S}^0(y))) \to_{\text{select}} \mathsf{A}(x,\mathsf{M}(x,y))$.

The main theorem is proven by employing an ingenious (constructive) proof technique due to Buchholz [5].[6]

Lemma 2. *If R is terminating, then $\mathcal{L}ex^\omega$ is terminating.*

---

[6] The technique has been discovered independently by Jouannaud and Rubio [13], who show that it combines well with the Tait–Girard reducibility technique (both are essentially based on induction on terms), leading to a powerful termination proof technique also applicable to higher-order term rewriting.

Proof. To make the notation uniform, in the sense that all function symbols will carry a label, we employ $f^\omega(t)$ to denote the term $f(t)$ for an *unmarked* symbol f. This allows us to write *any* $\mathcal{L}ex^\omega$-term uniquely as an ω-*marked term* of the form $f^\alpha(t)$ for some unmarked symbol f, some ordinal α and some vector of terms t. The ordinal α will be a natural number n or ω. In the crucial induction in this proof we will make use of the fact that in the ordering of the ordinals we have $\omega > n$ for each natural number n.

We prove by induction on the construction of terms that any ω-marked term is terminating. To that end it is sufficient to show that any ω-marked term $f^\alpha(t)$ is terminating under the assumption (the induction hypothesis) that its arguments t are terminating.

So assume that $t_1, \ldots, t_n$ are terminating, with n the arity of f. We prove that $f^\alpha(t)$ is terminating by a further induction on the triple consisting of f, t, and α in the lexicographic product of the relations R, $(\to_{\mathcal{L}ex^\omega} \lceil \mathcal{SN})^n$ and $>$. Here $(\to_{\mathcal{L}ex^\omega} \lceil \mathcal{SN})^n$ is the n-fold lexicographic product of the terminating part of $\to_{\mathcal{L}ex^\omega}$, with n the arity of f.

Clearly, the term $f^\alpha(t)$ is terminating if all its one-step $\to_{\mathcal{L}ex^\omega}$-reducts are.[7] The latter we prove by distinguishing cases on the type of the reduction step.

1. If the step is a head step, we perform a further case analysis on the rule applied.

   (put) The result follows by the IH for the third component of the triple, since $\omega > m$ for any natural number m.

   (select) The result follows by the termination assumption for the t.

   (copy) Then α is of the form $m + 1$ for some natural number m, and the reduct has shape $g^\omega(f^m(t), \ldots, f^m(t))$ for some g such that f R g. By the IH for the third component, each of the $f^m(t)$ is terminating. Hence, by the IH for the first component, the reduct is terminating.

   (lex) Then α is of the form $m+1$ for some natural number m and the reduct has shape $f^\omega(t_1, \ldots, t_{i-1}, g^m(s), f^m(t), \ldots, f^m(t))$, with $t_i = g^\omega(s)$. Each $f^m(t)$ is terminating by the IH for the third component. Hence, the reduct is terminating by the IH for the second component, since $g^m(s)$ is a one-step $\mathcal{L}ex^\omega$-reduct of t_i (for the put-rule).

2. If the step is a non-head step, then it rewrites some direct argument, and the result follows by the IH for the second component. □

Theorem 1. *ILPO-termination implies termination.*

Proof. Suppose the TRS $\mathcal{T}$ is ILPO-terminating for some terminating relation R on its signature, i.e. l R_{ilpo} r holds for each rule $l \to r$ in $\mathcal{T}$.

Since R_{ilpo} is defined as a restriction of $\to^+_{\mathcal{L}ex}$ which in turn coincides with $\to^+_{\mathcal{L}ex^\omega}$ by Lemma 1, it is a transitive relation that is closed under contexts and substitutions and, by Lemma 2, also terminating. Hence, R_{ilpo} is a reduction order, and therefore $\mathcal{T}$ must be terminating. □

---

[7] This observation can be used as an inductive characterization of termination: a term is terminating if and only if all its one-step reducts are. In a constructive rendering of the proof one can take this characterization as the definition of termination.

Hence termination of the example TRSs such as $\mathcal{D}$ed, follows from their ILPO-termination as established above.

Remark 4. It is worth noting that Buchholz's proof technique can also be applied to non-simplifying TRSs. For instance, for proving termination of the one-rule TRS $f(f(x)) \to f(g(f(x)))$ the technique boils down to showing that any instance $f(g(f(t)))$ of the right-hand side is terminating on the assumption that the direct subterm $f(t)$ of the corresponding instance of the left-hand side is. This follows by 'induction' on the right-hand side and cases on the shape of the left-hand side of the rule: $f(g(f(t)))$ can be rewritten neither at the head nor at position 1, hence it is terminating if $f(t)$ is.

5   Equivalence of ILPO with the Recursive Lexicographic Path Order

We show that ILPO is at least as powerful as the recursively defined lexicographic path order found in the literature, and is equivalent to it for transitive relations. The following definition of $>_{lpo}$ for a given strict order $>$ on the signature Σ, is copied verbatim from Definition 5.4.12 in the textbook by Baader and Nipkow [7].

Definition 4. *Let Σ be a finite signature and $>$ be a strict order on Σ. The **lexicographic path order** $>_{lpo}$ on $T(\Sigma, V)$ induced by $>$ is defined as follows: $t >_{lpo} s$ iff*

(LPO1) $s \in Var(t)$ *and* $t \neq s$, *or*
(LPO2) $t = f(t_1, \ldots, t_m)$, $s = g(s_1, \ldots, s_n)$, *and*
  (LPO2a) *there exists* i, $1 \leq i \leq m$, *with* $t_i \geq_{lpo} s$, *or*
  (LPO2b) $f > g$ *and* $t >_{lpo} s_j$ *for all* j, $1 \leq j \leq n$, *or*
  (LPO2c) $f = g$, $t >_{lpo} s_j$ *for* j, $1 \leq j \leq n$, *and there exists* i, $1 \leq i \leq m$, *such that* $t_1 = s_1$, ..., $t_{i-1} = s_{i-1}$ *and* $t_i >_{lpo} s_i$.

It is easy to see that this is still a correct recursive definition for $>$ being an arbitrary relation R, yielding R_{lpo}. We call a TRS $\mathcal{T} = \langle \Sigma, \mathcal{R} \rangle$ LPO-terminating for a terminating relation R, if $\mathcal{R} \subseteq R_{lpo}$.

Lemma 3. $R_{lpo} \subseteq R_{ilpo}$, *for any relation R.*

Proof. We show by induction on the definition of $t\ R_{lpo}\ s$ that $t^* \twoheadrightarrow_{\mathcal{L}ex} s$, where $(f(t))^* = f^*(t)$. This suffices, since $t \to_{\mathrm{put}} t^*$ and t, s are not marked.

(LPO1) If $s \in Var(t)$ and $t \neq s$, then the result follows by repeatedly selecting the subterm on a path to an occurrence of s in t.
(LPO2) Otherwise, let $t = f(t_1, \ldots, t_m)$, $s = g(s_1, \ldots, s_n)$.
  (LPO2a) Suppose there exists i, $1 \leq i \leq m$, with either $t_i = s$ or $t_i\ R_{lpo}\ s$. In the former case, the result follows by a single application of the select-rule for index i. In the latter case, this step is followed by an application of the put-rule after which the result follows by the IH.

(LPO2b) Suppose $f \mathrel{R} g$ and $t \mathrel{R_{lpo}} s_j$ for all j, $1 \leq j \leq n$. Then the result follows by a single application of the copy-rule and n applications of the IH.

(LPO2c) Suppose $f = g$, $t \mathrel{R_{lpo}} s_j$ for j, $1 \leq j \leq n$, and there exists i, $1 \leq i \leq m$, such that $t_1 = s_1$, ..., $t_{i-1} = s_{i-1}$ and $t_i \mathrel{R_{lpo}} s_i$. Then the result follows by a single application of the lex-rule, selecting the ith argument, and the IH for $t_i \mathrel{R_{lpo}} s_i$ and $t \mathrel{R_{lpo}} s_j$ for j, $i < j \leq n$. $\square$

We call the $\mathcal{L}$ex-strategy implicit in this proof the *wave* strategy. The idea is that the marked positions in a term represent the wave front, which moves downwards, i.e. from the root in the direction of the subtrees of a left-hand side, generating an ever growing prefix of the right-hand side behind it. This is visualised abstractly in Figure 3, and for the atomic $\mathcal{L}$ex-decomposition of $\mathsf{M}(x, \mathsf{S}(y)) \to \mathsf{A}(x, \mathsf{M}(x, y))$

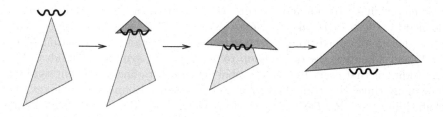

Fig. 3. Wave strategy

of Figure 1, in Figure 4. (In fact, all $\mathcal{L}$ex-reductions given above adhere to the wave strategy.) One can prove a converse to Lemma 3 by a detailed proof-

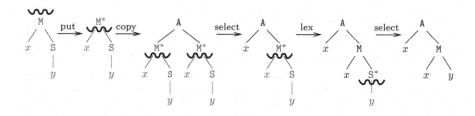

Fig. 4. Wave strategy for atomic $\mathcal{L}$ex-decomposition of $\mathsf{M}(x, \mathsf{S}(y)) \to \mathsf{A}(x, \mathsf{M}(x, y))$

theoretic analysis, showing that any $\mathcal{L}$ex-reduction can be transformed, into a wave reduction. The upshot is that in general $R_{ilpo} = (R_{lpo})^+$. As a corollary we then have that ILPO is equivalent with LPO for any strict order, and that

R_{ilpo} is decidable, in case R is a terminating relation for which reachability is decidable: simply 'try all waves up to the size of the right-hand side'.[8]

Remark 5. Note that $(R^+)_{ilpo}$ may differ from R_{ilpo}. Consider the signature consisting of nullary symbols a, b, and unary symbols f, g, and the terminating relation $f\ R\ b\ R\ g$. Then the one-rule TRS $f(a) \to g(a)$ is not ILPO-terminating, but it is ILPO-terminating for R^+. The problem is that making $f(a)$ smaller using R forces erasure of its argument a, because b is nullary.

A proof-theoretic analysis of the wave strategy is beyond the scope of this paper. Here we will be satisfied by giving a rather *ad hoc* proof of the converse of Lemma 3, for the case where we start with a transitive relation R.

Lemma 4. $R_{ilpo} \subseteq R_{lpo}$, *for any transitive relation R.*

Proof. Fix the relation R. By definition, if $t\ R_{ilpo}\ s$ then $t \to^+_{\mathcal{L}\text{ex}} s$. By Lemma 1, then also $t \to^+_{\mathcal{L}\text{ex}^\omega} s$. To show that this implies $t\ R_{lpo}\ s$, we employ a homomorphic embedding ϵ of ω-marked terms (as introduced in the proof of Lemma 2) defined by $f^\alpha(\boldsymbol{u}) \mapsto f(\epsilon(\boldsymbol{u}), \alpha)$. Here the terms in the range of ϵ are terms over the signature obtained from Σ by increasing the arity of every function symbol by 1, and adjoining the ordinals up to ω as nullary symbols. The idea of the embedding is that every function symbol gets an extra final argument signifying how many times the symbol may be 'used'. Initially (unmarked) it is set to ω. Embedding the TRS $\mathcal{L}\text{ex}^\omega$ (see Table 5) yields the TRS $\epsilon(\mathcal{L}\text{ex}^\omega)$ having rules given in Table 6.

$$f(\boldsymbol{x}, \omega) \to_{\text{put}} f(\boldsymbol{x}, n)$$
$$f(\boldsymbol{x}, n) \to_{\text{select}} x_i \quad (1 \le i \le |\boldsymbol{x}|)$$
$$f(\boldsymbol{x}, n+1) \to_{\text{copy}} g(f(\boldsymbol{x}, n), \ldots, f(\boldsymbol{x}, n), \omega) \quad (f\ R\ g)$$
$$f(\boldsymbol{x}, g(\boldsymbol{y}, \omega), \boldsymbol{z}, n+1) \to_{\text{lex}} f(\boldsymbol{x}, g(\boldsymbol{y}, n), \boldsymbol{l}, \omega) \quad (\boldsymbol{l} = f(\boldsymbol{x}, g(\boldsymbol{y}, \omega), \boldsymbol{z}, n))$$

Table 6. Rules of $\epsilon(\mathcal{L}\text{ex}^\omega)$

By definition of ϵ, $t \to^+_{\mathcal{L}\text{ex}^\omega} s$ implies $\epsilon(t) \to^+_{\epsilon(\mathcal{L}\text{ex}^\omega)} \epsilon(s)$. It is easy to verify that each of the $\epsilon(\mathcal{L}\text{ex}^\omega)$-rules is contained in $\epsilon(R)_{lpo}$, where $\epsilon(R)$ is obtained by taking the union of R and the natural greater than relation $>$ on the ordinal symbols, and relating every function symbol to any ordinal symbol. Note that $\epsilon(R)$ is transitive since R and $>$ are. Since the relation $\epsilon(R)_{lpo}$ is closed under contexts and substitutions and is transitive if $\epsilon(R)$ is (see e.g. [10,7]),[9] we conclude that

---

[8] This is somehow analogous to the way in which *rippling* guides the search for a proof of a goal from a given [14].

[9] Note that these properties need to be verified separately for the *recursive* lexicographic path order (or any variation on it). In case of the *iterative* lexicographic path order (or any variation on it), these properties hold automatically by it being given by a TRS, allowing one to focus on establishing termination.

$\epsilon(t)$ $\epsilon(R)_{lpo}$ $\epsilon(s)$. That this implies t R_{lpo} s, follows by an easy induction on the definition of the former. The crux of the proof is that the adjoined ordinal symbols do not relate to the original symbols from Σ. The only problematic cases (since then the IH could not be applied) are:

(LPO2a) holds since either $\omega = \epsilon(s)$ or ω $\epsilon(R)_{lpo}$ $\epsilon(s)$. Neither can be the case since ω is not related to symbols in Σ, in particular not to the head symbol of $\epsilon(s)$ (which is the same as the head of s).

(LPO2c) holds since ω $\epsilon(R)_{lpo}$ ω. This obviously cannot be the case.

In the other cases the IH does the job, e.g.

(LPO2a) holds since either $\epsilon(t_i) = \epsilon(s)$, or $\epsilon(t_i)$ $\epsilon(R)_{lpo}$ $\epsilon(s)$, for some i. Then either $t_i = s$ by injectivity of ϵ, or t_i $\epsilon(R)_{lpo}$ s by the IH, and we conlude t R_{lpo} s. □

Combining Lemmas 3 and 4 yields our second main result.

Theorem 2. $>_{ilpo} = >_{lpo}$, for any transitive relation (order) $>$.

6   Conclusion

We have shown that our iterative set-up of ILPO can serve as an independent alternative to the classical recursive treatment of LPO. It can be seen as being obtained by *decomposing* the recursive definition, extracting *atomic* rules from the inductive clauses. From this perspective it is only natural that we have taken an arbitrary terminating relation (instead of order) on the signature as our starting point, so one could speak, in the spirit of Persson's presentation of recursive path relations [15], of iterative lexicographic path *relations*.

We claim that the correspondence between recursive and iterative ways of specifying path orders is robust, i.e. goes through for variants of LPO like the embedding relation and recursive path orders. Substantiating the claim is left to future research.

Another direction for further investigation is suggested by Remark 4. It seems that an analogous argument can be used to yield soundness of Arts and Giesl's *dependency-pair* technique for proving termination. (See e.g. [10, Section 6.5.5].) Thus, whereas non-simplifying TRSs are traditionally out of the scope of the recursive path order method, by their termination proof being tied to Kruskal's Tree Theorem, Buchholz's technique will give us a handle on a uniform treatment of both path orders and the dependency-pair technique.

References

1. Dershowitz, N.: Orderings for term rewriting systems. TCS **17**(3) (1982) 279–301
2. Marcone, A.: Fine analysis of the quasi-orderings on the power set. Order **18** (2001) 339–347

554   Jan Willem Klop, Vincent van Oostrom, and Roel de Vrijer

3. Bergstra, J., Klop, J.: Algebra of communicating processes. TCS **37**(1) (1985) 171–199
4. Bergstra, J., Klop, J., Middeldorp, A.: Termherschrijfsystemen. Programmatuurkunde. Kluwer (1989)
5. Buchholz, W.: Proof-theoretic analysis of termination proofs. APAL **75**(1-2) (1995) 57–65
6. Kamin, S., Lévy, J.J.: Two generalizations of the recursive path ordering. University of Illinois (1980)
7. Baader, F., Nipkow, T.: Term Rewriting and All That. Cambridge University Press (1998)
8. Geser, A.: Relative Termination. PhD thesis, Universität Passau, Germany (1990)
9. Klop, J.: Term rewriting systems. In Abramsky, S., Gabbay, D., Maibaum, T., eds.: Handbook of Logic in Computer Science. Volume 2, Background: Computational Structures. Oxford University Press (1992) 1–116
10. Terese: Term Rewriting Systems. Volume 55 of Cambridge Tracts in Theoretical Computer Science. Cambridge University Press (2003)
11. Dedekind, R.: Was sind und was sollen die Zahlen?, Brunswick (1888)
12. Dershowitz, N., Jouannaud, J.P.: Rewrite systems. In van Leeuwen, J., ed.: Handbook of Theoretical Computer Science. Volume B, Formal Models and Semantics. Elsevier (1990) 243–320
13. Jouannaud, J.P., Rubio, A.: The higher-order recursive path ordering. In: 14th Annual IEEE Symposium on Logic in Computer Science, IEEE Computer Society (1999) 402–411
14. Bundy, A., Basin, D., Hutter, D., Ireland, A.: Rippling: Meta-Level Guidance for Mathematical Reasoning. Volume 56 of Cambridge Tracts in Theoretical Computer Science. Cambridge University Press (2005)
15. Persson, H.: Type Theory and the Integrated Logic of Programs. PhD thesis, Chalmers, Sweden (1999)

A Functorial Framework for Constraint Normal Logic Programming

Paqui Lucio[2], Fernando Orejas[1], Edelmira Pasarella[1], and Elvira Pino[1]

[1] Departament LSI
Universitat Politècnica de Catalunya,
Campus Nord, Mòdul C5, Jordi Girona 1-3, 08034 Barcelona, Spain
{orejas,edelmira,pino}@lsi.upc.es
[2] Departament LSI
Univ. Pais. Vasco,
San Sebastián, Spain
jiplucap@si.ehu.es

Abstract. The semantic constructions and results for definite programs do not extend when dealing with negation. The main problem is related to a well-known problem in the area of algebraic specification: if we fix a constraint domain as a given model, its free extension by means of a set of Horn clauses defining a set of new predicates is semicomputable. However, if the language of the extension is richer than Horn clauses its free extension (if it exists) is not necessarily semicomputable. In this paper we present a framework that allows us to deal with these problems in a novel way. This framework is based on two main ideas: a reformulation of the notion of constraint domain and a functorial presentation of our semantics. In particular, the semantics of a logic program P is defined in terms of three *functors*: $(O\mathcal{P}_P, \mathcal{ALG}_P, \mathcal{LOG}_P)$ that apply to constraint domains and provide the operational, the least fixpoint and the logical semantics of P, respectively. The idea is that the application of $O\mathcal{P}_P$ to a specific constraint solver, provides the operational semantics of P that uses this solver; the application of $\mathcal{ALG}_P$ to a specific domain, provides the least fixpoint of P over this domain; and the application of $\mathcal{LOG}_P$ to a theory of constraints provides the logic theory associated to P. We prove that these three functors are in some sense equivalent.

1 Introduction

Constraint logic programming was introduced in ([9]) as a powerful and conceptually simple extension of logic programming. Following that seminal paper, the semantics of definite (constraint) logic programs has been studied in detail (see, e.g. [10], [11]). However, the constructions and results for definite programs do not extend when dealing with negation. The main problem is related to a well-known problem in the area of algebraic specification: if we fix a constraint domain as a given model, its free extension by means of a set of Horn clauses defining a set of new predicates is semicomputable. However, if the language of the extension is richer than Horn clauses its free extension (if it exists) is not necessarily semicomputable ([8]). Now, when working without negation we are in the former case, but when working with negation we are in the latter case. In particular, this implies that the results about the soundness and completeness of the

K. Futatsugi et al. (Eds.): Goguen Festschrift, LNCS 4060, pp. 555–577, 2006.
© Springer-Verlag Berlin Heidelberg 2006

operational semantics with respect to the logical and algebraic semantics of a definite constraint logic program do not extend to the case of programs with negation, except when we impose some restrictions to these programs.

The only approach that we know dealing with this problem is ([19]). In that paper, Stuckey presents one of the first operational semantics which is proven complete for programs that include (constructive) negation. Although we use a different operational semantics, that paper has had an important influence in our work on negation. The results in ([19]) were very important when applied to the case of standard (non-constrained) logic programs because they provided some good insights about constructive negation. However, the general version (i.e., logic programs over an arbitrary constraint domain) is not so interesting (in our opinion). The reason is that the completeness results are obtained only for programs over *admissible* constraints. We think that this restriction on the constraints that can be used in a program is not properly justified.

In our opinion, the problem when dealing with negation is not on the class of constraints considered, but rather, in the notion of constraint domain used. In particular, the notion of constraint domain used in the context of definite programs is not adequate when dealing with negation. Instead, we propose a small reformulation of the notion of constraint domain. To be precise, we propose that a domain should be defined in terms of a class of elementarily equivalent models and not in terms of a single model. With this variation we show the equivalence of the logical, operational, and fixpoint semantics of programs with negation without needing to restrict the class of constraints.

The logical semantics that we have used is the standard Clark-Kunen 3-valued completion of programs (see, e.g. [19]). The fixpoint semantics that we are using is a variation of other well-known fixpoint semantics used to deal with negation ([5,19,6,15]). Finally, the operational semantics that we are using is an extension of a semantics called BCN that we have defined in ([16]) for the case of programs without constraints. The main reason for using this semantics and not Stuckey's semantics is that our semantics is simpler. This implies having simpler proofs for our results. In particular, we do not claim that our semantics is better than Stuckey's (nor that it is worse). A proper comparison of these two semantics and of others like [5,6] would need experimental work. We have a prototype implementation of BCN ([1]), but we do not know if the other approaches have been implemented. Anyhow, the pragmatic virtues of the various operational approaches to constructive negation are not a relevant issue in this paper.

Our semantics is functorial. We consider that a constraint logic program is a program that is parameterized by the given constraint domain, i.e., that the semantics of a program should be some kind of mapping. However, we also think that working in a categorical setting provides some additional advantages that are shown in the paper.

The paper is organized as follows. In the following section we give a short introduction to the semantics of (definite) constraint logic programs. In Section three, we discuss the inadequacy of the standard notion of constraint domain when dealing with negation and propose a new one. In Section four we study the semantics of programs when defined over a given arbitrary constraint domain. Then, in the following section we define several categories for defining the various semantic domains involved and define the functorial semantics of logic programs. Finally, in Section 6 we prove the equivalence of the logical, fixpoint and operational semantics.

2 Preliminaries

A signature Σ consists of a pair of sets (FS_Σ, PS_Σ) of function and predicates symbols, respectively, with some associated arity. $T_\Sigma(X)$ denotes the set of all *first-order Σ-terms* over variables from X, and T_Σ denotes the set of all ground terms. A literal is either an atom $p(t_1, \ldots, t_n)$ (namely a positive literal) or a negated atom $\neg p(t_1, \ldots, t_n)$ (namely a negative literal). The set *Form*$_\Sigma$ is formed by all *first-order Σ-formulas* written (from atoms) using connectives $\neg, \wedge, \vee, \rightarrow, \leftrightarrow$ and quantifiers $\forall, \exists$. We denote by $free(\varphi)$ the set of all free variables occurring in φ. $\varphi(\bar{x})$ specifies that $free(\varphi) \subseteq \bar{x}$. *Sent*$_\Sigma$ is the set of all $\varphi \in$ *Form*$_\Sigma$ such that $free(\varphi) = \emptyset$, called Σ-sentences. By $\varphi^{\forall \setminus \bar{z}}$ (resp. $\varphi^{\exists \setminus \bar{z}}$) we denote the formula $\forall x_1 \ldots \forall x_n(\varphi)$ (resp. $\exists x_1 \ldots \exists x_n(\varphi)$), where $x_1 \ldots x_n$ are the variables in $free(\varphi) \setminus \bar{z}$. In particular, the universal (resp. existential) closure, that is $\varphi^{\forall \setminus \emptyset}$ (resp. $\varphi^{\exists \setminus \emptyset}$) is denoted by $\varphi^\forall$ (resp. $\varphi^\exists$).

The semantics of normal logic programs is defined using a concrete *three-valued* extension of the classical two-valued interpretation of logical symbols. The connectives $\neg, \wedge, \vee$ and quantifiers ($\forall, \exists$) are interpreted as in Kleene's logic ([12]). However, $\leftrightarrow$ is interpreted as the identity of truth-values (hence, $\leftrightarrow$ is two-valued) Moreover, to make $\varphi \leftrightarrow \psi$ logically equivalent to $(\varphi \rightarrow \psi) \wedge (\psi \rightarrow \varphi)$, Przymusinski's interpretation ([17]) of $\rightarrow$ is required. It is also two-valued and gives the value $\underline{f}$ exactly in the following three cases: $\underline{t} \rightarrow \underline{f}$, $\underline{t} \rightarrow \underline{u}$ and $\underline{u} \rightarrow \underline{f}$. Equality is two-valued also. Following [3], it is easy to see that the above three-valued logic satisfies (as classical first-order logic does) all of the basic *metalogical properties*, in particular completeness and compactness.

A three-valued Σ-structure, $\mathcal{A}$, consists of a universe of values A, and an interpretation of each function symbol by a total function (of adequate arity), and of each predicate symbol by a total function on the set of the three boolean values $\{\underline{t}, \underline{f}, \underline{u}\}$ (i.e., a partial relation). Hence, terms cannot be undefined, but atoms can be interpreted as $\underline{u}$. Mod$_\Sigma$ denotes the set of all three-valued Σ-structures. A Σ-structure $\mathcal{A} \in$ Mod$_\Sigma$ is a model of (or satisfies) a set of sentences Φ if, and only if, $\mathcal{A}(\varphi) = \underline{t}$ for any sentence $\varphi \in \Phi$. This is also denoted by $\mathcal{A} \models \Phi$. We will denote by $\mathcal{A} \models_\sigma \Phi$ that $\mathcal{A}$ satisfies the sentence $\sigma(\Phi)$, resulting from the valuation $\sigma : free(\Phi) \rightarrow A$ of the formula Φ. Given a set Φ of Σ-sentences Mod$_\Sigma(\Phi)$ is the subclass of Mod$_\Sigma$ formed by the models of Φ. Logical consequence $\Phi \models \varphi$ means that $\mathcal{A} \models \varphi$ holds for all $\mathcal{A} \in$ Mod$_\Sigma(\Phi)$. Two Σ-structures $\mathcal{A}$ and $\mathcal{B}$ are *elementarily equivalent*, denoted $\mathcal{A} \simeq \mathcal{B}$ if $\mathcal{A}(\varphi) = \mathcal{B}(\varphi)$ for each first-order Σ-sentence φ. We denote by $EQ(\mathcal{A})$ the set of all Σ-structures that are elementarily equivalent to $\mathcal{A}$.

A Σ-*theory* is a set of Σ-sentences closed under logical consequence. A theory can be presented *semantically* or *axiomatically*. A semantic presentation is a class C of Σ-structures. Then, the theory semantically presented by C is the set of all Σ-sentences which are satisfied by C:

$$Th(C) = \{\varphi \in Sent_\Sigma \mid for\ all\ \mathcal{A} \in C\ \mathcal{A}(\varphi) = \underline{t}\}$$

An *axiomatic* presentation is a decidable set of axioms $Ax \subseteq Sent_\Sigma$. Then, the theory axiomatically presented by Ax is the set of all logical consequences of Ax:

$$Th(Ax) = \{\varphi \in Sent_\Sigma \mid Ax \models \varphi\}$$

A Σ-theory $\mathcal{T}$ is said to be *complete* if $\varphi \in \mathcal{T}$ or $\neg\varphi \in \mathcal{T}$ holds for each Σ-sentence φ.

2.1 Constraint Domains

A constraint logic program can be seen as a program where some function and predicate symbols have a predefined meaning on a given domain, called the constraint domain. In particular, according to the standard approach for defining the class of $CLP(X)$ programs ([10], [11]), a constraint domain X consists of five parts $(\Sigma_X, \mathcal{L}_X, Ax_X, \mathcal{D}_X, solv_X)$, where $\Sigma_X = (FS_X, PS_X)$ is the constraint signature, i.e., the set of symbols that are considered to be predefined; $\mathcal{L}_X$ is the constraint language, i.e., the class of Σ_X-formulas that can be used in programs; $\mathcal{D}_X$ is the domain of computation, i.e., a model defining the semantics of the symbols in Σ_X; Ax_X is an axiomatization of the domain, i.e., a decidable set of Σ_X-sentences such that $\mathcal{D}_X \models Ax_X$; and, finally, $solv_X$ is a constraint solver, i.e., an oracle that answers queries about constraints and that is used for defining the operational semantics of programs. In general, constraint solvers are expected to solve constraints, i.e., given a constraint c, one would expect that the solver will provide the values that satisfy the constraint or that it returns an equivalent constraint in *solved form*. However, in our case, we just need the solver to answer (un)satisfiability queries. We consider that, given a constraint c, $solv_X(c)$ may return F, meaning that c is not satisfiable or it may answer T, meaning that c is valid in the constraint domain, i.e., that $\neg c$ is unsatifiable. The solver may also answer U meaning that either the solver does not know the right answer or that the constraint is neither valid nor unsatifiable.

In addition, a constraint domain X must satisfy:

- T, F, $t_1 = t_2 \in \mathcal{L}_X$ (hence the equality symbol $=$ belongs to PS_X) and $\mathcal{L}_X$ is closed under variable renaming, existential quantification and conjunction. Moreover, the equality symbol $=$ is interpreted as the equality in $\mathcal{D}_X$, and Ax_X includes the equality axioms for $=$.
- The solver does not take variable names into account, that is, for all renamings ρ, $solv_X(c) = solv_X(\rho(c))$
- $Ax_X, \mathcal{D}_X$ and $solv_X$ agree in the sense that:
  1. $\mathcal{D}_X$ is a model of Ax_X.
  2. For all $c \in \mathcal{L}_X \cap Sent_{\Sigma_X}$: $solv_X(c) = \text{T} \Rightarrow Ax_X \models c$.
  3. For all $c \in \mathcal{L}_X \cap Sent_{\Sigma_X}$: $solv_X(c) = \text{F} \Rightarrow Ax_X \models \neg c$.

Moreover, $solv_X$ must be well-behaved, i.e., for any constraints c_1 and c_2:

1. $solv_X(c_1) = solv_X(c_2)$ if $\models c_1 \leftrightarrow c_2$.
2. If $solv_X(c_1) = \text{F}$ and $\models c_1 \leftarrow c_2^{\exists \setminus free(c_1)}$ then $solv_X(c_2) = \text{F}$.

In what follows, a constraint domain $X = (\Sigma_X, \mathcal{L}_X, Ax_X, \mathcal{D}_X, solv_X)$ will be called a $(\Sigma_X, \mathcal{L}_X)$-constraint domain.

2.2 Constraint Logic Programs

A constraint logic program over a $(\Sigma_X, \mathcal{L}_X)$-constraint domain X can be seen as a generalization of a definite logic program. In particular, a constraint logic program consists of rules $p : - q_1, ..., q_n \square c_1, ..., c_m$, where each q_i is an atom and each c_i is a constraint in

$\mathcal{L}_X$, and where atoms have the form $q(t_1, \ldots, t_n)$, where q is a user-defined predicate and $t_1, \ldots, t_n$ are terms over Σ_X. A program rule can be written, equivalently, in flat form

$$p(X_1, \ldots, X_n) : - q_1, \ldots, q_n \Box c_1, \ldots, c_m, X_1 = t_1, \ldots X_n = t_n$$

where $X_1, \ldots, X_n$ are fresh new variables. In what follows we will assume that constraint logic programs consist only of flat rules.

The semantics of a $(\Sigma_X, \mathcal{L}_X)$-logic program P can be also seen as a generalization of the semantics of a (non-constrained) logic program. In particular, in [10,11], the meaning of P is given in terms of the usual three kinds of semantics.

The *operational semantics* is defined in terms of finite or infinite derivations $S_1 \rightsquigarrow S_2 \rightsquigarrow \ldots \rightsquigarrow S_n \ldots$, where the states S_i in these derivations are tuples $G_i \Box C_i$, where G_i is a goal (i.e., a sequence of atoms) and C_i is a sequence of constraints (actually a constraint, since constraints are closed under conjunction). In particular, from a state $S = G \Box C$ we can derive the state $S' = G' \Box C'$ if there is a rule $p(X_1, \ldots, X_n) : - G_0 \Box C_0$, and an atom $p(t_1, \ldots, t_n)$ in G, where $X_1, \ldots, X_n$ are fresh new variables not occurring in $G \Box C$, such that $G' = < G_0, (G \backslash p(t_1, \ldots, t_n)) >$ and $C' = < C, C_0, X_1 = t_1, \ldots X_n = t_n >$ is satisfiable. Then, given a derivation $S_1 \rightsquigarrow S_2 \rightsquigarrow \ldots \rightsquigarrow S_n$, with $S_n = G_n \Box C_n$, we say that C_n is an answer to the query $S_1 = G_1 \Box C_1$ if G_n is the empty goal.

The *logical semantics* of P is defined as the theory presented by $P \cup Ax_X$.

Finally its *algebraic semantics*, $M(P, X)$, is defined as the least model of P extending $\mathcal{D}_X$, in the sense that this model agrees with $\mathcal{D}_X$ in the corresponding universe of values and in the interpretation of the symbols in Σ_X. It may be noted that Σ-structures extending $\mathcal{D}_X$ can be seen as subsets of $Base_P(\mathcal{D}_X)$, where $Base_P(\mathcal{D}_X)$ is the set of all atoms of the form $p(\alpha_1, \ldots, \alpha_n)$, where p is a user-defined predicate and $\alpha_1, \ldots, \alpha_n$ are values in $\mathcal{D}_X$. As in the standard case, the algebraic semantics of P can be defined as the least fixpoint of the immediate consequence operator $T_P^X : 2^{Base_P(\mathcal{D}_X)} \to 2^{Base_P(\mathcal{D}_X)}$ defined as follows:

$$T_P^X(I) = \{\sigma(p) \mid \sigma : free(p) \to \mathcal{D}_X \text{ is a valuation, } (p : - \bar{a} \Box c) \in P, I \models_\sigma \bar{a} \text{ and } \mathcal{D}_X \models_\sigma c\}$$

In [11] it is proved that the above three semantics are equivalent in the sense that:

- The operational semantics is sound with respect to the logical semantics. That is, if a goal G has answer c then $P \cup Ax_X \models c \to G$.
- The operational semantics is also sound with respect to the algebraic semantics. That is, if a goal G has answer c then $M(P, X) \models c \to G$.
- The operational semantics is complete with respect to the logical semantics. That is, if $P \cup Ax_X \models c \to G$, then G has answers $c_1, \ldots, c_n$ such that $Ax_X \models c \leftrightarrow c_1 \vee \ldots \vee c_n$.
- The operational semantics is complete with respect to the algebraic semantics. That is, if $M(P, X) \models_\sigma G$, where $\sigma : free(G) \to \mathcal{D}_X$ is a valuation, then G has an answer c such that $\mathcal{D}_X \models_\sigma c$

2.3   A Functorial Semantics for Constraint Logic Programs

The semantic definitions sketched in the previous subsection are, in our opinion, not fully satisfactory. On one hand, a constraint logic program can be seen as a logic program parameterized by the constraint domain. Then, we think that its semantics should also be parameterized by the domain. This is not explicit in the semantics sketched above. On the other hand, we think that the formulation of some of the previous equivalence results could be found to be, in some sense, not fully satisfactory. Let us consider, for instance, the last result, i.e., the completeness of the operational semantics with respect to the algebraic semantics. In our opinion, a fully satisfactory result would have said something like:

if $M(P,X) \models_\sigma G$ where $\sigma: free(G) \to \mathcal{D}_X$ is a valuation, then G has an answer c such that $solv_X(c) \neq \mathbf{F}$

However this property will not hold unless the constraint solver $solv_X$ is also complete with respect to the computation domain.

In our opinion, each of the three semantics (logical, algebraic and operational semantics) of a constraint logic program should be some kind of mapping. Moreover, we can envision that the parameters of the logical definitions would be constraint theories. Similarly, the parameters for algebraic definitions would be computation domains. Finally, the parameters for the operational definitions would be constraint solvers. In this context, proving the soundness and completeness of one semantics with respect to another one would mean comparing the corresponding mappings. In particular, a given semantics would be sound and complete with respect to another one if the two semantic mappings are in some sense equivalent. On the other hand, we believe that these mappings are better studied if the given domains and codomains are not just sets or classes but categories, which means taking care of their underlying structure. As a consequence, these mappings would be defined as functors and not just as plain set-theoretic functions.

In Section 5 the above ideas are fully developed for the case of constraint normal logic programs. The case of constraint logic programs can be seen as a particular case.

3 Domain Constraints for Constraint Normal Logic Programs

In this section, we provide a notion of constraint domain for constraint normal logic programming. The idea, as discussed in the introduction, is that this notion, together with a proper adaptation of the semantic constructions used for (unconstrained) normal logic programs, will provide an adequate semantic definition for constraint normal logic programs. In particular, the idea is that the logical semantics of a program should be given in terms of the (3-valued) Clark-Kunen completion of the program, the operational semantics in terms of some form of constructive negation [19,5,6], and the algebraic semantics in terms of some form of fixpoint construction (as in [19,6,15]).

The main problem is that a straightforward extension (as it may be just the inclusion of negated atoms in the constraint languages) of the notion of constraint domain introduced in Subsection 2.1 will not work, as the following example shows.

Example 1 *Let P be the CLP($\mathcal{N}$) program:*

$q(z) : - \Box z = 0$
$q(v) : - q(x) \Box v = x + 1$

and assume that its logical semantics is given by its completion:

$$\forall z(q(z) \leftrightarrow (z = 0 \lor \exists x(q(x) \land v = x+1))).$$

This means, obviously, that $q(n)$ should hold for each n. Actually, the model defined by the algebraic semantics seen in Subsection 2.1 would satisfy $\forall z q(z)$. Now consider that P is extended by the following definitions:

$r : - \neg q(x)$
$s : - \neg r$

whose completion is:

$$(r \leftrightarrow \exists x(\neg q(x))) \land (s \leftrightarrow \neg r).$$

Now, the operational semantics, and also the ω-iteration of the Fitting's operator [7], would correspond to a three-valued structure extending $I\!N$, where both r and s are undefined and where, as before, $q(n)$ holds for each n. Unfortunately, such a structure would not be a model of the completion of the program since this structure satisfies $\forall z q(z)$ but it does not satisfy either $\neg r$ or s. ∎

The problem with the example above is that, if the algebraic semantics is defined by means of the ω-iteration of an immediate consequence operator, then, in many cases, the resulting structure would not be a model of the completion of the program. Otherwise, if we define the algebraic semantics in terms of some least (with respect to some order relation) model of the completion extending $I\!N$, then, in many cases, the operational semantics would not be complete with respect to that model. Actually, in some cases this model could be non (semi-)computable ([2], [8]).

In our opinion, the problem is related to the following observation. Let us suppose that $X = (\Sigma, L, Ax, D, solv)$ and $X' = (\Sigma, L, Ax, D', solv)$ are two constraint domains that only differ in their domains of computation, D and D', which are elementarily equivalent. Now, a program defined over any of these domains would show exactly the same behaviour, since both algebras satisfy exactly the same constraints, i.e., we may consider that two structures that are elementarily equivalent should be considered indistinguishable as domains of computation for a given constraint domain. As a consequence, we may consider that the semantics of a program over two indistinguishable constraint domains should also be indistinguishable. However, if P is a (Σ, L)-program, then $M(P, X)$ and $M(P, X')$ are not necessarily elementarily equivalent. In particular if we consider the program P of Example 1 and we consider as constraint domain a non-standard model of the natural numbers $I\!N'$, then we would have that $M(P, I\!N) \models \forall z q(z)$ but $M(P, I\!N') \not\models \forall z q(z)$.

We think that this problem is caused by considering that the domain of computation, $\mathcal{D}_X$, of a constraint domain is a single structure. In the case of programs without negation this apparently works fine and it seems quite reasonable from an intuitive point of view. For instance, if we are writing programs over the natural numbers, it seems reasonable to think that the computation domain is the algebra of natural numbers. However, when dealing with negation, we think that the computation domain of a constraint domain should be defined in terms of the class of all the structures which are elementarily equivalent to a given one. To be precise, we reformulate the notion of constraint domain as follows:

Definition 2 *A $(\Sigma_X, \mathcal{L}_X)$-constraint domain X is a 5-tuple $(\Sigma_X, \mathcal{L}_X, Ax_X, Dom_X, solv_X)$, where $\Sigma_X = (FS_X, PS_X)$ is the constraint signature, $\mathcal{L}_X$ is the constraint language, $Dom_X = EQ(\mathcal{D}_X)$ is the domain of computation, i.e., the class of all Σ_X-structures which are elementarily equivalent to a given structure $\mathcal{D}_X$, Ax_X is a decidable set of Σ_X-sentences such that $\mathcal{D}_X \models Ax_X$, and $solv_X$ is a constraint solver, such that:*

- *T, F, $t_1 = t_2 \in \mathcal{L}_X$ (hence the equality symbol $=$ belongs to PS_X) and $\mathcal{L}_X$ is closed under variable renaming, existential quantification, conjunction and negation. Moreover, the equality symbol $=$ is interpreted as the equality in Dom_X and Ax_X includes the equality axioms for $=$.*
- *The solver does not take variable names into account, that is, for all variable renamings ρ, $solv_X(c) = solv_X(\rho(c))$*
- *Ax_X, Dom_X and $solv_X$ agree in the sense that:*
    1. *$\mathcal{D}_X$ is a model of Ax_X.*
    2. *For all $c \in \mathcal{L}_X \cap Sent_\Sigma$: $solv_X(c) = T \Rightarrow Ax_X \models c$.*
    3. *For all $c \in \mathcal{L}_X \cap Sent_\Sigma$: $solv_X(c) = F \Rightarrow Ax_X \models \neg c$.*

As before, we assume that $solv_X$ is well-behaved, i.e., for any constraints c_1 and c_2:

1. *$solv_X(c_1) = solv_X(c_2)$ if $\models c_1 \leftrightarrow c_2$.*
2. *If $solv_X(c_1) = F$ and $\models c_1 \leftarrow c_2^{\exists \setminus free(c_1)}$ then $solv_X(c_2) = F$.*

4   Semantic Constructions for Constraint Normal Logic Programs

Analogously to constraint logic programs, given a signature $\Sigma = (PS_\Sigma, FS_\Sigma)$, normal constraint logic Σ-programs over a constraint domain $X = (\Sigma_X, \mathcal{L}_X, Ax_X, Dom_X, solv_X)$, can be seen as a generalization of a normal logic programs. So, a Σ-program now consists of clauses of the form $a : - \ell_1, ..., \ell_m \Box c_1, ..., c_n$, where a and the ℓ_i, $i \in \{1, ..., m\}$, are a flat atom and flat literals, respectively, whose predicate symbols belong to $PS_\Sigma \setminus PS_X$ and the c_j, $j \in \{1, ..., n\}$ belong to $\mathcal{L}_X$. As before, we also assume that all clauses defining the same predicate p have exactly the same head $p(X_1, ..., X_m)$.

4.1   Logical Semantics

The standard logical meaning of a Σ-program P is its (generalized) Clark's completion $Comp_X(P) = Ax_X \cup P^*$, where P^* includes a sentence

$$\forall \bar{z}(q(\bar{z}) \leftrightarrow ((G_1 \wedge c_1)^{\exists \setminus \bar{z}} \vee ... \vee (G_k \wedge c_k)^{\exists \setminus \bar{z}}))$$

for each $q \in PS_\Sigma \setminus PS_X$, and where $\{(q(\bar{z}) : - G_1 \Box c_1), ..., (q(\bar{z}) : - G_k \Box c_k)\}$ is the set [3] of all the clauses in P with head predicate q. In what follows, this set will be denoted by $Def_P(q)$. Intuitively, in this semantics we are considering that $Def_P(q)$ is a *complete*

---

[3] If the set is empty, then the above sentence is simplified to $\forall \bar{z}(q(\bar{z}) \leftrightarrow F$

definition of the predicate q. A weaker logical meaning for the program P is obtained by defining its semantics as $Ax_X \cup P^\forall$, where $P^\forall$, is the set including a sentence

$$\forall \bar{z}(q(\bar{z}) \leftarrow ((G_1 \wedge c_1)^{\exists \setminus \bar{z}} \vee \ldots \vee (G_k \wedge c_k)^{\exists \setminus \bar{z}}))$$

for each $q \in PS_\Sigma \setminus PS_X$.

4.2 The *BCN* Operational Semantics

In this section we generalize the *BCN* operational semantics introduced in [16] and refined in [1] in such a way that it can be used for any constraint domain. The *BCN* operational semantics is based on two operators originally introduced by Shepherdson [18] to characterize Clark-Kunen's semantics in terms of satisfaction of (equality) constraints. Such operators exploit the definition of literals in the completion of programs and associate a constraint formula to each query. As a consequence, the answers are computed, on one hand, by a symbolic manipulation process that obtains the associated constraint(s) of the given query and, on the other hand, by a constraint checking process that deals with such constraint(s). In particular, the original version ([16]) of the *BCN* operational semantics works with programs restricted to the constraint domain of terms with equality.

Definition 3 *For any program P, the operators T_k^P and F_k^P associate a constraint to each query, as follows:*

Let $Def_P(q) = \{q(\bar{z}) : - \bar{\ell}_i \Box c_i \mid 1 \le i \le m\}$

$$T_0^P(q(\bar{z})) = \mathbf{F} \qquad T_{k+1}^P(q(\bar{z})) = \bigvee_{i=1}^m \exists \bar{y}^i (c_i \wedge T_k^P(\bar{\ell}_i))$$

$$F_0^P(q(\bar{z})) = \mathbf{F} \qquad F_{k+1}^P(q(\bar{z})) = \bigwedge_{i=1}^m \forall \bar{y}^i (\neg c_i \vee F_k^P(\bar{\ell}_i))$$

For all $k \in \mathbb{N}$:

$$T_k^P(\mathbf{T}) = \mathbf{T} \qquad\qquad F_k^P(\mathbf{T}) = \mathbf{F}$$

$$T_k^P(\neg q(\bar{z})) = F_k^P(q(\bar{z})) \qquad F_k^P(\neg q(\bar{z})) = T_k^P(q(\bar{z}))$$

$$T_k^P(\bigwedge_{j=1}^n \ell_j) = \bigwedge_{j=1}^n T_k^P(\ell_j) \qquad F_k^P(\bigwedge_{j=1}^n \ell_j) = \bigvee_{j=1}^n F_k^P(\ell_j)$$

For any $c \in L_X$, for any $k \in \mathbb{N}$:

$$T_k^P(c) = c \qquad F_k^P(c) = \neg c$$

Definition 4 *Let P be a program and $solv_X$ a constraint solver. A $BCN(P, solv_X)$-derivation step is obtained by applying the following derivation rule:*

(R) $\bar{\ell}_1, \bar{\ell}_2 \Box d$ is $BCN(P, solv_X)$-derived from $\bar{\ell}_1, \ell(\bar{x}), \bar{\ell}_2 \Box c$ if there exists $k > 0$ such that $d = T_k^P(\ell(\bar{x})) \wedge c$ and $solv_X(d^\exists) \ne \mathbf{F}$.

Definition 5 *Let P be a program and $solv_X$ a constraint solver.*

1. *A $BCN(P, solv_X)$-derivation from the query L is a succession of $BCN(P, solv_X)$-derivation steps of the form $L \leadsto_{(P, solv_X)} \cdots \leadsto_{(P, solv_X)} L'$. Then, $L \overset{n}{\leadsto}_{(P, solv_X)} L'$ means that the query L' is $BCN(P, solv_X)$-derived from the query L in n $BCN(P, solv_X)$-derivation steps.*

2. *A finite $BCN(P, solv_X)$-derivation $L \overset{n}{\leadsto}_{(P, solv_X)} L'$ is a successful derivation if $L' = \Box c$. In this case, $c^{\exists \backslash free(L)}$ is the corresponding $BCN(P, solv_X)$-computed answer.*

3. *A query $L = \bar{\ell} \Box c$ is a $BCN(P, solv_X)$-failed query if $solv_X((c \to F_k^P(\bar{\ell}))^\forall) = \mathbf{T}$ for some $k > 0$ such that $solv_X(F_k^P(\bar{\ell})^\forall) \neq \mathbf{F}$.*

A *selection rule* is a function selecting a literal in a query and, whenever *Solvx* is well-behaved, $BCN(P, solv_X)$ is independent of the selection rule used. To prove this assertion we follow the strategy used in [14,11], so we first prove the next lemma.

Lemma 6 (Switching Lemma) *Let P be a program and Solvx be a well-behaved solver. Let L be a query, ℓ_1, ℓ_2 be literals in L and let $L \leadsto_{(P, solv_X)} L_1 \leadsto_{(P, solv_X)} L'$ be a non-failed derivation in which ℓ_1 has been selected in L and ℓ_2 in L_1. Then there is a derivation $L \leadsto_{(P, solv_X)} L_2 \leadsto_{(P, solv_X)} L''$ in which ℓ_2 has been selected in L and ℓ_1 in L_2, and L' and L'' are identical up to reordering of their constraint component.*

Theorem 7 (Independence of the selection rule) *Let P be a program and $solv_X$ a well-behaved solver. Let L be a query and suppose that there exists a successful $BCN(P, solv_X)$-derivation from L with computed answer c. Then, using any selection rule R there exists another successful $BCN(P, solv_X)$-derivation from L of the same length with an answer which is a reordering of c.*

Next, we establish the basis for relating the $BCN(P, solv_X)$ operational semantics to the logical semantics of a particular class of constraint logic programs. The proposition below provides the basis for proving soundness and completeness of the semantics.

Proposition 8 *Let $\Sigma = (FS_X, PS_X \cup PS)$ be an extension of a given signature of constraints $\Sigma_X = (FS_X, PS_X)$ by a set of predicates PS, and let P be a Σ-program. For each Σ_X-theory of constraints Ax_X, each conjunction of Σ-literals $\bar{\ell}$ and each k in $\mathbb{N}$:*

$$P^* \cup Th(Ax_X) \models (T_k^P(\bar{\ell}) \to \bar{\ell})^\forall$$

4.3 Fixpoint Semantics

According to what is argued in Section 3, we consider the domain $(Dom_\Sigma/_\equiv, \preceq)$ for computing immediate consequences defined as follows: Let Dom_Σ be the class of three-valued Σ-interpretations which are extensions of models in Dom_X. Then, as it is done in [19] to extend [13] to the general constraint case, we consider the Fitting's ordering on Dom_Σ interpreted in the following sense: For all partial interpretations $\mathcal{A}, \mathcal{B} \in Dom_\Sigma$, for each Σ_X-constraint $c(\bar{x})$ and each Σ-literal $\ell(\bar{x})$:

$$\mathcal{A} \preceq \mathcal{B} \quad iff \quad \mathcal{A}((c \to \ell)^\forall) = \underline{t} \Rightarrow \mathcal{B}((c \to \ell)^\forall) = \underline{t}$$

It is quite easy to see that $(Dom_\Sigma, \preceq)$ is a preorder. Therefore, we consider the equivalence relation $\equiv$ induced by $\preceq$ ($\mathcal{A} \equiv \mathcal{B}$ if, and only if, $\mathcal{A} \preceq \mathcal{B}$ and $\mathcal{B} \preceq \mathcal{A}$), and the induced partial order

$$[\mathcal{A}], [\mathcal{B}] \in Dom_\Sigma/_\equiv : [\mathcal{A}] \preceq [\mathcal{B}] \ iff \ \mathcal{A} \preceq \mathcal{B}$$

to build a cpo $(Dom_\Sigma/_\equiv, \preceq)$ with a bottom class $[\bot_\Sigma]$ such that for each $\mathcal{A} \in [\bot_\Sigma]$ we have that $\mathcal{A}((c \to \ell)^\vee) \neq \underline{t}$ for all Σ_x-constraint $c(\bar{x})$ and all Σ-literal $\ell(\bar{x})$. That is, the set of goals of the form $(c \to \ell)^\vee$ satisfied by the models in $[\bot_\Sigma]$ is empty.

Proposition 9 $(Dom_\Sigma/_\equiv, \preceq)$ is a cpo with respect to $\preceq$, and, the equivalence class $[\bot_\Sigma]$ is its bottom element.

Definition 10 (Immediate consequence operator $T_P^{Dom_X}$**)** Let P be a Σ-program, then the immediate consequence operator $T_P^{Dom_X} : Dom_\Sigma/_\equiv \to Dom_\Sigma/_\equiv$ is defined for each $[\mathcal{A}] \in Dom_\Sigma/_\equiv$, as

$$T_P^{Dom_X}([\mathcal{A}]) = [\Phi_P^{\mathcal{D}_X}(\mathcal{A})]$$

where $\mathcal{D}_X$ is the distinguished domain model in the class Dom_X, $\mathcal{A}$ is any model in $[\mathcal{A}]$, and $[\Phi_P^{\mathcal{D}_X}(\mathcal{A})]$ is the $\equiv$-class of models such that for each Σ_x-constraint $c(\bar{x})$ and each Σ-atom $p(\bar{x})$,

(i) $\Phi_P^{\mathcal{D}_X}(\mathcal{A})((c \to p)^\vee) = \underline{t}$ if, and only if, there are (renamed versions of) clauses $\{p(\bar{x}) : -\ell_1^i, \ldots, \ell_{n_i}^i \Box d_i \mid 1 \le i \le m\} \subseteq Def_P(p)$ and $\mathcal{D}_X$-satisfiable constraints $\{c_j^i \mid 1 \le i \le m \wedge 1 \le j \le n_i\}$ such that

- $\mathcal{A}((c_j^i \to \ell_j^i)^\vee) = \underline{t}$
- $\mathcal{D}_X((c \to \bigvee_{1 \le i \le m} \exists \bar{y}_i (\bigwedge_{1 \le j \le n_i} c_j^i \wedge d_i))^\vee) = \underline{t}$

(ii) $\Phi_P^{\mathcal{D}_X}(\mathcal{A})((c \to \neg p)^\vee) = \underline{t}$ if, and only if, for each (renamed version) clause in $\{p(\bar{x}) : -\ell_1^i, \ldots, \ell_{n_i}^i \Box d_i \mid 1 \le i \le m\} = Def_P(p(\bar{x}))$ there is a $J_i \subseteq \{1, \ldots n_i\}$ and $\mathcal{D}_X$-satisfiable constraints $\{c_j^i \mid 1 \le i \le m \wedge j \in J_i\}$ such that

- $\mathcal{A}((c_j^i \to \neg \ell_j^i)^\vee) = \underline{t}$
- $\mathcal{D}_X((c \to \bigwedge_{1 \le i \le m} \forall \bar{y}_i (\bigvee_{j \in J_i} c_j^i \vee \neg d_i))^\vee) = \underline{t}$

where, for each $i \in \{1, \ldots, m\}$, $\bar{y}_i$ are the free variables in $\{\ell_1^i, \ldots, \ell_{n_i}^i, d_i\}$ not in $\bar{x}$.

Remark 11 In the definition of the operator $\Phi_P^{\mathcal{D}_X}$, we could choose any other model in Dom_X, instead of $\mathcal{D}_X$, since all of them are elementarily equivalent, and the domain is just used for constraint satisfaction checking. Similarly, $\mathcal{A}$ could be any other model in $[\mathcal{A}]$ since it is used for checking satisfaction of sentences of the form $(c \to \ell)^\vee$. Moreover, models in a $\equiv$-class $[\Phi_P^{\mathcal{D}_X}(\mathcal{A})]$ are elementarily equivalent in its restrictions to Σ_X. In fact, $[\Phi_P^{\mathcal{D}_X}(\mathcal{A})]|_{\Sigma_X} = Dom_X$ since, all classes in Dom_Σ are (conservative) predicative extensions of Dom_X and, the operator $T_P^{Dom_X}$ does not compute new consequences from $\mathcal{L}_X$.

In what follows we will prove that $\mathcal{T}_P^{Dom_X}$ is continuous in the cpo $Dom_\Sigma/_\equiv$. As a consequence, it has an effectively computable least fixpoint.

Theorem 12 $\mathcal{T}_P^{Dom_X}$ *is continuous in the cpo* $(Dom_\Sigma/_\equiv, \preceq)$, *so it has a least fixpoint* $\mathcal{T}_P^{Dom_X} \uparrow \omega = \bigsqcup [\Phi_P^{\mathcal{D}_X} \uparrow n]$.

However, it is important to notice that, as we will show in example 13,

$$\bigsqcup [\Phi_P^{\mathcal{D}_X} \uparrow n] \neq [\Phi_P^{\mathcal{D}_X} \uparrow \omega]$$

In fact, the operator $\Phi_P^{\mathcal{D}_X}$ can be considered a variant of the Stuckey's immediate consequence operator in [19], so, it inherits its drawbacks. On one hand, $\Phi_P^{\mathcal{D}_X}$ is monotonic but not continuous. On the other hand, it will have different behavior depending on the constraint domain in Dom_X that may be predicatively extended.

Example 13 *Consider the* $CNLP(\mathcal{N})$*-program from example 1:*

$q(z) : - \Box z = 0$
$q(v) : - q(x) \Box v = x + 1$
$r : - \neg q(x)$

First, let us look at the behaviour of the operator Φ:

- $\Phi_P^{\mathcal{N}} \uparrow \omega$ *would be the model extending* $\mathcal{N}$ *where* r *is undefined and all the sentences* $\{(z = n \to q(z))^\vee | n \geq 0\}$ *are true, so, the sentence* $\forall z.q(z)$ *will be evaluated as true in* $\Phi_P^{\mathcal{N}} \uparrow \omega$. *This is not a fixpoint since we can iterate once more, to obtain a different model* $\Phi_P^{\mathcal{N}} \uparrow (\omega + 1)$ *where* $\neg r$ *is true.*
- *In contrast, if we consider any non-standard model* $\mathcal{M}$ *elementarily equivalent to* $\mathcal{N}$, *the sentence* $\forall z.q(z)$ *will be evaluated as undefined in* $\Phi_P^{\mathcal{M}} \uparrow \omega$, *so, no more consequences will be obtained if we iterate once more.*

Now we can compare with the behaviour of $\mathcal{T}$:

Similar to the first case, $\mathcal{T}_P^{EQ(\mathcal{N})} \uparrow \omega$ *is the class of* $\equiv$-*equivalent models extending* $EQ(\mathcal{N})$, *where* r *is undefined and all the sentences*

$$\{(z = n \to q(z))^\vee | n \geq 0\}$$

are true. But now, this is a fixpoint in contrast to what happens with any other operator working over just one standard model. In particular, it is not difficult to see that the sentence $\forall z.q(z)$ *is never satisfied (by models) in* $[\Phi_P^{\mathcal{N}} \uparrow k]$ *for any* k. *This is because we are considering also non standard models (as the predicative extension of the above* $\mathcal{M}$) *at each iteration. Therefore, as a consequence of the definition of* $\bigsqcup$, *we have that* $\forall z.q(z)$ *is not satisfied in*

$$\mathcal{T}_P^{EQ(\mathcal{N})} \uparrow \omega = \bigsqcup [\Phi_P^{\mathcal{N}} \uparrow k]$$

Finally, as a consequence of the continuity of $\mathcal{T}_P^{Dom_X}$, we can extend a result from Stuckey [19] related to the satisfaction of the logical consequences of the completion in any ordinal iteration of $\Phi_P^{\mathcal{D}_X}$, until the ω iteration of $\mathcal{T}_P^{Dom_X}$ (its least fixpoint):

Theorem 14 (Extended Theorem of Stuckey)
Let $Th(Dom_X)$ be the complete theory of Dom_X. For each Σ-goal $\bar{\ell}_\Box c$:

1. $P^* \cup Th(Dom_X) \models_3 (c \rightarrow \bar{\ell})^\forall \Leftrightarrow \forall \mathcal{A} \in \mathcal{T}_P^{Dom_X} \uparrow \omega : \mathcal{A}((c \rightarrow \bar{\ell})^\forall) = \underline{t}$
2. $P^* \cup Th(Dom_X) \models_3 (c \rightarrow \neg\bar{\ell})^\forall \Leftrightarrow \forall \mathcal{A} \in \mathcal{T}_P^{Dom_X} \uparrow \omega : \mathcal{A}((c \rightarrow \neg\bar{\ell})^\forall) = \underline{t}$

5 Functorial Semantics

As introduced in Subsection 2.3, one basic idea in this work is to formulate the constructions associated to the definition of the operational, least fixpoint and logical semantics of constraint normal logic programs in functorial terms. This allows us to separate the study of the properties satisfied by these three semantic constructions, from the classic comparisons of three kinds of semantics of programs over a specific constraint domain. Moreover, once the equivalence of semantic constructions is (as intended) obtained, the classical *completeness* results that can be obtained depending on the relations among solvers, theories and domains, are just consequences of the functorial properties.

Comparing these semantic functors is not straightforward since, intuitively, their domains and codomains are different categories. We can see the logical semantics of a $(\Sigma_X, \mathcal{L}_X)$-constraint logic program P as a mapping (a functor), let us denote it by $\mathcal{LOG}_P$, whose arguments are logical theories and whose results are also logical theories. The algebraic semantics of P, denoted $\mathcal{ALG}_P$, can be seen as a functor that takes as arguments logical structures and returns as results logical structures. Finally, the operational semantics of P, denoted $\mathcal{OP}_P$ can be considered to take as arguments constraint solvers and return as results (for instance) interpretations of computed answers.

We solve this problem by representing all the semantic domains involved in terms of sets of formulas. This is a quite standard approach in the area of Logic Programming where, for instance, (finitely generated) models are often represented as Herbrand structures (i.e., as classes of ground atoms) rather than as algebraic structures. One could criticize this approach in the framework of constraint logic programming, since a class does not faithfully represents a single model (the constraint domain of computation Dom_X) but a class of models. However, we have argued previously that, when dealing with negation, a constraint domain of computation should not be a single model, but the class of models which are elementarily equivalent to Dom_X. In this sense, one may note that a class of elementarily equivalent models is uniquely represented by a complete theory. However, since we are dealing with three-valued logic, we are going to represent model classes, theories and solvers as pairs of sets of sentences, rather than just as single sets.

In what follows, we present the categorical setting required for our purposes. Being more precise, first of all, we need to define the categories associated to solvers, computation domains and theories (axiomatizable domains). Then, we will define the category which properly represents the semantics of programs. Finally, we will define the three functors that respectively represent the operational, logical and algebraic semantics of a constraint normal logic programs.

Definition 15 *Given a signature* Σ_X, *a* Σ_X-*pre-theory* $\mathcal{M}$ *is a pair of sets of* Σ_X-*sentences* $(\mathcal{M}_{\underline{t}}, \mathcal{M}_{\underline{f}})$.

Remarks and Definitions 16

1. *Given a solver* $solv_X$ *of a given language* $\mathcal{L}_X$ *of* Σ_X-*constraints, we will denote by* $\mathcal{M}_{solv_X}$ *the pre-theory associated to* $solv_X$, *i.e., the pair* $(\mathcal{M}_{\underline{t}}, \mathcal{M}_{\underline{f}})$ *where* $\mathcal{M}_{\underline{t}}$ *is the set of all constraints* $c \in \mathcal{L}_X$ *such that* $solv_X(c) = \textsc{t}$ *and* $\mathcal{M}_{\underline{f}}$ *is the set of all constraints* $c \in \mathcal{L}_X$ *such that* $solv_X(c) = \textsc{f}$.
2. *Similarly, given a set of axioms* Ax_X *of a given language* $\mathcal{L}_X$ *of* Σ_X-*constraints, we will denote by* $\mathcal{M}_{Ax_X}$ *the theory associated to* Ax_X.
3. *Finally, given a computation domain* Dom_X *of a given language* $\mathcal{L}_X$ *of* Σ_X-*constraints, we will denote by* $\mathcal{M}_{Dom_X}$ *the theory associated to* Dom_X, *i.e., the pair* $(\mathcal{M}_{\underline{t}}, \mathcal{M}_{\underline{f}})$ *where* $\mathcal{M}_{\underline{t}}$ *is the set of sentences satisfied by* Dom_X *and* $\mathcal{M}_{\underline{f}}$ *is the set of sentences which are false in* Dom_X. *Note that, since constraint domains are typically two-valued,* $\mathcal{M}_{\underline{t}}$ *would typically be a complete theory and, therefore,* $\mathcal{M}_{\underline{f}}$ *is the complement of* $\mathcal{M}_{\underline{t}}$.

For the sake of simplicity, given a pre-theory $\mathcal{M}$, *we will write* $\mathcal{M}(c) = \underline{t}$, *to mean* $c \in \mathcal{M}_{\underline{t}}$; $\mathcal{M}(c) = \underline{f}$, *to mean* $c \in \mathcal{M}_{\underline{f}}$; *and* $\mathcal{M}(c) = \underline{u}$, *otherwise.*

Now, according to the above ideas, we will define categories to represent constraint solvers, computation domains and domain axiomatizations. Also, following similar ideas we are going to define a category of semantic domains for programs. In this case, we will define the semantics in terms of sets of formulas. However, we will restrict ourselves to sets of answers, i.e., formulas with the form $c \to G$, where G is any goal.

Definition 17 (Categories for Constraint Domains and Program Interpretations)
Given a signature Σ_X *we can define the following categories:*

1. *The category of* Σ_X-*pre-theories,* $\underline{PreTh}_{\Sigma_X}$ *(or just* $\underline{PreTh}$ *if* Σ_X *is clear from the context) is defined as follows:*
   - *Its class of objects is the class of* Σ_X-*pre-theories.*
   - *For each pair of objects* $\mathcal{M}$ *and* $\mathcal{M}'$ *there is a morphism from* $\mathcal{M}$ *to* $\mathcal{M}'$, *noted just by* $\mathcal{M} \preceq_c \mathcal{M}'$, *if* $\mathcal{M}_{\underline{t}} \subseteq \mathcal{M}'_{\underline{t}}$ *and* $\mathcal{M}_{\underline{f}} \subseteq \mathcal{M}'_{\underline{f}}$
2. $\underline{Th}_{\Sigma_X}$ *(or just* $\underline{Th}$) *is the full subcategory of* $\underline{PreTh}_{\Sigma_X}$ *whose objects are theories.*
3. $\underline{CompTh}_{\Sigma_X}$ *(or just* $\underline{CompTh}$) *is the full subcategory of* $\underline{PreTh}_{\Sigma_X}$ *whose objects are complete theories*
4. *Given a constraint language* $\mathcal{L}_X$ *and a signature* Σ *extending* Σ_X, $\underline{ProgInt}^{\Sigma}_{(\Sigma_X, \mathcal{L}_X)}$ *(or just* $\underline{ProgInt}$ *if* Σ, Σ_X *and* $\mathcal{L}_X$ *are clear from the context) is the category where:*
   - *Its objects are sets of sentences* $(c \to \bar{\ell})^{\forall}$ *or* $(c \to \neg\bar{\ell})^{\forall}$, *where* $c \in \mathcal{L}_X$ *and* $\bar{\ell}$ *is a conjunction of* Σ-*literals.*

- *For each pair of objects A and A' there is a morphism from A to A', noted just by $A \preceq A'$ if $A \subseteq A'$*
  - $\Sigma_X \subseteq \Sigma_{X'}$ *and*
  - *for each $(FS_X, PS_X \cup PS)$-literal $\ell(\overline{x})$ and Σ_X-formula $c(\overline{x})$, $A((c \to \overline{\ell})^\vee) = \underline{t}$ implies $A'((c \to \overline{\ell})^\vee) = \underline{t}$, and $A((c \to \neg\overline{\ell})^\vee) = \underline{t}$ implies $A'((c \to \neg\overline{\ell})^\vee) = \underline{t}$.*

As pointed out before, this categorical formulation allows us to speak about relations among solvers, domains and theories by establishing morphisms among them in the common category *PreTh*, in such a way that the morphism between two objects represents the relation *"agrees with"* (or *completeness* if they are seen in the reverse sense). To be more precise, given a constraint domain $X = (\Sigma_X, L_X, Ax_X, Dom_X, solv_X)$, we can reformulate the conditions (in Section 2.1) required among $solv_X$, Dom_X and Ax_X as:

$$\mathcal{M}_{solv_X} \preceq_c \mathcal{M}_{Ax_X} \preceq_c \mathcal{M}_{Dom_X}$$

in *PreTh*. That is, since Dom_X must be a model of Ax_X, there is a morphism from $\mathcal{M}_{Ax_X}$ to $\mathcal{M}_{Dom_X}$. Moreover, since $solv_X$ must agree with Ax_X, there is a morphism from $\mathcal{M}_{solv_X}$ to $\mathcal{M}_{Ax_X}$. Then, by transitivity, $solv_X$ agrees with Dom_X, so there is a morphism $\mathcal{M}_{solv_X}$ to $\mathcal{M}_{Dom_X}$. In addition, we can also reformulate other conditions in these terms:

- $solv_X$ is Ax_X-complete (respectively, Dom_X-complete) if, and only if, $\mathcal{M}_{Ax_X} \preceq_c \mathcal{M}_{solv_X}$ (respectively, $\mathcal{M}_{Dom_X} \preceq_c \mathcal{M}_{solv_X}$).
- Ax_X completely axiomatizes Dom_X if, and only if, $\mathcal{M}_{Dom_X} \preceq_c \mathcal{M}_{Ax_X}$, so, as expected $\mathcal{M}_{Ax_X} = \mathcal{M}_{Dom_X}$.

Finally, we will define the three functors that represent, for a given program P, its operational, its algebraic or least fixpoint, and its logical semantics.

Definition 18 (Functorial semantics) *Let P be a Σ-program. We can define three functors $O\mathcal{P}_P : \underline{PreTh} \to \underline{ProgInt}$, $ALG_P : \underline{CompTh} \to \underline{ProgInt}$ and $LOG_P : \underline{Th} \to \underline{ProgInt}$ such that:*

a) *$O\mathcal{P}_P$, ALG_P and LOG_P assign objects $\mathcal{M}$ in its corresponding source category to objects in $\underline{ProgInt}$, in the following way*

1. *Operational Semantics:*

$$O\mathcal{P}_P(\mathcal{M}) = \{(c \to \overline{\ell})^\vee \mid (\mathcal{M}(c^\exists) \neq \underline{f}) \text{ and there is a } BCN(P, \mathcal{M}) - derivation$$
$$\text{for } \overline{\ell}_\Box T \text{ with computed answer } d \text{ such that } \mathcal{M}((c \to d)^\vee) = \underline{t}\} \cup$$
$$\{(c \to \neg\overline{\ell})^\vee \mid \overline{\ell}_\Box c \text{ is a } BCN(P, \mathcal{M}) - failed goal\}$$

2. *Least Fixpoint Semantics:*

$$ALG_P(\mathcal{M}) = \{(c \to \overline{\ell})^\vee \mid (\mathcal{M}(c^\exists) \neq \underline{f}) \wedge T_P^{\mathcal{M}} \uparrow \omega \models (c \to \overline{\ell})^\vee\} \cup$$
$$\{(c \to \neg\overline{\ell})^\vee \mid (\mathcal{M}(c^\exists) \neq \underline{f}) \wedge T_P^{\mathcal{M}} \uparrow \omega \models (c \to \neg\overline{\ell})^\vee\}$$

3. *Logical Semantics:*

$$LOG_P(\mathcal{M}) = \{(c \to \bar{\ell})^\vee \mid (\mathcal{M}(c^\exists) \neq \underline{f}) \wedge P^* \cup Th(\mathcal{M}) \models (c \to \bar{\ell})^\vee\} \cup$$
$$\{(c \to \neg\bar{\ell})^\vee \mid (\mathcal{M}(c^\exists) \neq \underline{f}) \wedge P^* \cup Th(\mathcal{M}) \models (c \to \neg\bar{\ell})^\vee\}$$

b) *To each pair of objects* $\mathcal{M}$ *and* $\mathcal{M}'$ *such that* $\mathcal{M} \preceq_c \mathcal{M}'$ *in the corresponding source category,* $\mathcal{F} \in \{ALG_P, LOG_P\}$ *assigns the morphism* $\mathcal{F}(\mathcal{M}) \preceq \mathcal{F}(\mathcal{M}')$ *in ProgInt. However,* OP_P *is contravariant, i.e.,* $\mathcal{M} \preceq_c \mathcal{M}'$ *in* <u>PreTh</u> *implies* $\mathcal{F}(\mathcal{M}') \preceq \mathcal{F}(\mathcal{M})$ *in ProgInt.*

It is easy to see that ALG_P and LOG_P are functors as a straightforward consequence of the fact that morphisms are partial orders and the monotonicity of the operator $T_P^{\mathcal{M}}$ and the logic, respectively. The contravariance of OP_P is a consequence of the fact that the *BCN*-derivation process only makes unsatisfiability queries to the solver to prune derivations. This means that when $\mathcal{M}_{\underline{f}}$ is larger the derivation process prunes more derivation sequences.

Now, given a $(\Sigma_X, \mathcal{L}_X)$-program P, we can define the semantics of P as

$$[\![P]\!] = (OP_P, ALG_P, LOG_P)$$

6   Equivalence of Semantics

In this subsection, we will first prove that the semantic constructions represented by the functors OP_P, ALG_P and LOG_P are equivalent in the sense that for each object $\mathcal{M}$ in CompTh, $OP_P(\mathcal{M})$, $ALG_P(\mathcal{M})$, and $LOG_P(\mathcal{M})$ are the same object in *ProgInt*.

Then, we will show the completeness of the operational semantics with respect to the algebraic and logical semantics just as a consequence of the fact that functors preserve the relations from its domains into its codomains.

Theorem 19 *Let P be a Σ-program. For each object $\mathcal{M}$ in* CompTh,

$$OP_P(\mathcal{M}) = ALG_P(\mathcal{M}) = LOG_P(\mathcal{M})$$

Finally, we present the usual completeness results of the operational semantics that can be obtained when the domains, theories and solvers are not equivalent. As we pointed out before, these results can be obtained just as a consequence of working with functors. In particular, since $\mathcal{M}_{solv_X} \preceq_c \mathcal{M}_{Dom_X}$ the contravariance of OP_P implies that $ALG_P(\mathcal{M}_{Dom_X}) \preceq_c OP_P(\mathcal{M}_{solv_X})$, and similarly for the logical semantics. That is:

Corollary 20   (Completeness of the operational semantics) *For any program P, OP_P is complete with respect to ALG_P and with respect to LOG_P. That is, for each constraint domain $(\Sigma_X, \mathcal{L}_X, Ax_X, Dom_X, solv_X)$:*

- $ALG_P(\mathcal{M}_{Dom_X}) \preceq_c OP_P(\mathcal{M}_{solv_X})$
- $LOG_P(\mathcal{M}_{Ax_X}) \preceq_c OP_P(\mathcal{M}_{solv_X})$

Acknowledgements: The authors would like to thank an anonymous referee for his work in improving this paper. This work has been partially supported by the Spanish CICYT project GRAMMARS (ref. TIN2004-07925-C03).

References

1. J. Álvez, P. Lucio, and F. Orejas. Constructive negation by bottom-up computation of literal answers. *Proc. 20004 ACM Symp. on Applied Computing*, pp. 1468–1475, 2004.
2. J. A. Bergstra, M. Broy, J. V. Tucker, and M. Wirsing. On the power of algebraic specifications. In Jozef Gruska and Michal Chytil, editors, *Math. Foundations of Computer Science 1981, Lecture Notes in Computer Science* 118: 193–204. Springer, 1981.
3. W. A. Carnielli. Sistematization of finite many-valued logics through the method of tableaux. *J. of Symbolic Logic*, 52(2):473–493, 1987.
4. K.L. Clark. Negation as failure. In H. Gallaire and J. Minker, editors, *Logic and Databases*, pages 293–322. Plenum Press. New York, 1978.
5. W. Drabent. What is a failure? An approach to constructive negation. *Acta Informática*, 32:27–59, 1995.
6. F. Fages. Constructive Negation by pruning. *J. of Logic Programming*, 32:85–118, 1997.
7. M. Fitting. A Kripke-Kleene semantics for logic programs. *J. of Logic Programming*, 4:295–312, 1985.
8. J. Goguen and J. Meseguer. *Initiality, Induction and Computability*. in *Algebraic Methods in Semantics*, (M. Nivat and J. Reynolds, eds.). Cambridge Univ. Press:459–540, 1985.
9. J. Jaffar and J.-L. Lassez. Constraint logic programming. In *POPL*, pages 111–119, 1987.
10. J. Jaffar and M. Maher. Constraint logic programming: a survey. *J. of Logic Programming*, (19/20):503–581, 1994.
11. J. Jaffar, M. Maher, K. Marriot, and P. Stukey. The semantics of constraint logic programs. *J. of Logic Programming*, (37):1–46, 1998.
12. S. C. Kleene. *Introduction to Metamathematics*. Van Nostrand, 1952.
13. K. Kunen. Signed data dependencies in logic programs. *J. of Logic Programming*, 7:231–245, 1989.
14. J. W. Lloyd. *Foundations of Logic Programming*. Springer-Verlag, 2nd edition, 1987.
15. P. Lucio, F. Orejas, and E. Pino. An algebraic framework for the definition of compositional semantics of normal logic programs. *J. of Logic Programming*, 40:89–123, 1999.
16. E. Pasarella, E. Pino, and F. Orejas. Constructive Negation without subsidiary trees. In *9th Int. Workshop on Functional and Logic Programming*, Benicassim, Spain. 2000.
17. T. Przymusinski. On the declarative semantics of deductive databases and logic programs. In J. Minker, editor, *Foundations of Deductive Databases and Logic Progamming*, pages 193–216. Morgan Kaufmann, 1988.
18. J.C. Shepherdson. Language and equality theory in logic programming. Technical Report PM-91-02, University of Bristol, 1991.
19. P. J. Stuckey. Negation and constraint logic programmming. *Information and Computation*, 118:12–23, 1995.

7   Appendix

Proof of Lemma 6 Let L be $\overline{\ell}_1, \ell_1, \overline{\ell}_2, \ell_2, \overline{\ell}_3 \square c$. Then, $L_1 = \overline{\ell}_1, \overline{\ell}_2, \ell_2, \overline{\ell}_3 \square c \wedge T_k^P(\ell_1), k > 0$, and $solv_X((c \wedge T_k^P(\ell_1))^\exists) \neq \mathbf{F}$, and, $L' = \overline{\ell}_1, \overline{\ell}_2, \overline{\ell}_3 \square c \wedge T_k^P(\ell_1) \wedge T_{k'}^P(\ell_2), k > 0, k' > 0$ and $solv_X((c \wedge T_k^P(\ell_1) \wedge T_{k'}^P(\ell_2))^\exists) \neq \mathbf{F}$.

Now, to construct the derivation $L \rightsquigarrow_{(P,solv_X)} L_2 \rightsquigarrow_{(P,solv_X)} L''$ in which ℓ_2 is select first in L_1 we choose $L_2 = \overline{\ell}_1, \ell_1, \overline{\ell}_2, \overline{\ell}_3 \square c \wedge T_{k'}^P(\ell_2)$ and $L'' = \overline{\ell}_1, \overline{\ell}_2, \overline{\ell}_3 \square c \wedge T_{k'}^P(\ell_2) \wedge T_k^P(\ell_1)$. Since $solv_X((c \wedge T_k^P(\ell_1) \wedge T_{k'}^P(\ell_2))^\exists) \neq \mathbf{F}$, by the well-behavedness property of $solv_X$, we know that $solv_X((c \wedge T_{k'}^P(\ell_2))^\exists) \neq \mathbf{F}$ and $solv_X((c \wedge T_{k'}^P(\ell_2) \wedge T_k^P(\ell_1))^\exists) \neq \mathbf{F}$. Hence, $L \rightsquigarrow_{(P,solv_X)} L_2 \rightsquigarrow_{(P,solv_X)} L''$ is a valid $BCN(P, solv_X)$-derivation. ∎

Proof of Theorem 7   The proof follows by induction on the length, n, of the $BCN(P, solv_X)$-derivation. The base step, $n = 0$, trivially holds. Assume that the statement holds for $n' < n$. Now, to prove the inductive step, consider the $BCN(P, solv_X)$-derivation

$$L \rightsquigarrow_{(P, solv_X)} L_1 \rightsquigarrow_{(P, solv_X)} \cdots \rightsquigarrow_{(P, solv_X)} L_{n-1} \rightsquigarrow_{(P, solv_X)} \Box c$$

Since this is a successful derivation, each literal in L is selected at some point of the derivation. Let us consider the literal ℓ in L and suppose that it is selected in L_i. By applying Lemma 6 i times we can reorder the above derivation to obtain the following one $L \rightsquigarrow_{(P, solv_X)} L'_1 \rightsquigarrow_{(P, solv_X)} \cdots \rightsquigarrow_{(P, solv_X)} L'_{n-1} \rightsquigarrow_{(P, solv_X)} \Box c'$, such that ℓ is selected in L and c' is a reordering of c. Assume that the selection rule R selects literal ℓ when considering the singleton derivation L. From the induction hypothesis, there is another $BCN(P, solv_X)$-derivation $L'_1 \overset{n-1}{\rightsquigarrow}_{(P, solv_X)} \Box c''$, using the selection rule R', where R' selects literals as they are selected by the rule R when considering the derivation $L \rightsquigarrow_{(P, solv_X)} L'_1 \overset{n-1}{\rightsquigarrow}_{(P, solv_X)} \Box c''$. So, c'' is a reordering of c' and hence of c. Thus, $L \rightsquigarrow_{(P, solv_X)} L'_1 \rightsquigarrow_{(P, solv_X)} \cdots \rightsquigarrow_{(P, solv_X)} L'_{n-1} \rightsquigarrow_{(P, solv_X)} \Box c''$ is the $BCN(P, solv_X)$-derivation we were looking for.   ∎

Proof of proposition 8   We are going to prove that $P^* \cup Th(Ax_X) \models (T_k^P(\ell) \to \ell)^\forall$ for each $k \in \mathbb{N}$, since it is easy to see that the general case is a straightforward consequence of Definition 3.

The proof follows by induction on k and it merely relies on standard syntactical properties of first-order logic. For the base case, $k = 0$, the proposition trivially holds. Assume that the statement holds for $k' < k$. Assume $T_k^P(\ell)$ is satisfiable (if it is not satisfiable the proposition trivially holds). There are two cases:

1. $\ell = p(\overline{x})$. Then, applying twice the definition of T_k^P, the first time for atoms and the second time for the conjunction of literals, we obtain the following:

$$T_k^P(p(\overline{x})) = \bigvee_{i=1}^{m} \exists \overline{y}^i (c^i \wedge T_{k-1}^P(\overline{\ell}^i)) = \bigvee_{i=1}^{m} \exists \overline{y}^i (c^i \wedge \bigwedge_{j=1}^{n_i} T_{k-1}^P(\ell_j^i))$$

Now, from the induction hypothesis we have that, for all $i \in \{1, \ldots, m\}$ and for all $j \in \{1, \ldots, n_i\}$:

$$P^* \cup Th(Ax_X) \models (T_{k-1}^P(\ell_j^i) \to \ell_j^i)^\forall$$

Then, it follows logically that,

$$P^* \cup Th(Ax_X) \models (\bigvee_{i=1}^{m} \exists \overline{y}^i (c^i \wedge \bigwedge_{j=1}^{n_i} T_{k-1}^P(\ell_j^i)) \to \bigvee_{i=1}^{m} \exists \overline{y}^i (c^i \wedge \bigwedge_{j=1}^{n_i} \ell_j^i))^\forall$$

And, again, applying the definition of T_k^P we obtain the following:

$$P^* \cup Th(Ax_X) \models (T_k^P(p) \to \bigvee_{i=1}^{m} \exists \overline{y}^i (c^i \wedge \bigwedge_{j=1}^{n_i} \ell_j^i))^\forall \tag{1}$$

In addition, by the completion of predicate $p(\bar{x})$, we have that,

$$P^* \cup Th(Ax_X) \models (\bigvee_{i=1}^{m} \exists \bar{y}^i (c^i \wedge \bigwedge_{j=1}^{n_i} \ell_j^i) \to p(\bar{x}))^{\forall} \qquad (2)$$

Hence, by (1) and (2), we can conclude that

$$P^* \cup Th(Ax_X) \models (T_k^P(p(\bar{x})) \to p(\bar{x}))^{\forall}, \quad k > 0$$

2. $\ell = \neg p(\bar{x})$. Then, $T_k^P(\neg p(\bar{x})) = F_k^P(p(\bar{x}))$, and applying the definition of F_k^P we obtain the following:

$$F_k^P(p(\bar{x})) = \bigwedge_{i=1}^{m} \forall \bar{y}^i (\neg c^i \vee F_{k-1}^P(\bar{\ell}^i)) = \bigwedge_{i=1}^{m} \forall \bar{y}^i (\neg c^i \vee \bigvee_{j=1}^{n_i} F_{k-1}^P(\ell_j^i))$$

Using the induction hypothesis we have that, for all $i \in \{1, \ldots, m\}$, $j \in \{1, \ldots, n_i\}$:

$$P^* \cup Th(Ax_X) \models (F_{k-1}^P(\ell_j^i) \to \neg \ell_j^i)^{\forall}$$

Therefore, it follows logically that,

$$P^* \cup Th(Ax_X) \models (\bigwedge_{i=1}^{m} \forall \bar{y}^i (\neg c^i \vee \bigvee_{j=1}^{n_i} F_{k-1}^P(\ell_j^i)) \to \bigwedge_{i=1}^{m} \forall \bar{y}^i (\neg c^i \vee \bigvee_{j=1}^{n_i} \neg \ell_j^i))^{\forall}$$

Again, applying the definition of F_k^P, we have that,

$$P^* \cup Th(Ax_X) \models (F_k^P(p(\bar{x})) \to \bigwedge_{i=1}^{m} \forall \bar{y}^i (\neg c^i \vee \bigvee_{j=1}^{n_i} \neg \ell_j^i))^{\forall} \qquad (3)$$

Finally, we use the completion of the predicate $p(\bar{x})$ to obtain:

$$P^* \cup Th(Ax_X) \models (\bigwedge_{i=1}^{m} \forall \bar{y}^i (\neg c^i \vee \bigvee_{j=1}^{n_i} \neg \ell_j^i) \to \neg p(\bar{x}))^{\forall} \qquad (4)$$

Hence, by (3) and (4), we can conclude that

$$P^* \cup Th(Ax_X) \models (F_k^P(p(\bar{x})) \to \neg p(\bar{x}))^{\forall}, \quad k > 0 \quad \blacksquare$$

Proof of proposition 9 To prove that $(Dom_{\Sigma}/_{\equiv}, \preceq)$ is a cpo, we show that each increasing chain $\{[\mathcal{A}_i]\}_{i \in I} \subseteq Dom_{\Sigma}/_{\equiv}$, $[\mathcal{A}_1] \preceq \ldots \preceq [\mathcal{A}_n] \preceq \ldots$, has a least upper bound $\bigsqcup [\mathcal{A}_n]$. Let $[\mathcal{A}]$ be such that $\mathcal{A}((c \to \ell)^{\forall}) = \underline{t}$ iff, for some n, $\mathcal{A}_n((c \to \ell)^{\forall}) = \underline{t}$. Then, it is almost trivial to see that

- for each n, $[\mathcal{A}_n] \preceq [\mathcal{A}]$
- for any other $[\mathcal{B}]$ such that $[\mathcal{A}_n] \preceq [\mathcal{B}]$ for each n, $[\mathcal{A}] \preceq [\mathcal{B}]$.

Finally, it is trivial to see that $[\perp_{\Sigma}] \preceq [\mathcal{A}]$ for all $[\mathcal{A}] \in Dom_{\Sigma}/_{\equiv}$. $\blacksquare$

Proof of Theorem 12 First of all, $T_P^{Dom_X}$ is monotonic as a consequence of the fact that $\Phi_P^{\mathcal{D}_X}$ is monotonic. Then, being $T_P^{Dom_X}$ monotonic, to prove that it is continuous it is enough to prove that is is finitary. That is: For each increasing chain $\{[\mathcal{A}_n]\}_{n \in I}$, $[\mathcal{A}_1] \preceq \ldots \preceq [\mathcal{A}_n] \preceq \ldots$

$$T_P^{Dom_X}\left(\bigsqcup[\mathcal{A}_n]\right) \preceq \bigsqcup T_P^{Dom_X}([\mathcal{A}_n])$$

Let $[\mathcal{A}] = \bigsqcup[\mathcal{A}_n]$ and $[\mathcal{B}] = T_P^{Dom_X}(\bigsqcup[\mathcal{A}_n]) = [\Phi_P^{\mathcal{D}_X}(\mathcal{A})]$. Let us assume $\mathcal{B}((c \to \ell)^\forall) = \underline{t}$. We have two cases:

(a) If $\ell = p(\bar{x})$ then, by the definition of the operator $\Phi_P^{\mathcal{D}_X}$, we know there are (renamed versions of) clauses $\{p(\bar{x}) : -\ell_1^i, \ldots, \ell_{n_i}^i \Box d_i \mid 1 \leq i \leq m\}$ in P and $\mathcal{D}_X$-satisfiable constraints $\{c_j^i \mid 1 \leq i \leq m \wedge 1 \leq j \leq n_i\}$ such that

- $\mathcal{A}((c_j^i \to \ell_j^i)^\forall) = \underline{t}$
- $\mathcal{A}((c \to \bigvee_{1 \leq i \leq m} \exists \bar{y}_i(\bigwedge_{1 \leq j \leq n_i} c_j^i \wedge d_i))^\forall) = \underline{t}$

In such a situation, by definition of $\bigsqcup$, we know that for each $1 \leq i \leq m$ and $1 \leq j \leq n_i$ there is a $[\mathcal{A}_k] \in \{[\mathcal{A}_n] \mid n \in I\}$ such that $\mathcal{A}_k((c_j^i \to \ell_j^i)^\forall) = \underline{t}$. Then, since $(Dom_\Sigma/_\equiv, \preceq)$ is a cpo, we know that each finite sub-chain has a least upper bound in $\{[\mathcal{A}_n]\}_{n \in I}$. Let it be $[\mathcal{A}_s]$. In addition, since all models in $\mathcal{D}_\Sigma$ are elementarily equivalent we can state that

- $\mathcal{A}_s((c_j^i \to \ell_j^i)^\forall) = \underline{t}$
- $\mathcal{A}_s((c \to \bigvee_{1 \leq i \leq m} \exists \bar{y}_i(\bigwedge_{1 \leq j \leq n_i} c_j^i \wedge d_i))^\forall) = \underline{t}$

Therefore, $\Phi_P^{\mathcal{D}_X}(\mathcal{A}_s)((c \to p(\bar{x}))^\forall) = \underline{t}$ so for all models $C \in [\Phi_P^{\mathcal{D}_X}(\mathcal{A}_s)]$ we have that $C((c \to p(\bar{x}))^\forall) = \underline{t}$. Thus, by definition of $\bigsqcup$, this implies that for all $C' \in \bigsqcup[\Phi_P^{\mathcal{D}_X}(\mathcal{A}_n)] = \bigsqcup T_P^{Dom_X}([\mathcal{A}_n])$ we have that $C'(c \to p(\bar{x})))^\forall = \underline{t}$.

(b) The proof for $\ell = \neg p(\bar{x})$ proceeds in the same way. That is, by the definition of the operator $\Phi_P^{\mathcal{D}_X}$, we know that for each (renamed version) clause in $\{p(\bar{x}) : -\ell_1^i, \ldots, \ell_{n_i}^i \Box d_i \mid 1 \leq i \leq m\} = Def_P(p(\bar{x}))$ there is a $J_i \subseteq \{1, \ldots n_i\}$ and $\mathcal{D}_X$-satisfiable constraints $\{c_j^i \mid 1 \leq i \leq m \wedge j \in J_i\}$ such that

- $\mathcal{A}((c_j^i \to \neg\ell_j)^\forall) = \underline{t}$
- $\mathcal{A}((c \to \bigwedge_{1 \leq i \leq m} \forall \bar{y}_i(\bigvee_{j \in J_i} c_j^i \vee \neg d_i))^\forall) = \underline{t}$

Again, by definition of $\bigsqcup$, we know that for each $j \in J$ there is a $[\mathcal{A}_j] \in \{[\mathcal{A}_n] \mid n \in I\}$ such that $\mathcal{A}_j((c_j \to \neg\ell_j)^\forall) = \underline{t}$. Then, as a consequence of $(Dom_\Sigma/_\equiv, \preceq)$ being a cpo, and all models in $\mathcal{D}_\Sigma$ being elementarily equivalent, there is a class $[\mathcal{A}_s]$ in the chain such that

- $\mathcal{A}_s((c_j^i \to \neg\ell_j)^\forall) = \underline{t}$
- $\mathcal{A}_s((c \to \bigwedge_{1 \leq i \leq m} \forall \bar{y}_i(\bigvee_{j \in J_i} c_j^i \vee \neg d_i))^\forall) = \underline{t}$

Therefore $\Phi_P^{\mathcal{D}_X}(\mathcal{A}_s)((c \to \neg p(\bar{x}))^\forall) = \underline{t}$ so, for all models $C \in [\Phi_P^{\mathcal{D}_X}(\mathcal{A}_s)]$ we have that $C((c \to \neg p(\bar{x}))^\forall) = \underline{t}$. And, finally, by definition of $\bigsqcup$, this implies that for all $C' \in \bigsqcup[\Phi_P^{\mathcal{D}_X}(\mathcal{A}_n)] = \bigsqcup T_P^{Dom_X}([\mathcal{A}_n])$ we have that $C'(c \to \neg p(\bar{x})))^\forall = \underline{t}$.   ∎

Proof of Theorem 14 We prove that 1 and 2 hold for a goal $\ell \Box c$. Then, the general case for $\bar{\ell} \Box c$ easily follows from the logical definition of the truth-value of $(c \to \bar{\ell})^\forall$ and $(c \to \neg \bar{\ell})^\forall$.

The Stuckey's result states that $P^* \cup Th(Dom_X) \models_3 (c \to \ell)^\forall$ if, and only if,

$$\Phi_P^{\mathcal{D}x} \uparrow k((c \to \ell)^\forall) = \underline{t}$$

for some finite k. So, by definition of $T_P^{Dom_X}$, this is equivalent to

$$\forall \mathcal{A} \in T_P^{Dom_X} \uparrow k : \mathcal{A}((c \to \ell)^\forall) = \underline{t}$$

for some finite k. And, by definition of $\bigsqcup$, to

$$\forall \mathcal{A} \in \bigsqcup T_P^{Dom_X} \uparrow k : \mathcal{A}((c \to \ell)^\forall) = \underline{t}$$

■

Proof of Theorem 19 First of all, we have that $\mathcal{LOG}_P(\mathcal{M}) = \mathcal{ALG}_P(\mathcal{M})$ as a direct consequence of Theorem 14 (Extended Theorem of Stuckey). We will prove that

- $\mathcal{ALG}_P(\mathcal{M}) \preceq \mathcal{OPP}_P(\mathcal{M})$ and
- $\mathcal{OPP}_P(\mathcal{M}) \preceq \mathcal{LOG}_P(\mathcal{M})$

(a) To prove that $\mathcal{ALG}_P(\mathcal{M}) \preceq \mathcal{OPP}_P(\mathcal{M})$ we use induction on the number of iterations of $T_P^{\mathcal{M}}$. We just consider goals such that $\bar{\ell} = p(\bar{x})$ and $\bar{\ell} = \neg p(\bar{x})$, since the general case follows from the properties of operators T_k^P and F_k^P and the fact that *BCN* is independent of the selection rule. The base case $n = 0$ is trivial, since $T_P^{\mathcal{M}} \uparrow 0 = [\bot_\Sigma]$ and $[\bot_\Sigma]((c \to \ell)^\forall) \neq \underline{t}$ for all Σ_x-constraint $c(\bar{x})$ and all Σ-literal $\ell(\bar{x})$.

Assume that for all $k \leq n$, $T_P^{\mathcal{M}} \uparrow k((c \to \ell)^\forall) = \underline{t}$ implies $\mathcal{OPP}_P(\mathcal{M})((c \to \ell)^\forall) = \underline{t}$. If $\bar{\ell} = p(\bar{x})$ then, by the definition of $T_P^{\mathcal{M}}$, we know that there are clauses $\{p(\bar{x}) : -\ell_1^i, \ldots, \ell_{n_i}^i \Box d_i \mid 1 \leq i \leq m\}$ in P and $\mathcal{M}$-satisfiable constraints $\{c_j^i \mid 1 \leq i \leq m \wedge 1 \leq j \leq n_i\}$ such that $T_P^{\mathcal{M}} \uparrow n((c_j^i \to \ell_j^i)^\forall) = \underline{t}$ and

$$\mathcal{M}((c \to \bigvee_{i=1}^{m} \exists \bar{y}_i (\bigwedge_{j=1}^{n_i} c_j^i \wedge d_i))^\forall) = \underline{t}$$

Then, by the induction hypothesis we have that $\mathcal{OPP}_P(\mathcal{M})((c_j^i \to \ell_j^i)^\forall) = \underline{t}$ for all $1 \leq i \leq m$ and $1 \leq j \leq n_i$. Thus, there exist successful $BCN(P, \mathcal{M})$-derivations for each $1 \leq i \leq m$ and $1 \leq j \leq n_i$:

$$\ell_j^i \Box d_i \rightsquigarrow_{(P,\mathcal{M})} \Box T_{k_j^i}^P(\ell_j^i) \wedge d_i$$

such that $\mathcal{M}((T_{k_j^i}^P(\ell_j^i)) \wedge d_i)^\exists) \neq \underline{f}$ and $\mathcal{M}(c_j^i \to T_{k_j^i}^P(\ell_j^i))^\forall) = \underline{t}$.

Let $k > 0$ be the largest number in $\{k_j^i \mid 1 \le i \le m \wedge 1 \le j \le n_i\}$. Then, as a consequence of the monotonicity of the operator T_-^P, we know $\mathcal{M}\left(\left(\bigwedge_{j=1}^{n_i} T_k^P(\ell_j^i)\right) \wedge d_i\right)^\exists) \ne \underline{f}$. And since

$$T_k^P\left(\bigwedge_{j=1}^{n_i} \ell_j^i\right) = \bigwedge_{j=1}^{n_i} T_k^P(\ell_j^i)$$

and

$$\mathcal{M}\left(\left(\bigwedge_{j=1}^{n_i} c_j^i \to T_k^P\left(\bigwedge_{j=1}^{n_i} \ell_j^i\right)\right)^\forall\right) = \underline{t}$$

we have that

$$\mathcal{M}\left(\left(c \to \bigvee_{i=1}^{m} \exists \bar{y}_i\left(T_k^P\left(\bigwedge_{j=1}^{n_i} \ell_j^i\right) \wedge d_i\right)\right)^\forall\right) = \underline{t}$$

That is, $\mathcal{M}\left(T_{k+1}^P(p(\bar{x}))^\exists\right) \ne \underline{f}$ and

$$\mathcal{M}\left(\left(c \to T_{k+1}^P(p(\bar{x}))\right)^\forall\right) = \underline{t}$$

Therefore, we can guarantee the existence of a successful $BCN(P, \mathcal{M})$-derivation:

$$p(\bar{x}) \square \underline{t} \rightsquigarrow_{(P, \mathcal{M})} \square T_{k+1}^P(p(\bar{x}))$$

such that $O\mathcal{P}_P(\mathcal{M})((c \to p(\bar{x}))^\forall) = \underline{t}$.

The proof for $\bar{\ell} = \neg p(\bar{x})$ proceeds in the same way. That is, according to the definition of the operator $T_P^{\mathcal{M}}$, we know that for each (possibly renamed) clause in $\{p(\bar{x}) :- \ell_1^i, \dots, \ell_{n_i}^i \square d_i \mid 1 \le i \le m\} = Def_P(p(\bar{x})))$ there is a $J_i \subseteq \{1, \dots n_i\}$ and $\mathcal{M}$-satisfiable constraints $\{c_j^i \mid 1 \le i \le m \wedge j \in J_i\}$ such that:

- $T_P^{\mathcal{M}} \uparrow n((c_j^i \to \neg \ell_j)^\forall) = \underline{t}$
- $\mathcal{M}\left(\left(c \to \bigwedge_{1 \le i \le m} \forall \bar{y}_i(\bigvee_{j \in J_i} c_j^i \vee \neg d_i)\right)^\forall\right) = \underline{t}$

Again, by the induction hypothesis we have that for all $1 \le i \le m$ and $j \in J_i$, $O\mathcal{P}_P(\mathcal{M})((c_j^i \to \neg \ell_j^i)^\forall) = \underline{t}$ so, for some $r_j^i > 0$

$$\mathcal{M}\left(\left(c_j^i \to F_{r_j^i}^P(\ell_j^i)\right)^\forall\right) = \underline{t}$$

Let $r > 0$ be the largest number in $\{r_j^i \mid 1 \le i \le m \wedge j \in J_i\}$. Then, as a consequence of the monotonicity of the operator F_-^P, we know $\mathcal{M}\left(\left(\bigvee_{j \in J_i} F_r^P(\ell_j^i)\right)^\exists\right) \ne \underline{f}$. And, since $F_r^P(\bigvee_{j \in J_i} \ell_j^i) = \bigvee_{j \in J_i} F_r^P(\ell_j^i)$ and $\mathcal{M}\left(\left(\bigvee_{j \in J_i} c_j^i \to F_r^P(\bigvee_{j \in J_i} \ell_j^i)\right)^\forall\right) = \underline{t}$ we have that

$$\mathcal{M}\left(\left(c \to F_{r+1}^P(p(\bar{x}))\right)^\forall\right) = \underline{t}$$

Therefore, we can guarantee that $p(\bar{x}) \square c$ is a $BCN(P, \mathcal{M})$-failure, so $O\mathcal{P}_P(\mathcal{M})((c \to \neg p(\bar{x}))^\forall) = \underline{t}$.

(b) Finally, we prove that $O\mathcal{P}_P(\mathcal{M}) \preceq \mathcal{L}O\mathcal{G}_P(\mathcal{M})$. Again we have two cases:

   (i) Suppose that $O\mathcal{P}_P(\mathcal{M})((c \to \neg \bar{\ell})^\forall) = \underline{t}$ so, $\bar{\ell} \square c$ is a $BCN(P, \mathcal{M})$-failed goal. Hence, $\mathcal{M}\left(\left(c \to F_k^P(\bar{\ell})\right)^\forall\right) = \underline{t}$, for some $k > 0$. Therefore, by Proposition 8, we can conclude that $P^* \cup Th(\mathcal{M}) \models (c \to \neg \bar{\ell})^\forall$.

(ii) Suppose now that $O\mathcal{P}_P(\mathcal{M})((c \to \bar{\ell})^\vee) = \underline{t}$. Again we will prove the case $\bar{\ell} = p(\bar{x})$ since the general case will follow from the properties of T_k^P and the fact that BCN is independent of the selection rule. So we assume $p(\bar{x})\square c$ has a $BCN(P,\mathcal{M})$-derivation

$$p(\bar{x})\square\underline{t} \rightsquigarrow_{(P,\mathcal{M})} \square T_k^P(p(\bar{x}))$$

such that $\mathcal{M}((c \to T_k^P(p(\bar{x})))^\vee) = \underline{t}$. Then, again as a consequence of Proposition 8, we can conclude that $P^* \cup Th(\mathcal{M}) \models (c \to p(\bar{x}))^\vee$.   ∎

A Stochastic Theory of
Black-Box Software Testing

Karl Meinke

School of Computer Science and Communication,
Royal Institute of Technology, 100-44 Stockholm, Sweden

Abstract. We introduce a mathematical framework for black-box software testing of functional correctness, based on concepts from stochastic process theory. This framework supports the analysis of two important aspects of testing, namely: (i) coverage, probabilistic correctness and reliability modelling, and (ii) test case generation. Our model corrects some technical flaws found in previous models of probabilistic correctness found in the literature. It also provides insight into the design of new testing strategies, which can be more efficient than random testing.

1 Introduction

Structural or glass-box testing of the functional correctness of software systems has been theoretically studied at least since the early 1950s (see for example [Moore 1956]). Although many useful structural test strategies have been developed, (see for example the survey [Lee and Yannakakis 1996]), theoretical studies clearly indicate the limitations of structural testing. In particular the complexity of structural testing techniques often grows exponentially with the size of the system.

To overcome this limitation, software engineers use black-box testing methods for large systems (see e.g. [Beizer 1995]). We will assume that a functional requirement on a system S can be modelled as a pair (p, q) consisting of precondition p on the input data and a postcondition q on the output data of S. The simplest strategy for black-box testing is random testing, in which input vectors satisfying p are randomly generated, and the output of each execution is compared with the postcondition q as a test oracle.

Efforts to improve on random testing, for example by careful manual design of test cases based on system knowledge and programming expertise, face the problem of proving their cost effectiveness. In fact to date, a general theoretical framework in which different black-box test strategies can be compared seems to be lacking. Constructing such a theory is a challenging problem for the theoretician. Not least because when structural features of a system S are hidden, much less remains on which to build a mathematical model. Essentially we have only the pre and postconditions p and q and the semantics or black-box behaviour of S.

K. Futatsugi et al. (Eds.): Goguen Festschrift, LNCS 4060, pp. 578–595, 2006.

In this paper we introduce a mathematical foundation for black-box testing of functional correctness based on the theory of stochastic processes. Our approach is sufficiently general to deal with both: (i) coverage, test termination and reliability modelling, and (ii) efficient test case generation. These two issues seem to be central to any improvement in software testing technology.

A system under test is invisible under the black-box methodology. Thus we can view its output as a source of stochastic behaviour under a sequence of tests. Our approach exploits the analogy between:

(i) a *black-box test* which can be viewed as an input/output measurement made on an unseen and randomly given program, and
(ii) a *random variable* which is a measurable function between two σ-fields.

Thus we can study how different strategies for black-box testing involve different assumptions about the finite-dimensional distributions (FDDs) of such random variables. While exact calculations of probabilities are generally difficult, it seems possible to identify heuristics which simplify and approximate these calculations. These can form the basis for new test strategies and tools.

The organisation of this paper is as follows. In Section 2 we formalise the concept of test success for a program S with respect to a functional correctness requirement $\{p\}S\{q\}$. In Section 3, we introduce the necessary concepts from measure theory and the theory of stochastic processes which are used to formalise probabilistic statements about testing. Our model corrects some technical flaws found in previous models of probabilistic correctness in the literature. In Section 4, we consider the coverage problem and termination criteria for testing. In Section 5 we consider efficient test case generation. Section 6 considers open problems and future research.

2   Logical Foundations of Functional Black-Box Testing

In this section we formalise functional black-box testing within the traditional framework of program correctness. The principle concept to be defined is that of a *successful black-box test* for functional correctness.

To simplify our exposition, we consider requirements specifications and computation over the ordered ring $\mathbb{Z}$ of integers. It should be clear that our approach can be generalised to any countable many-sorted data type signature Σ, (see for example [Loeckx et al. 1996]) and any minimal Σ algebra A.

The first-order language or signature Σ^{ring} for an ordered ring of integers consists of two constant symbols 0, 1, three binary function symbols $+$, $*$, $-$ and two binary relation symbols $=$, $\leq$. The ordered ring $\mathbb{Z}$ of integers is the first-order structure with domain $\mathbb{Z}$ where the constant, function and relation symbols are interpreted by the usual arithmetic constants, functions and relations.

Let X be a set of variables. The set $T(\Sigma^{ring}, X)$ of all *terms* is defined inductively in the usual way, and $T(\Sigma^{ring}) = T(\Sigma^{ring}, \varnothing)$ denotes the subset of all variable free or *ground terms*. If $\alpha : X \to \mathbb{Z}$ is any assignment then

$\overline{\alpha} : T(\Sigma, X) \to \mathbb{Z}$ denotes the *term evaluation mapping*. For any ground term t we may write $t_{\mathbb{Z}}$ for $\overline{\alpha}(t)$.

We assume the usual definition of the set $L(\Sigma^{ring}, X)$ of all *first-order formulas* over Σ^{ring} and X as the smallest set containing all atomic formulas (equations and inequalities) which is closed under the propositional connectives $\wedge$, $\neg$ and the quantifier $\forall$. The expression $\phi \vee \psi$ denotes $\neg(\neg\phi \wedge \neg\psi)$ while $\phi \to \psi$ denotes $\neg\phi \vee \psi$. The set $L_{\omega_1, \omega}(\Sigma^{ring}, X)$ of all *infinitary first-order formulas* over Σ^{ring} and X extends $L(\Sigma^{ring}, X)$ with closure under countably infinite conjunctions

$$\bigwedge_{i \in I} \phi_i,$$

for I a countable set. (See e.g. [Barwise 1968].) This infinitary language plays a technical role in Section 3 to translate first-order formulas into probabilistic statements about testing.

We assume the usual definitions of a free variable in a formula or term, and the substitution of a free variable by a term.

Next we recall how first-order formulas are used to define pre and postconditions for a program within the framework of the Floyd-Hoare theory of program correctness. For an overview of this theory see e.g. [de Bakker 1980].

Definition 1. *For any set X of variable symbols. define $X' = \{x' \mid x \in X\}$. A variable $x \in X$ is termed a prevariable while $x' \in X'$ is termed the corresponding postvariable.*

A precondition is a formula $p \in L(\Sigma^{ring}, X)$ with only prevariables, while a postcondition is a formula $q \in L(\Sigma^{ring}, X \cup X')$ that may have both pre and postvariables.

Let $\Omega(X)$ be an arbitrary programming language in which each program $\omega \in \Omega(X)$ has an *interface* $i(\omega) = \overline{x}$ where $\overline{x} = x_1, \ldots, x_k \in X^k$ is a finite sequence of length $k \geq 1$ of integer variables. The interface variables x_i are considered to function as both input and output variables for ω. If ω has the interface $\overline{x}$ we may simply write $\omega[\overline{x}]$. Note that, consistent with black-box testing, we assume no internal structure or syntax for $\Omega(X)$ programs. We will assume semantically that each program $\omega[\overline{x}] \in \Omega(X)$ has a simple *deterministic transformational action* on the initial state of $x_1, \ldots, x_k$. In the sequel, for any set B, we let $B_{\perp} = B \cup \{\perp\}$ where $\perp$ denotes an *undefined value*. We let $f : A \to B_{\perp}$ denote a *partial function* between sets A and B that may be undefined on any value $a \in A$, in which case we write $f(a) = \perp$. If $f(a)$ is defined and equal to $b \in B$ we write $f(a) = b$. We use $[A \to B_{\perp}]$ to denote the set of all partial functions from A to B.

Definition 2. *Let $\Omega(X)$ be a programming language. By a semantic mapping for $\Omega(X)$ we mean a mapping of programs into partial functions,*

$$[\![\cdot]\!] : \bigcup_{k \geq 1} [\, \mathbb{Z}^k \to \mathbb{Z}^k_{\perp} \,]$$

where for any program $\omega \in \Omega(X)$ with interface $i(\omega) = x_1, \ldots, x_k$

$$[\![\omega]\!] : \mathbb{Z}^k \to \mathbb{Z}_\perp^k$$

is a partial recursive function.

Intuitively, for any input $a = a_1, \ldots a_k \in \mathbb{Z}^k$, either ω fails to terminate when the interface variables $x_1, \ldots, x_k$ are initialised to $a_1, \ldots a_k$ respectively, and $[\![\omega]\!](a) = \perp$, or else ω terminates under this initialisation, and $[\![\omega]\!](a) = b_1, \ldots b_k$, where b_i is the state of the interface variable x_i after termination.

Let us collect together the definitions introduced so far to formalise the concepts of test success and failure. Recall the usual satisfaction relation $\mathbb{Z}, \alpha \models \phi$ in $\mathbb{Z}$ for a formula ϕ (finitary or infinitary) under an assignment $\alpha : X \to \mathbb{Z}$ in the Σ^{ring} structure $\mathbb{Z}$.

Definition 3. *Let $\omega \in \Omega(X)$ be a program with interface $i(\omega) = x_1, \ldots, x_k \in X^k$.*

(i) A functional specification $\{p\}\omega\{q\}$ is a triple, where $p \in L(\Sigma^{ring}, X)$ is a precondition and $q \in L(\Sigma^{ring}, X \cup X')$ is a postcondition.

(ii) A specification $\{p\}\omega\{q\}$ is said to be true in $\mathbb{Z}$ under assignments $a : X \to \mathbb{Z}$ and $b : X' \to \mathbb{Z}$ if, and only if, $\mathbb{Z}, a \models p$ and if $[\![\omega]\!](a(x_1), \ldots a(x_k)) = b(x_1'), \ldots, b(x_k')$ then $\mathbb{Z}, a \cup b \models q$. If $\{p\}\omega\{q\}$ is true in $\mathbb{Z}$ under a and b we write

$$\mathbb{Z}, a \cup b \models \{p\}\omega\{q\}.$$

We say that $\{p\}\omega\{q\}$ is valid in $\mathbb{Z}$ and write $\mathbb{Z} \models \{p\}\omega\{q\}$, if, and only if, for every $a : X \to \mathbb{Z}$ if there exists $b : X' \to \mathbb{Z}$ such that $[\![\omega]\!](a(x_1), \ldots a(x_k)) = b(x_1'), \ldots, b(x_k')$ then $\mathbb{Z}, a \cup b \models \{p\}\omega\{q\}$.

(iii) Let p be a precondition and q be a postcondition. For any $a : X \to \mathbb{Z}$ we say that ω fails the test a of $\{p\}\omega\{q\}$ if, and only if, there exists $b : X' \to \mathbb{Z}$ such that $\mathbb{Z}, a \models p$ and $[\![\omega]\!](a(x_1), \ldots a(x_k)) = b(x_1'), \ldots, b(x_k')$ and

$$\mathbb{Z}, a \cup b \not\models q.$$

We say that ω passes the test a of $\{p\}\omega\{q\}$ if ω does not fail a.

Intuitively ω fails the test $a : X \to \mathbb{Z}$ of $\{p\}\omega\{q\}$ if a satisfies the precondition p, and ω terminates on the input a but the resulting output assignment $b : X' \to \mathbb{Z}$ does not satisfy the postcondition q. This definition is consistent with the *partial correctness interpretation* of validity for $\{p\}\omega\{q\}$ used in Definition 3.(ii) (c.f. [de Bakker 1980]). Partial correctness (rather than the alternative *total correctness interpretation*) is appropriate if we require a failed test case to produce an observably incorrect value in finite time rather than an unobservable infinite loop. In particular, for this choice of Definitions 3.(ii) and 3.(iii) we have

$$\mathbb{Z} \not\models \{p\}\omega\{q\} \quad \Leftrightarrow \quad \omega \text{ fails some test } a \text{ of } \{p\}\omega\{q\}.$$

Thus we have formalised program testing as the search for counterexamples to program correctness (under the partial correctness interpretation).

3   A Stochastic Calculus for Program Correctness

Within the framework of program correctness outlined in Section 2, we wish to approach functional black-box testing in a quantitative way. The following question seems central. *Given a specification $\{p\}\omega\{q\}$ suppose that ω has passed n tests $a_1, \ldots, a_n$ of p and q: for any new test a_{n+1} what is the probability that ω will fail a_{n+1}?* Intuitively, we expect this failure probability to monotonically decrease as a function of n. For fixed n we also expect the failure probability to depend on the values chosen for $a_1, \ldots, a_n$. An optimal testing strategy would be to choose a_{n+1} with the maximal probability that ω fails a_{n+1}, for each $n \geq 0$. In this section we will introduce a probabilistic model of testing that can be used to answer this question.

Recall from probability theory the concept of a σ–*algebra* or σ–*field* $(\Omega, \, F)$ of events, where Ω is a non-empty set of elements termed *outcomes* and $F \subseteq \wp(\Omega)$ is a collection of sets, known as *events*, which is closed under countable unions and intersections. (See e.g. [Kallenberg 1997].) Importantly for us, $(A, \, \wp(A))$ is a σ-field, for any countable set A, including any set of partial recursive functions.

Definition 4. *Let $\overline{x} = x_1, \ldots, x_k \in X^k$, for $k \geq 1$, be an interface. By a sample space of programs over $\overline{x}$ we mean a pair*

$$\Omega[\overline{x}] = (\, \Omega[\overline{x}], \, eval \,),$$

where $\Omega[\overline{x}] \subseteq \Omega(X)$ is a subset of programs $\omega[\overline{x}]$ all having the same interface $\overline{x}$, and $eval : \Omega[\overline{x}] \times \mathbb{Z}^k \to \mathbb{Z}_\perp^k$ is the program evaluation mapping given by

$$eval(\, \omega, \, a_1, \ldots, a_k \,) = [\![\, \omega \,]\!](\, a_1, \ldots, a_k \,).$$

We say that $\Omega[\overline{x}]$ is extensional if, and only if, for all programs $\omega, \omega' \in \Omega[\overline{x}]$

$$(\, \forall a \in \mathbb{Z}^k \ eval(\omega, \, a) = eval(\omega', \, a) \,) \ \to \ \omega = \omega'.$$

We consider extensional sample spaces of programs only. It is important for distribution modeling that all programs $\omega \in \Omega[\overline{x}]$ have the same interface $\overline{x}$.

Recall that a *probability measure* $\mathbb{P}$ on σ-field $(\Omega, \, F)$ is a function $\mathbb{P} : F \to [0, 1]$ satisfying: $\mathbb{P}(\emptyset) = 0$, $\mathbb{P}(\Omega) = 1$, and for any collection $e_1, e_2, \ldots \in F$ of events which are pairwise disjoint, i.e. $i \neq j \Rightarrow e_i \cap e_j = \emptyset$,

$$\mathbb{P}(\, \bigcup_{i=1}^{\infty} e_i \,) = \sum_{i=1}^{\infty} \mathbb{P}(\, e_i \,).$$

The triple $(\Omega, \, F, \, \mathbb{P})$ is termed a *probability space*.

Let $(\Omega, \, F)$ and $(\Omega', \, F')$ be σ-fields. A function $f : \Omega \to \Omega'$ is said to be *measurable*, if, and only if, for each event $e \in F'$,

$$f^{-1}(e) \in F.$$

Let $P = (\Omega, F, \mathbb{P})$ be a probability space. A *random variable* $X : \Omega \to \Omega'$ is a measurable function. If $X : \Omega \to \Omega'$ is any random variable then X induces a probability function $\mathbb{P}_X : F' \to [0, 1]$ defined on any $e \in F'$ by

$$\mathbb{P}_X(e) = \mathbb{P}(\, X^{-1}(e)\,).$$

Then $(\Omega', F', \mathbb{P}_X)$ is a probability space. Thus one important role of a random variable that we will exploit in Definition 9 is to transfer a probability measure from F-events onto F'-events.

We may take more than one measurement on the outcome of any random experiment. We can consider any finite number or even an infinite number of measurements. This leads us naturally to the important concept of a stochastic process. Let I be any non-empty set. A *stochastic process* S over (Ω', F') is an I-indexed family of random variables $S = \langle S_i : \Omega \to \Omega' \mid i \in I \rangle$. For each $\omega \in \Omega$, the function $S(\omega) : I \to \Omega'$ defined by

$$S(\omega)(i) = S_i(\omega)$$

is termed a *path* of the process S. (See e.g. [Grimmet et al. 1982].)

Definition 5. *Let $\overline{x} = x_1, \ldots, x_k \in X^k$, for $k \geq 1$, be an interface and let $\Omega[\overline{x}]$ be a sample space of programs. Define*

$$I_{\overline{x}} = \{x_1, \ldots, x_k\} \times \mathbb{Z}^k$$

to be an indexing set for a family of random variables. For each index

$$(x_i, a_1, \ldots, a_k) \in I_{\overline{x}},$$

define the random variable $S_{(x_i, a_1, \ldots, a_k)} : \Omega[\overline{x}] \to \mathbb{Z}_\perp$ by

$$S_{(x_i, a_1, \ldots, a_k)}(\omega) = \begin{cases} \perp & \text{if } eval(\, \omega, a_1, \ldots, a_k\,) = \perp, \\ eval(\, \omega, a_1, \ldots, a_k\,)_i, & \text{otherwise.} \end{cases}$$

Thus $S_{(x_i, a_1, \ldots, a_k)}(\omega)$ gives the output obtained for the interface variable x_i by executing program ω on the input $a_1, \ldots, a_k$.

A *path* $S(\omega) : \{x_1, \ldots, x_k\} \times \mathbb{Z}^k \to \mathbb{Z}_\perp$ for the stochastic process $S = \langle S_i \mid i \in I_{\overline{x}} \rangle$ gives the entire input/output behaviour of the program ω.

Definition 5 exploits the analogy between: (i) a *black-box test* which can be viewed as an input/output measurement made on an unseen and therefore essentially randomly given program, and (ii) a *random variable* which is a measurable function between two σ-fields. A test sequence is just a sequence of such input/output measurements made on the same unseen program. Therefore, it is natural to model all possible tests on the same program as a path of a stochastic process. Hence we arrive at the model given by Definition 5.

Modelling programs as stochastic processes in this way now makes it possible to derive probabilistic statements about test success and failure.

In the remainder of this section, we let ($\Omega[\overline{x}]$, F, $\mathbb{P}$) denote an arbitrary probability space, where $\Omega[\overline{x}] = (\Omega[\overline{x}], eval)$ is a countable extensional sample space of programs, and $F = \wp(\Omega[\overline{x}])$. In order to assign probabilities to pre and postconditions on any program $\omega \in \Omega[\overline{x}]$ we need to be able to represent these as events (i.e. sets of programs) within the σ-field F. For this we begin by showing how first-order terms can be analysed in terms of the random variables introduced in Definition 5.

Now $\mathbb{Z}$ is a minimal Σ^{ring} structure. So by definition, for any integer $i \in \mathbb{Z}$ there exists a canonical numeral $\overline{i} \in T(\Sigma^{ring})$ which denotes i in $\mathbb{Z}$, i.e. $\overline{i}_{\mathbb{Z}} = i$.

Definition 6. *For any assignment* $a : X \to \mathbb{Z}$, *we define the translation mapping*

$$a^\sharp : T(\Sigma^{ring}, X \cup \{x'_1, \ldots, x'_k\}) \to T(\Sigma^{ring}, I_{\overline{x}})$$

by induction on terms.
(i) $a^\sharp(0) = 0$ *and* $a^\sharp(1) = 1$.
(ii) For any variable $x \in X$, $a^\sharp(x) = \overline{a(x)}$,
(iii) For any postvariable $x' \in \{x'_1, \ldots, x'_k\}$,

$$a^\sharp(x') = (x, a(x_1), \ldots, a(x_k)).$$

(iv) For any terms t_1, $t_2 \in T(\Sigma^{ring}, X \cup X')$, *and for any function symbol* $op \in \{+, *, -\}$,

$$a^\sharp(t_1 \ op \ t_2) = (a^\sharp(t_1) \ op \ a^\sharp(t_2)).$$

In essence $a^\sharp$ replaces each logical variable and prevariable with the name of its value under a. Also $a^\sharp$ replaces each postvariable with the index of its corresponding random variable under a.

In order to translate first-order formulas into events we extend $a^\sharp$ to all first-order formulas by mapping these into the quantifier-free fragment of $L_{\omega_1, \omega}(\Sigma, I_{\overline{x}})$, in which even bound variables have dissappeared.

Definition 7. *For any assignment* $a : X \to A$, *we define the translation mapping*

$$a^\sharp : L(\Sigma^{ring}, X \cup \{x'_1, \ldots, x'_k\}) \to L_{\omega_1, \omega}(\Sigma^{ring}, I_{\overline{x}})$$

by induction on formulas.
(i) For any terms $t_1, t_2 \in T(\Sigma^{ring}, X \cup \{x'_1, \ldots, x'_k\})$ *and any relation symbol* $R \in \{\leq, =\}$,

$$a^\sharp(t_1 \ R \ t_2) = (a^\sharp(t_1) \ R \ a^\sharp(t_2)).$$

(ii) For any formulas ϕ_1, $\phi_2 \in L(\Sigma^{ring}, X \cup \{x'_1, \ldots, x'_k\})$,

$$a^\sharp(\phi_1 \wedge \phi_2) = (a^\sharp(\phi_1) \wedge a^\sharp(\phi_2))$$

$$a^\sharp(\neg \phi_1) = \neg(a^\sharp(\phi_1))$$

(iii) For any variable $x \in X \cup \{x'_1, \ldots, x'_k\}$, *and any formula* $\phi \in L(\Sigma^{ring}, X \cup \{x'_1, \ldots, x'_k\})$,

$$a^{\sharp}(\forall x \; \phi) = \bigwedge_{i \in \mathbb{Z}} a[z \to i]^{\sharp}(\phi[x/z])$$

where $z \in X - \{x_1, \ldots, x_n\}$ *is the least (non-interface) variable (under a fixed enumeration) such that* z *is not free in* ϕ *and* $a[z \to i] : X \to A$ *agrees with* a *everywhere except on* z, *where* $a[z \to i](z) = i$.

Now we can easily associate an event consisting of a set of programs with every quantifier free formula $\phi \in L_{\omega_1, \omega}(\Sigma^{ring}, I_{\overline{x}})$ as follows.

Definition 8. *Define the event set* $\mathfrak{F}(\phi) \subseteq \Omega[\overline{x}]$ *for each quantifier free formula* $\phi \in L_{\omega_1, \omega}(\Sigma^{ring}, I_{\overline{x}})$ *by induction on formulas.*
(i) For any terms $t_1[i_1, \ldots, i_m], t_2[i_1, \ldots, i_m] \in T(\Sigma^{ring}, I_{\overline{x}})$, *and any relation symbol* $R \in \{\leq, =\}$,

$$\mathfrak{F}(t_1 \; R \; t_2)) =$$

$$\langle S_{i_1}, \ldots, S_{i_m} \rangle^{-1}(\{b \in \mathbb{Z}^m \mid \mathbb{Z}, b \models t_1 \; R \; t_2\}).$$

(ii) For any quantifier free formulas $\phi_1, \phi_2 \in L_{\omega_1, \omega}(\Sigma^{ring}, I_{\overline{x}})$,

$$\mathfrak{F}(\phi_1 \wedge \phi_2) = \mathfrak{F}(\phi_1) \cap \mathfrak{F}(\phi_2)$$

$$\mathfrak{F}(\neg \phi_1) = \Omega[\overline{x}] - \mathfrak{F}(\phi_1)$$

(iii) For any countable family of quantifier free formulas $\langle \phi_i \in L_{\omega_1, \omega}(\Sigma^{ring}, I_{\overline{x}}) \mid i \in I \rangle$,

$$\mathfrak{F}(\bigwedge_{i \in I} \phi_i) = \bigcap_{i \in I} \mathfrak{F}(\phi_i).$$

Notice in Definition 8.(i) we assume that $\{b \in \mathbb{Z}^m \mid \mathbb{Z}, b \models t_1 \; R \; t_2\}$ is indeed an event on $\mathbb{Z}^m$. In the case that we take the discrete σ-field $\wp(\mathbb{Z}^m)$ this requirement is trivially satisfied.

Using Definition 8 we can now translate the probability distribution $\mathbb{P}$ on programs into probability values for correctness statements. Thus we come to the central definitions of this section.

Definition 9. *Let* $\phi \in L(\Sigma^{ring}, X \cup \{x'_1, \ldots, x'_k\})$ *be any formula.*

(i) We define the probability that ϕ *is satisfiable under* $a : X \to \mathbb{Z}$ *by*

$$\mathbb{P}(Sat_a(\phi)) = \mathbb{P}(\mathfrak{F}(a^{\sharp}(\phi))).$$

(ii) We define the probability that ϕ *is satisfiable by*

$$\mathbb{P}(Sat(\phi)) = \mathbb{P}(\bigcup_{a : X \to \mathbb{Z}} \mathfrak{F}(a^{\sharp}(\phi))).$$

(i) We define the probability that ϕ is valid by

$$\mathbb{P}(\, \mathbb{Z} \models \phi \,) = \mathbb{P}(\, \bigcap_{a:X \to \mathbb{Z}} \mathfrak{F}(\, a^{\sharp}(\phi) \,) \,).$$

Definition 9.(i) provides a rigorous mathematical answer to the initial question this section. It is helpful to illustrate this definition with some simple examples.

Example 1. Let $x \in X$ be a single variable interface.
(i) Consider the specification

$$\{\ \}\omega\{x' = m * x + c\},$$

which asserts that $\omega[x]$ computes a linear function $f(x) = mx + c$ of its input variable x. For any input assignment of $a \in \mathbb{Z}$ to x

$$\mathfrak{F}(\, a^{\sharp}(x' = m * x + c)\,) =$$

$$\mathfrak{F}(\,(\,x,\,a\,) = m * a + c\,) =$$

$$\langle\, S_{(x,\,a)}\,\rangle^{-1}\{b \in \mathbb{Z} \mid \mathbb{Z},\, b \models (\,x,\,a\,) = m * a + c\} =$$

$$\{\omega \in \Omega[x] \mid eval(\omega,\,a) = m * a + c\}.$$

Thus

$$\mathbb{P}(\, Sat_a(\,\neg\, x' = m * x + c\,)\,) =$$

$$\mathbb{P}(\,\{\omega \in \Omega[x] \mid eval(\omega,\,a) \neq m * a + c\}\,)$$

is the probability that a randomly chosen single variable program $\omega[x] \in \Omega[x]$ fails the postcondition $x' = m * x + c$ on the test input $a \in \mathbb{Z}$.
(ii) More generally, the probability that a program $\omega[x]$ will fail a test $a_{k+1} \in \mathbb{Z}$ of p and q given that ω has already passed k tests $a_1, \ldots, a_k \in \mathbb{Z}$ with output $b_1, \ldots, b_k \in \mathbb{Z}$ is the conditional probability

$$\mathbb{P}(\, Sat_{a_{k+1}}(\, p \wedge \neg q\,) \mid \{\omega \in \Omega : eval(\omega,\,a_i) = b_i \text{ for } 1 \leq i \leq k\}\,).$$

Definition 9 satisfies several intuitive properties.

Proposition 1. *Let ϕ, $\psi \in L(\, \Sigma^{ring},\, X \cup \{x'_1,\, \ldots,\, x'_k\}\,)$ be any formulas.*

(i) If $\mathbb{Z} \models \phi$ then $\mathbb{P}(\, \mathbb{Z} \models \phi\,) = 1$.

(ii) If $\mathbb{Z} \models \neg\phi$ then $\mathbb{P}(\, \mathbb{Z} \models \phi\,) = 0$.

(iii) If $\mathbb{Z} \models \phi \to \psi$ then $\mathbb{P}(\, \mathbb{Z} \models \phi\,) \leq \mathbb{P}(\, \mathbb{Z} \models \psi\,)$.

(iv) $\mathbb{P}(\, \mathbb{Z} \models \phi\,) = 1 - \mathbb{P}(\, Sat(\neg\phi)\,)$.

(v) If $\mathbb{Z} \models \phi \leftrightarrow \psi$ then $\mathbb{P}(\, \mathbb{Z} \models \phi\,) = \mathbb{P}(\, \mathbb{Z} \models \psi\,)$.

Proof. Follows easily from Definition 9.

By Proposition 1.(v) the probability that a correctness formula is valid is independent of its syntactic structure and depends only on its semantics.

Although these properties are intuitive, they are not satisfied by any of the reliability models of [Hamlet 1987], [Miller et al. 1992] or [Thayer et al. 1978]. For example, these models all assign a probability $p < 1$ of satisfying a tautology. We believe this points to a significant conceptual flaw in existing models of probabilistic correctness in the literature.

4   Test Coverage and Software Reliability

Given that n tests of a program $\omega[\overline{x}]$ are unsuccessful in finding an error in ω, what is the probability that ω satisfies a specification $\{p\}\omega\{q\}$? If it is possible to calculate or even estimate this probability value, then we have a clearly defined stopping criterion for black-box testing: we may terminate when a desired probability of correctness has been achieved. Thus the concept of probability of correctness gives a formal model of black-box test coverage, where by coverage we mean the extent of testing. We shall apply the theoretical model introduced in Section 3 to consider the problem of estimating the probability of correctness.

An obvious technical problem is to find a distribution $\mathbb{P}$ on programs which is realistic. However, to begin to study coverage and the testing termination problem we can use very simple heuristical probability distributions, and examine how calculations can be made.

For simplicity, we consider programs with a single integer variable interface $x \in X$. Let

$$\Omega[x] = (\ \Omega[x], \ eval\ ).$$

Also, for simplicity, we assume that $\Omega[x]$ is a subrecursive language, i.e. each program $\omega[x] \in \Omega[x]$ terminates on all inputs. Thus $eval : \Omega[x] \times \mathbb{Z} \to \mathbb{Z}$ is also a total function, which allows us to work with totally defined random walk models of paths.

To calculate the probability of satisfying a formula ϕ, we need a probability distribution $\mathbb{P} : \wp(\ \Omega[x]\ ) \to [0, 1]$ Recalling Definition 5, one approach is to consider the associated family of random variables

$$S = \langle\ S_{(x,\ i)} : \Omega[x] \to \mathbb{Z}\ |\ i \in \mathbb{Z}\ \rangle.$$

A simple model of the FDDs of these random variables is to assume a *random walk hypothesis*. Writing S_i for $S_{(x,\ i)}$ we can relate S_n and S_{n+1} by a formula $S_{n+1} = S_n + X_n$ where

$$\langle\ X_i : \Omega[x] \to \mathbb{Z}\ |\ i \in \mathbb{Z}\ \rangle$$

is another family of random variables. A simple relationship is to define an *exponential distribution* on the X_i by

$$\mathbb{P}(X_i = +n) = \mathbb{P}(X_i = -n) = \frac{1}{3}\left(\frac{1}{2}\right)^n$$

for all $n \in \mathbb{N}$.

An exponential random walk over a finite interval can model any total function over that interval. This distribution captures the simple intuition that fast growing functions are increasingly unlikely. Furthermore it is easily analysed, and for simple formulas, we can estimate the probability of satisfiability. After n test passes on the inputs $a_1 \leq a_2 \leq \ldots \leq a_n \in \mathbb{Z}$ we have a sequence of $n-1$ intervals $[a_i, a_{i+1}]$. We can consider the probability of satisfying a formula $p \wedge \neg q$ (where p is a precondition and q is a postcondition) over each of these intervals separately. To perform such an analysis, we first note that the interval $[a_i, a_{i+1}]$ can be renormalised to the interval $[0, a_{i+1} - a_i]$ without affecting the probability values. (An exponential random walk is homogeneous along the x-axis.) Let us consider probabilities for the individual paths over an interval $[0, a]$.

The probability of a path following an exponential random walk is a function of its length and its volatility.

Definition 10. *Let* $y = y_0, y_1, \ldots, y_n \in \mathbb{Z}^{n+1}$ *be a path of length* $n \geq 1$. *We define the volatility* $\lambda(y) \in \mathbb{Z}$ *of* y *by*

$$\lambda(y) = \sum_{i=1}^{n} | y_i - y_{i-1} |.$$

Proposition 2. *Let* $y = y_0, y_1, \ldots, y_n \in \mathbb{Z}^{n+1}$ *be a path of length* $n \geq 1$. *Then*

$$\mathbb{P}(S_i = y_i \ for \ i = 1, \ldots, n \mid S_0 = y_0) = \left(\frac{1}{3}\right)^n \left(\frac{1}{2}\right)^{\lambda(y)}.$$

Proof. By induction on n.

Let us consider monotone paths.

Definition 11. *Let* $y = y_0, y_1, \ldots, y_n \in \mathbb{Z}^{n+1}$ *be a path of length* $n \geq 1$. *We say that* y *is monotone if* $y_0 \leq y_1 \leq \ldots \leq y_n$ *or* $y_0 \geq y_1 \geq \ldots \geq y_n$.

Proposition 3. *Let* $y = y_0, y_1, \ldots, y_n \in \mathbb{Z}^{n+1}$ *be a path of length* $n \geq 1$.
(i) If y *is monotone then* $\lambda(y) = |y_n - y_0|$.
(ii) If y *is non-monotone then* $\lambda(y) > |y_n - y_0|$.

Proof. By induction on the length of paths.

Thus it is easy to calculate the probability of a monotone path.

Corollary 1. *Let* $y = y_0, y_1, \ldots, y_n \in \mathbb{Z}^{n+1}$ *be a monotone path of length* n. *Then*

$$\mathbb{P}(S_i = y_i \ for \ i = 1, \ldots, n \mid S_0 = y_0) = \left(\frac{1}{3}\right)^n \left(\frac{1}{2}\right)^{|y_n - y_0|}.$$

Proof. Immediate from Propositions 2 and 3.

So the probability of a monotone path under the exponential distribution is independent of the steps taken, and depends only on the start and end points. In fact, by Proposition 3, this property characterises the monotone paths. Furthermore, any non-monotone path y from y_0 to y_n has exponentially lower probability than any monotone path from y_0 to y_n by a factor $\left(\frac{1}{2}\right)^{\lambda(y)-|y_n-y_0|}$.

Let $C^R(n, r)$ be the number of ways of selecting a collection of r objects from a total of n objects with repetitions. Then

$$C^R(n, r) = C(n+r-1, r)$$

where $C(n, r)$ is the binomial coefficient defined by

$$C(n, r) = \frac{n(n-1)\ldots(n-r+1)}{r!}.$$

Recall the well known *upper negation identity* (see e.g. [Graham et al. 1989])

$$C(n, r) = (-1)^r C(r-n-1, r),$$

from which we can infer $C^R(n, r) = (-1)^r C(-n, r)$.

Theorem 1. *For any $n \geq 1$ and y_0, $y_n \in \mathbb{Z}$,*

$$\mathbb{P}(S_n = y_n \mid S_0 = y_0) =$$

$$\left(\frac{1}{3}\right)^n \left(\frac{1}{2}\right)^{|y_n-y_0|} \left(C^R(n, |y_n-y_0|) + \right.$$

$$\left. \sum_{i>0} \sum_{k=1}^{min(i,\,n-1)} C(n, k)\, C^R(k, i-k)\, C^R(n-k, |y_n-y_0|+i)\left(\frac{1}{2}\right)^i \right).$$

Proof. Apply Proposition 2 and sum over all volatility values.

To illustrate the approach, we estimate the probability of a single variable program $w[x]$ failing a test of a simple linear equational specification

$$\{\ \}w\{x' = m * x + c\},$$

within an interval $[0, n]$. (Recall Example 1.) We may assume that w passes both tests of the endpoints 0 and n.

Theorem 2. *(i) For any $n > 1$ and any $c \in \mathbb{Z}$,*

$$\mathbb{P}(S_i \neq c \ \text{for some}\ 0 < i < n \mid S_0 = c,\ S_n = c) \approx \left(\frac{n-1}{n+1}\right).$$

(ii) For any $n > 1$ and any m, $c \in \mathbb{Z}$, where $m \neq 0$,

$$\mathbb{P}(S_i \neq mi+c \ \text{for some}\ 0 < i < n \mid S_0 = c,\ S_n = mn+c)$$

$$\approx 1 - \frac{1}{C^R(n, |nm|)}.$$

Proof. (i) Follows from Proposition 2 and Theorem 1 by considering non-monotone paths with highest probability satisfying $S_i \neq c$ for some $0 < i < n$.

(ii) Clearly, there is only one monotone path $y_0, \ldots, y_n$ of length n, satisfying $y = mx + c$, namely

$$y = y_0 = c, \; y_1 = m + c, \; \ldots, \; y_n = nm + c.$$

So for all monotone paths from c to $nm + c$ excluding y, using Corollary 1 and Theorem 1,

$$\mathbb{P}(\, S_n = nm + c, \; S_i \neq mi + c \; for \; some \; 0 < i < n \; \mid \; S_0 = c \,) =$$

$$(\, C^R(\, n, \, |nm| \,) - 1 \,) \left(\frac{1}{3}\right)^n \left(\frac{1}{2}\right)^{|nm|}.$$

Hence the result follows.

By a similar analysis of other types of correctness formulas, it becomes clear that closed form solutions to reliability estimation problems become intractable for anything other than simple kinds of formulas. For practical testing problems, Monte Carlo simulation (see e.g. [Bouleau 1994]) seems to be a necessary tool to estimate reliability after n tests, even for such a simple distribution as the exponential random walk.

Clearly, the results of this section depend on specific properties of the exponential random walk. This distribution model represents a naive but mathematically tractable model of reality. An open question for future research is to find more realistic models or program distributions. Of course, more accurate models would lead to slightly different results than those presented here.

5   Test Case Generation (TCG)

In section 4 we considered the problem of stopping the testing process with some quantitative conclusion about the reliability of a program after n tests have been passed. In this section we consider how to apply our stochastic model to the actual testing phase that precedes termination. How can we use the stochastic approach to efficiently generate test cases that can effectively uncover errors? We have already seen in Section 4 that calculations of correctness probabilities may be computationally expensive. However, for certain kinds of probability distributions an approach to TCG can be developed from outside probability theory, using classical *function approximation theory*, with the advantage of efficient speed.

For clarity of exposition, we will again deal with the case of a program interface consisting a single input/output variable $x \in X$. Furthermore, we will generalise from probability measures to arbitrary finite measures at this point. (Recall that every probability measure is a measure, but not vice-versa.)

Definition 12. *Let $M : [\,[m, n] \to D\,] \to \mathbb{R}^+$ be a measure for $D \subseteq \mathbb{Z}$. (If the codomain of M is $[0, 1]$ then M is an FDD.) Then M is elective if, and only if, for any $(a_1, b_1), \ldots, (a_k, b_k) \in [m, n] \times D$ the set*

$$F_{(a_1,\, b_1),\, \ldots,\, (a_k,\, b_k)} = \{\; g : [m, n] \to D \mid g(a_i) = b_i \; for \; 1 \leq i \leq k \;\}$$

has a unique maximum member under M, i.e. there exists $f \in F_{(a_1,\, b_1),\, \ldots,\, (a_k,\, b_k)}$ such that for all $g \in F_{(a_1,\, b_1),\, \ldots,\, (a_k,\, b_k)}$

$$g \neq f \;\Rightarrow\; M(g) < M(f).$$

To understand Definition 12, suppose that $b_1, \ldots, b_k \in D$ are the results of executing a program $w[x]$ on the test inputs $a_1, \ldots, a_k \in [m, n]$ respectively. Then an elective measure M gives for the input/output pairs

$$(a_1, b_1), \ldots, (a_k, b_k)$$

a unique "most likely candidate" for a function f extending these pairs to the entire interval $[m, n]$. This candidate function f, which is the maximum member $f \in F_{(a_1,\, b_1),\, \ldots,\, (a_k,\, b_k)}$ under M, represents a "best" guess of what the partially known system under test might look like in its entirety.

An elective measure $M : [\,[m, n] \to D\,] \to \mathbb{R}^+$ gives rise to an iterative test case generation procedure in the following way. Given k executed test cases $a_1, \ldots, a_k \in [m, n]$ for a program $w[x]$ with results $b_1, \ldots, b_k \in D$, we can consider the unique elected function $f_k \in F_{(a_1,\, b_1),\, \ldots,\, (a_k,\, b_k)}$ as a model of $w[x]$. By analysing f_k we may be able to locate a new test case $a_{k+1} \in [m, n]$ such that a_{k+1} satisfies a precondition p but $f_k(a_{k+1})$ does not satisfy a postcondition q. Then a_{k+1} is a promising new test case to execute on $w[x]$. If no such a_{k+1} exists we can use some other choice criteria (e.g. random) for the $k + 1$-th test, and hope for a more promising test case later as the sequence of elected functions $f_k : k \geq 1$ converges to the actual input/output behaviour of $w[x]$.

The fundamental technical problem for this approach to TCG is to find a suitable elective measure M. One pragmatic solution to this problem is introduced in [Meinke 2004] using function approximation theory. Specifically, an *interpolant* (usually a local interpolant) of $(a_1, b_1), \ldots, (a_k, b_k) \in [m, n] \times D$ is chosen as the elected function. In [Meinke 2004] piecewise polynomials were investigated as local interpolants. Many other classes of approximating functions are known in the literature such as splines, wavelets, radial basis functions, etc. Thus the technique gives a rich source of algorithms.

Our main result in this section is to show that for a large class of approximation methods, the function approximation approach (which is non-probabilistic, and fast to the extent that interpolants can be efficiently computed and evaluated) is equivalent to the measure theoretic approach.

Definition 13. *Let $D \subseteq \mathbb{Z}$ be any subset. An interpolation scheme is a mapping $I : \wp([m, n] \times D) \to [\,[m, n] \to D\,]$ such that for all $1 \leq i \leq k$*

$$I(\; \{(a_1, b_1), \ldots, (a_k, b_k)\} \;)(a_i) = b_i.$$

We will show that a large class of interpolation schemes, including polynomial interpolation, actually give rise to elective measures, and even elective probability measures. Thus the approximation approach can be seen as a special case of the stochastic approach, where the FDDs are implicit, but can be efficiently computed.

Definition 14. *Let* $\mu : 2^{2^{[m,n]}} \to \mathbb{R}^+$ *be a finite measure (not necessarily a probability measure). Let*

$$I : \wp([m,n] \times D) \to [\,[m,n] \to D\,]$$

be an interpolation scheme. Define the measure

$$\mu^I : [\,[m,n] \to D\,] \to \mathbb{R}^+$$

by

$$\mu^I(\,f\,) = \mu(\,\{\,\{a_{i_1},\ \ldots,\ a_{i_k}\,\} \subseteq [m,n]\,\mid$$
$$I[\,\{\,(a_{i_1},\ f(a_{i_1})),\ \ldots,\ (a_{i_1},\ f(a_{i_1}))\,\}\,] = f\,\}\,).$$

Proposition 4. *If* $D \subseteq \mathbb{Z}$ *is finite then* μ *can be defined so that* μ^I *is a probability measure for any interpolation scheme* I.

Proof. By construction.

Definition 15. *Let* $M : [\,[m,n] \to D\,] \to \mathbb{R}^+$ *be an elective measure. Define the interpolation scheme* $I^M : \wp([m,n] \times D) \to [\,[m,n] \to D\,]$ *by*

$$I^M[\,\{\,(a_0,\ b_0),\ \ldots,\ (a_k,\ b_k)\,\}\,] = f,$$

where $f \in \{\,g : [m,n] \to D \mid g(x_i) = y_i \text{ for } 1 \leq i \leq k\,\}$ *is the unique element for which* $M(f)$ *is maximum.*

Definition 16. *Let* $I : \wp([m,n] \times D) \to [\,[m,n] \to D\,]$ *be an interpolation scheme.*
(i) I is monotone if, and only if, $I[\,\{\,(a_0,\ b_0),\ \ldots,\ (a_k,\ b_k)\,\}\,] = f$ *and* $\{\,(a_0',\ b_0'),\ \ldots,\ (a_j',\ b_j')\,\} \subseteq f$ *and* $\{\,a_0,\ \ldots,\ a_k\,\} \subseteq \{\,a_0',\ \ldots,\ a_j'\,\}$ *imply*

$$I[\,\{\,(a_0',\ b_0'),\ \ldots,\ (a_j',\ b_j')\,\}\,] = f.$$

(ii) I is permutable if, and only if, for any $\{\,a_1,\ \ldots,\ a_k\,\} \subseteq [m,n]$ *and* $f : [m,n] \to D$ *if* $I[\,\{\,(a_1,\ f(a_1)),\ \ldots,\ (a_k,\ f(a_k))\,\}\,] = f$ *then for any* $\{\,a_1',\ \ldots,\ a_k'\,\} \subseteq [m,n]$

$$I[\,\{\,(a_1',\ f(a_1')),\ \ldots,\ (a_k',\ f(a_k'))\,\}\,] = f.$$

Example 2. Polynomial approximation is monotone and permutable.

Theorem 3. *Let $I : \wp([m, n] \times D) \to [\, [m, n] \to D\,]$ be a monotone permutable interpolation scheme. Then μ^I is elective and $I^{\mu^I} = I$.*

Proof. Consider any $\{(a_1, b_1), \ldots, (a_n, b_n)\} \subseteq [m, n] \times D$ and suppose

$$I[\; \{(a_1, b_1), \ldots, (a_n, b_n)\}\;] = f,$$

then we need to show that

$$I^{\mu^I}[\; \{(a_1, b_1), \ldots, (a_n, b_n)\}\;] = f.$$

It suffices to show that for any $g : [m, n] \to D$ such that $g(a_i) = b_i$ for $1 \leq i \leq n$ and $g \neq f$, $\mu^I(g) < \mu^I(f)$, i.e. μ^I is elective. Consider any such g and any $\{a'_1, \ldots, a'_k\} \subseteq [m, n] \times D$ and suppose that

$$I[\; \{(a'_1, g(a'_1)), \ldots, (a'_k, g(a'_k))\}\;] = g.$$

Since I is permutable we must have $k \geq n$.

Since I is an interpolation scheme

$$I[\; \{(a_1, f(a_1)), \ldots, (a_n, f(a_n))\}\;] = f.$$

Then since I is permutable

$$I[\; \{(a'_1, f(a'_1)), \ldots, (a'_n, f(a'_n))\}\;] = f,$$

Finally since I is monotone and $k \geq n$,

$$I[\; \{(a'_1, f(a'_1)), \ldots, (a'_k, f(a'_k))\}\;] = f.$$

Therefore

$$\{\{a_1, \ldots, a_k\} \subseteq [m, n] \mid I[\; \{(a_1, g(a_1)), \ldots, (a_k, g(a_k))\}\;] = g\} \subseteq$$

$$\{\{a_1, \ldots, a_k\} \subseteq [m, n] \mid I[\; \{(a_1, f(a_1)), \ldots, (a_k, f(a_k))\}\;] = f\}.$$

Thus since μ is a measure, $\mu^I(f) > \mu^I(g)$, i.e. μ^I is elective.

6   Conclusions

In this paper we have introduced a stochastic model for black-box testing of the functional correctness of programs. This model allows us to derive a probability value for the validity of a correctness formula of the form $\{p\}w\{q\}$ conditional on the results of any finite set of black-box tests on w. It corrects technical problems with similar models occuring previously in the literature. Our model provides a solution to the difficult problem of measuring coverage in black-box testing. It also suggests new approaches to the test case generation process itself.

Further research is necessary to establish accurate models of the probabilistic distribution of programs. Furthermore, we may generalise our model to consider how program distributions are influenced by the choice of the programming problem to be solved (the precondition p and postcondition q). This would give a theoretical model of the *competent programmer hypothesis* of [Budd 1980]. This also requires consideration of the difficult problem of non-termination. For example, it may be necessary to introduce a non-functional time requirement into specifications, in order to abort a test that can never terminate. Research into other abstract data types and concrete data structures also presents an important problem in this area.

Much of this research was carried out during a sabbatical visit to the Department of Computer Science Engineering at the University of California at San Diego (UCSD) during 2003. We gratefully acknowledge the support of the Department, and in particular the helpful comments and advice received from Joseph Goguen and the members of the Meaning and Computation group. We also acknowledge the financial support of TFR grant 2000-447.

References

[de Bakker 1980] J.W. de Bakker, *Mathematical Theory of Program Correctness*, Prentice-Hall, 1980.

[Barwise 1968] J. Barwise (ed), *The Syntax and Semantics of Infinitary Languages*, Lecture Notes in Mathematics 72, Springer-Verlag, Berlin, 1968.

[Bauer 1981] H. Bauer, *Probability Theory and Elements of Measure Theory*, Academic Press, London, 1981.

[Beizer 1995] B. Beizer, *Black-Box Testing*, John Wiley, 1995.

[Bouleau 1994] N. Bouleau, D. Lepingle, *Numerical Methods for Stochastic Processes*, John Wiley, New York, 1994.

[Budd 1980] Budd, T.A. DeMillo, R.A. Lipton, R.J. Sayward, F.G. Theoretical and Empirical Studies on Using Program Mutation to Test the Functional Correctness of Programs, Proc. 7th ACM SIGPLAN-SIGACT Symp. on Principles of Programming Languages, 220-223, 1980.

[Graham et al. 1989] R.L. Graham, D.E. Knuth and O. Patashnik, *Concrete Mathematics*, Addison-Wesley, Reading Mass., 1989.

[Grimmet et al. 1982] G. Grimmet, D. Stirzaker, *Probability and Random Processes*, Oxford University Press, 1982.

[Hamlet 1987] Hamlet, R.G. Probable Correctness Theory, Inf. Proc. Letters 25, 17-25, 1987.

[Kallenberg 1997] O. Kallenberg, *Foundations of Modern Probability*, Springer Verlag, 1997.

[Lee and Yannakakis 1996] D. Lee, M. Yannakakis, Principles and Methods of Testing Finite State Machines - a Survey, Proc. IEEE, **84** (8), 1090-1123, 1996.

[Loeckx et al. 1996] J. Loeckx, H-D. Ehrich, M. Wolf, *Specification of Abstract Data Types*, Wiley Teubner, Chichester 1996.

[Meinke 2004] K. Meinke, Automated Black-Box Testing of Functional Correctness using Function Approximation, pp 143-153 in: G. Rothermel (ed) *Proc. ACM SIG-SOFT Int. Symp. on Software Testing and Analysis, ISSTA 2004*, Software Engineering Notes 29 (4), ACM Press, 2004.

[Miller et al. 1992] Miller, K.W. Morell, L.J. Noonan, R.E. Park, S.K. Nicol, D.M. Murrill, B.W. Voas, J.M.: Estimating the Probability of Failure when Testing Reveals no Failures, IEEE Trans. Soft. Eng. 18 (1), 33-43, 1992.

[Moore 1956] E.F. Moore, Gedanken-experiments on Sequential Machines, Princeton Univ. Press, Ann. Math. Studies, 34, 129-153, Princeton NJ, 1956.

[Thayer et al. 1978] Thayer, T.A. Lipow, M. Nelson, E.C.: Software Reliability, North Holland, New York, 1978.

[Weiss and Weyuker 1988] Weiss, S.N. Weyuker, E.J.: An Extended Domain-Based Model of Software Reliability, IEEE Trans. Soft. Eng. 14 (10), 1512-1524, 1988.

Some Tips on Writing Proof Scores
in the OTS/CafeOBJ Method

Kazuhiro Ogata[1,2] and Kokichi Futatsugi[2]

[1] NEC Software Hokuriku, Ltd.
ogatak@acm.org
[2] Japan Advanced Institute of Science and Technology (JAIST)
kokichi@jaist.ac.jp

Abstract. The OTS/CafeOBJ method is an instance of the proof score approach to systems analysis, which has been mainly devoted by researchers in the OBJ community. We describe some tips on writing proof scores in the OTS/CafeOBJ method and use a mutual exclusion protocol to exemplify the tips. We also argue soundness of proof scores in the OTS/CafeOBJ method.

1   Introduction

The proof score approach to systems analysis has been mainly devoted by researchers in the OBJ community [10,8]. In the approach, an executable algebraic specification language is used to specify systems and system properties, and a processor of the language, which has a rewrite engine as one of its functionalities, is used as a proof assistant to prove that systems satisfy system properties. Proof plans called *proof scores* are written in the algebraic specification language to conduct such proofs and the proof scores are executed by the language processor by means of rewriting to check if the proofs are success.

Proof scores can be regarded as programs to prove that algebraic specifications satisfy system properties. While proof scores are being designed, constructed and debugged, we can understand algebraic specifications being analyzed more profoundly, which may even let us find flaws lurked in the specifications [15,14]. Our thought on proof is similar to that of the designers of LP [11]. Proof scripts written in a tactic language provided by proof assistants such as Coq [1] and Isabel/HOL [13] may be regarded as such programs, but it seems that such proof assistants rather aim for mechanizing mathematics.

We have argued that the proof score approach to systems analysis is an attractive approach to design verification in [6] thanks to (1) balanced human-computer interaction and (2) flexible but clear structure of proof scores. The former means that humans are able to focus on proof plans, while tedious and detailed computations can be left to computers; humans do not necessarily have to know what deductive rules or equations should be applied to goals to prove. The latter means that lemmas do not need to be proved in advance and proof scores can help humans comprehend the corresponding proofs; a proof that a

K. Futatsugi et al. (Eds.): Goguen Festschrift, LNCS 4060, pp. 596–615, 2006.

system satisfies a system property can be conducted even when all lemmas used have not been proved, and assumptions used are explicitly and clearly written in proof scores. To precisely assess the achievement of (1) and (2) in the proof score approach and compare it with systems analysis with other existing proof assistants, however, we need further studies.

The OTS/CafeOBJ method [17,4,7] is an instance of the proof score approach to systems analysis. In the OTS/CafeOBJ method, observational transition systems (OTSs) are used as models of systems and CafeOBJ [2], an executable algebraic specification language/system, is used; OTSs are transition systems, which are straightforwardly written as algebraic specifications. An older version of the OTS/CafeOBJ method is described in [17,4], and the latest version is described in [7]. We have conducted case studies, among which are [15,18,19,16,20], to demonstrate the usefulness of the OTS/CafeOBJ method. In this paper, we describe some tips on writing proof scores in the OTS/CafeOBJ method. A mutual exclusion protocol called *Tlock* using *atomicInc*, which atomically increments the number stored in a variable and returns the old number, is used as an example. We also argue soundness of proof scores in the OTS/CafeOBJ method.

The rest of the paper is organized as follows. Section 2 describes the OTS/CafeOBJ method. Section 3 describes tips on writing proof scores in the OTS/CafeOBJ method. Section 4 informally argue soundness of proof scores in the OTS/CafeOBJ method. Section 5 concludes the paper.

2   The OTS/CafeOBJ Method

In the OTS/CafeOBJ method, systems are analyzed as follows.

1. Model a system as an OTS $\mathcal{S}$.
2. Write $\mathcal{S}$ in CafeOBJ as an algebraic specification. The specification consists of sorts (or types), operators on the sorts, and equations that define (properties of) the operators. The specification can be executed by using equations as left-to-right rewrite rules by CafeOBJ.
3. Write system properties in CafeOBJ. Let P be the set of such system properties and let P' be the empty set. .
4. If P is empty, the analysis has been successfully finished, which means that $\mathcal{S}$ satisfies all the properties in P'. Otherwise, extract a property p from P and go next. .
5. Write a proof score in CafeOBJ to prove that $\mathcal{S}$ satisfies p. The proof may need other system properties as lemmas. Write such system properties in CafeOBJ and put them that are not in P' into P if any. .
6. Execute (or play) the proof score with CafeOBJ. If all the results are as expected, then the proof is discharged. Put p into P' and go to 4. If all the results are not as expected, rewrite the proof score and repeat 6. .

Tasks 5 and 6 may be interactively conducted together. A counterexample may be found in tasks 5 and 6.

In this section, we mention CafeOBJ, describe the definitions of basic concepts on OTSs, write on how to write OTSs in CafeOBJ and how to write proof scores that OTSs satisfy invariant properties in CafeOBJ.

2.1   CafeOBJ

CafeOBJ [2] is an algebraic specification language/system mainly based on order-sorted algebras and hidden algebras [9,3]. Abstract data types are specified in terms of order-sorted algebras, and abstract machines are specified in terms of hidden algebras. Algebraic specifications of abstract machines are called *behavioral specifications*. There are two kinds of sorts in CafeOBJ: *visible sorts* and *hidden sorts*. A visible sort denotes an abstract data type, while a hidden sort denotes the state space of an abstract machine. There are three kinds of operators (or operations) with respect to (wrt) hidden sorts: *hidden constants, action operators* and *observation operators*. Hidden constants denote initial states of abstract machines, action operators denote state transitions of abstract machines, and observation operators let us know the situation where abstract machines are located. Both an action operator and an observation operator take a state of an abstract machine and zero or more data. The action operator returns the successor state of the state wrt the state transition denoted by the action operator plus the data. The observation operator returns a value that characterizes the situation where the abstract machine is located.

Basic units of CafeOBJ specifications are modules. CafeOBJ provides built-in modules. One of the most important built-in modules is BOOL in which propositional logic is specified. BOOL is automatically imported by almost every module unless otherwise stated. In BOOL and its parent modules, declared are the visible sort Bool, the constants true and false of Bool, and operators denoting some basic logical connectives. Among the operators are not_, _and_, _or_, _xor_, _implies_ and _iff_ denoting negation ($\neg$), conjunction ($\wedge$), disjunction ($\vee$), exclusive disjunction (xor), implication ($\Rightarrow$) and logical equivalence ($\Leftrightarrow$), respectively. The operator if_then_else_fi corresponding to the if construct in programming languages is also declared. CafeOBJ uses the Hsiang term rewriting system (TRS) [12] as the decision procedure for propositional logic, which is implemented in BOOL. CafeOBJ reduces any term denoting a proposition that is always true (false) to true (false). More generally, a term denoting a proposition reduces to an exclusively disjunctive normal form of the proposition.

2.2   Observational Transition Systems (OTSs)

We suppose that there exists a universal state space denoted Υ and that each data type used in OTSs is provided. The data types include Bool for truth values. A data type is denoted D_*.

Definition 1 (OTSs). *An OTS $\mathcal{S}$ is $\langle \mathcal{O}, \mathcal{I}, \mathcal{T} \rangle$ such that*

- $\mathcal{O}$: *A finite set of observers. Each observer $o_{x_1:D_{o1},\ldots,x_m:D_{om}} : \Upsilon \to D_o$ is an indexed function that has m indexes $x_1, \ldots, x_m$ whose types are*

$D_{o1}, \ldots, D_{om}$. The equivalence relation $(v_1 =_S v_2)$ between two states $v_1, v_2 \in \Upsilon$ is defined as $\forall o_{x_1,\ldots,x_m} : \mathcal{O}. (o_{x_1,\ldots,x_m}(v_1) = o_{x_1,\ldots,x_m}(v_2))$, where $\forall o_{x_1,\ldots,x_m} : \mathcal{O}$ is the abbreviation of $\forall o_{x_1,\ldots,x_m} : \mathcal{O}. \forall x_1 : D_{o1} \ldots \forall x_m : D_{om}$.

- $\mathcal{I}$: The set of initial states such that $\mathcal{I} \subseteq \Upsilon$.
- $\mathcal{T}$: A finite set of transitions. Each transition $t_{y_1:D_{t1},\ldots,y_n:D_{tn}} : \Upsilon \to \Upsilon$ is an indexed function that has n indexes $y_1, \ldots, y_n$ whose types are $D_{t1}, \ldots, D_{tn}$ provided that $t_{y_1,\ldots,y_n}(v_1) =_S t_{y_1,\ldots,y_n}(v_2)$ for each $[v] \in \Upsilon/=_S$, each $v_1, v_2 \in [v]$ and each $y_k : D_{tk}$ for $k = 1, \ldots, n$. $t_{y_1,\ldots,y_n}(v)$ is called the successor state of v wrt S. Each transition $t_{y_1,\ldots,y_n}$ has the condition $c\text{-}t_{y_1:D_{t1},\ldots,y_n:D_{tn}} : \Upsilon \to$ Bool, which is called the effective condition of the transition. If $c\text{-}t_{y_1,\ldots,y_n}(v)$ does not hold, then $t_{y_1,\ldots,y_n}(v) =_S v$.   □

We note the following two points on transitions, which have something to do with writing proof scores. (1) Although transitions are defined as relations among states in some other existing transition systems, transitions are functions on states in OTSs. This is because transitions are represented by (action) operators in behavioral specifications and operators are functions in CafeOBJ. However, multiple transitions that are functions on states can be substituted for one transition that is a relation among states. (2) Basically there is no restriction on the form of effective conditions. But, effective conditions should be in the form $c_1\text{-}t_{y_1,\ldots,y_n}(v) \wedge \ldots \wedge c_M\text{-}t_{y_1,\ldots,y_n}(v)$, where each $c_k\text{-}t_{y_1,\ldots,y_n}(v)$ has no logical connectives or has one negation at head, so that proof scores can have clear structure. When an effective condition is not in this form, it is converted to a disjunctive normal form. If the disjunctive normal form has more than one disjunct, multiple transitions each of which has one of the disjuncts as its effective condition can be substituted for the corresponding transition.

Definition 2 (Reachable states). *Given an OTS S, reachable states wrt S are inductively defined:*

- *Each $v_{\text{init}} \in \mathcal{I}$ is reachable wrt S.*
- *For each $t_{y_1,\ldots,y_n} \in \mathcal{T}$ and each $y_k : D_{tk}$ for $k = 1, \ldots, n$, $t_{x_1,\ldots,x_n}(v)$ is reachable wrt S if $v \in \Upsilon$ is reachable wrt S.*

Let $\mathcal{R}_S$ be the set of all reachable states wrt S .   □

Predicates whose types are $\Upsilon \to$ Bool are called *state predicates*. All properties considered in this paper are *invariants*.

Definition 3 (Invariants). *Any state predicate $p : \Upsilon \to Bool$ is called invariant wrt S if p holds in all reachable states wrt S, i.e. $\forall v : \mathcal{R}_S. p(v)$.*   □

We suppose that each state predicate p considered in this paper has the form $\forall z_1 : D_{p1} \ldots \forall z_a : D_{pa}. P(v, z_1, \ldots, z_a)$, where $v, z_1, \ldots, z_a$ are all variables in p and $P(v, z_1, \ldots, z_a)$ does not contain any quantifiers.

A concrete example of how to model a system as an OTS is given.

Example 1 (Tlock). The pseudo-code executed by each process i can be written as follows:

Loop

 l1: $ticket[i] := atomicInc(tvm)$;
 l2: **repeat until** $ticket[i] = turn$;
   Critical section;
 cs: $turn := turn + 1$;

tvm and $turn$ are non-negative integer variables shared by all processes and $ticket[i]$ is a non-negative integer variable that is local to process i. Initially, each process i is at label l1, tvm and $turn$ are 0, and $ticket[i]$ for each i is unspecified. The value of tvm (which stands for a ticket vending machine) is the next available ticket. Each process i obtains a ticket, which is stored in $ticket[i]$, at label l1. A process is allowed to enter the critical section if its ticket equals the value of $turn$ at label l2. $turn$ is incremented when a process leaves the critical section at label cs.

Let Label, Pid and Nat be the types of labels (l1, l2 and cs), process IDs and non-negative integers (natural numbers). Tlock can be modeled as the OTS $\mathcal{S}_{\text{Tlock}}$ such that

- $\mathcal{O}_{\text{Tlock}} \triangleq \{\text{tvm} : \Upsilon \rightarrow \text{Nat}, \text{turn} : \Upsilon \rightarrow \text{Nat}, \text{ticket}_{i:\text{Pid}} : \Upsilon \rightarrow \text{Nat}, \text{pc}_{i:\text{Pid}} : \Upsilon \rightarrow \text{Label}\}$
- $\mathcal{I}_{\text{Tlock}} \triangleq \{v_{\text{init}} \in \Upsilon \mid \text{tvm}(v_{\text{init}}) = 0 \wedge \text{turn}(v_{\text{init}}) = 0 \wedge \forall i : \text{Pid}. (\text{pc}_i(v_{\text{init}}) = \text{l1})\}$
- $\mathcal{T}_{\text{Tlock}} \triangleq \{\text{get}_{i:\text{Pid}} : \Upsilon \rightarrow \Upsilon, \text{check}_{i:\text{Pid}} : \Upsilon \rightarrow \Upsilon, \text{exit}_{i:\text{Pid}} : \Upsilon \rightarrow \Upsilon\}$

The three transitions are defined as follows:

- get_i : c-$\text{get}_i(v) \triangleq \text{pc}_i(v) = \text{l1}$. If c-$\text{want}_i(v)$, then

  $\text{tvm}(\text{get}_i(v)) \triangleq \text{tvm}(v) + 1$, $\text{turn}(\text{get}_i(v)) \triangleq \text{turn}(v)$,
  $\text{ticket}_j(\text{get}_i(v)) \triangleq$ **if** $i = j$ $\text{tvm}(v)$ **else** $\text{ticket}_j(v)$, and
  $\text{pc}_j(\text{get}_i(v)) \triangleq$ **if** $i = j$ **then** l2 **else** $\text{pc}_j(v)$.

- check_i : c-$\text{check}_i(v) \triangleq \text{pc}_i(v) = \text{l2} \wedge \text{ticket}_i(v) = \text{turn}(v)$. If c-$\text{want}_i(v)$, then

  $\text{tvm}(\text{check}_i(v)) \triangleq \text{tvm}(v)$, $\text{turn}(\text{check}_i(v)) \triangleq \text{turn}(v)$,
  $\text{ticket}_j(\text{get}_i(v)) \triangleq \text{ticket}_j(v)$, and $\text{pc}_j(\text{check}_i(v)) \triangleq$ **if** $i = j$ **then** cs **else** $\text{pc}_j(v)$.

- exit_i : c-$\text{exit}_i(v) \triangleq \text{pc}_i(v) = \text{cs}$. If c-$\text{want}_i(v)$, then

  $\text{tvm}(\text{exit}_i(v)) \triangleq \text{tvm}(v) + 1$, $\text{turn}(\text{exit}_i(v)) \triangleq \text{turn}(v)$,
  $\text{ticket}_j(\text{get}_i(v)) \triangleq \text{ticket}_j(v)$, and $\text{pc}_j(\text{exit}_i(v)) \triangleq$ **if** $i = j$ **then** l1 **else** $\text{pc}_j(v)$.

Let $\text{MX}(v)$ be $\forall i, j : \text{Pid}. [(\text{pc}_i(v) = cs \wedge \text{pc}_j(v) = cs) \Rightarrow i = j]$. $\text{MX}(v)$ is invariant wrt $\mathcal{S}_{\text{Tlock}}$, i.e. $\forall v : \mathcal{R}_{\mathcal{S}_{\text{Tlock}}} . \text{MX}(v)$, although it may need to be verified. $\qquad \square$

2.3 Specifying OTSs in CafeOBJ

We suppose that a visible sort V_* corresponding to each data type D_* used in OTSs and the related operators are provided. X_k and Y_k are CafeOBJ variables corresponding to indexes x_k and y_k of observers and transitions, respectively.

The universal state space Υ is represented by a hidden sort, say H declared as *[H]* by enclosing it with *[and]*. Given an OTS $\mathcal{S}$, an arbitrary initial state is represented by a hidden constant, say init, each observer $o_{x_1,\ldots,x_m}$ is represented by an observation operator, say o, and each transition $t_{y_1,\ldots,y_n}$ is represented by an action operator, say t. The hidden constant init, the observation operator o and the action operator t are declared as follows:

```
op init : -> H
bop o : H V_{o1} ... V_{om} -> V_o
bop t : H V_{t1} ... V_{tn} -> H
```

The keyword bop or bops is used to declare observation and action operators.

We suppose that the value returned by $o_{x_1,\ldots,x_m}$ in an arbitrary initial state can be expressed as $f(x_1,\ldots,x_m)$. This is expressed by the following equation:

```
eq o(init,X_1,...,X_m) = f(X_1,...,X_m) .
```

$f(X_1,\ldots,X_m)$ is the CafeOBJ term corresponding to $f(x_1,\ldots,x_m)$.

Each transition $t_{y_1,\ldots,y_n}$ is defined by describing what the value returned by each observer $o_{x_1,\ldots,x_m}$ in the successor state becomes when $t_{y_1,\ldots,y_n}$ is applied in a state v. When $c\text{-}t_{y_1,\ldots,y_n}(v)$ holds, this is expressed generally by a conditional equation that has the form

```
ceq o(t(S,Y_1,...,Y_n),X_1,...,X_m) = e-t(S,Y_1,...,Y_n,X_1,...,X_m)
      if c-t(S,Y_1,...,Y_n) .
```

S is a CafeOBJ variable of H, corresponding to v. e-t(S,Y$_1$,...,Y$_n$,X$_1$,...,X$_m$) is the CafeOBJ term corresponding to the value returned by $o_{x_1,\ldots,x_m}$ in the successor state denoted by t(S,Y$_1$,...,Y$_n$). c-t(S,Y$_1$,...,Y$_n$) is the CafeOBJ term corresponding to $c\text{-}t_{y_1,\ldots,y_n}(v)$.

If $c\text{-}t_{y_1,\ldots,y_n}(v)$ always holds in any state v or the value returned by $o_{x_1,\ldots,x_m}$ is not affected by applying $t_{y_1,\ldots,y_n}$ in any state v (i.e. regardless of the truth value of $c\text{-}t_{y_1,\ldots,y_n}(v)$), then a usual equation is used instead of a conditional equation. The usual equation has the form

```
eq o(t(S,Y_1,...,Y_n),X_1,...,X_m) = e-t(S,Y_1,...,Y_n,X_1,...,X_m) .
```

e-t(S,Y$_1$,...,Y$_n$,X$_1$,...,X$_m$) is S if the value returned by $o_{x_1,\ldots,x_m}$ is not affected by applying $t_{y_1,\ldots,y_n}$ in any state.

When $c\text{-}t_{y_1,\ldots,y_n}(v)$ does not hold, $t_{y_1,\ldots,y_n}$ changes nothing, which is expressed by a conditional equation that has the form

```
ceq t(S,Y_1,...,Y_n) = S if not c-t(S,Y_1,...,Y_n) .
```

We give the CafeOBJ specification of $\mathcal{S}_{\text{Tlock}}$.

Example 2 (CafeOBJ specification of $\mathcal{S}_{\text{Tlock}}$). $\mathcal{S}_{\text{Qlock}}$ is specified in CafeOBJ as the module TLOCK:

```
mod* TLOCK {  pr(PNAT) pr(LABEL) pr(PID)
 *[Sys]*
-- an arbitrary initial state
  op init : -> Sys
-- observation operators
  bops tvm turn : Sys -> Nat  bop ticket : Sys Pid -> Nat
  bop  pc : Sys Pid -> Label
-- action operators
  bops get check exit : Sys Pid -> Sys
-- CafeOBJ variables
  var S : Sys  vars I J : Pid
-- init
  eq tvm(init) = 0 .  eq turn(init) = 0 .  eq pc(init,I) = 11 .
-- get
  op c-get : Sys Pid -> Bool {strat: (0 1 2)}
  eq c-get(S,I) = (pc(S,I) = 11) .
  --
  ceq tvm(get(S,I)) = s(tvm(S)) if c-get(S,I) .
  eq  turn(get(S,I)) = turn(S) .
  ceq ticket(get(S,I),J)
      = (if I = J then tvm(S) else ticket(S,J) fi) if c-get(S,I) .
  ceq pc(get(S,I),J) = (if I = J then 12 else pc(S,J) fi) if c-get(S,I) .
  ceq get(S,I) = S if not c-get(S,I) .
-- check
  op c-check : Sys Pid -> Bool {strat: (0 1 2)}
  eq c-check(S,I) = (pc(S,I) = 12 and ticket(S,I) = turn(S)) .
  --
  eq  tvm(check(S,I)) = tvm(S) .  eq  turn(check(S,I)) = turn(S) .
  eq  ticket(check(S,I),J) = ticket(S,J) .
  ceq pc(check(S,I),J)
      = (if I = J then cs else pc(S,J) fi) if c-check(S,I) .
  ceq check(S,I) = S if not c-check(S,I) .
-- exit
  op c-exit : Sys Pid -> Bool {strat: (0 1 2)}
  eq c-exit(S,I) = (pc(S,I) = cs) .
  --
  eq  tvm(exit(S,I)) = tvm(S) .
  ceq turn(exit(S,I)) = s(turn(S)) if c-exit(S,I) .
  eq  ticket(exit(S,I),J) = ticket(S,J) .
  ceq pc(exit(S,I),J)
      = (if I = J then 11 else pc(S,J) fi) if c-exit(S,I) .
  ceq exit(S,I) = S if not c-exit(S,I) .
}
```

A comment starts with -- and terminates at the end of the line. PNAT, LABEL
and PID are the modules in which natural numbers, labels and process IDs
are specified. The keyword pr is used to imports modules. The operator s of
s(tvm(S)) and s(turn(S)) is the successor function of natural numbers. The
keyword start: is used to specify local strategies to operators [5]. The local
strategy (0 1 2) given to c-get indicates that when CafeOBJ meets a term

whose top is c-get such as c-get(s,i), CafeOBJ should try to rewrite the whole term such as c-get(s,i). If CafeOBJ does not find any rules with which the term is rewritten, it evaluates the first and second arguments such as s and i in that order, and tries to rewrite the whole term such as c-get(s',i') again, where s' and i' are the results obtained by evaluating s and i.   □

2.4 Proof Scores of Invariants

Although some invariants may be proved by rewriting and/or case splitting only, we often need to use induction, especially *simultaneous induction* [7]. We then describe how to verify $\forall v : \mathcal{R}_\mathcal{S}.\, p(v)$ by simultaneous induction by writing proof scores in CafeOBJ based on the CafeOBJ specification of $\mathcal{S}$.

It is often impossible to prove $\forall v : \mathcal{R}_\mathcal{S}.\, p(v)$ alone. We then suppose that it is possible to prove $\forall v : \mathcal{R}_\mathcal{S}.\, p(v)$ together with $N - 1$ other state predicates[3], that is, we prove $\forall v : \mathcal{R}_\mathcal{S}.\, (p_1(v) \wedge \ldots \wedge p_N(v))$, where p_1 is p. We suppose that each p_k has the form $\forall z_k : D_{pk}.\, P_k(v, z_k)$ for $k = 1, \ldots, N$. Note that the method described here can be used when p_k has more than one universally quantified variable. Let v_{init}^c be an arbitrary initial state of $\mathcal{S}$, and then for the base case, all we have to do is to prove

$$\forall z_1 : D_{p1}.\, P_1(v_{\text{init}}^c, z_1) \wedge \ldots \wedge \forall z_N : D_{pN}.\, P_N(v_{\text{init}}^c, z_N) \tag{1}$$

For each induction case (i.e. each $t_{y_1,\ldots,y_n} \in T$), all we have to do is to prove

$$\begin{aligned} &\forall z_1 : D_{p1}.\, P_1(v^c, z_1) \wedge \ldots \wedge \forall z_N : D_{pN}.\, P_N(v^c, z_N) \\ &\Rightarrow \forall z_1 : D_{p1}.\, P_1(t_{y_1^c,\ldots,y_n^c}(v^c), z_1) \wedge \ldots \wedge \forall z_N : D_{pN}.\, P_N(t_{y_1^c,\ldots,y_n^c}(v^c), z_N) \end{aligned} \tag{2}$$

for an arbitrary state v^c and an arbitrary value y_k^c for $k = 1, \ldots, n$.

To prove (1), we can separately prove each conjunct

$$P_i(v_{\text{init}}^c, z_k^c) \tag{3}$$

where z_k^c is an arbitrary value of D_{pk} for $k = 1, \ldots, N$. To prove (2), assuming $\forall z_1 : D_{p1}.\, P_1(v^c, z_1), \ldots, \forall z_N : D_{pN}.\, P_N(v^c, z_N)$, we can separately prove each $P_k(t_{y_1^c,\ldots,y_n^c}(v^c), z_k^c)$, where z_k^c is an arbitrary value of D_{pk}, for $k = 1, \ldots, N$. $P_k(v^c, z_k^c)$ is often used as an assumption to prove $P_k(t_{y_1^c,\ldots,y_n^c}(v^c), z_k^c)$. Therefore, the formula to prove has the form

$$(P_\alpha(v^c, d_\alpha) \wedge P_\beta(v^c, d_\beta) \wedge \ldots) \Rightarrow [P_k(v^c, z_k^c) \Rightarrow P_k(t_{y_1^c,\ldots,y_n^c}(v^c), z_k^c)] \tag{4}$$

where $\alpha, \beta, \ldots \in \{1, \ldots, N\}$ and $d_\alpha, d_\beta, \ldots$ are some values of $D_{p\alpha}, D_{p\beta}, \ldots$ for $i = 1, \ldots, N$.

We next describe how to write proof plans of (3) and (4) in CafeOBJ. We first declare the operators denoting $P_1, \ldots, P_N$ and the equations defining the operators. The operators and equations are declared in a module, say INV (which imports the module where $\mathcal{S}$ is written), as follows:

---

[3] Generally, such $N - 1$ state predicates should be found while $\forall v : \mathcal{R}_\mathcal{S}.\, p(v)$ is being proved.

```
op invₖ : H Vₚₖ -> Bool
eq invₖ(S,Zₖ) = Pᵢ(S,Zₖ) .
```

for $k = 1, \ldots, N$. Z_k is a CafeOBJ variable of V_{pk} and $P_i(S,Z_k)$ is a CafeOBJ term denoting $P_k(v, z_k)$. In INV, we also declare a constant z_k^c denoting an arbitrary value of V_{pk} for $i = 1, \ldots, N$. We then declare the operators denoting basic formulas to prove in the induction cases and the equations defining the operators. The operators and equations are declared in a module, say ISTEP (which imports INV), as follows:

```
op istepₖ : Vₚₖ -> Bool
eq istepₖ(Zₖ) = invₖ(s,Zₖ) implies invₖ(s',Zₖ) .
```

for $i = 1, \ldots, N$. s and s', which are declared in ISTEP, are constants of H. s denotes an arbitrary state and s' denotes a successor state of the state.

The proof plan of (3), written in CafeOBJ, has the form

```
open INV
  red invₖ(init,zₖᶜ) .
close
```

for $i = 1, \ldots, N$. The command open makes a temporary module that imports a given module and the command close destroys it. The command red reduces a given term. CafeOBJ scripts like this constitute proof scores. Such fragments of proof scores are called *proof passages*. Feeding such a proof passage into the CafeOBJ system, if the CafeOBJ system returns true, the corresponding proof is successfully done.

The proof of (4) often needs case splitting. We suppose that the state space is split into L_k sub-spaces[4] in order to prove (4) and that each sub-space is characterized by a proposition case_{kl} for $l = 1, \ldots, L_k$ provided that $\text{case}_{k1} \vee \ldots \vee \text{case}_{kL_k}$. The proof of (4) can be then replaced with

$$\begin{aligned} &\text{case}_{kl} \Rightarrow \\ &[(P_\alpha(v^c,d_\alpha) \wedge P_\beta(v^c,d_\beta) \wedge \ldots) \Rightarrow [P_k(v^c, z_k^c) \Rightarrow P_k(t_{y_1^c,\ldots,y_n^c}(v^c), z_k^c)]] \end{aligned} \quad (5)$$

for $l = 1, \ldots, L_k$ and $k = 1, \ldots, N$.

We suppose that $d_\alpha, d_\beta, \ldots$ are CafeOBJ terms denoting $d_\alpha, d_\beta, \ldots$ Then the proof passage of (5) has the form

```
open ISTEP
  -- arbitrary objects
  op y₁ᶜ : -> V₁ .   ···   op y_Nᶜ : -> V_N .
  -- assumptions
  Declaration of equations denoting caseₖₗ.
  -- successor state
  eq s' = t(s,y₁ᶜ,...,y_Nᶜ) .
  -- check
  red (invα(s,dα) and invβ(s,dβ) and ...) implies istepₖ(zₖᶜ) .
close
```

for $l = 1, \ldots, L_k$ and $k = 1, \ldots, N$.

---

[4] Generally, such case splitting should be done while $\forall v : \mathcal{R}_S . p(v)$ is being proved.

Equations *available in a proof passage* "open M ⋯ close" are those declared in the module M and the modules imported by M plus those declared in the proof passage. We say that the lefthand side of an equation $l = r$ (a term t) is *(ir)reducible in a proof passage* if l (t) is (ir)reducible wrt $E \setminus \{l = r\}$ (E), where E is the set of all equations available in the proof passage.

We briefly describe the proof scores of $\forall v : \mathcal{R}_{\mathcal{S}_{\mathrm{Tlock}}}.\mathrm{MX}(v)$.

Example 3 (Proof socres of $\forall v : \mathcal{R}_{\mathcal{S}_{\mathrm{Tlock}}}.\mathrm{MX}(v)$). We need four more state predicates to prove $\forall v : \mathcal{R}_{\mathcal{S}_{\mathrm{Tlock}}}.\mathrm{MX}(v)$, which are found while proving it. The four state predicates are as follows: $p_2(v) \triangleq \forall i, j : \mathrm{Pid}. [(\mathrm{pc}_i(v) = \mathrm{cs} \wedge \mathrm{pc}_j(v) = 12 \wedge \mathrm{ticket}_j(v) = \mathrm{turn}(v)) \Rightarrow i = j]$, $p_3(v) \triangleq \forall i : \mathrm{Pid}. (\mathrm{pc}_i(v) = \mathrm{cs} \Rightarrow \mathrm{turn}(v) < \mathrm{tvm}(v))$, $p_4(v) \triangleq \forall i, j : \mathrm{Pid}. [(\mathrm{pc}_i(v) = 12 \wedge \mathrm{pc}_j(v) = 12 \wedge \mathrm{ticket}_i(v) = \mathrm{ticket}_j(v)) \Rightarrow i = j]$, and $p_5(v) \triangleq \forall i : \mathrm{Pid}. (\mathrm{pc}_i(v) = 12 \Rightarrow \mathrm{ticket}_i(v) < \mathrm{tvm}(v))$. The proof of $\forall v : \mathcal{R}_{\mathcal{S}_{\mathrm{Tlock}}}.\mathrm{MX}(v)$ needs p_2, that of $\forall v : \mathcal{R}_{\mathcal{S}_{\mathrm{Tlock}}}.p_2(v)$ needs MX, p_3 and p_4, that of $\forall v : \mathcal{R}_{\mathcal{S}_{\mathrm{Tlock}}}.p_3(v)$ needs MX and p_5, that of $\forall v : \mathcal{R}_{\mathcal{S}_{\mathrm{Tlock}}}.p_4(v)$ needs p_5, and that of $\forall v : \mathcal{R}_{\mathcal{S}_{\mathrm{Tlock}}}.p_5(v)$ needs no other state predicates.

The module INV is declared as follows:

```
mod INV {  pr(TLOCK)
  ops i j : -> Pid
  op inv1 : Sys Pid Pid -> Bool   op inv2 : Sys Pid Pid -> Bool
  op inv3 : Sys Pid -> Bool       op inv4 : Sys Pid Pid -> Bool
  op inv5 : Sys Pid -> Bool
  var S : Sys   vars I J : Pid
  eq inv1(S,I,J) = ((pc(S,I) = cs and pc(S,J) = cs) implies I = J) .
  eq inv2(S,I,J) = ((pc(S,I) = cs and pc(S,J) = 12
                     and ticket(S,J) = turn(S)) implies I = J) .
  eq inv3(S,I)   = (pc(S,I) = cs implies turn(S) < tvm(S)) .
  eq inv4(S,I,J) = ((pc(S,I) = 12 and pc(S,J) = 12
                     and ticket(S,I) = ticket(S,J)) implies I = J) .
  eq inv5(S,I)   = (pc(S,I) = 12 implies ticket(S,I) < tvm(S)) .
}
```

The module ISTEP is declared as follows:

```
mod ISTEP {  pr(INV)
  ops s s' : -> Sys
  op istep1 : Pid Pid -> Bool   op istep2 : Pid Pid -> Bool
  op istep3 : Pid -> Bool       op istep4 : Pid Pid -> Bool
  op istep5 : Pid -> Bool
  vars I J : Pid
  eq istep1(I,J) = inv1(s,I,J) implies inv1(s',I,J) .
  eq istep2(I,J) = inv2(s,I,J) implies inv2(s',I,J) .
  eq istep3(I)   = inv3(s,I) implies inv3(s',I) .
  eq istep4(I,J) = inv4(s,I,J) implies inv4(s',I,J) .
  eq istep5(I)   = inv5(s,I) implies inv5(s',I) .
}
```

Let us consider the following proof passage of $\forall v : \mathcal{R}_{\mathcal{S}_{\text{Tlock}}} . \text{MX}(v)$:

```
open ISTEP
-- arbitrary values
  op k : -> Pid .
-- assumptions
  -- eq c-check(s,k) = true .
  eq pc(s,k) = 12 .  eq ticket(s,k) = turn(s) .
  qq i = k .  eq (j = k) = false .  eq pc(s,j) = cs .
-- successor state
  eq s' = check(s,k) .
-- check
  red istep1(i,j) .
close
```

The proof passage corresponds to a (sub-)case obtained by splitting the induction case for check_k. The (sub-)case is referred as case 1.check.1.1.0.1. CafeOBJ returns false for the proof passage. From the five equations that characterize the (sub-)case, however, we can conjecture p_2. When inv2(s,j,i) implies istep1(i,j) is used instead of istep1(i,j), CafeOBJ returns true for the proof passage.

Let us consider the following proof passage of $\forall v : \mathcal{R}_{\mathcal{S}_{\text{Tlock}}} . p_2(v)$:

```
open ISTEP
-- arbitrary values
  op k : -> Pid .
-- assumptions
  -- eq c-exit(s,k) = true .
  eq pc(s,k) = cs .
  eq (i = k) = false .  eq (j = k) = false .  eq pc(s,i) = cs .
-- successor state
  eq s' = exit(s,k) .
-- check
  red istep2(i,j) .
close
```

The proof passage corresponds to a (sub-)case obtained by splitting the induction case for exit_k. The (sub-)case is referred as case 2.exit.1.0.0.1. Although CafeOBJ returns neither true nor false for the proof passage, we notice that inv1(s,i,k) reduces to false in the proof passage. Therefore, we use inv1(s,i,k) implies istep2(i,j) instead of istep2(i,j) and then CafeOBJ returns true for the proof passage.   □

3   Tips

What we should do to prove a state predicate invariant wrt an OTS is three tasks: (1) use of simultaneous induction, (2) case splitting and (3) predicate (lemma) discovery/use. We use the proof of $\forall v : \mathcal{R}_{\mathcal{S}_{\text{Tlock}}} . \text{MX}(v)$ to describe the three tasks.

3.1   Simultaneous Induction

The first thing to do is to use simultaneous induction to break the proof into the four (sub-)goals (one is the base case and the others are the three induction cases) and the four proof passages are written. The proof passage of the base case is as follows:

```
open INV
  red inv1(init,i,j) .
close
```

The proof passage of the induction case for check$_k$ is as follows:

```
open ISTEP
  op k : -> Pid .
  eq s' = check(s,k) .
  red istep1(i,j) .
close
```

The case is referred as case 1.check. The proof passages of the remaining two induction cases are written likewise.

CafeOBJ returns **true** for the base case but neither **true** nor **false** for each of the three induction cases. What to do for the three induction cases are case splitting and/or predicate discovery/use.

3.2   First Thing to Do for Each Induction Case

Each induction case for $t_{y_1,...,y_n}$ is split into two (sub-)cases: (1) c-$t_{y_1,...,y_n}$ and (2) $\neg c$-$t_{y_1,...,y_n}$ unless c-$t_{y_1,...,y_n}$ holds in every case. Case 1.check is split into the two (sub-)cases whose corresponding proof passages are as follows:

```
open ISTEP                      open ISTEP
  op k : -> Pid .                 op k : -> Pid .
  eq c-check(s,k) = true .        eq c-check(s,k) = false .
  eq s' = check(s,k) .           eq s' = check(s,k) .
  red istep1(i,j) .              red istep1(i,j) .
close                           close
```

The two (sub-)cases are referred as case 1.check.1 and 1.check.0. CafeOBJ returns **true** for case 1.check.0 but neither **true** nor **false** for case 1.check.1. CafeOBJ always returns **true** for the (sub-)case where $\neg c$-$t_{y_1,...,y_n}$ due to Definition 1 if the OTS concerned is correctly written in CafeOBJ.

3.3   Appropriate Equations Declared in Proof Passages

As shown, each (sub-)case is characterized by equations. Equational reasoning by rewriting is used to check if a proposition holds in each case, but full equational reasoning power is not used because CafeOBJ does not employ any completion facilities. Therefore, equations that characterize a case heavily affects the success in proving that a proposition holds in the case. We describe appropriate equations, which characterize a case, declared in a proof passage. If CafeOBJ returns true for a proof passage, nothing should be done. Otherwise, the equations in the proof passage should be appropriate as described from now.

- The lefthand side of each equation should be irreducible in a proof passage so that the equation can be used effectively as a rewrite rule. This is because the rewriting strategy adopted by CafeOBJ is basically an innermost strategy.
- Let $PP(E)$, where E is a set of equations, be a proof passage in which the equations in E are declared, and E_1 and E_2 be sets of equations. We suppose that $\bigwedge_{e_1 \in E_1} e_1$ is equivalent to $\bigwedge_{e_2 \in E_2} e_2$. If every equation in E_1 can be proved by rewriting from $PP(E_2)$ but every equation in E_2 cannot be proved by rewriting from $PP(E_1)$, then E_2 should be used instead of E_1. Some examples are given.

  1. Let E_1 be $\{\rho_1 \wedge \rho_2 = \text{true}\}$ and E_2 be $\{\rho_1 = \text{true}, \rho_2 = \text{true}\}$. We suppose that $\rho_1 \wedge \rho_2$, ρ_1 and ρ_2 are irreducible in $PP(\emptyset)$. Then, $\rho_1 \wedge \rho_2$ reduces to true in $PP(E_2)$ but l_1 (l_2) does not necessarily reduce to true in $PP(E_1)$. Therefore, E_2 should be used instead of E_1.
  2. Let c be a binary data constructor. We suppose that $c(a_1, b_1)$ equals $c(a_2, b_2)$ if and only if a_1 equals a_2 and b_1 equals b_2. Let E_1 be $\{c(a_1, b_1) = c(a_2, b_2)\}$ and E_2 be $\{a_1 = a_2, b_1 = b_2\}$. We suppose that $c(a_1, b_1)$, a_1 and b_1 are irreducible in $PP(\emptyset)$. Then, both $c(a_1, b_1)$ and $c(a_2, b_2)$ reduce to a same term in $PP(E_2)$ but a_1 and a_2 (b_1 and b_2) do not necessarily reduce to a same term in $PP(E_1)$. Therefore, E_2 should be used instead of E_1.
  3. Let n be a natural number, N be a constant denoting an arbitrary multiset of natural numbers, the juxtaposition operator be a data constructor of multisets. The juxtaposition operator is declared as op __ : Bag Bag -> Bag {assoc comm id: empty}, where Bag is the visible sort for multisets of natural numbers and is a supersort of Nat, assoc and comm specify that the operator is associative and commutative, and id: empty specifies that empty, which is the constant denoting the empty multiset, is an identity of the operator. We suppose that we want to specify that N includes n. One way is to use $n \in N = \text{true}$, and the other way is to use $N = n\ N'$, where N' is another constant denoting an arbitrary multiset of natural numbers[5]. Let E_1 be $\{n \in N = \text{true}\}$ and E_2 be $\{N = n\ N'\}$. We suppose that $n \in N$ and N are irreducible in $PP(\emptyset)$. Then, $n \in N$ reduces to true in $PP(E_2)$ if $\in$ is defined appropriately in equation, but N and $n\ N'$ do not necessarily reduce to a same term in $PP(E_1)$. Therefore, E_2 should be used instead of E_1.
  4. $\neg \rho$ is reducible in any proof passage because of the Hsiang TRS. If ρ is irreducible in a proof passage, $\neg \rho$ reduces to ρ xor true in the proof

---

[5] Since N is an arbitrary multiset and includes n, N must be $n'\ N'$, where (1) n' equals n or (2) $n \in N'$. We can select (1) because the juxtaposition operator is associative and commutative.

passage. Therefore, one way of making the equation $(\neg\rho) = \texttt{true}$ effective is to use $(\rho \texttt{ xor true}) = \texttt{true}$. But, $\rho = \texttt{false}$ is more appropriate.

5. This example is a variant. Let E_1 be $\{(l = r) = \texttt{true}\}$ and E_2 be $\{l = r\}$. We suppose that $l = r$ and l are irreducible in $PP(\emptyset)$. $l = r$ reduces to $\texttt{true}$ in both $PP(E_1)$ and $PP(E_2)$. It is often the case, however, that E_2 is more appropriate than E_1 because l reduces r in $PP(E_2)$ but l does not in $PP(E_1)$.

According to what has been described in this subsection, the proof passage of case 1.check.1 should be rewritten as follows:

```
open ISTEP
  op k : -> Pid .
  -- eq c-check(s,k) = true .
  eq pc(s,k) = 12 .   eq ticket(s,k) = turn(s) .
  eq s' = check(s,k) .
  red istep1(i,j) .
close
```

CafeOBJ still returns neither $\texttt{true}$ nor $\texttt{false}$ for this proof passage. Then, what we should do is further case splitting.

3.4 Further Case Splitting

For a proof passage for which CafeOBJ returns neither $\texttt{true}$ nor $\texttt{false}$, the case corresponding to the proof passage is split into multiple (sub-)cases in each of which CafeOBJ returns either $\texttt{true}$ or $\texttt{false}$. When CafeOBJ returns $\texttt{true}$ in a (sub-)case, nothing should be done for the case. When CafeOBJ returns $\texttt{false}$ in a (sub-)case, it is necessary to find a state predicate that does not hold in the case and is likely invariant wrt an OTS concerned.

There are some ways of splitting a case into multiple (sub-)cases.

- Based on a proposition ρ: A case is split into two (sub-)cases where (1) ρ holds and (2) ρ does not, respectively. As shown in Subsect. 3.2, case 1.check is split into the two (sub-)cases based on the proposition $\texttt{c-check(s,k)}$.
- Based on data constructors: We suppose that a data type has M data constructors. Then, a case is split into M (sub-)cases. Some examples are given.
  1. $\texttt{Nat}$ has the two data constructors 0 and $\texttt{s}$. Let x be a constant denoting an arbitrary natural number in a proof passage. The case corresponding to the proof passage is split into the two (sub-)cases where (1) $x = 0$ and (2) $x = \texttt{s}(y)$, where y is another constant denoting an arbitrary natural number. Case (1) means that x is zero and case (2) means that x is not zero.
  2. $\texttt{Bag}$ has the two data constructors $\texttt{empty}$ and $\texttt{\_\_}$. Let N be a constant denoting an arbitrary multiset in a proof passage. The case corresponding to the proof passage is split into the two (sub-)cases where (1) $N = \texttt{empty}$ and (2) $N = n'\ N'$, where n' is a constant denoting an arbitrary natural number and N' is a constant denoting an arbitrary multiset. Case (1) means that N is empty and case (2) means that N is not empty.

- Based on a tautology whose form is $\rho_1 \vee \ldots \vee \rho_M$: A case is split into M (sub-) cases where (1) ρ_1 holds, ..., (M) ρ_M holds. This case splitting generalizes the case splitting based on a proposition because $\rho \vee \neg\rho$ is a tautology.

In order to apply one of the three ways of splitting a case, we need to find a proposition, a constant denoting an arbitrary value of a data type, or a tautology whose form is $\rho_1 \vee \ldots \vee \rho_M$. There are usually multiple candidates based on which a case is split. A selection from such candidates affects how well a proof concerned is conducted. It is necessary to understand an OTS concerned and experience writing proof scores so as to select a better one among such candidates. There are some heuristic rules, however, to select one among such candidates.

- Select a proposition that directly affects the truth value of a proposition to prove such as istep(i,j). If i equals j, istep(i,j) reduces to true in case 1.check.1, the proposition i = j may be a good candidate.
- Select a proposition ρ if ρ appears in a result obtained by reducing a proposition to prove. If ρ appears at the conditional position of if_then_else_fi such as if ρ then a else b fi, ρ may be a good candidate.

We describe how to split case 1.check.1. CafeOBJ returns ((if (k = i) then cs else pc(s,i) fi) = cs) and ... for the corresponding proof passage. Then, we select the proposition k = i to split the case. The equation i = k is declared[6] in one proof passage whose corresponding case is referred as case 1.check.1.1, and the equation (i = k) = false is declared in the other proof passage whose corresponding case is referred as case 1.check.1.0.

Since CafeOBJ returns if (k = j) then cs else pc(s,j) fi = cs and ... for the proof passage corresponding case 1.check.1.1, we select the proposition k = j to split the case. The equation j = k is declared in one proof passage whose corresponding case is referred as 1.check.1.1.1, and the equation (j = k) = false is declared in the other proof passage whose corresponding case is referred as case 1.check.1.1.0. CafeOBJ returns true for the former proof passage, but pc(s,j) = cs xor true for the latter proof passage. Then, case 1.check.1.1.0 is also split based on pc(s,j) = cs. The equation pc(s,j) = cs is declared in one proof passage whose corresponding case is referred as case 1.check.1.1.0.1, and the equation (pc(s,j) = cs) = false is declared in the other proof passage whose corresponding case is referred as case 1.check.1.1.0.0. CafeOBJ returns false for the former proof passage and true for the latter proof passage. Case 1.check.1.0 can be split into four (sub-)cases in the same was as case 1.check1.1.

3.5   Predicate (Lemma) Discovery/Use

When CafeOBJ returns false for a proof passage, there are two possibilities: (1) if an an arbitrary state characterized by the case corresponding to the proof passage is not reachable wrt an OTS S concerned, the case can be discharged,

---

[6] Note that i = k is declared instead for k = i.

and (2) otherwise, a state predicate concerned is not invariant wrt $\mathcal{S}$. If a state predicate is invariant wrt $\mathcal{S}$ and does not hold in the case, then an arbitrary state characterized by the case is not reachable wrt $\mathcal{S}$. That is why we find a state predicate that does not hold in the case and is likely invariant wrt $\mathcal{S}$.

Let E is a set of equations that characterize a case such that CafeOBJ returns `false` for a proof passage corresponding to the case. We suppose that $\bigwedge_{e \in E} e$ is equivalent to a proposition whose form is $Q(v^c, z^c_\alpha)$. Let $q(v)$ be $\forall z_\alpha : D_{q\alpha}. \neg Q(v, z_\alpha)$. Since q surely does not hold in the case characterized by E, q is one possible candidate. Generally, q' such that $q' \Rightarrow q$ can be a candidate because q' does not hold in the case characterized by E,

Let us consider the proof passage corresponding to case 1.check.1.1.0.1 shown in Example 3. From the five equations that characterize the case, we obtain the proposition `pc(s,i) = 12 and pc(s,j) = cs and ticket(s,i) = turn(s) and not(j = i)` by concatenating them with conjunctions, substituting k with i because of the equation `i = k`, and deleting the tautology `i = i`. p_2 is obtained from the proposition,

Some contradiction may be found in a set of equations that characterize a case even when CafeOBJ does not return `false` in a proof passage corresponding to the case. If that is the case, a state predicate can be obtained from the contradiction such that the state predicate does not hold in the case and is likely invariant wrt an OTS concerned.

Let us consider the proof passage corresponding to case 2.exit.1.0.0.1 shown in Example 3. We notice that the three equations `pc(s,k) = cs`, `pc(s,i) = cs` and `(i = k) = false` contradict $\forall v : \mathcal{R}_{\mathcal{S}_{\text{Tlock}}}. \text{MX}(v)$ and `inv1(s,i,k)` can be used in the proof passage.

Even when any contradictions are not found in a set of equations that characterize a case and CafeOBJ does not return `false` in a proof passage corresponding to the case, a state predicate may be found such that the state predicate can be used to discharge the case and is likely invariant wrt an OTS concerned.

Let us consider the proof passage corresponding to case 1.check.1.1.0. CafeOBJ returns `pc(s,j) = cs xor true` for the proof passage, but `inv2(s,j, i)` also reduces to `pc(s,j) = cs xor true` in the proof passage. Therefore, `inv2(s,j,i)` can be used to discharge the case and it is not necessary to split the case anymore.

4   Soundness of Proof Scores

Let us consider the proof of $\forall v : \mathcal{R}_{\mathcal{S}}. (p_1(v) \land \ldots \land p_N(v))$ described in Subsect. 2.4 again. If CafeOBJ returns `true` for each proof passage in the proof scores, $p_1, \ldots, p_N$ are really invariant wrt $\mathcal{S}$ provided that

1. Needless to say, the computer (including the operating system, the hardware, etc.) on which CafeOBJ works is reliable,
2. Equational reasoning is sound and rewriting faithfully (partially though) implements equational reasoning [8]; the CafeOBJ implementation of rewriting is reliable,

3. The Hsiang TRS is sound [12]; the TRS is reliably implemented in CafeOBJ,
4. The built-in equality operator _==_ is not used,
5. S is specified in CafeOBJ in the way described in Subsect. 2.3, and
6. The proof scores of $\forall v : \mathcal{R}_S.(p_1(v) \land \ldots \land p_N(v))$ are written in the way described in Subsect. 2.4.

When CafeOBJ meets the term a == b, it first reduces a and b to a' and b', which are irreducible wrt a set of equations (rewrite rules) concerned, and returns true if a' is exactly the same as b' and false otherwise. The combination of _==_ and not_ can damage the soundness. Since the built-in inequality operator _=/=_ is the combination of _==_ and not_, it should not be used either. Let us consider the following module:

```
mod! DATA { [Data]
  ops d1 d2 : -> Data
}
```

We try to prove $\forall d : \mathtt{Data}. \neg(d = \mathtt{d2})$ by writing a proof score. A plausible proof score that consists of one proof passage is as follows:

```
open DATA
  op d : -> Data .    -- an arbitrary value of Data.
  red not(d == d2) . -- or red d =/= d2 .
close
```

CafeOBJ returns true for this proof passage, which contradicts the fact that there exists the counterexample d2. Therefore, users should declare an equality operator such as _=_ for each visible sort and equations defining it instead of _==_ and _=/=_.

Under the above six assumptions, the only thing that we should take care of on the soundness is whether all necessary cases are checked by rewriting for each proof passage. A possible source of damaging it is transitions. Since transitions are functions on states in OTSs, however, the source can be dismissed. Every operator is a function in CafeOBJ as well. Therefore, rewriting surely covers all necessary cases for each proof passage.

Note that we do not have to assume that the CafeOBJ specification of S, when it is regarded as a TRS, is terminating or confluent for the soundness. If the CafeOBJ specification is not terminating, CafeOBJ may not return any results for a proof passage forever. This causes the success in proofs, but does not affect the soundness.

We suppose that a term a has two irreducible forms a' and a'' in a proof passage because the CafeOBJ specification is not confluent and that a actually reduces to a' but not to a''. Although CafeOBJ ignores a rewriting sequence that starts with a and ends in a'', this does not affect the soundness because a' equals a'' from an equational reasoning point of view and it is enough to use either a' or a''. Whether the CafeOBJ specification is confluent, however, can affects the success in proofs. Let us consider the following module:

```
mod! DATA2 { [Data2]
  ops d1 d2 d3 : -> Data2
  op _=_ : Data2 Data2 -> Bool {comm}
  var D : Data2
  eq (D = D) = true .
  eq d1 = d2 .   eq d1 = d3 .
}
```

We try to prove d1 = d3 by writing a proof passage. The case is split into two (sub-)cases where (1) d2 = d3 and (2) d2 ≠ d3. Then, the proof score that consists of two proof passages is as follows:

```
open DATA2
  eq d2 = d3 .
  red d1 = d3 .
close
open DATA2
  eq (d2 = d3) = false .
  red d1 = d3 .
close
```

CafeOBJ returns true for the first proof passage and false for the second proof passage. We stuck for the second proof passage unless we notice the equation d1 = d3 in the module DATA2.

From what has been described, it is desirable that the CafeOBJ specification of S is terminating and confluent.

We can check if proof scores that state predicates are invariant wrt S conforms to what is described in Subsect. 2.4. We suppose that all proofs are conducted by simultaneous induction. Let P and P' be sets of state predicate such that P' is empty. A procedure that makes such a check is as follows:

1. If P is empty, the procedure successfully terminates, which means that the proof score of $\forall v : \mathcal{R}_S . p(v)$ for each $p \in P'$ conforms to what is described in Subsect. 2.4; otherwise, extract a predicate p from P and go next.
2. Check if a proof score of $\forall v : \mathcal{R}_S . p(v)q$ has been written. If so, go next; otherwise, the procedure reports that a proof score of $\forall v : \mathcal{R}_S . p(v)$ has not been written and terminates.
3. Check if the proof score of $\forall v : \mathcal{R}_S . p(v)q$ conforms to simultaneous induction. If so, go next; otherwise, the procedure reports that the proof score of $\forall v : \mathcal{R}_S . p(v)q$ does not conform to simultaneous induction and terminates.
4. Check if the proof score of $\forall v : \mathcal{R}_S . p(v)q$ covers all necessary cases. If so, put p into P', put other state predicates that are used in the proof score and that are not in P' into P, and go to 1; otherwise, the procedure reports that the proof score of $\forall v : \mathcal{R}_S . p(v)$ does not cover all necessary cases and terminates.

The procedure can increase the confidence in soundness of proof scores.

5   Conclusion

We have described some tips on writing proof scores in the OTS/CafeOBJ method and used Tlock, a mutual exclusion protocol using *atomicInc*, to exemplify the tips. We have also informally argued soundness of proof scores in the OTS/CafeOBJ method.

We have been developing a tool called Gateau [21] that takes propositions used for case splitting and state predicates used to strengthen the basic induction hypothesis, and generates the proof score of an invariant, which conforms to what is described in Subsect. 2.4.

Proof scores can also be considered proof objects, which can be checked as described in Sect. 4. We think that it is worthwhile to develop a tool, which is an implementation of the procedure in Sect. 4 that checks if a proof score conforms to what is described in Subsect. 2.4. Such a tool can be complementary to Gateau.

References

1. Y. Bertot and P. Castéran. *Interactive Theorem Proving and Program Development – Coq'Art: The Calculus of Inductive Constructions*. Springer, 2004.
2. R. Diaconescu and K. Futatsugi. *CafeOBJ Report*, volume 6 of *AMAST Series in Computing*. World Scientific, 1998.
3. R. Diaconescu and K. Futatsugi. Behavioural coherence in object-oriented algebraic specification. *J. UCS*, 6:74–96, 2000.
4. R. Diaconescu, K. Futatsugi, and K. Ogata. CafeOBJ: Logical foundations and methodologies. *Computing and Informatics*, 22:257–283, 2003.
5. K. Futatsugi, J. A. Goguen, J. P. Jouannaud, and J. Meseguer. Principles of OBJ2. In *12th POPL*, pages 52–66. ACM, 1985.
6. K. Futatsugi, J. A. Goguen, and K. Ogata. Verifying design with proof scores. In *VSTTE 2005*, 2005.
7. K. Futatsugi, J. A. Goguen, and K. Ogata. Formal verification with the OTS/Cafe-OBJ method. submitted for publication, 2006.
8. J. Goguen. *Theorem Proving and Algebra*. The MIT Press, to appear.
9. J. Goguen and G. Malcolm. A hidden agenda. *TCS*, 245:55–101, 2000.
10. J. Goguen and G. Malcolm, editors. *Software Engineering with OBJ: Algebraic Specification in Action*. Kluwer, 2000.
11. J. V. Guttag, J. J. Horning, S. J. Garland, K. D. Jones, A. Modet, and J. M. Wing. *Larch: Languages and Tools for Formal Specification*. Springer, 1993.
12. J. Hsiang and N. Dershowitz. Rewrite methods for clausal and nonclausal theorem proving. In *10th ICALP*, LNCS 154, pages 331–346. Springer, 1983.
13. T. Nipkow, L. C. Paulson, and M. Wenzel. *Isabelle/HOL: A Proof Assistant for Higher-Order Logic*. LNCS 2283. Springer, Berlin, 2002.
14. K. Ogata and K. Futatsugi. Flaw and modification of the *i*KP electronic payment protocols. *IPL*, 86:57–62, 2003.
15. K. Ogata and K. Futatsugi. Formal analysis of the *i*KP electronic payment protocols. In *1st ISSS*, LNCS 2609, pages 441–460. Springer, 2003.
16. K. Ogata and K. Futatsugi. Formal verification of the Horn-Preneel micropayment protocol. In *4th VMCAI*, LNCS 2575, pages 238–252. Springer, 2003.

17. K. Ogata and K. Futatsugi. Proof scores in the OTS/CafeOBJ method. In *6th FMOODS*, LNCS 2884, pages 170–184. Springer, 2003.
18. K. Ogata and K. Futatsugi. Equational approach to formal verification of SET. In *4th QSIC*, pages 50–59. IEEE CS Press, 2004.
19. K. Ogata and K. Futatsugi. Formal analysis of the NetBill electronic commerce protocol. In *2nd ISSS*, volume 3233 of *LNCS*, pages 45–64. Springer, 2004.
20. K. Ogata and K. Futatsugi. Equational approach to formal analysis of TLS. In *25th ICDCS*, pages 795–804. IEEE CS Press, 2005.
21. T. Seino, K. Ogata, and K. Futatsugi. A toolkit for generating and displaying proof scores in the OTS/CafeOBJ method. In *6th RULE*, ENTCS 147(1), pages 57–72. Elsevier, 2006.

Drug Interaction Ontology (DIO) and the Resource-Sensitive Logical Inferences

Mitsuhiro Okada[1], Yutaro Sugimoto[1],
Sumi Yoshikawa[2], and Akihiko Konagaya[2]

[1] Logic Group, Department of Philosophy, Keio University
2-15-45 Mita, Minato-ku, Tokyo, 108-8345, Japan
{mitsu,sugimoto}@abelard.flet.keio.ac.jp

[2] Genomic Science Center (GSC), RIKEN
1-7-22 Suehiro-cho, Tsurumi-ku, Yokohama City, Kanagawa, 230-0045, Japan
{sumi,konagaya}@gsc.riken.jp

Abstract. In this paper, we propose a formulation for inference rules in Drug Interaction Ontology (DIO). Our formulation for inference rules is viewed from the standpoint of process-description. The relations in DIO are now described as resource-sensitive linear logical implications. The compositional reasoning on certain drug-interactions discussed in our previous work on DIO is represented as a construction of a linear logical proof. As examples of our formulation, we use some anti-cancer drug interactions.[3]

1 Introduction

Ontology-oriented knowledgebases have been studied and developed in various fields, where knowledgebases are designed in accordance with the underlying ontological structures, such as structures of persistent objects, structures of functions, structures of processes, etc. Ontologies of the biomedical and bioinformatic domain have been studied and developed very intensively, as well as ontologies of other specific domains and domain-independent general ontologies.[4] Needless to say, in order to make an ontology-based database useful and practical, it is important to provide a suitable formal language and an inference engine, which will make the best use of the ontological structures of the relevant domains. For this purpose, various formal language frameworks and various inference engines have been proposed in the literature on biomedical and bioinformatic applications.

Some have employed tree and graph structures for the basic formal structures and retrieving-search engines on the tree/graphs are used. [5] Others have used the relation-based predicate logic language and its variants as the formal

---

[3] We would like to express our sincere thanks to the anonymous referee for invaluable comments on earlier versions of this paper.

[4] For some survey of ontology-methodology for knowledgebases, see e.g. [17] [22] [26].

[5] E.g. Gene Ontology (GO) Editorial Style Guide [1] [2].

K. Futatsugi et al. (Eds.): Goguen Festschrift, LNCS 4060, pp. 616–642, 2006.

framework, in which the logical engine based on first order predicate logic is often employed explicitly or implicitly (e.g. [8] [13]): B.Smith *et al.* [23] [24], for example, proposed the use of a somewhat limited number of primitive predicates/relations for a biological ontology in the setting of a fragment of first order predicate logic. Description Logics are also considered variants of predicate logic, which enhance the latter's expressive power (for concepts, for example) to some extent while preserving its effective computability.[6] The transformation and integration techniques among different ontology languages are also important, and some pioneering works have been done by Goguen and others (cf., e.g. [5] [6] [7]) using category theoretical tools.

In the domain of drug/pharmaceutical applications, knowledgebases developed on the basis of molecular level ontology, are particularly useful, and it would be important to design the reasoning of drug-interactions according to the ontology-based knowledgebases, as well as to the traditional static ontological structures of drug-related knowledgebases. Although drug interactions are represented by relations in a static manner, it is desirable that the interaction processes themselves be captured within the logical reasoning/inference framework.

The main objective of this paper is to design a logical inference engine for a process-based biological ontology. And for that purpose, we will herein adopt a molecular interaction-based process ontology modeling method for some drug interactions (Drug Interaction Ontology: DIO) from [27] [28] [29]. In our previous work on DIO, we proposed certain schematic or abstract inferences based on basic triadic relations in molecular-interactions, such as "Drug a facilitates the generation of c under the action of enzyme b in a situated environment" ($facilitate(a, b, c)$, for short), or "Drug a inhibits the generation of c under the action of enzyme b in a situated environment".

While the use of such relations for interaction-processes provides a schematic inference tool in Drug Interaction Ontology, and the relation-based reasoning often hides the processes level, the question concerning the logical consolidation for the inference engine of DIO has been left opened. In this paper, we introduce a variant of resource-sensitive logic, linear logic, to explain the basic inference engine used in our previous work on DIO, where the relational approach is reduced to a more basic process approach, and accordingly the basic triadic relation for molecular interactions, for example, is expressed as the logical description of interaction processes, using a resource-sensitive logical implication such as "The coexistence of resource a and of the environmental resource b implies product c (in the linear logical expression $(a, !b) \multimap c$)." Here, $!b$ expresses a relatively large amount of resources b in the sense that the resource consumption of b can be ignored in the context of the reasoning.

---

[6] For the applications of Description Logics to ontologies, see e.g. [11] [25].

Various multiple interaction processes can then be described as a (linear logical) formal deduction proof, where a composed interaction process could be identified with a structurally composed logical proof. (In other words, a certain part of the DIO process could be simulated by a formal deductive proof process.) This suggests that some inquiries for the DIO-based ontological knowledgebase, which could be treated by a logical proof-search engine of a variant of linear logic. And indeed we shall argue hereinafter that a suitable formulation can be given for the basic inferences on DIO (in [27] [28] [29] etc.) in a resource-sensitive logic. Logical representations have been used in various areas of computer science where theorem-proving approaches are applied in order to represent the processes (cf. e.g. [9]). In particular, in this paper, we shall show that:

1. The relational-level oriented approach described by Yoshikawa *et al.* [27] [28] is reduced to the process-level approach, by the use of a logical system adopted for the linear logical (resource-sensitive logical) process-descriptions of drug interactions;
2. A composed process-description (for drug interactions) can be formulated by means of a composed logical proof of the resource-sensitive "linear logic";
3. For a negative expression, such as "inhibiting", used in the reasoning of DIO, we introduced a quantitative modality, in addition to the usual modality of resource-sensitive logics, in order to adjust the standard resource-sensitive process-description logic (such as linear logic) to the specific domain-oriented inferences of DIO. The usual modality $!A$ represents the existence of a relatively large amount of resource A, hence the consumption of resource A can be ignored when $!A$ is used for some reusable environmental resources, such as enzymes (cf., e.g. [19]). The typical use of the standard modality appears as an environmental resource of an enzyme, where the consumption of an environmental enzyme during the reaction can be ignored, and the enzyme can be considered to exist before and after the reaction. On the other hand, we introduced a non-logical domain-specific modality ∇, a domain specific quantitative modality, which describes a significant decrease of resource A. This modality is used for describing inhibition.

While the triadic relation (holding among the drug, environmental enzyme, and product) was introduced as the basic relation in the DIO to derive drug-drug interactions [28], in this paper we present a formal break-out of this basic relation into the process level, where the triadic relation is now described as a resource-sensitive linear logical formula. Then one might be able to fit the inferences for the drug-drug interactions in a precise logical inference system. In particular, this logical inference level is regarded as the process description level (using the concurrent process description methodology of linear logic), while preserving the basic DIO ontological modeling framework. (see Table 1) This precise logical formalization shows a way for DIO to be designed/implemented for practical uses, based on a logical inference engine.

Table 1. The Correspondence between the basic DIO relations and our logical process descriptions

DIO Descriptions for Molecular Interactions	Meaning	Linear Logical Process Descriptions
$MI(a,b,c)$ also written $facilitate(a,b,c)$	An emerged molecule a interacts with b, and c emerges as a result. From this relation, "a facilitates the generation of c (under the presence of b)" is inferred.	$(a, !b) \multimap c$
$bind\_more(a,b,a*b)$ (previously written as $bind\_more(a,b,a{=}b)$ in the DIO literature)	A molecular binding of a and b, caused by the emergence of a, is relatively more frequently or durably formed than the other binding complexes with b in the scope of interest.	$(a,b) \multimap a \otimes b$
$inhibit(A,B)$	The emergence of a resource A may inhibit B.	$A \multimap \triangledown!B$

$a, b, c, \ldots$ denote molecular expressions, and A, B, C denote composed states of molecular expressions (cf. section 3).

2 Preliminaries to Drug Interaction Ontology (DIO) Modeling

2.1 Biomedical Ontologies

In the biomedical domain, several ontologies have been developed for different purposes, and many of them were built through the so-called "concept-centered" or "terminology-centered" approaches. To be sure, these ontologies and/or terminology systems would be useful as repositories for the tasks involving a large amount of complex technical terms and emerging new terms. However, the concept/terminology centered approaches do not seem too efficient when applied to computational inferences. As a solution to this problem, we might think of adopting basic ontological schemes in the hope that this would help us develop well-structured knowledgebases, which are applicable to biological pathway models, and which can deal with sophisticated medical information and so on.

One of such basic ontological schemes, BFO [21], the Basic Formal Ontology developed by IFOMIS [12] for application in the medical domain, provides two categories, SNAP and SPAN, both of which are formulated in predicate logic: the former corresponds typically to continuants, and the latter to occurrents or processes. Another scheme, "The Relation Ontology" which is reported in *Relations in Biomedical Ontologies* [23], is designed to provide basic relations that cover every granular level, namely as molecules to organisms, in the biomedical domain. Here, biological entities are largely divided into continuants and processes, and the relation between the two is also defined as an instance-level primitive one; P $has\_participant$ C (P: process, C: continuant). (Some examples of continuants and molecular level processes are shown in the upper part of Figure 1.)

We think it is important to deal with the relation between *continuant* and *process* in a more precise or ontological manner; because the manipulations of their inter-relation is an essential part of communication in the discipline. We have to keep in mind that the main topic in the discipline includes such questions as "How are new biological substances endogenously produced as a result of a certain stimulus (an emergence of continuants)?", or "How are certain biological reactions/phenomena brought about or regulated?". Causality expressions are often used in tandem with terms describing the inter-relation, and such terms as "facilitate" (a term indicating the positive direction), or "inhibit" (a term indicating the negative direction) are used to refer to their manipulations. However, as far as our investigation indicates, there is no ontology that defines "facilitate"/"inhibit" or "facilitator"/"inhibitor" in the context of relations between continuants and processes, beyond the terminological level.

2.2   Drug Interaction Ontology as Molecular Level Process Ontology

Basic Formula of Molecular Interaction   Our previous work on Drug Interaction Ontology (DIO) [27] [28] [29] can be regarded as an attempt to put forward an ontology for primitive processes at the molecular level, using three role relations between continuants participating in the process. As shown in Table 1, a molecular interaction is in general represented as a triadic (role) relation of continuants; $MolecularInteraction(a, b, c)$ or $MI(a, b, c)$. They can be read as "An emerged molecule a interacts with b, and c emerges as a result", where a denotes an *input* or *trigger*, b denotes an (environmentally situated) *object*, and c denotes an *output* or *resultant product*, respectively. In other words, a (*input*) and b (*object*) are necessary participants to bring about the process (*enablers*), while c (*output*) is an emerged continuant as a result of the process. A difference between *input* and *object* is that the latter is a relatively "situated" continuant in the field/place of interest (e.g. a pool of reactions such as inside the cell). In other words, the *output* continuant, by its semantic definition, may be a *trigger* or constitute an *input* of another molecular interaction process.

The triadic relation for molecular interaction, $MI(a, b, c)$, can also be read as "The emergence of a facilitates the emergence of c, which is mediated by situated b". It can also be written as $facilitate(a, b, c)$. The latter part "mediated by situated b" could be sometimes omitted. In such case, it may also be written as $facilitate\ (a, c)$.

In the Drug Interaction Ontology model, we defined different types of interactions by comparing the existing pattern (e.g. a change of molecular population, a change of location pattern) of participants. In this paper, we will deal with the enzymatic catalytic reaction (substrate-enzyme reaction) and the so-called "enzyme inhibition" reaction (inhibitor-enzyme reaction). Several instances of these relations, both of which are subclasses of $MI(a, b, c)$, will be used later.

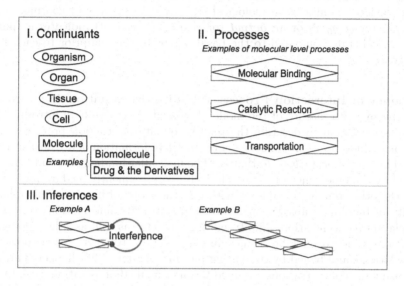

Fig. 1. Inferences Using Continuants and Processes

Transient Complex This triadic model encapsulates details of filling events during the time course of the process. For example, formation of transient complexes and (in some cases) their dissociation processes are encapsulated. In physico-chemical molecular interactions, a certain kind of transient binding complex is formed, but its lifetime may be very short. It might be represented by something like $MI\_transient(a, b, a * b)$, if we consider the formation of a transient complex is the end of the reaction process. Some binding complexes last for a relatively long time and are called "molecular bindings" or "formation of assemblies", which we deal with here as subclasses of the triadic molecular interaction.

The efficiency of the formation of transient complexes depends largely on the quantitative chances of encountering every enabler molecule and affinity property. An encounter of molecules may be largely influenced by the local molecular population (concentration) and by other factors, such as the existence of competitive interaction counterparts. The latter factor is influenced not only by intrinsic molecular affinities, but also by other environmental conditions, which might modulate the properties. Unlike chemical reactions, which can occur under artificially controlled settings, biological reactions usually take place in an environment filled with many concomitant continuants and processes, under physiological conditions.

As it would be almost impossible to describe every detail of an event including all continuants and in a given biological environment, what we represent as the triadic molecular interaction model, by its defined meanings, is a dynamic event which deviates from a basal biological state. The emergence of continuants as an *input* (or *trigger*) or an *output* (or *resultant product*) indicates a change of biological state compared to the basal one or to the one prior to each such emergence.

Efficiency of Interaction Processes: *bind_more* Relation In accordance with the semantic nature of the model discussed above, we here introduce the *bind_more* relation to describe the relative quality of reaction processes. The relation includes in it the influence of competitive formation of a transient complex. The $bind\_more(a, b, a * b)$ that is "The molecular binding of a and b, $(a * b)$, caused by the emergence of a", is relatively more frequently or durably formed than the other binding complexes with b in the scope of interest. This relative quality of binding is abstracted from the quantitative information concerning key players (concentration, affinity parameters, etc) as well as other biological conditions which cannot be defined in every detail. To some extent, however, this relation could be led by comparing pharmacokinetic / biochemical parameters, using the concentrations obtained by an ordinal administration dose. In our two examples below, one is known as a "mechanism-based inhibition" which is a tighter binding than any other ordinary substrate, and thus the *bind_more* relation is clearly manifest as reported in the literature. In the other example case too, the *bind_more* relation is adopted since the binding complex lasts longer than most of the ordinary substrates.

Inhibiting Relation In the literature of the bio-medical domain, narrative expressions such as "a inhibits b" are often used while ignoring the types of interactions. In Drug Interaction Ontology, direct molecular interaction is regarded as a subclass of triadic molecular interaction $MI(a, b, c)$, and represented as $MI\_xi(a, b, c)$.[7] Here, the relation $inhibit(a, b)$ means "The emergence of a decreases b". The inhibitory relations in other complex type of interactions, or in combination of more than two triadic interactions, are called "inferred inhibitions" in DIO.

2.3 Examples

Drug Interaction Between 5-FU and SRV [8]

Figure 2 (quoted from [27] with some modifications), is a pathway map which is manually created in order to explain the causal effect of the anticancer drug 5-*FU* (the upper map) and the effect of the concomitant use of *SRV* (the lower

---

[7] x is a variable, which indicates interaction modalities such as enzymatic reactions, transformations, etc.

[8] 5-FU: an anticancer drug, 5-fluorouracil, SRV: an antiviral drug, sorivudine

map). The arrows indicate the reactions, connecting node molecules (e.g. bio-transformation) corresponding to *input* and *output*. The names near the arrow lines indicate mediators of reactions (e.g. enzymes). The broken line arrows indicate aggregates of molecular interactions, while the solid line arrow indicate one unit of molecular interaction in the triadic molecular interaction model.

In this paper, we deal with two types of interactions only. Both occur in the lower map and involve the participation of the enzyme DPD. The first interaction, 5-FU ⟶ $H2FUra$ (mediated by DPD) is modelled by the expression MI_{es}(5-FU, DPD, $H2FUra$). $MI_{es}(a, b, c)$ is a subclass of $MI(a, b, c)$ and an abbreviation of "The emergence of a substrate a triggers a reaction catalyzed by a situated enzyme b and a is converted to c as a result".

The second interaction, shown in the lower map something like BVU ⟶■ DPD (down arrow), is represented by the expression $MI_{ei}(BVU, DPD, BVU *$ $DPD)$.

MI_{ei} (a, b, c) is another subclass of $MI(a, b, c)$. It can be also read as "The emergence of a triggers a reaction owing to which a situated enzyme b is less populated or less capable of reaction as a result, compared to the state prior to the reaction". In this example, an irreversible binding formation ($BVU *$ DPD) is confirmed. It can also be read as $bind\_more(BVU, DPD, BVU * DPD)$ in this example. It is based on the nature of the so-called "mechanism based inhibition" as opposed to the enzyme-substrate reaction in case of the usual oral administration dose under physiological conditions.

It is known that the former process is inhibited (or made less effective) when the latter reaction is also occurring. Intuitively "BVU inhibits (mediated by DPD) $H2FUra$ formation", that is it inhibits 5-FU's detoxification process. In section 4, we will provide a logical reasoning system representing the *inhibit* relation.

Drug Interaction Between CPT-11 and Ketoconazole The drug interaction between CPT-11 and *Ketoconazole* was treated by Yoshikawa *et al* as an example of the model [27]. An illustration of the pathway map of that interaction is shown in Section 4. This example is also adopted to explain of the drug-drug interaction behind the effect of the concomitant use of CPT-11 and *Ketoconazole*. A summary biochemical statement in this example may be expressed as "transformation of CPT-11 is interfered by *Ketoconazole* through the modulation of the enzyme $CYP3A4$". Its pharmacological semantics may be read as "The activity of the anticancer drug CPT-11 is elevated", or "The toxicity of CPT-11 is elevated" for short.

In this paper, we deal with local biochemical semantics by extracting only two molecular interactions that are conjoined to each other. One is CPT-11 ⟶ APC, which is mediated by $CYP3A4$, MI_{es} (CPT-11,$CYP3A4$, APC). The other is *Ketoconazole* ⟶■ $CYP3A4$, MI_{ei}(*Ketoconazole*, $CYP3A$, *Ketoconazole* $* CYP3A4$). Although the interaction modality of *Ketoconazole* with $CYP3A4$ differs from that in the above "mechanism-based inhibiton", the binding between *Ketoconazole* and $CYP3A4$ holds longer and tighter than the bind-

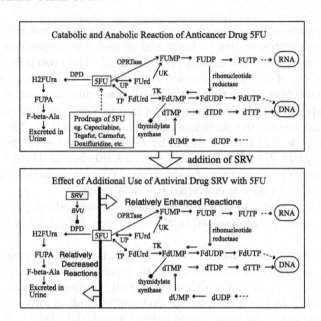

Fig. 2. 5-FU Associated Pathways and the Influence of Addition of SRV

ing between CPT-11 and $CYP3A4$. Therefore, the *bind_more* relation could also be applied in this case. These reactions will be taken up in Section 4 as examples of our logical representation of drug interactions.

3   Preliminaries to Linear Logical Inference Systems for Process Descriptions

In this section we introduce our process description language, a fragment of linear logic.[9] First, we introduce the vocabulary of our process description language as follows:

(1) Logical connectives: $A \otimes B$ ("the molecular binding of A and B"),
                                  (A , B) ("The co-existence of A and B"),
                                  $A \multimap B$ ("If A is added, B is generated"),
                                  $! A$ ("A exists as an environmental resource"),
(2) Additional non-logical
    modal connective: ∇A ("A decreases"),
(3) Molecular expressions: $A, B, C, \ldots, A_0, A_1, A_2, \ldots$
    Molecular variables: $a, b, c, \ldots, a_0, a_1, a_2, \ldots$

The outermost parentheses are often deleted. For example, $A \otimes B \multimap C$ is a molecular expression. $(((A, B), C), \ldots)$ can be abbreviated as $(A, B, C, \ldots)$ or

---

[9] We give a linear logical preliminary explanation, which would be minimal information for understanding this paper, more detailed and formal introduction may be found in e.g. [19],[4]. (See also [3],[18],[20].)

just $A, B, C, \ldots$. $A_1, \ldots, A_n \vdash B_1, \ldots, B_m$ is called a sequent, which may be paraphrased informally "If $A_1, \ldots, A_n$ co-exist then $B_1, \ldots, B_m$ are generated by consuming $A_1, \ldots, A_n$". A finite sequence of formulas (possibly the empty sequence) is denoted by $\Gamma, \Delta, \ldots$. Parentheses occurring in a formula may be deleted when this causes no ambiguity. A sequent is an expression of the form $\Gamma \vdash \Delta$.

There are two kinds of inference rules:

$$\frac{\Gamma_1 \vdash \Delta_1}{\Gamma' \vdash \Delta'}, \qquad\qquad \frac{\Gamma_1 \vdash \Delta_1 \quad \Gamma_2 \vdash \Delta_2}{\Gamma' \vdash \Delta'} .$$

The former has only one upper sequent $\Gamma_1 \vdash \Delta_1$, while the latter has two upper sequents $\Gamma_1 \vdash \Delta_1$ and $\Gamma_2 \vdash \Delta_2$. Both have only one lower sequent $\Gamma' \vdash \Delta'$. We also consider a special kind of inference rules for which there is no upper sequent. Such a special kind of inference rule is called an "axiom sequent".

In the traditional logics, including the classical predicate logic and constructive logic, the following logical inference is admitted as a valid inference:

$$\frac{C \longrightarrow A \quad C \longrightarrow B}{C \longrightarrow A \text{ and } B} .$$

Here, $A \longrightarrow B$ is read "If A then B". Then, the above inference thus says: From the two assumptions "If C then A" and "If C then B", one can reason "If C then A and B". This is obviously true for the usual mathematical reasoning. For example,

$$\frac{f(x) < a \longrightarrow b < x \quad f(x) < a \longrightarrow x < c}{f(x) < a \longrightarrow b < x \text{ and } x < c} .$$

However, when we try to apply this inference rule to the two premises; "If one has one dollar then one gets a chocolate package" and "If one has one dollar then one gets a candy package", then one may have the following inference:

$$\frac{\text{one has \$1} \longrightarrow \text{one gets a chocolate} \quad \text{one has \$1} \longrightarrow \text{one gets a candy}}{\text{one has \$1} \longrightarrow \text{one gets a chocolate and a candy}} .$$

A naive reading of this inference leads to a wrong conclusion "If one has one dollar then one gets both a chocolate package and a candy package". In fact, the following are implicitly assumed when the traditional logical rules are applied to some statements; (1) the statements are independent of temporality, i.e., the logic treats only "eternal" knowledge which is independent of time. (2) the logical implication "$\longrightarrow$" is independent of any consumption relation or any causal relation. These assumptions are appropriate when we treat the ordinary mathematics. Hence, the traditional logical inferences can be used for mathematical reasoning in general. However, when we would like to treat concurrency-sensitive matters or the resource-consumption relation we need to be careful with the application of the logical inference rules. The above example illustrates one such situation. In particular, when we would like to study the mathematical structures of information or computation in computer science, information science, etc., we often need to elaborate the traditional logical inferences since concurrency-sensitive setting and concepts for the consumption of computational resources or of resources for information processing are often essential in computer science and the related fields. Linear logic proposed by Girard is considered one of the basic logical systems which would provide a logical framework for such a new situation occurring in computer science and its related fields.[10] For example, instead of the traditional logical connective $\wedge$ ("and"), linear logic provides two different kinds of logical connectives $\otimes$ and $\&$, where $A \otimes B$ means "A and B hold in parallel (at the same time)" while $A \& B$ means "Either A or B can be chosen to hold (as you like) but only one of them at once".

The traditional logical implication $\longrightarrow$ is replaced by the linear implication $\multimap$, where $A \multimap B$ means "By the consumption of A, B is generated". With the explicit appearance of the resource consumption relation, the conjunction "A and B" naturally yields the co-existence of A and B. We use "comma" ("A, B") to denote the coexistence A and B. We also introduce a stronger notion of co-existence, namely, that of the molecular binding "$A \otimes B$".[11]

Following [27], we take as a basic primitive relation the specific triadic relation, called "the facilitate relation $(facilitate(A, B, C))$". ("A drug A under the environmental resource !B (such as enzymes) generates C" is expressed as $(A, !B) \multimap C$ in the logical sequent from the two transitions.) In this paper, we express this relation on a process description level as a linear logical $\vdash$ relation; The logical inference rule for "," is:

$$\frac{C \vdash A \quad D \vdash B}{(C, D) \vdash (A, B)}$$

When we apply "If one has \$1 then one gets a chocolate package" and "If one has \$1 then one gets a candy package" to the two premises of the left inference rule for "," , then we obtain as a conclusion "one has $\{\$1, \$1\} \multimap$ one gets (a chocolate , a candy)", which means "If one has two \$1's (namely, \$2) then one gets both a chocolate package and a candy package at the same time".

On the other hand, the infinite amount of a resource of A is expressed as !A, with the help of modal operator ! in linear logic. (!A is such a resource that one can consume A as many times as one wants without any loss of !A.) By using this modal operator ! one can express the traditional logical truth (i.e., eternal truth) within the framework of linear logic. Hence, linear logic contains the traditional logic (with the help of modal operator), and linear logic is considered a *refined* (or *fine-grained*) form of the traditional logic, rather than a logic *different* from the traditional logic.

---

[10] There are some other approaches in which the traditional first order logic is refined in order to capture actions and changes of states. Situation calculus proposed by J. McCarthy[15] is one such approach.

[11] Although in the original notation of linear logic, the symbol $\otimes$ is used for the parallel operator, we use a slight different symbolism for it in this paper. (See [19], [4].)

Note that the linear logical implication $A \multimap B$ means that "By consuming A, B is generated." Hence, when A exists and $A \multimap B$ holds, then B can be actually generated at the expense of the resource A.

On the other hand, the traditional logical implication, as in the one used in the above mathematical reasoning, does not consume the premise. That is to say, the traditional implication $A \to B$ means that when A holds and $A \multimap B$ holds, then B holds, where A continues to hold even after B is obtained from A and $A \to B$. (For a precise list of the formal inference rules, see the Appendix at the end of paper.)

The linear logical modal operator $!A$ usually stands for an infinite amount of a resource A. If an inference resource A is not resource-sensitive, it may be interpreted that such a resource can be repeatedly used without any loss, i.e., in the traditional logical sense "A holds" may be interpreted as "There are infinitely many amount of A available", that is, $!A$, in our symbolism.

The traditional implication may be expressed by the linear implication (the resource-consumptional implication) with the bang operator $!$; and thus $A \to B$, for example, may be represented by the linear logical formula $(!A) \to B$.

Accordingly, the standard rules for the bang-modal operator $!$ are formulated as follows:

$!$-left
(dereliction-left)

$$\frac{A, \Gamma \vdash \Delta}{!A, \Gamma \vdash \Delta}$$

(contraction-left)

$$\frac{!A, !A, \Gamma \vdash \Delta}{!A, \Gamma \vdash \Delta}$$

In the DIO-style reasoning, users sometimes wish to obtain certain inhibiting-information. For example, consider the following setting. The production of a is normally generated by a drug b. But, with the use of another drug c in the same environment, the production of a is inhibited, or in other words, the amount of the production of a is substantially decreased due to the use of c.

To deal with such a case, we introduce a new modal operator, which is called the quantitative modality, in symbol ∇.

It represents our thinking about the inhibiting-effects in the DIO-style reasoning.

∇-left   ∇-right

$$\frac{A, \Gamma \vdash \Sigma}{\nabla A, \Gamma \vdash \nabla \Sigma}$$ $$\frac{\Gamma \vdash \Sigma, !A}{\Gamma \vdash \Sigma, A, \nabla !A}$$

Now a careful reader might wonder about the consistency of such rules. The introduction of such inference rules, however, does not affect the consistency of the logical inference rules, as given by the following form of the Proposition.

Proposition: Modality rules for ∇ are consistent.

Proof
By deleting the weak modality symbol, the new rules are still derived rules of the fragment of the original linear logic. This means that the consistency problem of our logical inference system with the new quantitative modal operator is reduced to the consistency problem of the original linear logic. Since the original linear logic is known to be consistent, our new rules for the quantitative modal operator are consistent. □

4   A Linear Logical Formulation of Basic Relations in DIO

4.1   Basic Relations

The triadic relation $facilitate(a, b, c)$ which is explained in section 2.2 is considered as a consumption process (i.e. *input* or *object* may be consumed in this process and generate *output*). We have formulated this consumption relation by a linear logical consumption relation. For convenience, we used the following abbreviations:

a_1 : *input*
a_2 : *object*
a_3 : *output*

Now the triadic relation of $facilitate(a_1, a_2, a_3)$ is logically described as follows:

$$(a_1, !a_2) \multimap a_3$$

Then using linear logical inferences, one can obtain the following general abstract Lemma.

Lemma 1
$$!a_2, facilitate(a_1, a_2, a_3) \vdash a_1 \multimap a_3$$

This means that if there is an environmental resource $!a_2$ and if *facilitate*(a_1, a_2, a_3) holds, then if a_1 is added, a_3 is generated (addition of a_1 generates a_3).

Proof: the following is a formal linear logical proof of this Lemma.

$$
\frac{
\dfrac{a_1 \vdash a_1 \quad !a_2 \vdash !a_2}{a_1, !a_2 \vdash (a_1, !a_2)} \; parallel \qquad a_3 \vdash a_3
}{
\dfrac{a_1, !a_2, ((a_1, !a_2) \multimap a_3) \vdash a_3}{!a_2, ((a_1, !a_2) \multimap a_3) \vdash a_1 \multimap a_3} \; \multimap right
} \; \multimap left
$$

We show some examples of concrete applications of the above Lemma.

Example 4.1.1: *CPT-11 is catalyzed by CE and converted to SN38*
It is known that this actually holds in a certain part of the human liver. CPT-11 is an anti-cancer drug, which is also known as *Irinotecan*[12], CE[13] is an enzyme which exists ordinarily in the human liver, and $SN38$[14] is a drug-derived substance which is generated as a result of CE mediated catalytic reaction. We regard a_1 as CPT-11, a_2 as CE, and a_3 as $SN38$. And by this Lemma, if

1. $!a_2$ exists as an environmental resource; and
2. $facilitate(a_1, a_2, a_3)$ actually holds.

Therefore, the two premises of the Lemma $!a_2$ and $facilitate(a_1, a_2, a_3)$ hold. Hence, the Lemma tells that if CPT-11 is given, $SN38$ is generated.

Example 4.1.2: *5-FU is catalyzed by DPD and converted to H2FUR*
It is known that this actually holds in a certain part of the living human body. *5-FU* [15] is an anti-cancer drug, used to treat some types of cancer. And DPD [16] is an enzyme that exists mainly in the liver. On the other hand, $H2FUR$ is a drug-derived substance that is generated as a result of catalytic reaction mediated by DPD.
As in Example 4.1.1, we regard a_1 as *5-FU*, a_2 as DPD, and a_3 as $H2FUR$. By Lemma 1, if

1. $!a_2$ exists as an environmental resource; and
2. $facilitate(a_1, a_2, a_3)$ actually holds.

Then $facilitate(a_1, a_2, a_3)$ can be deduced in a logical proof.

---

[12] topoisomerase-I inhibitor,
    7-ethyl-10-[4-1(piperidino)-1-piperidino]-carbonyloxycamptothecin
[13] carboxylestherase
[14] 7-ethyl-10-hydroxycamptothecin
[15] 5-fluorouracil
[16] dihydropyrimidine dehydrogenase

4.2   Biomolecular Bindings Relations

To describe biomolecular bindings, the $\otimes$ (called *tensor*) symbol is used as a connective.

Rule: Biomolecular Binding
We express biomolecular bindings as follows:

$$(a_1, a_2) \multimap a_1 \otimes a_2$$

This means that if a_1 and a_2 exist, and it is known that they will actually bind together, then the consumption of a_1 and a_2 will generate the bound molecule $a_1 \otimes a_2$.

Example 4.2.1: *Ketoconazole binds to CYP3A4*
Ketoconazole, an anti-fungal drug, is known to be slowly metabolized by *CYP3A4* forming stable complexes. $CYP3A4^{17}$ is one of the so-called drug metabolizing enzymes which mainly exist in the human liver.
We express this as follows:

$$(Ketoconazole, CYP3A4) \multimap Ketoconazole \otimes CYP3A4$$

This means that if *Ketoconazole* and *CYP3A4* co-exist, *Ketoconazole* and *CYP3A4* will bind together.

Example 4.2.2: *BVU binds to DPD*
BVU (bromovinil uracil) is a drug-derived substance and binds to an enzyme *DPD*. We express this as follows:

$$(BVU, DPD) \multimap BVU \otimes DPD$$

This means that if *BVU* and *DPD* co-exist, *BVU* and *DPD* will bind together.

4.3   Inhibiting Relations

For inhibiting relations, we use the quantitative modality operator ∇ as introduced in Table 1.

Modality Rule I: $\nabla right$

$$\frac{\Gamma \vdash \Sigma, !a}{\Gamma \vdash \Sigma, a, \nabla !a} \ \nabla right$$

---

[17] cytochrome P-450 isoform 3A4

An environmental resource $!a$ can be considered as the sum of a and $\nabla!a$. This means that if a part of $!a$ is used in the environmental resource $!a$, then the environmental resource is consumed, and the amount of usable environmental resource $!a$ will decrease. This inference rule will be used to express the basic "inhibition" relation in DIO; a use of $!a$ which results in a product which may inhibit $!a$.

Modality Rule II: $\nabla left$

$$\frac{a, \Gamma \vdash b_1, b_2, ..., b_n}{\nabla a, \Gamma \vdash \nabla b_1, \nabla b_2, ..., \nabla b_n} \ \nabla left$$

If the effects of A decrease under the same context of Γ, the effect of Products of $\{b_1, b_2, ..., b_n\}$ may be affected. We define $inhibit(A, B)$ as $A \multimap B$, using ∇. Using the above rules, we can infer the inhibiting relations. For convenience, we also use the following abbreviations.

a_1 : Drug1
a_2 : Enzyme
a_3 : Product
a_4 : Drug2

Lemma 2:

$$(a_4, a_2) \multimap (a_4 \otimes a_2), !a_2 \vdash inhibit(a_4, !a_2)$$

This means that if there is an environmental resource $!a_2$ and if $bind(a, b)$ actually holds, then if a_4 is added, the bound molecule $a_4 \otimes a_2$ and the decreased a_2 are generated. (The addition of a_1 and a_4 generates the decreased a_2.)

Proof:

$$\cfrac{\cfrac{a_4 \vdash a_4 \quad \cfrac{\cfrac{!a_2 \vdash !a_2}{!a_2 \vdash \nabla!a_2, a_2} \ \nabla right}{a_4, !a_2 \vdash \nabla!a_2, (a_4, a_2)} \quad a_4 \otimes a_2 \vdash a_4 \otimes a_2}{\cfrac{\cfrac{(a_4, a_2) \multimap (a_4 \otimes a_2), a_4, !a_2 \vdash a_4 \otimes a_2, \nabla!a_2}{(a_4, a_2) \multimap (a_4 \otimes a_2), a_4, !a_2 \vdash \nabla!a_2} \ weakening\text{-}right}{(a_4, a_2) \multimap (a_4 \otimes a_2), !a_2 \vdash a_4 \multimap \nabla!a_2} \ \multimap right}}{} \multimap left$$

Example 4.2.1: *Ketoconazole may decrease the amount of CYP3A4*

We use the following abbreviations.

a_2 : $CYP3A4$
a_4 : $Ketoconazole$

We can see,

1. $!a_2$ really exists as an environmental resource; and
2. a_2 and a_4 are actually bound together, namely $bind(a_4, a_2)$ holds.

Therefore, the two premises of Lemma 2 hold. Then it can be proved that "a_4 may decrease the amount of a_2".

Example 4.2.2: *BVU may decrease the amount of H2FUR*

In the same way as in Example 4.2.1, we use the following correspondence Table.

$a_2 : DPD$
$a_4 : BVU$

We can see,

1. $!a_2$ really exists as an environmental resource; and
2. a_2 and a_4 are actually bound together, namely $bind(a_4, a_2)$

holds. Then, the two premises of Lemma 2 hold. Therefore, it can be proved that "a_4 may decrease the amount of a_2".

By using Lemma 2, we can infer Lemma 3 below.

Lemma 3:

$$bind(a_4, a_2), !a_2, facilitate(a_1, a_2, a_3) \vdash inhibit((a_1, a_4), \nabla a_3)$$

This means that if $bind(a, b)$ actually holds, and if there is an environmental resource $!a_2$, and if $facilitate(a_1, a_2, a_3)$ actually holds, then if a_1 and a_4 are added, the decreased a_3 is generated (the addition of a_1 and a_4 generates the decreased a_3).

Proof:

$$
\cfrac{
  \cfrac{
    \cfrac{From\ Lemma2}{bind(a_4,a_2), a_4, !a_2 \vdash \nabla !a_2}\ \multimap left
    \quad
    \cfrac{!a_2, a_1, facilitate(a_1,a_2,a_3) \vdash a_3}{\nabla !a_2, a_1, facilitate(a_1,a_2,a_3) \vdash \nabla a_3}\ \nabla left
  }{bind(a_4,a_2), a_1, a_4, !a_2, facilitate(a_1,a_2,a_3) \vdash \nabla a_3}\ cut
}{bind(a_4,a_2), !a_2, facilitate(a_1,a_2,a_3) \vdash (a_1, a_4) \multimap \nabla a_3}\ \multimap right
$$
,

where $bind(a_4, a_2)$ is the abbreviation of $(a_4, a_2) \multimap (a_4 \otimes a_2)$, $facilitate(a_1, a_2, a_3)$ stands for $(a_1, !a_2) \multimap a_3$.

Example 4.3.1: *Ketoconazole may inhibit the generation of APC*

We use the following correspondence Table.

$a_1 : CPT\text{-}11$
$a_2 : CYP3A4$
$a_3 : APC$
$a_4 : Ketoconazole$

APC^{18} is a drug-derived substance which is generated in this bio-molecular process. We can see,

1. $!a_2$ really exists as an environmental resource; and
2. $facilitate(a_1, a_2, a_3)$ actually holds; and
3. a_4 and a_2 are actually bound together, namely $bind(a_4, a_2)$ holds.

Then the premises of Lemma 3 hold. Therefore, it can be proved that "a_4 may inhibit the generation of a_3 in the presence of a_1".

Remark I: Non-monotonic Reasoning. Notice that non-monotonic reasoning is used in these inferences regarding the presence of a_4 (*Ketoconazole* in the above example) in the resource. The proof of a_3 is replaced by a proof of $\triangledown a_3$ under the assumptions of $facilitate(a_1, a_2, a_3)$ and giving a_1 with $!a$. This is one of the essential features of inhibition-related reasoning in DIO.

Remark II: Domain Specificity of the Quantitative Modality $\triangledown A$. Although the introduction of the quantitative modality $\triangledown A$ keeps the consistency of the logical proof system (as was shown in Proposition 1 in Section 3), this new modality is very domain-specific and destroys the basic universal structure of logical syntax, namely the cut-eliminability. In fact, the above proof of Lemma 2 serves as a counter-example of the cut-elimination theorem.

Example 4.3.2: *BVU may inhibit the generation of H2FUR*

In the same way as in Example 4.3.1, we use the following correspondence Table.

$a_1 : 5\text{-}FU$
$a_2 : DPD$
$a_3 : H2FUR$
$a_4 : BVU$

We can see,

1. $!a_2$ really exists as environmental resource; and
2. $facilitate(a_1, a_2, a_3)$ actually holds; and
3. a_2 and a_4 actually bind together; namely, $bind(a_4, a_2)$

actually holds. Then the premises of Lemma 3 hold. Therefore, it can be proved that "a_4 may inhibit the generation of a_3 in the presence of a_1".

---

[18] 7-ethyl-10-[4-N-(5-aminopentianoic acid)-1-piperidino]-carbonyloxycamptothecin

4.4 Remark on Logically Higher-Level Reasoning

For the implementation of highly complicated pharmacological relations, we claim that meta-level inferences on proofs, rather than assertions are required. Our logical formalism could be used as a tool to classify relations into different levels, in particular, we could clarify that some reasoning of DIO is not a reasoning at the assertion level, but that at the meta-logical level, namely at the level of inferences on proofs. Here, we show an example of such a meta-level reasoning in DIO.

Using the proof of Lemma 1 and the proof of Lemma 2, we can obtain the Meta-reasoning I given below by a meta-level-reasoning. (Notice: meta-level inference is shown by the bold line, and the inference requires two proofs (or, under the linear logical) proofs-as-composed processes identification, two composed processes as the premises of the inference.)

Meta-reasoning I

$$
\cfrac{
\cfrac{
a_6 \vdash a_6 \quad
\cfrac{
\cfrac{!a_4 \vdash !a_4}{!a_4 \vdash \nabla!a_4, a_4}\nabla\text{right}
}{}
}{
\cfrac{a_6, !a_4 \vdash \nabla!a_4, (a_6, a_4)}{a_6, !a_4 \vdash a_6 \otimes a_4, \nabla!a_4}
}
\quad
\cfrac{
(a_6, a_4) \vdash a_6 \otimes a_4 \quad !a_4, a_1, ((a_1, !a_4) \multimap a_5) \vdash a_5
}{
\cfrac{
a_1, a_6, !a_4, ((a_1, !a_4) \multimap a_5) \vdash a_6 \otimes a_4, \nabla a_5
}{
\cfrac{a_1, a_6, !a_4, (a_1, !a_4 \multimap a_5) \vdash \nabla a_5}{!a_4, (a_1, !a_4 \multimap a_5) \vdash (a_1, a_6) \multimap \nabla a_5}
}
}
}{}
$$

$$
\cfrac{
\cfrac{a_1 \vdash a_1 \quad !a_2 \vdash !a_2}{\cfrac{a_1, !a_2 \vdash (a_1, !a_2)}{\cfrac{a_1, !a_2, ((a_1, !a_2) \multimap a_3) \vdash a_3}{!a_2, ((a_1, !a_2) \multimap a_3) \vdash a_1 \multimap a_3}} \quad a_3 \vdash a_3}
}{}
$$

$$
\mathbf{!a_2, !a_4, ((a_1, !a_2) \multimap a_3), ((a_6, a_4) \multimap (a_6 \otimes a_4)), ((a_1, !a_4) \multimap a_5) \vdash (a_1, a_6) \multimap \triangle a_3}
$$

For example, we take the $Ketoconazole$ - CPT-11 interaction process as one such example. We use the following correspondence Table.

$a_1 : CPT\text{-}11$
$a_2 : CE$
$a_3 : SN\text{-}38$
$a_4 : CYP3A4$
$a_5 : APC$
$a_6 : Ketoconazole$

And here, we explain this situation more briefly. CPT-11 generates SN-38 in the situated environment of CE-enzyme, while CPT-11 generates APC in the situated environment $CYP3A4$, in some region of the liver. Those facilitation processes are described as linear-logical proofs in subsection 4.1, (which is represented by the left-upper proof in the Meta-reasoning I). On the other hand, the existence of $Ketoconazole$ inhibits $CYP3A4$, which we have formally described as a linear logical proof with the quantitative modality ∇ in the previous subsection, (which is represented by the right-upper proof in the Meta-reasoning I). Then, one can reason about the interaction between the CPT-11 - CE - SN-38 facilitation process and the CPT-11 - $CPY3A4$ - $Ketoconazole$ - APC inhibition process, which results in a further facilitation of CPT-11 - CE - SN-38 process (due to the fact that the resource CPT-11 can be used for generating SN-38 because of the assumption of CPT-11 for generating APC is inhibited).

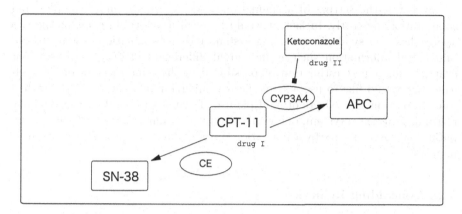

Fig. 3. CPT-11 and $Ketoconazole$ Interaction Process

5 Discussion and Conclusion

5.1 Discussion on the Symbolic Inference Methods and the Simulation Methods for Drug-Interaction Knowledge

We introduced a fragment of a logical inference system for the symbolic reasoning of drug interaction knowledge, where the usual treatment using the relations-based (or equivalently predicates-based) reasoning is analyzed into more primitive process description-based reasoning, using a variant of resource-sensitive logic.

A symbolic logical reasoning encapsulates concrete numerical values and the detailed (e.g. chemical) levels of processes, and it reasons about the drug-interaction on a certain abstract level, which could make the computational processing less costly and which efficiently provides important information. Of course, such abstract and symbolic approaches have some drawbacks: the results of inquiries depend on the way of abstracting the real concrete situations in the organic cells, and it might sometimes cause the validation problem concerning an abstract modeling process.

Here in DIO, we take the ontological methodology proposed by Yoshikawa *et al.* [27] [28] [29], where some specific way of abstraction and some symbolic way of reasoning/inferring are claimed to be ontologically essential and useful for the knowledge of drug-interaction processes. On the other hand, following a different approach, such as the simulation method based on numerical data from verious *in vitro* experiments, it would be possible to realize a wide range of interaction processes, e.g. in a cell. However, for human response to drugs (including drug-drug interactions), the results obtained with the simulation model using such numerical data have not been satisfactory so far, in view of quantitative prediction.

Moreover, the setting of a simulation model is usually more complex and computationally costly. In fact, it would be ideal if one could combine the two approaches, the symbolic reasoning system and the simulation system, to obtain useful information such as individual differences of drug responses. For example, for a first estimation of possible drug-drug interactions or drug effects, one would like to use a symbolic reasoning/inference system, while when some important side-effects of multiple use of drugs are found by the symbolic reasoning/inference system, one would like to re-examine them with more computational cost/resource by a simulation system taking into account quantitative matters.

5.2   Concluding Remarks

We have utilized a variant of linear logic, namely a resource-sensitive logic, to formulate some of the basic inferences used in our previous work on Drug Interaction Ontology (DIO). In particular, we have obtained the following results:

1. The original relational approach of DIO could be logically grounded in terms of the logical description level of the interaction processes, by the use of linear logical process descriptions.
2. The informal arguments for the basic information on *facilitation*, *inhibition, molecular binding*, etc. used in the former work of DIO are now described at the logical proof level.
3. A linear logical proof has a direct meaning in DIO as an interaction process. A complex process corresponds to a composite proof-structure.

In the course of our formulation of the logical language for the interaction-process descriptions, we introduced a new modality, the quantitative modality, in order to reason about a certain negative effect (inhibition, i.e., decreasing tendency of a product due to the interference of another co-existent component). Our logic for representing process descriptions has the characteristics of non-monotonic reasoning and resource- sensitive reasoning. And it seems both characteristics are necessary for the reasoning about the process description level of knowledge/information based on a drug-interaction ontology (DIO) model.

5.3   Future Work

We list here some items on which we plan to work in the future.

1. We plan to extend the basic part of our reasoning/inference system to a further range of drug-interaction ontology and related biomedical ontologies.

2. We also plan to build a combined framework of the symbolic inference method proposed in this paper and the simulation method mentioned earlier in this section, to obtain an integrated framework for the drug-interaction knowledge/information tool.

3. We plan to apply our linear logical process description framework to further ranges of process-based ontologies. It would be useful for the setting of process ontology modeling to investigate the relationship between the traditional relation/predicate-based descriptions (whose reasoning basically follows the traditional first order logic) and our process-based descriptions (whose reasoning follows linear logic and its variants, as explained in this paper). In fact, the formal (logical) process descriptions introduced in our linear logical language framework would be useful to define process-related interactive relations, such as *facilitate, bind more, inhibit,* etc. , while, once precisely defined, the precise formal definition and the precise operational meaning of such relations may be encapsulated for some simple queries on DIO. In fact, the statement of Lemma 1,2,3 has the form of horn clauses once those relations are defined by a linear logical formula. In such cases, one could make use of the traditional predicate logical inference engine, while it is necessary to return to the precise resource-sensitive (linear) logical inference engine for more resource-sensitive queries. So, interconnecting the traditional relational (and predicate-logical) approach and the linear logical approach would be useful for the practical purpose of DIO. Logic programming language frameworks have been well investigated in former work (by Dale Miller's group and others [10] [16]). Such logic programming frameworks might be useful for this direction of research.

4. We have not developed semantics for our system yet, except for the underlying operational semantics of our proof-syntax (of [19]). We plan to develop phase semantics by introducing the semantic denotation of $\triangledown A$ ("significantly small amount of A").

References

1. Gene Ontology Consortium: Creating the Gene Ontology resource: design and implementation. *Genome Res 2001,* (2001) 11:1425-1433
2. Gene Ontology Consortium: The Gene Ontology Editorial Style Guide. http://www.geneontology.org/GO.usage.shtml
3. Girard, J.Y.: Linear Logic, In: Theoretical Computer Science 50 (1987) 1-102
4. Girard, J.Y.: Linear Logic: its syntax and semantics. In J.Y. Girard, Y. Lafont, and L. Regnier, editors, Advances in Linear Logic. London Mathematical Society Lecture Note Series, Cambridge University Press (1995)
5. Goguen, J: Data, Schema, Ontology, and Logic Integration.In Logic Journal of the IGPL, edited by Walter Carnielli, Miguel Dionisio, and Paulo Mateus. volume 13, no. 6, (2006, to appear)
6. Goguen, J: Information Integration in Institutions. In: Jon Barwise memorial volume edited by Larry Moss, Indiana University Press, (2006, to appear)
7. Goguen, J: Ontology, Society, and Ontotheology. In: Formal Ontology in Information Systems, edited by Achille Varzi and Laure Vieu, IOS Press, (2004) 95-103

638   Mitsuhiro Okada et al.

8. Grenona, P., Smith, B.: SNAP and SPAN: Towards Dynamic Spatial Ontology. SPATIAL COGNITION AND COMPUTATION, vol(issue), start-end. Lawrence Erlbaum Associates, Inc.
9. Hasebe, K., Jouannaud J.P., Kremer, A., Okada, M, Zumkeller, R.: Formal Verification of Dynamic Real-Time State-Transition Systems Using Linear Logic. In: Proc. Software Science Conference, Nagoya, Sept. (2003) 5 pages
10. Hodas, J.S., Miller, D.: Logic Programming in a Fragment of Intuitionistic Linear Logic. In: Journal of Information and Computation, 110(2): 1 May (1994) 327-365
11. Horrocks, I.: Description logics in ontology applications. In: B. Beckert, editor, Proc. of the 9th Int. Conf. on Automated Reasoning with Analytic Tableaux and Related Methods (TABLEAUX 2005), no 3702 in Lecture Notes in Artificial Intelligence. Springer-Verlag (2005) 2-13
12. Institute for Formal Ontology and Medical Information Science. (IFOMIS). http://www.ifomis.uni-saarland.de/
13. Masolo, C., Borgo, S., Gangemi, A., Guarino, N., Oltramari, A.: WonderWeb Deliverable D18 - Ontology Library (final). http://wonderweb.semanticweb.org/deliverables/D18.shtml
14. Mathijssen, R.H.J., van Alpen, R.J., Verweij, J., Walter, J.L., Nooter, K., Stoter, G., Sparreboom, A.: Clinical Pharmacokinetics and Metabolism of Irinotecan (CPT-11). Clinical Cancer Research. Vol.7, August (2001) 2182-2194
15. McCarthy, J., Hayes, P.J.: Some Philosophical Problems from the Standpoint of Artificial Intelligence. Machine Intelligence 4 (1969) 463-502
16. Dale Miller. A Survey of Linear Logic Programming *Computational Logic: The Newsletter of the European Network in Computational Logic*, Volume 2, No.2, pp.63-67, December (1995)
17. OBO. Open Biomedical Ontologies. http://obo.sourceforge.net/
18. Okada, M.: Linear Logic and Intuitionistic Logic. In: La revue internationale de philosophie. no. 230, special issue "Intuitionism" (2004) 449-481
19. Okada, M.: An Introduction to Linear Logic: Phase Semantics and Expressiveness. In: Theories of Types and Proofs, eds. M.Takahashi-M.Dezani-M.Okada, Memoirs of Mathematical Society of Japan, vol.2 (1998) 255-295, second edition (1999)
20. Okada, M.: Girard's Linear Logic and Applications (in Japanese, partly in English). In: a JSSS Tutorial Lecture Note (41 pages), Software Science Society of Japan, (1993)
21. Smith, B., et al.: Basic Formal Ontology http://ontology.buffalo.edu/bfp
22. Smith, B., Kumar, A.: On Controlled Vocabularies in Bioinformatics: A Case Study in the Gene Ontology, BIOSILICO: Drug Discovery Today, 2 (2004) 246?252.
23. Smith, B., Ceusters, W., Klagges, B., Köhler, J., Kumar, A., Lomax, J., Mungall, C., Neuhaus, F., Rector, A.L., Rosse,C.: Relations in biomedical ontologies. Genome Biology (2005) 6:R46
24. Smith, B., Rosse,C.: The Role of Foundational Relations in the Alignment of Biomedical Ontologies. In: M. Fieschi, et al. (eds.), Medinfo 2004. Amsterdam: IOS Press, (2004): 444-448
25. Sattler, U.: Description Logics for Ontologies. In: Proc. of the International Conference on Conceptual Structures (ICCS 2003), volume 2746 of Lecture Notes in Artificial Intelligence. Springer Verlag (2003)
26. Stevens, R., Wroe, C., Lord P., Goble, C.: Ontologies in Bioinformatics. In: Handbook on Ontologies in Information Systems. Springer, (2003): 635-657
27. Yoshikawa, S., Satou, K., Konagaya, A.: Drug Interaction Ontology (DIO) for Inferences of Possible Drug-drug Interactions. In: MEDINFO 2004, M. Fieschi et al. (Eds) :IOS Press; (2004) : 454-458

28. Yoshikawa, S., Konagaya, A.: DIO: Drug Interaction Ontology - Application to Inferences in Possible Drug-drug Interactions 2003, In: Proceedings of The 2003 International Conference on Mathematics and Engineering Techniques in Medicine and Biological Sciences (METMBS '03), June 23 - 26, (2003). Las Vegas, Nevada, USA: 231-236

29. Yoshikawa, S., Satou, K., Konagaya, A.: Application of Drug Interaction Ontology (DIO) for Possible Drug-drug Interactions. In: Proceedings of Chem-BioInformatics Society (2003) : 320

Appendix I

Formal Rules for a Fragment of Linear Logic with Quantitative Modality (LLQ)

The following are formal rules for a fragment of Linear Logic with Quantitative Modality (LLQ), which is used in this paper. For further basic backgrounds of linear logic and process-descriptions with linear logic, see [19]

Definition 1 (Inference rules for LLQ). Below, A and B represent arbitrary molecular expressions and Γ, Δ, Σ, Π represent arbitrary (finite) sequences of molecular expressions, including the case of an empty sequence. A sequent $A_1, ..., A_n \vdash B_1, ..., B_m$ means informally $(A_1, ..., A_n) \multimap (B_1, ..., B_m)$, namely, if $A_1, ..., A_n$ are given at once, then $B_1, ..., B_n$ are generated by consuming $A_1, ..., A_n$.

- **Axiom sequent**
  Logical axiom sequent

$$A \vdash A$$

- **Cut-rule**

$$\frac{\Gamma \vdash \Delta, A \quad A, \Sigma \vdash \Pi}{\Gamma, \Sigma \vdash \Delta, \Pi}$$

- **Multiplicative (Parallel)**

"," (parallel)-right

$$\frac{\Gamma \vdash \Delta, A \quad \Sigma \vdash \Pi, B}{\Gamma, \Sigma \vdash \Delta, \Pi, A, B}$$

- **Linear Implication**

$\multimap$-left

$$\frac{\Gamma \vdash \Delta, A \quad B, \Sigma \vdash \Pi}{A \multimap B, \Gamma, \Sigma \vdash \Delta, \Pi}$$

$\multimap$-right

$$\frac{A, \Gamma \vdash \Delta, B}{\Gamma \vdash \Delta, A \multimap B}$$

- **Weakening**

Weakening-right

$$\frac{\Gamma \vdash \Delta, A}{\Gamma \vdash \Delta}$$

(Note that the right commas are the parallel conjunctions in our sequent calculus formulation, and the above weakening rule is a derived rule in the standard weakening rule of linear logic (in cf. [19])).

- **Modality**

!-left
(dereliction-left)

$$\frac{A, \Gamma \vdash \Delta}{!A, \Gamma \vdash \Delta}$$

(contraction-left)

$$\frac{!A, !A, \Gamma \vdash \Delta}{!A, \Gamma \vdash \Delta}$$

- **Quantitative Modality**

∇-left

$$\frac{A, \Gamma \vdash \Sigma}{\nabla A, \Gamma \vdash \nabla \Sigma}$$

∇-right

$$\frac{\Gamma \vdash \Sigma, !A}{\Gamma \vdash \Sigma, A, \nabla !A}$$

Appendix II

Scope of the DIO Model and Its Limitation

Pathway Model and Phenomenon in the Real World Our model for Drug Interaction Ontology, the triadic molecular interaction model, can be considered as an atomic component of the pathway model, which is used to describe not only biological phenomena but also the mechanisms of drug action. It is often used to explain the causality of drug action, side effects and other inducible phenomena. A biological reaction in reality, however, is very complex, and a pathway model itself is a kind of abstraction from information in nature. They both constitute only part of the full stream of events, being provided to explain phenomena of particular interest. In a pathway model for a given phenomenon, disregarded reactions are those with unknown associations, or with less influential effect. Obviously, undiscovered reactions are not included.

Usually, the time scale of reactions in a given pathway is implicit. The time scale in general, corresponds to the ones in experiments, by which the possibilities of reactions are identified and verified. These are mostly shorter than a monthly or yearly range. Long term reactions such as the accumulation of injury in mitochondrial DNA during the normal aging process are disregarded. Likewise, very short-time events in quantum level (e.g. atomic, electronic) are disregarded.

This model can be used for the (re)construction of a pathway model, by providing conjunction (inference) rules. This approach is different from the pathway decomposition approach or the pathway first approach, where a molecular interaction is tightly bound to its parent pathway model. An arbitrary molecular interaction represented by a triadic relation could be potentially integrated as a sub-process of different pathways, which would also be the case in the real world. Another aspect of the molecular interaction network in real world phenomena is its dynamic nature and complexity influenced by organism level regulations. There would be multiple feedback regulations and loops in the network map. Inferences using relative relations such as *inhibit* or *facilitate*, without using quantitative information and a time scale, would have limitations in such complicated network models.

Scope of the Molecular Binding Model Our triadic molecular interaction model, described above, reflects the process of binding-based interactions, mediated by a transient complex. There also exist non-binding type processes such as a movement by natural diffusion, a bio-transformation without mediators, and so on. These are outside the scope of this paper.

Molecular Level Granularity This interaction model is based on molecular interactions while our relational database schema include location information for each process participant. The relation is molecule *part_of* or *located_in* a subcellular component, and/or molecule *part_of* or *located_in* tissue/organ components. Some reactions, such as a transporter mediated process, are location sensitive. On the other hand, the type of reactions examplified in this article is not location sensitive. In our examples it is presumed that the participants and the events are all allocated in the same field, and thus we disregard the attributes of location. When we deal with location-sensitive reactions, however, inferences including different biological granularity would need further formalism.

Application for Prediction of Real World Event As was pointed out above, the pathway model itself has certain limitations in view of real world events. In addition, it does not deal with numerical data, and is not capable of a quantitative estimation of molecular events. We made some abstraction from such information by introducing *relative relation* in terms of "*bind_more*". The *relative relations* are also embedded in the semantics of triadic molecular interaction itself: The *emergence* of an input *triggers* the execution of a process and causes the *emergence* of a new product (*output*).

642   Mitsuhiro Okada et al.

Instead of dealing with numerical information, this model deals with a qualitative change of amount by providing semantics of a relative change to the relations, such as *activation* (relatively increased level of execution) or *inhibit* (relatively decreased level of execution). For a more precise prediction and for more complicated pathway network models (such as those of loops), the use of numerical data would be important. A practical solution might be a cooperative inference between these methods and simulation methods.

Author Index

Lecture Notes in Computer Science

For information about Vols. 1–3955

please contact your bookseller or Springer